Y0-BYK-874

THE ALKALOIDS

Chemistry and Physiology

VOLUME XVII

THE ALKALOIDS
Chemistry and Physiology

Edited by
R. H. F. MANSKE

Department of Chemistry, University of Waterloo
Waterloo, Ontario, Canada

R. G. A. RODRIGO

Wilfrid Laurier University
Waterloo, Ontario, Canada

VOLUME XVII

1979

ACADEMIC PRESS • NEW YORK • SAN FRANCISCO • LONDON

A Subsidiary of Harcourt Brace Jovanovich, Publishers

ACADEMIC PRESS, INC.
111 Fifth Avenue, New York, New York 10003

United Kingdom Edition published by
ACADEMIC PRESS, INC. (LONDON) LTD.
24/28 Oval Road, London NW1 7DX

Library of Congress Cataloging in Publication Data

Manske, Richard Helmuth Fred, Date
The alkaloids : chemistry and physiology.

Includes bibliographical references.
1. Alkaloids. 2. Alkaloids--Physiological effect.
I. Holmes, Henry Lavergne, joint author. II. Title.
QD421.M3 547'.72 50-5522
ISBN 0-12-469517-5 (v. 17)

PRINTED IN THE UNITED STATES OF AMERICA

79 80 81 82 9 8 7 6 5 4 3 2 1

CONTENTS

Chapter 4. Papaveraceae Alkaloids. II

F. Šantavý

Chapter 5. Monoterpene Alkaloid Glycosides

R. S. Kapil and R. T. Brown

LIST OF CONTRIBUTORS

Numbers in parentheses indicate the pages on which the authors' contributions begin.

R. T. BROWN, Department of Chemistry, The University, Manchester M13 9PL, England (545)

GEOFFREY A. CORDELL, College of Pharmacy, University of Illinois at the Medical Center, Chicago, Illinois 60612 (199)

M. F. GRUNDON, School of Physical Sciences, New University of Ulster, Coleraine, Northern Ireland (105)

R. S. KAPIL, Central Drug Research Institute, Lucknow 226001, India (545)

NARESH V. MODY, Institute for Natural Products Research and the Department of Chemistry, University of Georgia, Athens, Georgia 30602 (1)

S. WILLIAM PELLETIER, Institute for Natural Products Research and the Department of Chemistry, University of Georgia, Athens, Georgia 30602 (1)

F. ŠANTAVÝ, Institute of Chemistry, Medical Faculty, Palacký University, 775 15, Olomouc, Czechoslovakia (385)

PREFACE

The first volume of "The Alkaloids" appeared in 1950 and since then sixteen volumes of this series of reviews have been favorably received. We record here with deep regret the untimely death of the series' founding editor, Dr. R. H. F. Manske, and pay a brief tribute to the man and his work. The reviews now offered in Volume XVII were commissioned by him and it is our hope that this and subsequent volumes will continue to maintain the rigorous standards of excellence set by the late Dr. Manske.

R. G. A. RODRIGO

RICHARD HELMUTH FREDERICK MANSKE (1901–1977)

Richard H. F. Manske was born in Germany in 1901 and came with his parents to Canada in 1906. He died in Guelph, Ontario in September 1977 from injuries sustained in an automobile accident early in the same year.

Manske began his education in the public schools of Saskatchewan where his family had settled after arriving in Canada. He later journeyed east to Ontario to attend Queen's University in Kingston, Ontario, where he was awarded the bachelor of science degree in 1923 and the master of science degree during the following year. In his graduate studies at Queen's he held a bursary awarded by the National Research Council of Canada. At Queen's he met his first wife, Jean Gray, whom he married in 1924. There were two daughters from this marriage, Barbara and Cory. On completion of his master's degree he moved to Manchester, England as an 1851 Exhibition Scholar where he studied in the laboratories of Robert Robinson. It was there that he began his work on alkaloids for which he will always be remembered.

As a graduate student at Manchester, Manske determined the structures of harmine and harmaline and synthesized both alkaloids. He also synthesized rutaecarpine, as he liked to recall, by accident. Another fortuitous experiment was his discovery of the use of hydrazine in the hydrolysis of phthalimides. He also collaborated at this time with A. Lapworth in one of the pioneering studies of physical organic chemistry. He was awarded the Ph.D. by Manchester in 1926.

Upon his return to North America, Manske worked briefly with the General Motors Company in the United States as a research chemist and then for three years as a research fellow at Yale University. In collaboration with T. B. Johnson at Yale he developed a new synthesis of ephedrine and related compounds. In 1929 he returned to Canada to assume a post with the National Research Council.

In Ottawa he began a systematic investigation of Fumariaceous plants for their alkaloid content. By developing new methods of separation of the alkaloids he found many new bases of the aporphine, benzophenanthridine, phthalideisoquinoline, protopine, and protoberberine ring systems. Three new classes of isoquinoline alkaloids also owe their discovery to this work. In his examination of *Dicentra cucullaria* he found the alkaloid cularine, determined its structure, and thereby demonstrated that it belonged to a new ring system. In the same period he isolated many of the alkaloids that are now known as the spirobenzylisoquinolines. Some 25 years later he was co-author of the paper announcing the structure of ochotensimine, the first representative of this ring system to yield its structural secrets. The cancentrine alkaloids were the most complicated structurally of all the alkaloids that he isolated. They were obtained from *Dicentra canadensis* Walp. in 1932, and nearly 40 years elapsed before the structure of cancentrine itself was finally resolved.

Many of the alkaloids that he isolated in the 1930's and 1940's were found in small amounts, insufficient for structural examination by the methods available at that time. He had the foresight to put them away, carefully labeled and purified, until a time should come when science would reach the stage where they might be profitably examined. With the advent of refined spectroscopic methods in the last several decades the samples were removed from storage, usually from small brown bottles, and their structures elucidated, often in collaboration with younger and grateful colleagues.

Besides his classic studies on the isoquinoline family of alkaloids, Manske also undertook, in collaboration with Leo Marion, an examination of the Lycopodiaceae native to Canada. From this work some thirty alkaloids were characterized, and though he did not himself participate in a major way in their structural elucidation, he always followed the work with interest and insight. Other alkaloids with which he was concerned were members of the Senecio and Lobelia families, and in the latter group he discovered lobinaline, the first dimeric member of this family to be isolated.

In 1943 while still a relatively young man he left the National Research Council to assume the directorship of the newly established Research Laboratories of the Dominion Rubber Co. Ltd., now Uniroyal Ltd. He remained in this position for 23 years and retired in 1966. Under his direction the laboratory soon became a leading center of industrial research in Canada. At Guelph, alkaloids, of necessity, were relegated to a secondary role, but they were not entirely neglected. During this period "The Alkaloids" was begun and whenever there was time available from his other duties Manske would be found at the bench. He also encouraged his younger colleagues in the laboratory to carry on alkaloid research on a part-time basis. In this way some of us who worked there as young chemists acquired a taste for pure research and after a few years of Manske's tutelage moved on to university positions.

Upon his retirement from Uniroyal in 1966 he assumed the position of Adjunct Professor at the University of Waterloo where he applied himself once again to alkaloid chemistry. There he continued to be actively involved in research and teaching until his untimely accident.

During his career Manske was the recipient of many honors and awards. Among other things he was made a Fellow of the Royal Society of Canada in 1935, received a D.Sc. from Manchester in 1937, was Centenary Lecturer of the Chemical Society, London, in 1954, was awarded the Chemical Institute of Canada Medal in 1959 (the highest distinction of the Institute), was the recipient of several honorary degrees from Canadian universities, received the Morley Medal of the Cleveland Section of the American Chemical Society in 1972, and inaugurated the A. C. Neish lectures at the National Research Council of Canada, Halifax, Nova Scotia. He was active for many years in the Chemical Institute of Canada and served as its president from 1963 to 1964.

His other interests were many and varied and he approached them with the same enthusiasm and commitment as his chemistry. He loved music and was himself a violinist. Growing orchids was a passion with him. It began as a hobby, developed into a business, and in later years became a hobby once again. There was always an orchid blooming in his greenhouse which he was happy and pleased to show to visitors. On his extensive property in Guelph he had many varieties of trees, shrubs, and flowers, some of which were used by him and by others for scientific ends, but they were there mainly for his own enjoyment and that of others. Birds, stars, cooking, martinis, wines, politics, economics, and philosophy were all subjects about which he was knowledgeable and willing and eager to discourse upon.

One of his main concerns was the role of the scientist in modern society and he lectured extensively on this subject during his tenure as vice-president and president of the CIC. His views are succinctly expressed in the following passage taken from one of his lectures: "He who professes science is truly a scientist only if he strives to achieve an awareness of his place in society as a whole. If he buries himself in the confines of his discipline and neither knows nor cares about the broad vista of the world about him he has failed as a man; and if he does not apply the objective, that is scientific, method to matters other than to those of his narrow discipline he has failed as a man. Indeed the scientist has failed as a man if he does not make it a *sine qua non* of his life to question authority, be it of Mohamet or of Darwin."

He was a man with a tremendous appetite for life. He was always busy with his chemistry, his hobbies, or his family. He lived his life to the fullest, enjoyed it all, and transmitted his zest for living and learning to all who knew him well.

He is survived by his second wife, Doris, his two daughters, and five grandchildren. His death is mourned not only by his family but also by his many colleagues and friends.

D. B. MacLean
McMaster University
Hamilton, Ontario, Canada

CONTENTS OF PREVIOUS VOLUMES

Contents of Volume I

Contents of Volume II

Contents of Volume III

Contents of Volume IV

Contents of Volume V

Contents of Volume VI

Contents of Volume VII

Contents of Volume VIII

Contents of Volume IX

Contents of Volume X

Contents of Volume XI

Contents of Volume XII

CHAPTER

Contents of Volume XIII

Contents of Volume XIV

Contents of Volume XV

Contents of Volume XVI

—CHAPTER 1—

THE STRUCTURE AND SYNTHESIS OF C_{19}-DITERPENOID ALKALOIDS

S. WILLIAM PELLETIER AND NARESH V. MODY

Institute for Natural Products Research
University of Georgia, Athens, Georgia

ISBN 0-12-469517-5

I. Introduction

Since the early 1800's, the alkaloids (nitrogenous organic bases) isolated from plants of the *Delphinium* and *Aconitum* genera (Ranunculaceae) and more recently the *Garrya* genus (Garryaceae) and *Inula royleana* (Compositae) have been of interest because of their pharmacological properties, complex structures, and interesting chemistry. Biogenetically, these diterpenoid alkaloids are possibly formed in nature from tetracyclic or pentacyclic diterpenes in which the nitrogen atom of β-aminoethanol, methylamine, or ethylamine is linked to C-19 and C-20 in the C_{20}-diterpenoid skeleton and C-17 and C-19 in the C_{19}-diterpenoid skeleton to form a substituted piperidine ring.

These diterpenoid alkaloids may be divided into two broad groups: those based on a hexacyclic C_{19}-skeleton, and those based on a C_{20}-skeleton. The C_{19}-alkaloids are commonly called *aconitines*, and all possess either the aconitine, the lycoctonine, or the heteratisine skeleton. Usually in the literature, the C_{19}-diterpenoid alkaloids are referred to as either *aconitine-type* or *lycoctonine-type* alkaloids without structural differentiation. Because this practice sometimes creates confusion, we have divided the C_{19}-diterpenoid alkaloids into three categories, defined as follows:

i. *Aconitine-type.* These alkaloids possess the skeleton of aconitine, in which the C-7 position is not oxygenated or substituted by any other functional group.

ii. *Lycoctonine-type.* These alkaloids possess the skeleton of lycoctonine, in which the C-6 and C-7 positions are always oxygenated by hydroxyl or methylenedioxyl or methoxyl groups.

iii. *Heteratisine-type.* These alkaloids possess the skeleton of heteratisine, in which a lactone moiety is always present.

The numbering system for the alkaloids with basic aconitine (i), lycoctonine (ii), and heteratisine (iii) skeletons used in this chapter is in accord with proposals initiated by Dr. J. W. Rowe and cosponsored by the author.*

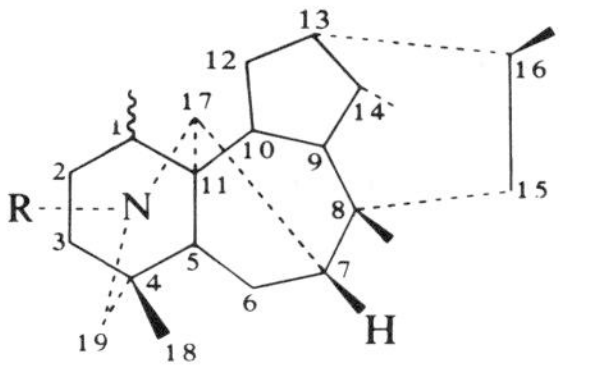

(i) Aconitine skeleton

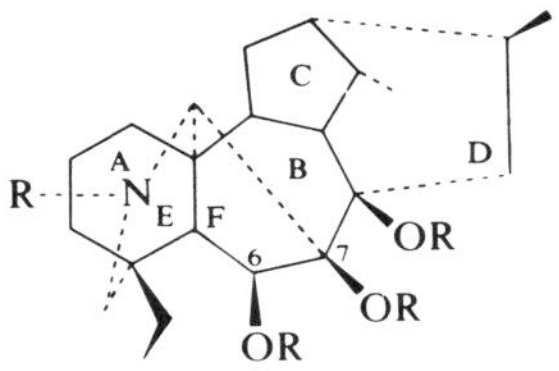

(ii) Lycoctonine skeleton

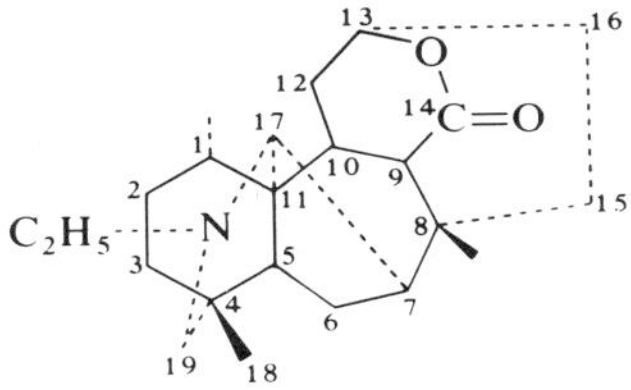

(iii) Heteratisine skeleton

The earlier work on the chemistry of C_{19}-diterpenoid alkaloids has been reviewed in Volumes IV (*1*), VII (*2*), and XII (*3*) of this series, in books on alkaloids by Boit (*4*) and Pelletier (*5*), in *The Alkaloids, Specialist Periodical Reports*, (*6–8*), and in other reviews (*9, 10*). This chapter deals with the recent developments in C_{19}-diterpenoid alkaloid chemistry reported in the literature available to us since July 1968, as well as certain unpublished work from our laboratory.

Because previously published books (*11, 12*) in the field of natural products have reported structures now known to be incorrect of several well-known C_{19}-diterpenoid alkaloids, we have included in this chapter a catalog of all the known C_{19}-diterpenoid alkaloids, showing the correct structures, physical properties, plant sources, and key references. This catalog should be very useful, for it presents in a single place important structural information on the C_{19}-diterpenoid alkaloids, which has been scattered through hundreds of papers and dozens of reviews.

* John W. Rowe, *The Common and Systematic Nomenclature of Cyclic Diterpenes*, Third revision, Forest Products Laboratory, Forest Service, U.S. Department of Agriculture, Madison, Wisconsin, 1968, p. 44.

II. Aconitine-Type Alkaloids

A. Aconitine

Aconitine, the major alkaloid of *Aconitum napellus L.*, has been known for its pharmacological and medicinal properties since 1833 (*13*). It is one of the best known alkaloids and possesses one of the most complicated structures of the C_{19}-diterpenoid alkaloids. This alkaloid occurs, together with the alkaloid mesaconitine, in *A. napellus, A. fauriei* Léveillé and Vaniot, *A. grossedentatum* Nakai, *A. hakusanense* Nakai, *A. mokchangense* Nakai, and *A. zuccarini* Nakai (*1*). It also has been isolated from the Chinese crude drugs Chuan-Wu, Fu-Tzu (*A. carmichaeli* Debaux = *A. fischeri* Reichb) (*14*) and Hye-shang-yi-zhi-hao (*A. bullatifolium* Léveillé var. homotrichum) (*15*). More recently, aconitine has been isolated (*16*) from the tubers of *A. karakolicum*, which were collected in the Terskei Ala-Tau ranges of the Kirghiz S.S.R.

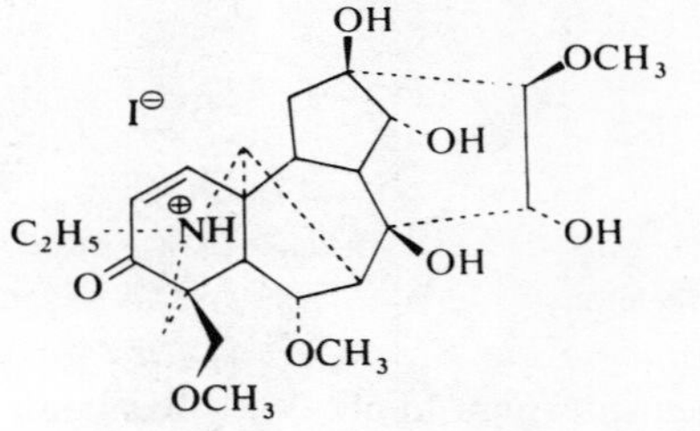

1

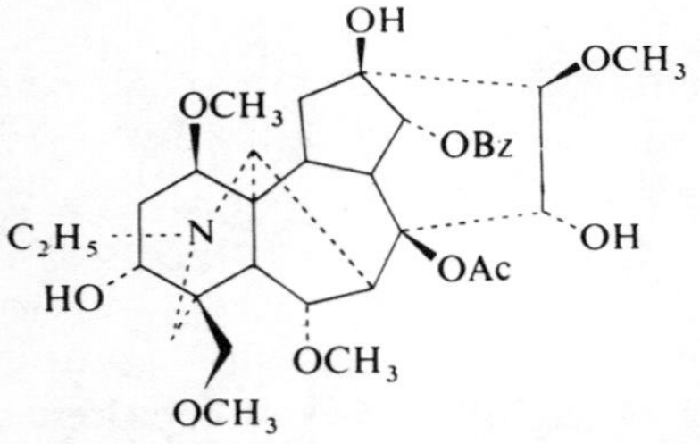

2

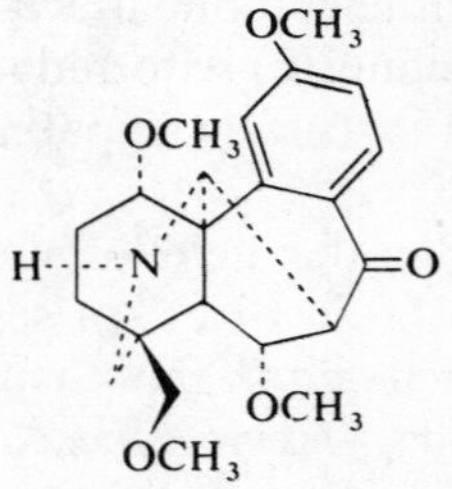

3

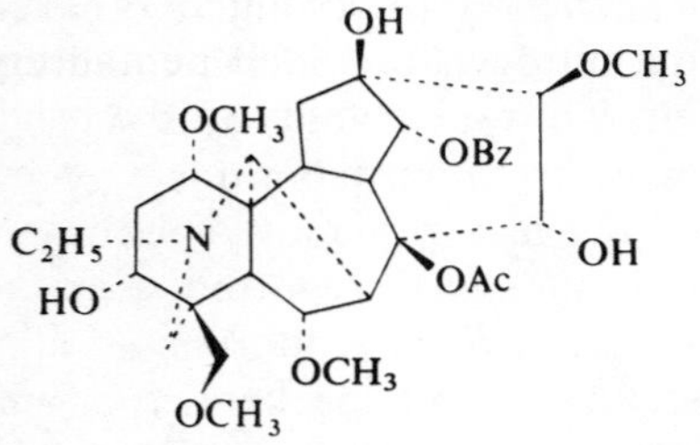

4 Aconitine

Early work on the chemistry of aconitine has been reviewed by Stern (*1, 2*) and recently by Pelletier and Keith (*3, 5*). On the basis of the X-ray structure (*17, 18*) of demethanolaconinone hydroiodide trihydrate (**1**), chemical data on a permanganate oxidation product of aconitine known as oxonitine, and conformational arguments (*19*), structure **2** was assigned to aconitine. Recently, Wiesner and co-workers (*20*) demonstrated that the C-1 methoxyl group in the related alkaloid, delphinine, has an α-equatorial configuration (cis to the nitrogen bridge) by an X-ray crystallographic study of the acid oxalate salt of a degradation product (**3**) of delphinine (see Section II,G). Delphinine has been correlated with aconitine and other C_{19}-diterpenoid alkaloids; thus, Wiesner and co-workers indicated that the structure of aconitine and related alkaloids must be revised to include an α-equatorial methoxyl group at the C-1 position. The complete structure of aconitine as **4** was finally elucidated after 40 years of extensive chemical and X-ray crystallographic studies.

B. Oxonitine

Much of the earlier chemical work on the structure elucidation of aconitine was based on oxonitine, a permanganate oxidation product of aconitine. The nature of the *N*-acyl group of oxonitine has been a disputed point in papers on the structure of oxonitine (**5**).

In 1954 Jacobs and Pelletier (*21*) at the Rockefeller Institute had proposed that oxonitine contains an *N*-acetyl group. Six years later they, as well as Turner and his co-workers (*22*), demonstrated that oxonitine actually contains an *N*-formyl group. They showed that the *N*-formyl group does not arise from the direct methanolic permanganate oxidation of the *N*-alkyl group. Thus, compound **6** was prepared by a series of reactions involving removal of the *N*-ethyl group from aconitine (**4**), acetylation, and formation of an *N*-formyl group. Comparison of an authentic sample of triacetyloxonitine with compound **6** confirmed identity. The most important evidence was that the permanganate oxidation of pentaacetylaconine (**7**) containing an *N*-ethyl group labeled at the carbon adjacent to the nitrogen provided pentaacetyl-*N*-formyl oxonitine (**8**) with only 6% residual radioactivity of the original activity (*22*). This tracer experiment clearly indicated that the *N*-formyl group of oxonitine did not derive from the *N*-ethyl group of aconitine.

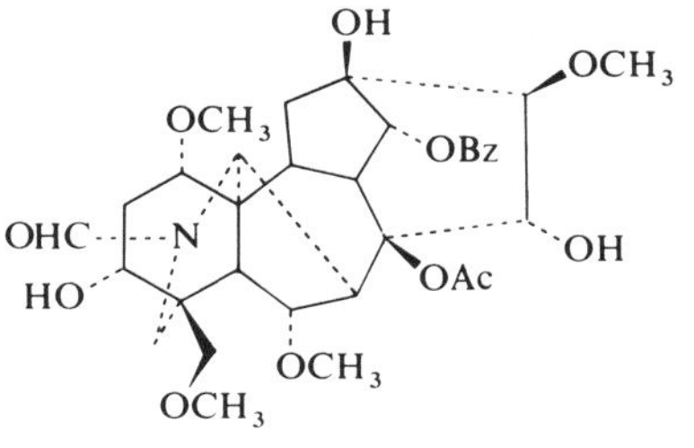

5 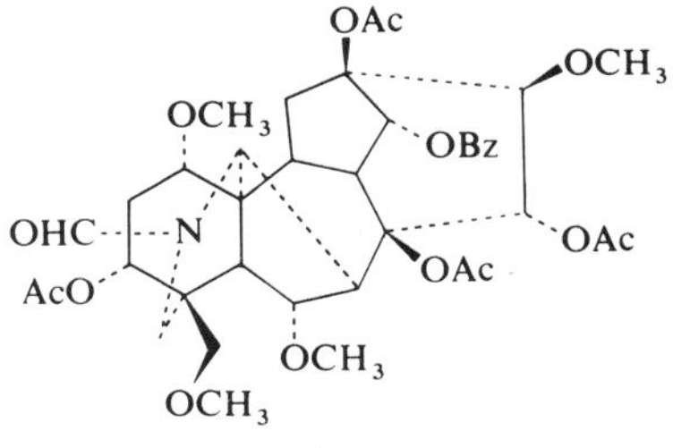 **6**

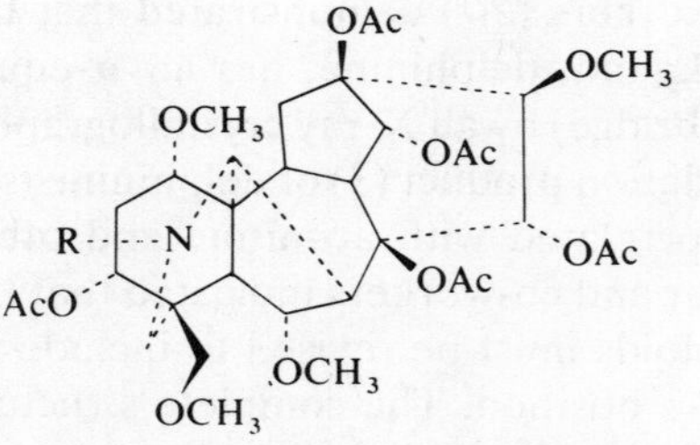

7 R = $^{14}CH_2{-}CH_3$
8 R = CHO

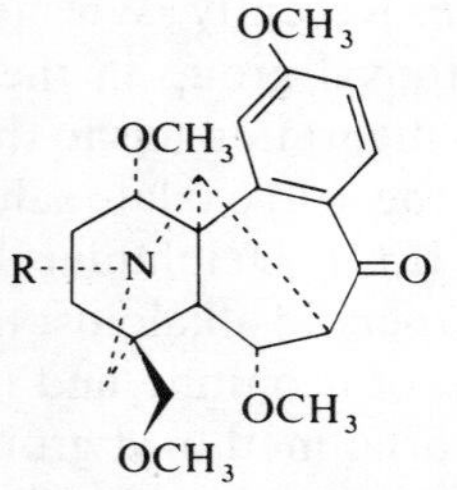

9 R = mixture of CHO and Ac
3 R = H

The question of whether oxonitine contains an *N*-formyl or an *N*-acetyl group has been studied recently by Wiesner and Jay (*23*). They have provided an explanation about the origin of oxonitine. The mass and ^{1}H NMR spectra of the *N*-acyl aromatization product (**9**), prepared from oxonitine, proved it to be a mixture of the *N*-acetyl and *N*-formyl derivatives, even though it was homogeneous on a TLC plate. Hydrolysis of this mixture (**9**) with hydrochloric acid in methanol afforded compound **3**. The latter was identical with samples prepared from delphinine and by total synthesis (see Section II,G). The authors indicated that, because aconitine is known to contain varying amounts of mesaconitine (**13**), the differing *N*-acyl groups can result from the oxidation of the *N*-methyl group of mesaconitine and the *N*-ethyl group of aconitine, present in the mixture.

It had been demonstrated earlier that the *N*-formyl group did not arise from the *N*-ethyl group during the oxidation; therefore, some of the conflicting data can be explained on the basis of the mesaconitine impurity in aconitine.

C. Jesaconitine and Mesaconitine

Jesaconitine was first isolated in 1909 (*24*) from a plant native to Hokkaido, *A. fischeri* Reichb. Since then, this alkaloid also has been isolated from different East Asian species, *A. subcuneatum* Nakai (*25, 26*), *A. sachalinense* F. Schmidt (*25, 26*), and *A. mitakense* Nakai, (*27*). Jesaconitine differs from aconitine in being an ester of anisic acid (4-methoxyl benzoate) instead of benzoic acid. The detailed chemical work on jesaconitine has been reviewed earlier (*3*). On the basis of 100 MHz ^{1}H NMR analysis of jesaconitine and its pyrolysis product, the structure **10** was assigned to jesaconitine (*28*). Since the structure of aconitine has been revised on the basis of an X-ray crystallographic study of a degradation product of delphinine, the structure of jesaconitine must be revised to **11** to indicate an α-equatorial methoxyl group at C-1.

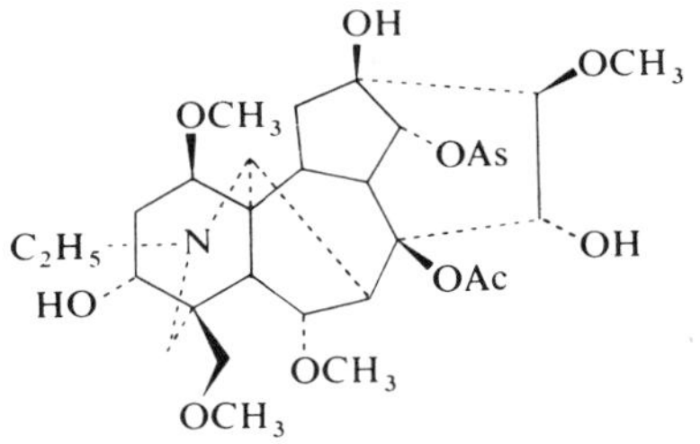

10

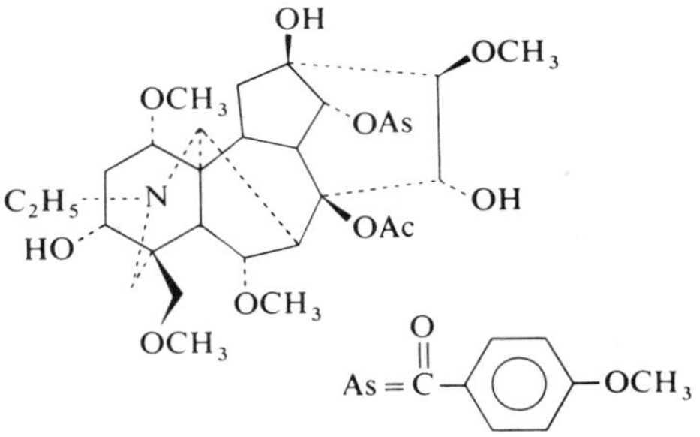

11 Jesaconitine

Mesaconitine was first isolated in 1929 (*29*). This alkaloid occurs in a large number of *Aconitum* species reported in the earlier review (*1*). It has also been isolated from *A. japonicum* (*29*), *A. mitakense* (*27*), *A. altaicum* Steinb. (*30*), and the Chinese drug known as Chuan-Wu (*14*) (*A. carmichalei*). The structure of mesaconitine as **12** was based on the structure of aconitine. Because mesaconitine differs from aconitine only in possessing an *N*-methyl instead of an *N*-ethyl group, the structure of mesaconitine must now be revised to **13**.

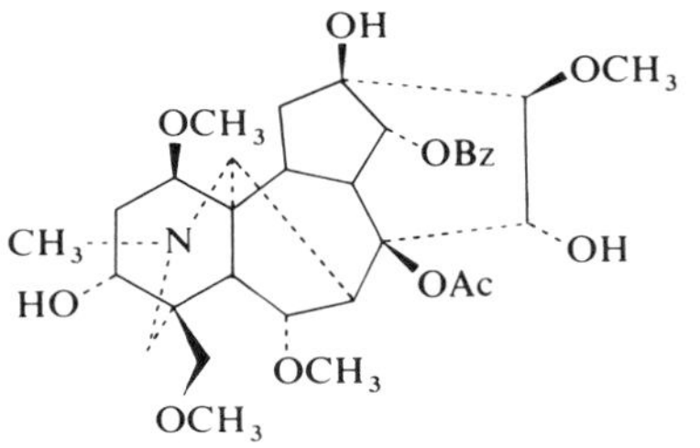

12

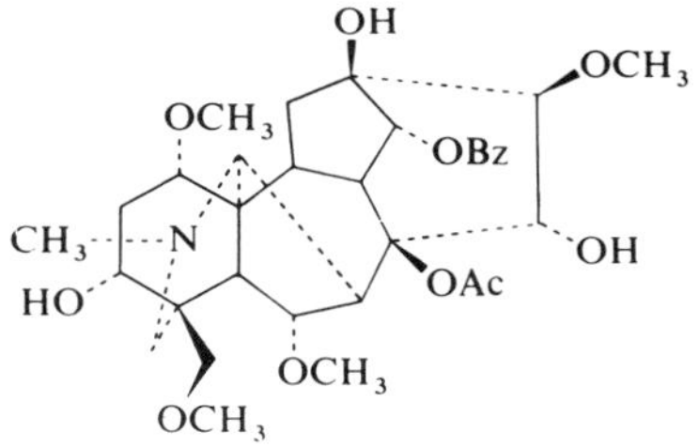

13 Mesaconitine

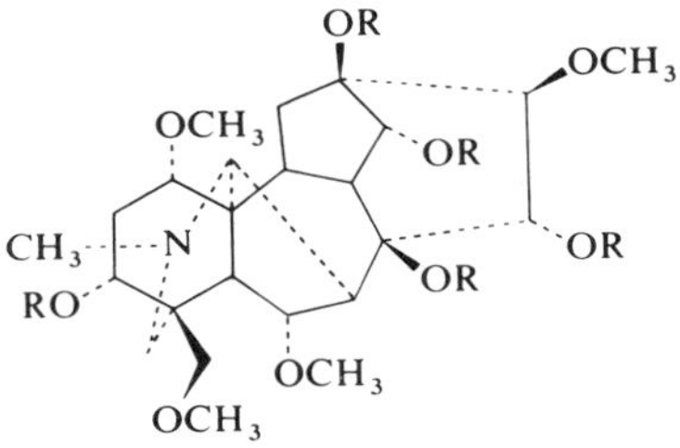

14 R = H
15 R = Ac

Recently, Ichinohe and Yamaguchi have reported the isolation of a new aconitine-type alkaloid from the roots of *A. sachalinense* Fr. Schmidt (*31*).

Chemical and spectral studies indicate the presence of an acetoxyl group, an anisic ester group (4-methoxyl benzoate), an *N*-methyl group, four aliphatic methoxyl groups, and three hydroxyl groups. This alkaloid undergoes the "pyro" reaction which is characteristic of aconitine-type alkaloids. The alkaloid's hydrolysis product has the same functional groups and molecular formula as mesaconine (**14**), but its pentaacetyl derivative was not identical with pentaacetylmesaconine (**15**). Therefore, the Japanese workers suggested that the hydrolysis product is an epimer with respect to a methoxyl or a secondary hydroxyl group of mesaconine.

D. Hypaconitine and Deoxyaconitine

Hypaconitine was isolated for the first time in 1929 (*32*) and frequently occurs with the closely related alkaloids aconitine (**4**) and mesaconitine (**13**). This alkaloid has been isolated from *A. sanyoense* Nakai var. *sanyoense* (*33*), *A. carmichaeli* (*14, 34*), *A. bullatifolium* var. *homotrichum* (the Chinese drug Hye-shang-yi-zhi-hao) (*15*), and *A. koreanum* (the Chinese crude drug Guan-bai-Fu-tsu) (*35*).

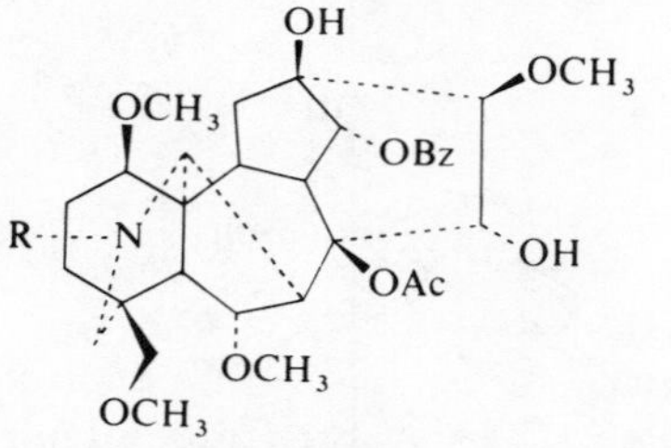

16 R = CH_3
17 R = CH_2—CH_3

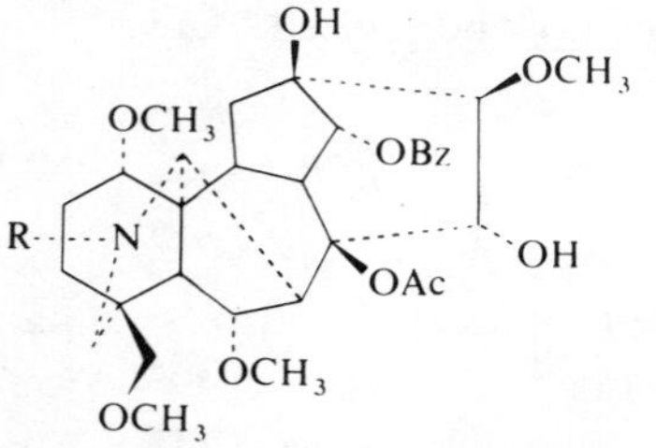

18 R = CH_3 Hypaconitine
19 R = CH_2—CH_3 Deoxyaconitine

Structure **16** was proposed originally for hypaconitine on the basis of the conversion of aconitine into hypaconitine via the intermediate deoxyaconitine (**17**), as described in the last review (*3*). Deoxyaconitine has been reported as a contaminant in crude aconitine (*36*). The structures of hypaconitine and deoxyaconitine were recently revised to **18** and **19** respectively, to indicate an α-equatorial methoxyl group at the C-1 position (*20*).

E. Pseudaconitine, Indaconitine, Veratroylpseudaconine, and Diacetylpseudaconitine

Pseudaconitine was isolated (*37, 38*) for the first time in 1877 from the roots of *A. ferox* Wall, a plant indigenous to the lower reaches of the

Himalayas in India. The roots of this plant are known as *Bish* and were used by the native Indians for medicinal purposes. Later, this alkaloid was isolated from the roots of *A. spictatum* Stapf (*39*). Czechoslovakian (*40*) and Indian (*41*) workers reexamined the alkaloidal constituents of *A. ferox* and have reported the isolation of pseudaconitine along with other known and uncharacterized alkaloids. Recently, pseudaconitine was isolated (*42*) as a minor constituent of *A. falconeri* Stapf. The structure of pseudaconitine (**20**) was assigned earlier and is covered in detail in the last review (*3*) in this series. Because pseudaconitine has been chemically correlated with delphinine, the structure of this alkaloid must now be revised to **21**.

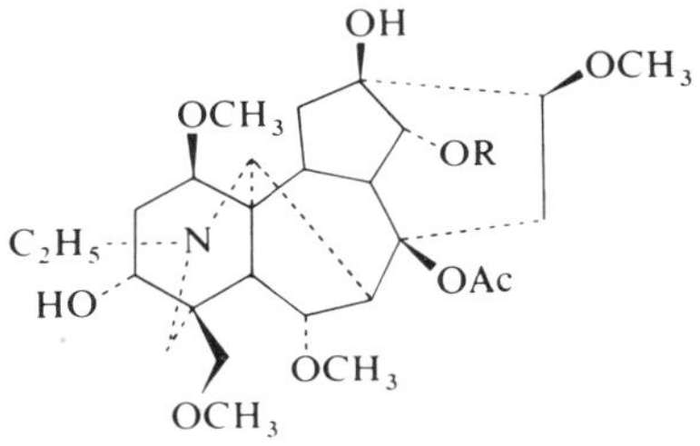

20 R = Vr
22 R = Bz

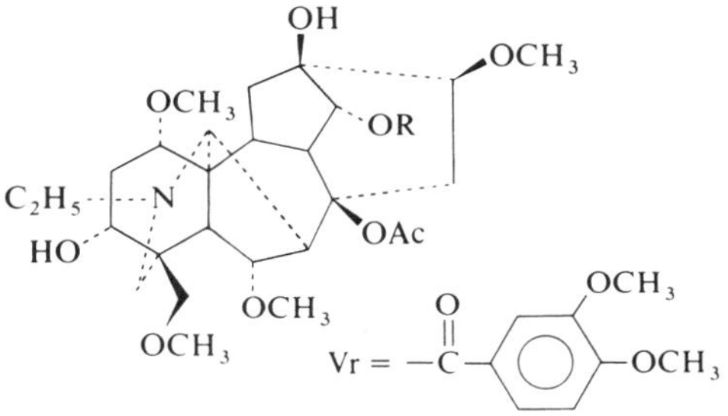

21 R = Vr Pseudaconitine
23 R = Bz Indaconitine

The alkaloid indaconitine was isolated (*43*) from *A. chasmanthum* Stapf for the first time in 1905. Recently, along with pseudaconitine, this alkaloid has been isolated as a minor alkaloid from the roots of *A. ferox* (*40*) and *A. falconeri* (*42*). Structure **22** was originally assigned to indaconitine, on the basis of chemical correlation with delphinine. The structure of delphinine has been revised; thus, indaconitine must be represented by structure **23** to indicate the α-configuration at the C-1 position.

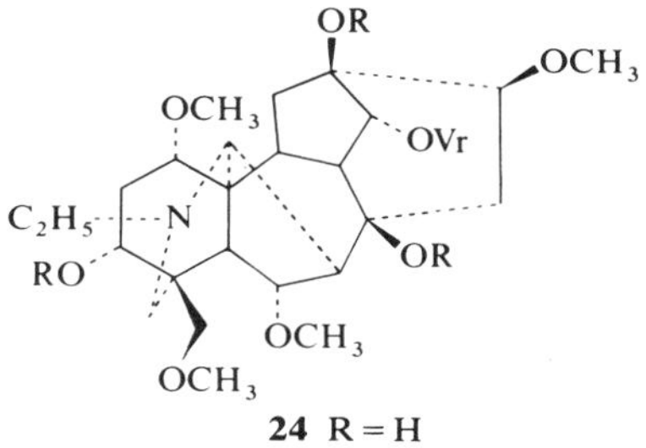

24 R = H
25 R = Ac

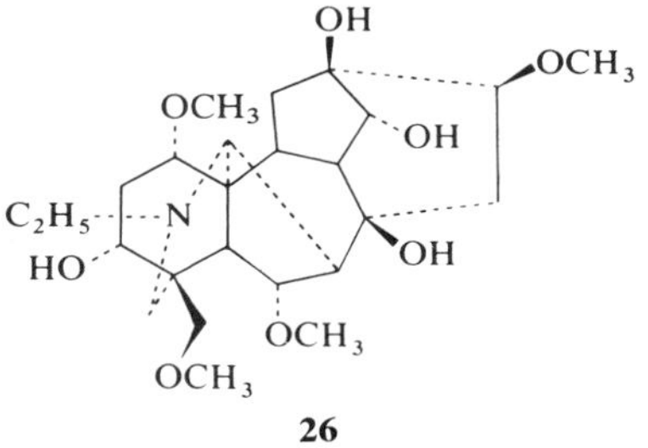

26

During a reexamination of alkaloidal constituents of the roots of *A. ferox*, Indian workers (*41*) isolated two new alkaloids, veratroylpseudaconine (**24**) and diacetylpseudaconitine (**25**) as minor components.

Recently, veratroylpseudaconine (**24**) was isolated by a combination of pH gradient, thin layer, and column chromatographic techniques as one of the major alkaloids of the methanolic extracts of the roots of *A. falconeri* (*42*). The basic hydrolysis of **24** yielded veratric acid and the crystalline parent amino alcohol, pseudaconine (**26**). Comparison of alkaloid **24** with an authentic sample of veratroylpseudaconine, prepared by heating pseudaconitine (**21**) with 0.1 *N* H_2SO_4 in a sealed tube, showed identity. The structure of veratroylpseudaconine as **24** was also established independently by a ^{13}C NMR analysis (*44*). Alkaline hydrolysis of **25** yielded acetic acid, veratric acid, and pseudaconine (**26**). Finally, structure **25** was established for diacetylpseudaconitine by comparison with an authentic sample, which was prepared by acetylation of **21** with acetic anhydride and *p*-toluenesulfonic acid.

F. Falaconitine and Mithaconitine

In 1966 Singh and his co-workers (*45*) reported the isolation and preliminary study of two new diterpenoid alkaloids, designated as bishaconitine and bishatisine from the indigenous crude drug known as *Bish, Bikh*, or *Mitha telia*. This drug was identified as the roots of *A. falconeri*. Recently, we reported (*46*) the isolation and structure determination of two alkaloids, falaconitine (**27**) and mithaconitine (**28**), from the methanolic extracts of the roots of *A. falconeri*. We found that Singh's data for "bishaconitine" are consistent with it being a mixture of several closely related alkaloids, with the principal constituent being falaconitine. Neither any atisine-type alkaloids nor the "bishatisine" reported earlier by Singh, Bajwa, and Singh (*45*) was encountered.

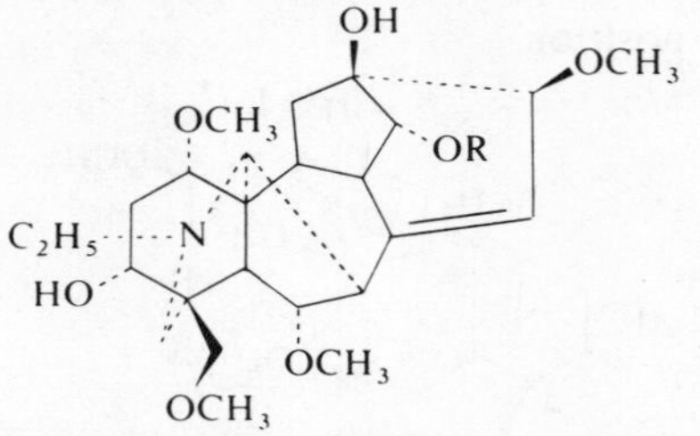

27 R = Vr Falaconitine
28 R = Bz Mithaconitine

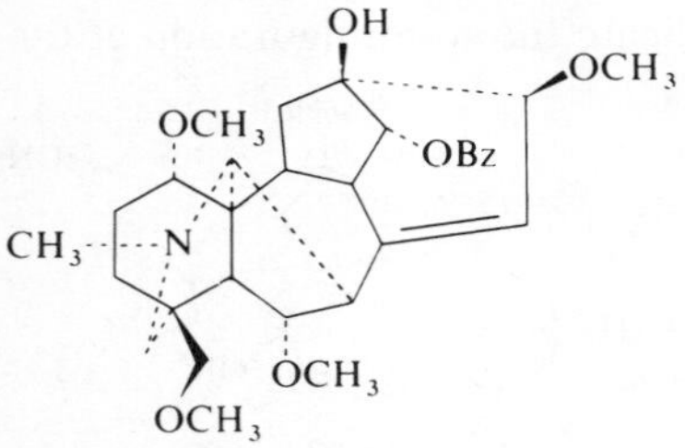

29 Pyrodelphinine

Falaconitine and mithaconitine as the major and minor alkaloids, respectively, were isolated as noncrystalline compounds by a combination of pH gradient, thin layer, and column chromatographic techniques. The structures of falaconitine (**27**) and mithaconitine (**28**) were established on the basis of ^{1}H and ^{13}C NMR analysis. The ^{1}H NMR spectrum of falaconitine was identical with mithaconitine except for the presence of the veratroyl group instead of a benzoyl group. Comparison of these alkaloids to pseudaconitine (**21**), indaconitine (**23**), and pyrodelphinine (**29**) was made through a study of their ^{13}C NMR spectra. Comparison of the ^{13}C chemical shifts of C-8 (singlet at 146.6 and 146.5 ppm) and C-15 (doublet at 116.1 and 116.4 ppm) in falaconitine and mithaconitine, respectively, with those of pyrodelphinine (**29**) (C-8 singlet at 146.6 ppm and C-15 doublet at 116.3 ppm) afforded evidence for the presence of a double bond between C-8 and C-15 in these new alkaloids. Further, the ^{13}C NMR data indicated that these new alkaloids were similar to pseudaconitine (**21**) and indaconitine (**23**) except for the presence of the double bond between the C-8 and C-15 positions. Finally, the structures of falaconitine and mithaconitine as **27** and **28** were confirmed by comparison with the pyrolysis products of alkaloids **21** and **23**, respectively.

Biogenetically, it is interesting to note that falaconitine and mithaconitine are the first naturally occurring alkaloids with a pyrodelphinine-type structure. These alkaloids may be biogenetic precursors for pseudaconitine (**21**), veratroylpseudaconine (**24**), and indaconitine (**23**), as well as other C_{19}-diterpenoid alkaloids containing a C-8 acetate and C-15 hydroxyl group, e.g., aconitine and hypaconitine.

G. Delphinine

Early in 1800 the seeds of *Delphinium staphisagria* L., commonly known as stavesacre, were extracted to furnish the oil and alkaloids in pure form in order to determine their insecticidal activity. The seeds of *D. staphisagria* have been found on extraction with ligroin to yield an appreciable amount of an alkaloid fraction, which consists mainly of the alkaloid delphinine. The latter was first isolated in 1819 by Lassaigne and Feneulle (*47*). Walz (*48*) revised the formula reported for delphinine by earlier workers to $C_{34}H_{47}NO_9$ on the basis of the analysis of the free base and its oxalate. This formula was also supported by Keller (*49*) and later by Markwood (*50*).

In 1939 Jacobs and Craig (*51*) modified the molecular formula of delphinine to $C_{33}H_{45}NO_9$ and assigned the nine oxygen atoms to four methoxyl groups, one hydroxyl group, one acetyl group, and a benzoyl group. In addition to this, they reported the presence of an *N*-methyl group in delphinine. Since 1939, delphinine became the subject of intensive chemical studies (*52–60*), and structure **30** with a β-methoxyl group at the C-1 position was assigned (*61*) to delphinine on the basis of a direct structural correlation with aconitine and what appeared to be a sound conformational argument (*19*). The detailed chemistry of delphinine

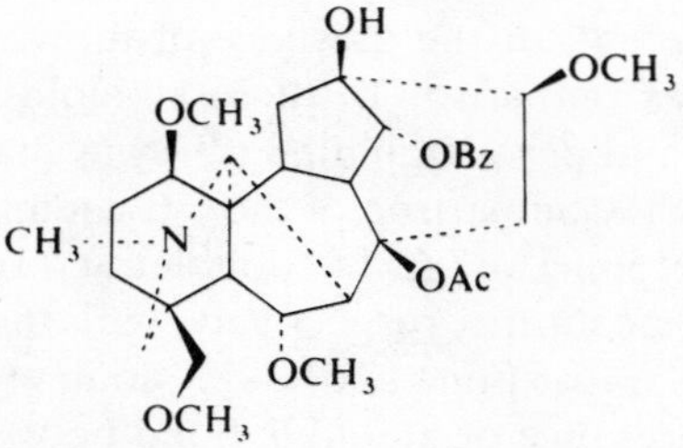

30

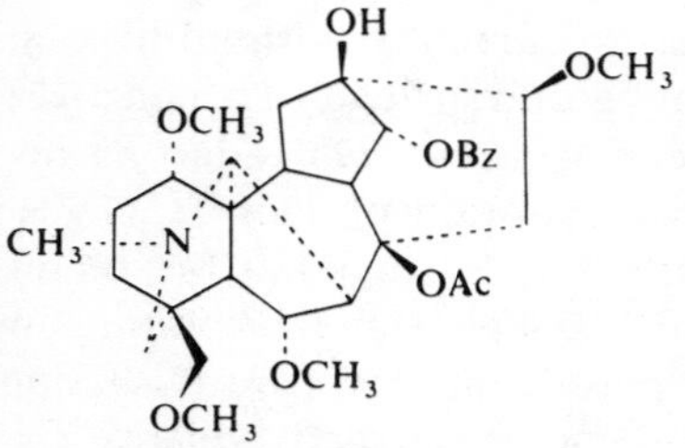

31 Delphinine

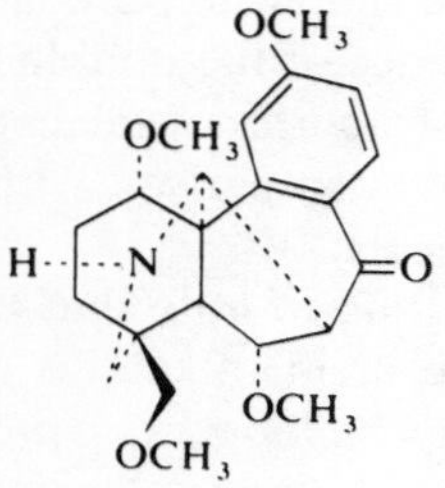

3

is covered in the last review article (*3*) in this series. Recently, Wiesner's research group at New Brunswick has assigned (*20*) a revised structure (**31**) for delphinine, on the basis of the X-ray crystallographic study of acid oxalate of the synthetic degradation product (**3**) of delphinine. (See Section VII,A.)

H. The Pyrodelphonine Chromophore

The pyrolysis product of delphinine (**31**), pyrodelphinine (**29**), on alkaline hydrolysis gives a parent pyroamino alcohol known as pyrodelphonine (**32**). In 1952 Edwards and Marion (*62*) observed the apparent diene chromophore in the "pyro" compounds derived from the aconitine-type alkaloids with no hydroxyl group at the C-15 position. Wiesner and co-workers (*63*) reported an unexpected UV absorption (λ_{max} 245 nm, ϵ_{max} 6300) for pyrodelphonine (**32**) and pyroneoline (**33**), which disappears upon acidification. To explain the unusual behavior of these pyro compounds, they postulated the participation of the lone-pair electrons of the nitrogen with the C-7–C-17 σ bond and the π system of C-8–C-15 double bond in an excited state resembling structure **32A** in pyrodelphonine and **33A** in pyroneoline. Later, the same phenomenon was also explained using valence bond language by Cookson and his co-workers (*64*).

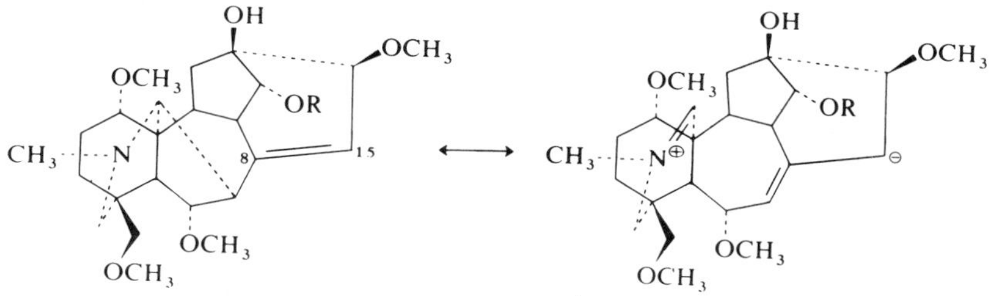

29 R = Bz Pyrodelphinine
32 R = H Pyrodelphonine

29A R = Bz
32A R = H

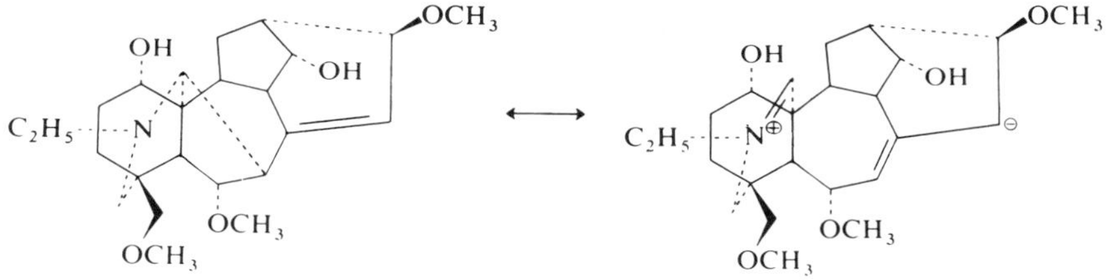

33 Pyroneoline

33A

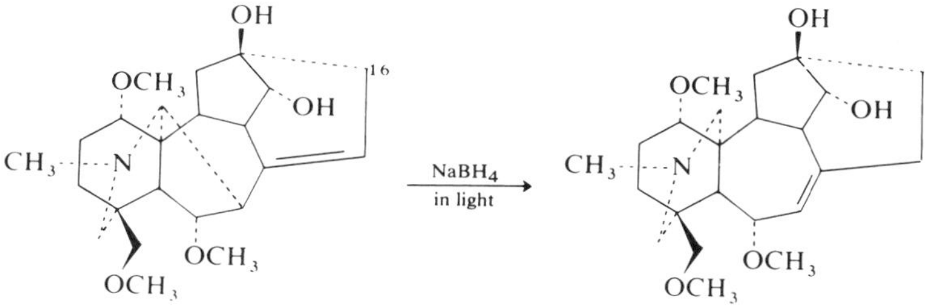

$NaBH_4$ in light

34

35

In 1969 Wiesner and Inaba (*65*) reported an unusual photoreduction of the C-7–C-17 bond in 16-demethoxylpyrodelphonine (**34**) to compound **35** by sodium borohydride in methanol. They also repeated the same reaction using sodium borodeuteride to confirm the postulated nature of the pyrochromophore. The high-resolution mass spectrum of the reaction product indicated that one deuterium was incorporated in the reaction product. This result suggested that one of the new hydrogen atoms in compound **35** came from the reagent and the other from the solvent. These results confirm the postulated nature of the pyro chromophore.

Recently, Pelletier and Djarmati (*66*) examined the pyrodelphinine chromophore in the electronic ground state of the molecule by ^{13}C and

^{1}H NMR spectroscopy. Their results were based on the differences in the ^{13}C and ^{1}H NMR spectra of pyrodelphinine effected by protonation of the nitrogen atom. The ^{13}C and ^{1}H NMR spectra were recorded during the stepwise protonation of the nitrogen atom by deuterated acetic acid, as well as N-oxidation by *m*-chloroperbenzoic acid. A downfield shift of the signal for C-8 (4.1 ppm) and a large upfield shift of the signal for C-15 (5.5 ppm) in compound **29** were observed upon protonation or N-oxidation. On the basis of these results, they postulated that the lone-pair electrons of the nitrogen, the C-7–C-17 σ bond, and the π electrons of the C-8–C-15 double bond are conjugated in the electronic ground state of the molecule. In other words, the structure of pyrodelphinine can be considered a resonance hybrid of the contributing structures **29** and **29A** and that of pyrodelphonine as a resonance hybrid of **32** and **32A**. Delocalization of the free electron pair of the nitrogen will give a negative charge at the C-15 position, and this results in an upfield shift of its resonance which can be observed upon acidification (protonation) or N-oxidation. Similar results were also obtained by observing the change in shift of the C-15 proton in the 100 MHz ^{1}H NMR upon protonation or N-oxidation.

I. Bikhaconitine, Chasmaconitine, and Chasmanthinine

Bikhaconitine was isolated for the first time by Dunstan and Andrews (*67*) in 1905 from the poisonous roots of *A. spictatum*. The roots of this plant are known as *Bikh* or *Bish* by the Indian people. Recently, this alkaloid was also isolated as a minor constituent by Czechoslovakian (*40*) and Indian (*41*) workers during the reexamination of alkaloidal constituents of another Indian plant, *A. ferox*. The structure of bikhaconitine (**36**) was established by the direct comparison with deoxypseudaconitine, which was prepared from pseudaconitine. Because the structure of pseudaconitine has been revised, the structure of bikhaconitine must now be represented by structure **37** indicating a C-1 α-methoxyl group in the A ring.

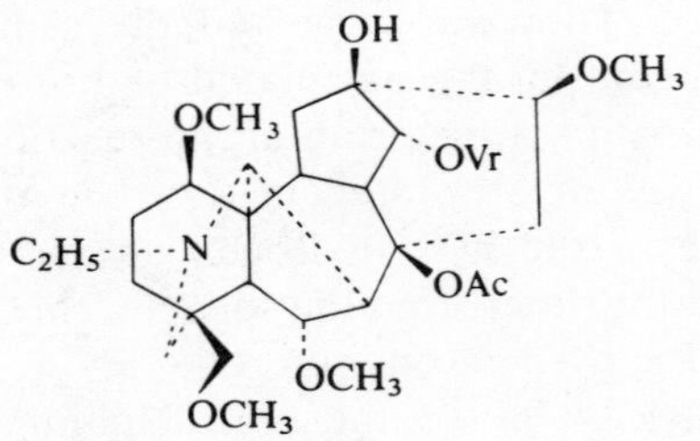

36

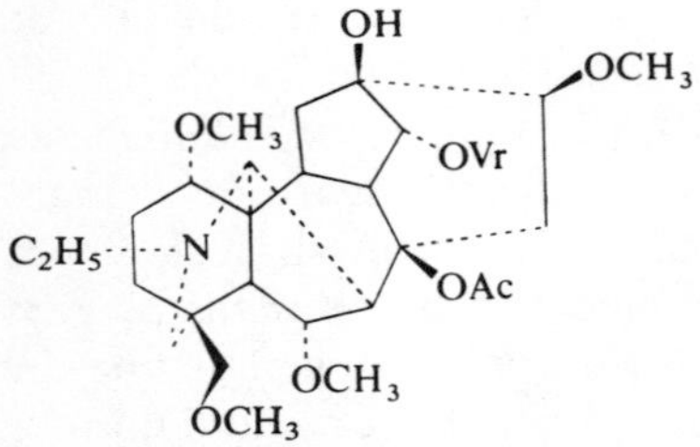

37 Bikhaconitine

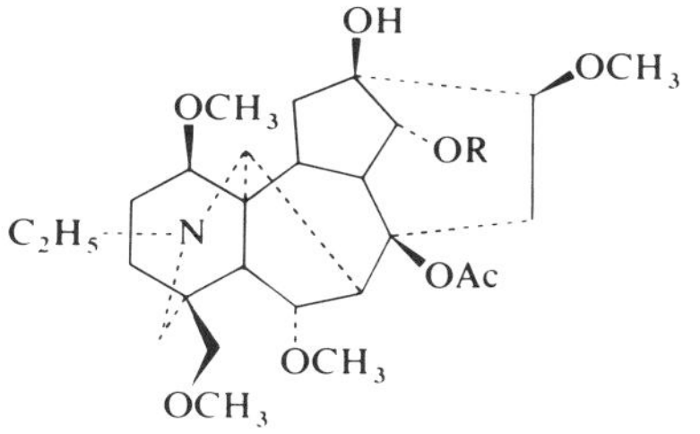

38 R = Bz
39 R = Cn

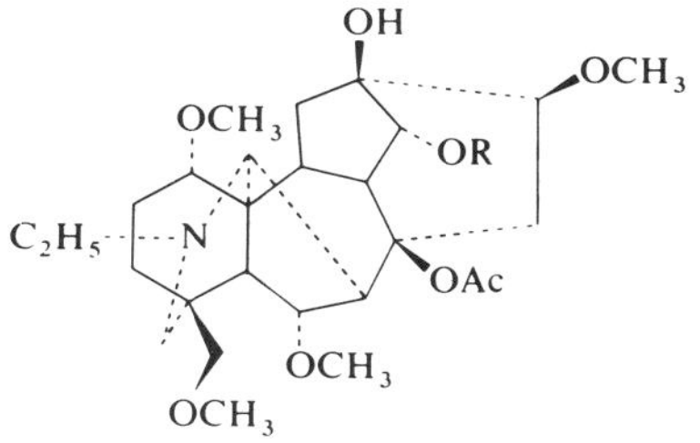

40 R = Bz Chasmaconitine
41 R = Cn Chasmanthinine

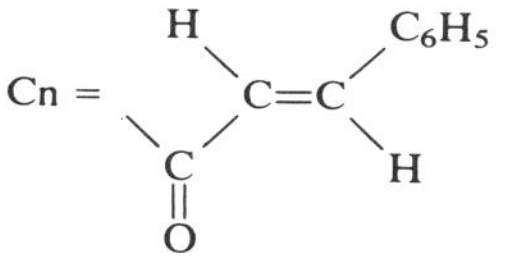

Cn = $-C(=O)-CH=CH-C_6H_5$

Chasmaconitine and chasmanthinine, two closely related alkaloids, were isolated (*68*) during reexamination of the roots of *A. chasmanthum*. The structure and stereochemistry of chasmaconitine (**38**) were assigned on the basis of the conversion of this alkaloid to delphinine. Since the structure of delphinine has been revised, chasmaconitine must be represented by structure **40**. Chasmanthinine was also correlated with chasmaconitine and other alkaloids, and structure **39** was originally assigned to it. With the revision of the structure of chasmaconitine, the structure for chasmanthinine must be revised to **41**. Thus far, neither of these two alkaloids has been isolated from any other plants.

J. Isodelphinine (Base D)

In 1959 Katsui and his co-workers (*69, 70*) isolated a diterpenoid alkaloid, designated initially as *base D* and later as *isodelphinine*, from the mother liquors accumulated during the isolation of miyaconitine and miyaconitinone from the roots of *A. miyabei* Nakai, a plant native to Hokkaido, Japan. On the basis of chemical studies, they reported that base D contains an *N*-methyl, a benzoxyl, an acetoxyl, a hydroxyl, and four methoxyl groups. On treatment with acetyl chloride, base D gave a monoacetate, mp 188–190°. The data indicated that base D has the same empirical formula and functional groups as delphinine (**31**), but differs from delphinine by the mixture-melting point (mmp) and infrared spectra. On the basis of an infrared analysis and chemical studies, Katsui postulated that isodelphinine is a C-14 epimer of delphinine.

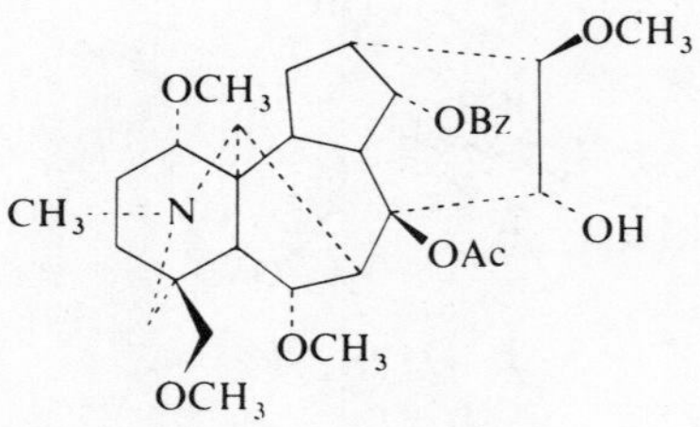

OCH$_3$ OBz OCH$_3$ CH$_3$ N OAc OH OCH$_3$ OCH$_3$

42 Isodelphinine

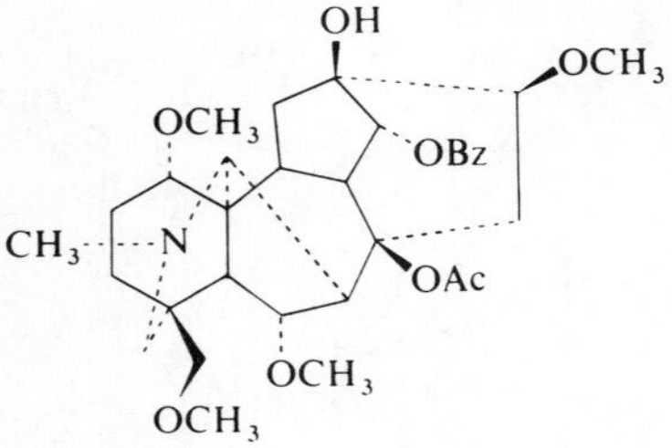

OH OCH$_3$ OCH$_3$ OBz CH$_3$ N OAc OCH$_3$ OCH$_3$

31 Delphinine

Recently, on the basis of ^{1}H and ^{13}C NMR analysis, we have assigned (*71*) structure **42** to isodelphinine. The ^{1}H NMR spectrum of isodelphinine was largely similar to that of delphinine. The presence of an 8-acetate–14-benzoate substitution pattern was established by observing the highly shielded acetate singlet at δ 1.45 in isodelphinine. The pattern of ^{13}C chemical shifts in isodelphinine was similar to that of delphinine, deoxyaconitine, and 8-acetyl-14-benzoylneoline. The characteristic shifts of isodelphinine indicated that the hydroxyl group is present at the C-15 position (78.9 ppm, doublet) and that no group is present at the C-13 position. The presence of hydroxyl at C-15 in isodelphinine was also confirmed by observing the downfield singlets at 92.3 and 172.3 ppm of C-8 and the carbonyl group of the C-8 acetate, respectively. The ^{13}C chemical shifts of all the carbons were in agreement with the assigned structure (**42**) for isodelphinine.

It is interesting to note that isodelphinine is the first naturally occurring example of an alkaloid possessing a C-15 hydroxyl in the absence of the C-13 hydroxyl group. Thus far, all the naturally occurring diterpenoid alkaloids bearing a C-15 hydroxyl have also possessed an hydroxyl at C-13, e.g., aconitine, jesaconitine, mesaconitine, deoxyaconitine, and hypaconitine. Thus, isodelphinine is an interesting subject for biogenetic speculation.

K. Delphisine, Neoline, Chasmanine, and Homochasmanine

Pelletier and co-workers have isolated (*72, 73*) a new alkaloid, designated as delphisine, from the mother liquors accumulated during the isolation of delphinine from the seeds of *D. staphisagria*. An X-ray crystallographic analysis of the hydrochloride salt of delphisine using direct phasing methods established its structure and stereochemistry as **43**. The absolute configuration of delphisine was determined as *1S, 4S, 5R, 6R, 7R, 8S, 9R, 10R, 11S, 13R, 14S, 16S,* and *17R* by Hamilton's method and confirmed by examination of sensitive Friedel pairs.* Ring D of delphisine exists in a boat conformation which is flattened at C-15. Ring A of delphisine exists in a boat conformation as shown in structure **44**, stabilized by intramolecular N · · · HO hydrogen bonding. The 100 MHz ^{1}H NMR study of delphisine also indicated that ring A retains its boat conformation in deuterochloroform solution at room temperature. The infrared spectrum of delphisine showed a broad absorption between 3640 and 3000 cm^{-1} with small peaks at 3600 and 3235 cm^{-1}, indicative of a hydrogen-bonded hydroxyl group and a boat form for ring A. When the C-1 α-hydroxyl group of delphisine was acetylated (**45**), the absence of intramolecular hydrogen bonding and the existence of ring A in a chair conformation as shown in structure **46** was observed.

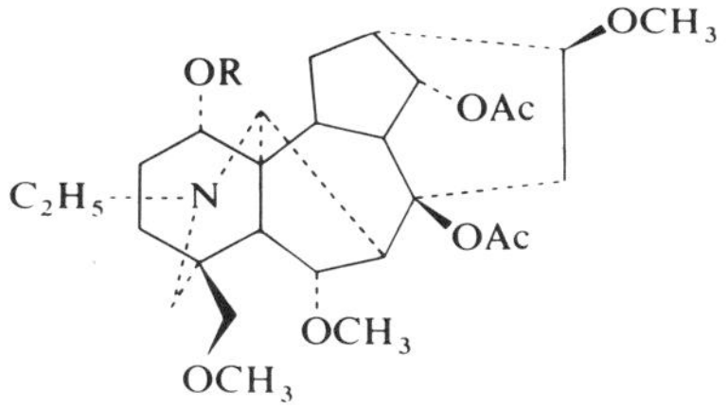

43 R = H Delphisine
45 R = Ac

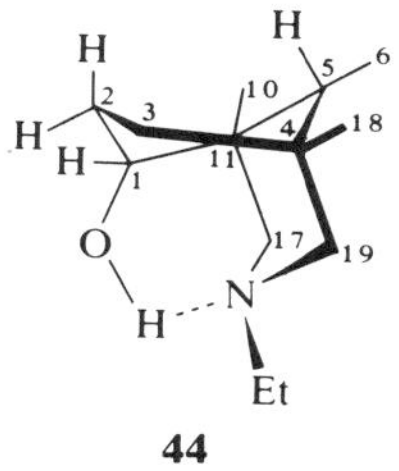

44

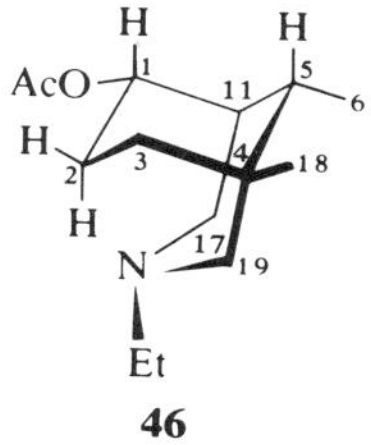

46

Upon alkaline hydrolysis, delphisine yielded the parent amino alcohol ($C_{24}H_{39}NO_6$; mp 160–161°), which was found to be identical with neoline. The latter was isolated as a minor constituent from *A. napellus* by Freudenberg and Rogers (*74*) in 1937. On the basis of chemical studies described in the previous review (*3*) in this series, Wiesner and his

* The absolute configuration at C-8 is incorrectly stated in the original articles for both delphisine (*72, 73*) and chasmanine (*73a*). In both cases, the correct Cahn–Ingold–Prelog notation is *8S* for the molecule, corresponding to our published coordinates and drawings. The absolute configuration at the remaining chiral centers is correctly stated in both cases. We express our appreciation to Prof. E. F. Meyer of Texas A & M University for bringing this error to our attention.

colleagues (*63*) assigned structure **47** to neoline (with the C-1 α-hydroxyl group). However, Marion and co-workers (*75*) correlated neoline with chasmanine, an alkaloid isolated during reexamination of the constituents of the roots of *A. chasmanthum*. Subsequently, they assigned structure **50** to chasmanine (*76, 77*) based on a correlation with browniine. On the basis of the neoline–chasmanine correlation, they assigned structure **48** with a C-1 β-hydroxyl group to neoline.

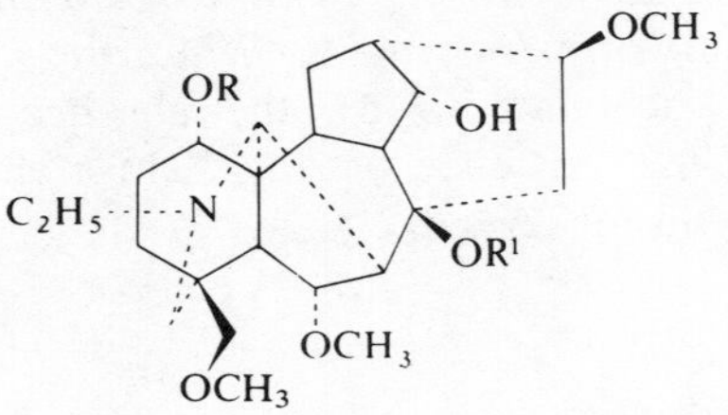

47 $R = R^1 = H$ Neoline
49 $R = CH_3, R^1 = H$ Chasmanine
51 $R = R^1 = CH_3$ Homochasmanine

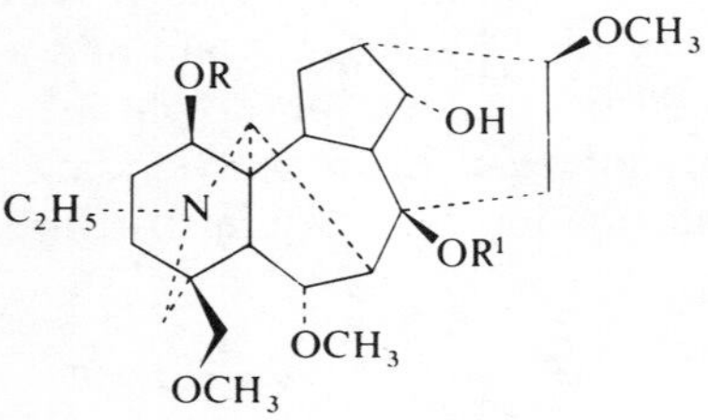

48 $R = R^1 = H$
50 $R = CH_3, R^1 = H$
52 $R = R^1 = CH_3$

The question about the different structures assigned to neoline has recently been resolved (*73, 78*). Delphisine was converted to a pair of C-1 hydroxyl epimers, **47** and **48**, by different routes. Mild alkaline hydrolysis of delphisine gave epimer **47**, which was shown to be identical (mmp, IR, ^1H and ^{13}C NMR) with an authentic sample of natural neoline. This chemical correlation with delphisine, a compound whose structure has been determined by X-ray analysis, proves that Wiesner's original structure for neoline (**47**) is correct and that Marion's revised structure for neoline is in error.

On oxidation with Cornforth reagent, delphisine (**43**) gave 1-ketodelphisine (**53**). On hydrolysis with aqueous methanolic potassium carbonate, compound **53** afforded the 1-keto-8,14-diol derivative (**54**). The latter was also prepared from delphisine by the reverse sequence. Hydrolysis of delphisine with mild base yielded the amino alcohol **47**, which on oxidation gave the 1-monoketo derivative (**54**) as well as the 1,14-diketo derivative (**55**). The latter was reduced selectively to **54** with 1 equivalent of sodium borohydride. The stereospecific reduction of the 14-keto group to the C-14 α-hydroxyl group was expected because of the less hindered β-side of the 14-keto group. On reduction with sodium borohydride in 10% aqueous methanol, compound **54** gave a mixture of epimers **47** and **48** in a ratio of 1:2, respectively. These epimers were separated using preparative silica gel thick-layer chromatographic plates. The less polar epimer (**48**), mp 100°, proved to be not identical with neoline, while the more polar epimer

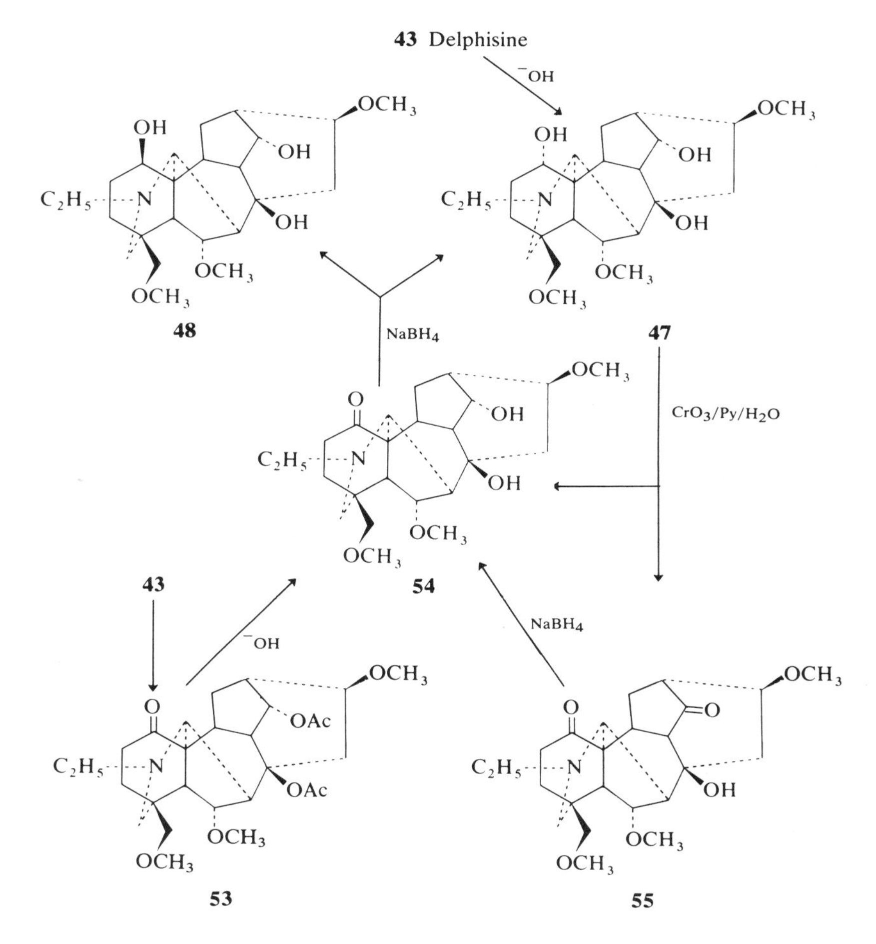

(**47**), mp 160–161°, was shown to be identical with neoline. The hydrolysis product of delphisine was also found to be identical with neoline. The triacetates prepared from epimer **47** by treatment with acetic anhydride and *p*-toluenesulfonic acid at 100° as well as by treating delphisine with acetic anhydride and pyridine at room temperature were found to be identical. On the basis of this chemical correlation, structure **47** has been assigned to neoline (*73, 78*).

Since Marion and colleagues have correlated chasmanine with neoline and homochasmanine with chasmanine, the structures previously assigned for chasmanine (*77*) and homochasmanine (*79*) (**50** and **52**, respectively), must be reconsidered. Chasmanine and neoline were converted to 8,14-di-*O*-methyl chasmanine and 1,8,14-tri-*O*-methylneoline (**56**), respectively,

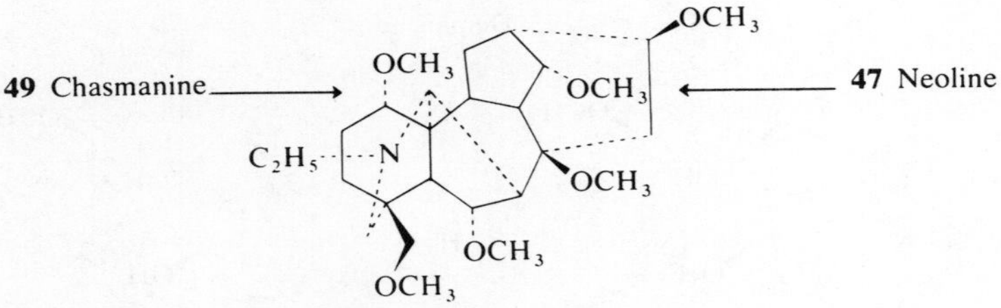

49 Chasmanine → ← **47** Neoline

56

by treatment of each alkaloid with sodium hydride and methyl iodide in refluxing dioxane. The reaction products from neoline and chasmanine were found to be identical. This simple chasmanine–neoline correlation indicated that chasmanine must also have a C-1 α-substituent, as present in neoline, and therefore structure **49** has been assigned to chasmanine (*73, 78*). Because chasmanine diacetate had been converted to homochasmanine by treatment with methanol under pressure, followed by basic hydrolysis, the structure of homochasmanine must be revised to **51**.

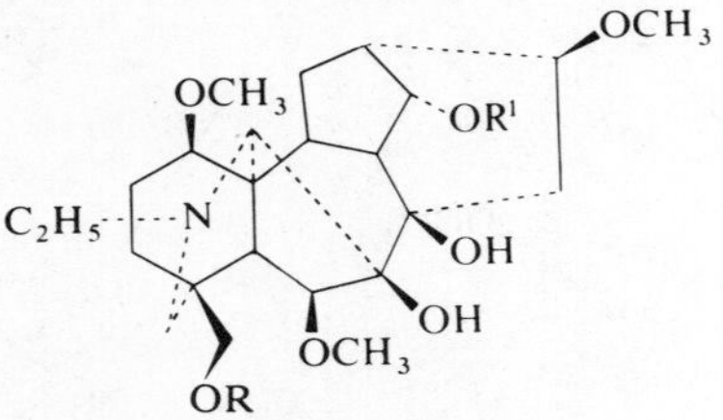

57 $R = CH_3, R^1 = H$ Browniine
58 $R = H, R^1 = CH_3$ Lycoctonine

These revised structures for neoline and chasmanine, (**47** and **49**, respectively) were also supported by a recent ^{13}C NMR study of 26 diterpenoid alkaloids and their derivatives (*66, 80*). Finally, the structure of chasmanine as **49** was confirmed by an X-ray crystallographic analysis (*73a, 81*) of chasmanine 14α-benzoate hydrochloride. With the structure of chasmanine now definitely established, attention was directed to the reported chemical correlation of chasmanine and browniine. This supposed correlation was found to be in error by a ^{13}C NMR study of browniine (**57**), lycoctonine (**58**), and certain of their derivatives. All but two ^{13}C chemical shifts for browniine and lycoctonine, those of C-14 and C-18, are essentially identical. The ^{13}C NMR study indicated identical stereochemistry of ring A in browniine and lycoctonine. This analysis provided evidence that the reaction sequence used in the reported correlation of browniine and chasmanine did not proceed as expected.

L. Delphidine

Recently, a new alkaloid, delphidine ($C_{26}H_{41}NO_7$; mp 98–100°) has been isolated (*82*) by a combination of gradient pH extractions and chromatographic techniques from the mother liquors accumulated during the isolation of delphinine from the seeds of *D. staphisagria*. The structure of delphidine has been established as 8-acetylneoline (**59**) by a chemical correlation and ^{13}C NMR analysis.

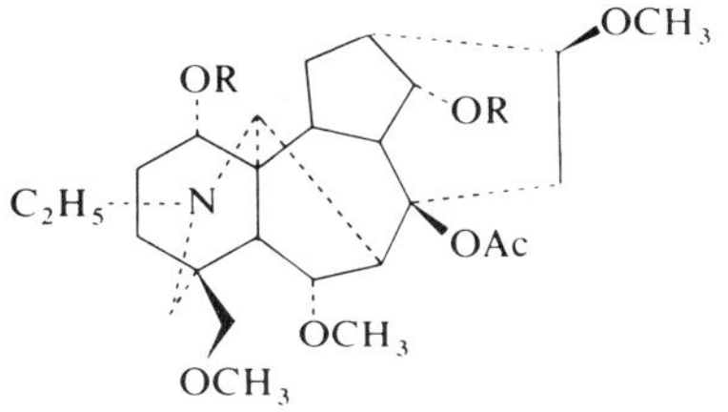

59 R = H Delphidine
60 R = Ac

Alkaline hydrolysis of delphidine (**59**) afforded an alkamine ($C_{24}H_{39}NO_6$; mp 159–161°), which was identical with neoline (**47**) by the usual criteria (mmp, IR, ^{1}H and ^{13}C NMR). Treatment of delphidine with acetic anhydride and pyridine at room temperature yielded neoline triacetate (**60**) (delphisine C-1 α-monoacetate) ($C_{32}H_{45}NO_9$; mp 149–151°). This compound was also obtained by the hydrolysis of delphidine to neoline (**47**), followed by acetylation of the latter with acetic anhydride and *p*-toluenesulfonic acid at 100° for 1 hr. These chemical studies indicated that the acetyl group in delphidine is present at the C-8 position. The ^{13}C NMR spectrum of delphidine was in complete agreement with the assigned structure (**59**).

Delphidine was also obtained by the selective hydrolysis of the 14-acetyl group of delphisine (**43**) over an alumina column (Activity III) as well as by treatment with mild base, e.g., potassium carbonate or potassium bicarbonate in aqueous methanol. The selective hydrolysis of the 14-acetate of delphisine by an alumina column suggested that delphidine might be an artifact formed during chromatographic separation of the alkaloids mixture. Recently, this possibility has been ruled out by isolating delphidine without use of an alumina or silica gel column.

M. Delphirine

In 1976 Pelletier and Bhattacharyya (*83*) reported the isolation of a minor alkaloid designated as delphirine ($C_{24}H_{39}NO_6$; mp 95–100°) from the seeds of *D. staphisagria*. The structure of delphirine was established as 1-epineoline (**48**) by ^{13}C NMR analysis and confirmed by comparison with a synthetic sample prepared from delphisine as shown earlier.

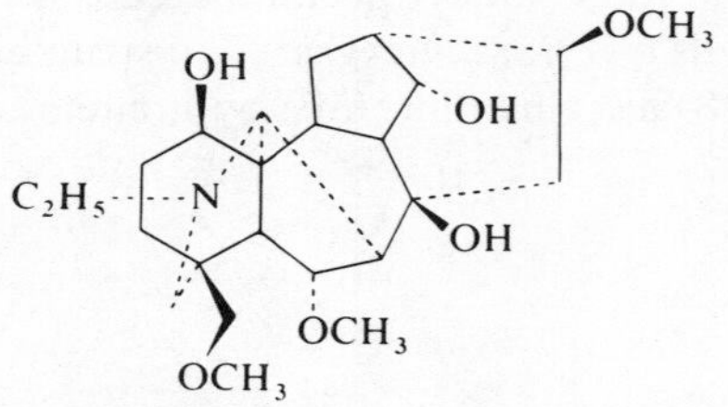

48 Delphirine

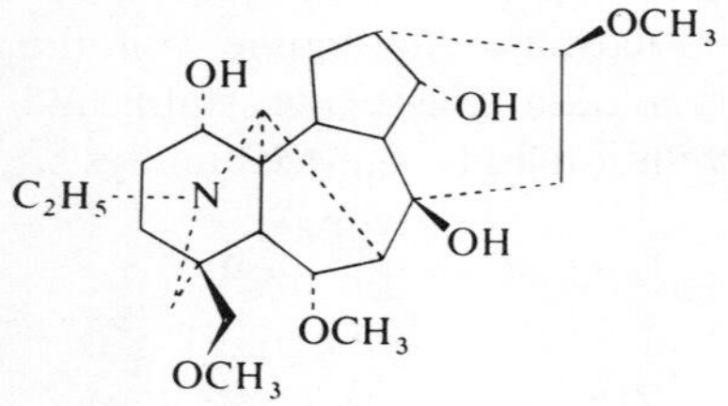

47 Neoline

An interesting feature of delphirine is that it reacts very slowly with Draggendorf's reagent, giving a precipitate after several minutes while neoline reacts instantly with the same reagent. The reduced basicity of the N atom of delphirine (pK_a 6.7) compared with that of neoline (pK_a 7.6) is in agreement with the reactivity of Draggendorf's reagent toward these alkaloids. On the other hand, the pK_a values of these alkaloids are in reverse of the order which would be predicted from hydrogen-bonding arguments; i.e., the C-1 α-hydroxyl group in neoline can form a hydrogen bond with the N atom, while the C-1 β-hydroxyl group in delphirine cannot. The difference in the basicity of delphirine and neoline may be explained by the orientation of the C-1 hydroxyl group and the conformation of ring A.

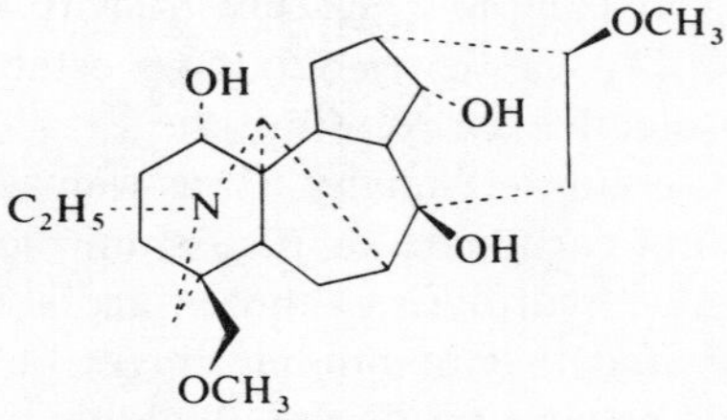

61 Isotalatizidine

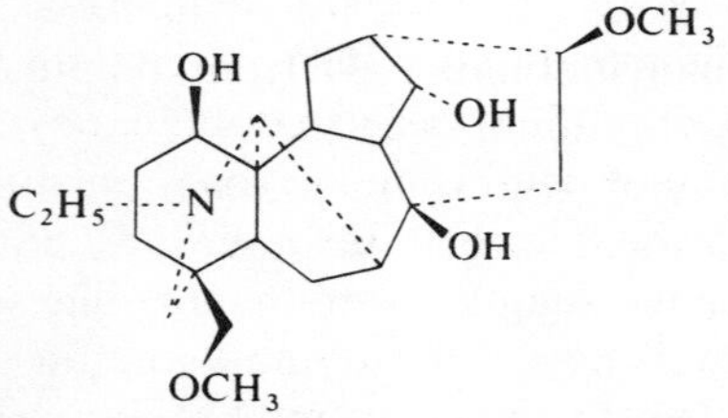

62 Talatizidine

Biogenetically, it is interesting to observe that the delphirine–neoline pair is the second example reported so far of C-1 hydroxyl epimers among the C_{19}-diterpenoid alkaloids. The isotalatizidine (**61**)–talatizidine (**62**) pair (*84, 85*) is the first example in this series.

N. NEOPELLINE

"Neopelline" was reported in 1924 by Schulze and Berger (*86*) as an amorphous alkaloid (mp ~ 80°) which was isolated as an impurity in crude aconitine from *A. napellus*. They reported the preparation of a few salts and assigned the molecular formula as $C_{32}H_{45}NO_8$. Saponification of neopelline gave acetic acid, benzoic acid, and the parent amino alcohol neoline. Freudenberg and Rogers (*74*) in 1937 failed to isolate neopelline from *A. napellus* following the method of Schulze and Berger (*86*) but did isolate neoline from the fraction which was supposed to provide neopelline. Because neoline was isolated from an extract which had been made basic with only 0.1 *N* sodium hydroxide solution, they suggested that neoline was possibly an artifact formed during a hydrolysis of neopelline. Until recently, no further work appears to have been done on neopelline.

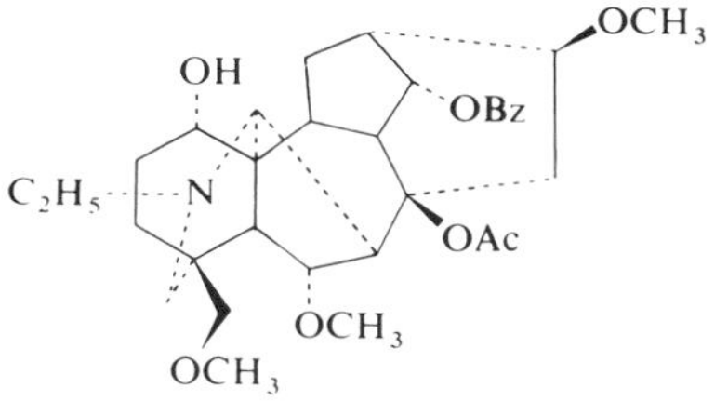

63

On the basis of a new molecular formula for neoline, $C_{24}H_{39}NO_6$, the molecular formula of neopelline has been revised to $C_{33}H_{45}NO_6$. Pelletier and Keith (*3*) recently suggested 8-acetyl-14-benzoylneoline (**63**) as a probable structure for neopelline. A partial synthesis (*87*) of 8-acetyl-14-benzoylneoline (**63**) from delphisine (**43**) was carried out to establish whether neopelline indeed has the suggested structure **63**.

Delphisine (**43**) was oxidized with Cornforth's reagent (CrO_3–Py–H_2O) to yield 1-ketodelphisine (**53**) in 93% yield. Compound **53** was hydrolyzed with 5% methanolic potassium hydroxide solution at room temperature to give 1-ketoneoline (**54**) in 76% yield. The latter compound was converted into the corresponding benzoate (**64**) by treatment with benzoyl chloride and pyridine for 3 hr. Compound **64** was converted to 8-acetyl-14-benzoyl-1-ketoneoline (**65**) by treatment with acetic anhydride and catalytic amounts of *p*-toluenesulfonic acid on a steam bath for 1 hr. Sodium borohydride reduction of compound **65** yielded the desired product, 8-acetyl-14-benzoylneoline (**63**), and its C-1 epimer (**66**). The structures of these compounds were confirmed by their ^{1}H and ^{13}C NMR analysis.

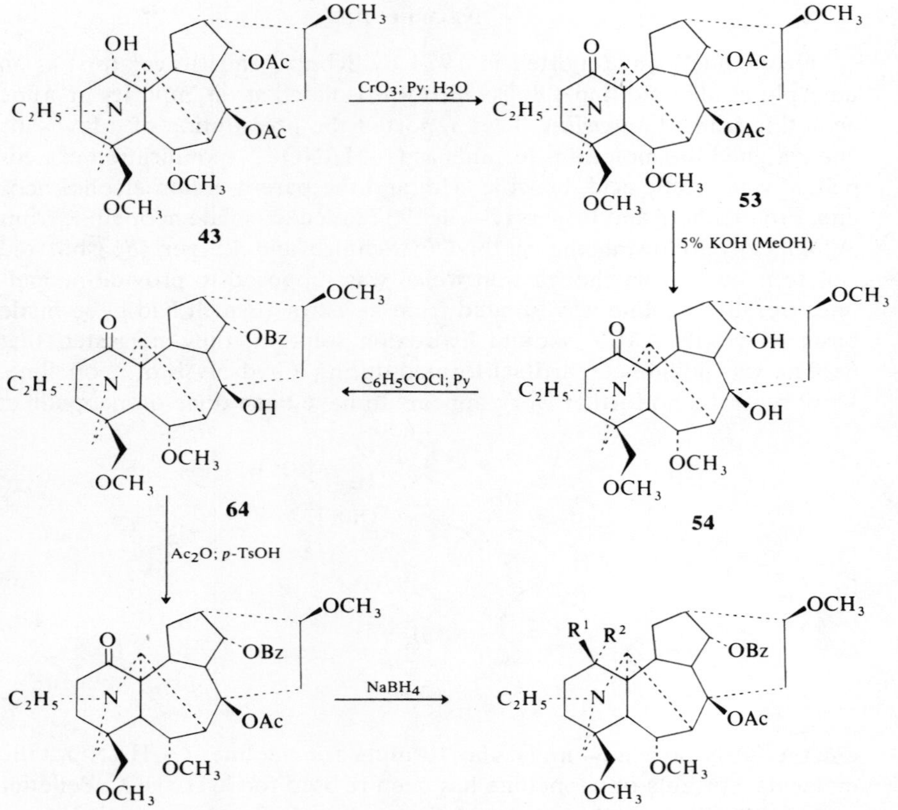

43 → (CrO_3; Py; H_2O) → 53

53 → (5% KOH (MeOH)) → 54

54 → (C_6H_5COCl; Py) → 64

64 → (Ac_2O; *p*-TsOH) → 65

65 → ($NaBH_4$) → 63, 66

63 $R^1 = H; R^2 = OH$
66 $R^1 = OH; R^2 = H$

On the basis of the physical properties of the synthetic 8-acetyl-14-benzoylneoline, the proposed structure (**63**) for neopelline has been ruled out. However, biogenetic considerations suggest the possibility of the existence of 8-acetyl-14-benzoylneoline in nature.

O. Absolute Configuration of Condelphine, Talatizidine, and Isotalatizidine

Condelphine, an alkaloid first isolated from *D. confusum* Popov in 1942 (*88*) and later from *D. denudatum* Wall (*85*), has been assigned structure **67** on the basis of extensive chemical studies and spectroscopic analysis. Condelphine is an *O*-acetate derivative of isotalatizidine (**61**). The latter alkaloid, along with its C-1 hydroxyl epimer, talatizidine (**62**), was isolated in 1940 from *A. talassicum* Popov, (*89, 90*). In 1961, Russian workers (*91*) postulated partial structures for these three alkaloids; subsequently, the structures for condelphine (**67**), isotalatizidine (**61**), and talatizidine (**62**) were established (*85*). This work is summarized in the previous review (*3*) in this series.

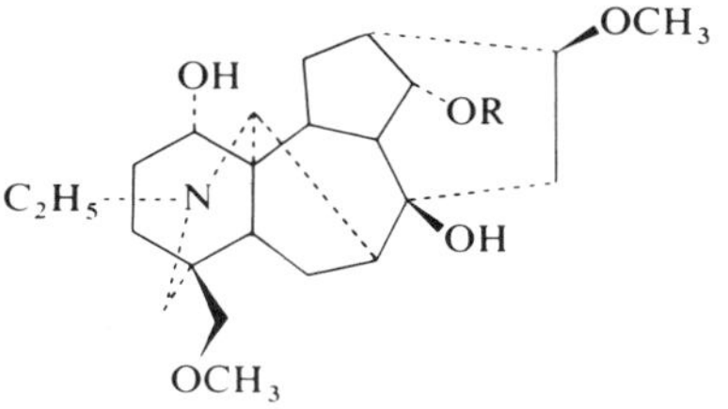

67 R = Ac Condelphine
61 R = H Isotalatizidine

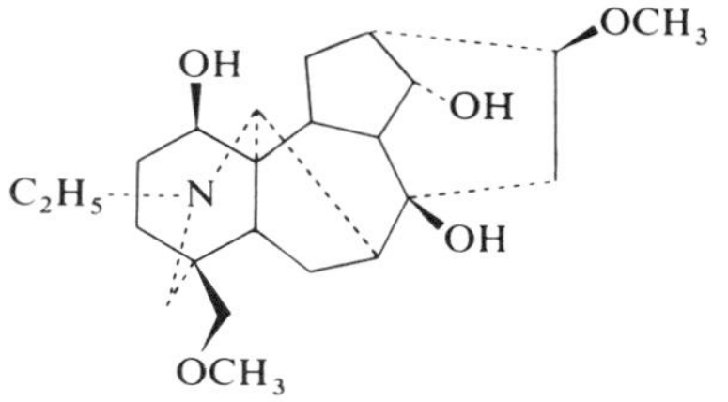

62 Talatizidine

A recent X-ray crystallographic study (*92*) of condelphine hydroiodide confirmed the assigned structure of condelphine as **67**. The absolute configuration was established as *1S, 4S, 5R, 7S, 8S, 9R, 10R, 11S, 13R, 14S, 16S,* and *17R* by examination of Friedel pairs of reflections. The structure has been refined to $R = 0.100$ and $R_w = 0.111$, based on 3828 observed reflections. Because condelphine has been correlated with isotalatizidine (**61**), talatizidine (**62**), talatizamine (**68**), and cammaconine (**69**), this work confirms the molecular structure and establishes the absolute configuration of the latter four alkaloids as well.

P. Talatizamine (Talatisamine)

Talatizamine ($C_{24}H_{39}NO_5$; mp 145–146°) was first isolated in 1940 from the roots of *A. talassicum* (*89, 93*). It has also been isolated from *A. nemorum* Popov (*93*) and *A. carmichaeli* (*14*) and, in 1967, was found in *A. variegatum* (*94*) along with closely related alkaloid, cammaconine (**69**). The structure of talatizamine (**68**) has been established (*95*) by chemical

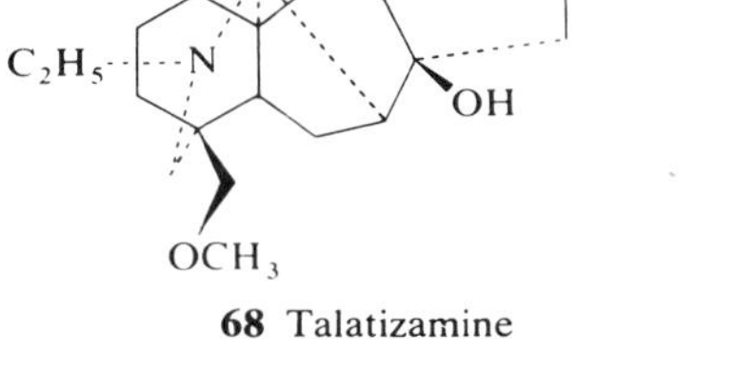

68 Talatizamine

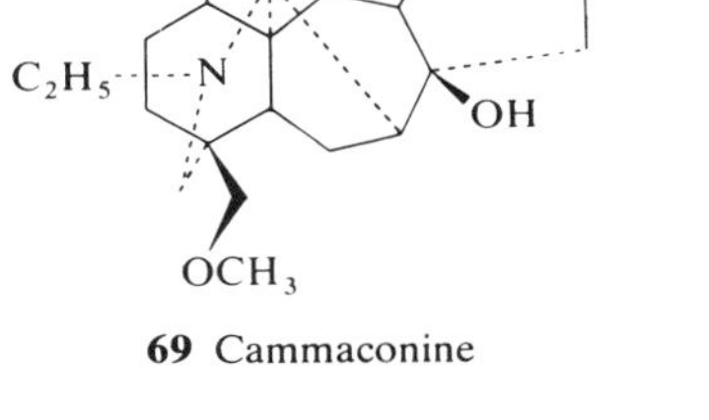

69 Cammaconine

correlation with isotalatizidine (**61**). The structure has been confirmed by a total synthesis of the alkaloid by Wiesner and his co-workers (*96*).

Talatizamine had been shown to contain an *N*-ethyl, one secondary and one tertiary hydroxyl, and three methoxyls on an aconitine-type skeleton. Analysis of the mass spectrum indicated that talatizamine is similar to that of the other aconitine-type alkaloids and that it contains a C-1 methoxyl group (*97*). Pyrolysis of diacetyltalatizamine (**70**) afforded the normal pyrolysis product, pyroacetyltalatizamine (**71**), and the rearranged product, isopyroacetyltalatizamine (**72**). Formation of these types of compounds during pyrolysis is characteristic of the aconitine-type alkaloids with a C-8 hydroxyl *or* acetate group and a C-16 methoxyl group. Reduction of the pyroderivative with lithium aluminum hydride confirmed the presence of the methoxyl group at the C-16 position in talatizamine, and on the basis of ^{1}H NMR analysis, a β-configuration was assigned to the

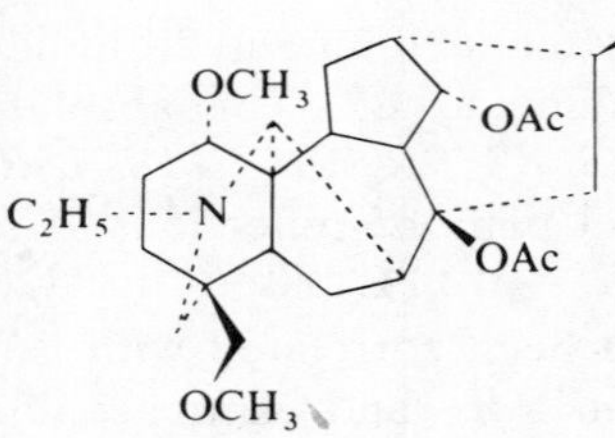

70

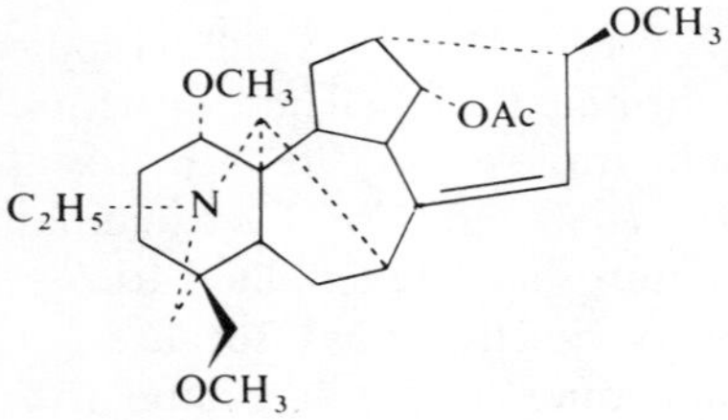

71

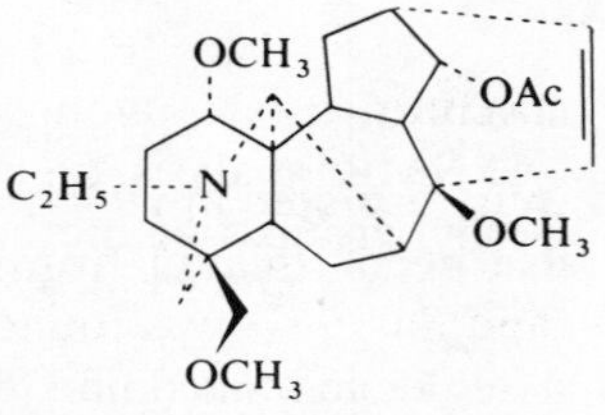

72

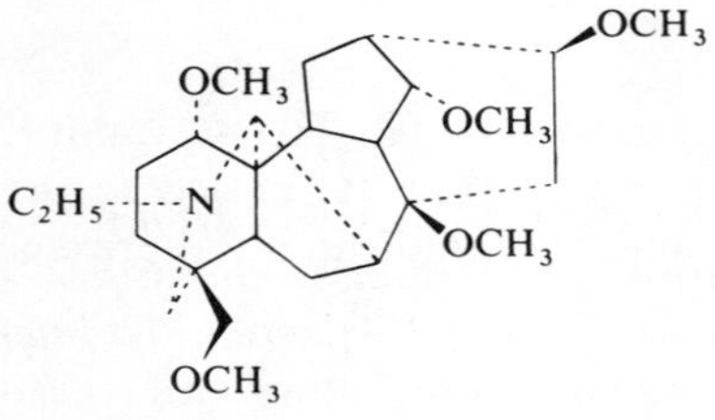

73

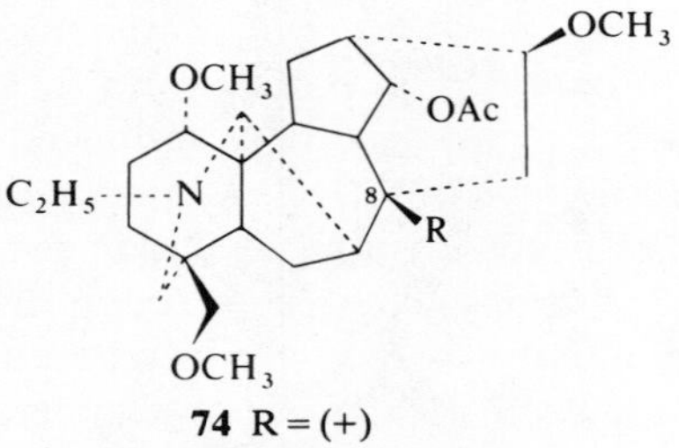

74 R = (+)
75 R = H

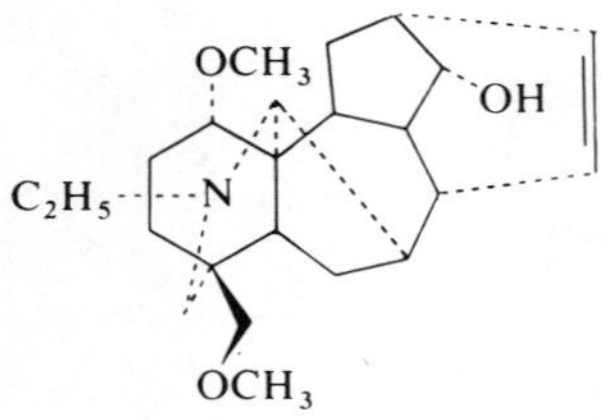

76

C-16 methoxyl group. Further ^{1}H NMR studies indicated the presence of a C-14 α-hydroxyl group and a C-18 methoxyl group.

Upon treatment with methyl iodide and sodium hydride, talatizamine yielded 1,8,14-tri-*O*-methylisotalatizidine (**73**) (*95*). Under the same reaction conditions, isotalatizidine (**61**) afforded two products. One of these was identified as 1,8,14-tri-*O*-methylisotalatizidine (**73**) and was found to be identical with di-*O*-methyltalatizamine. As the complete structure of isotalatizidine is known, this chemical correlation with isotalatizidine established the structure of talatizamine as **68**.

Pyrolysis of diacetyltalatizamine (**70**) in glycerine afforded a compound differing from the starting material by the loss of acetic acid, an acetyl group, and methanol (*98*). The ^{1}H NMR spectrum showed that the compound contains two methoxyl singlets at $\delta 3.20$ and $\delta 3.24$, and N–CH_2–CH_3 triplet at $\delta 0.97$, and no acetate group. The presence of one hydroxyl group detected by formation of a monoacetate and the position of this hydroxyl was fixed at C-14 by ^{1}H NMR analysis. On the basis of the above data, structure **76** was assigned to the pyrolysis product. Soviet workers have speculated that the final pyrolysis product may be formed by an ionic hydrogenation process. The generation of the tertiary carbonium at C-8 (**74**) is succeeded by the abstraction of hydride from the solvent (glycerine) to form compound **75**, which is then transformed to the final product **76** (*98*) by the loss of methanol and hydrolysis of the C-14 acetate group.

Q. Cammaconine

Cammaconine ($C_{23}H_{37}NO_5$; mp 135–137°), has been isolated along with talatizamine from *A. variegatum* (*94*). The ^{1}H NMR analysis of cammaconine (*95*) indicated the presence of an *N*-ethyl group, two methoxyl groups, and three hydroxyl groups. On refluxing with methyl iodide and sodium hydride, cammaconine afforded 1,8,14-tri-*O*-methylisotalatizidine (**73**). This observation confirmed that cammaconine has the same skeleton, oxygen substitution pattern, and stereochemistry as isotalatizidine, but only differs in the O-methylation pattern. The structure of cammaconine (**69**) was established on the basis of its oxidation products.

On oxidation with Sarett reagent, cammaconine produced two compounds, a monoketooxocammaconine (**77**) and a diketooxocammaconine (**78**). The IR and ^{1}H NMR spectra of both oxidation products revealed the presence of an *N*-ethyl group, two methoxyl groups, a hydroxyl group, a cyclopentanone moiety, and a six-membered ring lactam. The second product (**78**) also showed IR absorption characteristic of a six-membered ring ketone. There was no evidence for the presence of

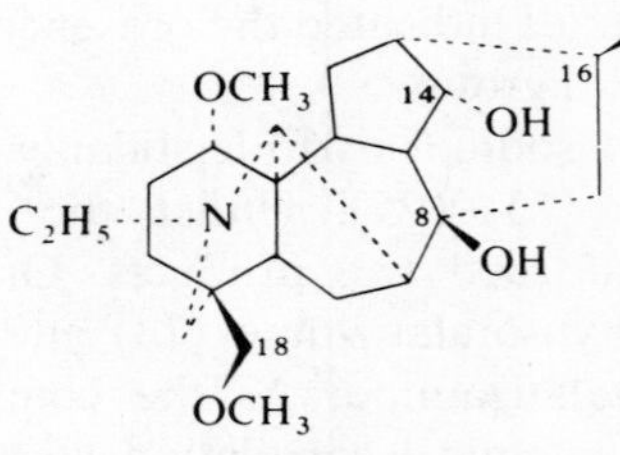

69 Cammaconine

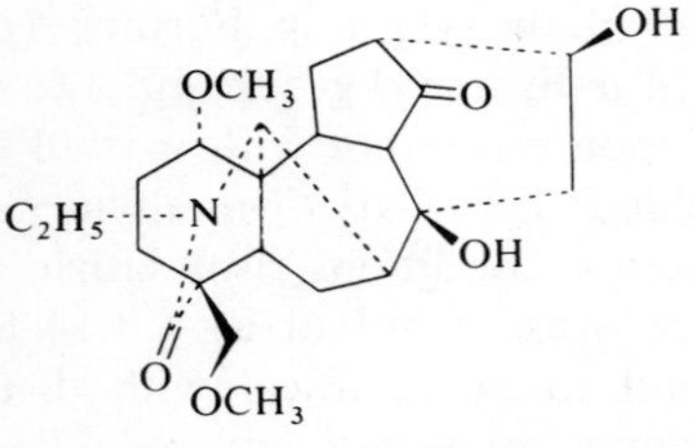

77

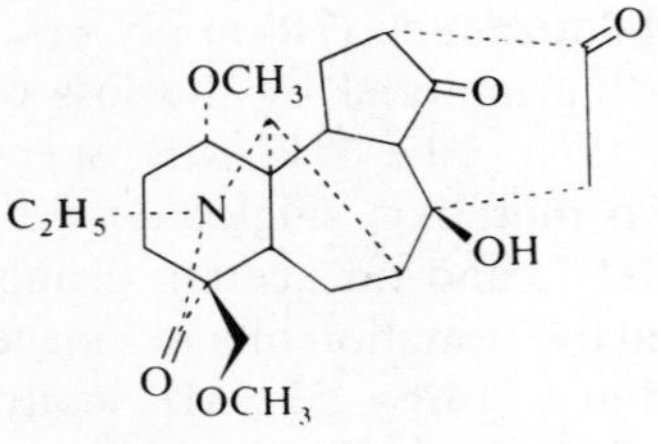

78

an aldehyde or an acidic proton in the ^{1}H NMR spectra of either oxidation product. This observation indicated that a hydroxyl group is not present at the C-18 position in cammaconine. The chemical correlation of cammaconine with isotalatizidine demonstrated the presence of a tertiary hydroxyl group at C-8 and a secondary hydroxyl group at C-14 in cammaconine. Thus, because it had been established that the hydroxyl group is not present at C-18, and to avoid identity with isotalatizidine, cammaconine must be represented by structure **69** (*95*).

It is interesting to note that cammaconine is the first example of an aconitine-type alkaloid with a β-hydroxyl group at the C-16 position instead of the usual β-methoxyl group.

R. Aconorine

In 1975 Soviet workers reported (*99*) the isolation and structure elucidation of a new amorphous alkaloid designated as aconorine ($C_{32}H_{44}N_2O_7$) from the roots of *A. orientale* Mill. It formed a perchlorate salt, mp 237°. On the basis of chemical studies and spectral analysis, they assigned (*99*) structure **79** to aconorine. This paper reported (*99*) that an alkaloid, columbianine, having the same structure as aconorine, had been isolated from *A. columbianum* by the Edwards research group at the National Research Council of Canada.

Upon basic hydrolysis, aconorine afforded the parent amino alcohol (**80**) and an *N*-acetylanthranilic acid. On treatment with acetyl chloride,

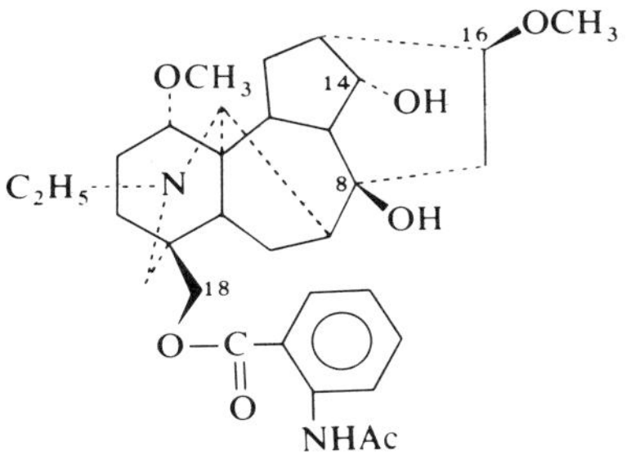

79 Aconorine

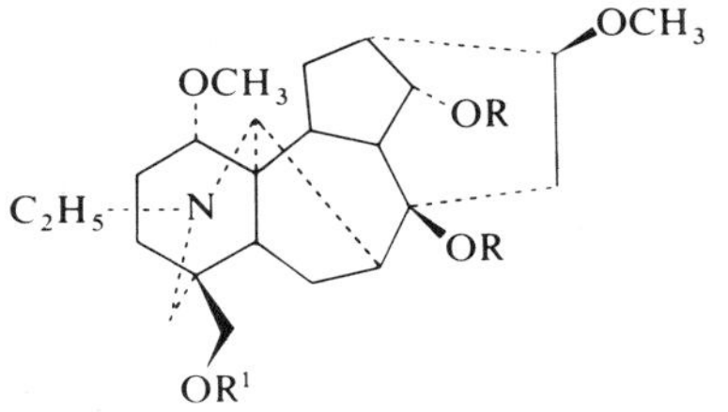

80 $R = R^1 = H$
81 $R = Ac$; $R^1 = -C(=O)-C_6H_4-NH-C(=O)-CH_3$

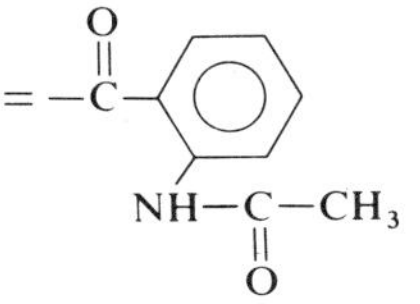

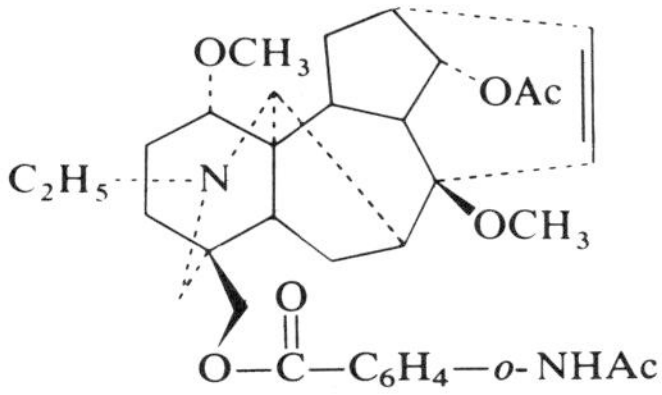

82

aconorine formed a diacetate derivative (**81**), which on pyrolysis yielded the rearranged pyrolysis product, isopyroacetylaconorine (**82**). This pyrolysis reaction is analogous to the pyrolytic rearrangement of talatizamine and other C_{19}-diterpenoid alkaloids bearing C-8 hydroxyl (or acetate) and C-16 methoxyl groups. On the basis of the pyrolysis product, the positions of hydroxyl and methoxyl groups were assigned to C-8 and C-16, respectively. The ^{1}H NMR analysis of aconorine (**79**) and its hydrolysis product (**80**) established the presence of the second hydroxyl group at C-14 and the point of attachment of the anthranilic acid derivative at the C-18 position.

S. Alkaloids A and B

The Canadian workers, Jones and Benn, reported the isolation of two new diterpenoid alkaloids A and B in very small amounts from *D. bicolor* Nutt (*100, 101*). Structures **83** and **84** were assigned to alkaloids A and B, respectively, primarily on the basis of the correlation with the ^{13}C NMR spectra of a variety of model C_{19}-diterpenoid alkaloids, e.g., deoxylycoctonine and isotalatizidine, and the results obtained from the pyrolysis study of alkaloid A. Alkaloid A ($C_{25}H_{39}NO_6$), an amorphous base, formed a hydroiodide salt (mp $>$ 240°). The IR and ^{1}H and ^{13}C NMR spectra revealed the presence of a tertiary methyl, an N-ethyl, an acetoxyl, two methoxyls, and two hydroxyl groups in alkaloid A. The presence of an acetoxyl group at the C-8 position was demonstrated by a pyrolysis study of alkaloid A in glycerol. The formation of acetic acid during pyrolysis was monitored by the ^{1}H NMR spectrum according to the published procedure (*28*).

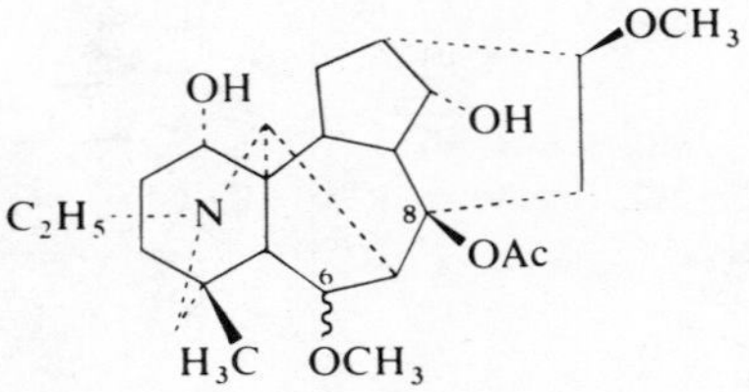

83

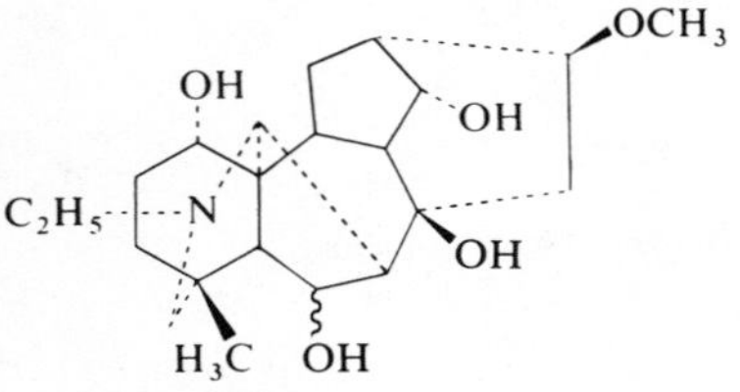

84 Alkaloid B

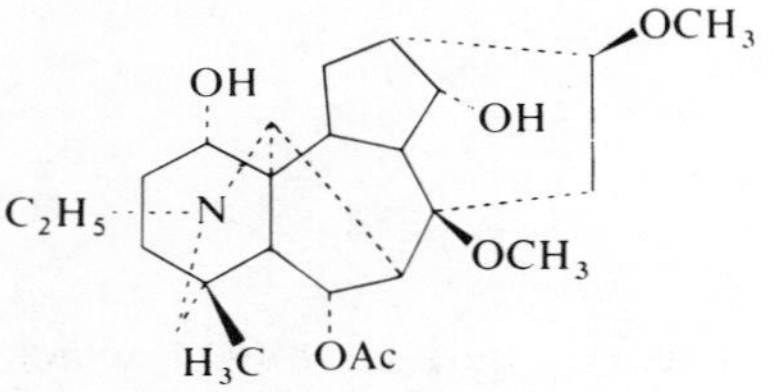

85 Alkaloid A

Recently, the structure of alkaloid A has been revised (*102*) from **83** to **85**, mainly on the basis of ^{13}C NMR spectral correlation of alkaloids A and B with those of closely related neoline-type alkaloids. The reinvestigation of the structure of alkaloid A by the ^{13}C NMR analysis was carried out because the position of acetoxyl group at C-8 in alkaloid A was rendered doubtful by the data from the pyrolysis study. The ^{13}C NMR analysis reveals that an α-acetyl group is present at the C-6 position instead of at C-8 and that a methoxyl group is present at the C-8 position. We have also revised the ^{13}C chemical shift assignments of alkaloid A (**85**).

Alkaloid B ($C_{22}H_{35}NO_3$; mp 190–191°) was also isolated in small amounts from *D. bicolor* (*100, 101*). It was shown to contain an *N*-ethyl, a tertiary methyl, a methoxyl, and four hydroxyl groups. Structure **84** was assigned to alkaloid B on the basis of comparisons of the ^{13}C chemical shifts for alkaloids A and B and their hydrochloride salts with those of deoxylycoctonine and isotalatizidine. Recently, the ^{13}C chemical shift assignments for alkaloid B have been revised (*102*). Because alkaloids A and B are closely related and the structure of alkaloid A was revised to **85** with the C-6 α-acetyl group, perhaps the C-6 hydroxyl group in alkaloid B is also present in the α-configuration.

T. LAPPACONITINE AND LAPPACONINE

Lappaconitine ($C_{32}H_{44}N_2O_8$) has been isolated from *A. septentrionale* Koelle (*103*), *A. orientale* (*104*), and *A. excelsum* Popov (*105*). Alkaline hydrolysis of lappaconitine yielded the parent alkamine, lappaconine ($C_{23}H_{37}NO_6$), and the *N*-acetylanthranilic acid. On the basis of extensive chemical studies, in 1969 Marion and his co-workers (*106*) assigned structure **86** to lappaconine. Meanwhile, Birnbaum reported (*107, 108*) an X-ray analysis of lappaconine hydrobromide and confirmed the proposed structure for lappaconine. Subsequently, the Soviet workers assigned (*109*) the *N*-acetylanthranilic ester moiety of lappaconitine (**87**) to the C-4 position.

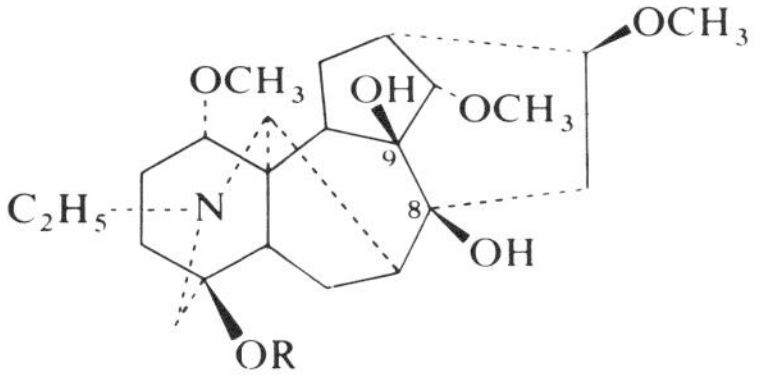

86 R = H Lappaconine
87 R = COC_6H_4—*o*-NHAc Lappaconitine

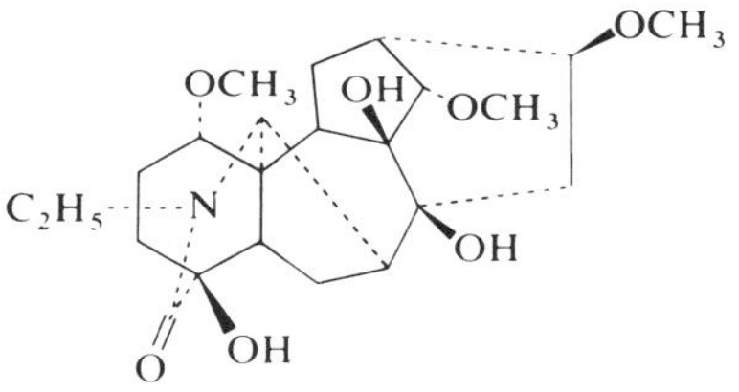

88

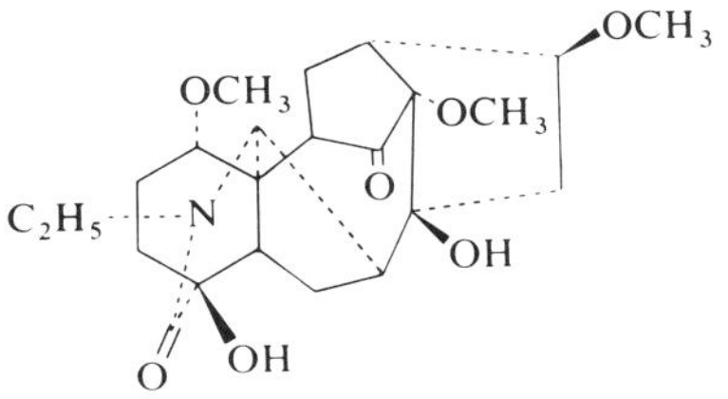

89

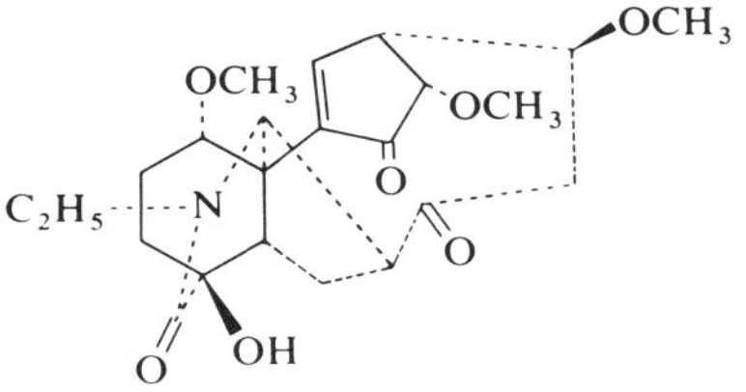

90

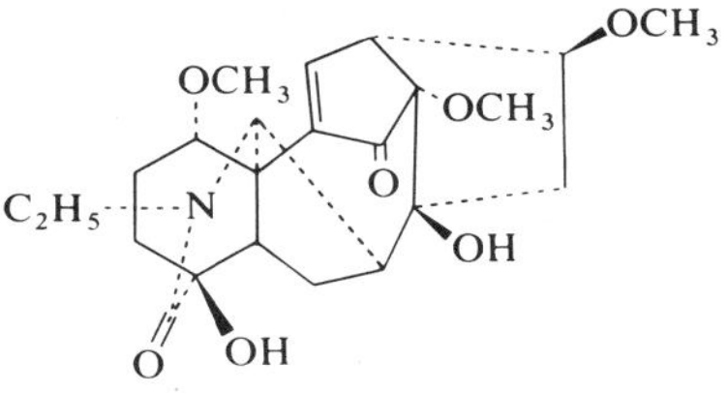

91

Potassium permanganate oxidation of lappaconine (**86**) yielded oxolappaconine (**88**). Further oxidation of oxolappaconine with periodic acid produced secooxolappaconine (**89**). However, some controversy has developed about the lead tetracetate cleavage of the C-8–C-9 vicinal diol in oxolappaconine to dehydrosecooxolappaconine (**91**). The Canadian workers (*106*) proposed that the intermediate **90** undergoes an aldol condensation to yield **91**, whose structure was assigned on evidence from its IR spectrum. The IR absorptions at 1645 and 1710 cm^{-1} were assigned to the lactam and α,β-unsaturated five-membered ring ketone in **91**. No other absorption due to the ketone group was present in the IR spectrum. In contrast, Soviet workers (*109*) have reported that there is no characteristic absorption due to the α,β-unsaturated ketone in the UV spectrum of compound **91**. However, hydrogenation of dehydrosecooxolappaconine (**91**) over palladium–carbon yielded secooxolappaconine (**89**) with the absorption of 1 mole of hydrogen. This controversy about the assignment of these structures can probably be resolved by an analyses of their ^{13}C NMR spectra.

Oxidation of lappaconitine with chromium trioxide in acetone followed by basic hydrolysis afforded a lactam–cyclopentanone. The identical lactam was also obtained by the oxidation of lappaconine with chromium trioxide in acetone. Only a vicinal diol (C-8 and C-9) could give rise to such an oxidation product; thus, Yunusov and his colleagues (*109*) assigned the *N*-acetylanthranilic ester moiety to the C-4 position in lappaconitine. Recently, we have confirmed (*110*) this assignment by ^{13}C NMR analysis of lappaconitine (**87**) and lappaconine (**86**).

Lappaconitine represents the first example of a C_{19}-diterpenoid alkaloid with one of the carbon atoms, which is usually attached to C-4, replaced by an oxygen atom. The presence of the C-8–C-9 vicinal diol system in lappaconitine is also unusual.

U. Lapaconidine

Lapaconidine ($C_{22}H_{35}NO_6$; mp 206–207°) has been isolated from *A. leucostomum* (*excelsum*) (*111, 112*). Its structure (**92**) was established on the basis of a chemical correlation with lappaconine. The ^{1}H NMR spectral analysis indicated the presence of an *N*-ethyl group and two methoxyl groups in lapaconidine. Upon treatment with acetyl chloride, lapaconidine formed a tetraacetate derivative, which indicated the presence of four hydroxyl groups in this alkaloid. On methylation with methyl iodide and sodium hydride, lapaconidine yielded tetra-*O*-methyllapaconidine, which was found to be identical with tri-*O*-methyllappaconine (**93**). This methylation reaction established that the positions of three out of four hydroxyl groups are identical in lappaconine and lapaconidine, and that the same oxygen substitution pattern and basic skeleton are present in both alkaloids. On the basis of mass spectral studies, the remaining hydroxy group in lapaconidine was assigned to the C-1 position.

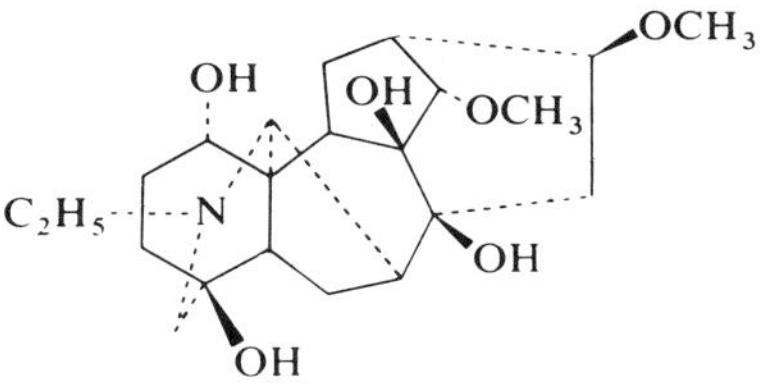

92 Lapaconidine

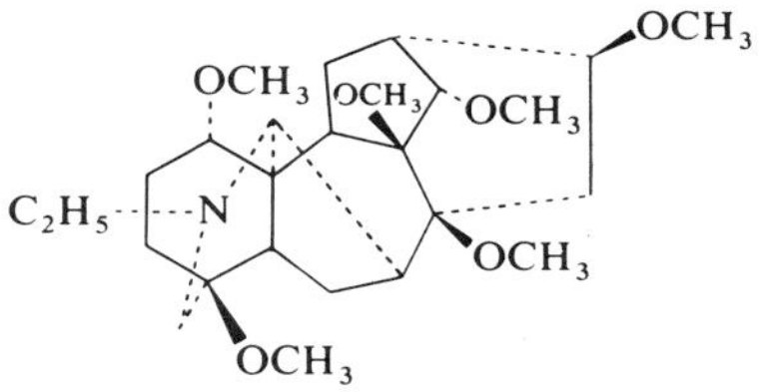

93

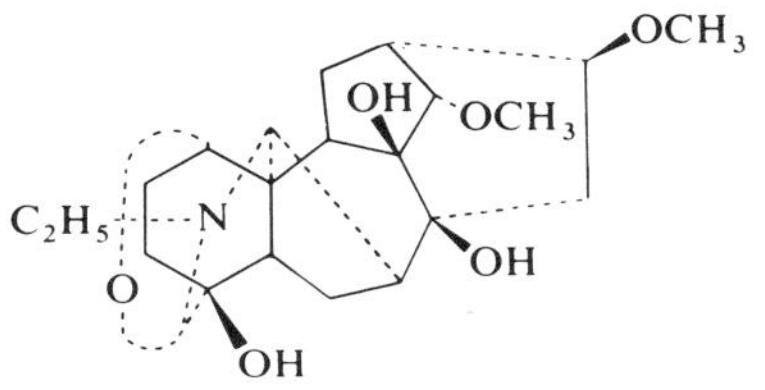

94

The presence of a C-1 hydroxyl group was confirmed by observing the formation of an inner carbinolamine ether (**94**) by potassium permanganate oxidation of lapaconidine. Reduction of compound **94** with Adams' catalyst regenerated the original alkaloid. The structure of the inner carbinolamine ether as **94** was also supported by observing a peak at $M^+ - 56$ (loss of acrolein) in the mass spectrum of **94**.

V. Rearrangements of Lappaconine and Lapaconidine

During their chemical studies of lappaconitine, Yunusov and his colleagues (*109*) observed an interesting rearrangement of lappaconine (**86**) upon treatment with sulfuric acid. Lappaconine produced an amorphous solid whose molecular formula differed from lappaconine by the loss of water and methanol. On the basis of spectral evidence, they proposed structure **96** for the reaction product, and suggested that the reaction may proceed via a pinacolic-type rearrangement.

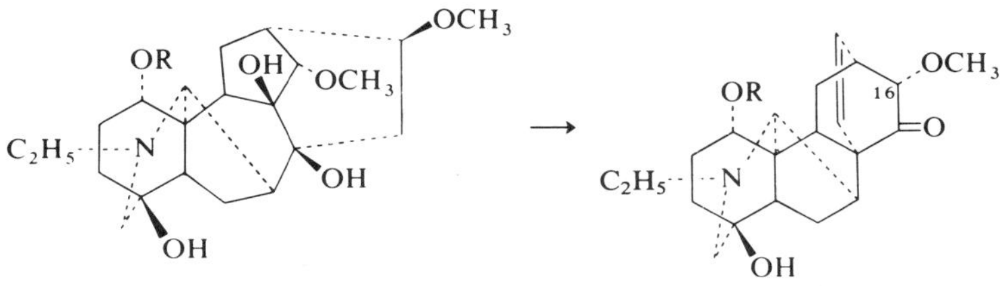

86 R = CH_3 Lappaconine
92 R = H Lapaconidine

96 R = CH_3
97 R = H
100 R = Ac

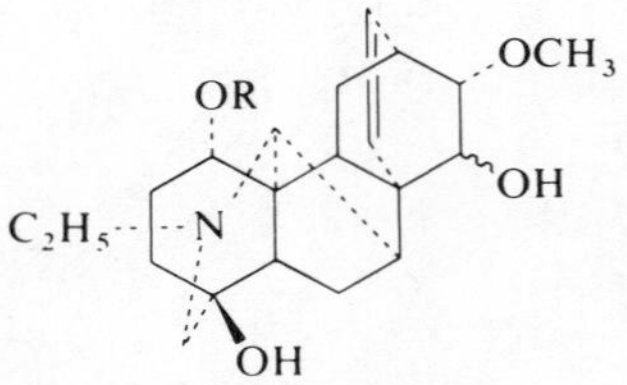

98 R = CH_3
101 R = H

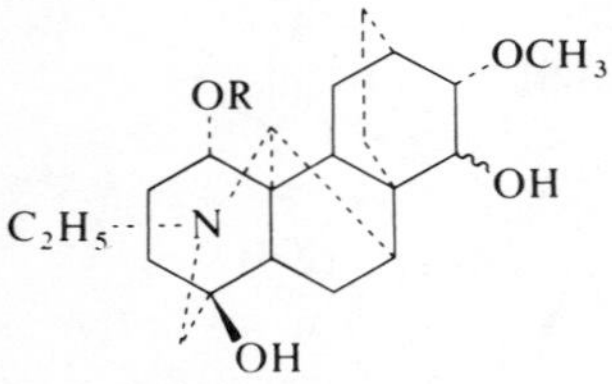

99 R = CH_3
102 R = H

On treatment with sulfuric acid, the closely related alkaloid lapaconidine (**92**) also yielded (*112*) a compound similar to **96**. This compound (**97**) probably also resulted via a pinacolic-type rearrangement. These rearrangement products contain the basic denudatine-type skeleton (atisine skeleton with an additional C-7–C-20 bond) and suggest the possible biogenetic interrelationship between the denudatine-type and aconitine-type alkaloids.

When reduced with sodium amalgam in absolute alcohol, compound **96** yielded a crystalline product with molecular formula $C_{22}H_{33}NO_4$. On the basis of mass spectral and IR data as well as the formation of a diacetate derivative, structure **98** was assigned to this reduction product. Hydrogenation of compound **96** with Adams' catalyst afforded compound **99**.

Subsequent work on these rearrangement products confirmed the structures assigned to them (*113*). The attempted cleavage of the methoxyl group at C-16 in compound **97** with zinc powder and glacial acetic acid resulted in the acetylation of the C-1 hydroxyl (**100**). Sodium borohydride reduction of **97** yielded derivative **101**. Similarly, hydrogenation of **97** with Adams' catalyst resulted in the reduction of ketone to hydroxyl as well as the reduction of the double bond to give compound **102**. The structures of these compounds were confirmed by their mass spectral analysis as well as comparison of mass spectral data with those of compounds prepared from lappaconine. The Soviet investigators reported that the $M^+ - 31$ peak in the mass spectrum of compound **96** originated from the C-16 methoxyl group and not from the C-1 methoxyl as previously suggested.

W. Excelsine

Excelsine ($C_{22}H_{33}NO_6$; mp 103–105°) has been isolated (*114*) from the roots of *A. excelsum* (*leucostomum*). On the basis of chemical and spectral studies, structure **103** was proposed for this alkaloid. Recently, the Soviet investigators have revised (*115*) the structure of excelsine to **104** and determined (*116*) its absolute configuration by an X-ray crystallographic study of the monohydrate–hydroiodide salt of (+)-excelsine. The structure was solved by the heavy-atom method with a final $R = 0.068$ based on 1817 reflections. The absolute configuration of excelsine was established by the examination of 20 Friedel pairs as *1S, 3S, 4R, 5S, 7S, 8S, 9S, 10S, 11R, 13S, 16S,* and *17R*. Ring A exists in a boat conformation with hydrogen bonding between the C-1 hydroxyl group and the nitrogen atom. Rings B and F were shown to be in chair conformations, while rings C and E exist in envelope forms. Ring D also exists in a boat form. All positional parameters, bond lengths, angles, and other pertinent intra- and inter-molecular atomic distances were reported.

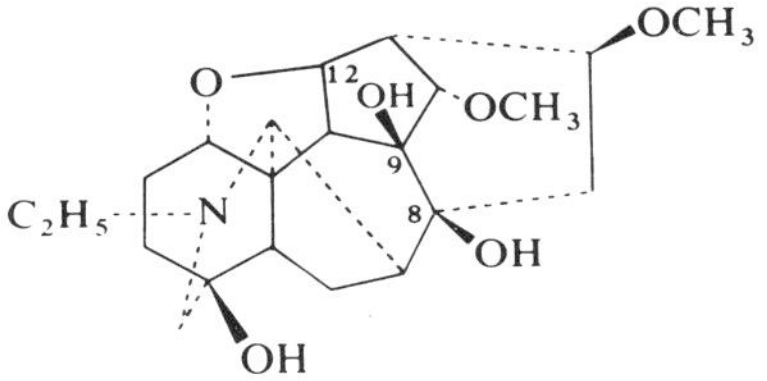

103

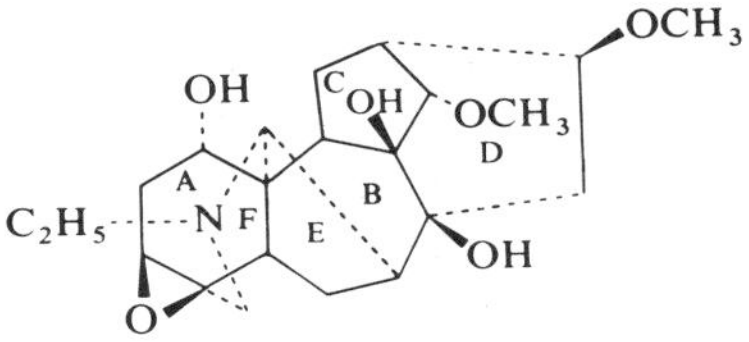

104 Excelsine

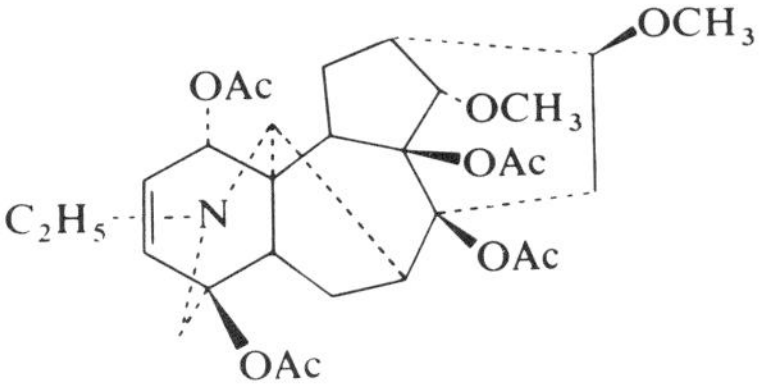

105

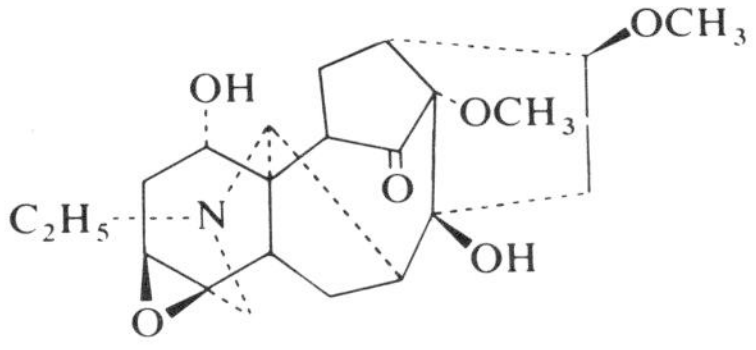

106

The presence of the epoxide moiety at C-3 and C-4 in excelsine explained the interesting chemical reactions observed earlier. On treatment with acetic anhydride and *p*-toluenesulfonic acid, excelsine yielded a triacetate derivative, while treatment with acetyl chloride afforded a tetraacetate derivative. On reduction with Raney nickel in methanolic base, excelsine yielded lapaconidine (**92**), but was inert toward other reducing agents, e.g., lithium aluminum hydride, sodium borohydride, and Adams' catalyst. Treatment of excelsine with boiling aqueous hydrochloric acid yielded an epimeric mixture of chlorohydrins with molecular formula $C_{22}H_{34}NO_6Cl$. These epimers were hydrolyzed to the crystalline compound $C_{22}H_{33}NO_6$ when treated with aqueous sulfuric acid. This compound formed a tetraacetate derivative for which structure **105** was proposed on the basis of spectral data.

Oxidation of excelsine with Kiliani reagent gave a compound ($C_{22}H_{31}NO_6$) containing a five-membered ring ketone. Under similar conditions, lappaconine and lapaconidine also gave five-membered ring ketones. This oxidation of excelsine proceeds via a cleavage of the C-8–C-9 diol system to give a product of structure **106**. Oxidation of excelsine with periodic acid at room temperature furnishes another compound ($C_{22}H_{33}NO_6$) with an oxocyclopentyl moiety. A definite structure for the periodate oxidation product was not proposed because of the unavailability of spectral and chemical data.

X. Karakolidine

Karakolidine ($C_{22}H_{35}NO_5$; mp 222–224°) has been isolated (*16, 119*) from the roots of *A. karakolicum* collected in the Kungei Ala-Tan range of the Kirghiz S.S.R. On the basis of chemical studies and comparison of karakolidine with the related alkaloid karakoline (**108**), karakolidine has been assigned the structure of 10β-hydroxykarakoline (**107**) (*16, 118*).

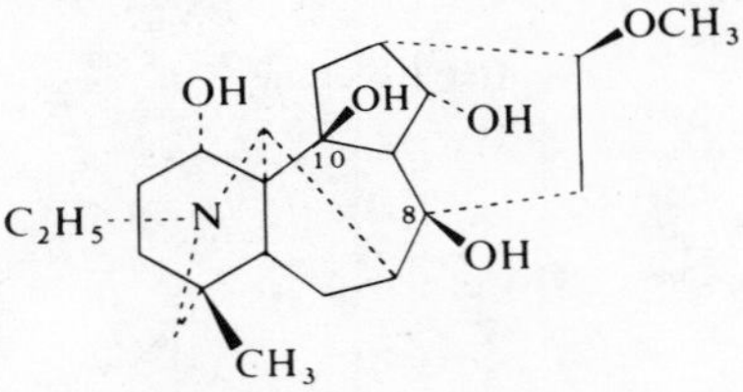

107 Karakolidine

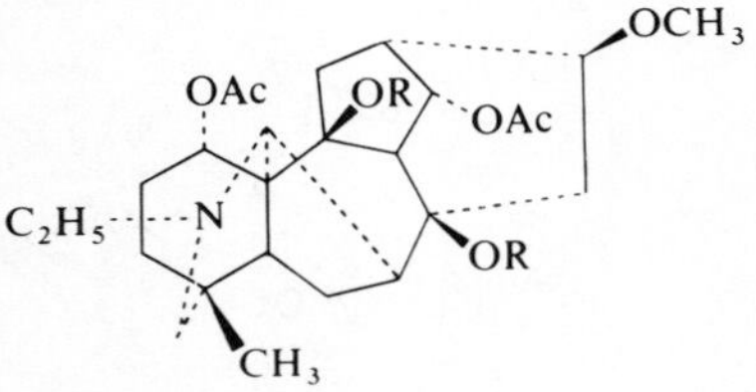

109 R = H
110 R = Ac

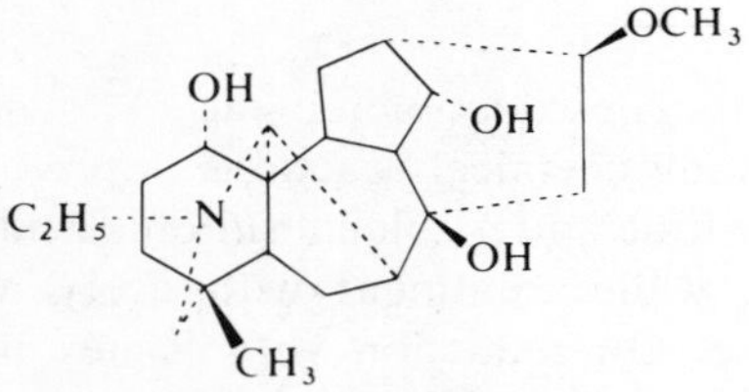

108 Karakoline

On acetylation of karakolidine with acetic anhydride and pyridine at room temperature, a diacetate derivative (**109**) was formed, while treatment with acetyl chloride yielded a tetraacetate derivative (**110**). Pyrolysis of compound **110** followed by alkaline hydrolysis yielded the normal pyrolysis product, pyrokarakolidine (**111**), and a rearranged product, isopyrokarakolidine (**112**). The latter compound was obtained from pyrokarakolidine (**111**) by treatment with methanolic hydrochloric acid. This pyrolysis reaction confirmed the presence of the C-8 acetoxyl and C-16 methoxyl groups in karakolidine. Oxidation of karakolidine with Kiliani

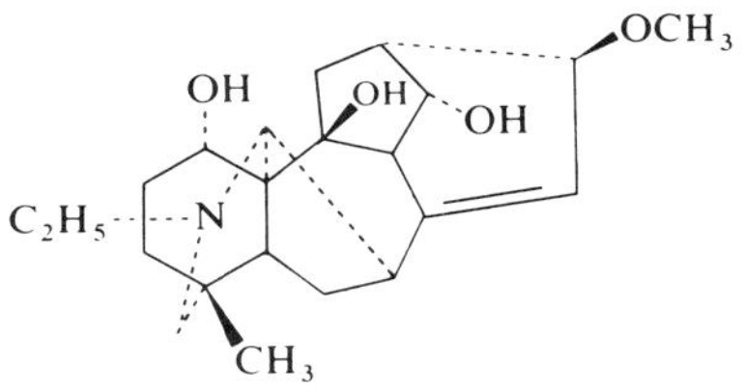

111 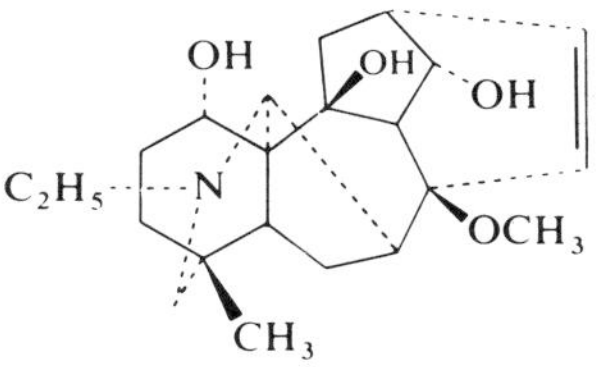 **112**

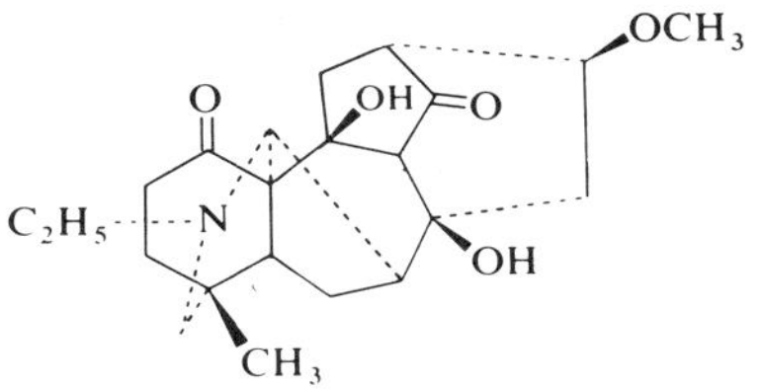

113 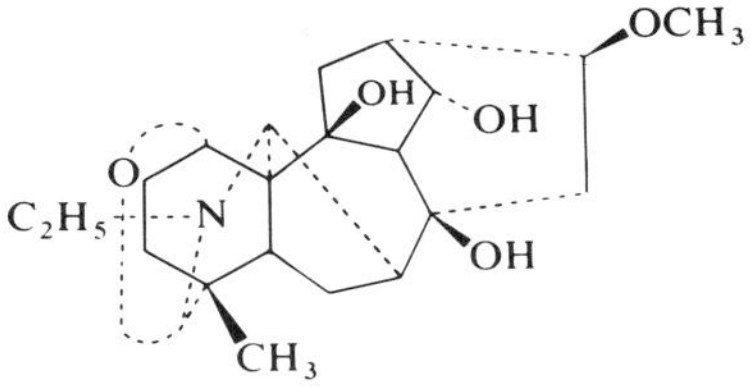 **114**

reagent, yielded 1,14-diketokarakolidine (**113**). The IR spectrum of the latter compound showed absorption for five- and six-membered ring ketones. When oxidized with potassium permanganate, karakolidine afforded an internal α-carbinolamine ether (**114**), which indicated the presence of a C-1 α-hydroxyl group. From these chemical transformations and spectral data, the Soviet chemists assigned structure **107** to karakolidine. There was no direct evidence reported by the Soviet workers for the presence of a C-10 hydroxyl group in karakolidine.

Y. Karakoline and Sachaconitine

Karakoline (**108**; $C_{22}H_{35}NO_4$; mp 183–184°), along with the related alkaloid, karakolidine, has been isolated (*117, 119*) from the roots of *A. karakolicum* growing in the Kungei Ala-Tan range of the Kirghiz S.S.R. Chemical transformations and spectral analyses of karakoline have led to the assignment of its structure as **108** (*117*). Karakoline yielded a diacetate (**115**) on acetylation with acetic anhydride and pyridine at room temperature, while treatment with acetyl chloride afforded a triacetylkarakoline (**116**). Pyrolysis of compound **116** followed by saponification of the reaction products afforded the rearranged product, acetyldemethylisopyrokarakoline (**117**). The structure of the latter was confirmed by chemical and spectral analyses. Hydrogenation of compound **117** with platinum catalyst yielded the dihydro derivative (**118**), which on acetylation formed a triacetate derivative (**119**). The stability of the C-1 acetate toward saponification will be discussed later. (See Section IV.) On

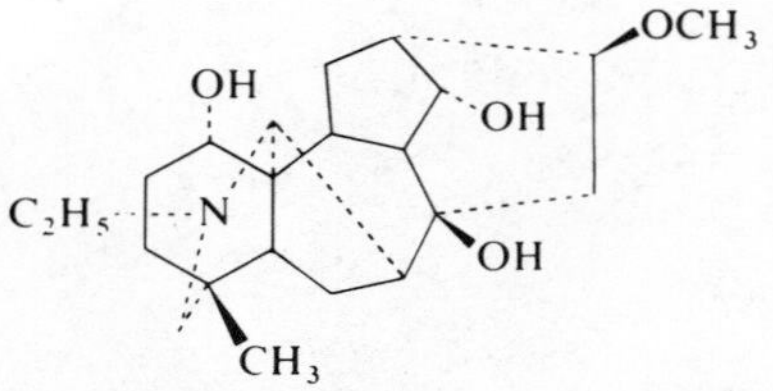

108 Karakoline

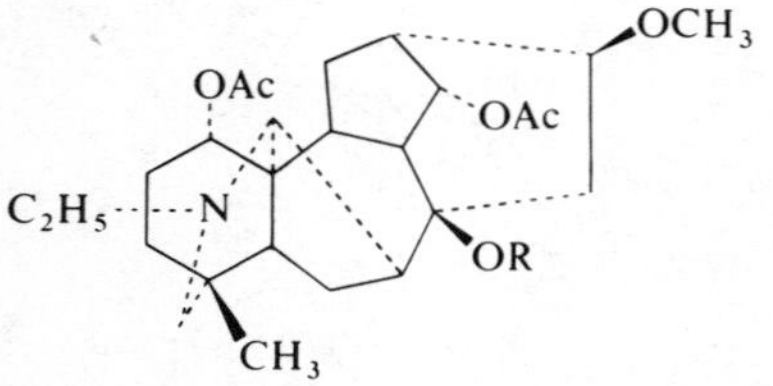

115 R = H
116 R = Ac

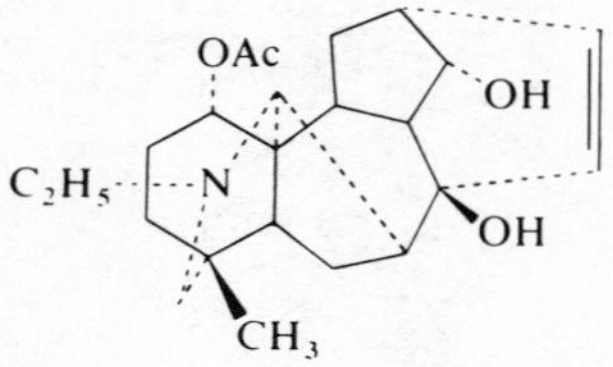

117

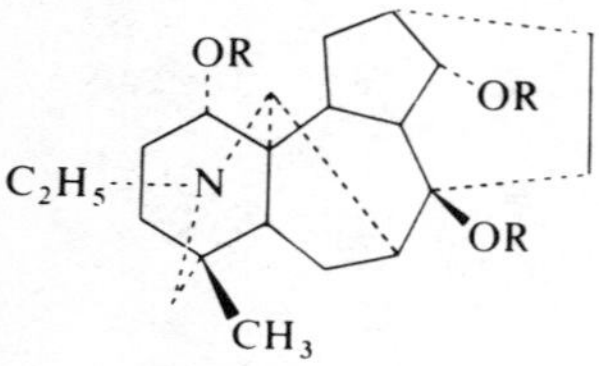

118 R = H
119 R = Ac

oxidation with chromic anhydride, karakoline formed a diketo derivative (**120**) containing cyclopentanone and cyclohexanone rings. Permanganate oxidation of karakoline yielded a C-1 α-internal carbinolamine ether (**121**) which on hydrogenation over Adams' catalyst was converted to the original alkaloid. These oxidation reactions confirmed the presence of a C-1 α-hydroxyl group in karakoline. From these classic chemical transformations and spectral studies, structure **108** was deduced for karakoline.

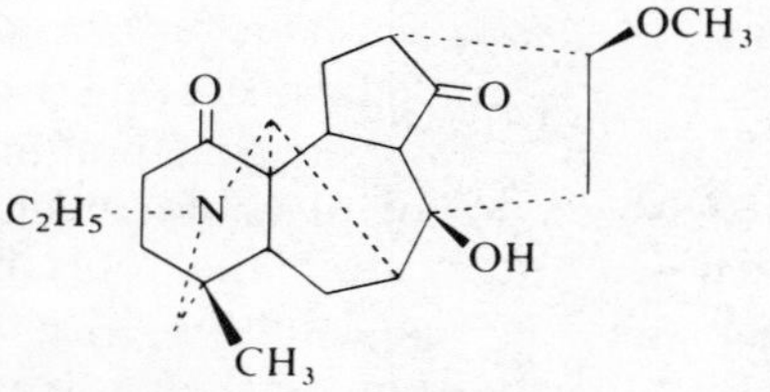

120

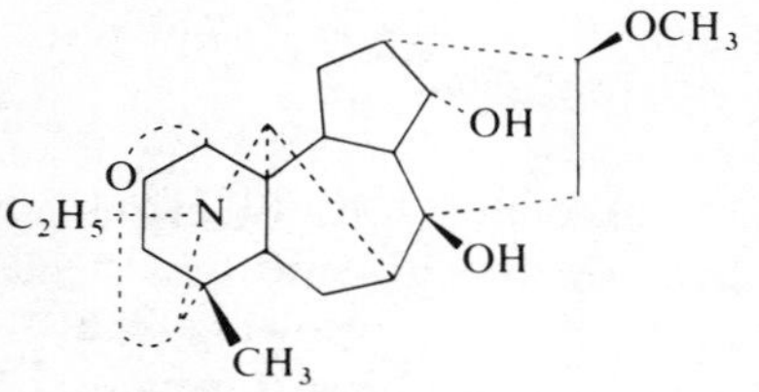

121

Sachaconitine ($C_{23}H_{37}NO_4$; mp 129–130°) has been isolated (*69, 120*) from the mother liquors accumulated during the isolation of two major C_{20}-diterpenoid alkaloids, miyaconitine and miyaconitinone, from the roots of *A. miyabei*, a plant native to Hokkaido, Japan. Structure **122** was deduced for sachaconitine.

On the basis of chemical studies, Katsui and Hasegawa (*120*) reported that sachaconitine contained an *N*-ethyl, two hydroxyl, and two methoxyl groups on an aconitine-type skeleton ($C_{19}H_{29}N$). On acetylation with acetyl chloride, sachaconitine formed a diacetate (**123**; mp 114–116°). The

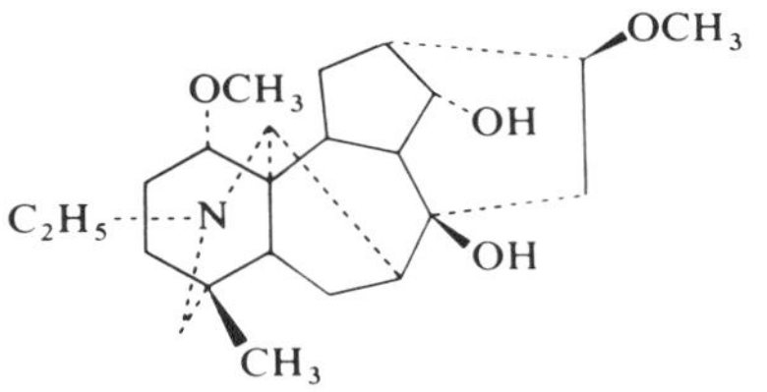

122 Sachaconitine

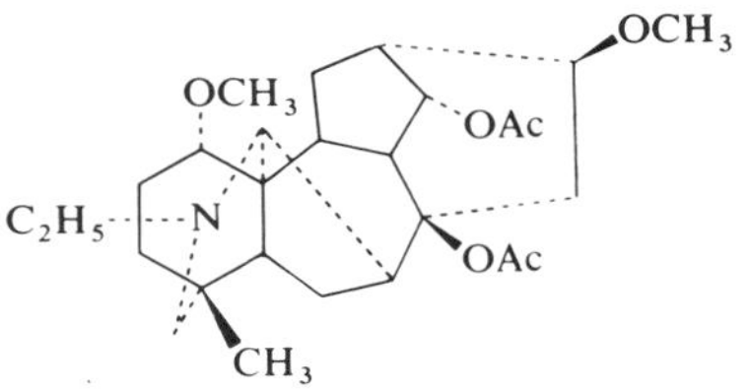

123

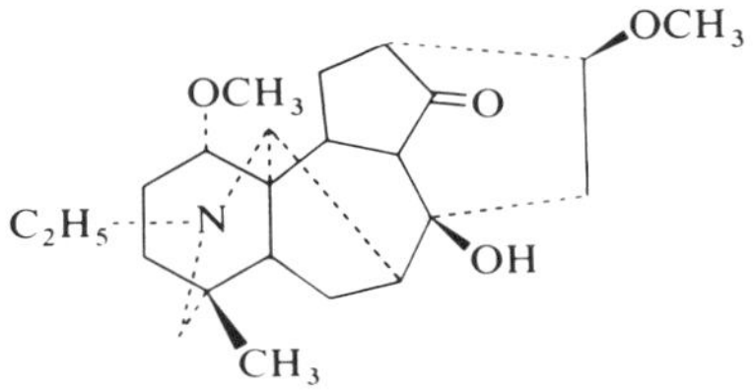

124

IR and UV characterization of the permanganate or chromic acid–pyridine oxidation products of sachaconitine indicated that one hydroxyl group in **122** had been oxidized to a cyclic five-membered ring ketone (**124**). Finally, the structure of sachaconitine as **122** was elucidated by correlation of ^{1}H and ^{13}C NMR spectra of sachaconitine with those of the closely related alkaloids talatizamine, karakoline, talatizidine, and chasmanine (*71*). A comparison of ^{13}C and ^{1}H NMR spectra of **122** with those of the alkaloids mentioned earlier afforded evidence for the presence of a C-4 methyl, *N*-ethyl, and two hydroxyl groups at the C-8 (singlet at 72.8 ppm) and C-14 (doublet at 75.7 ppm) positions in sachaconitine. The presence of two methoxyl groups at C-1 and C-16 was also confirmed by observing two doublets at 86.7 and 82.3 ppm and two quartets at 56.3 and 56.9 ppm in **122**, respectively. All other remaining signals were in complete agreement with the assigned structure (**122**) for sachaconitine.

Sachaconitine may be a biogenetic intermediate between the alkaloids aconosine (**125**) and talatizamine (**68**). Sachaconitine and karakoline (**108**) constitute an additional example of a C-1 α-hydroxyl–methoxyl pair along with the other known pairs, neoline–chasmanine, isotalatizidine–talatizamine, and lapaconidine–lappaconine. This additional example suggests the possible existence of other such pairs in nature.

Z. Aconosine

Aconosine ($C_{22}H_{35}NO_4$; mp 148°) has been isolated by Soviet chemists from *A. nasutum* Fisch and Rchb (*121*). Aconosine was found in all parts of the plant. Infrared and ^{1}H NMR spectra revealed the presence of an

N-ethyl, two hydroxyl, and two methoxyl groups in aconosine. Mass spectral analysis indicated that this alkaloid contained a basic aconitine-type skeleton. Oxidation of aconosine with chromic anhydride in acetone afforded a monoketoaconosine ($C_{22}H_{33}NO_4$), and structure **126** was assigned on the basis of IR absorption for a five-membered ring ketone. On

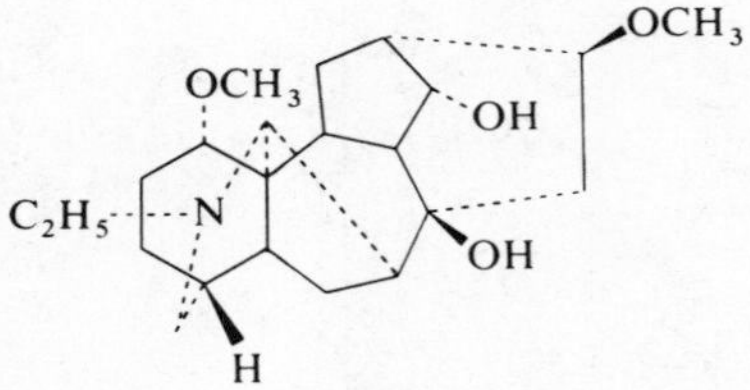

125 Aconosine

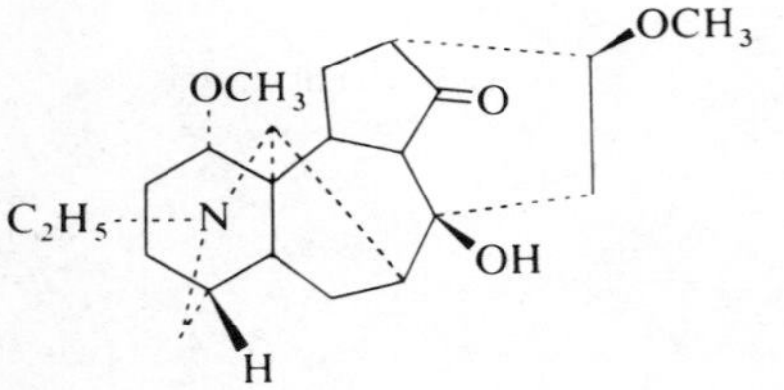

126

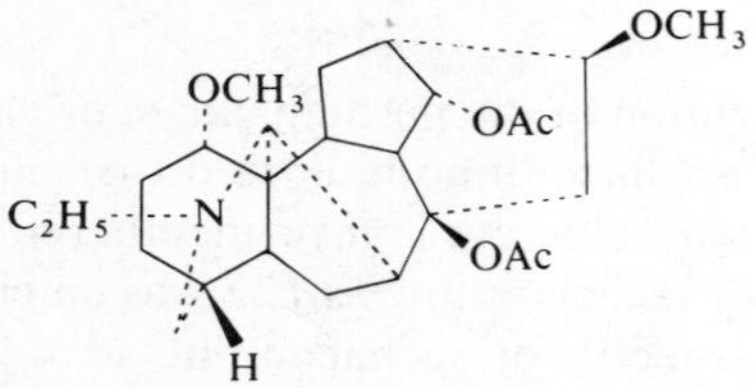

127

acetylation, aconosine formed a diacetate derivative (**127**). On the basis of the above chemical transformations and the mass spectral comparison of aconosine, ketoaconosine, and the diacetate derivative with talatizamine and its derivatives, structure **125** has been assigned to aconosine. It is interesting to note that karakoline, sachaconitine, and aconosine are the least oxygenated C_{19}-diterpenoid alkaloids known so far.

AA. Aconifine

The tubers of *A. karakolicum* collected from the Terskei Ala-Tan ranges of the Kirghiz S.S.R. yielded a new alkaloid (*16*) designated as aconifine ($C_{34}H_{47}NO_{12}$; mp 195–196°). Along with this new alkaloid, the tubers of *A. karakolicum* also afforded the well-known C_{19}-diterpenoid alkaloid aconitine (**4**) and the two known C_{20}-diterpenoid alkaloids, songorine and napelline.

On the basis of IR, ^{1}H NMR, and mass spectral data, a partial structure (**128**) has been proposed for aconifine. No further work related to this alkaloid has been reported. From the molecular formula and the known functional groups (4 methoxyl groups, 4 hydroxyl groups, a benzoate, and

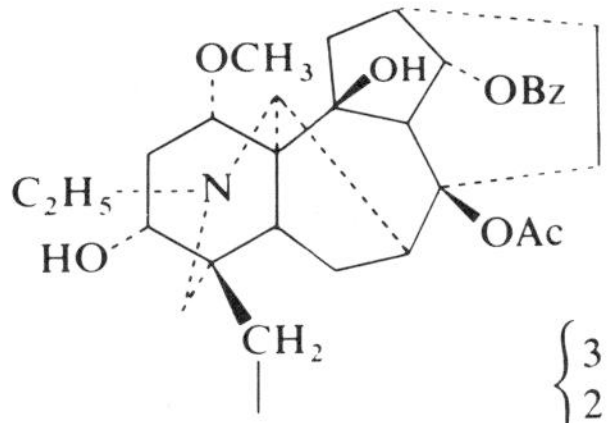

3 OCH_3
2 OH

128 Aconifine

an acetoxyl group), it is interesting to note that aconifine is more highly oxygen substituted than any other aconitine- or lycoctonine-type alkaloid.

III. Lycoctonine-Type Alkaloids

A. Acomonine

Acomonine ($C_{25}H_{41}NO_7$; mp 208–210°) has been recently isolated by Soviet chemists (*122*) from the roots of *A. monticola* along with three known C_{20}-diterpenoid alkaloids, songorine, norsongorine, and songoramine. They also reported the isolation of three closely related unknown alkaloids from the same plant, an amorphous alkaloid ($C_{22}H_{35}NO_6$) and two crystalline alkaloids ($C_{22}H_{33}NO_6$, mp 166–167°; $C_{22}H_{33}NO_5$, mp 161–164°). The structure of acomonine (**129**) was established (*123*) on the basis of chemical transformations and spectral analyses.

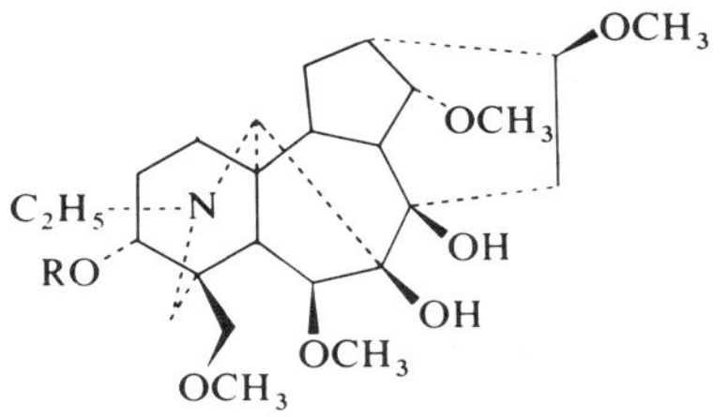

129 R = H Acomonine
130 R = Ac

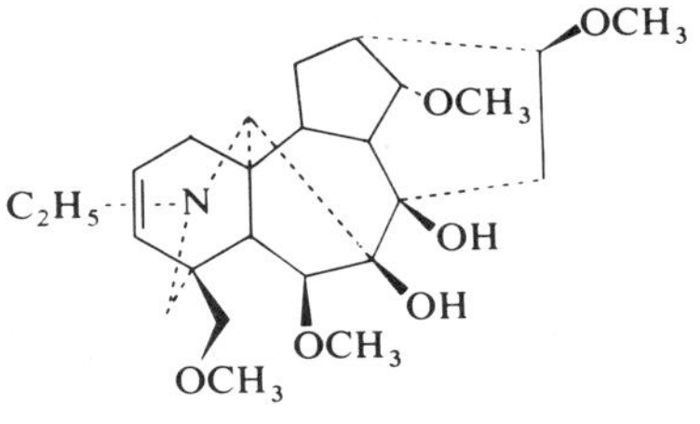

131

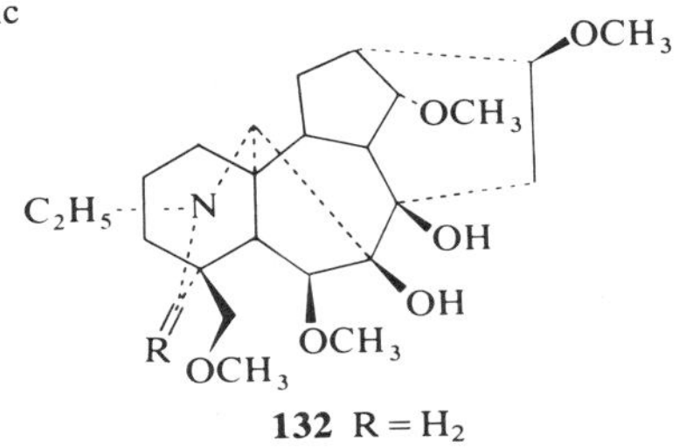

132 R = H_2
133 R = O

The spectral data revealed the presence of an *N*-ethyl, three hydroxyl, and four methoxyl functions in acomonine. Acetylation of acomonine with acetic anhydride in pyridine yielded the monoacetate derivative (**130**). Treatment of acomonine with *p*-toluenesulfonyl chloride yielded anhydroacomonine (**131**). On hydrogenation with Adams' catalyst **131** afforded deoxyacomonine (**132**). Oxidation of **132** with potassium permanganate yielded the lactam (**133**). On treatment with potassium permanganate in aqueous acetone, acomonine gave the inner ether (**134**), which was further oxidized to compound **135** with sodium metaperiodate. Similarly, periodate oxidation of anhydroacomonine (**131**) resulted in the formation of compound **136**. The spectral data for compounds **135** and **136** indicated

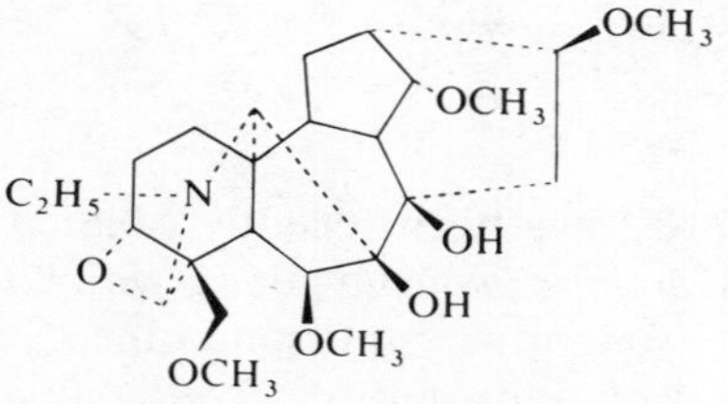

OCH_3 OCH_3 C_2H_5 N O OH OH OCH_3 OCH_3

134

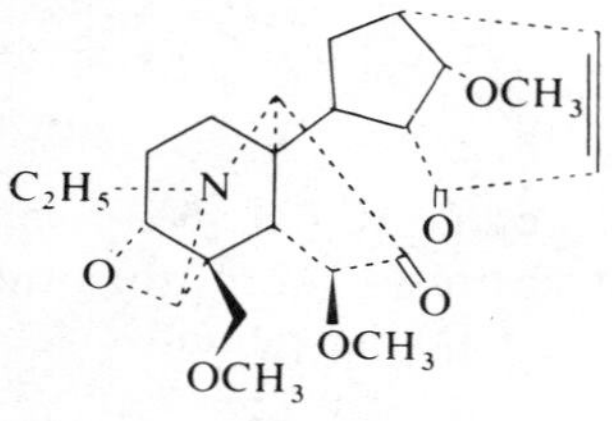

OCH_3 C_2H_5 N O O O OCH_3 OCH_3

135

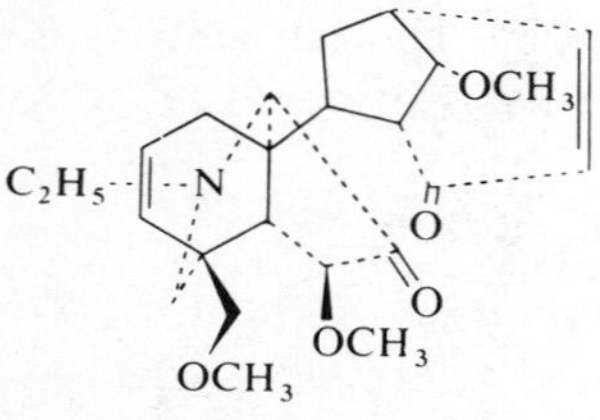

OCH_3 C_2H_5 N O O OCH_3 OCH_3

136

the presence of a cyclic α,β-unsaturated carbonyl function. These data are accommodated by an α-glycol system at C-7 and C-8 and a methoxyl group at C-16. Sodium borohydride reduction of the inner ether **134** resulted in the regeneration of acomonine. The presence of the secondary hydroxyl at C-3 was assigned on the basis of ^1H NMR and mass spectral data of acomonine and its monoacetate. The α-orientation of the C-3 hydroxyl group in acomonine was confirmed by observing the formation of the inner ether **134**. On the basis of ^1H NMR and mass spectral data, two methoxyl groups were assigned to C-14 and C-18. The remaining methoxyl group was assigned to C-6. Thus, acomonine is the first lycoctonine-type alkaloid which contains no oxygen function at the C-1 position.

B. Iliensine

Iliensine ($C_{24}H_{39}NO_9$; mp 201–203°) has been isolated by Soviet workers from the aerial parts of *D. biternatum* Huth (*124, 125*). The structure of iliensine (**137**) was established by chemical correlation with acomonine (**129**). Iliensine represents the second example of a lycoctonine-type alkaloid with no oxygen substituent at C-1.

From the spectral data, iliensine was shown to contain an *N*-ethyl, four hydroxyl, and three methoxyl groups. On acetylation with acetic anhydride in pyridine, iliensine yielded a diacetate (**138**), which indicated the presence of two secondary hydroxyl groups. Treatment of iliensine with potassium permanganate in aqueous acetone afforded the deethyl carbinol ether (**139**). The inner ether **139** yielded an *N,O*-diacetyl derivative (**140**)

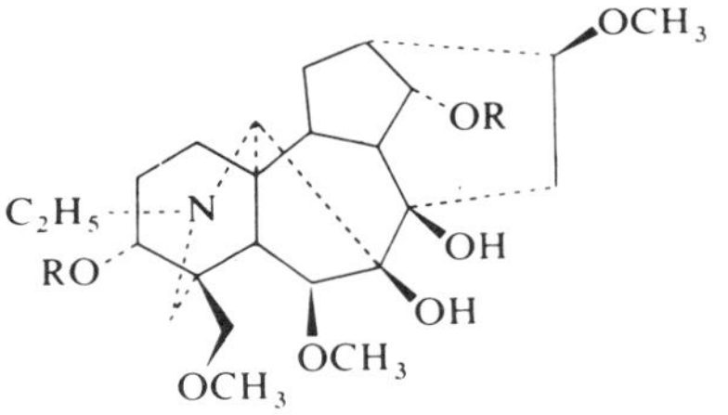

137 R = H Iliensine
138 R = Ac

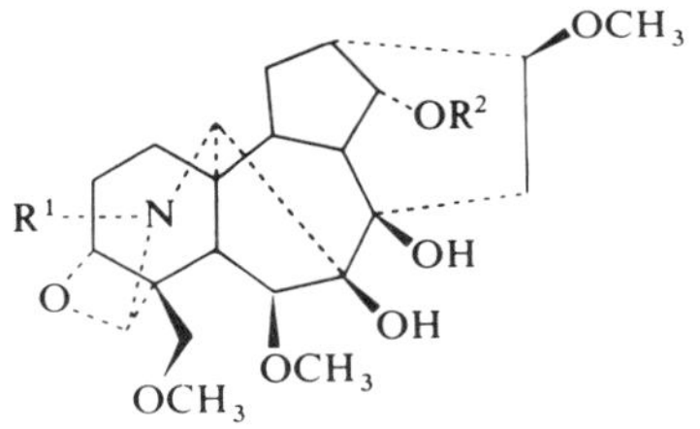

139 $R^1 = R^2 = H$
140 $R^1 = R^2 = Ac$
141 $R^1 = Ac$ $R^2 = H$

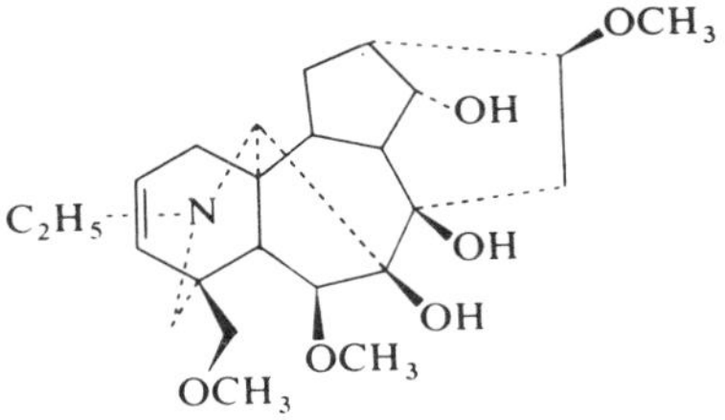

142

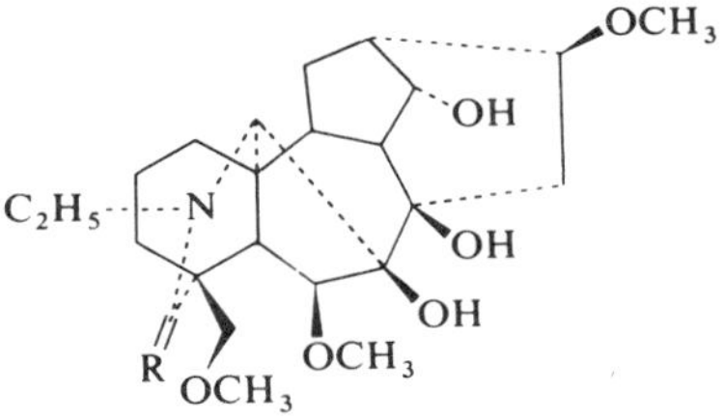

143 $R = H_2$
144 R = O

on acetylation. Upon alkaline hydrolysis with methanolic sodium hydroxide, diacetate **140** afforded the monoacetate derivative, **141**. Treatment of iliensine with *p*-toluenesulfonyl chloride yielded anhydroiliensine (**142**), which was hydrogenated with Adams' catalyst to deoxyiliensine (**143**). On the basis of the above chemical transformations, one of two hydroxyls was assigned to C-3 in an α-configuration. Permanganate oxidation of deoxyiliensine formed the lactam (**144**). Further oxidation of **144** with sodium

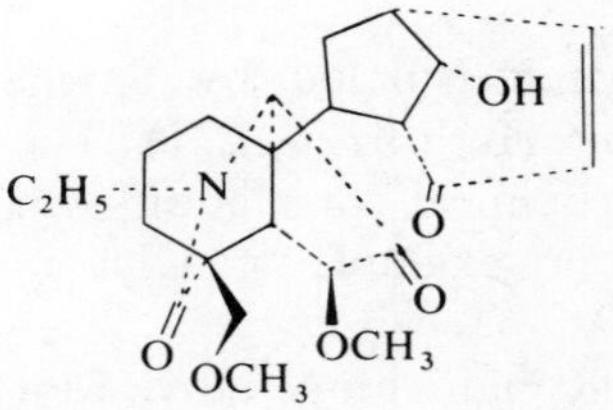

145

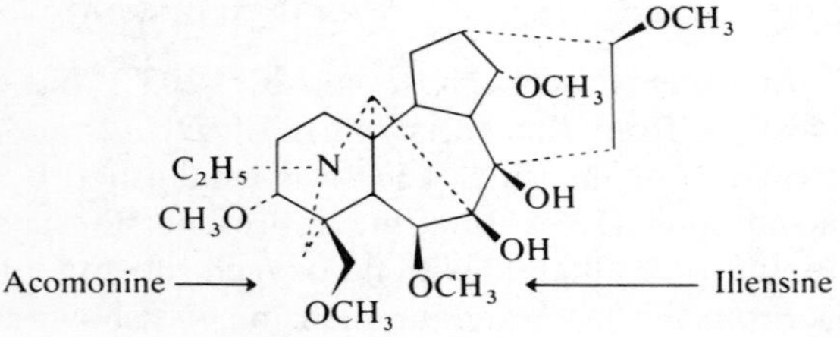

Acomonine ⟶ 146 ⟵ Iliensine

metaperiodate afforded the seco product (**145**). Methylation of iliensine with methyl iodide and sodium hydride yielded *O*-dimethyliliensine (**146**). The latter was found to be identical with *O*-methylacomonine (**146**). This methylation reaction and other chemical transformations led to the assignment of structure **137** to iliensine.

C. Delphatine

Delphatine ($C_{26}H_{43}NO_7$; mp 106°) has been isolated from the seeds of *D. biternatum* (*126, 127*). Structure **147** was established for delphatine (18-*O*-methyllycoctonine) by chemical correlation with the well-known alkaloid lycoctonine (**58**). The only difference between these two alkaloids is that delphatine has a methoxyl group at C-18, while lycoctonine has a hydroxy group at that position.

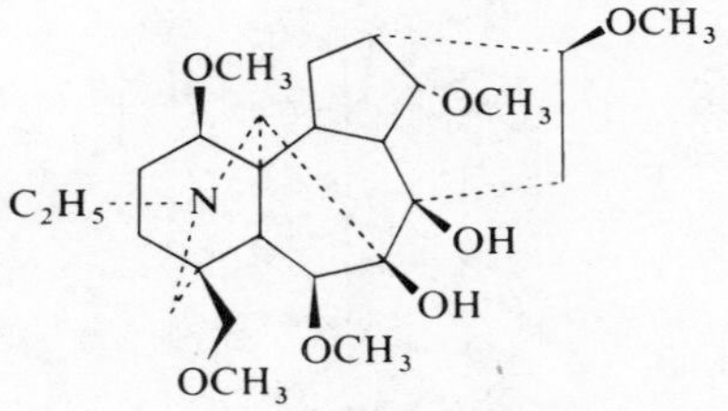

147 Delphatine

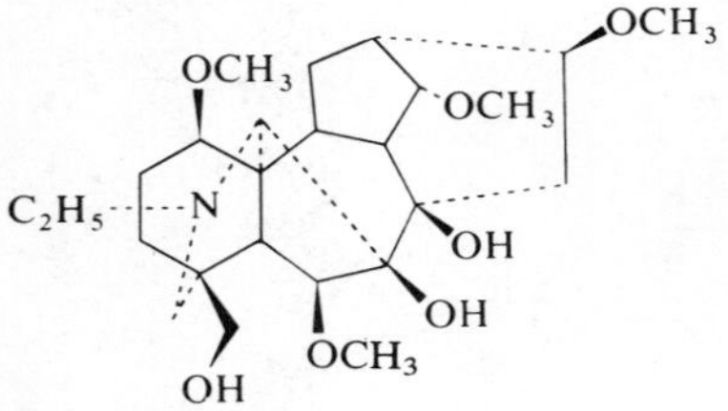

58 Lycoctonine

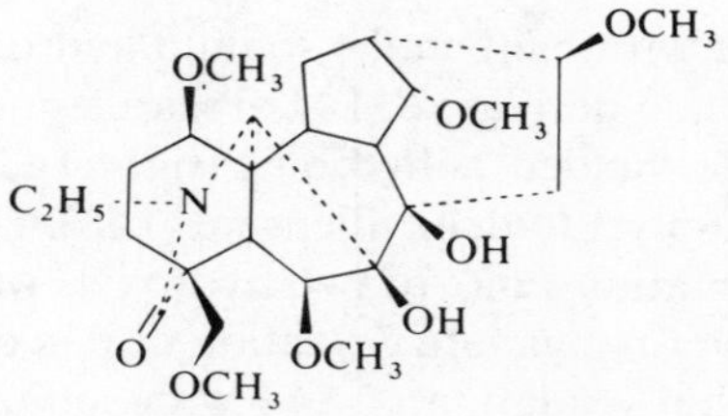

148

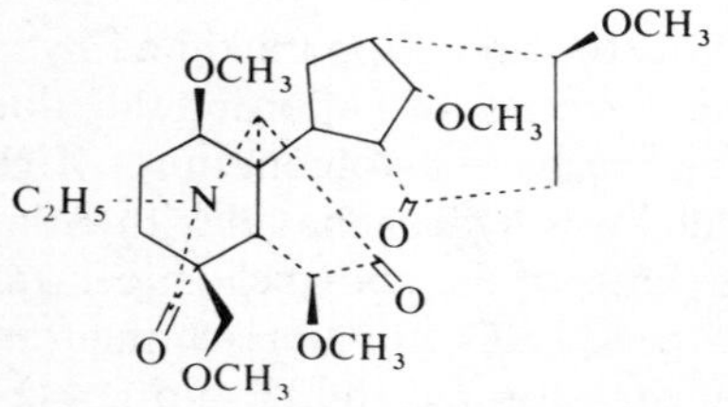

149

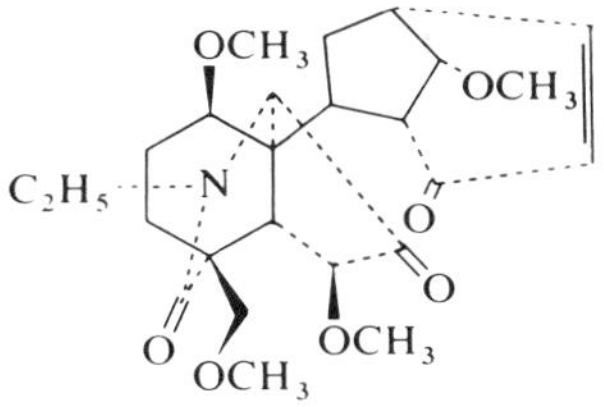

150 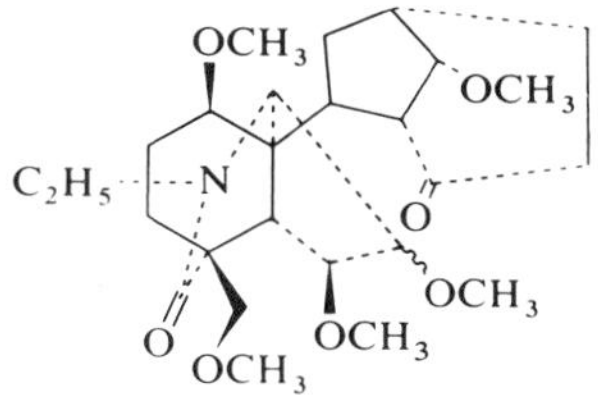 **151**

Permanganate oxidation of delphatine yielded the lactam (**148**), which was further oxidized with periodic acid to oxosecodelphatine (**149**). Upon treatment of compound **149** with sulfuric acid, a demethanol secodiketone (**150**) was formed. Hydrogenation of the latter over platinum, followed by methylation with methyl iodide and sodium hydride, yielded a compound (**151**) which was identical with the product obtained by a similar procedure from lycoctonine. The presence of two methoxyl groups at C-6 and C-18 in delphatine was detected by mass spectral comparison of delphatine and its derivative with lycoctonine and other related alkaloids and their derivatives.

D. Tricornine and Lycoctonine

Tricornine ($C_{27}H_{43}NO_8$; mp 187–189°) has been isolated (*128*) for the first time in our laboratory from the aerial parts and roots of *D. tricorne* Michaux (dwarf larkspur), a relatively rare plant collected from the mountains of North Carolina in the United States. Tricornine is accompanied by lycoctonine (**58**), and the latter also occurs in the aerial parts of *D. dictyocarpum* (*129*) along with some known and new alkaloids. The structure of tricornine (**152**) was assigned on the basis of ^{13}C NMR analysis of tricornine and lycoctonine and was confirmed by conversion of lycoctonine to tricornine (18-acetyllycoctonine).

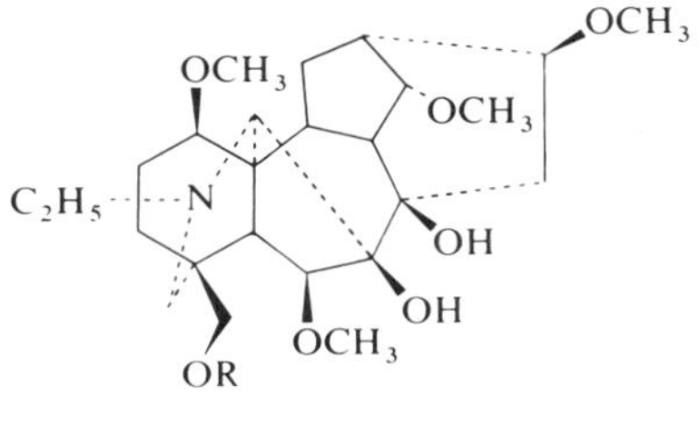

58 R = H Lycoctonine
152 R = Ac Tricornine
153 R = COC_6H_4—*o*-NH—$COCH_2CH_2CONH_2$ Avadharidine
154 R = COC_6H_4—*o*-$NHCOCH_3$ Ajacine

The ^{1}H and ^{13}C NMR data revealed the presence of an *N*-ethyl, an acetyl, four methoxyl, and two hydroxyl groups in tricornine. The ^{13}C NMR spectrum of tricornine closely resembled that of lycoctonine except for the presence of two additional signals at 20.8 (quartet) and 170.9 (singlet) ppm for the acetyl group. The above data indicated that tricornine is an acetyllycoctonine. Alkaline hydrolysis of tricornine yielded lycoctonine and acetic acid. On treatment with acetic anhydride in pyridine at room temperature, lycoctonine formed an 18-acetyllycoctonine which was identical with tricornine. It is interesting to note that although several C-18 N-substituted anthranilic acid esters of lycoctonine (*3*), e.g., avadharidine (**153**) and ajacine (**154**), are known, tricornine represents the only simple ester isolated so far in nature.

E. Methyllycaconitine and Delsemine

In 1973, methyllycaconitine (**155**) was reported to be present in *D. grandiflorum* L., *D. triste* Fisch, and *D. crassifolium* Schrad by Mats (*130*). He identified methyllycaconitine by paper and thin-layer chromatography during the study of curarelike neuromuscular effects of the total alkaloids from the above three species. Soviet chemists also isolated (*129*) methyllycaconitine along with other new alkaloids from the aerial parts of *D.*

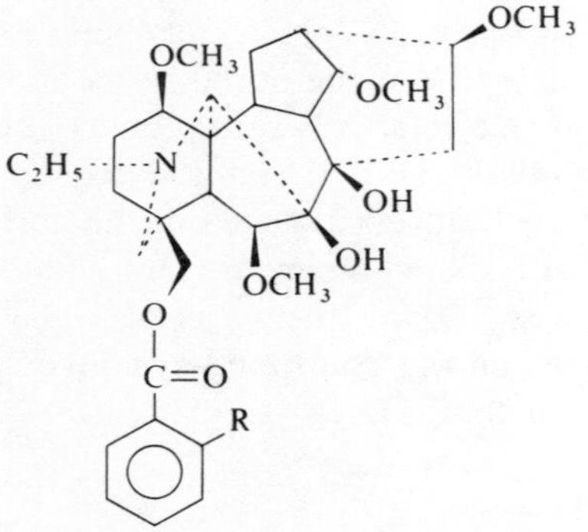

155 R = N-(3-methylsuccinimido) Methyllycaconitine

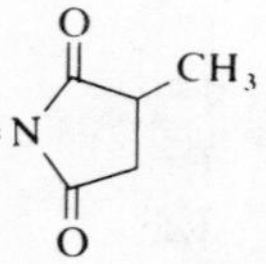

156 $R = NHCOCH_2CH(CH_3)CONH_2$ "Delsemine"

157 $R = NHCOCH(CH_3)CH_2CONH_2$ "Delsemine"

158 $R = NH_2$ Anthranoyllycoctonine (Inuline)

corumbosum (also known as *D. corymbosum* Regel and *D: dictyocarpum*). Recently, this same alkaloid has been isolated (*131*), along with tricornine and lycoctonine, from the whole plants of *D. tricorne*.

The alkaloid "delsemine" isolated by Yunusov and Abubakirov (*132*) from *D. semibarbatum* has been assigned alternative structure **156** or **157**. Treatment of methyllycaconitine with ammonia yielded a mixture of alkaloids **156** and **157** which proved to be identical with delsemine (*131*). The components of this mixture were detected by ^{13}C NMR analysis. This mixture was converted to anthranoyllycoctonine (**158**) by acidic hydrolysis to confirm the existence of **156** and **157** in the mixture. It is obvious that delsemine is an apparently inseparable mixture of the closely related compounds **156** and **157**, which are artifacts resulting from the use of NH_4OH during isolation.

F. DELECTINE AND *O,O*-DIMETHYLLYCOCTONINE

Delectine ($C_{31}H_{44}N_2O_8$; mp 107–109°) has been isolated (*133*) recently from *D. dictyocarpum*. This new alkaloid was shown to contain an anthranilic acid moiety on a lycoctonine-type skeleton. On the basis of chemical reactions and spectral studies, structure **159** was established for delectine.

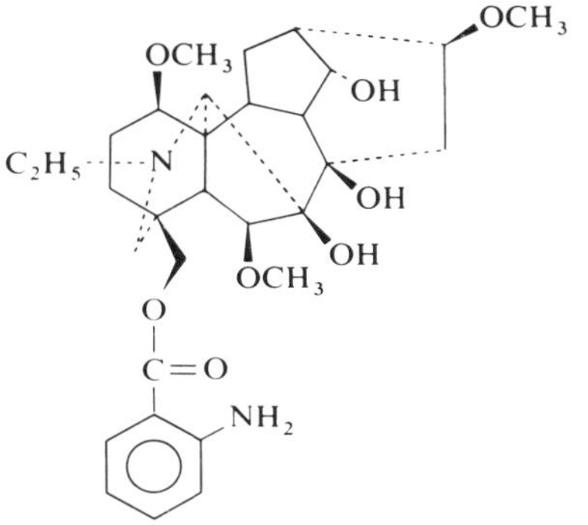

159 Delectine

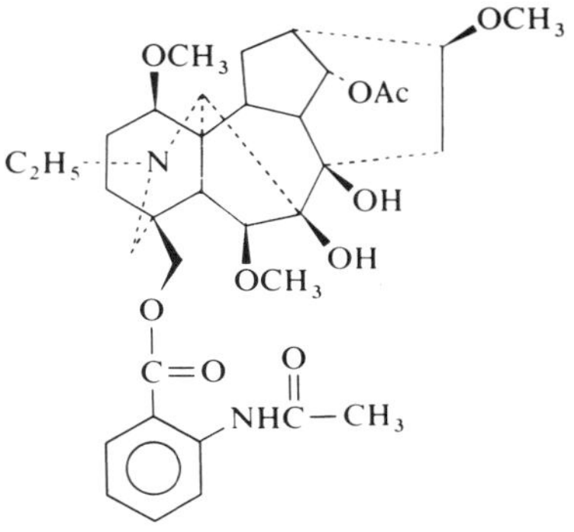

160

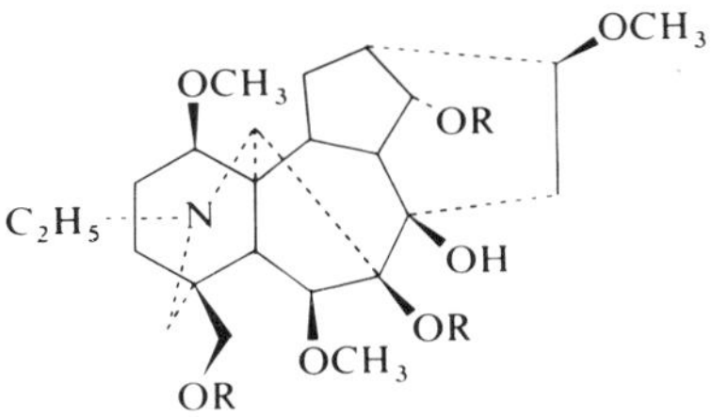

161 R = H
162 R = CH_3

Acetylation of delectine yielded an *N,O*-diacetyldelectine (**160**). Basic hydrolysis of delectine gave a parent amino alcohol (**161**) and anthranilic acid. Treatment of alkamine (**161**) with methyl iodide and sodium hydride afforded derivative **162**, which was identical with *O,O*-dimethyllycoctonine. Soviet chemists also reported the isolation of *O,O*-dimethyllycoctonine (**162**) from the same plant. The complete assignments of all functional groups and their stereochemistry were made on the basis of comparisons of ^{1}H NMR and mass spectral data with those of other lycoctonine-type alkaloids.

G. Delcosine, Acetyldelcosine, and Delsoline

During the mass spectral studies of diterpenoid alkaloids of the seeds of garden larkspur (*Consolida ambigua**), Waller and his co-workers (*134*) reported the presence of delcosine (**163**), acetyldelcosine (**164**), and delsoline (**165**). The mass fragmentation patterns of these alkaloids will be discussed in Section VI.

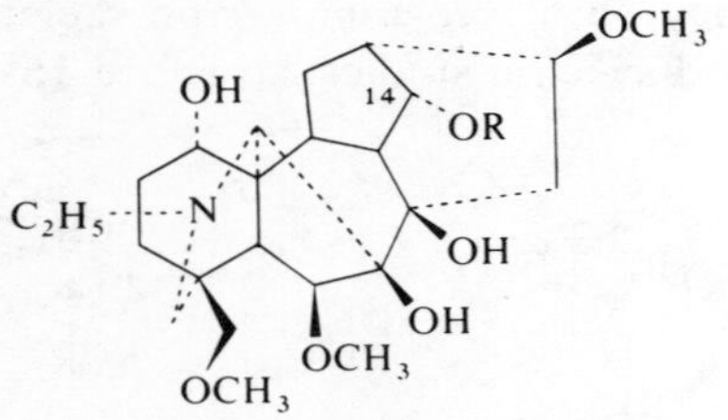

163 R = H Delcosine
164 R = Ac Acetyldelcosine
165 R = CH_3 Delsoline

H. Browniine and Acetylbrowniine

Recently, the presence of browniine has been observed (*135*) in the aerial parts of *D. carolinianum* as well as in *D. virescens* Nutt., a relatively rare plant collected from the mountains of northern Georgia in the United States. Browniine previously was isolated only from *D. brownii* Rydb., a plant native to Canada. Structure **57** was established for browniine by Edwards and his co-workers (*136*) in 1963. The ^{13}C NMR spectral analysis of browniine was consistent with the reported structure (**57**).

* This plant (garden larkspur) is designated in the older literature as *D. ajacis*. It is now correctly named as *Consolida ambigua* (L.) P. W. Ball & Heywood (C. S. Keener, *Castanea*, **41**, 15 (1976)). In our review the name *Consolida ambigua* will be used to designate the garden larkspur.

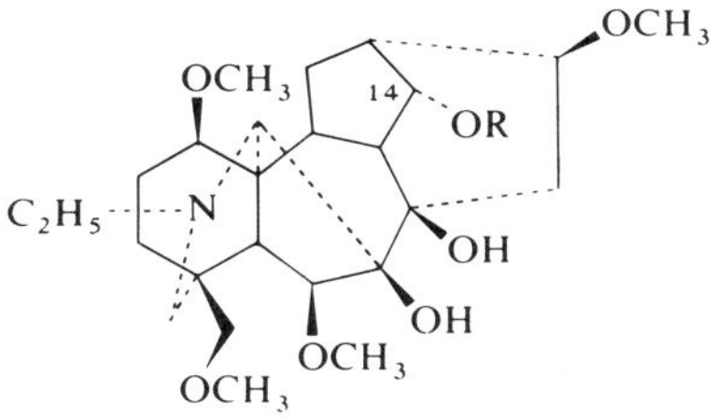

57 R = H Browniine
166 R = Ac Acetylbrowniine

A new alkaloid ($C_{27}H_{43}NO_8$; mp 123–124°) has been isolated (*137*) by a combination of gradient pH extraction and chromatographic techniques during the reexamination of alkaloidal constituents of the seeds of *Consolida ambigua*. The structure of this new alkaloid has been established as 14-acetylbrowniine (**166**) by a chemical correlation and ^{13}C NMR analysis. Basic hydrolysis of **166** afforded the parent alkamine, which was shown to be identical with the known alkaloid browniine. Acetylation of browniine with acetic anhydride and pyridine at room temperature yielded 14-acetylbrowniine, which was found to be identical with the new alkaloid. These hydrolysis and acetylation reactions confirmed the presence of acetyl group at C-14. The ^{13}C NMR spectrum of **166** was also in agreement with the assigned structure.

I. Delcorine

Delcorine ($C_{26}H_{41}NO_7$; mp 200–202°) has been isolated (*129*) from the chloroform extracts of the aerial parts of *D. corumbosum* (*D. corymbosum* Regel) along with an uncharacterized alkaloid (mp 93–95°) and the known alkaloid methyllycaconitine (**155**). From chemical and spectral studies, delcorine was shown to have structure **167**.

Preliminary spectral data indicated the presence of an *N*-ethyl, a methylenedioxyl, one hydroxyl, and four methoxyl groups in delcorine. On

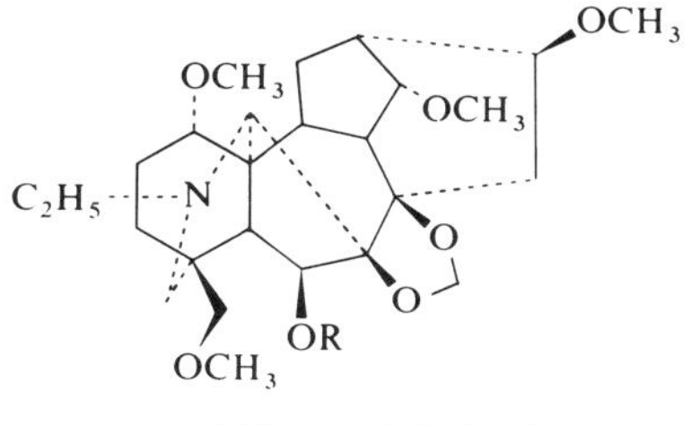

167 R = H Delcorine
168 R = Ac

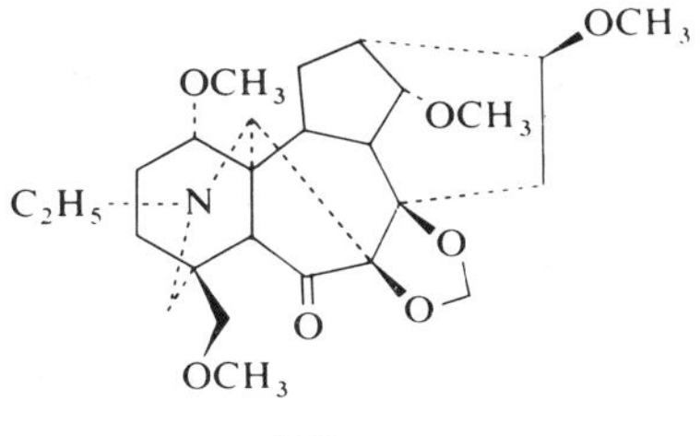

169

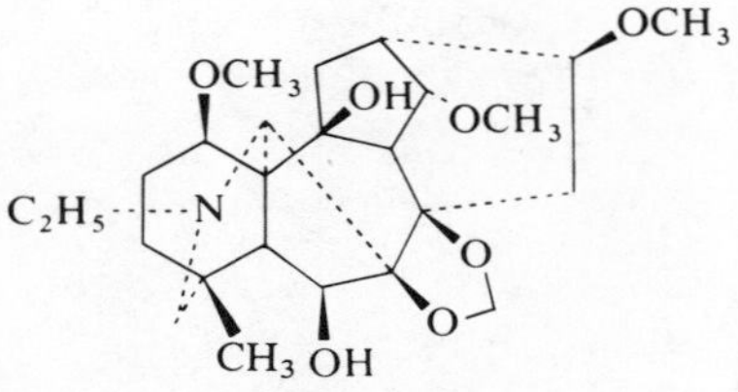

170 Deltamine

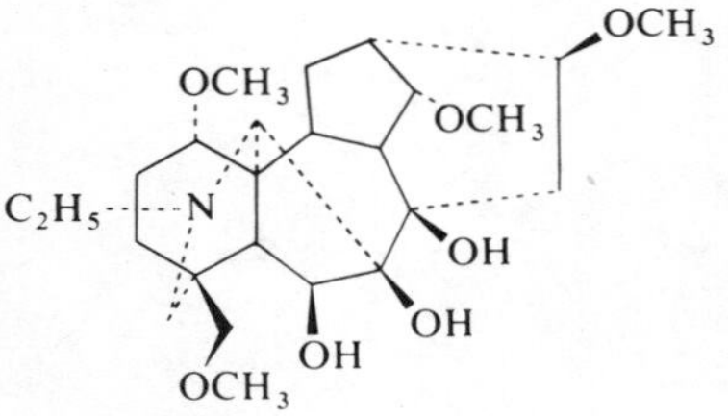

171

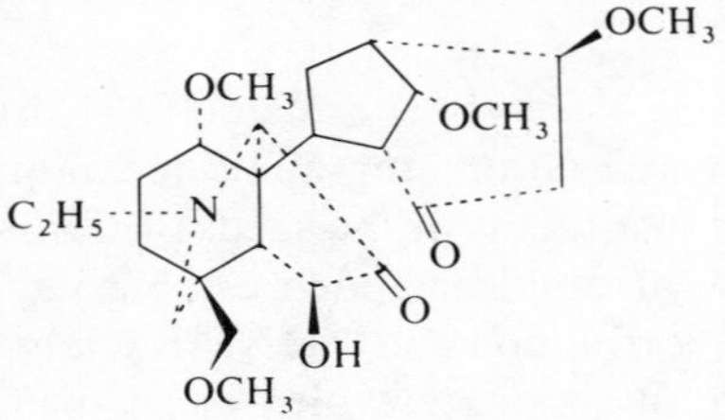

172

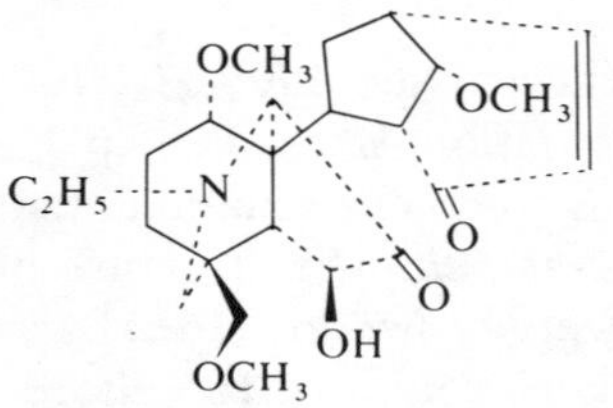

173

treatment with acetyl chloride, delcorine yielded a monoacetyl derivative (**168**). Delcorine formed a monoketodelcorine (**169**) on oxidation with chromium trioxide. The presence of the hydroxyl group at C-6 and the methylenedioxy group at C-7 and C-8 was detected by comparisons of the ^{1}H NMR and IR spectra of **167** and **169** with those of deltamine (**170**) and related alkaloids and their derivatives. The acetal function of delcorine was hydrolyzed by heating with sulfuric acid to give compound **171**. The latter was oxidized with periodic acid to afford compound **172**. The presence of a β-methoxyl group at C-16 was established by the formation of compound **173** from compound **172** on treatment with sulfuric acid. From the results obtained from the chemical and spectral studies, a methoxyl group was assigned to C-18. On the basis of mass spectral and ^{1}H NMR data, the remaining two methoxyl groups were assigned to C-1 and C-14 in an α-configuration. It is interesting to note that, with the exception of delcorine, all the lycoctonine alkaloids isolated to date from nature contain a C-1 β-methoxyl group.

The base (mp 93–95°) was partially characterized by preliminary spectral data which indicated the presence of an ethyl, a methylenedioxyl, and four methoxyl groups. Mass spectral analysis determined its molecular weight to be 463.

J. Dictyocarpine

Dictyocarpine ($C_{26}H_{39}NO_8$; mp 210–212°) has been isolated (*138*) from the chloroform extracts of the aerial parts of *D. dictyocarpum* collected in the Dzhungarskii Alatau (U.S.S.R.). The known alkaloids methyllycaconitine (**155**), deltaline (eldeline) (**174**), and deltamine (eldelidine) (**170**) were also isolated from this plant. The structure of dictyocarpine (**175**) was established (*139*) on the basis of ^{1}H NMR and mass spectral studies.

175 Dictyocarpine

176 R = H Dictyocarpinine
177 R = Ac

174 R^1 = Ac, R^2 = H Deltaline (Eldeline)
170 $R^1 = R^2$ = H Deltamine (Eldelidine)
179 $R^1 = R^2$ = Ac

178 R = H Delpheline
181 R = Ac Acetyldelpheline

180

Dictyocarpinine ⟶ ⟵ Deltamine

182

The presence of a C-4 methyl, an *N*-ethyl, an acetoxyl, a methylenedioxyl, and two methoxyl groups in dictyocarpine was detected by ^{1}H NMR analysis. Alkaline hydrolysis of dictyocarpine yielded the alkamine designated as dictyocarpinine (**176**) and acetic acid. Acetylation of this new alkaloid with acetyl chloride formed a triacetate derivative (**177**). On the basis of ^{1}H NMR studies, the acetoxyl group was assigned to C-6 in a β-configuration, and finally this assignment was confirmed by observing the $M^+ - 59$ ion peak in the mass spectrum of **175**. The hydroxyl group was assigned to C-10 in a β-configuration on the basis of the observed downfield shift of the C-14 proton signal. This downfield shift of the C-14 proton was explained by the deshielding influence of the C-10 hydroxyl group. For further ^{1}H NMR studies, deltaline (**174**) was converted into delpheline (**178**) by the following routes.

Pyrolysis of acetyldeltaline (**179**) at 210–220° for 30 min gave dehydroacetyldelpheline (**180**). Catalytic hydrogenation of **180** yielded acetyldelpheline (**181**), which was hydrolyzed with mild alkali to delpheline (**178**). ^{1}H NMR analysis of delpheline indicated that removal of the C-10 hydroxyl group led to a displacement of C-14 proton signal to the normal value. Thus, the deshielding effect of the C-10 hydroxyl group on the C-14 proton, as well as the presence of the C-10 and C-14 hydroxyl groups, was confirmed. One of the two methoxyl groups was also assigned to C-1 in a β-configuration by mass spectral studies of dictyocarpine.

The structure of dictyocarpine as **175** was confirmed by a methylation reaction. Methylation of dictyocarpinine (**176**) and deltamine (**170**) with methyl iodide in the presence of sodium hydride yielded the identical product (**182**). This reaction also indicated that the remaining methoxyl group was present at C-16.

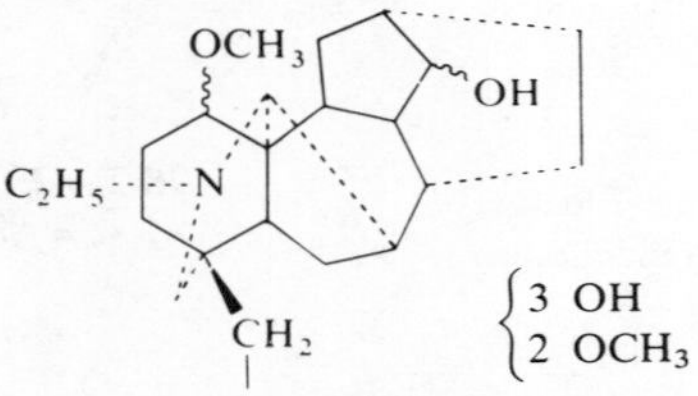

OCH_3 OH C_2H_5 N CH_2

$\left\{\begin{array}{l}3\ OH\\ 2\ OCH_3\end{array}\right.$

183

The chloroform extracts of the roots of *D. dictyocarpum* (*138*) yielded 1.83% total alkaloids consisting of the known alkaloids methyllycaconitine (**155**) and lycoctonine (**58**) and two new bases. A partial structure (**183**) has been suggested for the amorphous alkaloid with molecular weight 453 on the basis of chemical and extensive spectral data. The presence of two secondary hydroxyl groups was confirmed by acetylation with acetic anhydride in pyridine. The IR, ^{1}H NMR, and mass spectral data for the second amorphous alkaloid with molecular weight 541 indicated the presence of an *N*-ethyl, three methoxyl, and a benzoyl ester function.

K. Structure of Lycoctamone

Among the most unusual aspects of lycoctonine chemistry is the intriguing transformation of the lactam (**184**) to an α,β-unsaturated carbonyl compound, designated as lycoctamone, by vigorous treatment with acid. In

1971 Marion and his co-workers (*140*) established the structure of lycoctamone as **186** on the basis of extensive chemical and ^1H NMR studies.

Mild acid treatment of the **184** afforded the pinacolic dehydration product anhydrooxolycoctonine (**185**). A more vigorous treatment of compound **184** or **185** with acid gave lycoctamone, $C_{23}H_{31}NO_6$, with the overall loss of methanol, water, and the methyl of one methoxyl group. From the chemical and spectral data, the presence of an aldehyde group conjugated with trisubstituted double bond, a tertiary hydroxyl group, an exocyclic methylene group, and a δ-lactam group was confirmed. By studying several lycoctonine-type alkaloids and their derivatives in this rearrangement, the Canadian chemists concluded that the C-18 primary hydroxyl group and the C-1 and C-6 methoxyl groups are not involved in this rearrangement. By a process of elimination, the authors confirmed that the only methoxyl groups involved are those at the C-14 and C-16 positions.

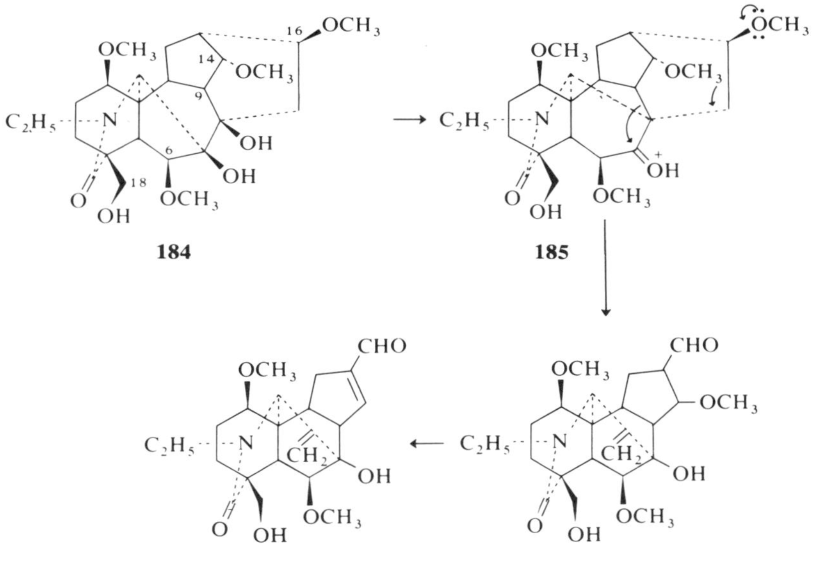

184

185

186 Lycoctamone

187

On the basis of extensive chemical and spectral studies, the authors suggested that the pinacolic dehydration product (**185**) and compound **187** are the most likely intermediates in the transformation to lycoctamone (**186**). They postulated that the repulsive interactions between the C-18 substituent and the C-6 hydrogen and between the C-9 hydrogen and the C-6 methoxyl group are the main driving forces in this transformation.

IV. Alkaline Hydrolysis and Acetylation Studies of C_{19}-Diterpenoid Alkaloids

The apparently anomalous failure of the basic hydrolysis of the C-1 acetyl group in **117** prompted Soviet chemists to examine the alkaline hydrolysis of diacetylkarakoline (*117*) (**115**), diacetyltalatizidine (**188**), and tetraacetyllapaconidine (**189**). In the time necessary for the complete hydrolysis of compounds **188** and **189**, diacetylkarakoline formed equal amounts of the C-1 monoacetate (**190**) and the starting material. This observation was explained by steric hindrance.

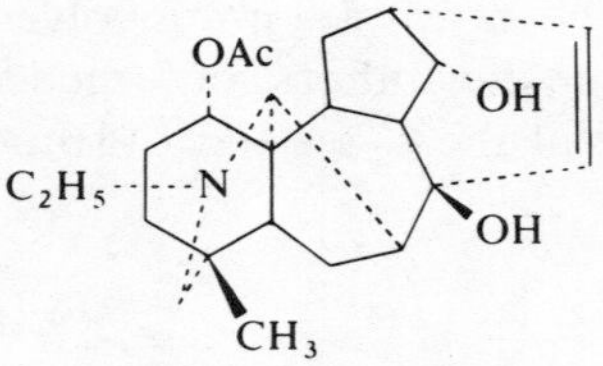

117

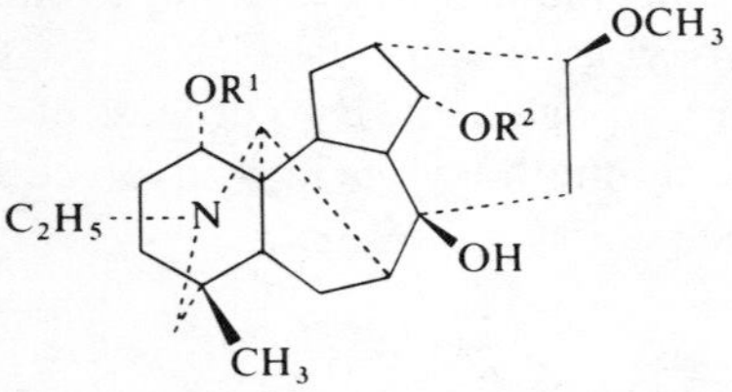

115 $R^1 = R^2 = Ac$
190 $R^1 = Ac; R^2 = H$

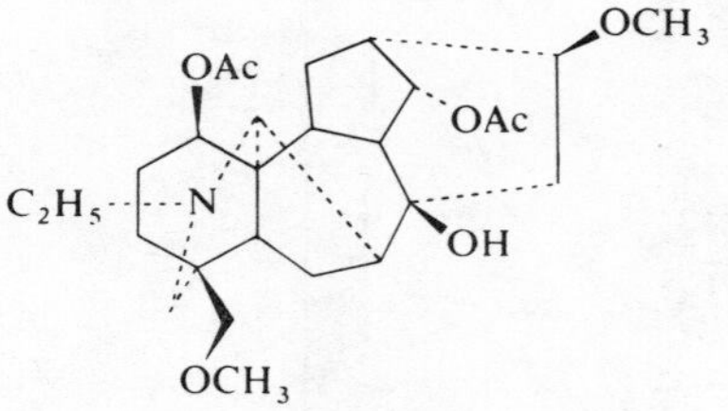

188

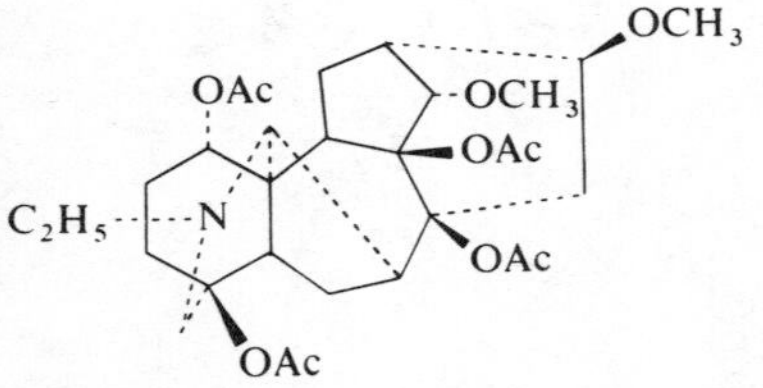

189

The relative reactivities of the C-1 and C-14 acetyl groups in a series of aconitine-type diterpenoid alkaloids have also been investigated (*141*). Alkaloids with C-1 methoxyl groups, e.g., diacetyltalatizamine (**70**) and triacetyllappaconine (**191**), were saponified to the parent alkamines approximately three times faster than the alkaloids with the C-1 acetoxyl groups, e.g., diacetylkarakoline (**115**), diacetyltalatizidine (**188**), and tetraacetyllapaconidine (**189**). However, karakoline (**108**) and karakolidine (**107**) were selectively acetylated at the C-1 position. On the basis of this acetylation experiment, Soviet workers have proposed that the C-1 hydroxyl group is sterically less hindered to acetylation than the C-14 hydroxyl which has cis–axial interactions with the hydroxyl and methoxyl groups at C-8 and C-16, respectively. During saponification, the carbonyl of the C-14 acetyl group would be removed from the ring plane so that

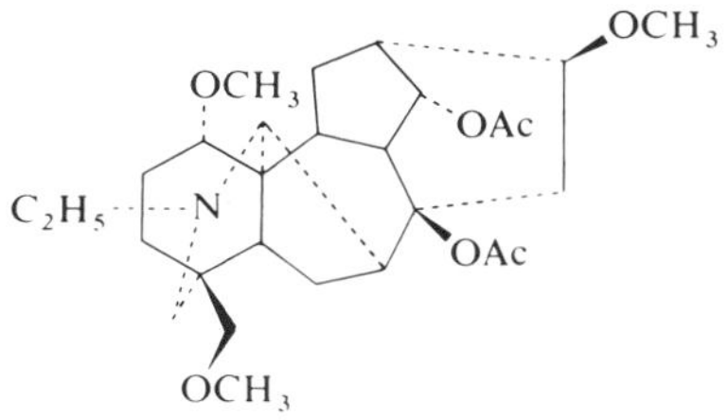

70

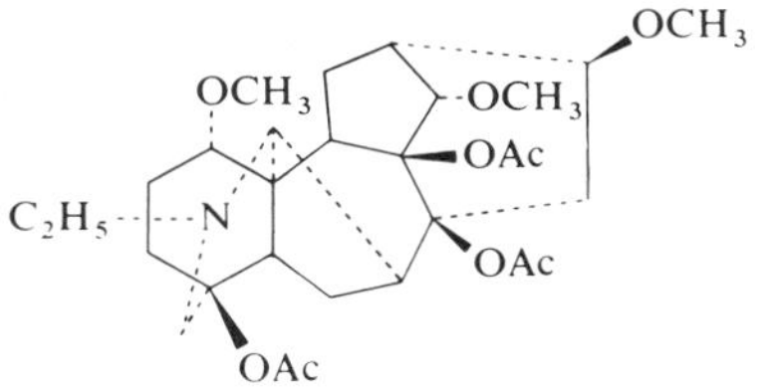

191

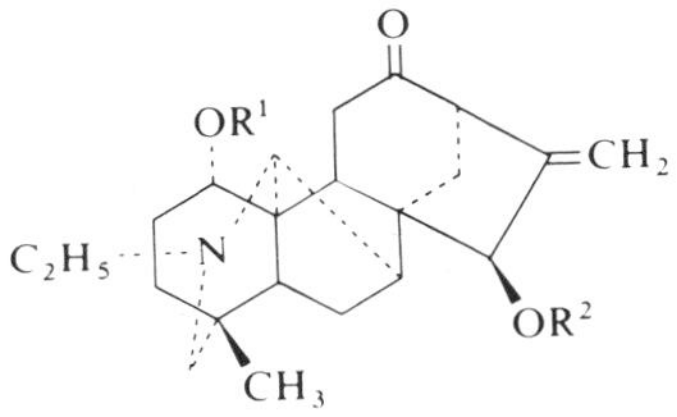

192 $R^1 = R^2 = Ac$
193 $R^1 = Ac; R^2 = H$

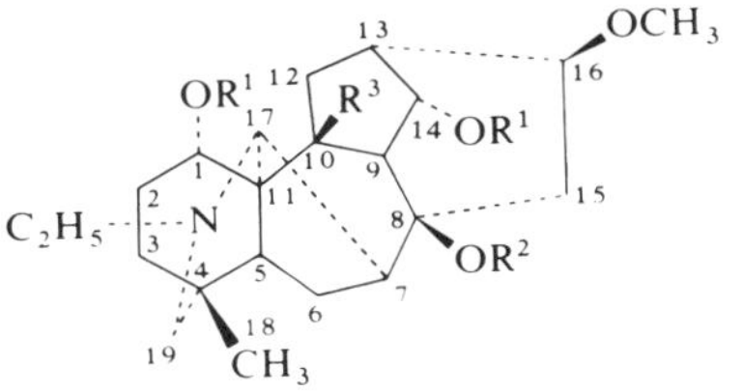

107 $R^1 = R^2 = H; R^3 = OH$
108 $R^1 = R^2 = R^3 = H$
194 $R^1 = R^2 = Ac; R^3 = OAc$
109 $R^1 = Ac; R^2 = H; R^3 = OH$

these interactions would be reduced. However, model studies indicate that the carbonyl of the C-1 acetyl group would also be hindered by the C-10, C-17, and C-12 hydrogens.

Songorinediacetate (**192**), in which the C-1 oxygen functionality is in a steric environment similar to that in the aconitine-type alkaloids, was also hydrolyzed partially to the C-1 monoacetyl derivative (**193**) in 45 min. On the other hand, the anamolous facile hydrolysis of the C-1 acetyl group in di- and tetraacetylkarakolidine (**109** and **194**) was explained by the steric influence of the C-10 hydroxyl group in these compounds.

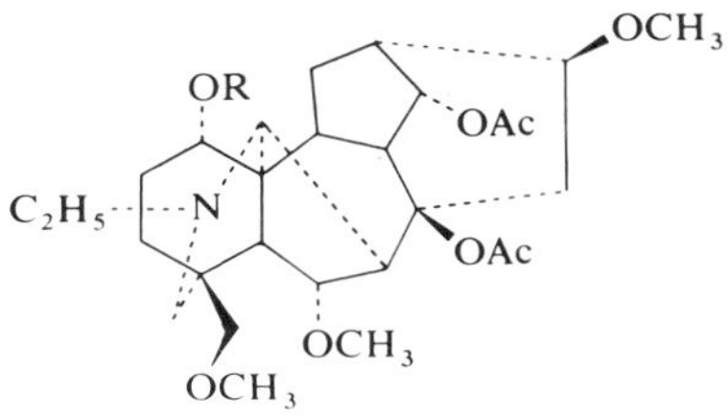

43 $R = H$
60 $R = Ac$

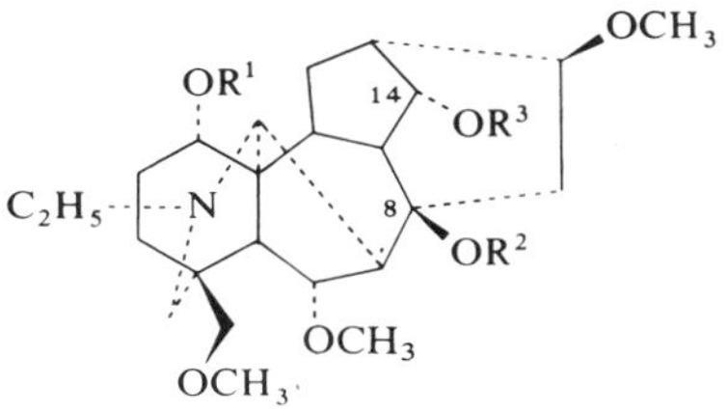

47 $R^1 = R^2 = R^3 = H$
59 $R^1 = R^3 = H; R^2 = Ac$
195 $R^1 = R^2 = Ac; R^3 = H$
196 $R^1 = Ac; R^2 = R^3 = H$

The Soviet hydrolysis studies are in agreement with the results of hydrolysis studies reported from our laboratory (*73*) for delphisine (**43**) and C-1 monoacetyldelphisine (**60**) (triacetylneoline). Hydrolysis of delphisine with aqueous methanolic potassium bicarbonate over two weeks at 35° gave the parent alkamine (**47**) and the partially hydrolyzed product, 8-acetylenoline (**59**). Under similar conditions of hydrolysis, delphisine monoacetate (**60**) formed 1,8-diacetyl neoline (**195**). Hydrolysis of delphisine acetate (**60**) with potassium carbonate in aqueous methanolic solution afforded the monoacetate (**196**) as the major product and the diacetate (**195**) and the parent alkamine (**47**) as minor products. These hydrolysis studies indicate that the rate of hydrolysis of the three acetyl groups in compound **60** is C-14 > C-8 > C-1.

It should be noted that steric or electronic effects may not be the only factors playing a role in these hydrolysis studies. Recent ^{1}H and ^{13}C NMR data from our laboratory (*66*) show that in the C_{19}-diterpenoid alkaloids with the C-1 α-hydroxyl group, ring A exists in a boat form stabilized by intramolecular hydrogen bonding (See Section II,K). On methylation *or* acetylation of the C-1 hydroxyl group, ring A is converted to the chair form. Nevertheless, the hydrolysis rate of the C-1 α-acetyl as well as the C-1 β-acetyl group in the different alkaloids was the same, a fact which indicates that the configuration at C-1 evidently has little effect on the hydrolysis rate.

V. ^{13}C NMR Spectroscopy of C_{19}-Diterpenoid Alkaloids

Interest in the application of ^{13}C NMR spectroscopy in the area of natural products during the last seven years has generated an abundance of fundamental substituent and additivity data for a series of simple and complex organic molecules. The ^{13}C NMR spectra of diterpenoid alkaloids not only reveal the number and type of carbon atoms in the system, but indicate close structural and family resemblances and give detailed information about the degree and sites of oxygen substitution. It is anticipated that ^{13}C NMR data will be of critical importance in determining the structure of newly isolated alkaloids from *Aconitum*, *Consolida*, *Delphinium*, and *Garrya* species. The importance of the ^{13}C NMR technique in solving difficult structural problems of complex diterpenoid alkaloids is discussed in this section.

In 1972, Jones and Benn (*100, 101*) made a significant contribution in the application of ^{13}C NMR spectroscopy by demonstrating the use of ^{13}C NMR studies in elucidating the structures of two new C_{19}-diterpenoid alkaloids. The ^{13}C NMR spectra of the several aconitine- and lycoctonine-type alkaloids and their derivatives, e.g., lycoctonine, deoxylycoctonine, deoxymethylenelycoctonine, browniine, isotalatizidine, delphonine, lycoctonal, and their corresponding hydrochloride *or* perchlorates salts, were examined in the Fourier mode at 22.63 MHz. With the help of proton decoupling techniques, additivity relationships, and the effects induced by specific structural changes, self-consistent assignments of nearly all the carbon resonances in these alkaloids were made and summarized in correlation diagrams. They observed particularly the general pattern of the ^{13}C chemical shifts of the quarternary carbons in diterpenoid alkaloids. The ^{13}C NMR data were utilized in the structure determination of two new C_{19}-diterpenoid alkaloids (alkaloids A and B, see Section II,S) isolated in small amounts from *D. bicolor*. Recently, we have revised (*102*) the ^{13}C chemical shift assignments of alkaloids A and B as well as the structure of alkaloid A mainly on the basis of ^{13}C NMR analysis of neoline (**47**), delphidine (**59**), delphisine (**43**), trimethoxylneoline (**56**), heteratisine (**197**), and 6-acetylheteratisine.

A comprehensive ^{13}C NMR study of nine aconitine-type alkaloids and 17 related derivatives has been also reported (*66*). The ^{13}C NMR spectra of aconitine (**4**), mesaconitine (**13**), deoxyaconitine (**19**), delphinine (**31**), chasmanine (**44**), delphisine (**43**), neoline (**47**), condelphine (**67**), isotalatizidine (**61**), as well as their derivatives have been analyzed. Certain corrections in some previously published ^{13}C chemical shift assignments for delphonine and isotalatizidine have also been reported. The differences in the oxygen substituents in all alkaloids greatly facilitated assignment of these chemical shifts. In addition, the oxygen functionalities were shown to cause chemical shifts of the carbons α and β to the carbon of direct attachment. This observation assisted in the calculation of ^{13}C chemical shifts using additivity relationships. The nature of the "pyrodelphinine chromophore" was also examined by the aid of ^{13}C and ^{1}H NMR spectroscopy. (See Section II,H.)

The problem of the incorrect assignment of the stereochemistry of the C-1 methoxyl group in chasmanine based on the chasmanine–browniine–lycoctonine correlations was investigated (*80, 81*) by ^{13}C NMR spectroscopy. The ^{13}C NMR spectra of lycoctonine and browniine and their corresponding acetates indicated that lycoctonine and browniine had identical stereochemistry at the C-1 position (β C-1 methoxyl), as do neoline and chasmanine (α C-1 functional group). Because the chemical reactions involved in the browniine–chasmanine correlation would not be expected to epimerize the oxygen function at C-1, it was concluded that the browniine–chasmanine correlation work must be in error.

The recent studies (*66, 100, 101, 102, 102a*) on ^{13}C NMR spectra of diterpenoid alkaloids have established a ^{13}C NMR data bank for the aconitine- and lycoctonine-type alkaloids which greatly facilitated structure elucidation of the alkaloid mentioned below.

The structures of two new C_{19}-diterpenoid alkaloids designated as falaconitine (**27**) and mithaconitine (**28**) and several known alkaloids isolated from the roots of *A. falconeri* have been elucidated with the aid of ^{13}C NMR spectroscopy. (*102a*) (See Section II,F.) Similarly, the structures of isodelphinine (**42**) and sachaconitine (**122**), alkaloids known for almost two decades, have been assigned (*71*) by the help of ^{13}C NMR spectroscopy.

This technique has also proven valuable in showing that "delsemine" is a mixture of two closely related alkaloids which differ only in the side chain of the anthranilic acid moiety (*131*).

Carbon 13 NMR spectroscopy has proven to be an important tool for solving structural and conformational problems in diterpenoid alkaloids. In the future, we may expect an increasing use of this technique for solving difficult structural problems in alkaloid chemistry. Unambiguous assignments of ^{13}C chemical shifts in diterpenoid alkaloids will also be useful in future biogenetic studies of these compounds.

VI. Mass Spectral Analysis of C_{19}-Diterpenoid Alkaloids

The mass spectra of C_{19}-diterpenoid alkaloids are complex, and there is a paucity of information concerning their fragmentation patterns. The first application of mass spectral data to the structure elucidation of complicated C_{19}-diterpenoid alkaloids involved a study of heteratisine and related alkaloids (*142*). On the basis of the mass spectral analysis of heteratisine (**197**), the structures of three related minor alkaloids, heterophyllidine (**198**), heterophyllisine (**199**), and heterophylline (200) were elucidated.

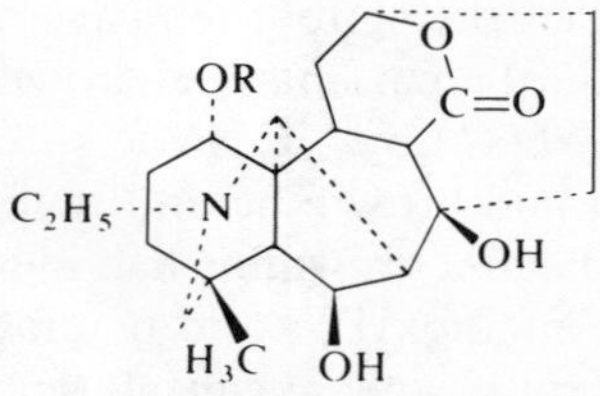

197 R = CH_3 Heteratisine
198 R = H Heterophyllidine

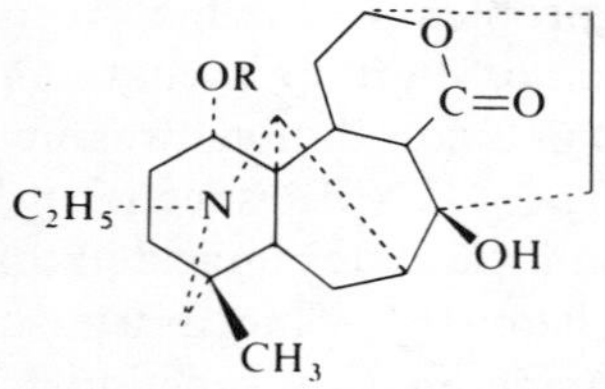

199 R = CH_3 Heterophyllisine
200 R = H Heterophylline

The main fragmentation involved the loss of the C-1 methoxyl group to give an ion at 360 (**202**). The transition state (**201**) was confirmed by accurate mass measurement and observation of a metastable peak. The

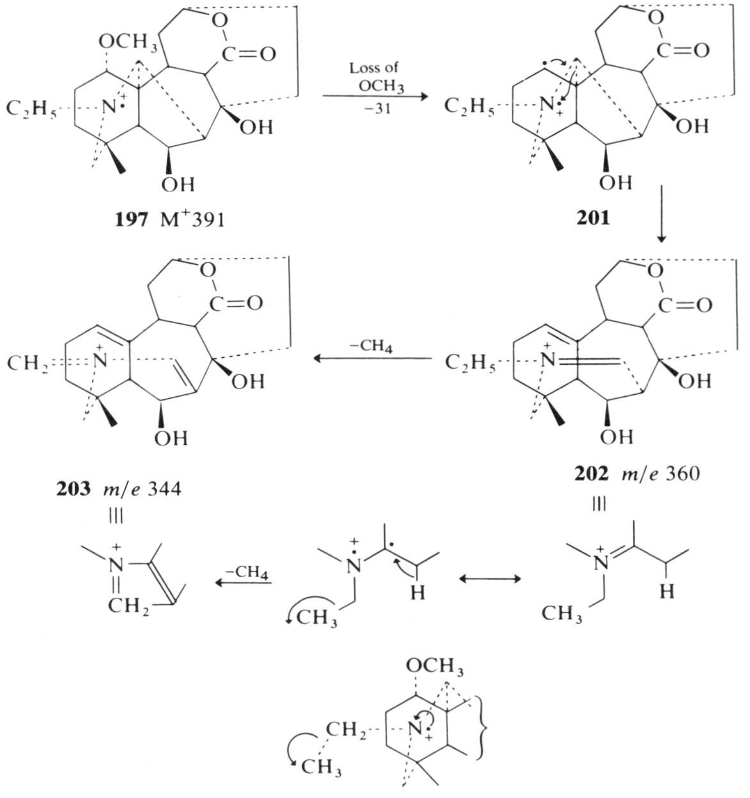

197 M^+391

201

202 *m/e* 360

203 *m/e* 344

204

loss of methane from **202** to give an ion at 344 (**203**) was also confirmed. The ion resulting from the loss of CH_4 from **202** was postulated to arise by the rearrangement shown above. The loss of methyl group from the parent molecular ion was also observed and explained by fragmentation of the N–CH_2–CH_3 group as shown in structure **204**.

In 1969 Yunusov and his co-workers (*143*) reported on their mass spectral studies of aconitine- and lycoctonine-type alkaloids. On the basis of mass spectral analysis of isotalatizidine (**61**), condelphine (**67**), talatizamine (**68**), neoline (**47**), aconitine (**4**), aconine, lycoctonine (**58**), and some of their derivatives, they reported that the main ionization center in these alkaloids is the nitrogen atom. During the study of lycoctonine-type alkaloids, they observed that the base peak is usually derived by the loss of the C-1 substituents as a radical. If a C-3 substituent was present, as in many of the aconitine-type alkaloids, the heteroring fragmented in a different manner. These fragmentation patterns were supported by the presence of metastable peaks in the mass spectra. The presence of one of most abundant ions ($M^+ - 18$) in the high-mass region of these spectra was explained by the loss of hydroxyl at the C-8 position. If the C-8 substituent was methoxyl or acetyl, the base peak corresponded to loss of the C-1 substituent and methanol *or* acetic acid, respectively. On the basis of their large mass spectra data bank, the Soviet scientists proposed on valid grounds fragmentation pathways for some of the observed ions which differed from those previously postulated.

Soviet chemists have examined (*144*) the pyrolyses of aconitine (**4**), benzoylacetyltalatizamine (**205**), diacetyltalatizamine (**206**), and 8-acetyl-14-ketotalatizamine (**207**) by mass spectral analyses at low (70–80°) and elevated (105–125°) temperatures. At temperatures above 100°, pyrolytic elimination of acetic acid occurred, while at 70–80° elimination was not observed. On the basis of observed steric interactions of the axial substituents at C-8, C-14, and C-16, they proposed that the relative ease of elimination of a C-8 acetate in these alkaloids is: (**4**) = (**205**) ≥ (**206**) > (**207**). Finally, they confirmed this postulate by observing the intensities of $M^+ - AcOH$ and $M^+ - AcO$ peaks and their molecular ion peaks in the mass spectra of these alkaloids.

At lower temperatures, pyrolytic elimination of the C-8 acetate was not observed in the mass spectra of these alkaloids. The Soviet chemists proposed that the formation of $M^+ - AcOH$ results from electron impact as shown in path A and with formation of the acetoxyl radical competing as in path B. The Soviet chemists suggested that compound **205** fragmented primarily according to path A at 125°, whereas at lower temperatures (70–80°) it fragmented mainly according to path B.

Yunusov and co-workers have further examined (*145*) the mass spectra of aconitine- and lycoctonine-type alkaloids to determine the orientation of the C-1 substituents in these alkaloids. They reported the mass spectral analysis of neoline, condelphine, isotalatizidine, talatizidine, acetylcondelphine, diacetyltalatizamine, aconine, lycoctonine, delphatine, and browniine. On the basis of comparison made of the mass spectra of the alkaloids bearing a C-1 α-hydroxyl group with that of talatizidine, which contains C-1 β-hydroxyl group, they observed that alkaloids with C-1 β-OH group exhibited a significant stable molecular ion peak with significant decrease in the intensity of the $M^+ - 17$ peak and a significant increase in the intensity of the $M^+ - 15$ peak. When comparison of the mass spectra of talatizamine and aconine (C−1 α-methoxyl group) was made with those of lycoctonine (**58**), delphatine (**147**), and browniine (**57**) (C-1 β-methoxyl group), no significant stable molecular ion peak was observed. But a sharp increase in the intensity of the $M^+ - 15$ peak was observed with

205 R = Bz
206 R = Ac

Pathway A

205 $R^1 = H; R^2 = OBz$
206 $R^1 = H; R^2 = OAc$
207 $R^1R^2 = O$

Pathway B

the change from the α- to the β-configuration. The ratios of $M^+ - OCH_3/M^+$ and $M^+ - OCH_3/M^+ - 15$ were found significantly larger in most, though not all, alkaloids with the C-1 α-configuration in comparison with the C-1 β-configuration. This relationship is valid in the case of the C-1 methoxyl-substituted alkaloids. These conclusions were supported by the presence of metastable peaks in the mass spectra of these alkaloids.

Because the methoxymethylene group attached at C-4 normally is not involved in the fragmentation of these alkaloids, the mass spectra of various carbinolamine ethers were utilized to determine the nature of the C-4 substituents. From the published mass spectral data, no definite conclusion about the nature of C-4 groups can be predicted.

In 1973, Waller and his colleagues (*134, 146*) reported the mass spectral analysis of the lycoctonine-type alkaloids delcosine (**163**), acetyldelcosine (**164**), and delsoline (**165**). They mentioned that these alkaloids showed a similar fragmentation pathway. As an example, the mass spectrum of delcosine is discussed here. Waller and his group also studied the mass spectra of the tetramethylsilane (TMS) derivatives of delcosine, acetyldelcosine, and delsoline, and found that the results obtained were in agreement with the proposed fragmentation outlined for delcosine.

Delcosine exhibited a molecular ion peak at 453 and a base peak of 438 ($M^+ - 15$). The mass spectrum also gave intense peaks corresponding to the loss of the CH_3 (*m/e* 438), hydroxyl (*m/e* 436), methoxyl (*m/e* 422), methanol (*m/e* 421), and both methyl and water (*m/e* 420). The common

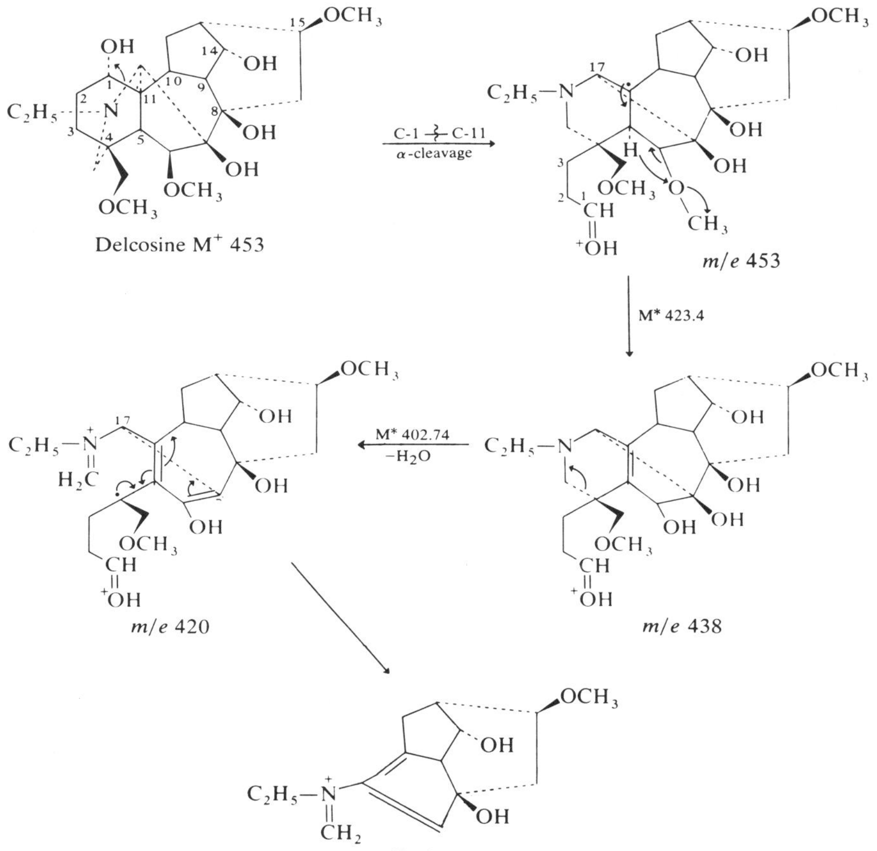

Delcosine M^+ 453

C-1 ⫻ C-11 α-cleavage

m/e 453

M* 423.4

m/e 438

M* 402.74 $-H_2O$

m/e 420

Pathway C

pathway of fragmentation of these three alkaloids was the formation of an M^+-189 ion peak. A mechanism postulated for the formation of the M^+-189 fragment ion for delcosine is shown in path C. The proposed structure for the m/e 264 fragment was confirmed by the accurate mass measurements for the M^+-189 ion.

VII. Synthetic Investigations

A. The Synthesis of Delphinine

Over the last ten years Professor Wiesner and his co-workers at the University of New Brunswick have made important contributions to the synthesis of delphinine (**31**) and related diterpenoid alkaloids (*147–150*).

They have achieved their goals in an elegant manner and proved that these complex alkaloids can be synthesized. In 1969 they reported the synthesis of compound **208**, and with it the solution of many of the key problems involved in the synthesis of the partially aromatic degradation product (**3**) of delphinine. After the synthesis of compound **208**, they effected (*151*) the synthesis of compound **3** by a similar route and correlated it with a degradation product of delphinine. This correlation resulted in a revision of the configuration of the C-1 methoxyl group in delphinine and its relatives. Because compound **208** is similar, except for the presence of the C-1 methoxyl group in the A ring, to degradation product **3**, only the synthesis of **3** will be presented in detail. As this partially aromatic product (**3**) can be obtained in relatively high yield from delphinine, it is an ideal advance relay compound for the total synthesis of delphinine.

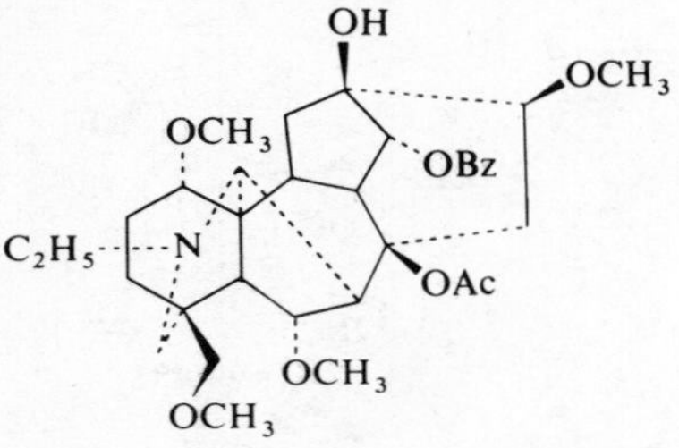

31 Delphinine

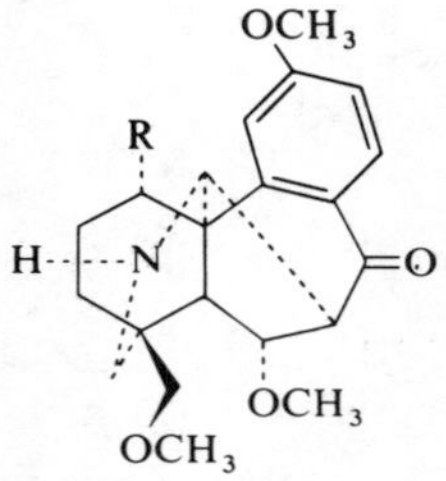

3 $R = OCH_3$
208 $R = H$

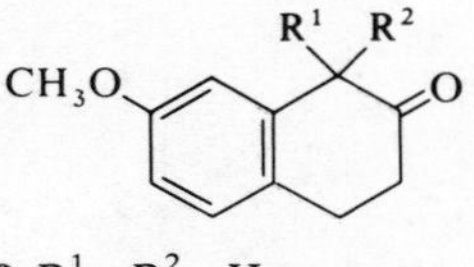

209 $R^1 = R^2 = H$
210 $R^1 = H; R^2 = CH_2{-}CH{=}CH_2$
211 $R^1 = CH_2{-}O{-}CH_2{-}C_6H_5$; $R^2 = CH_2{-}CH{=}CH_2$
212 $R^1 = CH_2{-}O{-}CH_2{-}C_6H_5$; $R^2 = CH_2{-}CH(OH){-}CH_2OH$
213 $R^1 = CH_2{-}O{-}CH_2{-}C_6H_5$; $R^2 = CH_2{-}CHO$

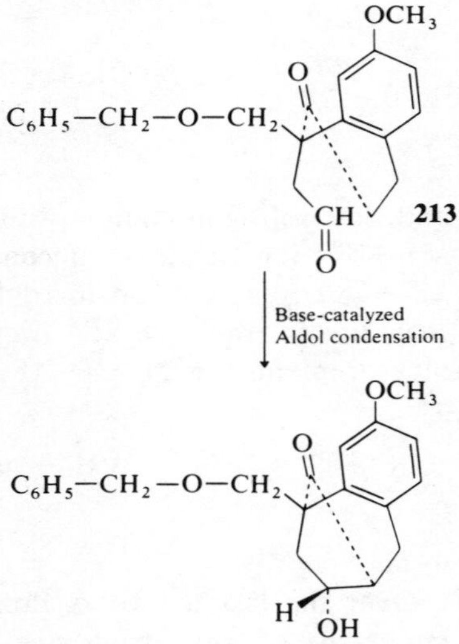

213

Base-catalyzed Aldol condensation

214

The starting methoxytetralone (**209**) was converted by the Stork pyrrolidine–enamine procedure to the alkyltetralone (**210**). On further alkylation with benzyl chloromethyl ether and sodium hydride in benzene, the latter yielded compound **211**, which was oxidized with osmic acid and sodium chlorate in tetrahydrofuran (THF) to give two diastereoisomeric diols (**212**). The aldehyde **213** was obtained in quantitative yield from compound **212** by treatment with excess metaperiodate in aqueous THF.

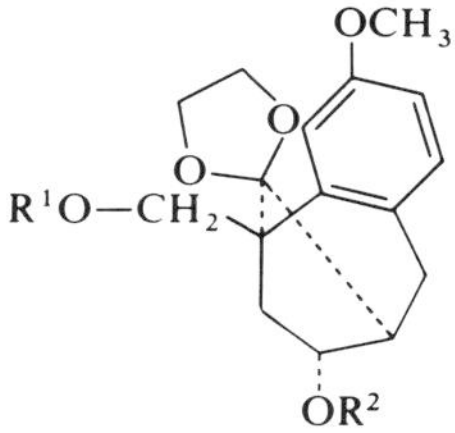

OCH3
R1O—CH2
OR2

215 $R^1 = CH_2{-}C_6H_5$; $R^2 = H$
216 $R^1 = CH_2{-}C_6H_5$; $R^2 = Ac$
217 $R^1 = H$; $R^2 = Ac$
218 $R^1 = THP$; $R^2 = Ac$
219 $R^1 = THP$; $R^2 = H$
220 $R^1 = THP$; $R^2 = CH_2{-}C_6H_5$
221 $R^1 = H$; $R^2 = CH_2{-}C_6H_5$

The crucial step in this synthesis, a base-catalyzed aldol condensation, furnished the hydroxyketone **214** in quantitative yield. Thus, three of the five rings of target compound **3** were constructed by a simple reaction sequence. The hydroxyketone was converted to the ketal **215** by a standard method. The benzyl blocking group was then transferred from the primary to the secondary alcoholic function in 65% overall yield by conventional reaction sequences, as shown above. At this point, the authors suggested that a much more direct conversion of **215** to **221** may be possible, using the acid-sensitive *p*-methoxybenzyl group as R^1 in compound **215**. This operation was not carried out, however, for the overall yield from **215** to **221** was satisfactory.

The primary alcohol **221** was oxidized to aldehyde **222** by chromium trioxide in pyridine in 85% yield. The alcohol **223** was obtained from compound **222** by treatment with Grignard reagent (**225**). The bromo compound (**224**) required for preparation of the Grignard reagent was prepared as outlined above in overall low yield.

Alcohol **223** was found to be in the configuration opposite to that of natural delphinine. Therefore, alcohol **223** was oxidized by Jones reagent to ketone **226** and the latter reduced with lithium aluminum hydride in dioxane to a mixture of alcohols **223** and **227** in a ratio of 7 : 3. The desired

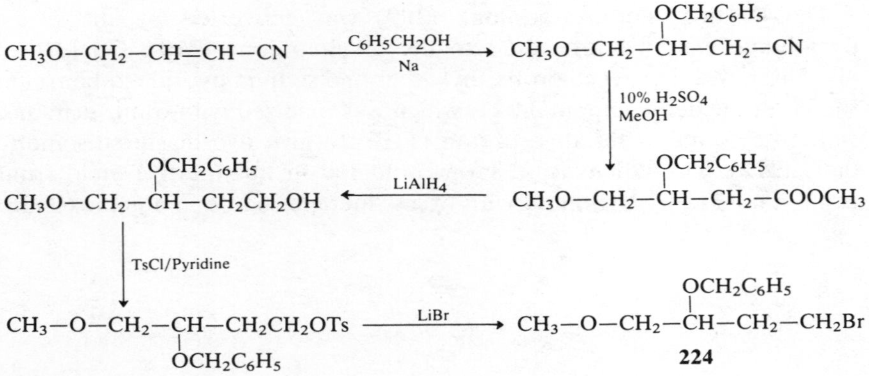

$CH_3O{-}CH_2{-}CH{=}CH{-}CN \xrightarrow[Na]{C_6H_5CH_2OH} CH_3O{-}CH_2{-}CH(OCH_2C_6H_5){-}CH_2{-}CN$

$\xrightarrow[MeOH]{10\%\ H_2SO_4} CH_3O{-}CH_2{-}CH(OCH_2C_6H_5){-}CH_2{-}COOCH_3$

$\xrightarrow{LiAlH_4} CH_3O{-}CH_2{-}CH(OCH_2C_6H_5){-}CH_2CH_2OH$

$\xrightarrow{TsCl/Pyridine} CH_3{-}O{-}CH_2{-}CH(OCH_2C_6H_5){-}CH_2CH_2OTs$

$\xrightarrow{LiBr} CH_3{-}O{-}CH_2{-}CH(OCH_2C_6H_5){-}CH_2{-}CH_2Br$

224

OCH_3, O, O, OHC, $OCH_2{-}C_6H_5$

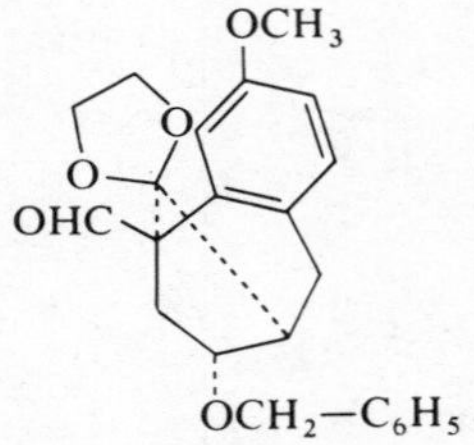

222

$CH_3{-}O{-}CH_2{-}CH(OCH_2C_6H_5){-}CH_2{-}CH_2{-}MgBr$

225

alcohol (**227**) was selectively acetylated in the mixture and separated by silica gel chromatography. The recovered epimer (**223**) was recycled in the oxidation–reduction sequence to yield an additional quantity of epimer **227**. The pure epimer (**227**) obtained by saponification of the acetate was methylated with an excess of methyl iodide and sodium hydride in refluxing dioxane to yield ketal **228**. The latter was hydrolyzed under reflux with aqueous acetic acid to give compound **229** in quantitative yield. Compound **229** was subjected to amination with Raney nickel in methanolic ammonia,

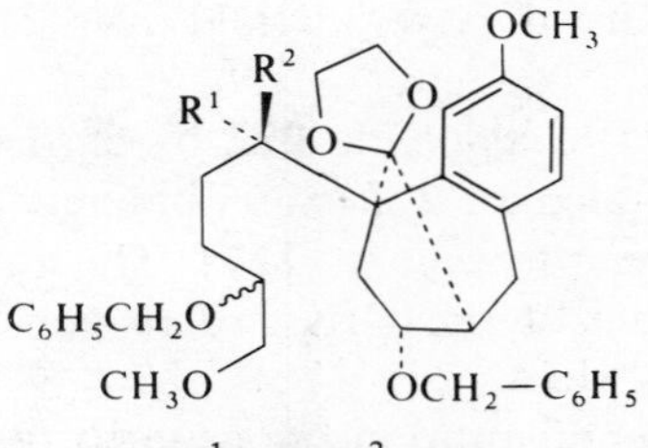

OCH_3, R^2, R^1, O, O, $C_6H_5CH_2O$, CH_3O, $OCH_2{-}C_6H_5$

223 $R^1 = H$; $R^2 = OH$
226 $R^1R^2 = O$
227 $R^1 = OH$; $R^2 = H$
228 $R^1 = OCH_3$; $R^2 = H$

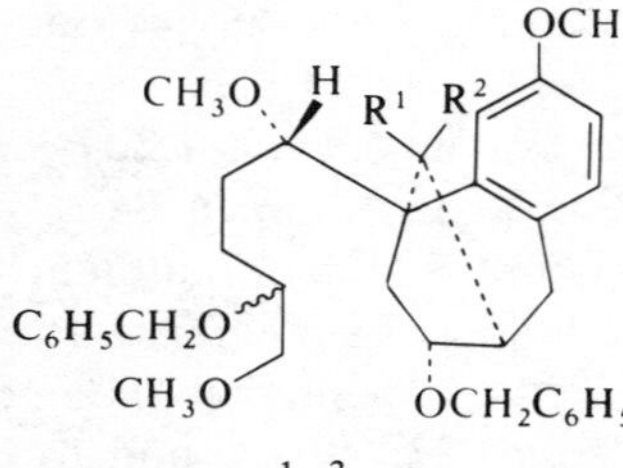

OCH_3, CH_3O, H, R^1, R^2, $C_6H_5CH_2O$, CH_3O, $OCH_2C_6H_5$

229 $R^1R^2 = O$
230 $R^1 = NH_2$; $R^2 = H$

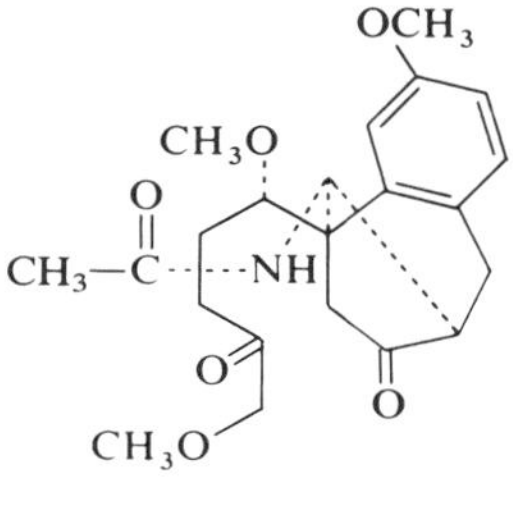

231

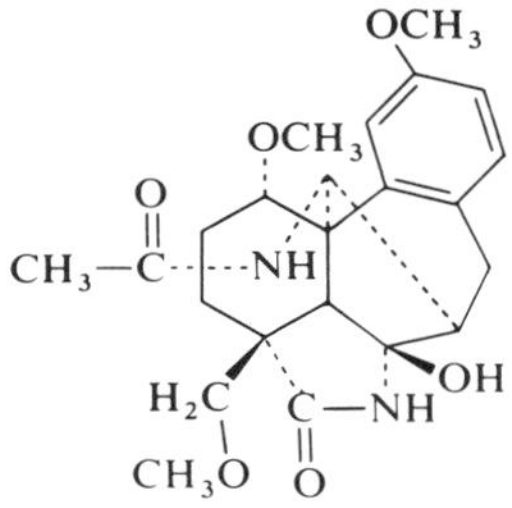

232

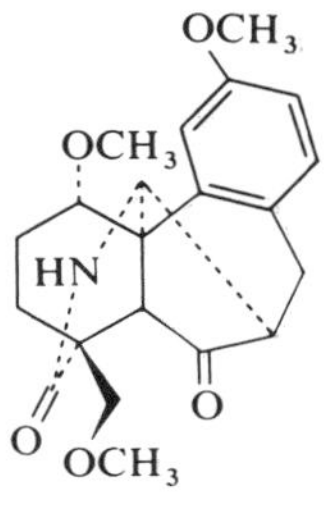

233

and the resulting crude amine (**230**) was acetylated with acetic anhydride in pyridine. The protecting benzyl groups were removed by hydrogenolysis, and the resulting hydroxyl groups were oxidized with chromium trioxide in pyridine to obtain the diketone (**231**). The latter was heated for 6 hr with potassium cyanide in aqueous ethanol to afford the desired lactamol (**232**). This reaction is believed to proceed via a base-catalyzed aldol condensation of the diketone **231** to an α,β-unsaturated ketone, which then adds the cyanide moiety. The nitrile group is subsequently hydrolyzed to a primary amide, which tautomerizes to the desired lactamol (**232**).

The ketolactam (**233**) was prepared from the lactamol (**232**) by heating in a mixture of methanol and concentrated hydrochloric acid. The ketolactam was reduced with lithium aluminum hydride in refluxing dioxane to afford the desired exoalcohol (**234**) and its epimer (**235**) in a 1 : 1 ratio. These epimers were separated by chromatography on alumina. The undesired epimer (**235**) was converted by Jones oxidation to the corresponding ketone (**236**), which was reduced with sodium in refluxing absolute alcohol to yield alcohols **234** and **235** in a more favorable 7 : 3 ratio. The exoalcohol **234** was acetylated with acetic anhydride in pyridine, and the product was hydrolyzed by heating with dilute methanolic potassium hydroxide to obtain the *N*-acetate alcohol (**237**). The latter was

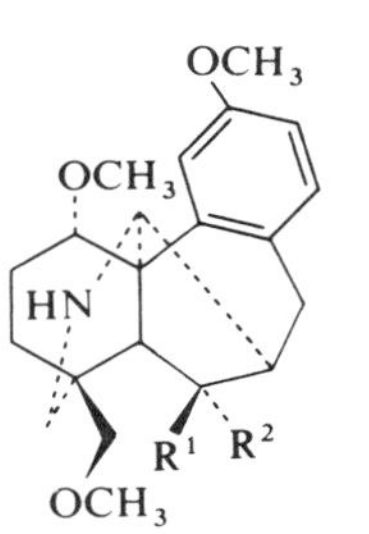

234 $R^1 = H; R^2 = OH$
235 $R^1 = OH; R^2 = H$
236 $R^1R^2 = O$

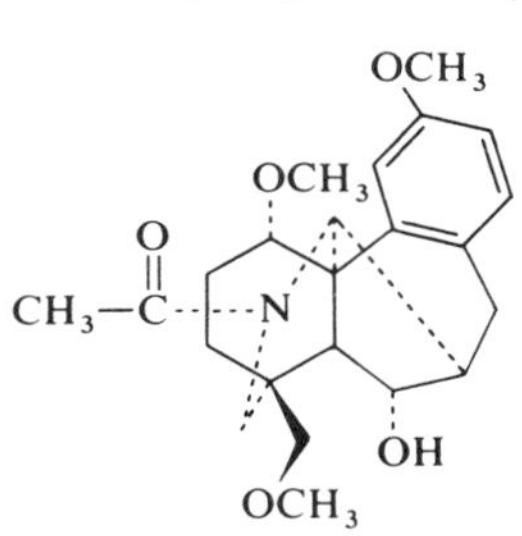

237

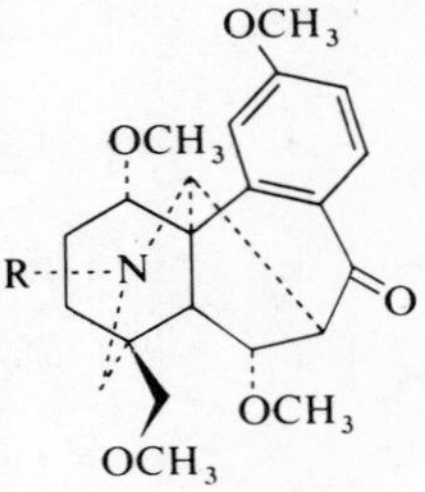

OCH_3 OCH_3 R N O OCH_3 OCH_3

238 R = C(=O)—CH_3
241 R = CHO
3 R = H

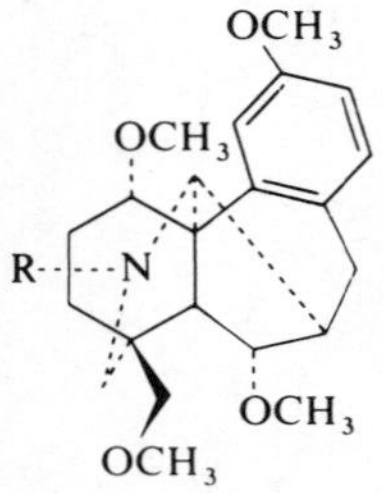

OCH_3 OCH_3 R N OCH_3 OCH_3

239 R = CH_3
240 R = CHO

methylated by a standard method, and the resulting product was converted to the desired intermediate (**238**) by Jones oxidation. The synthetic racemate (**238**) and the "natural" degradation product of delphinine were identical. Because compound **238** was exceedingly difficult to hydrolyze, the authors developed a second route for the final stages of the synthesis of compound **3**.

The exo-alcohol **234** was methylated with sodium hydride and methyl iodide in refluxing dioxane. The resulting tetramethoxyl *N*-methyl derivative (**239**) was converted to the *N*-formyl derivative (**240**) by oxidation with potassium permanganate in acetone at room temperature. Compound **240** was converted by Jones oxidation to compound **241**. The latter was also obtained from **239** by prolonged oxidation with excess potassium permanganate in acetone–acetic acid. The synthetic racemate (**241**) and the "natural" delphinine derivative gave identical IR, NMR, and mass spectra. The formyl derivative was easily hydrolyzed to the desired product (**3**) by refluxing with dilute hydrochloric acid. The synthetic racemate **3** was found to be indistinguishable from the optically active "natural" derivative obtained from delphinine. It should be noted that this stereoselective synthesis of **3** did not provide any evidence for the configuration of the C-1 methoxyl group.

To establish the configuration of the methoxyl group in ring A, an X-ray crystallographic analysis of the acid oxalate of the synthetic racemate **3** was performed. The X-ray structure determination corroborated all the features of the molecule and revealed the configuration of the C-1 methoxyl to be as shown in structure **3** (α, and cis to the nitrogen bridge). To obtain the synthetic optically active compound **3**, the racemate, *dl*-**3**, was resolved into its optical antipodes by selective reaction of the unwanted antipode with L-camphor sulfonyl chloride in pyridine. The unreacted free base was converted to the acid oxalate and crystallized to constant melting point.

This acid salt was shown to be identical with the oxalate derived from the "natural" degradation product. The totally synthetic optically active base obtained from the oxalate showed IR, NMR, and mass spectra which were superimposable with those of the "natural" compound (**3**).

This synthesis represents an outstanding achievement and provides a clarification of the structure of delphinine and related aconitine-type alkaloids.

B. The Total Synthesis of Talatizamine (Talatisamine)

In 1974 Wiesner and his co-workers reported (*96*) the first formal total synthesis of talatisamine (**68**) starting from an atisine-type intermediate (**242**). A key step in this synthesis involved a rearrangement of the atisine skeleton to the aconitine-type skeleton, a reaction previously suggested (*152*) for the biogenesis of the aconitine-type alkaloids. This type of rearrangement, which was reported by Johnston and Overton (*153, 154*) and also by Ayer and Deshpande (*155*) will be discussed later in Section VII,E.

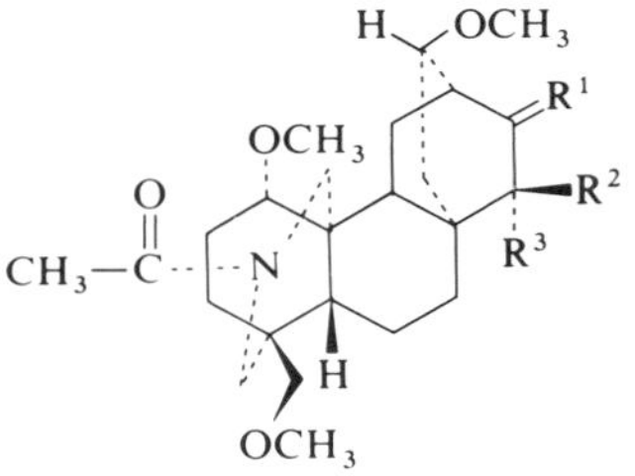

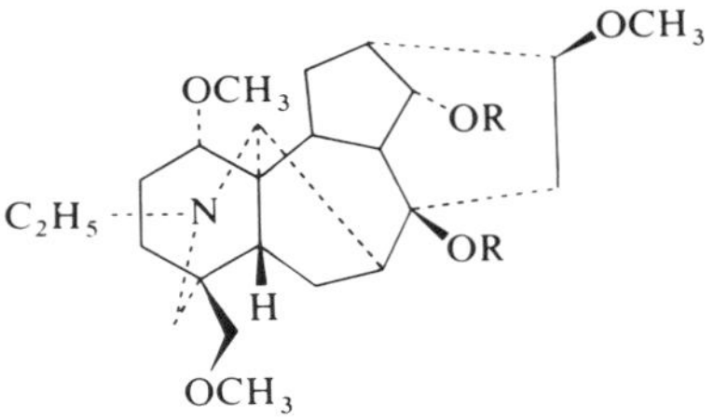

242 $R^1 = O$; $R^2 = R^3 = H$
243 $R^1 = O$; $R^2 = OCOC_6H_5$; $R^3 = H$
244 $R^1 =$ (ethylenedioxy ketal); $R^2 = OCOC_6H_5$; $R^3 = H$
245 $R^1 =$ (ethylenedioxy ketal); $R^2 = OH$; $R^3 = H$
246 $R^1 =$ (ethylenedioxy ketal); $R^2 = R^3 = O$

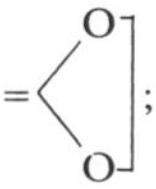

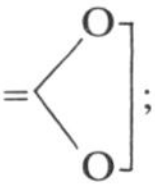

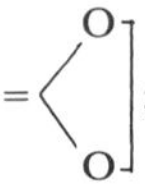

68 R = H
70 R = Ac

Benzoylation of compound **242** with sodium hydride and benzoyl peroxide gave the benzoate (**243**) in 65% yield. Ketalization of **243** with ethylene glycol and trimethyl orthoformate in the presence of sulfuric acid

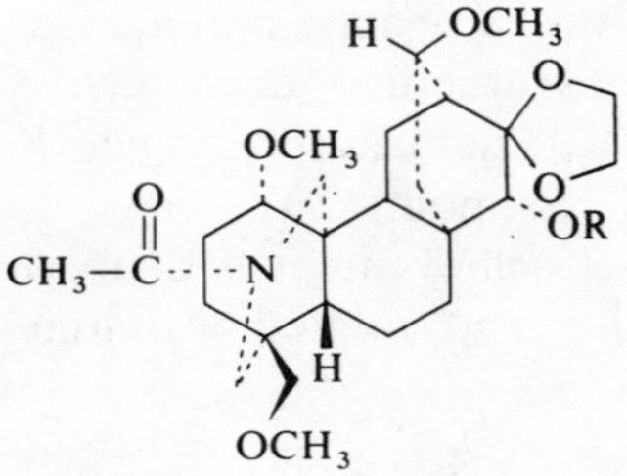

247 R = H
248 R = Ts

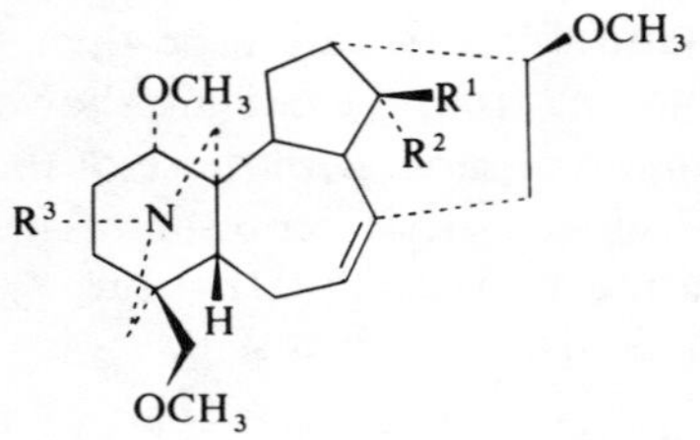

249 R^1R^2 = (O–CH₂CH₂–O ketal); R^3 = Ac

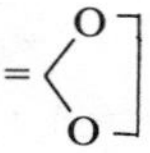

250 R^1R^2 = (O–CH₂CH₂–O ketal); $R^3 = C_2H_5$

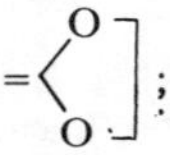

251 R^1R^2 = O; $R^3 = C_2H_5$
252 R^1 = H; R^2 = OH; $R^3 = C_2H_5$

afforded ketal **244**. Alkaline hydrolysis of **244** and oxidation of the resulting hydroxyketal **245** with chromium trioxide–pyridine in methylene chloride afforded ketone **246**. Sodium borohydride reduction of **246** in aqueous THF–methanol gave alcohol **247**, which was converted into the corresponding tosylate (**248**). The structure of compound **248** has been confirmed by an X-ray analysis (*156*). The racemic tosylate **248** was converted to the desired rearranged product (**249**) in 40% yield by heating in a 1 : 1 mixture of dimethyl sulfoxide and tetramethylguanidine at 180° for 24 hr. Reduction of compound **249** by lithium aluminum hydride in dioxane yielded the corresponding racemic *N*-ethyl compound (**250**). The latter compound was identical with the corresponding optically active derivative prepared from talatisamine as described below.

Talatizamine was converted to the diacetate (**70**), which was reductively cleaved to compound **252**. Jones oxidation of **252** afforded ketone **251**, which was converted to ketal **250** with ethylene glycol and *p*-toluenesulfonic acid in benzene.

Deketalization of compound **250** in aqueous methanolic hydrochloric acid yielded ketone **251** in 70% yield. Sodium borohydride reduction of the latter in THF–methanol afforded stereospecifically alcohol **252**. Oxidation of **252** with mercuric acetate yielded talatizamine in 40% yield.

C. Syntheses Directed toward Chasmanine

Recently, Wiesner and co-workers have reported (*157*) the stereospecific total synthesis of compound **253**, an intermediate to be used in the total synthesis of chasmanine (**49**). The starting material, the methoxyindanone (**254**), was converted to enol ether **255**, which was carboxylated with carbon dioxide–*n*-butyllithium in THF at −70° to afford the acid (**256**). The latter was converted into methyl ester **257** by treatment with HCl gas and methanol. The methyl ester (**257**) was converted to the tricyclic ester (**260**) via intermediates **258** and **259**, as described in earlier work (*158*). Reduction of the ester (**260**) with $LiAlH_4$ and oxidation of the resulting primary alcohol (**261**) with dicyclohexylcarbodiimide in dimethyl sulfoxide afforded aldehyde **262** in 89% yield. The side chain (**269**) was attached to **262** by a Grignard reaction to give the diastereomeric mixture (**270**) in 90% yield, as reported in previous cases (*151, 159*).

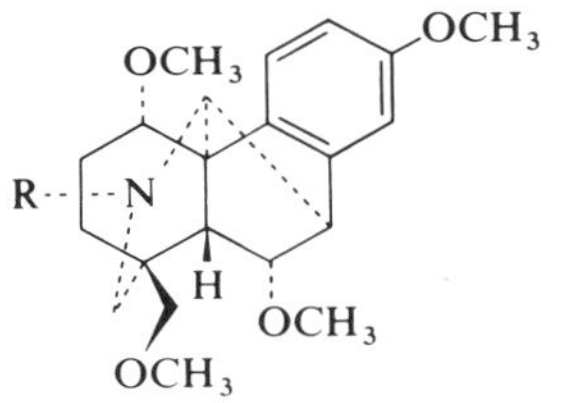

253 R = CHO
276 R = CH_3

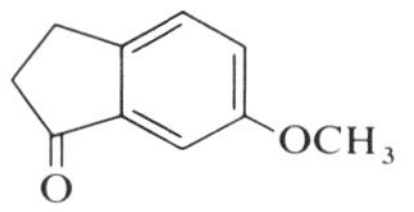

254

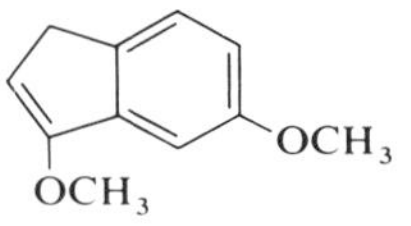

255

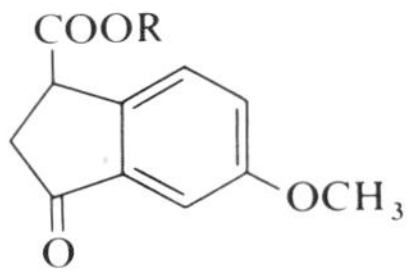

256 R = H
257 R = CH_3

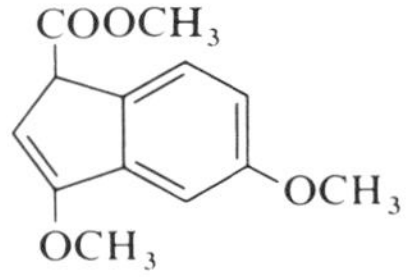

258

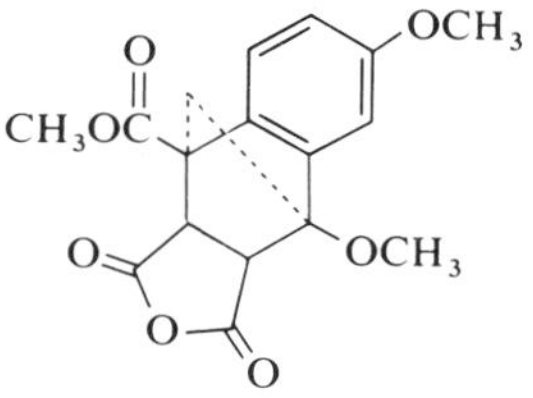

259

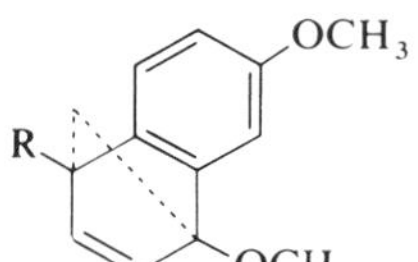

260 R = $COOCH_3$
261 R = CH_2OH
262 R = CHO

263 R = H
264 R = CPh_3

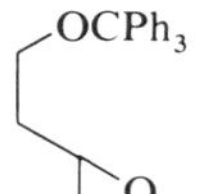

265

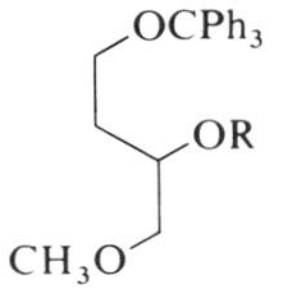

266 R = H
267 R = CH_2Ph

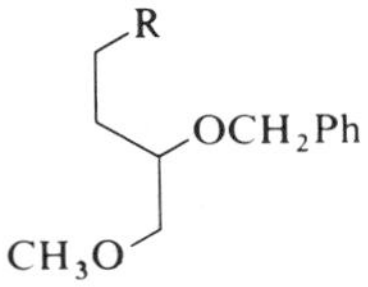

268 R = OH
269 R = Br

The bromo compound **269** was prepared from the commercially available homoallyl alcohol (**263**) as follows: Tritylation of **263** gave the ether (**264**), which was converted by treatment with perbenzoic acid to epoxide **265**. The latter was converted to methoxyl alcohol **266** by opening the epoxide ring in methanolic sodium methoxide. The primary alcohol **268** was prepared by benzylation of **266** with sodium hydride and benzyl chloride in dioxane, followed by treatment of the product (**267**) with 90% acetic acid at 45° for 24 hr. The alcohol (**268**) was converted to the bromide (**269**) as reported earlier (*151*).

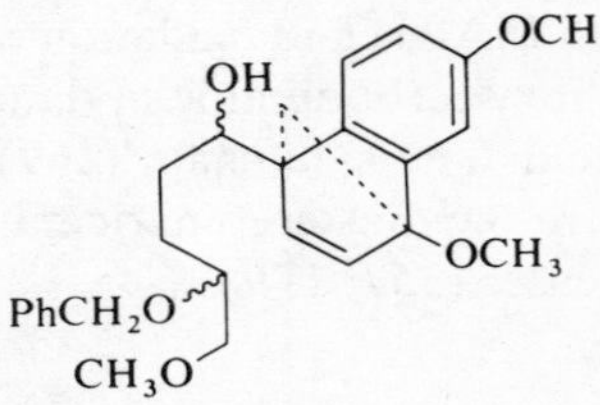

270 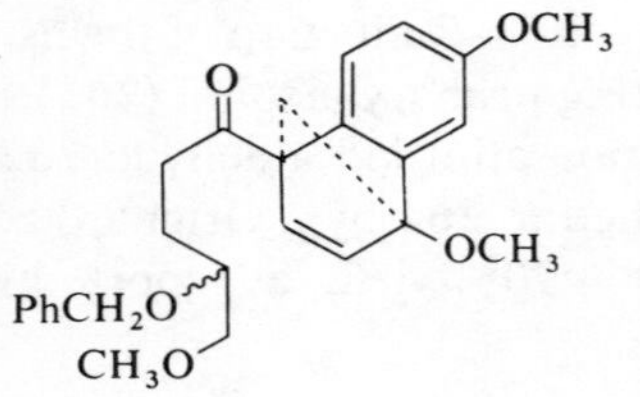 **271**

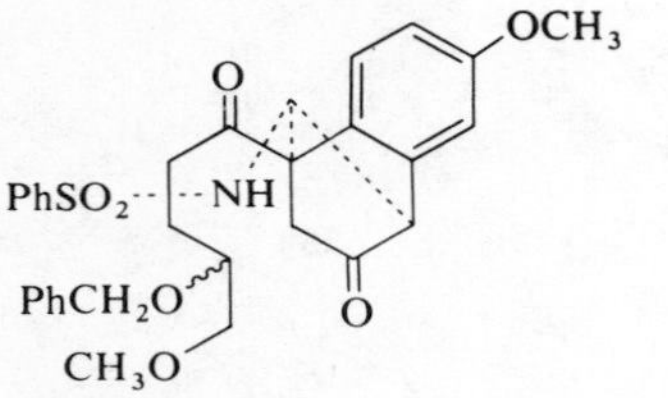

272

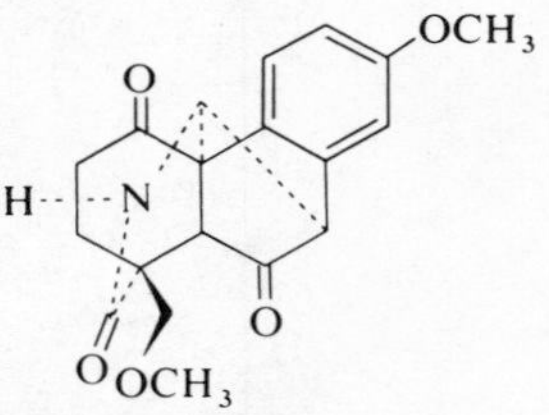

273

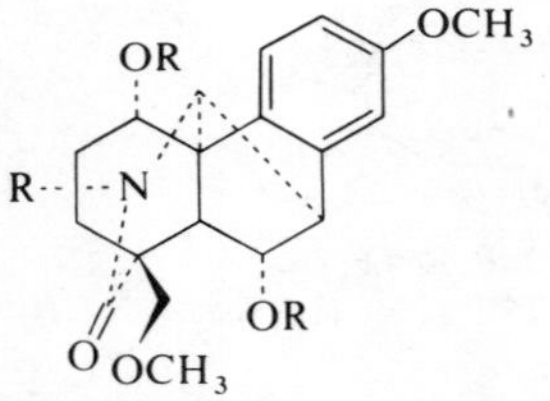

274 R = H
275 R = CH_3

Oxidation of compound **270** with chromium trioxide in pyridine gave the ketone (**271**). The latter was converted to diketone **272** in one step by treatment with a large excess of benzenesulfonylazide in glacial acetic acid containing a small amount of *p*-toluenesulfonic acid. The diketone **272** was converted to the diketolactam (**273**) by the methods worked out in the synthesis of napelline (*159*). Reduction of compound **273** with lithium tri-*tert*-butoxyaluminohydride yielded stereospecifically the dihydroxyl-lactam (**274**) in quantitative yield. Treatment of compound **274** with methyl iodide and sodium hydride in dry dioxane gave the tetramethoxyl-*N*-methyllactam (**275**) in 62% yield. Reduction of compound **275** with lithium aluminum hydride afforded amine **276**. The latter was oxidized with potassium permanganate in acetic acid to yield the *N*-formyl derivative (**253**).

OCH₃ O O O H H H CH₂

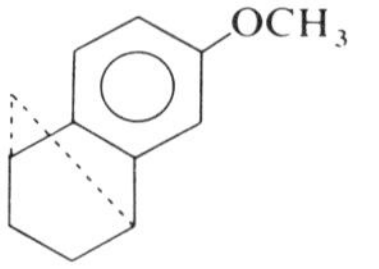

277 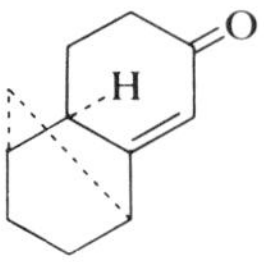 **278** 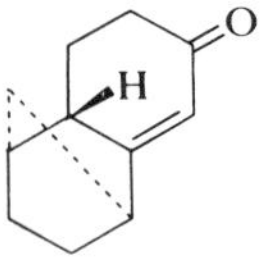 **279** 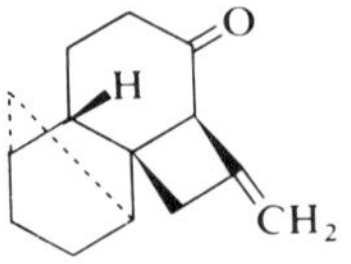 **280**

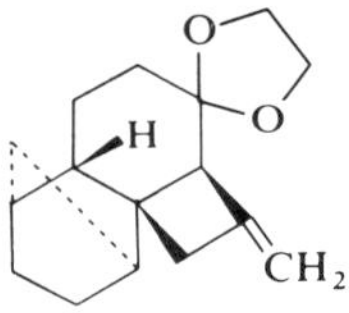

281 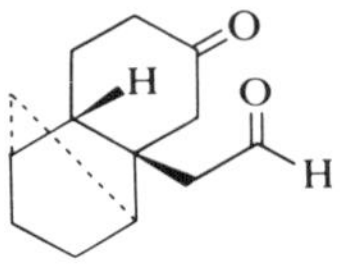 **282** 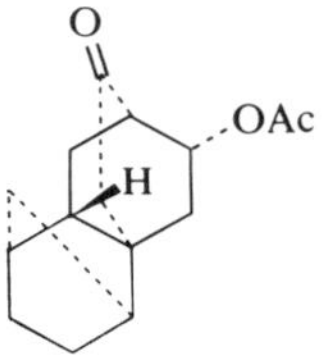 **283**

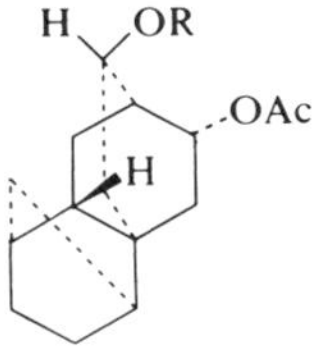

284 R = H
285 R = CH_3

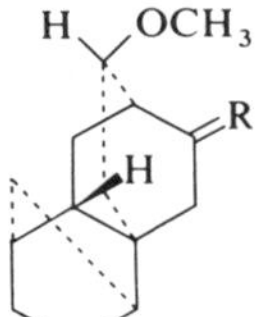

286 R = O
287 R = (ethylenedioxy)

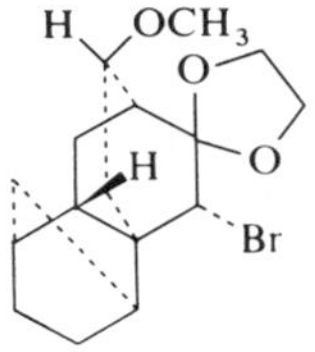

288

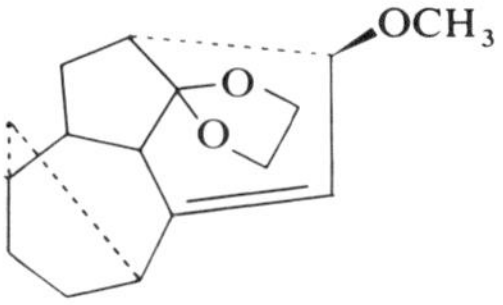

289

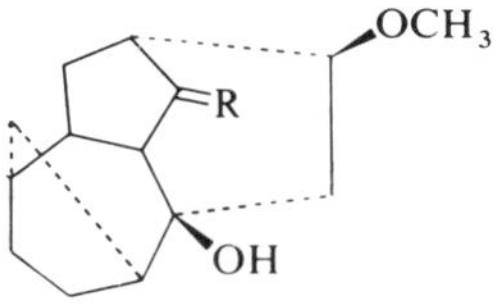

290 R = (ethylenedioxy)
291 R = O

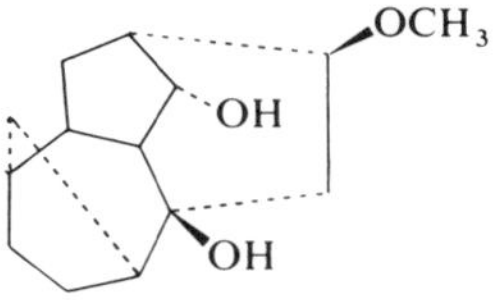

292

The reactions for the conversion of the *N*-formyl derivative (**253**) to chasmanine were studied (*160*) in a model system starting with compound **277**. Birch reduction of **277** afforded compound **278**, which was subjected to equilibration in methanolic hydrochloric acid–dioxane under reflux for six days to give epimers **278** and **279** in a ratio of 2.7 : 1. These epimers were separated by careful column chromatography on silica gel. The photoaddition of allene to the α,β-unsaturated ketone **279** afforded the stereospecific adduct **280**. The latter was converted to ketal **281**, and this compound gave by ozonolysis, sodium borohydride reduction, and acid treatment the ketoaldehyde (**282**) in 92% yield. Cyclization of compound **282** with BF_3–etherate in acetic anhydride and acetic acid yielded acetoxyl ketone **283**. The equilibrium in this condensation favored the retroaldol reaction, and neither base nor acid treatment effected ring closure. This condensation could be performed only under acetylating conditions. Compound **283** was reduced with sodium borohydride to afford the acetoxyl alcohol (**284**). The latter was methylated with diazomethane in BF_3–etherate to give the methoxyl derivative (**285**). The acetoxyl group of **285** was hydrolyzed, and the resulting alcohol was oxidized with chromium trioxide in pyridine to the ketone (**286**). The latter was converted to the ketal (**287**) by refluxing with ethylene glycol and *p*-toluenesulfonic acid in benzene. Bromination of **287** with pyridine hydrobromide perbromide in THF gave bromoketal **288**. Compound **288** was rearranged to **289** in 90% yield by refluxing in benzene with 1,5-diazabicyclo[3.4.0]nonene for several days. Treatment of **289** with mercuric acetate in 50% aqueous acetone at room temperature yielded compound **290**. Deketalization of **290** by heating with 50% acetic acid afforded ketone **291**. The structure of **291** was confirmed by single-crystal X-ray crystallography *(157)*. Reduction of ketone **291** with sodium borohydride in methanol gave the alcohol (**292**) in a yield of 80%.

Because all the steps in this model system proceeded under mild conditions and in high yield, it was mentioned that this process could be applied without major modification to compound **253** with a suitably blocked nitrogen. It appears that the total synthesis of chasmanine is now feasible.

D. The Construction of the C/D Ring System of the C_{19}-Diterpenoid Alkaloids

In 1973 Wiesner and his colleagues reported (*161*) a new synthetic route for the construction of the substituted C/D ring system of delphinine-type alkaloids. This process was based on a previous photochemical synthesis of the atisine skeleton (*162*).

The starting material, methoxytetralin (**293**), was converted into the α,β-unsaturated ketone (**294**) by Birch reduction. The allene adduct (**295**) was prepared from ketone **294** and was then rearranged to the hydroxyl ketone (**298**) via intermediates **296** and **297**. Compound **298** was methylated with sodium hydride and methyliodide to give the methoxyketone (**299**) in quanitative yield. Compound **299** was rearranged to **300** by refluxing with *tert*-butyl perbenzoate and cuprous bromide in absolute

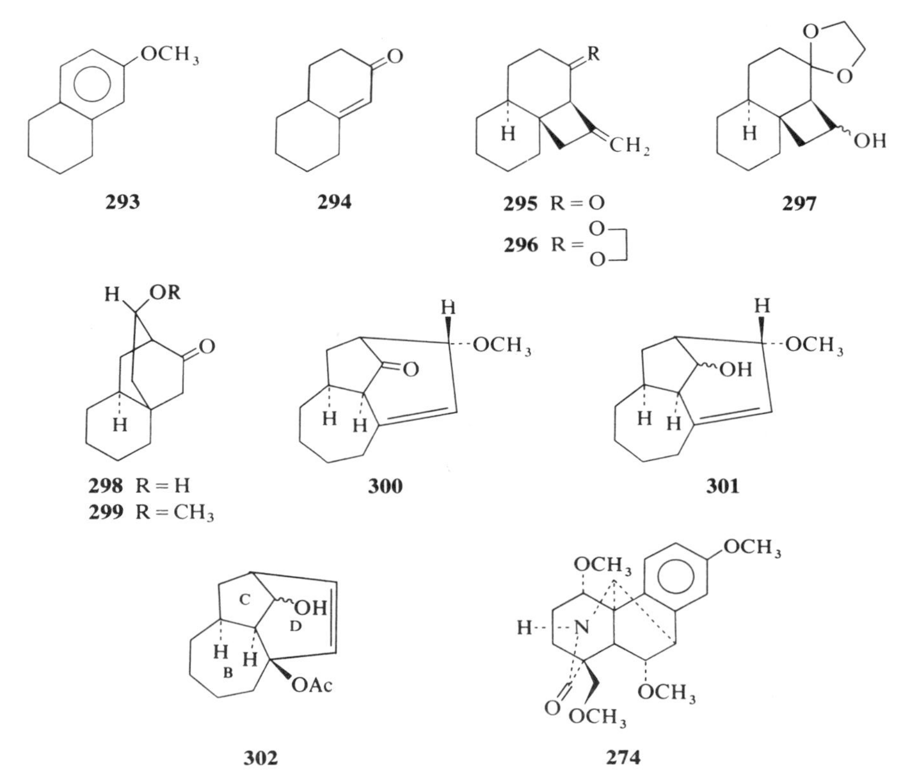

benzene for 24 hr. Reduction of **300** with lithium aluminum hydride yielded one epimeric alcohol (**301**). The latter underwent the typical pyro–isopyro rearrangement when heated with glacial acetic acid and *p*-toluenesulfonic acid to give the isopyrocompound **302**. Compound **302** contains the B, C, and D ring system of the C_{19}-diterpenoid alkaloids. Using these synthetic methods for construction of the C/D ring system, it should be possible to convert the previously prepared intermediate (**274**) to chasmanine. A similar construction of the C/D ring system is reviewed in Section VII,C.

E. Atisane-Aconane Conversions and Stereospecific Skeletal Rearrangements

Johnston and Overton have reported *(153, 154)* the conversion of the atisane skeleton to the aconitine skeleton, thus effecting the key step in the commonly accepted (*152*) route for the biogenesis of the aconitine-type alkaloids.

303 → A. HCl; B. Ac_2O → **304**

304 → C. KOH, $CHCl_3$; D. $NaBH_4$; E. Ac_2O/Py → **305**

305 → F. OsO_4, $NaIO_4$ in aq. dioxane; G. Na_2CO_3 in MeOH–H_2O; H. Ethylene glycol, TsOH → **306**

306 → I. CrO_3, Pyridine; J. $NaBH_4$; K. Aq. AcOH → **307**

307 $R^1 = OH; R^2 = H$
308 $R^1 = H; R^2 = OH$
309 $R^1 = OTs; R^2 = H$
310 $R^1 = H; R^2 = OTs$

311 **312**

The alkaloid atisine (**303**) was converted into the epimeric tosylates **309** and **310**, as shown above. On acetolysis, either **309** or **310** yielded the same olefin, **311**. However, preparative gas-phase pyrolysis of the α-tosylate (**310**) gave the desired ketoolefin (**312**) in a yield of 77%. Under the same pyrolytic conditions, the β-tosylate (**309**) afforded only the olefin

(**311**). The authors speculated that each conversion takes place stereospecifically via a seven-membered transition state. The structure of the ketoolefin **312** was confirmed (*163*) by an X-ray analysis of the ethylene ketal hydroiodide salt.

Ayer and Despande (*155*) have reported the rearrangement of tosylate **313** into diene **319** by solvolysis or even by passage of a benzene solution through silica gel. This rearrangement serves as a model for the proposed (*152*) biosynthetic transformation of the atisine-type alkaloids to the aconitine-type alkaloids.

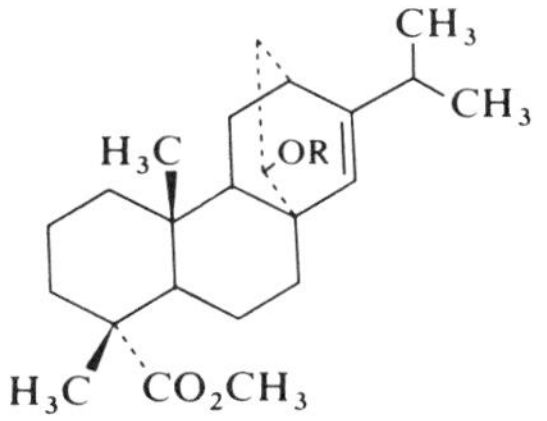

313 R = Ts
316 R = Ac
318 R = H

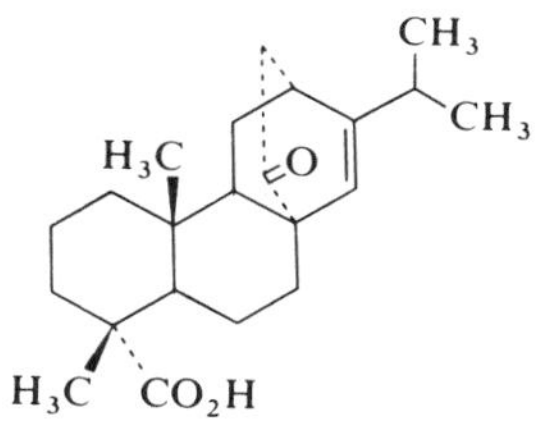

314

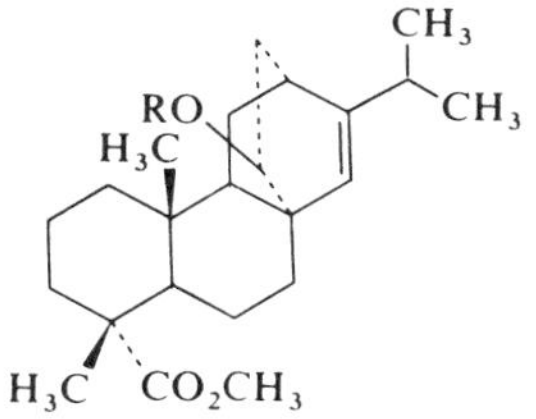

315 R = Ac
317 R = H

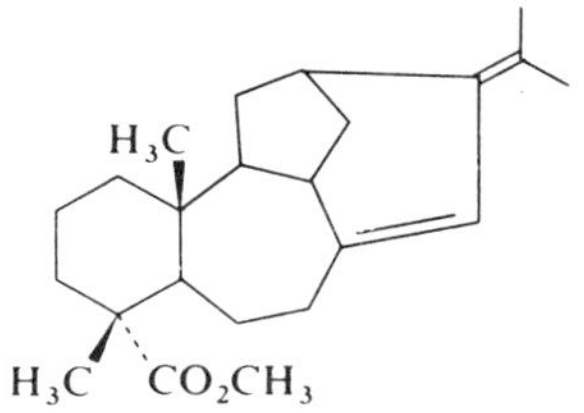

319

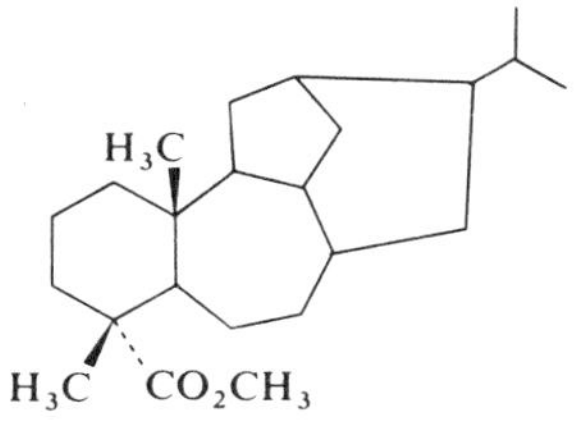

320

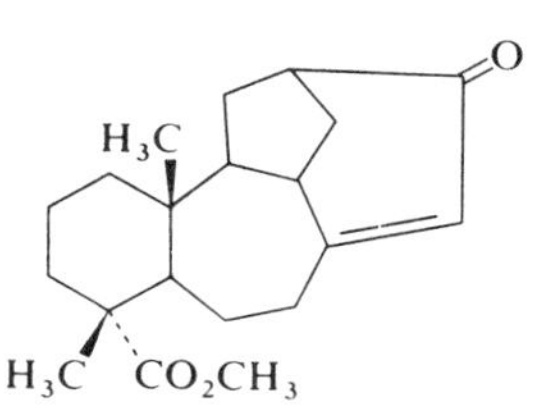

321

Levopimaric acid was converted into ketoacid **314** by the Diels–Alder addition of α-acetoxyacrylonitrile followed by basic work-up. Reduction of **314** with sodium borohydride afforded the crude product, which was treated with diazomethane and then acylated to give a mixture of acetoxylesters **315** and **316** in a 3:4 ratio. These compounds were separated by chromatography over alumina. Basic hydrolysis of compounds **315** and **316** afforded alcohols **317** and **318**, respectively. The alcohol **318** was converted to tosylate **313** by treatment with *p*-toluenesulfonyl chloride in pyridine. Chromatography of **313** over silica gel in benzene produced the desired diene (**319**) in a yield of 70% and the hydrolyzed product **318** in 15–20% yield. Solvolysis of **313** in hot acetic acid–sodium acetate solution yielded compound **319** in 40% yield and acetate **316** in 12% yield. Catalytic hydrogenation of the diene **319** over Adams' catalyst afforded the tetrahydro compound (**320**). Treatment of **319** with ozone in methylene chloride at −70° yielded the α,β-unsaturated ketone (**321**) in about 10% yield. These hydrogenation and ozonolysis reactions confirm the structure of the diene as **319**.

Recently, Wiesner and co-workers reported (*164*) the conversion of atisine (**303**) into a lycoctonine-type ketone (**322**). Atisine was converted into the ketal–ketone (**323**) as reported earlier by Johnston and Overton (*153*) (see above). Sodium borohydride reduction of **323** gave a mixture of alcohols **324** and **325**. The α-isomer (**324**) was separated from this mixture by preparative thin-layer chromatography and was tosylated with *p*-toluenesulfonyl chloride in pyridine to give the amorphous tosylate **326**. The latter was treated with tetramethylguanidine and dimethyl sulfoxide at 180° for 24 hr to afford the rearranged products **327** and **328** in 85% yield.

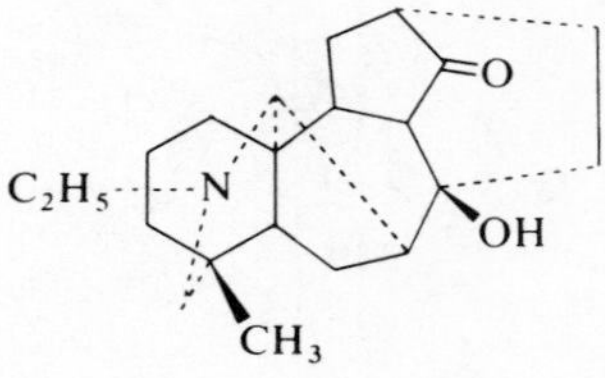

322

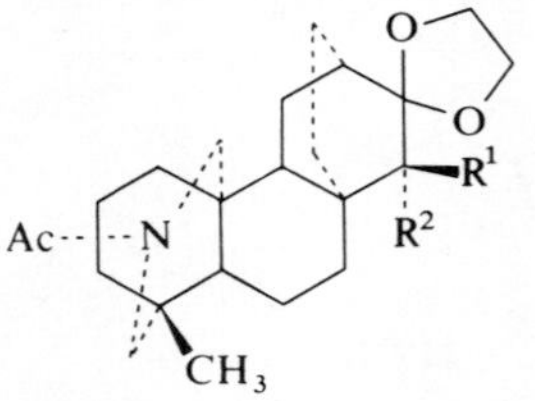

323 $R^1R^2 = O$
324 $R^1 = H$; $R^2 = OH$
325 $R^1 = OH$; $R^2 = H$
326 $R^1 = H$; $R^2 = OTs$

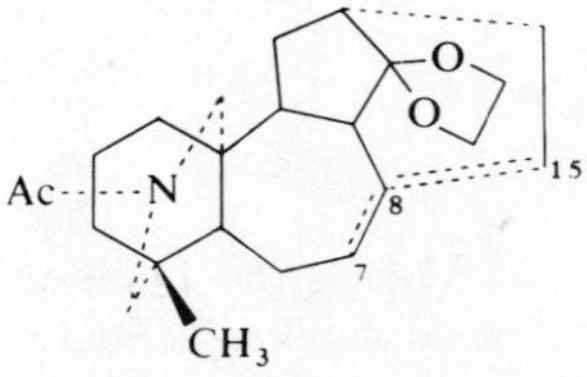

327 $\Delta^{8,15}$ compound
328 $\Delta^{7,8}$ compound

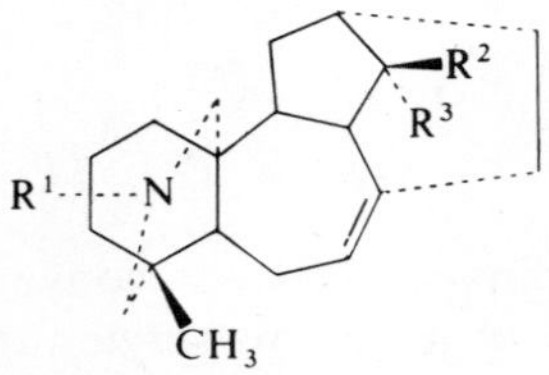

329 $R^1 = Ac$; $R^2 = R^3 = O$
330 $R^1 = Et$; $R^2 = H$; $R^3 = OH$
331 $R^1 = Et$; $R^2 = OH$; $R^3 = H$

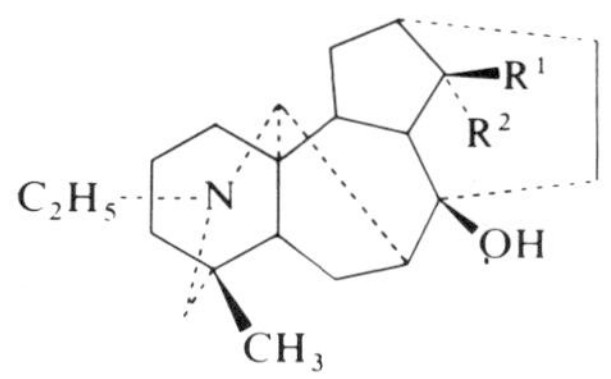

332 $R^1 = H; R^2 = OH$
333 $R^1 = OH; R^2 = H$

Compound **327** is identical with the product earlier prepared by pyrolytic rearrangement by the English workers (*153*). Hydrogenation of compounds **327** and **328** gave the same dihydro derivative. Deketalization of **328** yielded the corresponding ketone (**329**), which was reduced with lithium aluminum hydride to give a mixture of the epimeric hydroxyamines **330** and **331**. Treatment of **330** and **331** with mercuric acetate afforded the epimeric products **332** and **333**. Jones oxidation of this epimeric mixture yielded compound **322**, whose structure was confirmed by an X-ray crystallographic analysis.

In 1969 Tahara and Ohsawa (*165*) reported transformation of a diterpene acid (**337**) into an aconitine-type skeleton (**340**). Earlier, they described the cyclization of dioxoester **335** (obtained from enantiomeric deoxypodocarpic acid, **334**) into the c-homofluorene-type compound (**336**) by an aldolcondensation. Later they redirected the cyclization of a similar type compound (**339**) by introducing an oxygen bridge into the aconitine-type compound (**340**) as described below.

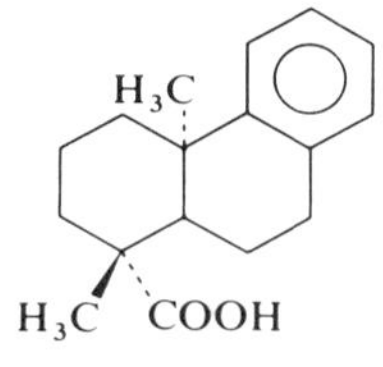

334 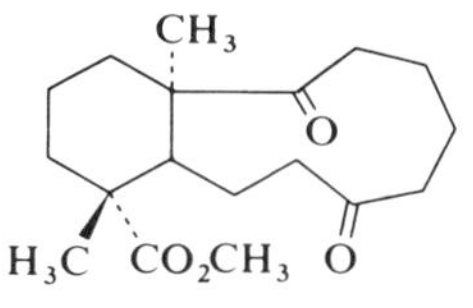 **335** 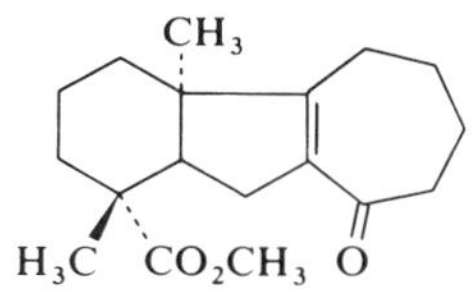 **336**

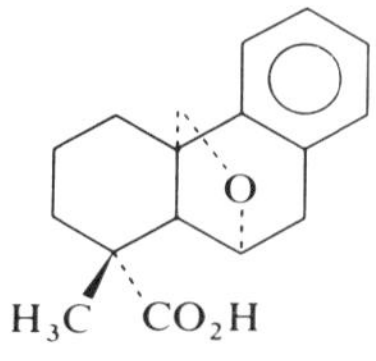

337 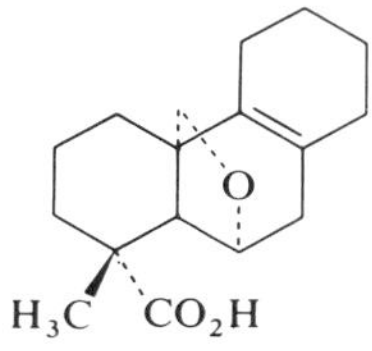 **338** 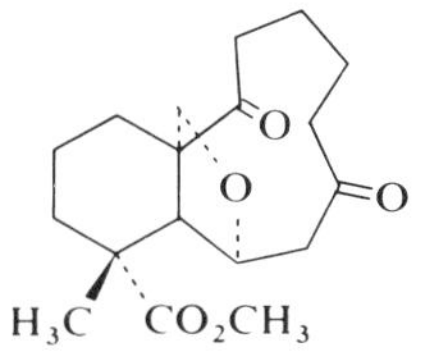 **339** 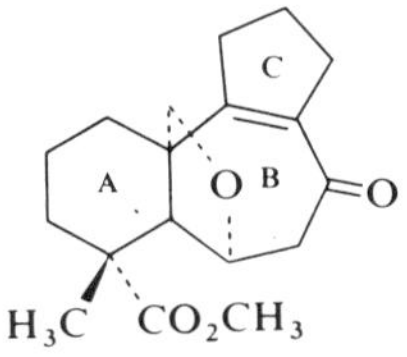 **340**

The acid (**337**) prepared from (−)-abietic acid was reduced by lithium ethylamine–*tert*-amyl alcohol to compound **338**. The methyl ester of **338** was hydroxylated and the resulting diol cleaved to give diketone **339**. The latter was cyclized by treatment with acid to the α,β-unsaturated ketone (**340**). Although rings A and B can be easily substituted by appropriate reactions to derive the corresponding aconitine-type alkaloid, the major problem with this route is the introduction of the ring C substituents.

VIII. A Catalog of C_{19}-Diterpenoid Alkaloids

14-Acetylbrowniine

$C_{27}H_{43}NO_8$; 123–124°
$[\alpha]_D + 27.8°$ (Chf.)
Consolida ambigua
^{13}C, ^{1}H NMR and chemical data
Refs. *102a, 137*

14-Acetyldelcosine

$C_{26}H_{41}NO_8$; 193–195°
$[\alpha]_D + 34°$ (EtOH)
Consolida ambigua
Correlated with delcosine; mass spectral analysis
Refs. *3, 134, 166, 167*

N-Acetyldelectine

$C_{33}H_{46}N_2O_9$; 116–118°
$[\alpha]_D$ ——
Delphinium dictyocarpum
Chemical and spectral analysis
Ref. *168*

Acomonine

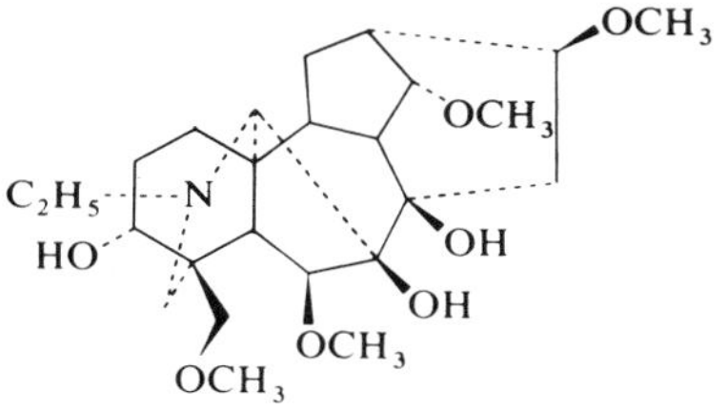

$C_{25}H_{41}NO_7$; 208–210°
$[\alpha]_D$ ——
Aconitum monticola
Chemical and spectral analysis
Refs. *122, 123*

Aconitine

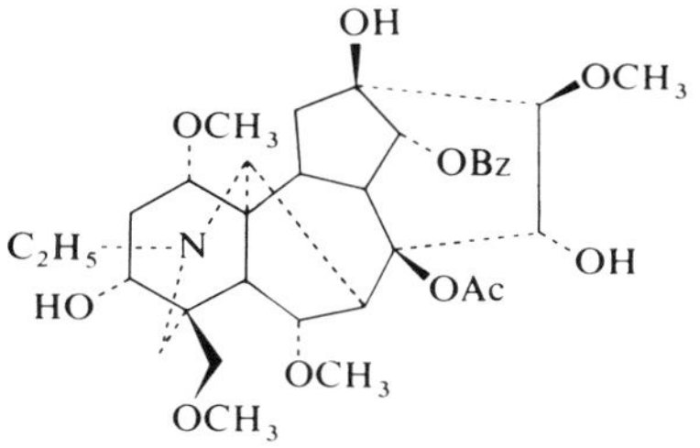

$C_{34}H_{47}NO_{11}$; 202–205°
$[\alpha]_D + 19°$ (Chf.)
Aconitum napellus; *A. fauriei*; *A. karakolicum* and others
Chemical and X-ray analysis
Refs. *1, 3, 17–20*

Aconorine

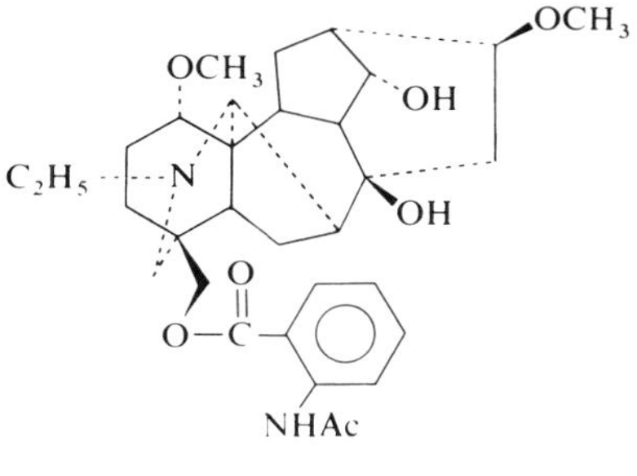

$C_{32}H_{44}N_2O_7$; amorphous; 237° (perchlorate)
$[\alpha]_D$ ——
Aconitum orientale; *A. columbianum*
Spectral and chemical data
Ref. *99*

Aconosine

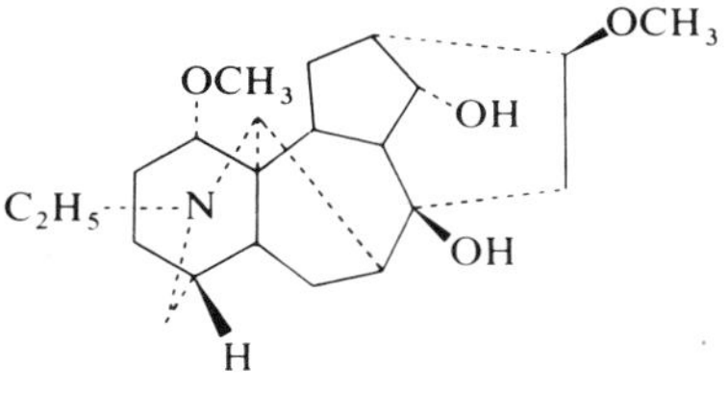

$C_{22}H_{35}NO_4$; 148°
$[\alpha]_D - 21°$ (MeOH)
Aconitum nasutum
Chemical and spectral data
Ref. *121*

Ajacine

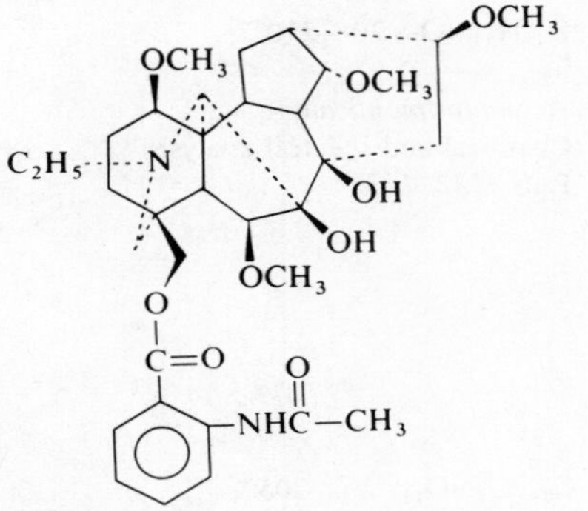

$C_{34}H_{48}N_2O_9$; 154° (hydrate)
$[\alpha]_D + 50°$ (EtOH) + 53° (Chf.)
Consolida ambigua; *D. orientale*
Correlated with lycoctonine
Refs. *3, 4, 132, 169*

Ajacusine

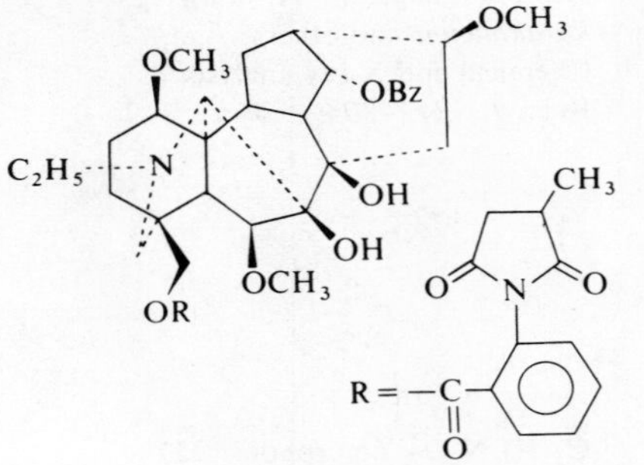

$C_{43}H_{52}N_2O_{11}$; 158–161°
$[\alpha]_D + 65.2°$ (abs. EtOH)
Consolida ambigua
^{13}C, ^{1}H NMR and chemical data
Ref. *170*

Ajadine

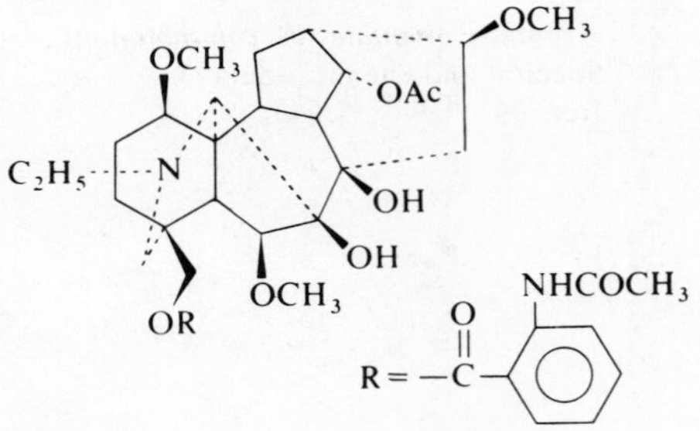

$C_{35}H_{48}N_2O_{10}$; 134–136° (d)
$[\alpha]_D + 43.9°$ (abs. EtOH)
Consolida ambigua
^{13}C, ^{1}H NMR and chemical data
Ref. *170*

Alkaloid A

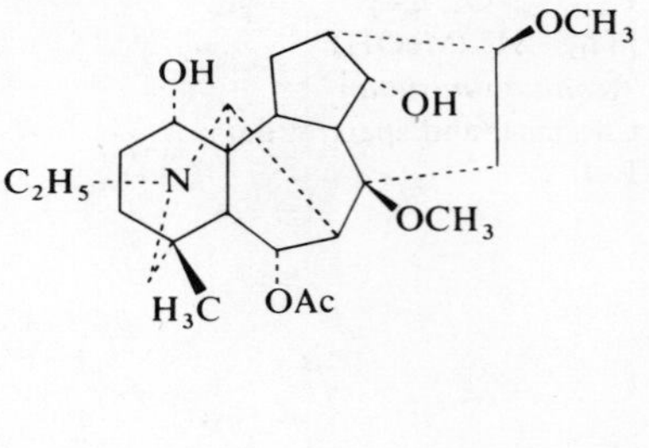

$C_{25}H_{39}NO_6$; amorphous
$[\alpha]_D + 10°$ (Chf.)
Delphinium bicolor
^{1}H and ^{13}C NMR analysis
Refs. *100–102*

Alkaloid B

$C_{22}H_{35}NO_5$; 190–191°
$[\alpha]_D + 16°$ (Chf.)
Delphinium bicolor
^1H and ^{13}C NMR analysis
Refs. *100–102*

Ambiguine

$C_{28}H_{45}NO_8$; 106–108°
$[\alpha]_D + 38°$ (Chf.)
Consolida ambigua
^{13}C, ^1H NMR, MS and chemical data
Ref. *171*

Anthranoyllycoctonine (Inuline)

$C_{32}H_{46}N_2O_8$; 165°
$[\alpha]_D + 51°$ (EtOH)
Inula royleana; *Delphinium consolida*; *D. barbeyi*
Correlated with lycoctonine
Refs. *3, 4, 132, 172, 173*

Avadharidine

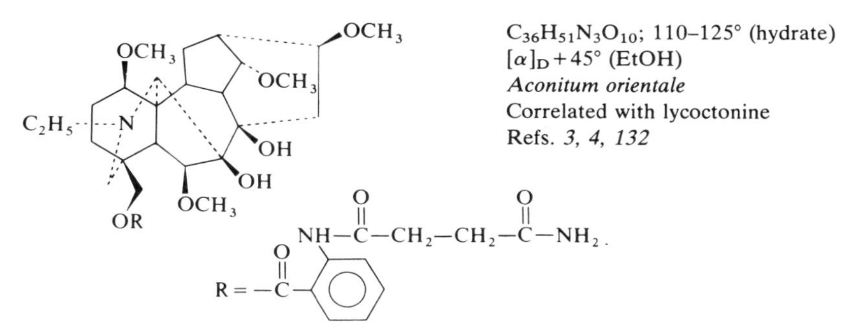

$C_{36}H_{51}N_3O_{10}$; 110–125° (hydrate)
$[\alpha]_D + 45°$ (EtOH)
Aconitum orientale
Correlated with lycoctonine
Refs. *3, 4, 132*

Bikhaconitine

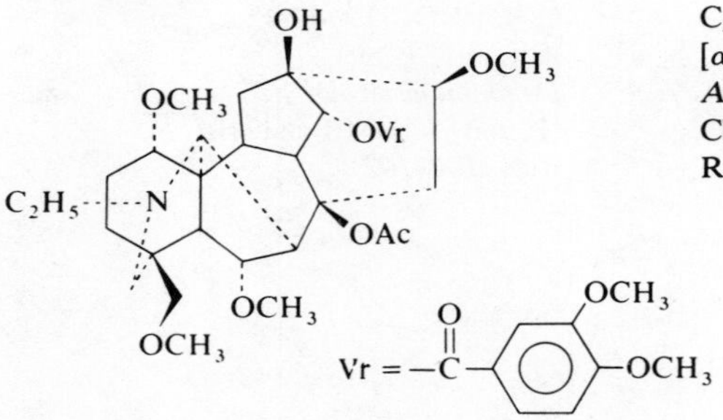

$C_{36}H_{51}NO_{11}$; 105–110° (hydrate)
$[\alpha]_D + 16°$ (EtOH)
Aconitum spicatum; *A. ferox*
Chemical and spectral analysis
Refs. *40, 41, 67, 174*

Browniine

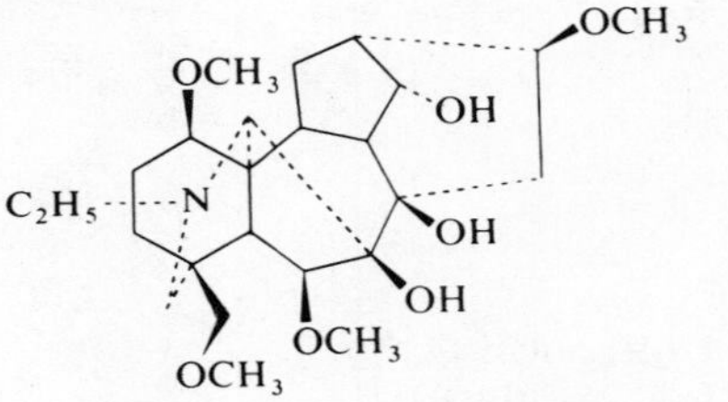

$C_{25}H_{41}NO_7$, amorphous
$[\alpha]_D + 25°$ for perchlorate (H_2O)
Consolida ambigua, Delphinium brownii; *D. carolinianum*; *D. virescens.*
Chemical and spectral analysis
Refs. *135–137*

Cammaconine

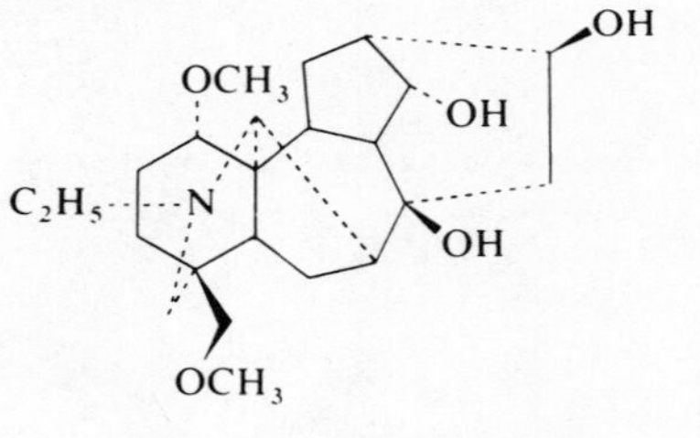

$C_{23}H_{37}NO_5$; 135–137°
$[\alpha]_D \pm 0°$
Aconitum variegatum
Correlated with isotalatizidine and talatisamine
Refs. *94, 95*

Chasmaconitine

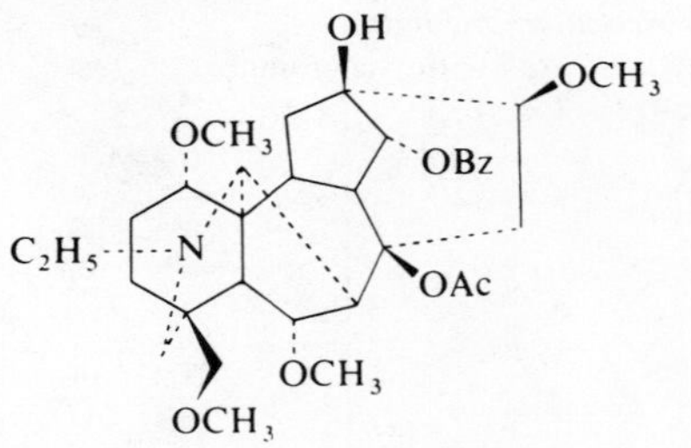

$C_{34}H_{47}NO_9$; 181–182° (Hexane), 165–167° (Ether)
$[\alpha]_D + 10.3°$ (EtOH)
Aconitum chasmanthum
Chemical and spectral analysis
Refs. *20, 68*

Chasmanine

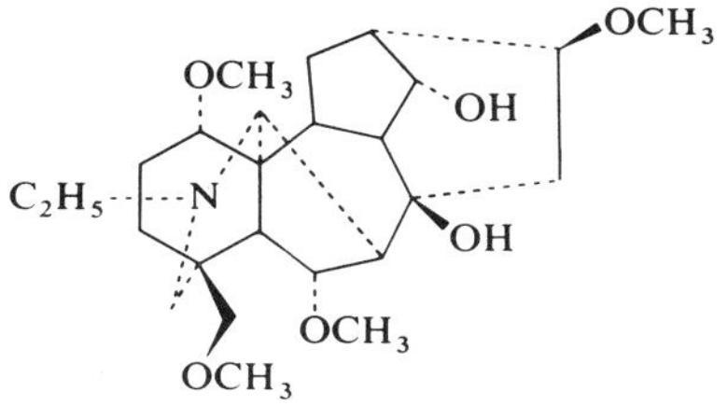

$C_{25}H_{41}NO_6$; 90–91°
$[\alpha]_D + 23.6°$ (EtOH)
Aconitum chasmanthum
^{13}C NMR, X-ray, and chemical analysis
Refs. *66, 73, 77, 78, 80, 81*

Chasmanthinine

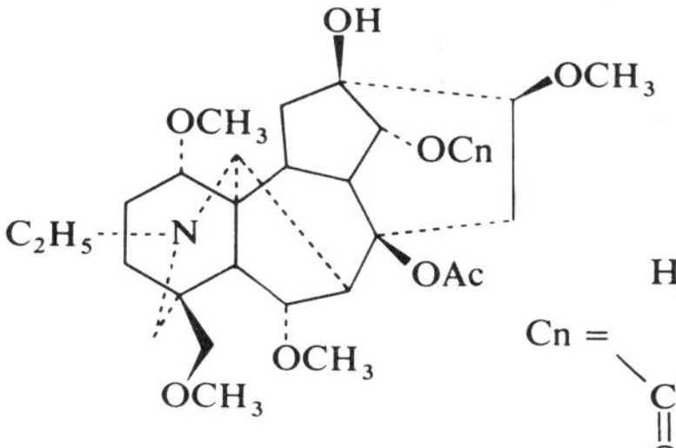

$C_{36}H_{49}NO_9$; 160–161°
$[\alpha]_D + 9.6$ (EtOH)
Aconitum chasmanthum
Chemical and spectral analysis
Refs. *20, 68*

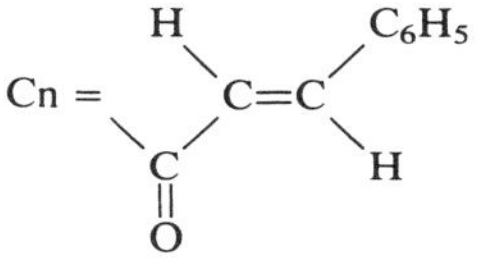

Condelphine

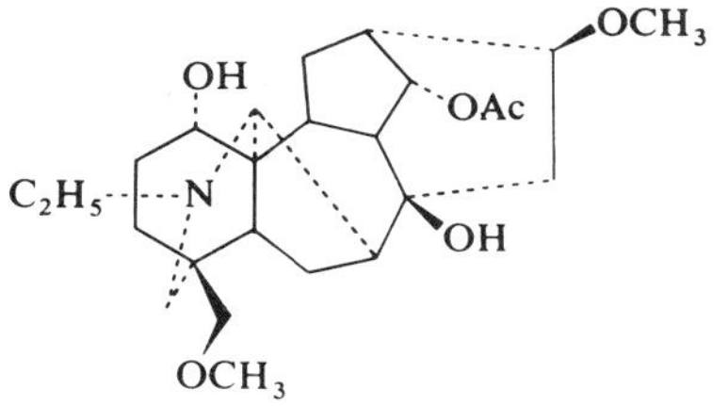

$C_{25}H_{39}NO_6$; 158–159°
$[\alpha]_D + 21.3°$
Delphinium denudatum; *D. confusum*; *D. oreophilum*
Chemical and spectral analysis
Refs. *85, 92*

14-Dehydrobrowniine

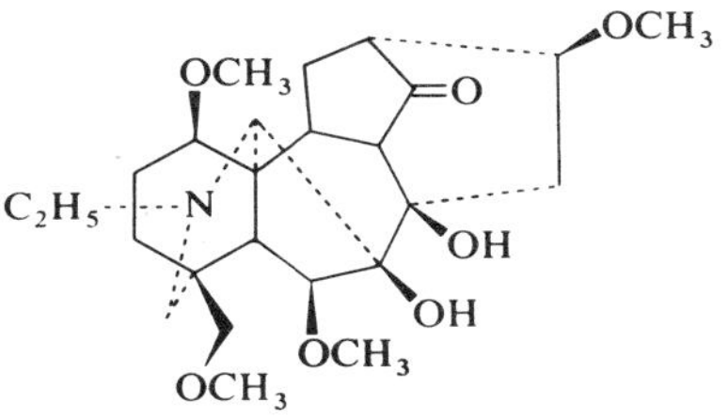

$C_{25}H_{39}NO_7$; 161–163°
$[\alpha]_D + 19°$ (EtOH)
Delphinium cardinale
Chemical and spectral analysis
Ref. *175*

14-Dehydrodelcosine (Shimoburo Base II)

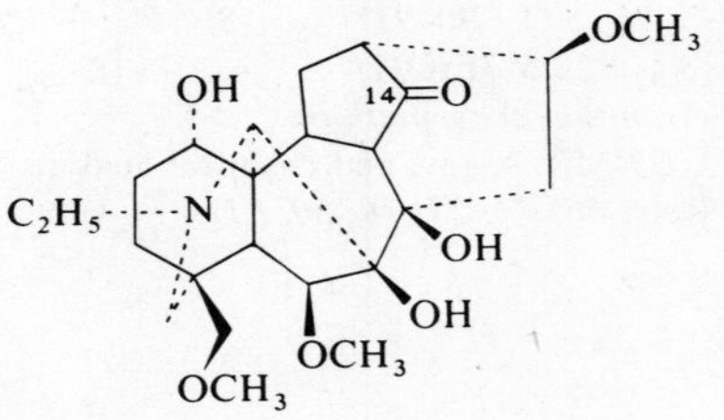

$C_{24}H_{37}NO_7$; 212.5–213.5°
$[\alpha]_D + 25.2°$ (Chf.)
Aconitum japonicum
Correlated with delcosine
Refs. *176, 177*

Delcorine

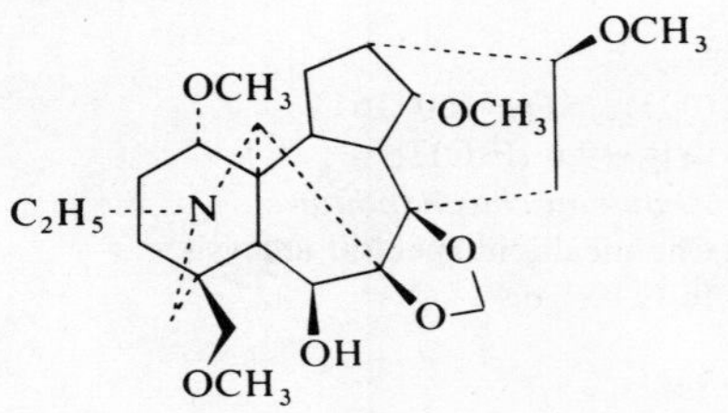

$C_{26}H_{41}NO_7$; 200–202°
$[\alpha]_D - 18°$ (Chf.)
Delphinium corumbosum (*corymbosum*)
Chemical and spectral analysis
Ref. *129*

Delcosine

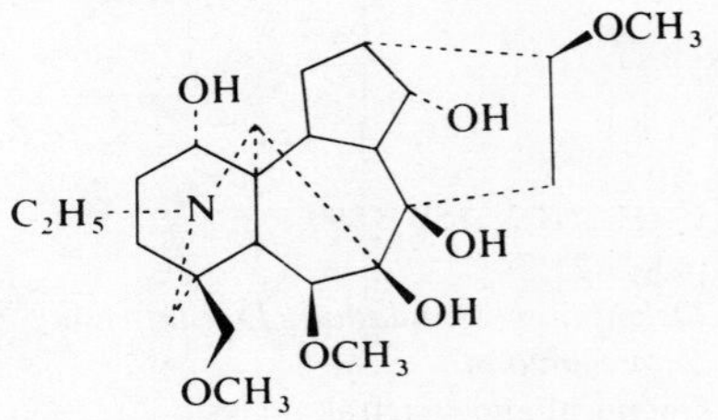

$C_{24}H_{39}NO_7$; 203–204°
$[\alpha]_D + 57°$ (Chf.)
Delphinium consolida; *Consolida ambigua*, and others
Chemical and spectral analysis
Refs. *3, 4, 134, 166, 167, 176*

Delectine

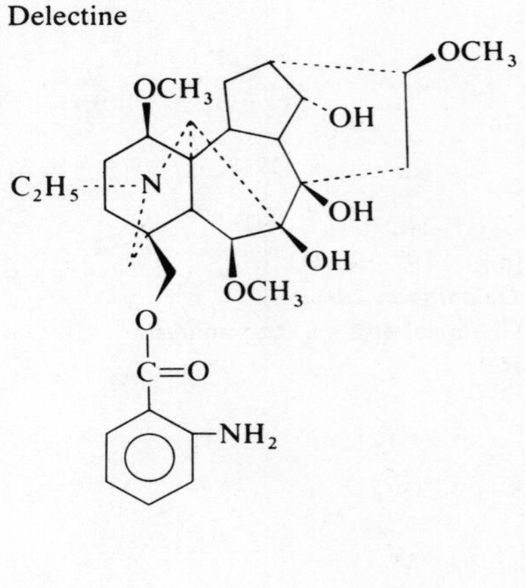

$C_{31}H_{44}N_2O_8$; 107–109°
$[\alpha]_D$ ——
Delphinium dictyocarpum
Chemical and spectral analysis
Ref. *133*

Delphatine

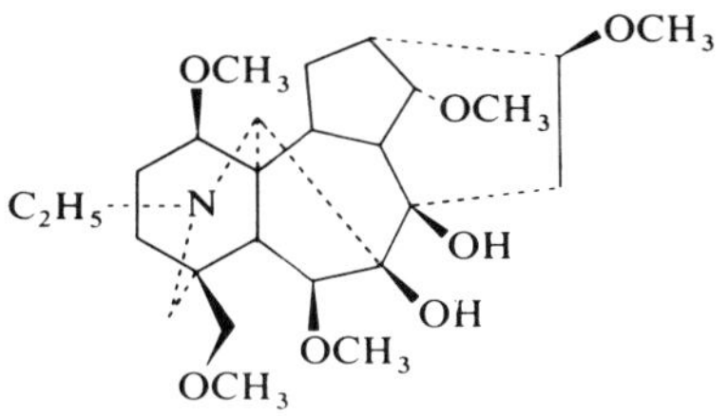

$C_{26}H_{43}NO_7$; 106°
$[\alpha]_D + 39°$
Delphinium biternatum
Correlated with lycoctonine
Refs. *126, 127*

Delpheline

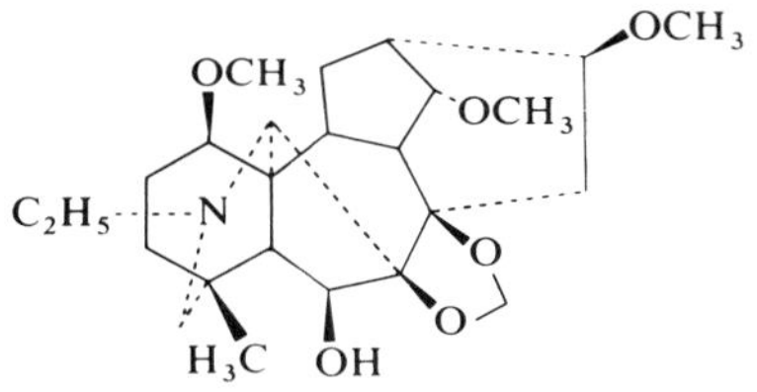

$C_{25}H_{39}NO_6$; 227°
$[\alpha]_D - 26°$ (Chf.)
Delphinium elatum
Correlated with lycoctonine
Refs. *2, 3, 178–180*

Delphidine

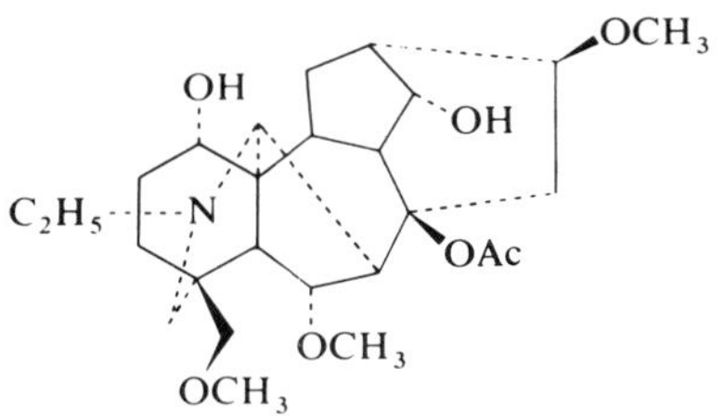

$C_{26}H_{41}NO_7$; 98–100° (dec.)
$[\alpha]_D + 16.6°$ (EtOH)
Delphinium staphisagria
^1H, ^{13}C NMR and chemical analysis
Refs. *82, 102*

Delphinine

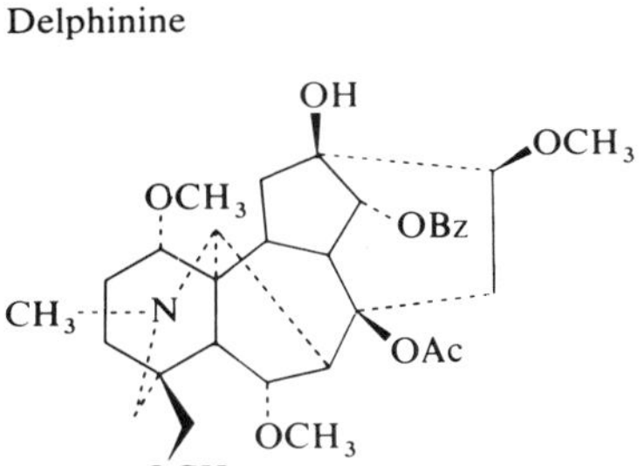

$C_{33}H_{45}NO_9$; 191–192°
$[\alpha]_D + 25°$ (EtOH)
Delphinium staphisagria
Chemical and X-ray analysis
Refs. *19, 20, 51–61*

Delphirine

OCH$_3$, OH, OH, C_2H_5, N, OH, OCH$_3$, OCH$_3$

$C_{24}H_{39}NO_6$; 95–100°
$[\alpha]_D + 3.8°$ (EtOH)
Delphinium staphisagria
Spectral and chemical analysis
Ref. *83*

Delphisine

OCH$_3$, OH, OAc, C_2H_5, N, OAc, OCH$_3$, OCH$_3$

$C_{28}H_{43}NO_8$; 122–123°
$[\alpha]_D + 7.1°$ (EtOH)
Delphinium staphisagria
X-ray and chemical analysis
Refs. *72, 73*

Delsemine

OCH$_3$, OCH$_3$, OCH$_3$, C_2H_5, N, OH, OH, OCH$_3$, O, C=O, R

$C_{37}H_{53}N_3O_{10}$; 125° (hydrate)
$[\alpha]_D + 43°$ (EtOH)
Delphinium semibarbatum; *D. oreophilum*; *D. tricorne* and others
Refs. *131, 132*

R = mixture of $NH-\overset{O}{\overset{\|}{C}}-\underset{}{\overset{CH_3}{\overset{|}{CH}}}-CH_2-\overset{O}{\overset{\|}{C}}-NH_2$

and

$NH-\overset{O}{\overset{\|}{C}}-CH_2-\underset{CH_3}{\underset{|}{CH}}-\overset{O}{\overset{\|}{C}}-NH_2$

Delsoline

OCH$_3$, OH, OCH$_3$, C_2H_5, N, OH, OH, OCH$_3$, OCH$_3$

$C_{25}H_{41}NO_7$; 215–216°
$[\alpha] + 53.4$ (Chf.)
Delphinium consolida; *D. orientale*; *Consolida ambigua* and others
Correlated with delcosine
Refs. *3, 133, 176, 181, 182*

Deltaline (Eldeline)

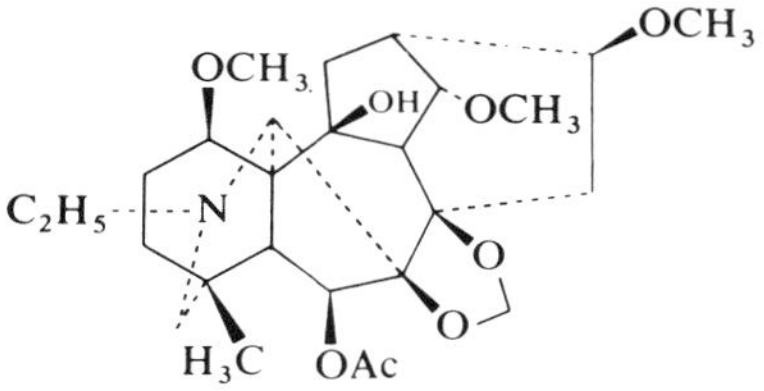

$C_{27}H_{41}NO_8$; 182–184°
$[\alpha]_D - 28°$ (MeOH)
Delphinium occidentale; *D. elatum*; *D. barbeyi* and others
Chemical data
Refs. *2, 3, 180*

Deltamine (Eldelidine)

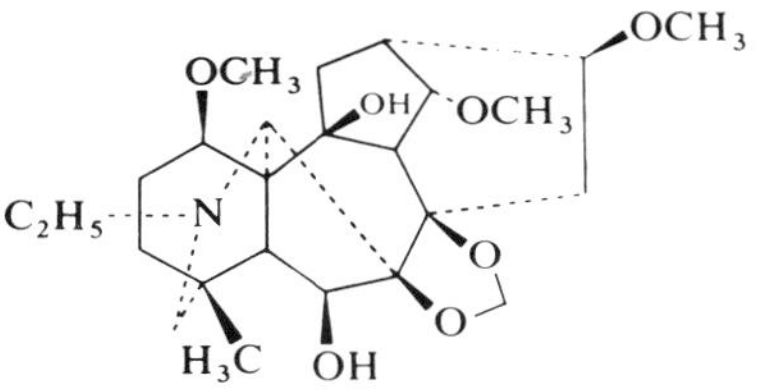

$C_{25}H_{39}NO_7$; 226–228°
$[\alpha]_D - 17°$ (MeOH)
Hydrolysis Product of Deltaline (Eldeline)
Chemical and spectral analysis
Refs. *2, 3, 180*

Demethyleneeldelidine

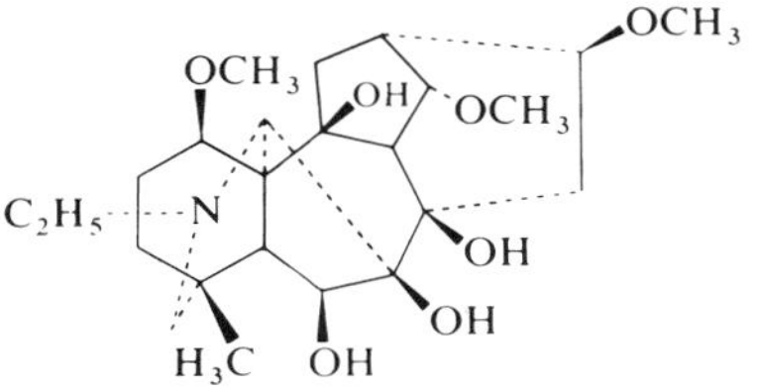

$C_{24}H_{39}NO_7$; 98–100°
$[\alpha]_D$ ——
Delphinium dictyocarpum
^{1}H NMR, MS and chemical data; Correlated with eldelidine
Ref. *168*

Deoxyaconitine

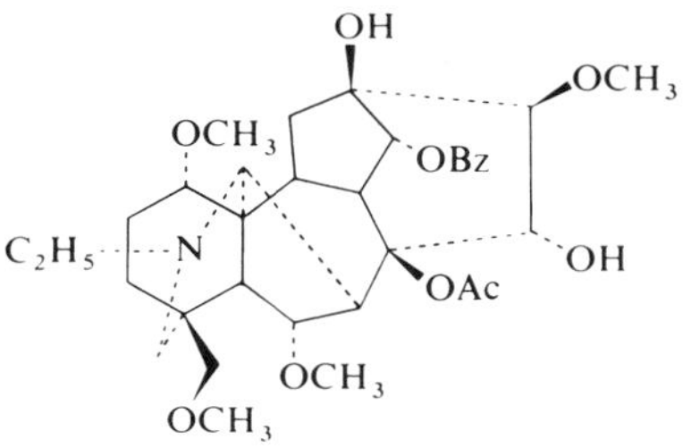

$C_{34}H_{47}NO_{10}$; 176–178°
$[\alpha]_D + 16°$ (EtOH)
Aconitum napellus
Correlated with aconitine. ^{13}C NMR analysis
Refs. *20, 36*

Diacetylpseudaconitine

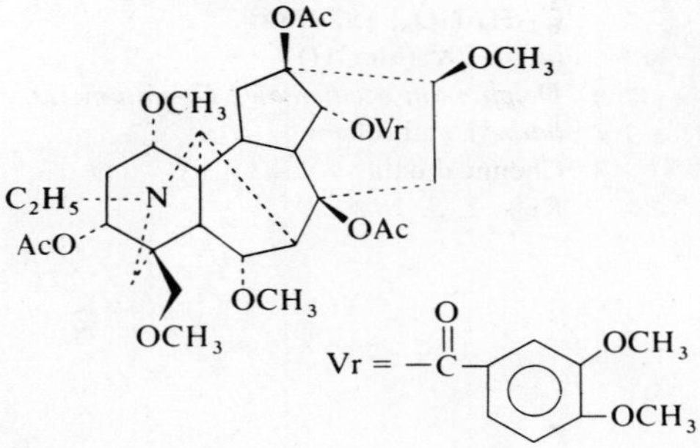

$C_{40}H_{55}NO_{14}$; 228–230°
$[\alpha]_D + 24°$ (EtOH)
Aconitum ferox
Correlated with pseudaconitine
^{1}H NMR analysis
Ref. *41*

Dictyocarpine

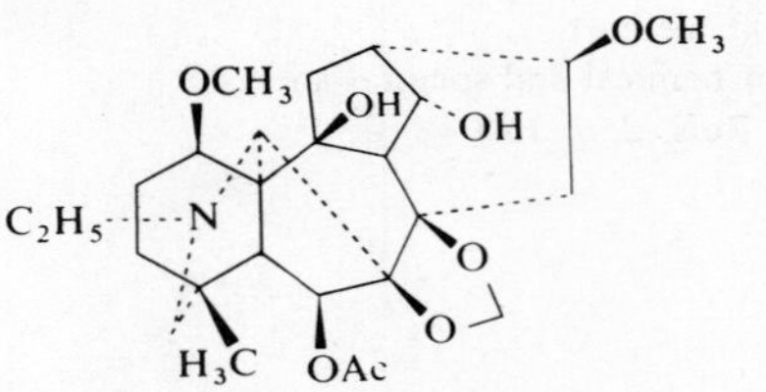

$C_{26}H_{39}NO_8$; 210–212°
$[\alpha]_D$ ——
Delphinium dictyocarpum
Chemical and spectral analysis
Refs. *138, 139*

Dictyocarpinine

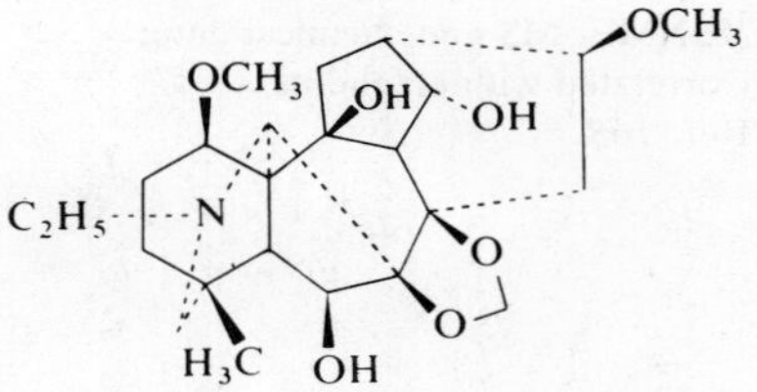

$C_{25}H_{37}NO_7$; 204–205°
$[\alpha]_D$ ——
Hydrolysis product of dictyocarpine
Refs. *138, 139*

O,O-Dimethyllycoctonine

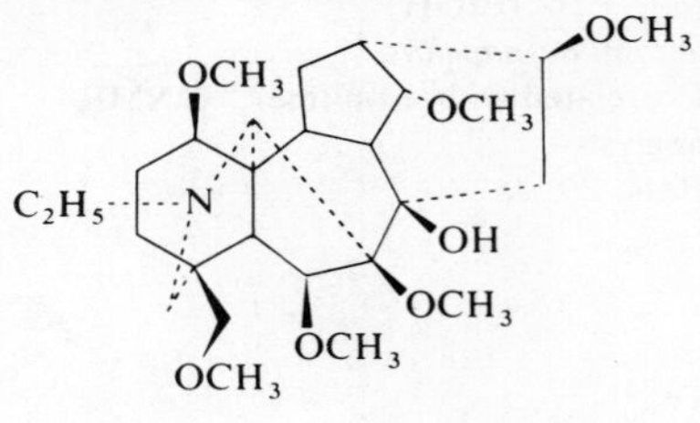

$C_{27}H_{45}NO_7$; ——
$[\alpha]_D$ ——
Delphinium dictyocarpum
Chemical analysis
Ref. *133*

Excelsine

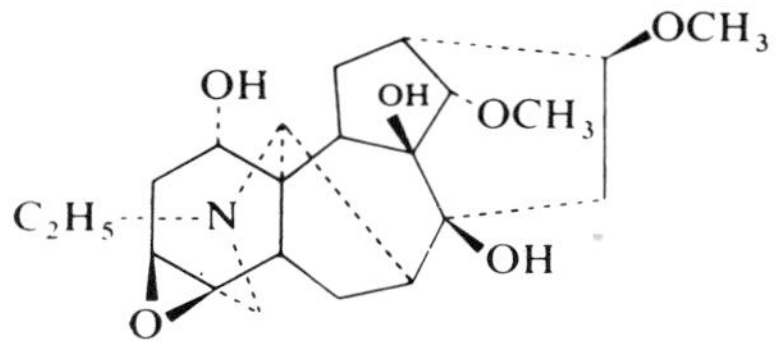

$C_{22}H_{33}NO_6$; 103–105°
$[\alpha]_D$ ——
Aconitum excelsum
Correlated with lapaconidine, X-ray analysis
Refs. *114–116*

Falaconitine

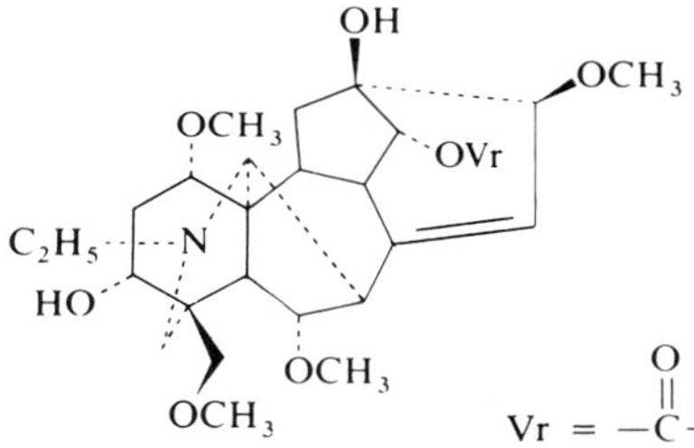

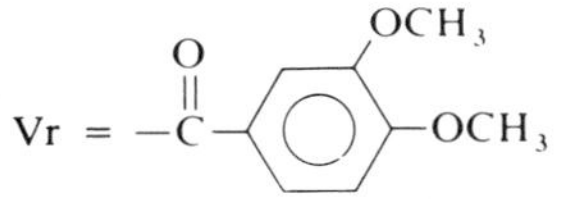

$C_{34}H_{47}NO_{10}$; amorphous
$[\alpha]_D + 111.5°$ (EtOH)
Aconitum falconeri
^1H, ^{13}C NMR and chemical analysis
Refs. *45, 46*

Heteratisine

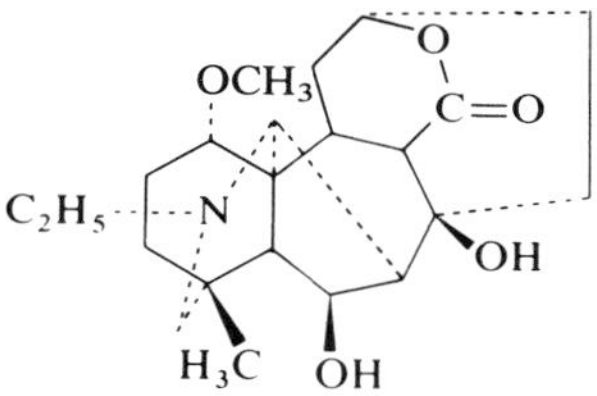

$C_{22}H_{33}NO_5$; 261–265° (dec)
$[\alpha]_D + 40°$ (MeOH)
Aconitum heterophyllum
Chemical, spectral, and X-ray analysis
Refs. *142, 183*

Heterophyllidine

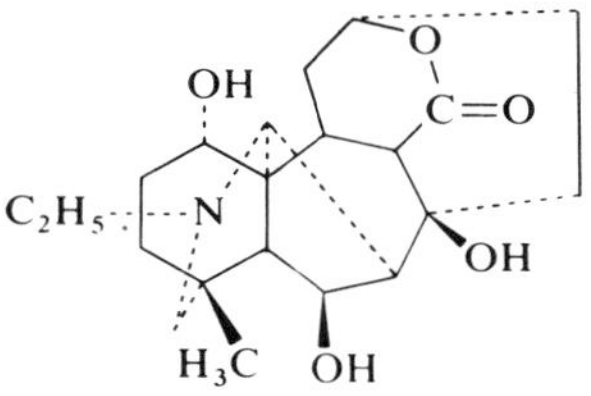

$C_{21}H_{31}NO_5$; 269–272°
$[\alpha]_D + 42.3°$ (MeOH)
Aconitum heterophyllum
IR, ^1H, and mass spectral analysis
Ref. *142*

Heterophylline

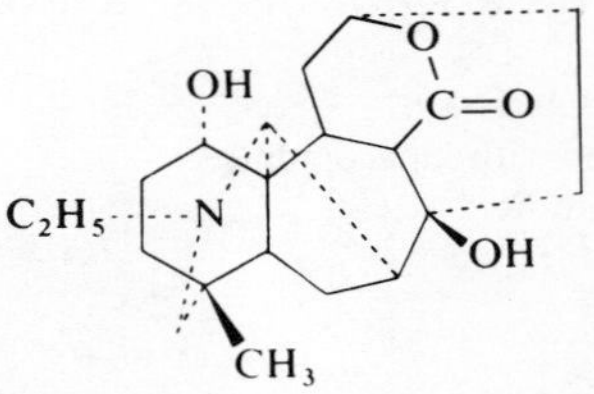

$C_{21}H_{31}NO_4$; 221.5–223°
$[a]_D + 10.5°$ (MeOH)
Aconitum heterophyllum
IR, ^{1}H, and mass spectral analysis
Ref. *142*

Heterophyllisine

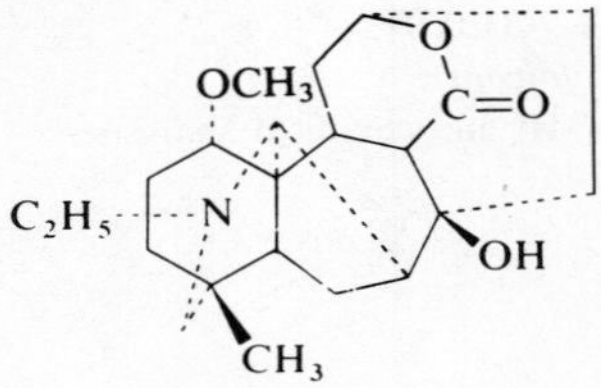

$C_{22}H_{33}NO_4$; 178–179°
$[\alpha]_D + 15.5°$ (MeOH)
Aconitum heterophyllum
IR, ^{1}H and mass spectral analysis
Ref. *142*

Homochasmanine

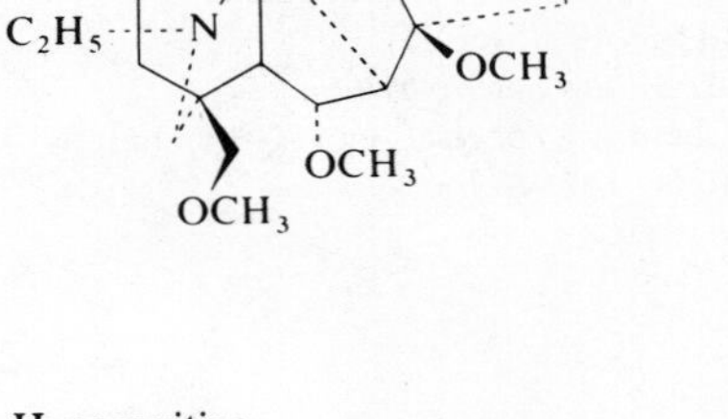

$C_{26}H_{43}NO_6$; 105–107°
$[\alpha]_D + 19.2°$ (EtOH)
Aconitum chasmanthum
Correlated with chasmanine
Refs. *73, 78, 79*

Hypaconitine

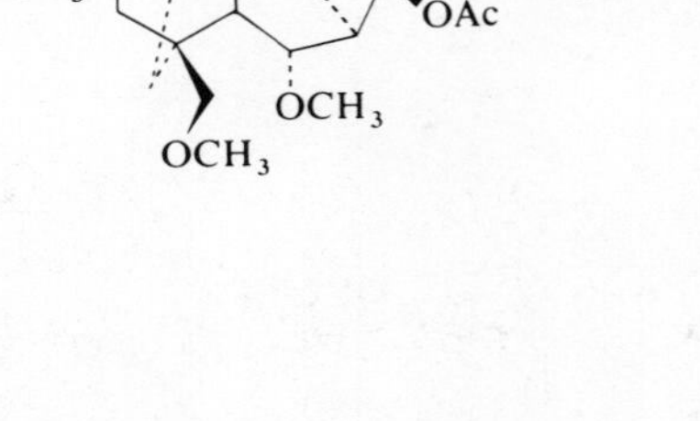

$C_{33}H_{45}NO_{10}$; 197–198°
$[\alpha]_D + 22°$ (Chf.)
Aconitum carmichaeli; *A. napellus*; *A. callianthum* and others
Correlated with aconitine
Refs. *14, 20, 32–36*

Iliensine

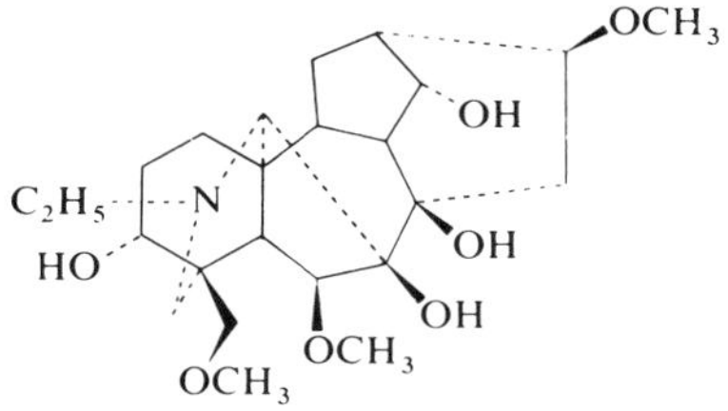

$C_{24}H_{39}NO_9$; 201–203°
$[\alpha]_D$ ——
Delphinium biternatum
Correlated with acomonine
Refs. *124, 125*

Indaconitine

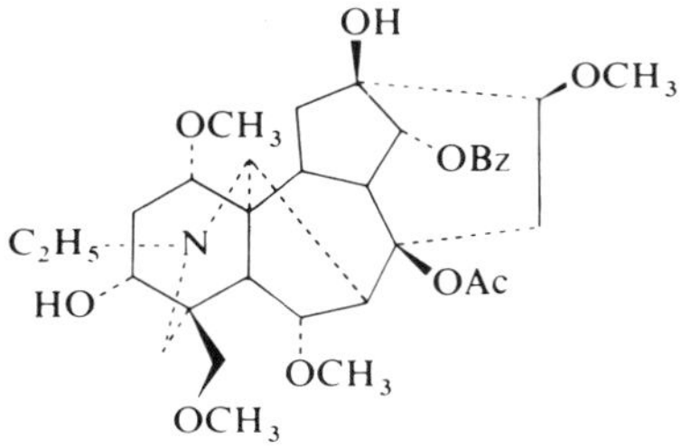

$C_{34}H_{47}NO_{10}$; 203–204°
$[\alpha]_D + 18°$ (EtOH)
Aconitum chasmanthum; *A. ferox*; *A. falconeri*
Correlated with delphinine
Refs. *20, 40, 42, 43*

Isodelphinine

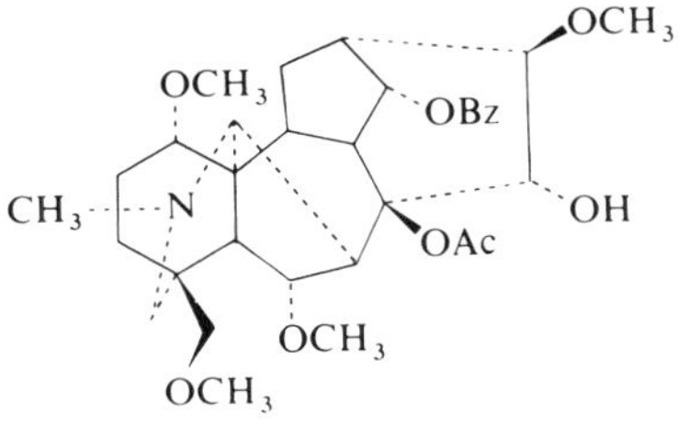

$C_{33}H_{45}NO_9$; 167–168°
$[\alpha]_D + 20.1°$ (EtOH)
Aconitum miyabei
^1H, ^{13}C NMR and chemical analysis
Refs. *69–71*

Isotalatizidine

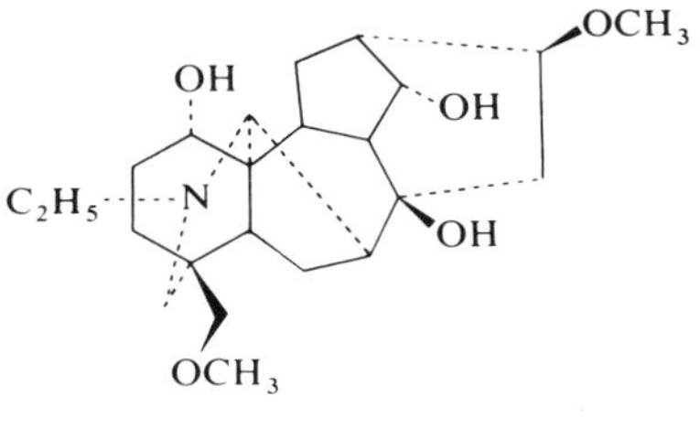

$C_{23}H_{37}NO_5$; 116–117°
$[\alpha]_D \pm 0°$
Aconitum talassicum; *Delphinium denudatum*
Correlated with condelphine
Refs. *85, 92*

Jesaconitine

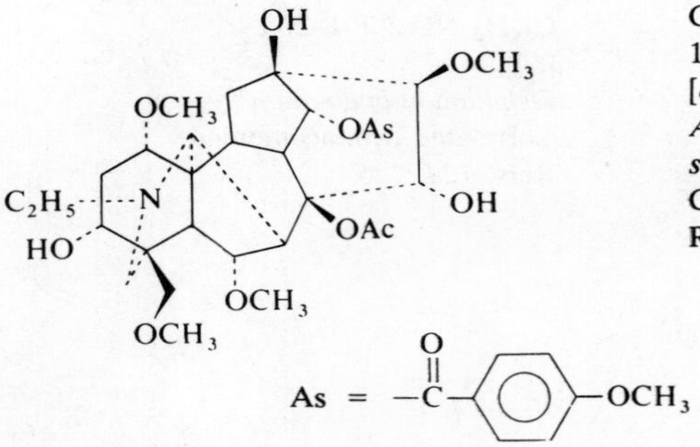

$C_{35}H_{49}NO_{12}$; amorphous (melts at 128–131°)
$[\alpha]_D - 17°$ for perchlorate (H_2O)
Aconitum fischeri; *A. subcuneatum*; *A. sachalinense* and others
Chemical, ^1H, ^{13}C NMR analysis
Refs. *20, 24–28*

Karakolidine (Karacolidine)

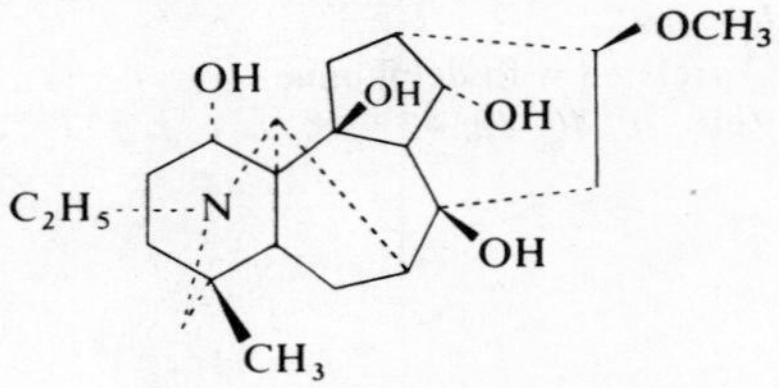

$C_{22}H_{35}NO_5$; 222–224°
$[\alpha]_D$ ——
Aconitum karakolicum
Chemical and spectral analysis
Refs. *16, 117, 118*

Karakoline (Karacoline)

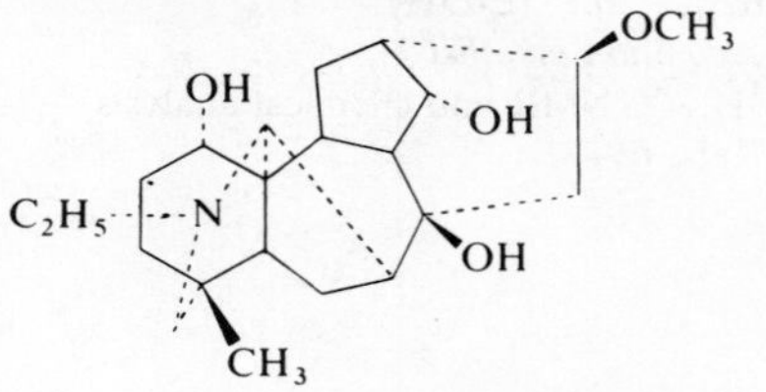

$C_{22}H_{35}NO_4$; 183–184°
$[\alpha]_D - 10°$ (MeOH)
Aconitum karakolicum
Chemical and spectral analysis
Refs. *16, 117, 119*

Lapaconidine

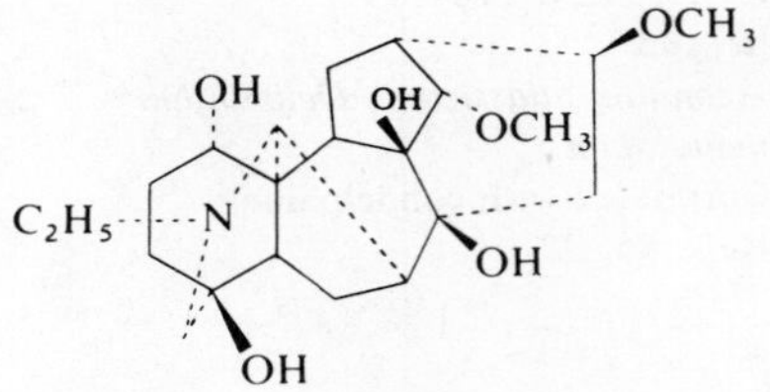

$C_{22}H_{35}NO_6$, 206–207°
$[\alpha]_D + 12.9°$ (Chf.)
Aconitum leucostomum (*excelsum*)
Correlated with lappaconine
Refs. *111, 112*

Lappaconitine

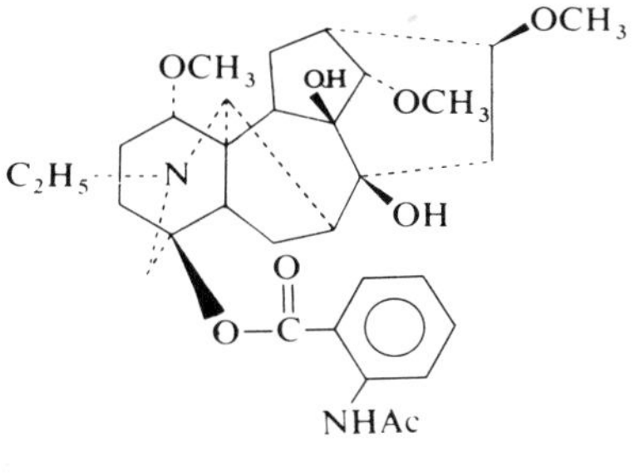

$C_{32}H_{44}N_2O_8$; 229°
$[\alpha]_D + 27°$ (Chf.)
Aconitum septentrionale; *A. orientale*; *A. excelsum*
Chemical, X-ray, and ^{13}C NMR analysis
Refs. *103–110*

Lycaconitine

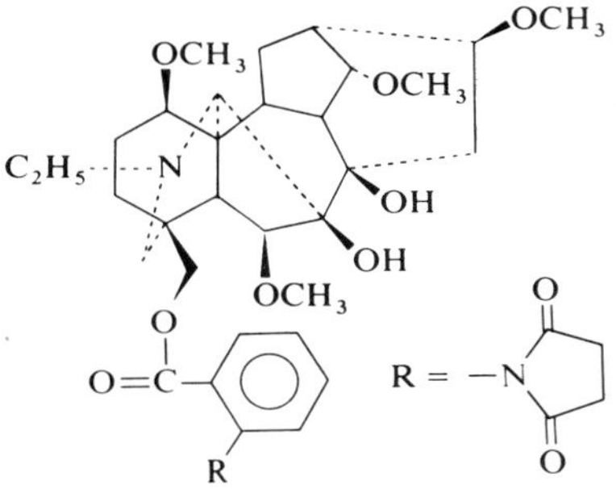

$C_{36}H_{48}N_2O_{10}$; amorphous
$[\alpha]_D + 43°$ (EtOH)
Aconitum gigas; *A. lycoctonum* and others
Correlated with lycoctonine
Refs. *3, 4*

Lycoctonine

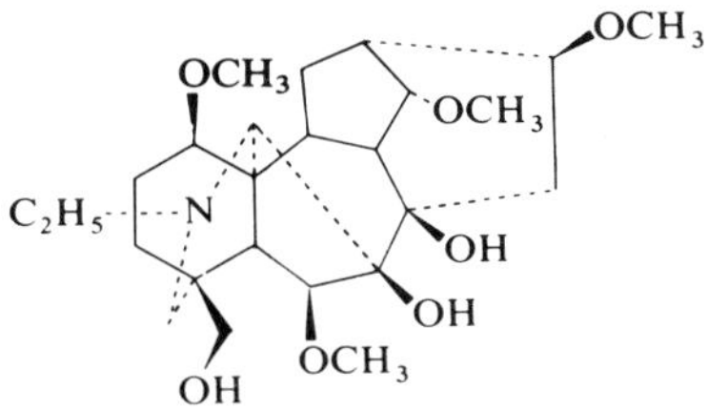

$C_{25}H_{41}NO_7$; 151–153°
$[\alpha]_D + 53°$ (EtOH)
Consolida ambigua; *D. barbeyi*; *D. tricorne*; *Inula royleana* and others
X-ray and chemical analysis
Refs. *128, 129, 184–186*

Mesaconitine

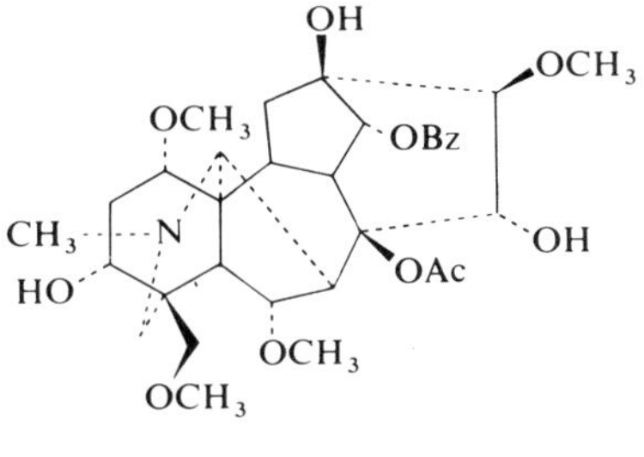

$C_{33}H_{45}NO_{11}$; 208–209°
$[\alpha]_D + 25°$ (Chf.)
Aconitum napellus; *A. fauriei*; *A. japonicum* and others
Correlated with aconitine
Refs. *1, 3, 4, 20, 27, 28, 30*

Methyllycaconitine

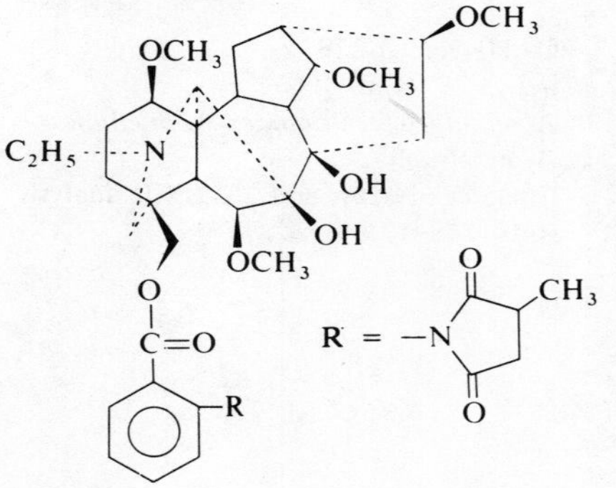

$C_{37}H_{50}N_2O_{10}$; 130° (amorphous)
$[\alpha]_D + 49°$ (EtOH)
D. dyctiocarpum; *D. elatum*; *D. tricorne* and others
Correlated with lycoctonine
Refs. *3, 4, 129–131*

Mithaconitine

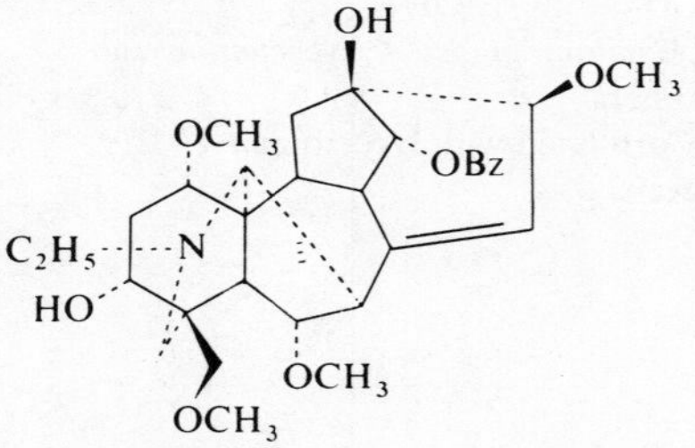

$C_{32}H_{43}NO_8$; amorphous
$[\alpha]_D + 94°$ (EtOH)
Aconitum falconeri
^1H, ^{13}C NMR and chemical analysis
Ref. *46*

Neoline

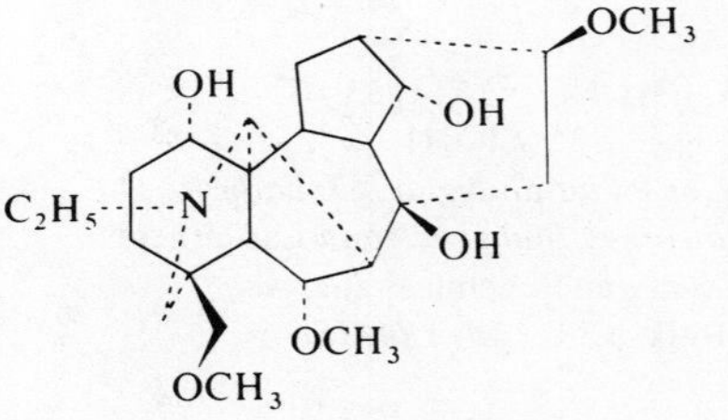

$C_{24}H_{39}NO_6$; 159–161°
$[\alpha]_D + 22°$ (EtOH)
Aconitum napellus; *A. stoerckianum*
Correlated with delphisine; ^1H and ^{13}C NMR analysis
Refs. *63, 73–75, 78*

Pseudaconitine

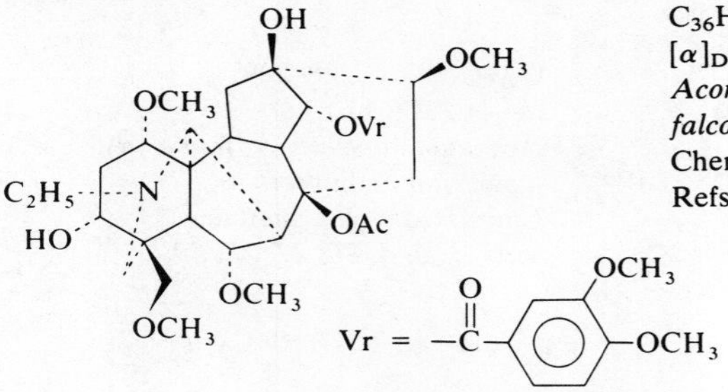

$C_{36}H_{51}NO_{12}$; 205–208°
$[\alpha]_D + 24°$ (Chf.)
Aconitum ferox; *A. spictatum*; *A. falconeri*
Chemical and spectral analysis
Refs. *3, 37–42*

Sachaconitine

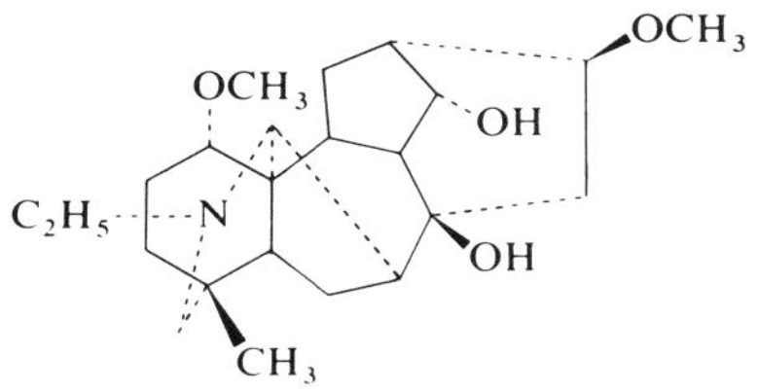

$C_{23}H_{37}NO_4$; 129–130°
$[\alpha]_D - 13.1°$ (EtOH)
Aconitum miyabei
^1H, ^{13}C NMR and chemical analysis
Refs. *69, 71, 120*

Talatizamine (Talatisamine)

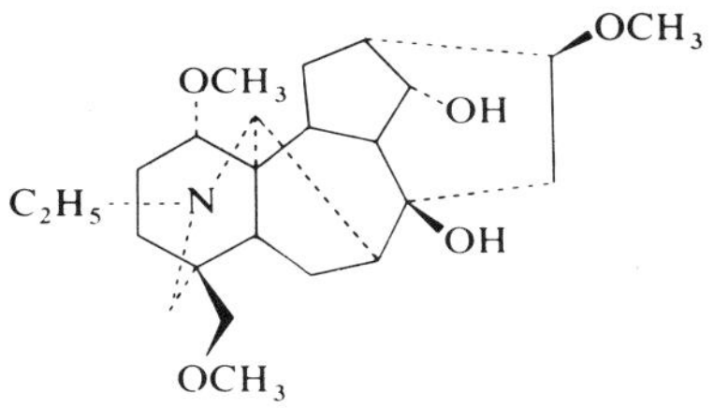

$C_{24}H_{39}NO_5$; 145–146°
$[\alpha]_D \pm 0°$
Aconitum talassicum; *A. nemorum*; *A. carmichaeli* and others
Correlated with isotalatizidine; synthesis
Refs. *14, 89, 93–98*

Talatizidine

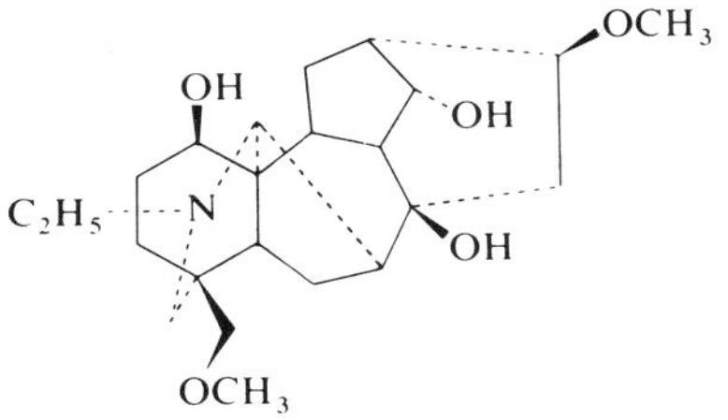

$C_{23}H_{37}NO_5$; 220–221°
$[\alpha]_D - 20°$ (MeOH)
Aconitum talassicum
Correlated with isotalatizidine
Refs. *85, 89–92*

Tricornine

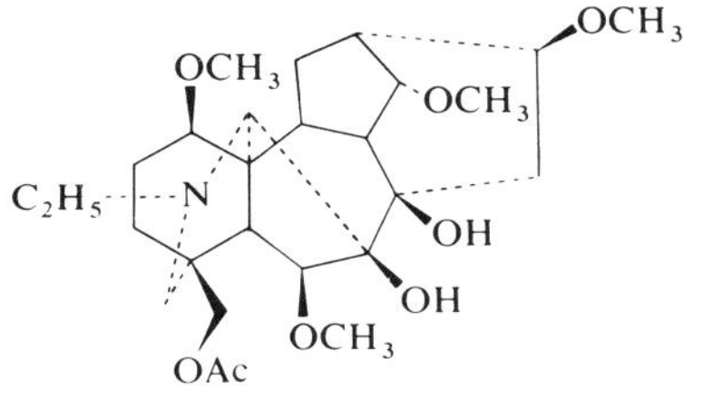

$C_{27}H_{43}NO_8$; 187–189°
$[\alpha]_D + 47.3°$ (EtOH)
Delphinium tricorne
^1H, ^{13}C NMR and chemical analysis
Ref. *128*

Umbrosine

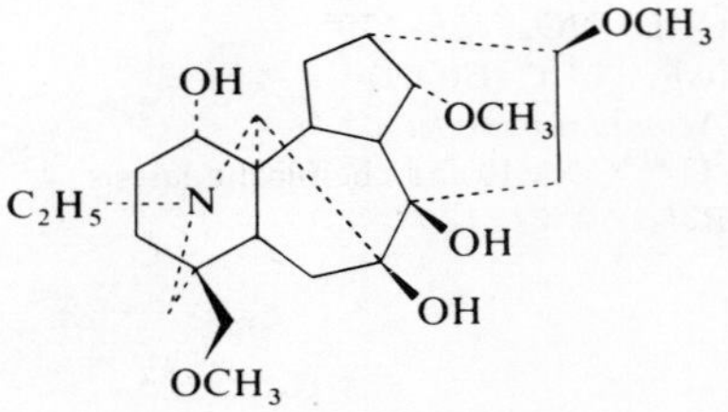

$C_{24}H_{39}NO_6$; 150–151°
$[\alpha]_D$ ——
Aconitum umbrosum
Chemical and spectral data
Ref. *187*

Veratroylpseudaconine

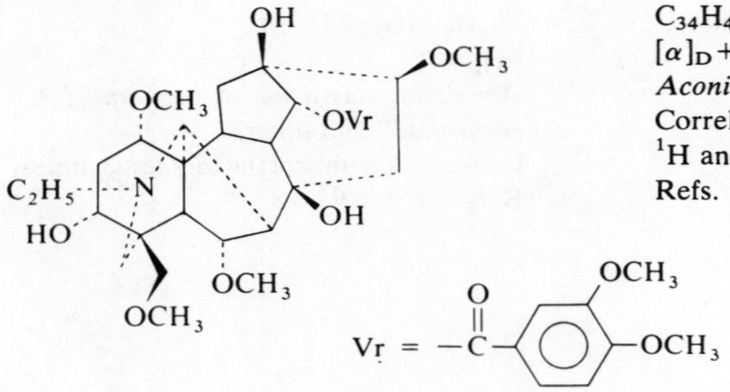

$C_{34}H_{49}NO_{11}$; 211–213°
$[\alpha]_D + 36.8°$ (EtOH)
Aconitum falconeri; *A. ferox*
Correlated with pseudaconitine;
^1H and ^{13}C NMR analysis
Refs. *41, 42, 44*

References

1. E. S. Stern, *in* "The Alkaloids" (R. H. F. Manske and H. L. Holmes, eds.), Vol. 4, Chapter 37. Academic Press, New York, 1954.
2. E. S. Stern, *in* "The Alkaloids" (R. H. F. Manske, ed.), Vol. 7, Chapter 22. Academic Press, New York, 1960.
3. S. W. Pelletier and L. H. Keith, *in* "The Alkaloids" (R. H. F. Manske, ed.), Vol. 12, Chapter 1. Academic Press, New York, 1970.
4. H. G. Boit, "Ergebnisse der Alkaloid-Chemie bis 1960," pp. 851–905, 1008–1011. Akademie-Verlag, Berlin, 1961.
5. L. H. Keith and S. W. Pelletier, *in* "Chemistry of the Alkaloids" (S. W. Pelletier, ed.), Chapter 18. Van Nostrand-Reinhold, Princeton, New Jersey, 1970.
6. O. E. Edwards, *Alkaloids (London)* **1**, 343–381 (1971).
7. S. W. Pelletier and L. H. Wright, *Alkaloids (London)* **2**, 247–258 (1972).
8. S. W. Pelletier and S. W. Page, *Alkaloids (London)* **3**, 232–257 (1973); **4**, 323–345 (1974); **5**, 230–241 (1975); **6**, 256–271 (1976); **7**, 247–267 (1977).
9. S. W. Pelletier and S. W. Page, *MTP International Review of Science: Alkaloids, Org. Chem., Ser. One* **9**, 319–346 (1973).
10. S. W. Pelletier and N. V. Mody, *Heterocycles* **5**, 771 (1976).
11. K. Nakanishi, T. Goto, S. Itô, S. Natori, and S. Nozol, eds., "Natural Products Chemistry," Vol. 2, pp. 271–272, Academic Press, New York, 1975. J. S. Glasby, "Encyclopedia of the Alkaloids." Plenum, New York, 1975.

12. T. K. Devon and A. I. Scott, "Handbook of Naturally Occurring Compounds," Vol. 2, pp. 244–246. Academic Press, New York, 1972.
13. P. L. Geiger, *Ann.* **7**, 269 (1833).
14. Y. Chen, Y. L. Chu, and J. H. Chu, *Yao Hsueh Hsueh Pao* **12**, 435 (1965); *CA* **63**, 16400a (1965).
15. J. H. Chu and S. T. Fang, *Hua Hsueh Hsueh Pao* **31**, 222 (1965); *CA* **63**, 16400a (1965).
16. M. N. Sultankhodzhaev, M. S. Yunusov, and S. Yu. Yunusov, *Khim. Prir. Soedin.* **9**, 127 (1973); *CA* **78**, 159946 (1973).
17. M. Przybylska and L. Marion, *Can. J. Chem.* **37**, 1116 (1959).
18. M. Przybylska and L. Marion, *Can. J. Chem.* **37**, 1843 (1959).
19. F. W. Bachelor, R. F. C. Brown, and G. Buchi, *Tet. Lett.* **10**, 1 (1960).
20. K. B. Birnbaum, K. Wiesner, E. W. K. Jay, and L. Jay, *Tet. Lett.* 867 (1971).
21. W. A. Jacobs and S. W. Pelletier, *J. Am. Chem. Soc.* **76**, 4048 (1954).
22. R. B. Turner, J. P. Yeschke, and M. S. Gibson, *J. Am. Chem. Soc.* **82**, 5182 (1960); W. A. Jacobs and S. W. Pelletier, *Chem. Ind. (London)* 591 (1960).
23. K. Wiesner and L. Jay, *Experientia* **27**, 758 (1971).
24. K. Markoshi, *Arch. Pharm. (Weinheim, Ger.)* **247**, 243 (1909).
25. R. Majima, H. Suginome, and S. Morio, *Ber. B* **57**, 1486 (1924).
26. H. Suginome and S. Imato, *J. Fac. Sci., Hokkaido Univ., Ser 3* **4**, 33 (1950).
27. E. Ochiai, T. Okamoto, and S. Sakai, *J. Pharm. Soc. Jpn.* **75**, 545 (1955).
28. L. H. Keith and S. W. Pelletier, *J. Org. Chem.* **33**, 2497 (1968).
29. E. Ochiai, T. Okamoto, S. Sakai, M. Kaneko, K. Fiyisowa, U. Nagai, and H. Tani, *J. Pharm. Soc. Jpn.* **76**, 550 (1956); S. I. Morio, *Ann.* **476**, 181 (1929).
30. T. E. Monakhova, T. F. Plantonova, A. D. Kuzovkov, and A. I. Shreter, *Khim. Prir. Soedin.* **2**, 113 (1965); *CA* **63**, 7347 (1965).
31. Y. Ichinohe and M. Yamaguchi, *Bull. Chem. Soc. Jpn.* **42**, 3038 (1969).
32. R. Majima and S. I. Morio, *Ann.* **476**, 171 (1929); *CA* **24**, 619 (1930).
33. E. Ochiai, T. Okamoto, S. Sakai, and A. Saito, *J. Pharm. Soc. Jpn.* **76**, 1414 (1956); *CA* **51**, 6661 (1957).
34. J. Iwasa and S. Narato, *Yakugaku Zasshi* **86**, 585 (1966); *CA* **65**, 10629 (1966).
35. K. Kao, F. Yo, and J. Chu, *Yao Hsueh Hsueh Pao* **13**, 186 (1966); *CA* **65**, 3922 (1966).
36. Y. Tsuda, O. Achmatowicz, and L. Marion, *Ann.* **680**, 88 (1964).
37. C. R. A. Wright and A. P. Luff, *J. Chem. Soc.* **31**, 143 (1877).
38. C. R. A. Wright and A. P. Luff, *J. Chem. Soc.* **33**, 151 (1878).
39. Y. Tsuda and L. Marion, *Can. J. Chem.* **41**, 1485 (1963).
40. A. Klásek, V. Simánek, and F. Santary, *Lloydia* **35**, 55 (1972).
41. K. K. Purushothaman and S. Chandrasekharan, *Phytochemistry* **13**, 1975 (1974).
42. S. W. Pelletier, N. V. Mody, and H. S. Puri, *Phytochemistry* **16**, 623 (1977).
43. W. R. Dunstan and A. E. Andrews, *J. Chem. Soc.* **87**, 1620 (1905).
44. S. W. Pelletier, N. V. Mody, R. S. Sawhney, and J. Bhattacharyya. *Heterocycles* **7**, 327 (1977).
45. N. Singh, G. S. Bajwa, and M. G. Singh, *Indian J. Chem.* **4**, 39 (1966).
46. S. W. Pelletier, N. V. Mody, and H. S. Puri, *Chem. Commun.* **12** (1977).
47. J. L. Lassaigne and H. Feneulle, *Ann. Chim. Phys.* **12**, 358 (1819).
48. T. Walz, *Arch. Pharm. (Weinheim, Ger.)* **260**, 9 (1922).
49. O. Keller, *Arch. Pharm. (Weinheim, Ger.)* **263**, 274 (1925).
50. L. N. Markwood, *J. Am. Pharm. Assoc.* **16**, 928 (1927).
51. W. A. Jacobs and L. C. Craig, *J. Biol. Chem.* **127**, 361 (1939).

52. W. A. Jacobs and S. W. Pelletier, *J. Am. Chem. Soc.* **76**, 161 (1954), and references cited therein.
53. W. A. Jacobs and S. W. Pelletier, *Chem. Ind.* (*London*) 948 (1955).
54. W. A. Jacobs and S. W. Pelletier, *J. Am. Chem. Soc.* **78**, 3542 (1956).
55. S. W. Pelletier, W. A. Jacobs, and P. Rathgeb, *J. Org. Chem.* **21**, 1514 (1956).
56. W. A. Jacobs and S. W. Pelletier, *J. Org. Chem.* **22**, 1428 (1957).
57. K. Wiesner, F. Bickelhaupt, and Z. Valenta, *Tetrahedron* **4**, 418 (1958).
58. K. Wiesner, F. Bickelhaupt, D. R. Babin, and M. Gotz, *Tet. Lett.* No. 3, 11 (1959).
59. K. Wiesner, D. L. Simmons, and L. R. Fowler, *Tet. Lett.* No. 18, 1 (1959).
60. K. Wiesner, F. Bickelhaupt, D. R. Babin, and M. Gotz, *Tetrahedron* **9**, 254 (1960).
61. K. Wiesner, D. L. Simmons, and R. H. Wightman, *Tet. Lett.* No. 15, 23 (1960).
62. O. E. Edwards and L. Marion, *Can. J. Chem.* **30**, 627 (1952).
63. K. Wiesner, H. W. Brewer, D. L. Simmons, D. R. Babin, F. Bickelhaupt, J. Kallos, and T. Bogri, *Tet. Lett.* No. 3, 17 (1960).
64. R. C. Cookson, J. Henstock, and J. Hudec, *J. Am. Chem. Soc.* **88**, 1060 (1966).
65. K. Wiesner and T. Inaba, *J. Am. Chem. Soc.* **91**, 1036 (1969).
66. S. W. Pelletier and Z. Djarmati, *J. Am. Chem. Soc.* **98**, 2626 (1976).
67. W. R. Dunstan and A. E. Andrews, *J. Chem. Soc.* **87**, 1636 (1905).
68. O. Achmatowicz and L. Marion, *Can. J. Chem.* **42**, 154 (1964).
69. H. Suginome, N. Katsui, and G. Hasegawa, *Bull. Chem. Soc. Jpn.* **32**, 604 (1959).
70. N. Katsui, *Bull. Chem. Soc. Jpn.* **32**, 774 (1959).
71. S. W. Pelletier, N. V. Mody, and N. Katsui, *Tet. Lett.* 4027 (1977).
72. S. W. Pelletier, W. H. De Camp, S. Lajšić, Z. Djarmati, and A. H. Kapadi, *J. Am. Chem. Soc.* **96**, 7815 (1974).
73. S. W. Pelletier, Z. Djarmati, S. Lajšić, and W. H. De Camp, *J. Am. Chem. Soc.* **98**, 2617 (1976).
73a. W. H. De Camp and S. W. Pelletier, *Acta Cryst allogr., Sect. B* **33**, 722 (1977).
74. W. Freudenberg and E. F. Rogers, *J. Am. Chem. Soc.* **59**, 2572 (1937).
75. L. Marion, J. P. Boca, and J. Kallos, *Tetrahedron, Suppl.* **8**, Part 1, 101 (1966).
76. O. Achmatowicz, Jr., Y. Tsuda, L. Marion, T. Okamoto, M. Natsume, H. H. Chang, and K. Kajima, *Can. J. Chem.* **43**, 825 (1965).
77. O. E. Edwards, L. Fonzes, and L. Marion, *Can. J. Chem.* **44**, 583 (1966).
78. S. W. Pelletier, Z. Djarmati, and S. Lajšić, *J. Am. Chem. Soc.* **96**, 7817 (1974).
79. O. Achmatowicz, Jr. and L. Marion, *Can. J. Chem.* **43**, 1093 (1965).
80. S. W. Pelletier, Z. Djarmati, and W. H. De Camp, *Acta Crystallogr., Sect. A* **31**, S111 (1976).
81. S. W. Pelletier, W. H. De Camp, and Z. Djarmati, *Chem. Commun.* 253 (1976).
82. S. W. Pelletier, J. K. Thakkar, N. V. Mody, Z. Djarmati, and J. Bhattacharyya, *Phytochemistry* **16**, 404 (1977).
83. S. W. Pelletier and J. Bhattacharyya, *Tet. Lett.* 4679 (1976).
84. A. D. Kuzovkov and T. F. Platonova, *J. Gen. Chem. USSR* (*Engl. Transl.*) **31**, 1286 (1961).
85. S. W. Pelletier, L. H. Keith, and P. C. Parasarathy, *J. Am. Chem. Soc.* **89**, 4146 (1967).
86. H. Schulze and G. Berger, *Arch. Pharm.* (*Weinheim, Ger.*) **262**, 553 (1924).
87. S. W. Pelletier, J. Bhattacharyya, and N. V. Mody, *Heterocycles* **6**, 463 (1977).
88. M. S. Rabinovich and R. A. Konovalova, *Zh. Obshch Khim.* **12**, 329 (1942).
89. R. A. Konowalowa and A. P. Orekhoff, *Bull. Soc. Chim. Fr., Mem.* [*5*] **7**, 95 (1940), *CA* **34**, 5450 (1940).
90. R. A. Konovalova and A. P. Orekhov, *Zh. Obshch. Khim.* **10**, 745 (1940).

91. A. D. Kuzovkov and T. F. Platonova, *J. Gen. Chem. USSR* (*Engl. Transl.*) **31**, 1286 (1961).
92. S. W. Pelletier, W. H. De Camp, D. L. Herald, Jr., S. W. Page, and M. G. Newton, *Acta Crystallogr., Sect. B* **33**, 716 (1977).
93. T. F. Platonova, A. D. Kuzovhov, and P. S. Massagetov, *J. Gen. Chem. USSR* (*Engl. Transl.*) **28**, 3157 (1958); *Zh. Obshch. Khim.* **28**, 3126 (1958).
94. M. A. Khaimova, M. D. Palamareva, L. G. Grozdanova, N. M. Mollov, and P. P. Panov, *C. R. Acad. Bulg. Sci.* **20**, 193 (1967); *CA* **67**, 54296c (1967).
95. M. A. Khaimova, M. D. Palamareva, N. M. Mollov, and V. P. Krestev, *Tetrahedron* **27**, 819 (1971).
96. K. Wiesner, T. Y. R. Tsai, K. Huber, S. E. Bolton, and R. Vlahov, *J. Am. Chem. Soc.* **96**, 4990 (1974); K. Wiesner, *Pure Appl. Chem.* **41**, (1–2), 93 (1975).
97. M. S. Yunusov and S. Yu. Yunusov, *Khim. Prir. Soedin.* **6**, 90 (1970).
98. M. S. Yunusov, V. A. Telnov, and S. Yu. Yunusov, *Khim. Prir. Soedin.* **6**, 774 (1970).
99. V. A. Telnov, M. S. Yunusov, S. Yu. Yunusov, and B. Sh. Ibragimov, *Khim. Prir. Soedin.* **11**, 814 (1975).
100. A. J. Jones and M. H. Benn, *Tet. Lett.* 4351 (1972).
101. A. J. Jones and M. H. Benn, *Can. J. Chem.* **51**, 486 (1973).
102. S. W. Pelletier, N. V. Mody, A. J. Jones, and M. H. Benn, *Tet. Lett.* 3025 (1976).
102a. S. W. Pelletier, N. V. Mody, R. S. Sawhney, and J. Bhattacharyya, *Heterocycles* **7**, 327 (1977).
103. H. Schulze and F. Ulfert, *Arch. Pharm.* (*Weinheim, Ger.*) **260**, 230 (1922); H. V. Rosendahl, *Arb. Pharmakol. Inst. Dorpat.* **11**, 1 (1895); G. Weidemann, *Arch. Exp. Pathol. Pharmakol.* **95**, 166 (1922).
104. A. D. Kuzovkov, P. S. Massagetov, *J. Gen. Chem. USSR* (*Engl. Transl.*) **25**, 161 (1955).
105. T. F. Platonova, A. D. Kuzovkov, and P. S. Massagetov, *J. Gen. Chem. USSR* (*Engl. Transl.*) **28**, 259 (1958).
106. N. Mollov, M. Tada, and L. Marion, *Tet. Lett.* No. 26, 2189 (1969); N. Mollov, M. Haimova, P. Tscherneva, N. Pecigargova, I. Ognjanov, and P. Panov, *C. R. Acad. Bulg. Sci.* **17**, 251 (1964); K. Khaimova, N. Mollov, P. Cerneva, A. Antonova, and V. Ivanova, *Tet. Lett.*, No. 38, 2711 (1964); L. Marion, L. Fonzes, C. K. Wilkins, Jr., and J. P. Boca, *Can. J. Chem.* **45**, 969 (1967).
107. G. I. Birnbaum, *Tet. Lett.* No. 26, 2193 (1969).
108. G. I. Birnbaum, *Acta Cryst allogr., Sect. B* **26**, 755 (1970).
109. V. A. Telnov, M. S. Yunusov, and S. Yu. Yunusov, *Khim. Prir. Soedin.* **6**, 583 (1970).
110. S. W. Pelletier, N. V. Mody, and A. J. Aasen, unpublished results.
111. V. A. Telnov, M. S. Yunusov, and S. Yu. Yunusov, *Khim. Prir. Soedin.* **6**, 639 (1970).
112. V. A. Telnov, M. S. Yunusov, Ya. V. Rashkes, and S. Yu. Yunusov, *Khim. Prir. Soedin.* **7**, 622 (1971).
113. V. A. Telnov, M. S. Yunusov, and S. Yu. Yunusov, *Khim. Prir. Soedin.* **9**, 130 (1973).
114. V. A. Telnov, M. S. Yunusov, and S. Yu. Yunusov, *Khim. Prir. Soedin.* **9**, 129 (1973).
115. S. M. Nasirov, V. G. Andrianov, Yu. T. Struchkov, V. A. Telnov, M. S. Yunusov, and S. Yu. Yunusov, *Khim. Prir. Soedin.* **10**, 812 (1974).
116. S. M. Nasirov, V. G. Andrianov, Yu. T. Struchkov, and S. Yu. Yunusov, *Khim. Prir. Soedin.* **12**, 206 (1976).
117. M. N. Sultankhodzhaev, M. S. Yunusov, and S. Yu. Yunusov, *Khim. Prir. Soedin.* **9**, 199 (1973).
118. M. N. Sultankhodzhaev, M. S. Yunusov, and S. Yu. Yunusov, *Khim. Prir. Soedin.* **11**, 481 (1975).

119. M. N. Sultankhodzhaev, M. S. Yunusov, and S. Yu. Yunusov, *Khim. Prir. Soedin.* **8**, 399 (1972).
120. N. Katsui and G. Hasegawa, *Bull. Chem. Soc. Jpn.* **33**, 1037 (1960).
121. D. A. Muravjeva, T. I. Plekhanova, and M. S. Yunusov, *Khim. Prir. Soedin.* **8**, 128 (1972).
122. V. E. Nezhevenko, M. S. Yunusov, and S. Yu. Yunusov, *Khim. Prir. Soedin.* **10**, 409 (1974).
123. V. E. Nezhevenko, M. S. Yunusov, and S. Yu. Yunusov, *Khim. Prir. Soedin.* **11**, 389 (1975).
124. M. S. Yunusov, V. E. Nezhevenko, and S. Yu. Yunusov, *Khim. Prir. Soedin.* **11**, 107 (1975).
125. M. S. Yunusov, V. E. Nezhevenko, and S. Yu. Yunusov, *Khim. Prir. Soedin.* **11**, 770 (1975).
126. M. S. Yunusov and S. Yu. Yunusov, *Dokl. Akad. Nauk SSSR* **188**, 1077 (1969).
127. M. S. Yunusov and S. Yu. Yunusov, *Khim. Prir. Soedin.* **6**, 334 (1970).
128. S. W. Pelletier and J. Bhattacharyya, *Phytochemistry* **16**, 1464 (1977).
129. A. S. Narzullaev, M. S. Yunusov, and S. Yu. Yunusov, *Khim. Prir. Soedin.* **9**, 497 (1973).
130. M. N. Mats, *Rastit. Resur.* **8**, 249 (1972).
131. S. W. Pelletier and J. Bhattacharyya, *Tet. Lett.* No. 32, 2735 (1977).
132. S. Yunusov and N. K. Abubakirov, *Dokl. Akad. Nauk Uzb. SSR* **8**, 21 (1949); A. D. Kuzovkov and T. F. Platonova, *J. Gen. Chem. USSR* (*Engl. Transl.*) **29**, 2746 (1959).
133. B. T. Salimov, M. S. Yunusov, S. Yu. Yunusov, and A. S. Narzullaev, *Khim. Prir. Soedin.* **11**, 665 (1975).
134. G. R. Waller, S. D. Sastry, and K. F. Kinneberg, *Proc. Okla. Acad. Sci.* **53**, 92 (1973).
135. S. W. Pelletier and N. V. Mody, unpublished results.
136. M. H. Benn, M. A. M. Cameron, and O. E. Edwards, *Can. J. Chem.* **41**, 477 (1963).
137. S. W. Pelletier and R. S. Sawhney, unpublished results on the alkaloids of *Consolida ambigua*.
138. A. S. Narzullaev, M. S. Yunusov, and S. Yu. Yunusov, *Khim. Prir. Soedin.* **8**, 498 (1972).
139. A. S. Narzullaev, M. S. Yunusov, and S. Yu. Yunusov, *Khim. Prir. Soedin.* **9**, 443 (1973).
140. M. H. Benn, J. D. Connolly, O. E. Edwards, L. Marion, and Z. Stojanac, *Can. J. Chem.* **49**, 425 (1971).
141. M. N. Sultankhodzhaev, M. S. Yunusov, and S. Yu. Yunusov, *Khim. Prir. Soedin.* **11**, 381 (1975).
142. S. W. Pelletier and R. Aneja, *Tet. Lett.* 557 (1967).
143. M. S. Yunusov, Ya. V. Rashkes, V. A. Telnov, and S. Yu. Yunusov, *Khim. Prir. Soedin.* **5**, 515 (1969).
144. M. S. Yunusov, Ya. V. Rashkes, and S. Yu. Yunusov, *Khim. Prir. Soedin.* **7**, 626 (1971).
145. M. S. Yunusov, Ya. V. Rashkes, and S. Yu. Yunusov, *Khim. Prir. Soedin.* **8**, 85 (1972).
146. S. D. Sastry, *in* "Biochemical Applications of Mass Spectrometry" (G. R. Waller, ed.), Chapter 24, p. 662, and references cited therein. Wiley (Interscience), New York, 1972.
147. K. Wiesner, W.-L. Kao, and J. Šantroch, *Can. J. Chem.* **47**, 2431 (1969).
148. K. Wiesner, W.-L. Kao, and E. W. K. Jay, *Can. J. Chem.* **47**, 2734 (1969).
149. K. Wiesner, E. W. K. Jay, C. Demerson, T. Kanno, J. Krepinsky, L. Poon, T. Y. R. Tsai, A. Vilim, and C. S. Wu, *Experientia* **26**, 1030 (1970).
150. K. Wiesner, E. W. K. Jay, and L. Jay (Poon), *Experientia* **27**, 363 (1971).

151. K. Wiesner, E. W. K. Jay, T. Y. R. Tsai, C. Demerson, L. Jay, T. Kanno, J. Krepinsky, A. Vilim, and C. S. Wu, *Can. J. Chem.* **50**, 1925 (1972).
152. K. Wiesner and Z. Valenta, *Prog. Chem. Org. Nat. Prod.* **16**, 50–53 (1958).
153. J. P. Johnston and K. H. Overton, *J. Chem. Soc., Perkin Trans. I* 1490 (1972).
154. J. P. Johnston and K. H. Overton, *Chem. Commun.* 329 (1969).
155. W. A. Ayer and P. D. Deshpande, *Can. J. Chem.* **51**, 77 (1973).
156. F. R. Ahmed, *Acta Crystallogr., Sect. B* **30**, 2558 (1974).
157. S-F. Lee, G. M. Sathe, W. W. Sy, P.-T. Ho, and K. Wiesner, *Can. J. Chem.* **54**, 1039 (1976).
158. K. Wiesner, P. T. Ho, R. C. Jain, S. F. Lee, S. Oida, and A. Philipp, *Can. J. Chem.* **51**, 1448 (1973).
159. K. Wiesner, P. T. Ho, D. Chang, Y. K. Lam, C. S. J. Pan, and W. Y. Ren, *Can. J. Chem.* **51**, 3978 (1973).
160. K. Wiesner, P. T. Ho, W. C. Liu, and M. N. Shanbhag, *Can. J. Chem.* **53**, 2140 (1975).
161. K. Wiesner, T. Y. R. Tsai, K. Huber, and S. Bolton, *Tet. Lett.* 1233 (1973).
162. R. W. Guthrie, Z. Valenta, and K. Wiesner, *Tet. Lett.* 4645 (1966).
163. G. Ferguson and J. P. Johnston, *Chem. Commun.* 330 (1969).
164. M. Przybylska, T. Y. R. Tsai, and K. Wiesner, *Chem. Commun.* 297 (1975).
165. A. Tahara and T. Ohsawa, *Tet. Lett.* 2469 (1969).
166. W. I. Taylor, W. E. Walles, and L. Marion, *Can. J. Chem.* **32**, 780 (1954).
167. L. Marion, *Pure Appl. Chem.* **6**, 621 (1963).
168. B. T. Salimov, M. S. Yunusov, and S. Yu Yunusov, *Khim. Prir. Soedin*, (1), 128 (1977).
169. T. F. Platonova and A. D. Kuzovkov, *Med. Pharm SSSR* **17**, 19 (1963); *CA* **59**, 6723 (1963).
170. S. W. Pelletier and R. S. Sawhney, *Heterocycles*, **9**, 463 (1978).
171. S. W. Pelletier, R. S. Sawhney, and N. V. Mody, *Heterocycles*, in press (1978).
172. O. E. Edwards and M. N. Rodger, *Can. J. Chem.* **37**, 1187 (1959).
173. S. Talapatra and A. Chatterjee, *J. Indian Chem. Soc.* **36**, 437 (1959).
174. Y. Tsuda and L. Marion, *Can. J. Chem.* **41**, 3055 (1963).
175. M. H. Benn, *Can. J. Chem.* **44**, 1 (1966).
176. V. Skaric and L. Marion, *Can. J. Chem.* **38**, 2433 (1960).
177. E. Ochiai, T. Okamoto, and M. Kaneko, *Chem. Pharm. Bull.* **6**, 730 (1958), and references cited therein.
178. R. C. Cookson and M. E. Trevett, *J. Chem. Soc.* 2689, 3121 (1956).
179. O. E. Edwards, L. Marion, and K. H. Palmer, *Can. J. Chem.* **36**, 1097 (1958).
180. M. Carmack, J. P. Ferris, J. Harvey, P. L. Magat, E. W. Martin, and D. W. Mayo, *J. Am. Chem. Soc.* **80**, 497 (1958).
181. F. Sparatore, R. Greenhalgh, and L. Marion, *Tetrahedron* **4**, 157 (1958).
182. L. Marion and V. Skaric, *J. Am. Chem. Soc.* **80**, 4434 (1958).
183. R. Aneja, D. M. Locke, and S. W. Pelletier, *Tetrahedron* **29**, 3297 (1973).
184. M. Przybylska and L. Marion, *Can. J. Chem.* **34**, 185 (1965).
185. O. E. Edwards, L. Marion, and D. K. R. Stewart, *Can. J. Chem.* **34**, 1315 (1956).
186. M. Przybylska and L. Marion, *Can. J. Chem.* **37**, 1843 (1959).
187. V. A. Tel'nov, N. M. Golubev, and M. S. Yunusov, *Khim. Prir. Soedin*, (5), 675 (1976).

—CHAPTER 2—

QUINOLINE ALKALOIDS RELATED TO ANTHRANILIC ACID

M. F. GRUNDON

School of Physical Sciences, New University of Ulster, Coleraine, Northern Ireland

I. Introduction

In the ten years since quinoline alkaloids were last reviewed by Openshaw in Volume IX of this treatise, the constituents of a considerable number of rutaceous species have been studied for the first time and more than 70 new quinoline alkaloids have been isolated. The most notable structural work includes the study of the phenolic alkaloids of the dihydrofuro- and dihydropyranoquinoline group found in *Balfourodendron* and

ISBN 0-12-469517-5

Ruta species, and the identification from a *Ptelea* species of a new group of quinoline alkaloids containing a terminal double bond in the prenyl portion of the molecule. New types of quinoline alkaloids are represented by the 1,2-dimethylallyl derivatives of a *Ravenia* species and the alkaloids of *Haplophyllum bucharicum* Litv. incorporating a C_{10} geranyl side chain.

By 1966 synthetic routes to the main groups of quinoline alkaloids had been well established, and studies during the last decade have been directed to the improvement of existing methods, to the elaboration of new high-yield procedures for the furoquinolines, and to the synthesis of new types, for example, the 1,1- and 1,2-dimethylallylquinolines and the group based on 3-prenylquinolines with a terminal double bond. The asymmetric synthesis of dihydrofuro- and dihydropyranoquinolines and related compounds has been accomplished, and the products have been used to establish the absolute configuration of the alkaloids.

The important first experiments on the biosynthesis of quinoline alkaloids were reported in Volume IX, Chapter 6. Extensive ^{14}C and ^{3}H tracer feeding experiments have now led to the elucidation of the main biosynthetic pathway to typical furoquinoline and hydroxyisopropyldihydrofuroquinoline alkaloids. In fact, the plan adopted for this chapter is based on the biosynthetic sequence.

Since publication of Volume IX of this treatise, thin-layer chromatography has been used extensively to detect the presence of quinoline alkaloids, and the subject has been reviewed (*1*). Gas chromatography has been introduced for the estimation of quinoline alkaloids (*2, 3*).

The application of spectroscopic methods for the elucidation of the structures of quinoline alkaloids has developed to the point where the constitutions of many new alkaloids have been established solely by UV, IR, NMR, and mass spectrometry. The NMR spectra of quinoline alkaloids have been discussed (Volume IX of this treatise) (*4, 5*), and the application of mass spectrometry has also been reviewed (*6*). Schemes 1 and 22 summarize NMR data for a selection of typical quinoline alkaloids. In discussing structure work spectroscopic data will be given only when they are of special significance.

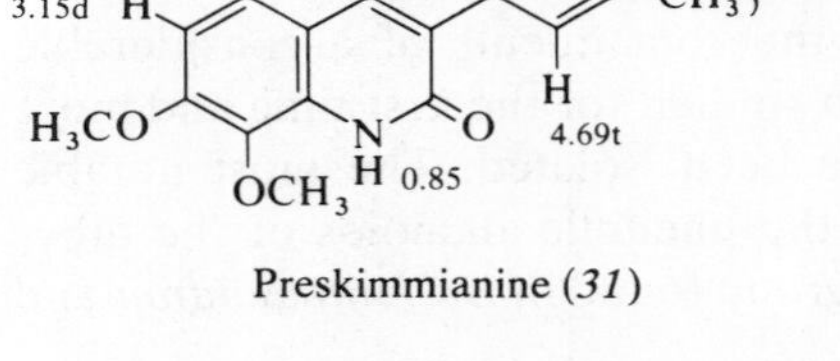

Preskimmianine (*31*)

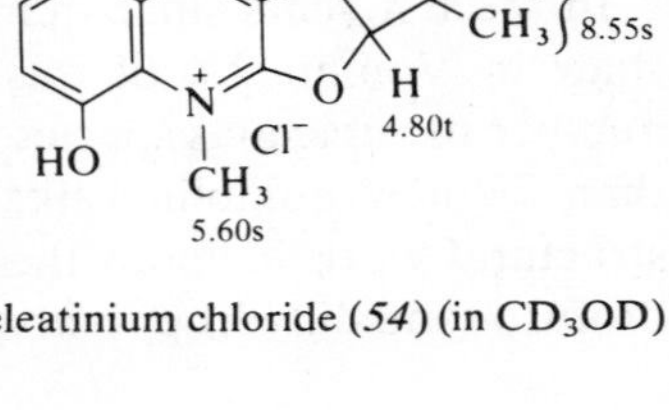

Pteleatinium chloride (*54*) (in CD_3OD)

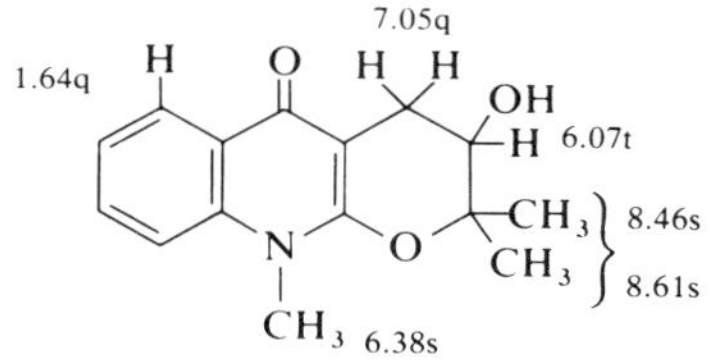

Ribalinine (*27*)

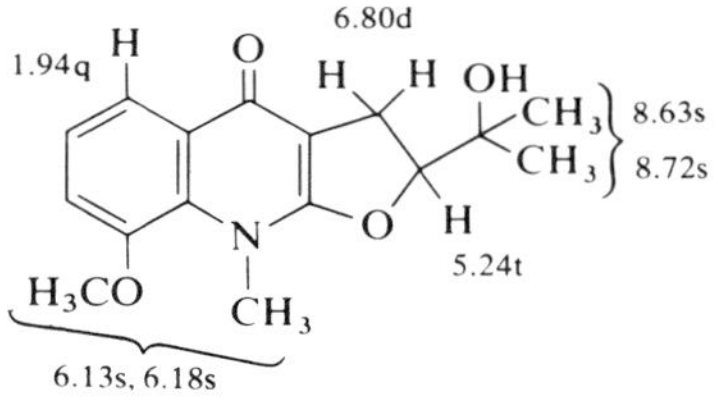

Balfourodine (*27*)

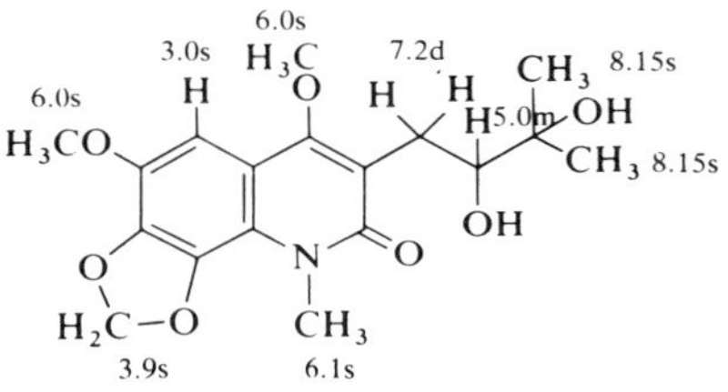

6-Methoxyhydroxylunidine (*57*)

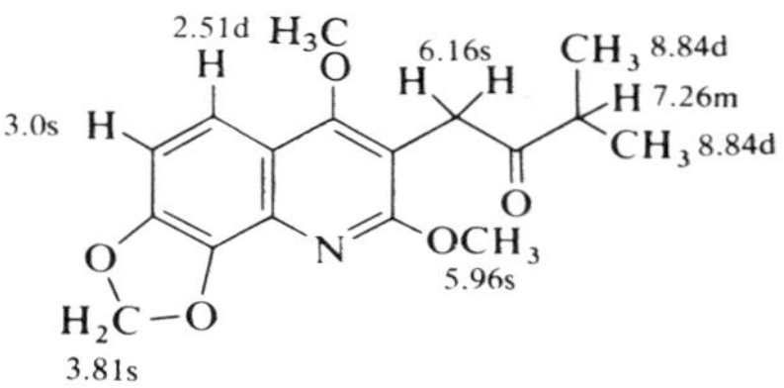

Orixinone (*59*)

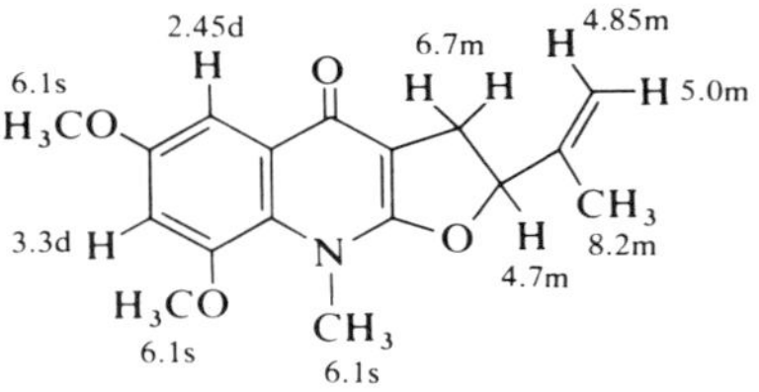

Ptelefolone (*72*)

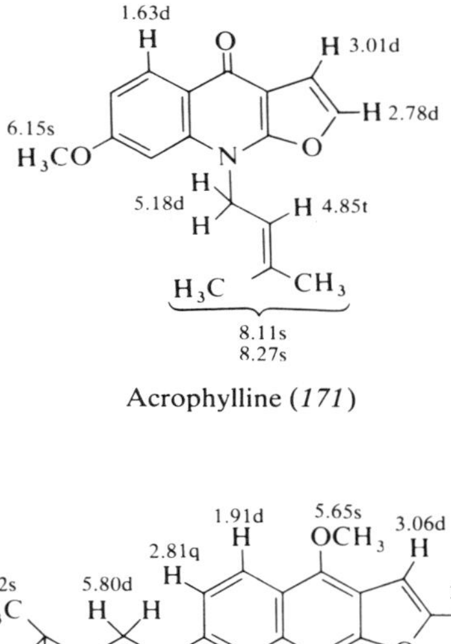

Acrophylline (*171*)

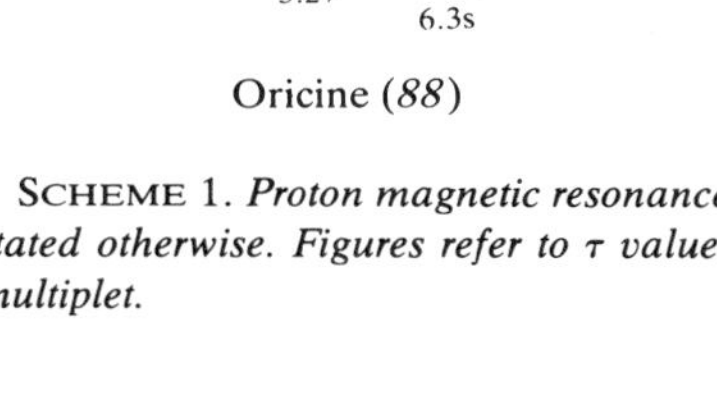

Oricine (*88*)

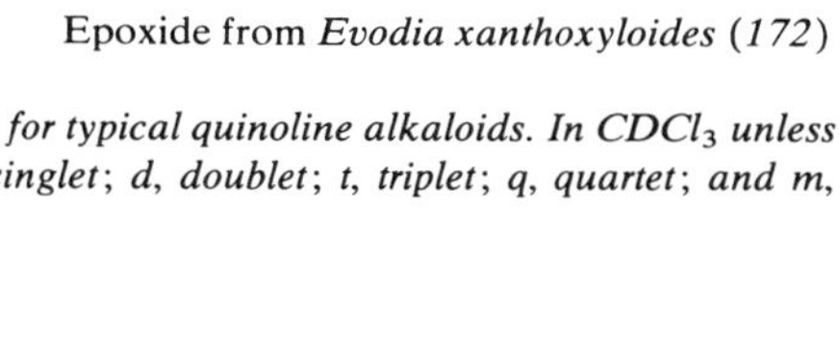

Epoxide from *Evodia xanthoxyloides* (*172*)

SCHEME 1. *Proton magnetic resonance data for typical quinoline alkaloids. In* $CDCl_3$ *unless stated otherwise. Figures refer to* τ *values; s, singlet; d, doublet; t, triplet; q, quartet; and m, multiplet.*

Concise general reviews of quinoline alkaloids appeared in 1970 (*7*) and in 1974 (*8*), and annual coverage is provided by the Chemical Society, London (*9*). A review listing in convenient form the occurrence of rutaceous alkaloids up to April 1973 has been published (*10*). A valuable account of the synthesis of 2,4-dioxyquinoline alkaloids is now available (*11*).

This review includes work published during the period 1966–mid 1976.

II. 4-Methoxy-2-quinolones

4-Hydroxy-2-quinolone is a biosynthetic precursor of quinoline alkaloids (see Section VII), and it is not surprising that compounds of this type occur in rutaceous plants (Volume IX, p. 225). Additional 4-methoxy-2-quinolones with or without *N*-methyl groups have now been reported.

4-Methoxy-1-methyl-2-quinolone (**1**) was isolated from *Hesperethusa crenulata* M. Roem. (*12*) and from the wood of *Fagara boninensis* (*13*). The alkaloid folimine from *Haplophyllum foliosum* Vved. (*14*) was identified as 4,8-dimethoxy-1-methyl-2-quinolone (**2**), which had been obtained earlier during the degradation of foliosine (Volume IX, p. 225). Folifidine from *H. dubium* Eug. Kor (*15*) and from *H. foliosum* (*16*) is the 8-hydroxy-2-quinolone **3**.

The alkaloid edulitine from *Casimiroa edulis* Llave and Lex was assigned structure **5** on the basis of spectroscopic studies (*17*), and this was confirmed by several syntheses (*18–20*). For example, Piozzi and coworkers cyclized the anilide **4**, obtained from *o*-methoxyaniline and

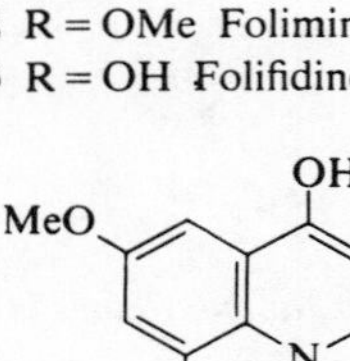

OMe; N; O; R; Me

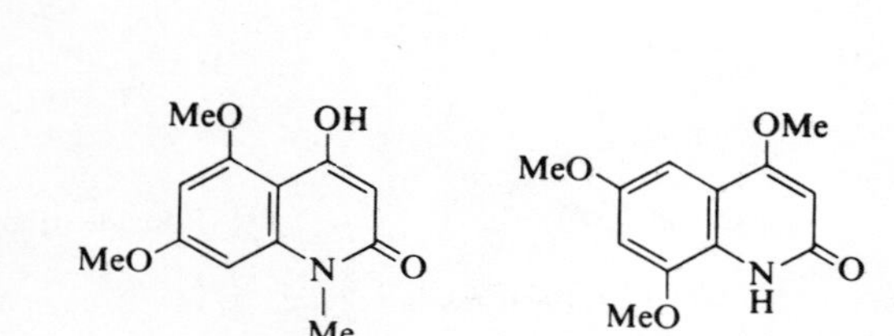

1 R = H
2 R = OMe Folimine
3 R = OH Folifidine

MeO; $NHCOCH_2CO_2H$

4

(i) PPA
(ii) CH_2N_2

OMe; MeO; N; H; O

5 Edulitine

OH; MeO; MeO; N; O; Me

6

MeO; OH; MeO; N; O; Me

7

OMe; MeO; MeO; N; H; O

8 Halfordamine

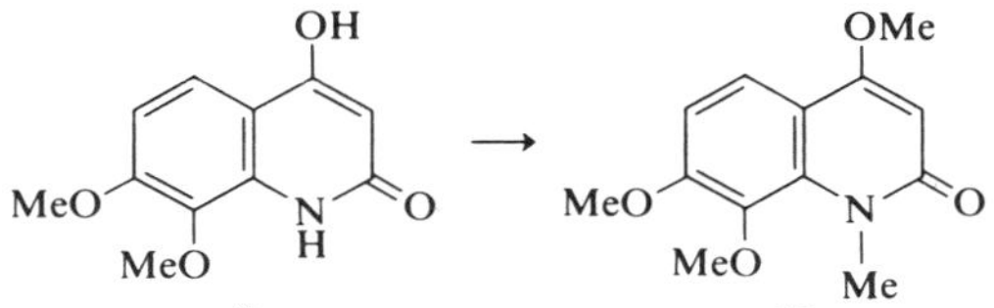

OH → OMe

MeO, MeO, N H, O — 9 → MeO, MeO, N Me, O — 10

malonic acid, with polyphosphoric acid to 4-hydroxy-8-methoxy-2-quinolone, which with diazomethane furnished edulitine (*18*). The alkaloid, robustinine, from *Haplophyllum bungei* Trautv is identical with edulitine (*21*).

The bark of *Halfordia scleroxyla* F. Muell. was shown by Crow and Hodgkin (*22*) to contain a 2-quinolone ($C_{12}H_{13}NO_4$) named halfordamine. The NMR spectrum of halfordamine indicated the presence of three methyl groups and two meta-coupled aromatic protons. A resonance at $\tau 0.32$, which disappeared on the addition of deuterium oxide, was attributed to a 4-hydroxyl group. Hence the alkaloid was formulated as a dimethoxy 4-hydroxy-1-methyl-2-quinolone (**6** or **7**). Halfordamine was reinvestigated independently by Storer and Young (*23*) and by Piozzi and co-workers (*24*). Both groups concluded that halfordamine was 4,6,8-trimethoxy-2-quinolone (**8**) and synthesized the alkaloid by a method similar to that described above for edulitine. It was pointed out that the NMR resonance at $\tau 3.02$ is consistent with the presence of a C-5 proton in structure **8** and that the absence of a shift in the UV spectrum in basic solution indicated that the alkaloid is a 4-methoxy-2-quinolone rather than a 1-methyl-4-hydroxy-2-quinolone as supposed previously.

Another 4-methoxy-1-methyl-2-quinolone (**10**) was isolated from the roots of *Spathelia sorbifolia* L. (*25*). The constitution of the alkaloid was established by spectroscopy and confirmed by dimethylation of quinolone **9** with dimethyl sulfate in dimethylformamide.

III. 3-Prenyl-2-quinolones and Related Tricyclic Alkaloids

A. Introduction

Investigation of the alkaloids of *Lunasia amara* Blanco, *Balfourodendron riedelianum* Engl., and *Platydesma campanulata* Mann (Volume IX, p. 236) resulted in the isolation and characterization of hydroxyisopropyldihydrofuroquinolines, for example, *O*-methylbalfourodinium salt (**11**), balfourodine (**12**), and platydesmine (**13**). The dihydropyranoquinoline, isobalfourodine (**14**), isomeric with balfourodine, was also obtained from *B. riedelianum.* A number of characteristic reactions were explored, particularly the interconversions of the quaternary alkaloid **11** and the 4-quinolone (**12**), and the cleavage of the quaternary salt with base to the diol, balfourolone (**15**). The structures of *O*-methylbalfourodinium salt, balfourodine, isobalfourodine, and platydesmine were confirmed by biomimetic synthesis from 3-prenylquinolines.

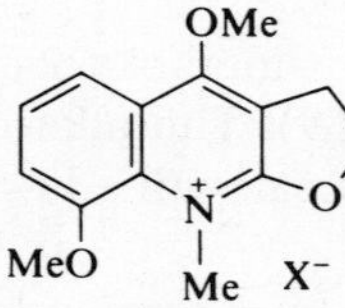

11 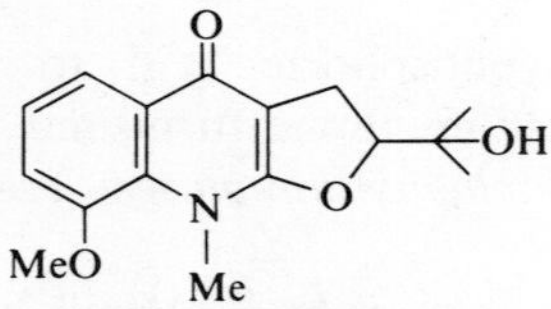 **12** Balfourodine 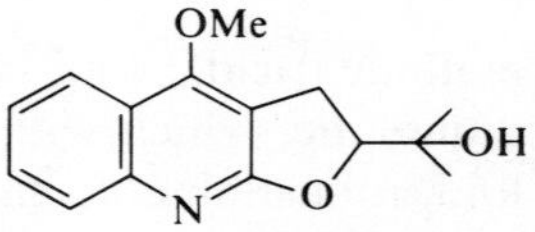 **13** Platydesmine

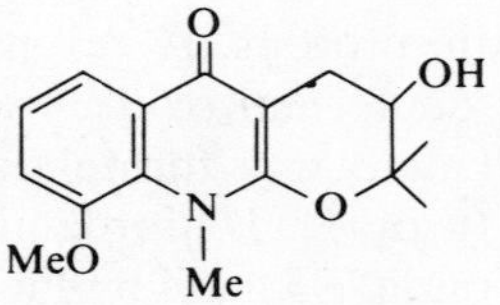

14 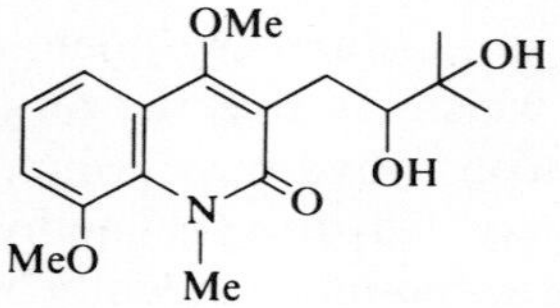 **15**

During the last decade, new alkaloids of this type have been isolated, methods of asymmetric synthesis have been developed, the absolute stereochemistry of a number of alkaloids has been established, and the mechanisms of certain rearrangement reactions of furo- and pyranoquinoline alkaloids of the group have been studied. Additional pyranoquinolines of the flindersine type have been discovered, and new methods of synthesis have been developed. Isopropyldihydrofuroquinoline alkaloids without a hydroxy group in the side chain are characteristic components of *Lunasia* species, but not much new work on this group has appeared recently.

B. 3-Prenyl-2-quinolones

4-Methoxy-3-prenyl-2-quinolones were first prepared in connection with the synthesis and biosynthesis of quinoline alkaloids (Volume IX, pp. 247, 257), but soon afterward prenyl derivatives were identified as constituents of rutaceous plants (Table I). *O*-Prenyl and *O*-geranyl ethers without substituents in the 3-position of the quinoline nucleus will be discussed in Section V of this chapter.

Eshiett and Taylor isolated atanine (**16**) from *Fagara xanthoxyloides* Lam. (*26*). The structure was established by NMR spectroscopy, by

TABLE I
3-PRENYL-2-QUINOLONES

Compound	Structure number	Molecular formula	Melting point (°C)	Ref.
Atanine	**16**	$C_{15}H_{17}NO_2$	130	*26, 28*
3-Isopentenyl-4-methoxy-7,8-methylenedioxy-2-quinolone	**24**	$C_{16}H_{17}NO_4$	159–162	*32*
N-Methylatanine	**17**	$C_{16}H_{19}NO_2$	Oil	*29*
Preskimmianine	**23**	$C_{17}H_{21}NO_4$	151–152	*31*
Ptelecortine (Pt/30)	**25**	$C_{18}H_{21}NO_5$	126–128	*33*
3-Isopentenyl-1-methyl-4,6,8-trimethoxy-2-quinolone (Pt/46)	**27**	$C_{18}H_{23}NO_4$	69–71	*34*
3-Isopentenyl-4-isopentenyloxy-2-quinolone	**28**	$C_{19}H_{23}NO_2$	114–115	*36*

cyclization with hydrogen iodide to dihydroflindersine (**18**), and by catalytic reduction to 4-methoxy-3-(3-methylbutyl)-2-quinolone. The alkaloid had been synthesized earlier (*27*), and later was identified as a constituent of *Ravenia spectabilis* Engl. (*28*).

16 R = H Atanine
17 R = Me *N*-methylatanine

18 Dihydroflindersine

N-Methylatanine (**17**) was obtained as an oil from the roots of *Ailanthus giraldii* Dode and identified by spectroscopic methods and by synthesis from *N*-methylaniline and diethyl(3-methylbut-2-enyl)malonate followed by reaction of the product with diazomethane (*29*).

Mitscher and co-workers (*30*) have explored alternative methods of synthesizing 3-prenyl-2-quinolones (Scheme 2) whereby the formation of by-products characteristic of the prenylmalonate route is avoided by blocking the 3-position. Thus, allylation of the 3-bromo-4-hydroxy-*N*-methyl-2-quinolone yielded ketone **19** which, with zinc and acid, was converted into the 3-prenyl-2-quinolone (**20**). The ketone **22**, obtained from ester **21**, was transformed into the 2-quinolone (**20**) by heating with copper acetate in hexamethylphosphorus triamide.

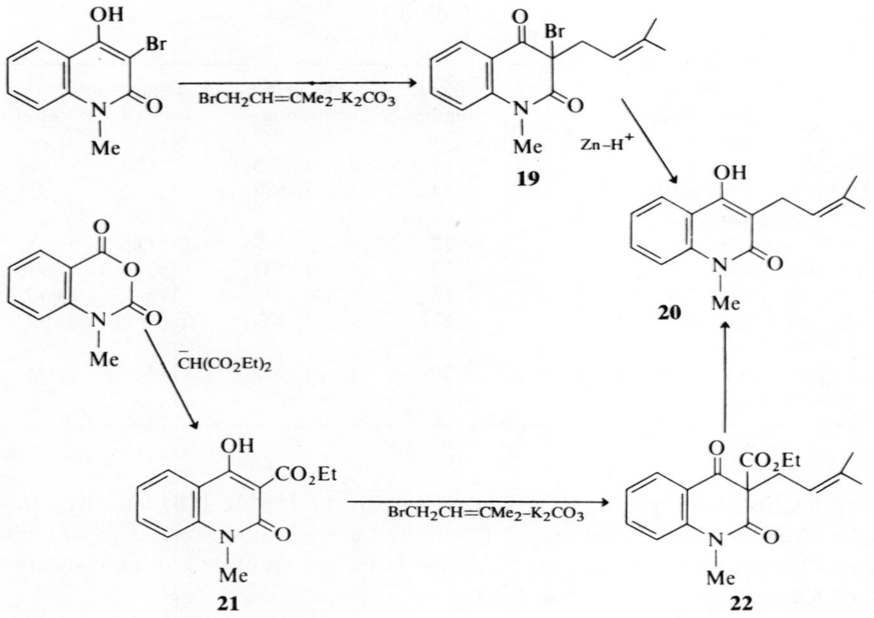

OH Br N Me O $BrCH_2CH{=}CMe_2–K_2CO_3$ O Br N Me O 19 Zn–H^+ OH N Me O 20

O O N Me O $\bar{C}H(CO_2Et)_2$ OH CO_2Et N Me O 21 $BrCH_2CH{=}CMe_2–K_2CO_3$ O CO_2Et N Me O 22

SCHEME 2. *Synthesis of 4-hydroxy-3-prenyl-2-quinolones.*

Storer and Young (*31*) isolated the 3-prenyl-2-quinolone (**23**) from the roots of *Dictamnus albus* L. The alkaloid was called preskimmianine on the assumption that it is a precursor of skimmiamine, but biosynthetic studies (see Section VII) now make this suggestion less likely. The structure of the alkaloid was established by both spectroscopy and synthesis.

Three prenylquinolones and an isopentyl derivative have been isolated from *Ptelea trifoliata* L.: compound **24** from the aerial parts of the plant growing in Arizona (*32*), ptelecortine (**25**) and pteleoline (**26**) from the European species (*33*), and compound **27** from the root bark (*34*). The structures of these alkaloids were identified by spectroscopic methods and, in the case of the 6,8-dimethoxy derivative (**27**), by synthesis (see Scheme 4, Section III,D,3) (*35*).

Haplophyllum tuberculatum A. Jass. contains the unusual diprenyl alkaloid **28** (*36*). The structure of this compound was apparent from the NMR

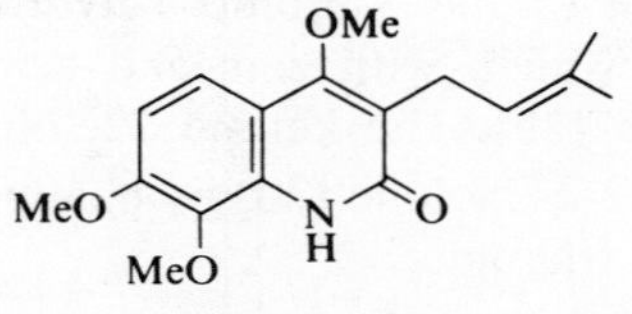

OMe MeO MeO N H O

23 Preskimmianine

OMe O O N H O

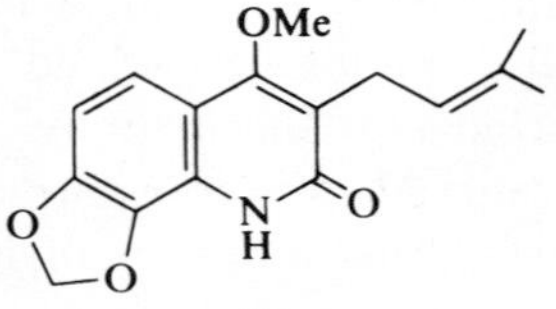

24

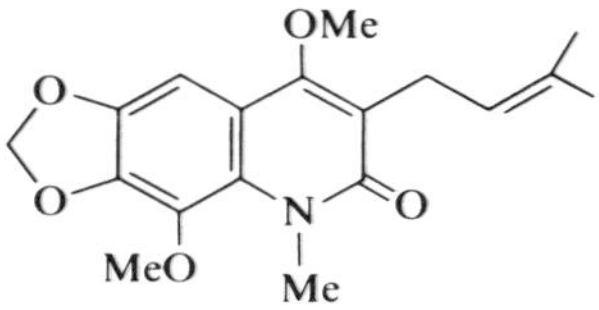

25 Ptelecortine

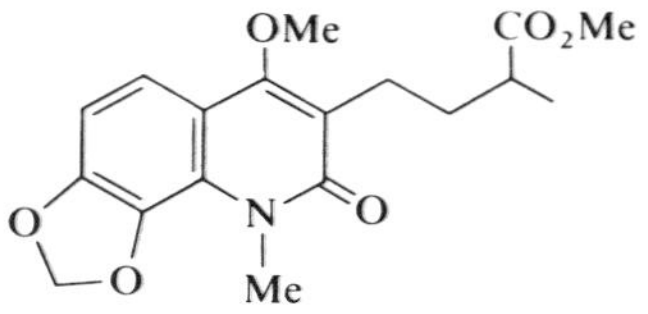

26 Pteleoline

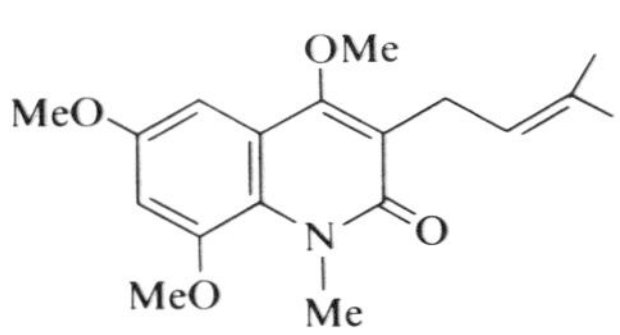

27

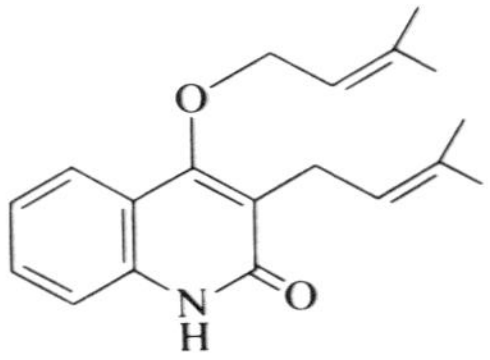

28

and mass spectra and from hydrogenolysis to the known 4-hydroxy-3-prenyl-2-quinolone; reaction of the latter compound with 3,3-dimethylallyl bromide afforded alkaloid **28** and the isomeric 3,3-diprenyl derivative.

C. DIHYDROFUROQUINOLINES, DIHYDROPYRANOQUINOLINES, AND RELATED ALKALOIDS

Data on dihydrofuroquinolines, dihydropyranoquinolines, and related alkaloids are given in Table II (*16, 17, 37–60*).

TABLE II
DIHYDROFUROQUINOLINES, DIHYDROPYRANOQUINOLINES, AND RELATED ALKALOIDS[a]

Compound	Structure number	Melting point (°C)	$[\alpha]_D$(Solvent)	Ref.
Unsubstituted in aromatic ring				
N-Methylplatydesminium perchlorate	**51**	200–202	+33° (MeOH) +38° (MeOH)[b]	*37* *38*
N-Methylplatydesminium chloride (R_{16})		108–111	—	*39*
Platydesmine	**49**	138–139	+47° ($CHCl_3$)	*40*
		136–137	$[\alpha]_{589}$ +44° (MeOH) $[\alpha]_{400}$ +90° (MeOH)	*41*
		135–137	0°	*42*
Platydesmine acetate	**50**	126–127	+23° ($CHCl_3$)	*40*
Isoplatydesmine	**52**	208–210	—	*43*
		191	+60° ($CHCl_3$)	*44*
		187	+48° ($CHCl_3$)	*45*

(*Continued*)

TABLE II (*Continued*)

Compound	Structure number	Melting point (°C)	$[\alpha]_D$ (Solvent)	Ref.
Araliopsine	**53**	152	+40° ($CHCl_3$)	*44*
Ribalinine (folifine)	**42**	233–234	0°	*46*
		232	−10° ($CHCl_3$)	*44*
		225–226	+14° (MeOH)	*16*
		222	+8° ($CHCl_3$)	*45*
Edulinine	**54**	114–117	−15°	*17, 47*
		140–141	−15°	*48*
		138–141	0°	*49*
		136–138	+27.2°	*43*
		138	—	*50*
Haplobucharine	**75**	126		*51*
6-Hydroxy and 6-methoxy compounds				
Ribalinium chloride	**30**	188–190	+40° (MeOH)	*52*
Ribalinium chloride		200–202 (R_{13})	+61° (H_2O)	*53*
Ribaline	**45**	259–260	0°	*46*
		>210	+86° (MeOH)	*46*
Rutalinium chloride (R_{15})	**46**	246–248	0°	*53*
Ribalinidine	**43**	257–258	−15° (MeOH)	*46*
		255–258 (R_{17})		*53*
8-Hydroxy compounds				
Pteleatinium chloride	**38**	267–270	*c*	*54*
7,8-Methylenedioxy compounds				
Hydroxylunine (Pt/12)	**55**	224–227	+60° (EtOH)	*55*
O-Methylhydroxyluninium chloride	**56**	133–135	−17.5° (MeOH)	*54*
Pteleflorine (Pt/47)	**47**	93–96		*56*
Neohydroxylunine	**48**	228–231	—	*57*
N-Demethyllunidonine	**69**			*58*
Hydroxylunidonine	**62**	145–149		*57*
		202–205		*56*
Orixinone	**72**	102–103		*59*
O-Methylluninium perchlorate	**65**	208–209	−23.8° (MeOH)	*60*
O-Methylluninium iodide		216–220	−27.4° (EtOH)	*60*
6-Methoxy-7,8-methylenedioxy compounds				
6-Methoxyhydroxylunidine	**59**	181–183		*57*
6-Methoxylunidonine	**71**	123–125		*57*
6-Methoxylunidine	**70**	145–147		*57*

[a] Supplementary to Volume IX, Chapter 6, Table II.

[b] The optical activity of the chloride from *R. graveolens* (*39*) was not recorded; a later isolation from this source is mentioned in Grundon and McColl (*38*, ref 26).

[c] CD data reported.

1. Ribalinium and Pteleatinium Salts

O-Methylbalfourodinium salt (**11**) is the major quaternary alkaloid of young Brazilian trees of the *Balfourodendron riedelianum* Engl. species (*61*). Working with older trees growing in Argentina, Corral and Orazi (*62*) showed that a new quaternary alkaloid, the ribalinium salt (**30**), is the principal alkaloid of the bark, and they elucidated its structure.

29

32

HCO_3^-

pyridine

MeI

30 Ribalinium salt

31

pH 10.6

NaH in DMF

^-OH

33 R = H
34 R = Me

35

^-OH

37

36

The presence of a phenolic group was indicated by a positive ferric chloride test and by methylation to ether **31**, which showed reactions typical of quaternary salts of the *O*-methylbalfourodinium type. Thus, heating *O*-methylribalinium iodide in pyridine furnished the 4-quinolone (**32**), which was reconverted into the quaternary salt by methyl iodide. The structures of **30** and **32** were confirmed by NMR and mass spectrometry.

Corral and Orazi (*52*) carried out a most interesting study of the reactions of ribalinium with base. At pH 10.6, a yellow solution was formed from which ribalinium was recovered. Addition of one equivalent of sodium bicarbonate to an aqueous solution of ribalinium and evaporation furnished the yellow crystalline zwitterion **29**, which could also be obtained by the use of a basic ion-exchange resin. Treatment of ribalinium chloride with N-sodium hydroxide for 3 days furnished the 1,2-diol (**33**). The 6-methoxy derivative **31** gave the corresponding diol **34** at pH 10.6, but not stereospecifically; fractional crystallization of the product gave first the (±)-diol and then the (−)-diol, representing a 1.9 : 1 ratio of (−) and (+) enantiomers. It was suggested that nucleophilic attack at the 2-position resulting in retention of configuration at the chiral center was in competition with a second route, **31** → **35** → **36** → **37** → **34**.

Evidence for an intermediate epoxide was obtained by reacting the salt **31** in dimethylformamide with sodium hydride to give an optically inactive compound shown by NMR spectroscopy to be epoxide **37**; racemization was assumed to occur via the zwitterion **36**. Treatment of the epoxide with sodium hydroxide furnished the optically inactive diol **34** by a process thought to involve direct attack of hydroxide ion on the epoxide ring. An alternative mechanism for this reaction is proposed later (see Section III, F), and other features of the epoxide route are also discussed. The diol **33** from ribalinium salt was also obtained as a mixture of the (−) and (+) enantiomers (ratio, 4.4 : 1), and a corresponding epoxide pathway may also be involved.

Ribalinium salt was also isolated by Reisch *et al.* from *Ruta graveolens* L. (*53*).

As a result of the reported antimicrobial action of extracts of the hop tree, *Ptelea trifoliata*, Mitscher *et al.* (*54*) investigated the aboveground constituents and found that the sole active agent was a new phenolic quinolinium derivative pteleatinium salt. The alkaloid was isolated as the chloride, $C_{16}H_{20}NO_4Cl$, and was shown to be the 8-hydroxy-*N*-methylplatydesminium salt (**38**) on the basis of IR, UV, NMR, and mass spectrometry and of correlation with known quinoline alkaloids. Thus, pyridine effected demethylation to the 4-quinolone **39** which, with diazomethane, was converted into (+)-balfourodine (**12**), while methylation of pteleatinium chloride with methyl iodide and sodium carbonate gave *O*-methyl-

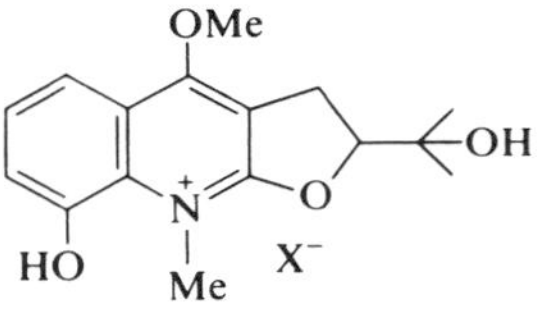

38 Pteleatinium salt

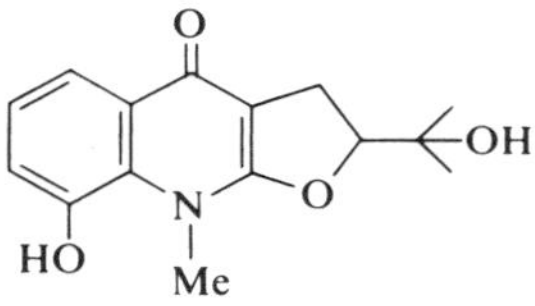

39

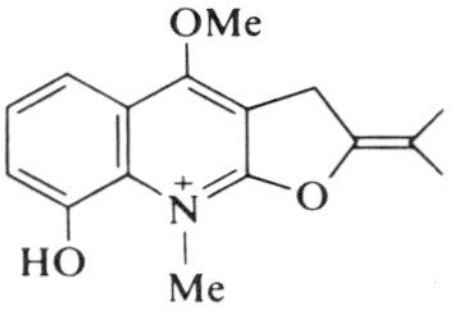

40

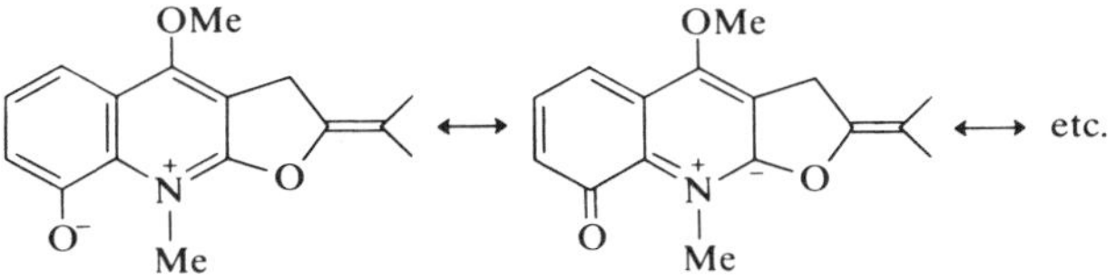

41

balfourodinium iodide (**11**). Dehydration of pteleatinium chloride with concentrated sulfuric acid and recovery of the product from a basic solution furnished a yellow compound, $C_{16}H_{18}NO_3$, which was formulated as the 8-hydroxy derivative **40**. By analogy with compound **29** obtained by treatment of ribalinium salt with base, the dehydration product of pteleatinium salt is more likely to be the zwitterion **41**, $C_{16}H_{17}NO_3$; the UV spectrum and analytical data are consistent with this structure. The presence of a tetrasubstituted double bond is apparent from the singlets at 6.50 (2H, $-CH_2-$), 8.50 (3H, $-CH_3$), and 8.55τ (3H, $-CH_3$), and is of some interest, because dehydration of such hydroxyisopropyldihydrofuroquinolines as balfourodine (**12**) (Volume IX, p. 245) and platydesmine (**13**) (*63*) usually furnishes isopropylfuroquinolines.

(+)-*O*-Methylbalfourodinium salt (**11**) has been isolated from two new sources, *Choisya ternata* H.B and K (*64*) and *Orixa japonica* Thunb. (*59*).

2. Ribalinine, Folifine, Ribalinidine, Ribaline, Rutalinium Salt, Pteleflorine, and Neohydroxylunine

Corral and Orazi (*46*) isolated a number of new tertiary bases from Argentinian *Balfourodendron riedelianum*. The phenolic alkaloids ribaline and ribalinidine were not found in the tertiary base fraction but were precipitated as reineckates with the quaternary alkaloids, presumably because of their zwitterion character and their low solubility in chloroform.

Ribalinine, which is optically inactive, was shown to be the pyranoquinolone **42** mainly by NMR spectroscopy and by synthesis. Bowman and Grundon (*27*) synthesized ribalinine before the isolation of the alkaloid by the method developed for isobalfourodine (**14**), and Corral and Orazi prepared ribalinine by rearrangement of compound **52** with acetic anhydride and pyridine (*46*). A key problem in this group of alkaloids is the distinction between furo and pyrano isomers, and this is best solved by NMR spectroscopy. The methine proton of ribalinine acetate resonates at 4.88τ, 1.2 ppm lower than in ribalinine, and typical of the secondary acetate group in the pyrano isomer (*46*). Further, the hydroxyl resonance of ribalinine appears as a doublet in dimethyl sulfoxide because of spin–spin coupling with the adjacent CH proton but as a singlet in the furo isomer, cf. **12** (*27*). In the same year, Yunusov and co-workers (*16*) investigated the optically active alkaloid folifine, from *Haplophyllum bucharicum*, and showed by spectroscopic studies of the alkaloid and its acetate that folifine was the (+) enantiomer corresponding to ribalinine (**42**). The same enantiomer occurs in *Araliopsis tabouensis* Aubrev et Pellegr. (*45*), and (−)-ribalinine has been found in the root bark of *A. soyauxii* Engl. (*44*).

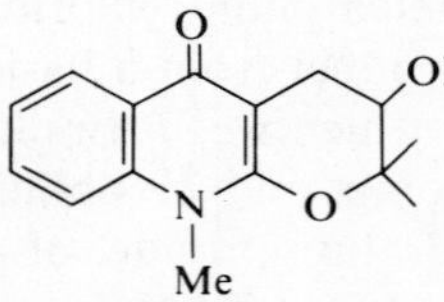

42 Ribalinine (Folifine)

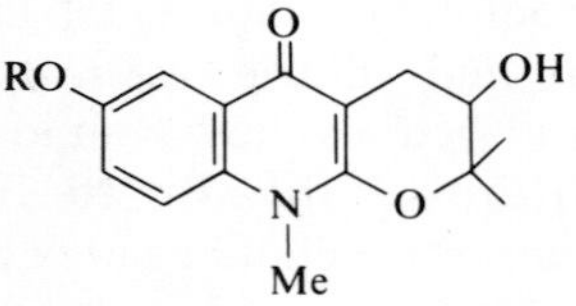

43 R = H Ribalinidine
44 R = Me

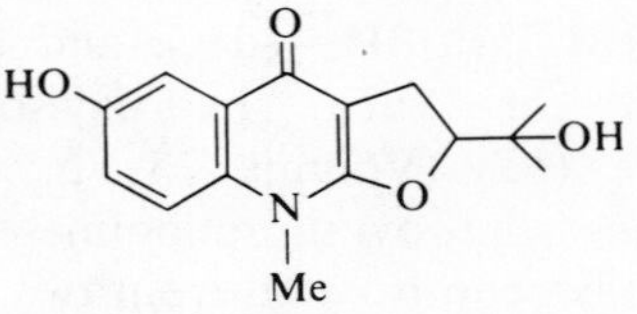

45 Ribaline

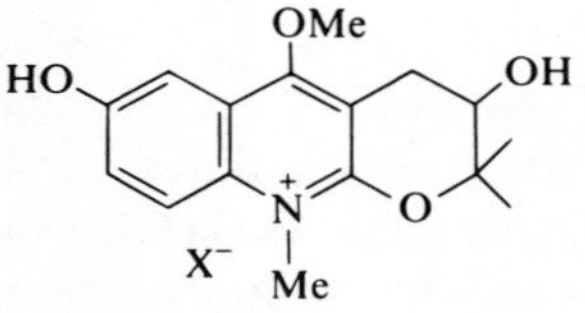

46 Rutalinium salt

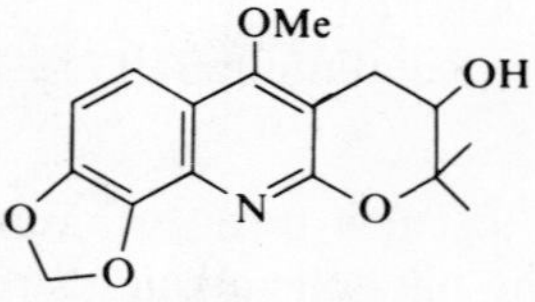

47 Pteleflorine

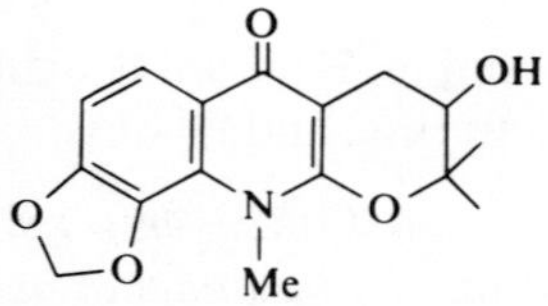

48 Neohydroxylunine

(−)-Ribalinidine is a phenolic alkaloid giving a monomethyl derivative with diazomethane, and the presence of a second hydroxyl group is indicated by the formation of a diacetate. The NMR spectra of ribalinidine and its derivatives show that the alkaloid has structure **43**, the C-5 proton being deshielded by the neighboring carbonyl group and producing a quartet through para- and meta-coupling (*46*).

The structure of ribaline (**45**) was established by correlation with ribalinium salt and with ribalinidine (*46*). Heating (+)-ribalinium salt (**30**) with pyridine afforded (+)-ribaline; this reaction occurred with partial racemization, and (±)-ribaline was obtained by crystallization. The furoquinolone–pyranoquinolone rearrangement (see Section III,F) was applied to (±)-ribaline which, with acetic anhydride in pyridine at 125°, furnished ribalinidine diacetate, yielding ribalinidine (**43**) on hydrolysis. Similarly, (+)-methylribaline was transformed into (+)-methylribalinidine acetate and then by hydrolysis into (−)-methylribalinidine (**44**).

Ribalinidine and ribaline were also obtained from *Ruta graveolens* (*53*). This species contains the quaternary rutalinium salt (**46**), corresponding to ribalinidine (**43**). Rutalinium, the first example of a quaternary salt in the pyranoquinoline series of alkaloids, was converted into the tertiary base on heating in pyridine (*53*).

The flowers of *Ptelea trifoliata* were shown by Reisch and co-workers (*56*) to contain the 4-methoxyhydroxydimethyldihydropyranoquinoline, pteleflorine (**47**); the isomeric *N*-methyl-4-quinolone, neohydroxylunine (**48**), was isolated by Mitscher *et al.* (*57*) from the stem tissue of this species. The constitutions of the alkaloids were determined by spectroscopic methods.

3. Platydesmine, *N*-Methylplatydesminium Salt, Isoplatydesmine, Araliopsine, and Edulinine

The 4-methoxyquinoline alkaloid, (+)-platydesmine (**49**), was first isolated from *Platydesma campanulata* (see Volume IX of this treatise), and its structure was confirmed by synthesis of the racemic base (*27*). It has since been obtained from both *Zanthoxylum parviflorum* Benth. (*41*) and, together with its acetate **50**, from *Geijera salicifolia* Schott. (*40*). (±)-Platydesmine has recently been identified as a constituent of *Melicope perspicuinervia* (*42*), and a sample of unspecified optical rotation was isolated from *Zanthoxylum belizense* Lundell (*65*). The alkaloid is a key intermediate in the biosynthesis of quinoline alkaloids (see Section VII of this chapter).

A quaternary alkaloid extracted by Boyd and Grundon (*37*) from leaves of *Skimmia japonica* Thunb. was shown to be the (+)-*N*-methylplatydesminium salt (**51**) by comparison with the racemate that had been synthesized earlier (*27*). The alkaloid has subsequently been isolated from *Ruta graveolens* (*39*) and from *Araliopsis tabouensis* (*66*).

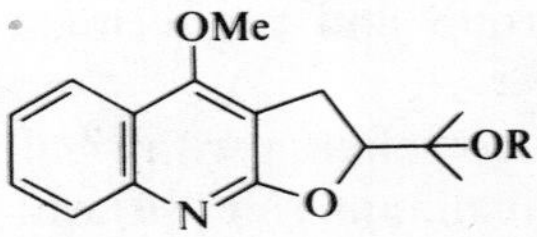

49 R = H Platydesmine
50 R = Ac

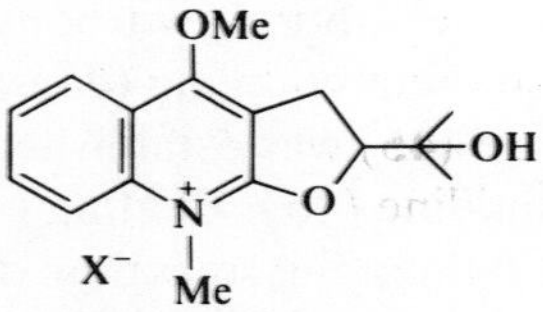

51 *N*-Methylplatydesminium salt

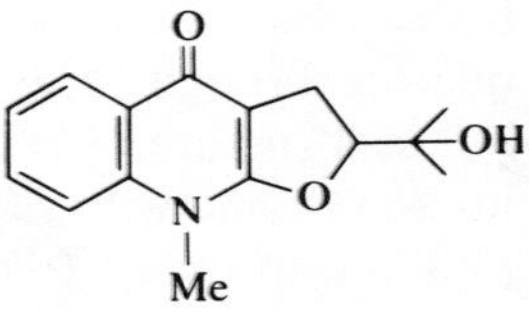

52 Isoplatydesmine

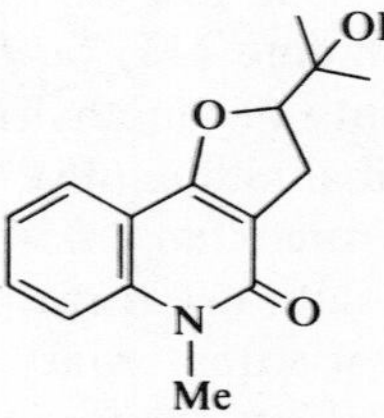

53 Araliopsine

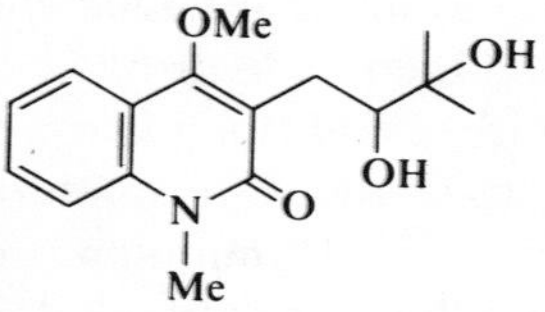

54 Edulinine

When heated briefly above its melting point, (+)-*N*-methylplatydesminium iodide gave the 4-quinolone **52**, $[\alpha]_D + 82°$ (MeOH) (*37*), identified by comparison with the synthetic racemate (*27*). Higa and Scheuer (*43*) have now isolated from *Pelea barbigera* (Gray) Hillebrand quinolone **52**, named isoplatydesmine. *Araliopsis soyauxii* has been shown to contain (+)-isoplatydesmine as the major alkaloid of the root bark (*44*); the same enantiomer is a constituent of *A. tabouensis* (*45*).

The most interesting constituent of *A. soyauxii* is the optically active alkaloid araliopsine, shown by spectroscopic studies to have structure **53** (*44*). This alkaloid, therefore, is the angular isomer of isoplatydesmine and is the first compound of this type to have been obtained from natural sources, although ψ-balfourodine (**126**), the 8-methoxy derivative of araliopsine is a well-known rearrangement product of balfourodine (**123**) (Section III,F). (+)-Isoplatydesmine (**52**), obtained by asymmetric synthesis (Section III,E), is similarly converted into (+)-araliopsine by treatment with sodium methoxide (*67*).

The alkaloid edulinine was first obtained from *Casimiroa edulis* Llave and Lex., and it was assigned the diol structure **54** on the basis of NMR and mass spectra (*17*). An optical rotation was not recorded, but the alkaloid from this source was later described as being optically active, $[\alpha]_D - 15°$ (*47*). The constitution was confirmed by comparison with (±)-edulinine obtained from synthetic *N*-methylplatydesminium iodide by treatment with aqueous ammonia at 20° (*37*) and with (−)-edulinine, $[\alpha]_D - 20°$ ($CHCl_3$), resulting from a similar reaction on (+)-*N*-methylplatydesminium salt (*37*). The alkaloid has been obtained from a number of other sources:

(−)-edulinine $[\alpha]_D - 15°$ ($CHCl_3$) from *Citrus macroptera* Montr. (*48*); the racemate from cell suspensions of *Ruta graveolens* (*49*); (+)-edulinine, $[\alpha]_D + 27.2°$ (C_6H_6), from *Pelea barbigera* (*43*); and a sample of unspecified optical properties from *Eriostemon trachyphyllus* F. Muell. (*50*). The melting points recorded for edulinine, as well as the specific rotations, show considerable variation.

(−)-Edulinine from *Casimiroa edulis* may be formed from (+)-*N*-methylplatydesminium salt during isolation. (+)-Edulinine from *Pelea barbigera* is certainly an artifact, and was obtained only when base was employed during isolation; the precursor, presumably (−)-*N*-methylplatydesminium salt, could not be detected (*43*). (±)-Edulinine is conveniently prepared by reaction of 4-methoxy-1-methyl-3-(3-methylbut-2-enyl)-2-quinolone with *m*-chloroperbenzoic acid and then with aqueous base, without isolating intermediates (*2*).

4. *O*-Methylhydroxyluninium Salt, 6-Methoxyhydroxylunidine, 6,8-Dimethoxyedulinine, and Hydroxylunidonine

The 7,8-methylenedioxyhydroxyisopropyldihydrofuroquinolone alkaloid, (−)-hydroxylunine (**55**), was isolated from *Lunasia amara* (Volume IX, p. 245), and the (+) enantiomer has now been shown to be a constituent of *Ptelea trifoliata* (*55*). Mitscher *et al.* (*54*) obtained the corresponding quaternary salt, *O*-methylhydroxyluninium salt (**56**), from the stems of *P. trifoliata*, and they determined the structure of the alkaloid by spectroscopic analysis and by conversion with pyridine into hydroxylunine.

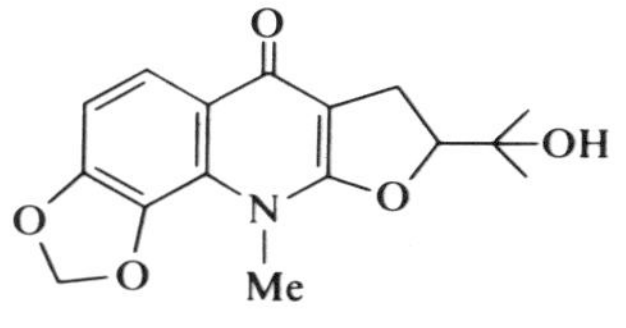

55 Hydroxylunine

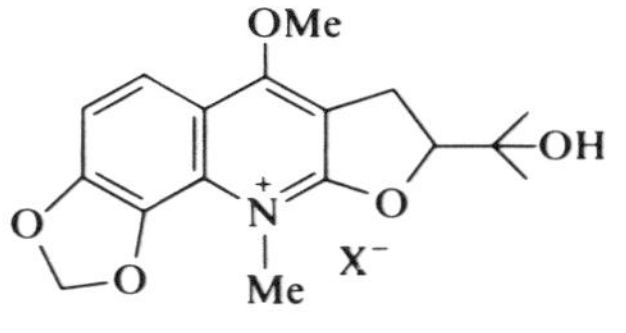

56 *O*-Methylhydroxyluninium salt

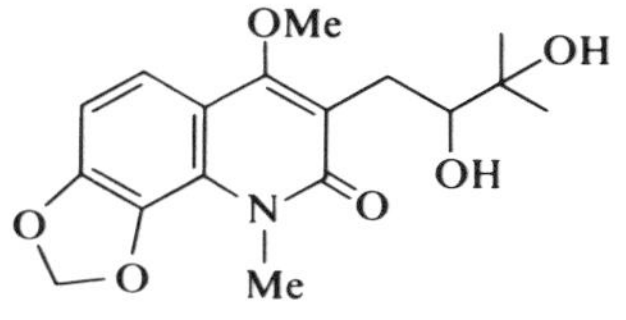

57 Hydroxylunidine

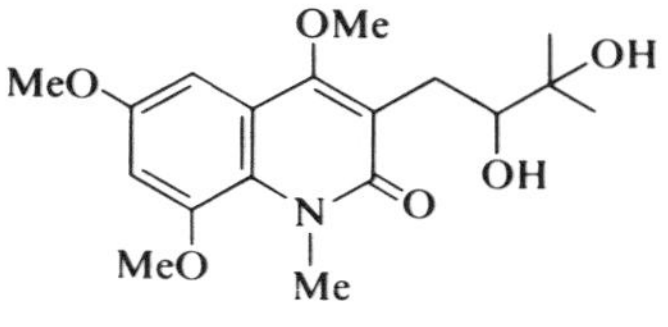

58 6,8-Dimethoxyedulinine

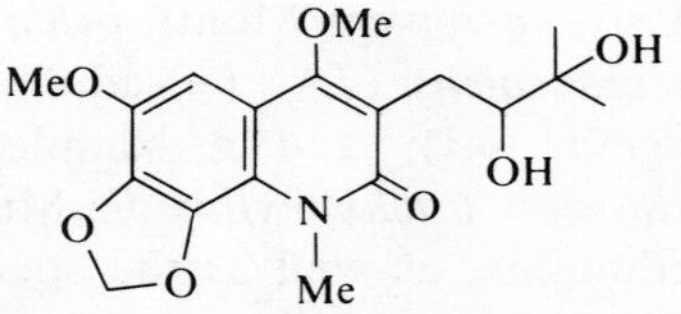

59 6-Methoxyhydroxylunidine

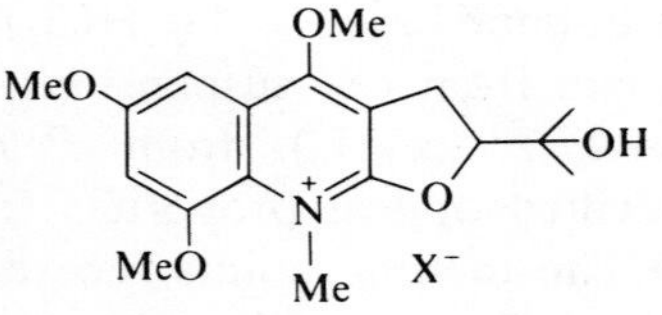

60

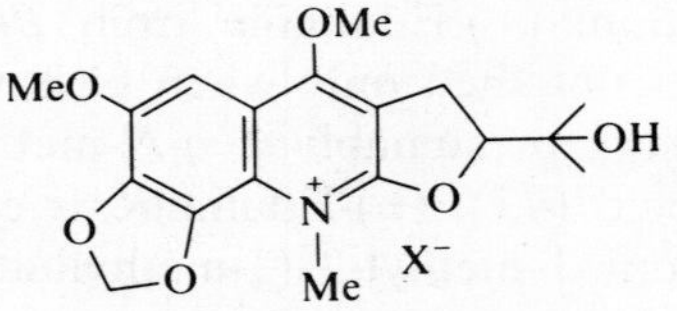

61

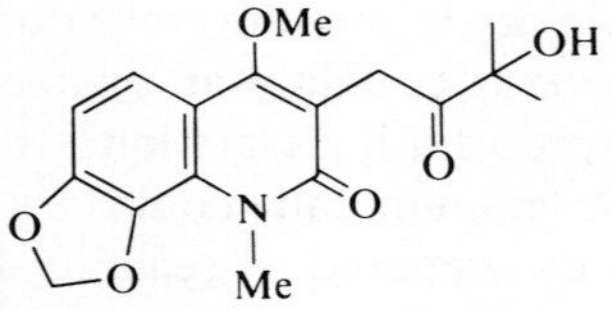

62 Hydroxylunidonine

A recent report by Reisch and his co-workers (*68*) indicated that treatment of an unresolved mixture of quinolinium salts from *P. trifoliata* with base yielded a group of hydrolysis products that was in any case isolated directly from the tertiary base fraction of the extract. One of these compounds, hydroxylunidine (**57**), was obtained previously from *L. amara* (Volume IX, p. 245), and two new diols, 6,8-dimethoxyedulinine (**58**) and 6-methoxyhydroxylunidine (**59**), were identified. The latter compound was also isolated from the stems of *P. trifoliata*, and its structure was established by spectroscopy (*57*). The formation of the three diols from the mixture of salts suggested that the latter contained *O*-methylhydroxyluninium salt (**56**) and the hitherto unknown quinolinium derivatives **60** and **61**, although the presence of pyrano isomers, cf. rutalinium salt (**46**), which would give the same diols on treatment with base, cannot be excluded.

A new alkaloid (mp 145–149°) of *P. trifoliata* was named hydroxylunidonine and assigned structure **62** (*57*). The presence of a hydroxyl function and ketonic and 2-quinolone carbonyl groups was indicated by IR absorption at 3300, 1710, and 1640 cm^{-1}. The constitution of the alkaloid was apparent from the NMR spectrum, which showed singlets at 7.2 (2H, Ar–CH_2–$\overset{\overset{\displaystyle O}{\|}}{C}$–), 8.5 (3H, –$CH_3$), and 8.65$\tau$ (3H, –CH_3), and from reaction with zinc and acetic acid to give lunidonine (**68**), presumably by a process of reduction–elimination. Hydroxylunidonine (mp 202–205°) was obtained from the flowers of *P. trifoliata* (*56*).

5. *O*-Methylluninium Salt, *N*-Demethyllunidonine, 6-Methoxylunidine, 6-Methoxylunidonine, Orixinone, and Haplobucharine

Isopropyldihydrofuroquinoline alkaloids, such as (−)-lunasine (**63**) and (−)-lunine (**64**), are constituents of *Lunasia* species (Volume IX, p. 239)

and can be formally regarded as cyclization products of 3-prenyl-2-quinolones. The bark of *Lunasia quercifolia* (Warb.) Lauterb. and K. Schum contains (−)-lunasine as principal quaternary alkaloid, but a study of the quaternary salts from the leaves showed that (−)-*O*-methylluninium salt (**65**) is the main constituent (*60, 69*). By standard methods, the salt furnished (−)-lunine (**64**) and (+)-lunidine (**66**). The latter compound was previously cyclized with inversion of configuration to the enantiomeric salt (+)-*O*-methylluninium perchlorate (Volume IX, p. 243). Compound **65** was subsequently isolated from *Ptelea trifoliata* together with (+)-lunine (**64**) (*55*). Reaction of (+)-lunidine (**66**) with acid gave the angular furoquinolone **67** (*69*).

The ketone lunidonine (**68**), which can be regarded either as an oxidation product of lunidine (**66**) or as a dehydration product of the diol

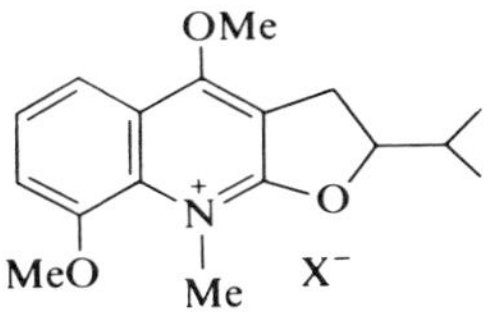

63 Lunasine

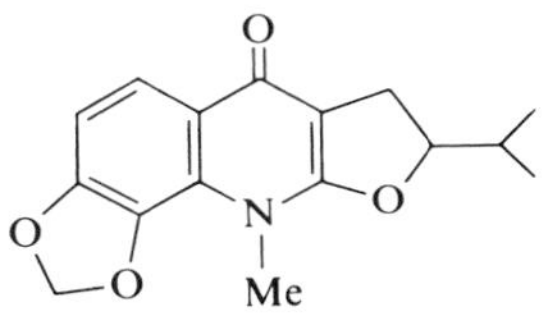

64 Lunine

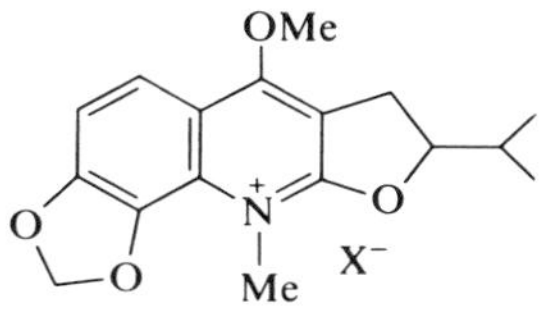

65 *O*-Methylluninium salt

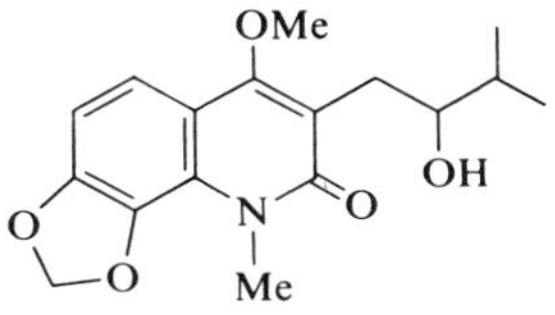

66 Lunidine

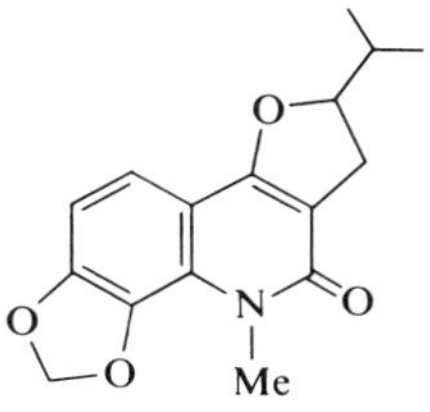

67

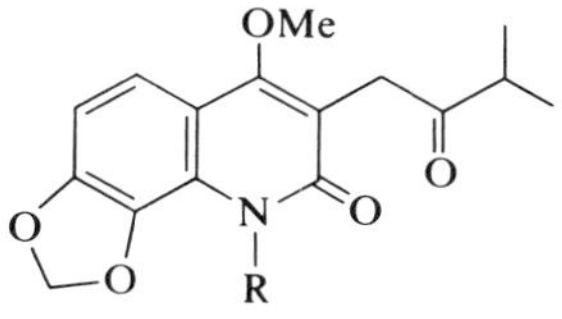

68 R = Me Lunidonine
69 R = H *N*-Demethyllunidonine

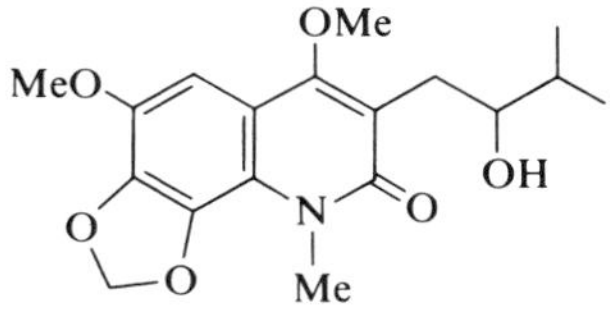

70 6-Methoxylunidine

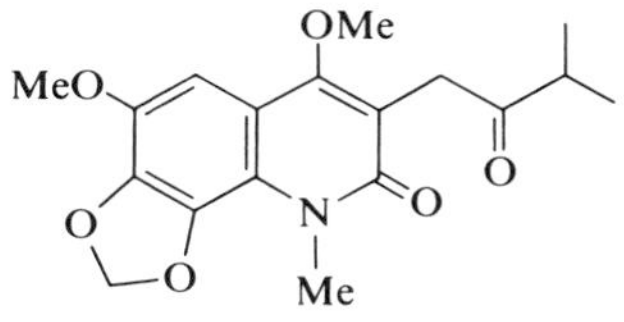

71 6-Methoxylunidonine

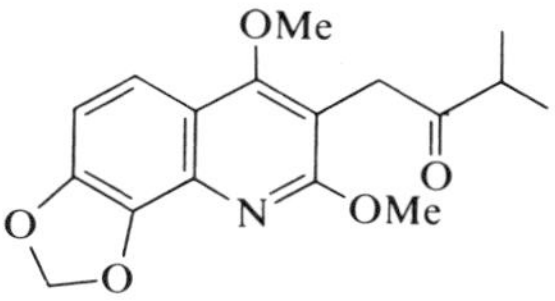

72 Orixinone

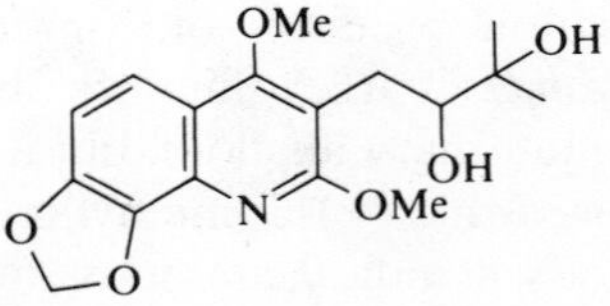

73 Orixine

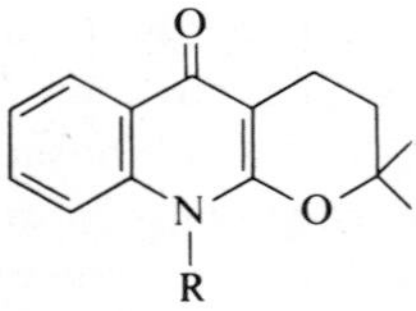

74 R = H Khaplofoline
75 R = $CH_2CH{=}CMe_2$ Haplobucharine

hydroxylunidine (**57**), is a constituent of *Lunasia amara* (Volume IX, p. 243); it has also been isolated from *P. trifoliata*, together with the new alkaloid, *N*-demethyllunidonine (**69**) (*58*).

In the 6-methoxy-7,8-methylenedioxy series, 6-methoxylunidine (**70**) and 6-methoxylunidonine (**71**) have been obtained from the stems of *P. trifoliata* (*57*). The structures of these alkaloids were established by spectroscopy and by interconversions. Thus, oxidation of 6-methoxylunidine (**70**) with Jones' reagent furnished the ketone 6-methoxylunidonine (**71**), and the reverse reaction was effected with borohydride. Dehydration of the α-glycol 6-methoxyhydroxylunidine (**59**) with *p*-toluenesulfonic acid in boiling toluene gave 6-methoxylunidonine (**71**).

A reexamination of the aerial parts of *Orixa japonica* resulted in the isolation of a new ketone, orixinone (**72**), which was present in the stems only (*59*). The structure of orixinone was established by spectroscopy and by its formation from the diol (±)-orixine (**73**) by means of dehydration with aqueous acid. The probability that the isoprenoid diol evoxine is formed from the epoxide alkaloid **174** (see Section IV,A,3) during isolation suggested that orixine, obtained from the root bark of *O. japonica*, and orixinone might be artifacts derived from epoxide **106** (see Section III,E,1). The availability of the synthetic epoxide prompted a thin-layer chromatographic study of the constituents of the leaves and stems of the plant, but neither the epoxide nor the orixine was detected.

The *Haplophyllum* alkaloid, khaplofoline (**74**) (Volume IX, p. 255) (*70*), is a dimethyldihydropyranoquinolone corresponding to the isopropyldihydrofuroquinolines of the type represented by lunasine (**63**) and lunine (**64**); haplobucharine (**75**), the *N*-prenyl derivative of khaplofoline, has now been isolated from *H. bucharicum* and its structure has been determined by spectroscopic methods (*51*).

D. Terminal Olefins and Related Alkaloids

1. Alkaloids of *Ptelea trifoliata*

An extensive investigation of the constituents of *Ptelea trifoliata* by Reisch and his co-workers (*33, 55, 56, 71–73*) led to the isolation of a new

TABLE III
TERMINAL OLEFINS OF *Ptelea trifoliata* L.

Compound	Structure number	Melting point (°C)	$[\alpha]_D$	Ref.
Isoptelefoline (Pt/6)	**77**	145–146	0°	*71*
Ptelefolidine (Pt/10)	**78**	118–119	0°	*71*
Ptelefoline (Pt/13)	**79**	91–93	0°	*55*
Ptelefructine (Pt/3)	**76**	146–149	0°	*71*
Ptelefoline methyl ether (KPt/5)	**81**	Oil	$[\alpha]_{578}-14°$ ($CHCl_3$)	*33*
Ptelefolidine methyl ether (Pt/23)	**80**	133–135	0°	*71*
Ptelefolone (Pt/22)	**82**	70–71	+1.9° ($CHCl_3$)	*72*
Ptelefolidone (Pt/40)	**83**	152–154	—	*56*
O-Methylptelefolonium chloride	**84**	90–93	—	*73*
O-Methylptelefolonium chloride	**84**	94–96	—	*54*

group of quinoline alkaloids (**76–84**) in which a 3-prenyl group incorporates a terminal double bond (see Table III). Four alkaloids, **76–79**, are 3-(2-hydroxy-3-methylbut-3-enyl)-4-methoxy-*N*-methyl-2-quinolones differing only in the nature of the substituents in the aromatic ring and are thus formally dehydro derivatives of edulinine (**54**). These four alkaloids together with the methyl ether **80** are racemates, but ptelefoline methyl ether (**81**) is optically active. The structures were established entirely by spectroscopic methods, NMR and mass spectrometry proving to be particularly useful. Taking ptelefoline (**79**) as an example (*55*), IR absorption at 3410 and at 1635 cm^{-1} indicates the presence of the hydroxyl and a 2-quinolone carbonyl group, probably linked through intramolecular hydrogen bonding. In the NMR spectrum the methylene and methine protons constitute an ABX system, the olefinic protons resonate at 5.0 and 5.2τ, and a three-proton singlet at 8.7τ is attributed to the side-chain methyl group; meta-coupled one-proton doublets at 3.2 and 3.35τ suggest that ptelefoline is a 6,8-dimethoxyquinolone. The mass spectra of alkaloids **76–79** show peaks due to the molecular ions. Fragmentation peaks are characteristic of the side chain and are interpreted as indicated in Scheme 3. The only major peaks in the mass spectrum of ptelefolidine methyl ether (**80**) are due to the molecular ion and to the M^+-85 fragment (*71*).

An optically active alkaloid, ptelefolone, isolated from *P. trifoliata*, was shown by spectroscopic methods to be the 4-quinolone **82** (*72*); extraction of the flowers of this species yielded the analogous compound ptelefolidone (**83**) (*56*). The metho salt of ptelefolone, *O*-methylptelefolonium salt (**84**), is reported to be the major quaternary salt of the leaves of *P.*

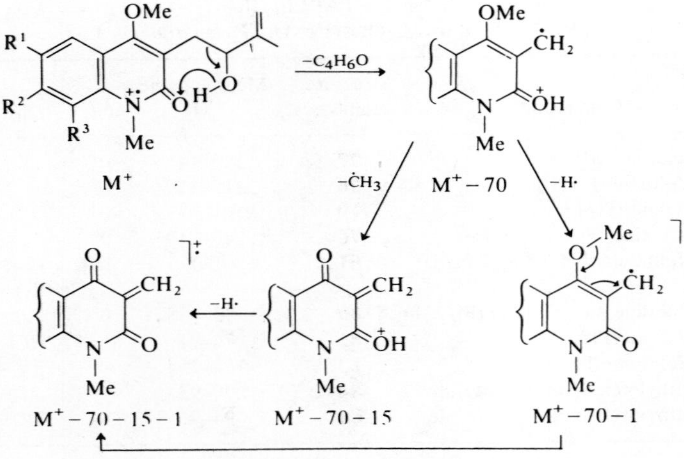

SCHEME 3. *Mass spectra of* 3-(2-*hydroxy*-3-*methylbut*-3-*enyl*)-2-*quinolones.*

trifoliata (*73*); *O*-methylptelefolidonium salt (**85**) is probably a constituent as well, for treatment of the quaternary alkaloid fraction with base afforded ptelefolidine (**78**) (*68*). Interconversions characteristic of *O*-methylbalfourodinium salt and balfourodine were not described for the furoquinolines containing terminal double bonds, although the mass spectrum of *O*-methylptelefolonium salt (**84**) indicates that a primary process involves conversion into alkaloid **82** by loss of an *O*-methyl group (*73*). It seems probable that the allylic alcohols **76–79** are artifacts arising from reaction of the corresponding quaternary alkaloids with base during isolation. In analogous cases the base-catalyzed ring cleavage involves only

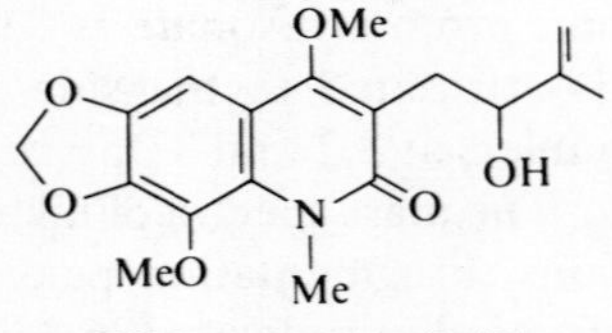

76 Ptelefructine (Pt/3)

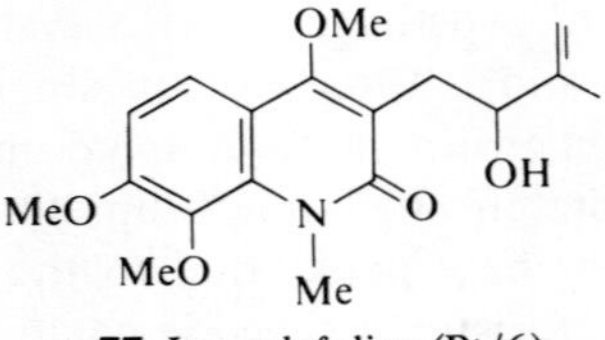

77 Isoptelefoline (Pt/6)

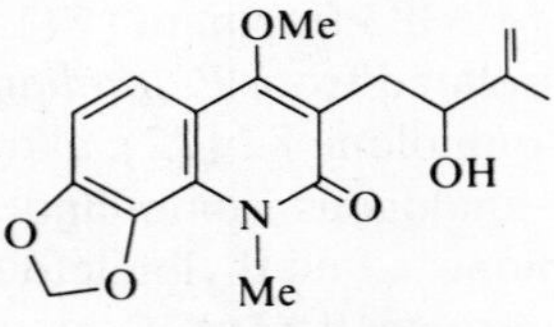

78 Ptelefolidine (Pt/10)

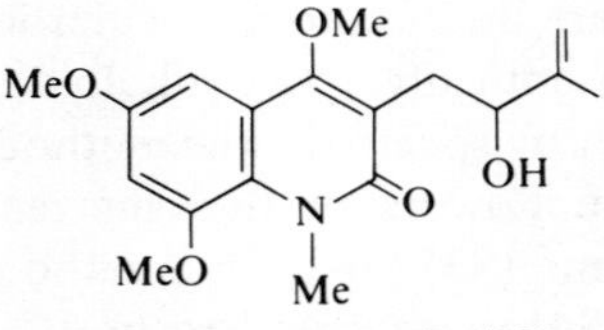

79 Ptelefoline (Pt/13)

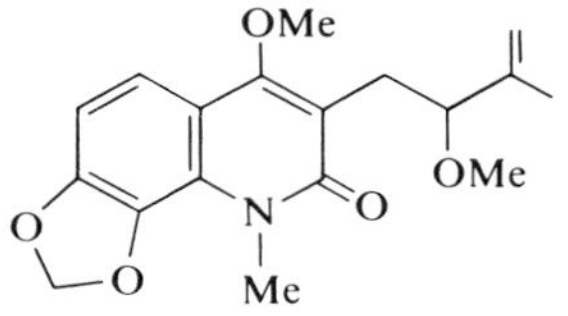

80 Ptelefolidine methyl ether (Pt/23)

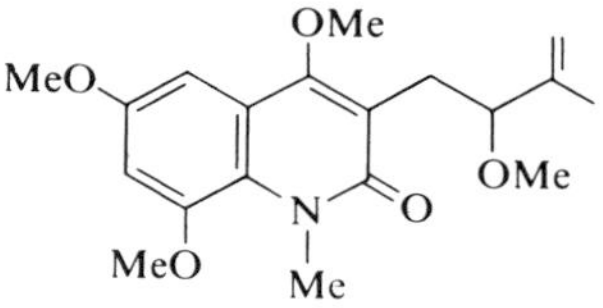

81 Ptelefoline methyl ether (KPt/5)

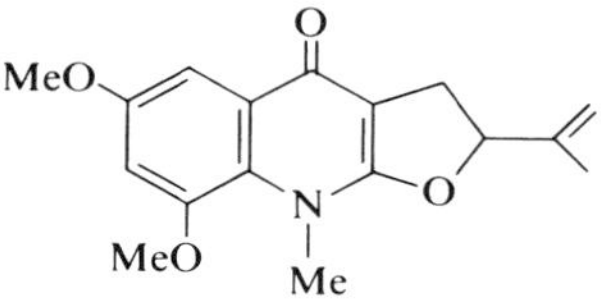

82 Ptelefolone (Pt/22)

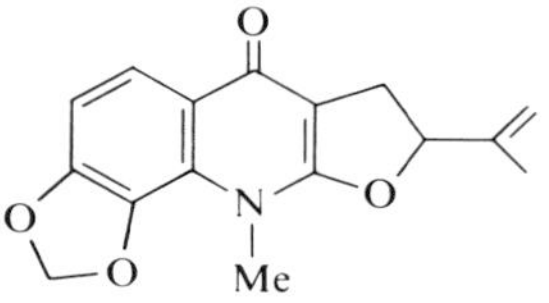

83 Ptelefolidone (Pt/40)

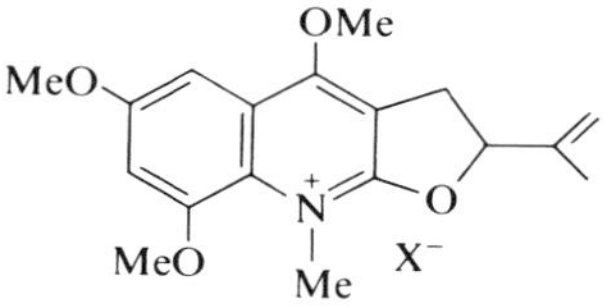

84 *O*-Methylptelefolonium salt

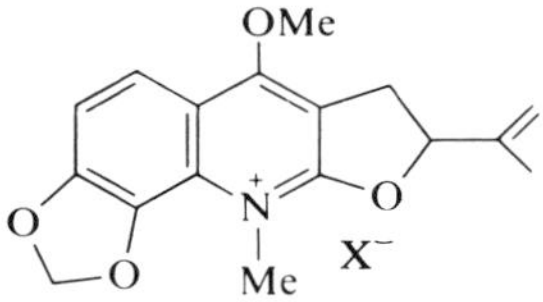

85 *O*-Methylptelefolidonium salt

partial racemization (see Section III,C,1); the lack of optical activity of alkaloids **76–80** may arise from preferential crystallization of the racemates or, more likely, from base-catalyzed racemization of the allylic system during isolation.

2. Dubinidine, Dubinine, and Folisine

This group of alkaloids found in *Haplophyllum* species can be regarded as hydroxylated derivatives of dihydrofuroquinolines containing terminal double bonds.

Yunusov *et al.* showed that dubinidine is an optically active tertiary base, and structure **86** was proposed, particularly on the basis of permanganate oxidation to dictamnic acid and formation of an aldehyde, dubinidal, by reaction with periodic acid. In discussing the results Openshaw pointed out (Volume IX, p. 254) that further evidence was required before structure **86** could be accepted; application of NMR and mass spectrometry has now led to revised structures **87** and **88**, respectively, for dubinidine and dubinine (*74, 75*). Dubinidal is clearly ketone **89**.

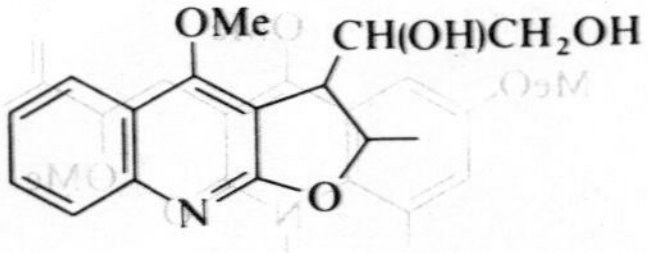
OMe
CH(OH)CH₂OH
N
O

86

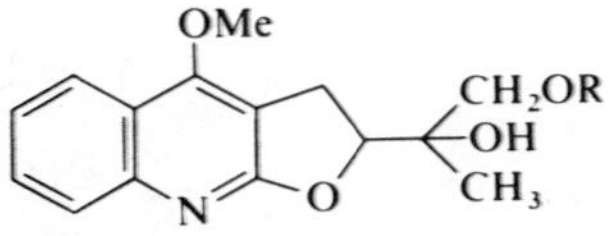
OMe
CH₂OR
OH
CH₃
N
O

87 R = H Dubinidine
88 R = Ac Dubinine

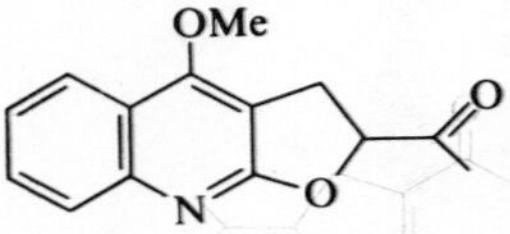
OMe
O
N
O

89 Dubinidal

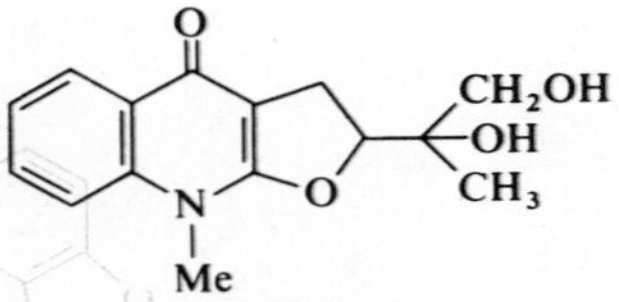
O
CH₂OH
OH
CH₃
N
O
Me

90 Folisine

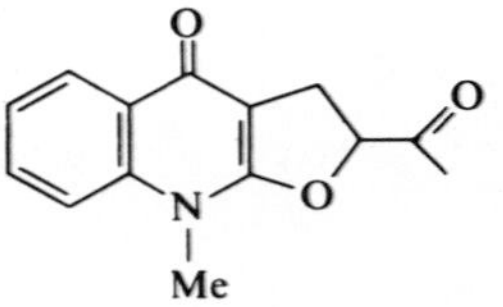
O
O
N
O
Me

91

Bessonova and Yunusov (*76*) isolated folisine {$C_{15}H_{17}NO_4$; mp 236–237°; $[\alpha]_D -123°$ (MeOH)} from *H. foliosum* and established structure **90** for the alkaloid by spectroscopic studies and by periodate oxidation to ketone **91**. Dubinidine methiodide was converted into folisine by reaction with pyridine.

3. Synthesis

A synthetic study by Grundon *et al.* (*35, 77*) is in progress and has resulted so far in the preparation of the terminal olefins *O*-methylptelefolonium salt and ptelefolone and the 1,2-diol, dubinidine (Scheme 4).

The approach used depends on the selective dehydration of hydroxyisopropyldihydrofuroquinolines to produce terminal olefins. Acid-catalyzed dehydration resulted in the exclusive formation of the more stable isopropylfuroquinolines, e.g. **95**. Although other conventional procedures were also largely unsuccessful, reaction of the hydroxyisopropyldihydrofuroquinoline **93** with thionyl chloride and pyridine gave the terminal olefin **94**, which was converted into *O*-methylptelefolonium salt (**84**). Triphenylphosphite dihalides in the presence of base, however, proved to be effective reagents. Thus, the phosphite dibromide and potassium carbonate reacted with platydesmine to give olefin **97** predominantly, which was converted into dubinidine (**87**) by treatment with osmium tetroxide. Ptelefolone was the only product formed when the 6,8-dimethoxy-4-quinolone **92** was treated successively with triphenylphosphite dichloride and with pyridine.

The hydroxyisopropyldihydrofuroquinoline intermediates **92** and **93** were prepared by established procedures (see Scheme 4).

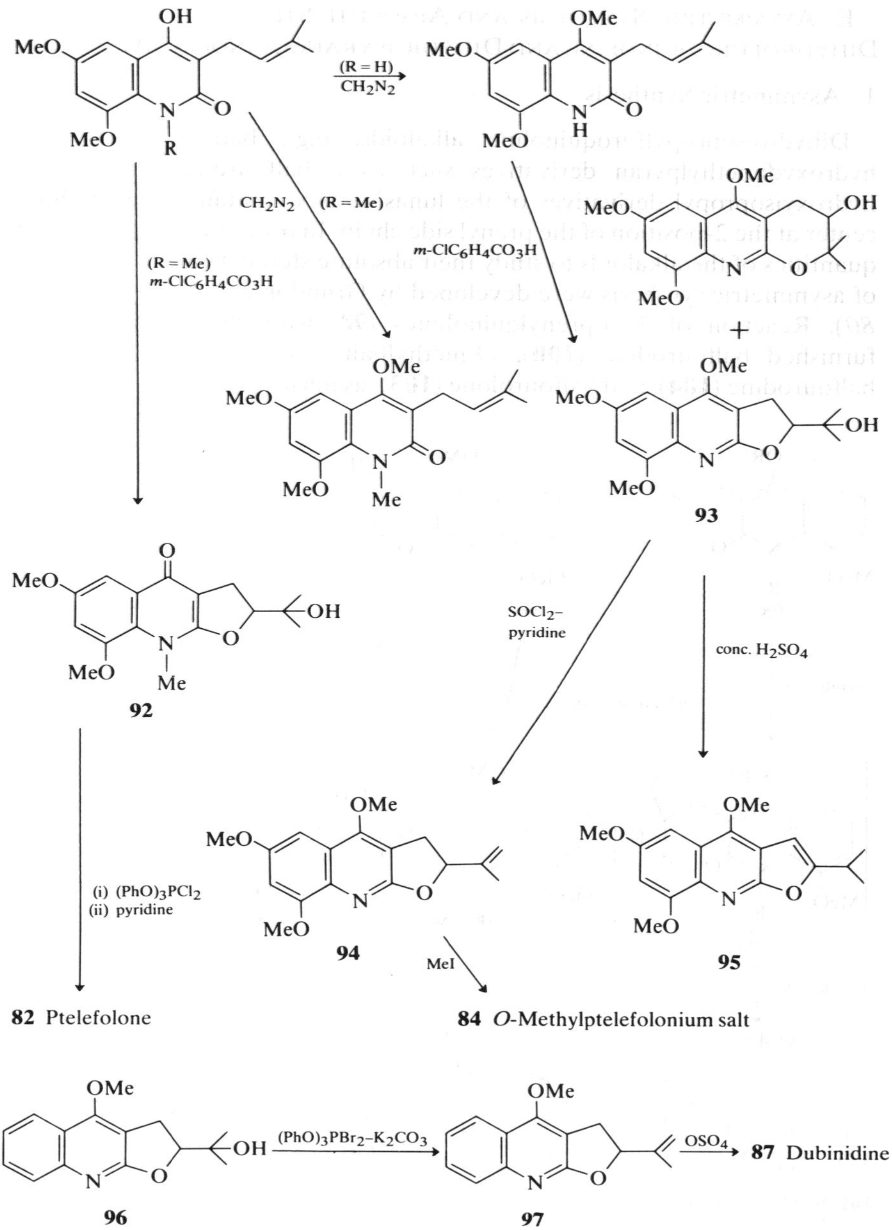

(R = H) CH_2N_2

CH_2N_2 (R = Me)

m-$ClC_6H_4CO_3H$

(R = Me) *m*-$ClC_6H_4CO_3H$

93

92

$SOCl_2$–pyridine

conc. H_2SO_4

(i) $(PhO)_3PCl_2$
(ii) pyridine

94

MeI

95

82 Ptelefolone

84 *O*-Methylptelefolonium salt

96

$(PhO)_3PBr_2$–K_2CO_3

97

OsO_4

87 Dubinidine

SCHEME 4. *Synthesis of terminal olefins and dubinidine.*

E. Asymmetric Synthesis and Absolute Stereochemistry of Dihydrofuroquinoline and Dihydropyranoquinoline Alkaloids

1. Asymmetric Synthesis

Dihydroisopropylfuroquinoline alkaloids, e.g., balfourodine (**100**), hydroxydimethylpyran derivatives such as isobalfourodine (**104**), and hydroxyisopropyl derivatives of the lunasine type contain a single chiral center at the 2-position of the prenyl side chain. In order to obtain sufficient quantities of the alkaloids to study their absolute stereochemistry, methods of asymmetric synthesis were developed by Grundon and co-workers (*78–80*). Reaction of 3-isoprenylquinolones (**98**) with chiral peroxy acids furnished balfourodine (**100**), *O*-methylbalfourodinium salt (**101**) isobalfourodine (**104**), and balfourolone (**103**), as illustrated in Scheme 5, with

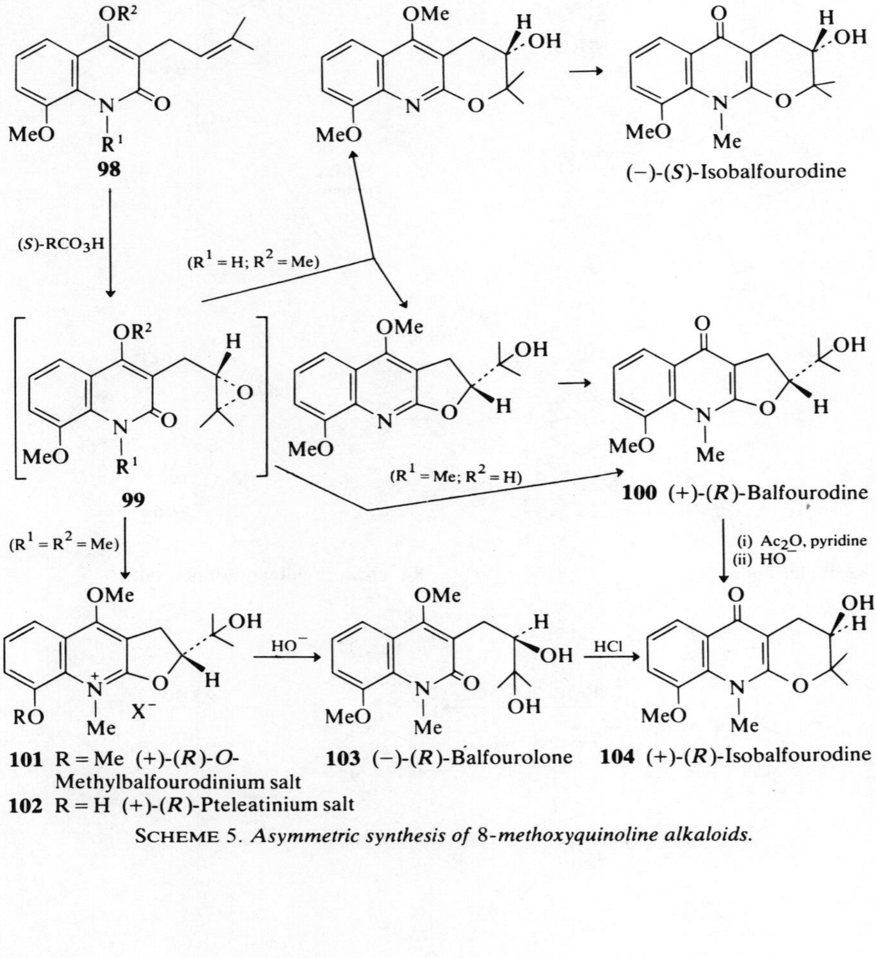

98

(*S*)-RCO_3H

(R^1 = H; R^2 = Me)

(−)-(*S*)-Isobalfourodine

99

(R^1 = Me; R^2 = H)

100 (+)-(*R*)-Balfourodine

(R^1 = R^2 = Me)

(i) Ac_2O, pyridine
(ii) HO^-

101 R = Me (+)-(*R*)-*O*-Methylbalfourodinium salt
102 R = H (+)-(*R*)-Pteleatinium salt

103 (−)-(*R*)-Balfourolone

104 (+)-(*R*)-Isobalfourodine

SCHEME 5. *Asymmetric synthesis of 8-methoxyquinoline alkaloids.*

2–10% optical induction. A range of chiral acids was employed in the formation of balfourodine from quinolone **98**, and it was found that (*S*)-peroxy acids gave (+)-balfourodine, while a preponderance of (−)-balfourodine was formed from (*R*)-peroxy acids. Crystallization of partially optically active balfourodine from ethyl acetate resulted in preferential separation of the racemate; balfourodine of up to 40% optical purity could be obtained from the mother liquors.

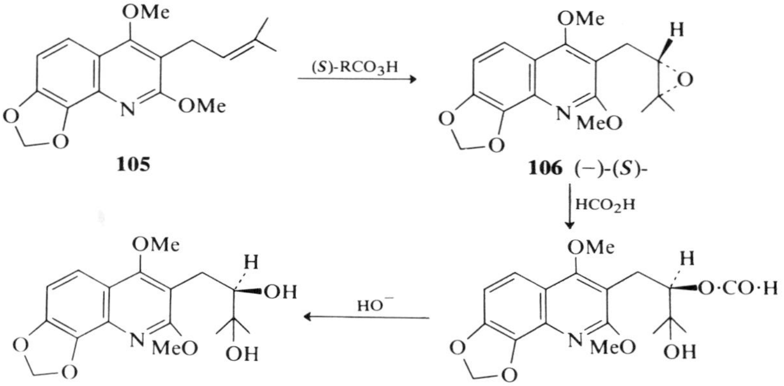

OMe, OMe, N, O, O — **105** —(S)-RCO_3H→ **106** (−)-(*S*)- (OMe, H, O, N, MeO) ↓HCO_2H **108** (+)-(*R*)-Orixine formate (OMe, H, O·CO·H, N, MeO, OH) —HO^-→ **107** (+)-(*R*)-Orixine (OMe, H, OH, N, MeO, OH)

SCHEME 6. *Asymmetric synthesis of orixine.*

In these reactions, epoxide **99** was not isolated, presumably because of rapid cyclization, but when the 2- and 4-oxygen functions were methylated, an epoxide (**106**) was obtained from compound **105** and converted into optically active orixine (**107**) via the formate ester **108** (Scheme 6) (*79*). This represents the first synthesis of orixine.

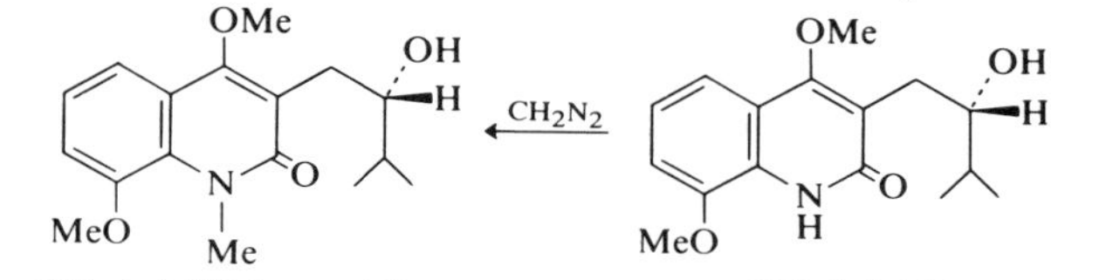

109 (−)-(*S*)- (OMe, H, O, N, MeO, MeO) —B_2H_6–$LiBH_4$→ **110** (+)-(*S*)- (OMe, OH, H, N, MeO, MeO) ↓HCl **112** (−)-(*S*)- (OMe, OH, H, N, H, O, MeO) —CH_2N_2→ **111** (−)-(*S*)-Lunacridine (OMe, OH, H, N, O, Me, MeO)

SCHEME 7. *Asymmetric synthesis of lunacridine.*

In the 8-methoxy series, the (−)-epoxide **109** was treated with diborane–lithium borohydride, and the product (**110**) of this reaction was cleaved selectively to the 2-quinolone **112**; methylation then afforded lunacridine (**111**) containing a preponderance of the (−) enantiomer (Scheme 7) (*80*).

2. Absolute Configurations

(+)-*N*-Methylplatydesminium salt (**113**) was shown (*81*) to have the (*R*) configuration by ozonolysis of the optically pure alkaloid to the (+)-(*R*)-hydroxylactone (**114**), of established stereochemistry (*81, 82*). Similarly, synthetic (+)-balfourodine, (+)-isobalfourodine, and (−)-balfourolone each gave the hydroxylactone containing a preponderance of the (+) enantiomer (*78*). Thus, the three alkaloids, which were isolated as (+), (+), and (−) enantiomers, respectively, from *Balfourodendron riedelianum*, have the (*R*) configurations **100**, **104**, and **103**. (+)-*O*-Methylbalfourodinium salt from *B. riedelianum* clearly has the (*R*) configuration **101**, as it is formed from (+)-(*R*)-balfourodine (**100**) with methyl iodide. Correlation of (+)-ptelea-tinium salt (**102**) with (+)-balfourodine (see Section III,C,1) indicates that this quaternary alkaloid also has an (*R*) configuration.

The assignment of configurations supports the mechanisms proposed for some well-known reactions of furo- and pyranoquinoline alkaloids. Thus, balfourolone (**103**) is believed to be an artifact formed from *O*-methyl-balfourodinium salt with aqueous base, and the (*R*) configurations of both compounds are consistent with a mechanism involving nucleophilic reaction at the 2-position of the quinoline ring. Conversion of (−)-(*R*)-balfourolone (**103**) into (+)-(*R*)-isobalfourodine (**104**) with aqueous acid occurs at the tertiary center without affecting the chiral center, as expected.

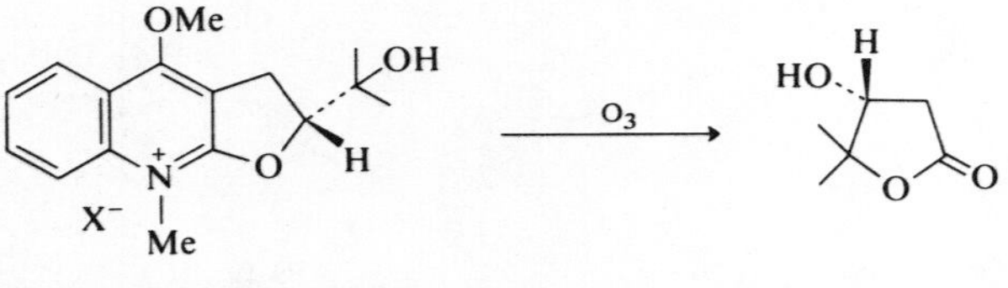

OMe OH N O H X⁻ Me $\xrightarrow{O_3}$ H HO O O

113 (+)-(*R*)-*N*-Methylplatydesminium salt

114 (+)-(*R*)-

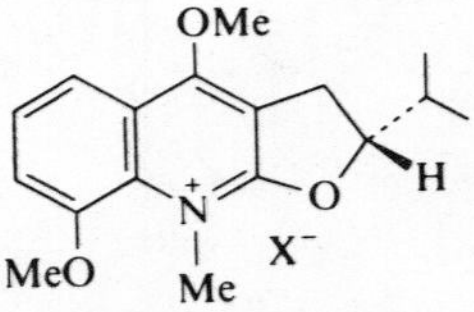

OMe N O H MeO Me X⁻

115 (−)-(*R*)-Lunasine

O N O H MeO Me

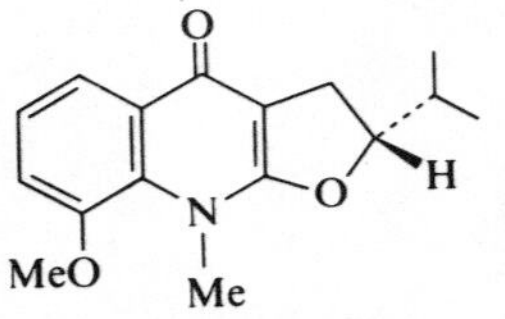

116 (−)-(*R*)-Lunacrine

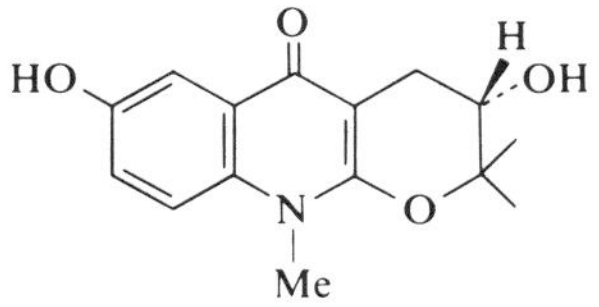

117 (−)-(*S*)-Ribalinidine

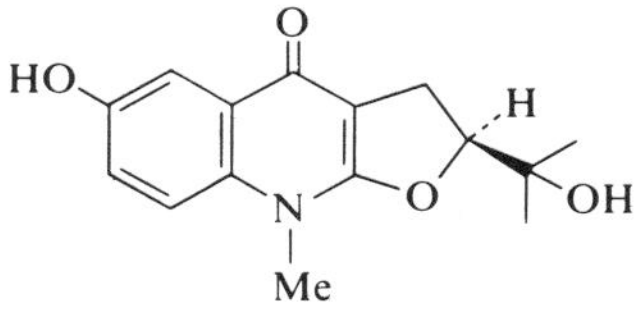

118 (+)-(*S*)-Ribaline

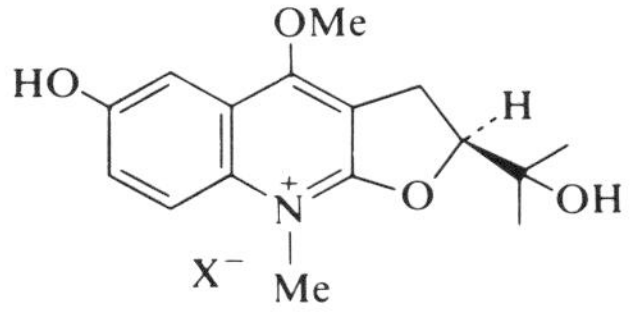

119 (+)-(*S*)-Ribalinium salt

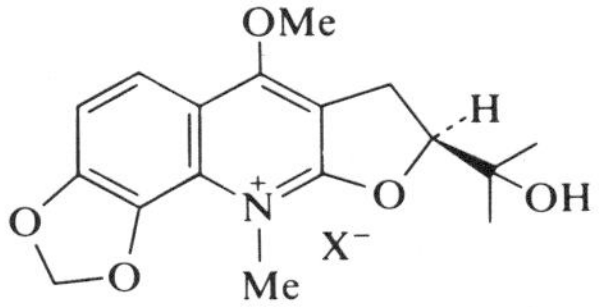

120 (−)-(*S*)-*O*-Methylhydroxyluninium salt

The sequence of reactions leading to (+)-(*R*)-balfouridine (**100**) and to (−)-(*S*)-isobalfourodine (Scheme 5) indicates that reaction of 3-prenylquinolines with (*S*)-peroxy acids yields (*S*)-epoxides (**99**). On this basis, the 2,4-dimethoxyquinoline epoxides obtained from (+)-(*S*)-peroxycamphoric acid also have the (*S*) configuration (**106** and **109**). Because the (−)-(*S*)-epoxide **106** is converted via reaction with formic acid (inversion) followed by hydrolysis of the formate **108** (retention) into (+)-orixine (*79*), the dextrorotatory alkaloid of *Orixa japonica* is assigned the (*R*) configuration (**107**). The accepted mechanism of the Brown–Yoon reaction (**109** → **110**) implies retention of configuration at the chiral center, and subsequent reactions (**110** → **112** → **111**) leading to the (−)-lunacridine (**111**) are also unlikely to affect the configuration; consequently, (+)-lunacridine from *Lunasia amara* is thought to be the (*R*) enantiomer (*80*). Lunasine (**115**) and lunacrine (**116**) were also assigned (*R*) configurations as a result of known transformations. The absolute stereochemistries of orixine, lunacridine, and related quinoline alkaloids are indicated tentatively in Table IV; confirmation by more direct methods of assignment is clearly desirable in some cases.

The pyranoquinoline alkaloid (−)-ribalinidine was shown to have the (*S*) configuration **117** by application of Horeau's method to (−)-*O*-methylribalinidine (*46*). The latter alkaloid is the predominant product from the rearrangement of (+)-methylribaline, which consequently also has an (*S*) configuration if, by analogy with the balfourodine–isobalfourodine conversion (see Section III,F), the rearrangement does not involve the chiral center; hence, the alkaloids (+)-ribaline (**118**) and (+)-ribalinium salt (**119**) are (*S*) enantiomers. It appears, therefore, that the balfourodinium series of alkaloids isolated from *B. riedelianum* from Brazil and the ribalinium series

TABLE IV
ABSOLUTE STEREOCHEMISTRY OF QUINOLINE ALKALOIDS

Compound	Structure number	Sign of $[\alpha]_D$	Method[a]	Configuration	Ref.
Unsubstituted in aromatic ring					
N-Methylplatydesminium salt	**113**	+	A	*R*	*81*
Isoplatydesmine	**124**	+	B and C	*R*	*37, 67*
Araliopsine	**127**	+	B	*R*	Section III,F
Edulinine	cf. **54**	−	B	*R*	*37*
8-Hydroxy and 8-methoxy compounds					
Balfourodine	**100**	+	A	*R*	*78*
Isobalfourodine	**104**	+	A	*R*	*78*
Balfourolone	**103**	−	A	*R*	*78*
O-Methylbalfourodinium salt	**101**	+	B	*R*	*78*
Pteleatinium salt	**102**	+	B	*R*	*54*
ψ-Balfourodine	**126**	+	A	*R*	*83*
ψ-Isobalfourodine	**130**	−	A	*S*	*83*
Lunacridine	cf. **111**	+	C	*R*[b]	*80*
Lunasine	**115**	−	B	*R*[b]	*80*
Lunacrine	**116**	−	B	*R*[b]	*80*
7,8-Methylenedioxy compounds					
Methylhydroxyluninium salt	**120**	−	D	*S*[b]	*54*
Orixine	**107**	+	C	*R*[b]	*78*
6-Hydroxy and 6-methoxy compounds					
Ribalinidine	**117**	−	E	*S*	*46*
Ribaline	**118**	+	B	*S*[b]	Section III, E,2
Ribalinium salt	**119**	+	B	*S*[b]	Section III, E,2

[a] A, Ozonolysis to lactone **114** or **129** of known configuration; B, correlation studies; C, asymmetric synthesis with peroxy acids of known configuration; D, circular dichroism studies; E, Horeau's method.
[b] Configuration by direct methods desirable.

from Argentine *B. riedelianum*, differing only in the nature of substituents present in the aromatic ring, have "opposite" configurations. This is contrary to the general stereochemical correlation of structurally related and cooccurring coumarins and quinoline alkaloids (*38*), and raises the possibility that two species of *Balfourodendron* are involved; further study of the stereochemistry of the ribaline group of alkaloids is clearly required.

Another apparent anomaly arises from the assignment of an (*S*) configuration (*54*) to (−)-*O*-methylhydroxyluninium salt (**120**) by comparison of the circular dichroism spectrum with (+)-(*R*)-pteleatinium salt (**102**); these alkaloids cooccur in *Ptelea trifoliata*, and the stereochemical result implies that they are derived biosynthetically from epoxides of "opposite" configuration.

(+)-(*R*)-*N*-Methylplatydesminium salt (**113**) is converted by a reaction not involving the chiral center into (+)-isoplatydesmine (**124**) (*37*). Thus the latter, which cooccurs with the (+)-quaternary salt in *Araliopsis tabouensis* also has an (*R*) configuration, an assignment supported by the asymmetric synthesis of (+)-isoplatydesmine by the reaction of the 3-prenyl-2-quinolone **20** with (+)-*S*-peroxycamphoric acid (*67*). The absolute stereochemistry of (+)-ribalinine (**42**), another constituent of this species, has not been established, but by analogy with the balfourodine–isobalfourodine pair found in *B. riedelianum* would be expected to be (*R*) as well.

F. Rearrangement Reactions of Dihydrofuroquinoline and Dihydropyranoquinoline Alkaloids

Rapoport and Holden (*61*) studied rearrangements of balfourodine (**123**) and isobalfourodine (**131**), and their work has already been reviewed briefly (Volume IX, p. 246). After the stereochemistry of the two alkaloids had been established (*78*), reexamination of the reactions by Grundon *et al.* (*78, 83*) using alkaloids obtained by asymmetric synthesis revealed apparent experimental contradictions and led to a revision of the mechanisms proposed earlier. It is appropriate, therefore, to discuss these interesting reactions, which are summarized in Table V in detail.

(−)-Balfourodine was reported to rearrange with acetic anhydride and pyridine into (+)-isobalfourodine acetate (**122**), which was hydrolyzed to (+)-isobalfourodine (overall 48% racemization), but it is now known that these products are derived from (+)-balfourodine rather than from the levorotatory enantiomer. Because (+)-balfourodine occurs in *Balfourodendron riedelianum*, and there is no indication that (−)-balfourodine was available from another source, it appears probable that (−)-balfourodine is a misprint for (+)-balfourodine in the earlier paper. Alternative mechanisms involving inversion at the chiral center were proposed for the rearrangement of balfourodine effected by acetic anhydride (*61*), but now that it is known that the two alkaloids have the (*R*) configuration these mechanisms are no longer tenable for the major reaction. A plausible mechanism implying retention of configuration is indicated in

TABLE V

REARRANGEMENTS OF DIHYDROFURO- AND DIHYDROPYRANOQUINOLINES[a]

Reactant	Structure number	Reagent[b]	Product	Ref.
(−)-Balfourodine		A	(+)-Isobalfourodine acetate	*61*
(+)-(*R*)-Balfourodine	**123**	A	(+)-(*R*)-Isobalfourodine acetate[c] (**122**)	*78*
(+)-(*S*)-Methylribaline	cf. **118**	A	(+)-(*S*)-Methylribalinidine acetate,[d] cf. (**117**)	*46*
(−)-Balfourodine		B	(+)-ψ-Balfourodine and (−)-ψ-Isobalfourodine	*61*
(+)-(*R*)-Balfourodine	**123**	C or D	(+)-(*R*)-ψ-Balfourodine (**126**)	*83*
(+)-(*R*)-Isoplatydesmine	**124**	D	(+)-(*R*)-Araliopsine (**127**)	*67*
(+)-(*R*)-Balfourodine	**123**	B	(−)-(*S*)-ψ-Isobalfourodine (**130**)	*83*
(+)-Isobalfourodine		B	(+)-ψ-Isobalfourodine	*61*
(+)-(*R*)-Isobalfourodine	**131**	B	(+)-(*R*)-ψ-Isobalfourodine	*84*
(+)-(*R*)-Isobalfourodine	**131**	D	(−)-(*S*)-ψ-Balfourodine	*84*
(+)-(*R*)-ψ-Balfourodine	**126**	B	(−)-(*S*)-ψ-Isobalfourodine (**130**)	*83*
(−)-(*S*)-ψ-Isobalfourodine	**130**	C or D	No reaction	*84*
Dubinidine diacetate	**136**	B	Pyrano-2-quinolone (**138**)	*75*
(±)-Balfourodine	**12**	Aq. HBr at 75°	Isopropylfuroquinolone (**140**)	*84*

[a] Supplementary to Volume IX, pp. 246, 251–252.

[b] A, Refluxing in Ac_2O–pyridine; B, refluxing in 20–25% NaOH or KOH in methanol, ethanol, or aqueous ethanol; C, NaH in DMF at 50°; and D, NaOMe in DMF at 15°.

[c] Hydrolysis gives (+)-(*R*)-isobalfourodine.

[d] Hydrolysis gives (−)-(*S*)-methylribalinidine.

Scheme 8. The presence of a 4-carbonyl group appears to be necessary for rearrangement, as the 4-methoxyquinoline acetate **121** is unaffected when heated with acetic anhydride. The rearrangement has also been observed with methylribaline, cf. **118** (see Section III,C,2).

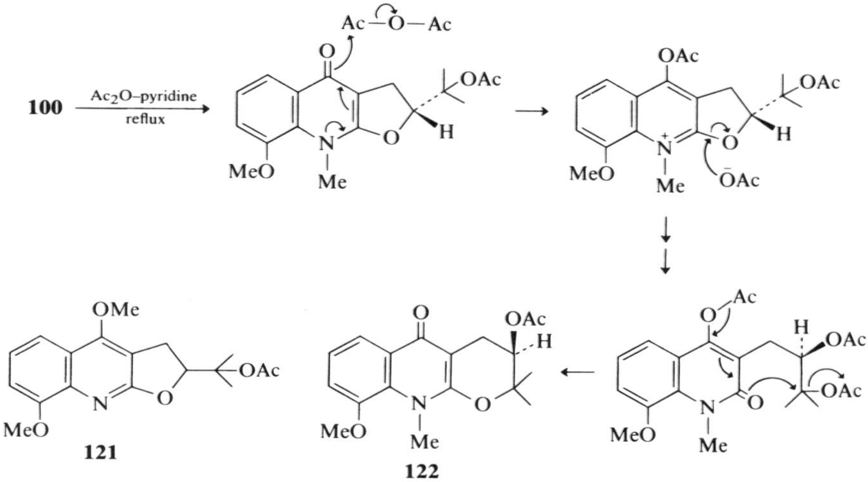

SCHEME 8. *Mechanism of the balfourodine–isobalfourodine acetate rearrangement.*

An interesting series of base-catalyzed rearrangements (Table V) was discovered by Rapoport and Holden (*61*). For instance, (−)-balfourodine, on refluxing in concentrated alkali, was reported to yield a mixture of the angular isomers (+)-ψ-balfourodine (**126**) and (−)-ψ-isobalfourodine (**130**), while in the same conditions (+)-isobalfourodine (**131**) afforded (+)-ψ-isobalfourodine. A mechanism was proposed involving substitution by hydroxyl ion at the carbon adjacent to nitrogen. According to this mechanism, the chiral centers are not affected (*61*), and because optically antipodal ψ-isobalfourodines were obtained from balfourodine and isobalfourodine, it was concluded that the alkaloids had "opposite" absolute configurations. The knowledge that the alkaloids are both (+)-(*R*) enantiomers (*78*) prompted a new study of these rearrangements (Table V) (*83, 84*).

Reaction of (+)-(*R*)-balfourodine (**123**) in dimethylformamide with sodium hydride at 50° or with sodium methoxide at 15° gave (+)-ψ-balfourodine (**126**) almost quantitatively, but when refluxed with methanolic potassium hydroxide, afforded (−)-ψ-isobalfourodine (**130**) as the only product. Again, the different results can be explained if it is assumed that (−)-balfourodine is a misprint for (+)-balfourodine in the earlier paper.

(+)-(*R*)-Isobalfourodine was converted with sodium methoxide into (−)-ψ-balfourodine, and, under more strenuous conditions, into (+)-ψ-isobalfourodine. It appeared that ψ-balfourodine (**126**) was formed under kinetic control, while the more stable ψ-isobalfourodine (**130**) resulted from an equilibrium controlled reaction at elevated temperatures. The structures of the rearrangement products were confirmed by NMR spectroscopy, and their absolute stereochemistry was established by exhaustive ozonolysis; (+)-ψ-balfourodine gave the (+)-(*R*)-lactone (**125**) and is thus an (*R*) enantiomer, and (−)-ψ-isobalfourodine (**130**) has the (*S*) configuration because it gives the (−)-(*S*)-lactone (**129**). The stereochemical relationships cannot be accommodated in the earlier mechanism, and an alternative was proposed (Scheme 9) involving intermediate epoxides. According to this mechanism, (+)-balfourodine is converted via an (*S*)-epoxide **128** into (+)-ψ-balfourodine with two inversions of configuration and into (−)-isobalfourodine with one inversion. A similar scheme involving an (*R*)-epoxide **132** can be applied to the rearrangement of (+)-isobalfourodine (**131**).

Support for the new mechanism was provided by trapping the intermediate with methyl iodide to give epoxide **133**. The structure of this compound was indicated by IR absorption at 1646 cm^{-1} (2-quinolone carbonyl) and by the NMR spectrum. The epoxide was converted readily into the diol balfourolone (**135**) on alumina chromatography or by treatment with 2 *N* sodium hydroxide at 20°; the mild conditions and the failure of 2,4-dimethoxyquinoline epoxides (e.g., **109**) to react under similar conditions suggest that this reaction occurs through formation of *O*-methylbalfourodinium salt (**134**) and subsequent nucleophilic attack at the 2-position rather than by direct reaction of hydroxide ion on the epoxide ring. Epoxide **133** is the presumed intermediate in reaction of 4-methoxy-*N*-methyl-3-prenyl-2-quinolone (**98**) with peroxy acid to give *O*-methylbalfourodinium salt, and treatment of the epoxide (**133**) in ether with hydriodic acid gave an immediate precipitate of the quaternary iodide (**134**), as expected. The reaction **128** ⇄ **126** is regarded as reversible, because (+)-ψ-balfourodine (**126**) is isomerized to (−)-ψ-isobalfourodine (**130**), and in the presence of methyl iodide the same epoxide (**133**) was obtained.

An analogous epoxide (**37**) was obtained from ribilinium salt (see Section III,C,1), and in a similar way *O*-methylbalfourodinium iodide gave epoxide **133**. The epoxide **37** was unexpectedly isolated as a racemate; it was hoped that this anomaly would be resolved by a study of epoxide **133**, but not enough of the pure compound was available for an accurate determination of its optical activity to be made.

(+)-(*R*)-Isoplatydesmine (**124**) is converted by base into its angular isomer, (+)-araliopsine (**127**) (*67*). The latter enantiomer occurring in

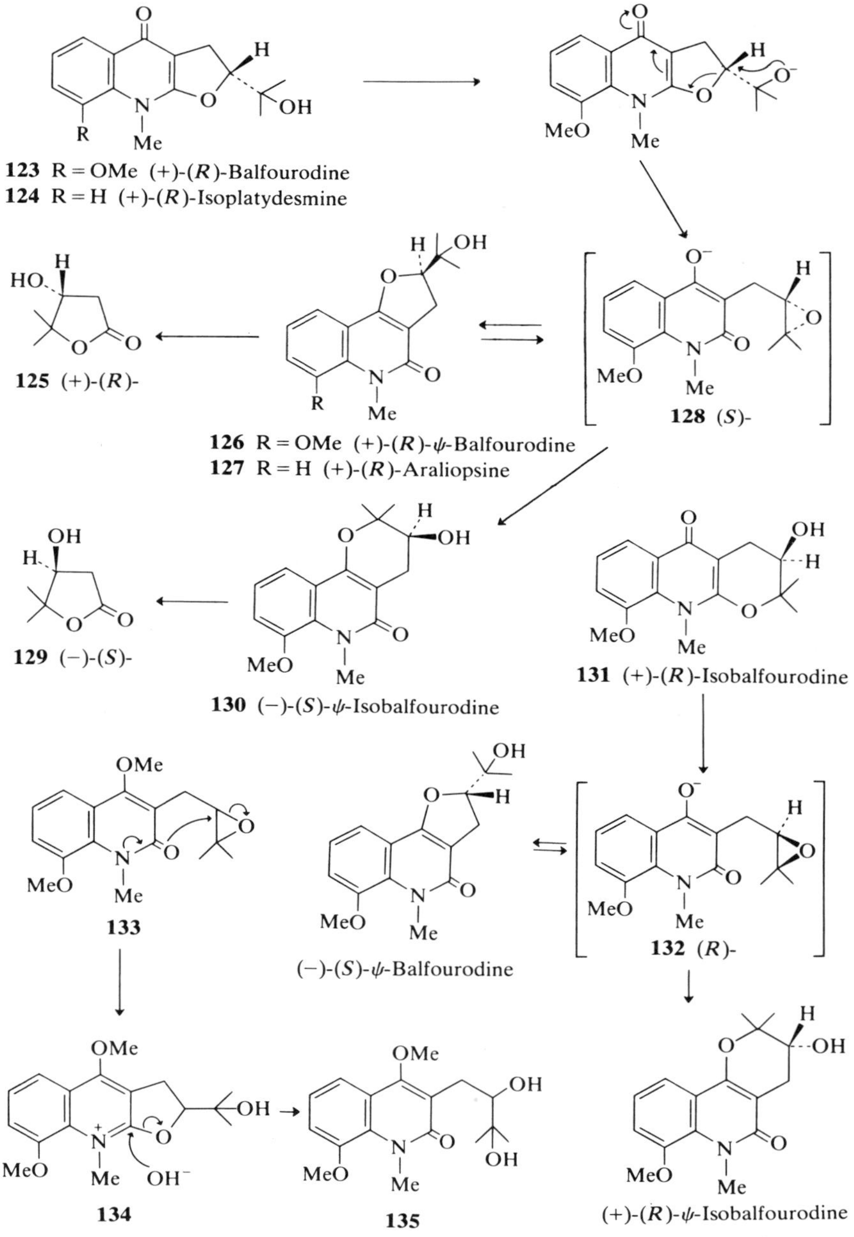

123 R = OMe (+)-(*R*)-Balfourodine
124 R = H (+)-(*R*)-Isoplatydesmine

125 (+)-(*R*)-

126 R = OMe (+)-(*R*)-ψ-Balfourodine
127 R = H (+)-(*R*)-Araliopsine

128 (*S*)-

129 (−)-(*S*)-

130 (−)-(*S*)-ψ-Isobalfourodine

131 (+)-(*R*)-Isobalfourodine

133

(−)-(*S*)-ψ-Balfourodine

132 (*R*)-

134

135

(+)-(*R*)-ψ-Isobalfourodine

SCHEME 9. *Base-catalyzed rearrangements of balfourodine, isobalfourodine, and isoplatydesmine.*

Araliopsis soyauxii (Section III,C,3) thus has the (*R*) configuration, assuming that the rearrangement reaction follows the same stereochemical pathway as the balfourodine–ψ-balfourodine transformation.

Yunusov *et al.* (*75*) observed that dubinidine diacetate (**136**), on heating with methanolic sodium hydroxide, gave the *N*-methylpyranoquinoline **138**. A possible explanation of this reaction (Scheme 10) is that hydrolysis and isomerization to the *N*-methyl-4-quinolone **137** are followed by a rearrangement analogous to the balfourodine–ψ-isobalfourodine transformation; the formation of an *N*-methyl derivative in the absence of a conventional methylating reagent is unexpected, and the reaction requires further investigation.

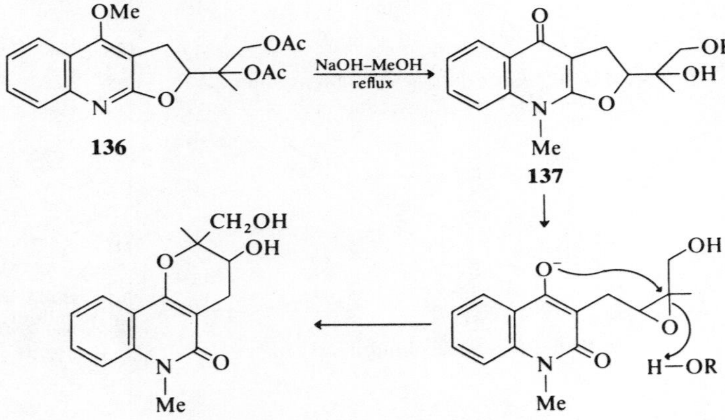

OMe OAc OAc N O **136** NaOH–MeOH reflux O OH OH N O Me **137**

CH₂OH OH O N O Me **138** O⁻ OH O N O Me H–OR

SCHEME 10. *Base-catalyzed rearrangement of dubinidine diacetate.*

A preliminary study has been carried out of acid-catalyzed reactions of balfourodine (*84*). Heating balfourodine with hydrobromic acid gave a complex mixture, but the major component was identified by spectroscopic methods as the angular furoquinoline **140**. The reaction may occur by dehydration to the enol ether **139** which then rearranges to its angular isomer by means of hydrolysis and then cyclization (Scheme 11).

G. 2,2-Dimethylpyranoquinoline Alkaloids of the Flindersine Group

Flindersine (**141**) has been known for many years and provides an intriguing exception to the general experience that furo- and pyranoquinoline alkaloids have a linear arrangement of the three rings. During the

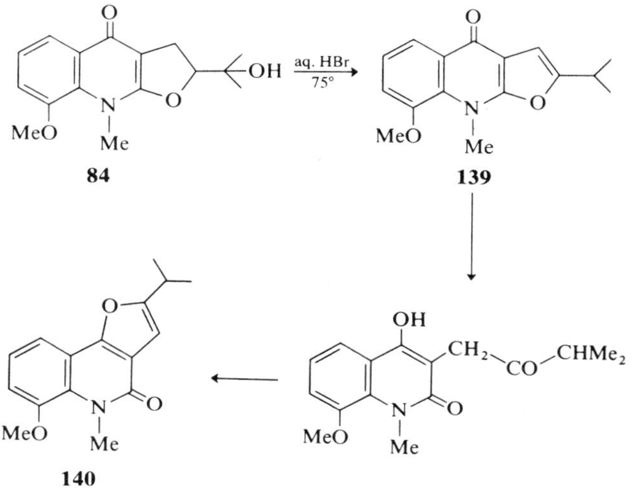

SCHEME 11. *Acid-catalyzed rearrangement of balfourodine.*

last decade, additional members of the flindersine group of alkaloids have been isolated and new syntheses of the ring system have been devised.

1. Occurrence and New Alkaloids

Flindersine, originally isolated from *Flindersia australis* R.Br., has now been obtained from *Haplophyllum tuberculatum* (*36*) and *Geijera parvifolia* Lidl. (*85*).

An alkaloid, $C_{15}H_{15}NO_2$, from *Spathelia sorbifolia* was shown to be *N*-methylflindersine (**142**) on the basis of its NMR spectrum and by comparison with a sample obtained by methylation of flindersine (*86*). *N*-Methylflindersine has also been isolated from the root bark of *Ptelea trifoliata* (*34*). The NMR spectra of flindersine and *N*-methylflindersine show one-proton doublets in the regions τ3.2–3.3 and 4.4–4.5 (–CH=CH–), which are particularly useful for the identification of 2,2-dimethylpyranoquinolines of the flindersine type.

Yunusov and his collaborators (*87*) obtained the alkaloid haplamine from *Haplophyllum perforatum*. Spectroscopic data indicated that haplamine was a methoxyflindersine, and it was assigned structure **143** on the basis of its reaction with base, followed by methylation, to give 4,6-dimethoxy-1-methyl-2-quinolone, although evidence for the structure of this degradation product was not recorded. The constitution of the alkaloid has since been confirmed by synthesis (see Section III,G,2).

Abe and Taylor isolated a new quinoline alkaloid, $C_{17}H_{19}NO_4$, named oricine, from timber of the West African tree *Oricia suaveolens* (Engl.) Verdoon. The UV, IR, and NMR spectra (Scheme 1, Section I) indicated that oricine was 6,7-dimethoxy-*N*-methylflindersine (**144**), and this was confirmed by synthesis (*88*).

2. Synthesis

Flindersine was synthesized some time ago (see Volume III of this treatise), but Piozzi (*89*) has introduced a new and convenient method involving reaction of 4-hydroxy-3-prenyl-2-quinolone with dichlorodicyanobenzoquinone (DDQ). Application of the procedure to the prenylquinolone **146** led to the formation of haplamine (**143**) (*90*). The pyranoquinolone **145** was prepared similarly and converted by methylation into oricine (**144**) (*88*). Treatment of the linear dihydropyranoquinoline alkaloid khaplofoline (**74**) with DDQ also gives flindersine (*91*), probably through ring cleavage to a prenylquinolone.

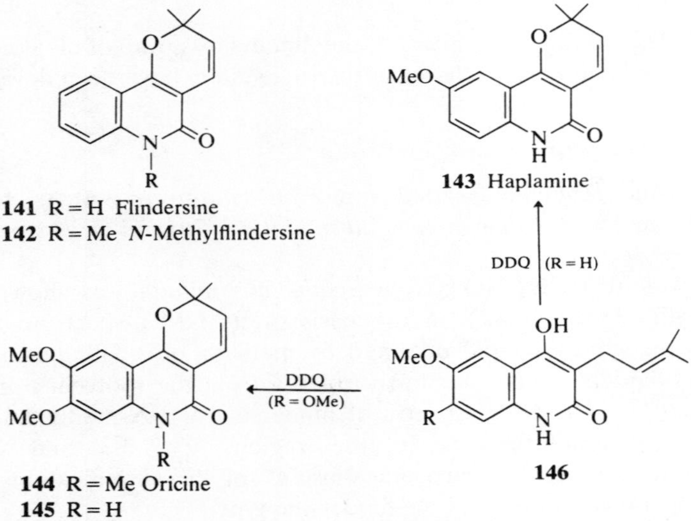

141 R = H Flindersine
142 R = Me *N*-Methylflindersine

143 Haplamine

DDQ (R = H)

144 R = Me Oricine
145 R = H

DDQ (R = OMe)

146

A new synthesis of 2,2-dimethylpyranoquinolines was developed by Bowman, Grundon, and James (*92*) (Scheme 12). Reaction of 3-isoprenyl-2,4-dimethoxyquinoline epoxides (**147**) with potassium hydroxide in aqueous dimethyl sulfoxide (DMSO) at 100°, followed by dilution with water, gives a clear solution which on neutralization furnishes 2-methoxy-(2,2-dimethyl)-pyranoquinolines **151** and **152** in high yield; the latter compounds are insoluble in base and so are not the initial products from the reaction. An NMR study of the reaction in deuterated solvents indicated that the first product was the allylic alcohol **148**, which could be isolated. The allylic alcohol was then converted slowly into base-soluble compound **149**; treatment of the solution with dimethyl sulfate resulted in regeneration of the 4-methoxyallylic alcohol, and neutralization gave the chromene **152**. A mechanism was proposed (Scheme 12) involving cyclization of a quinone methide **150**. Reaction of chromene **151** with hydrobromic acid gave flindersine (**141**).

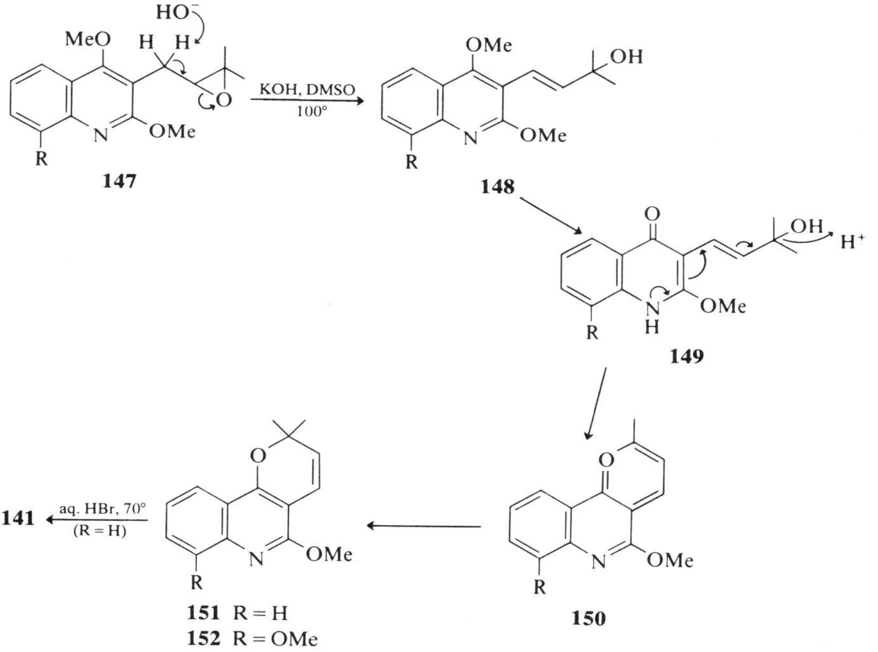

SCHEME 12. *Synthesis of flindersine alkaloids from prenyl epoxides.*

Huffman and Hsu (*93*) prepared flindersine and its 8-methoxy derivative from thallous salts of 4-hydroxy-2-quinolones and 3-chloro-3-methylprop-1-yne; the reaction apparently occurs by means of a rearrangement of the 4-ethers (**154**) (Scheme 13). The by-product **153** was isolated in one case and presumably arises by substitution at the 3-position and subsequent cyclization.

A simple and efficient synthesis of flindersine results from reaction of 4-hydroxy-2-quinolone with 3-methylbut-2-enal in refluxing pyridine (Scheme 13) (*94*).

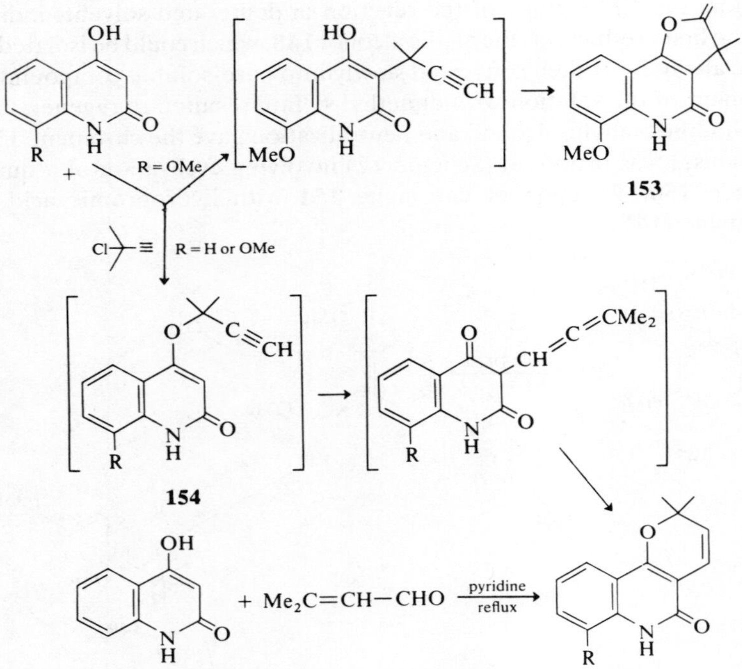

SCHEME 13. *Synthesis of flindersine alkaloids from* 4-*hydroxy*-2-*quinolones.*

3. Reactions

Reduction of the pyranoquinolone alkaloids gives dihydro derivatives, cf. **156**, which can be synthesized by acid-catalyzed cyclization of 3-prenylquinolones **155** or by reaction of monoanilides **157** with polyphosphoric acid (PPA) (*89*). Bromination of dihydroflindersine, previously thought to yield the monobromo derivative **160** (*89*), has been reinvestigated by Grundon and James (*95*). The NMR spectrum of the compound obtained from dihydroflindersine with *N*-bromosuccinimide (NBS) shows a singlet at $\tau 2.87$ due to a vinylic proton and is more consistent with structure **161**. As observed by Piozzi *et al.* (*89*), the same compound is obtained by reaction of flindersine (**158**) with bromine. When flindersine was heated with NBS in ethanol-free chloroform, no succinimide precipitated; chromatography afforded the vinyl bromide **161**, and treatment of the initial product with ethanol gave the *trans*-bromo ether **162**. It is apparent that an initial adduct **159** undergoes ready elimination or solvolysis. The pathway (Scheme 14) from dihydroflindersine to the vinyl bromide

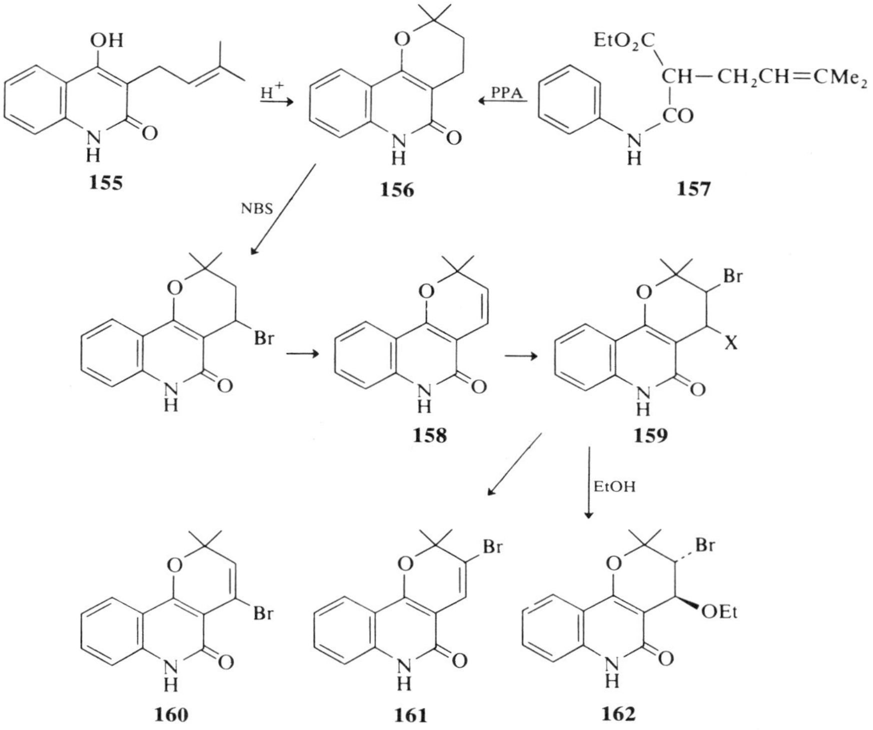

SCHEME 14. *Bromination of flindersine and dihydroflindersine.*

161 instituted by NBS is believed to involve normal benzylic bromination and then elimination to give flindersine followed by addition of NBS (or Br_2) and finally a second elimination. The abnormal reaction of benzyl derivatives with NBS seems likely to occur when an initial benzylic bromination product undergoes rapid elimination (cf. lemobiline, Section V,C,2).

IV. Furoquinoline Alkaloids

Known members of the furoquinoline group of alkaloids, particularly the ubiquitous skimmianine, have been obtained from a large number of new sources. Reported occurrences are listed in Table VI, which supplements earlier data (Volume IX, p. 227, Table I) and includes furoquinolines first obtained during the period 1957–1966 (Volume IX). The compounds in Table VI can be regarded as derivatives of dictamine (**163**).

TABLE VI
OCCURRENCE OF FUROQUINOLINES OF THE DICTAMNINE GROUP[a]

Plant source	Ref.
Dictamnine (**163**)	
Afraegle paniculata	*96*
Chorilaena quercifolia Endl.	*97*
Decatropsis bicolor (Zucc.) Radlk.	*98*
Dictamnus angustifolius	*99*
D. caucasicus Fisch.	*100*
Evodia belahe Baillon	*101*
Fagara mayu	*102*
Flindersia pimenteliana	*103*
Glycosmis pentaphylla (Retz.) Correa	*104*
Halfordia kendack (Mntr.) Giull.	*22*
Haplophyllum bucharicum Litv.	*105*
H. bungei Trautv.	*106*
H. ramossisium Vved.	*21, 106*
H. suaveolens (DC) G. Don.	*107*
Helietta longifoliata Britton	*108*
Medicosma cunninghamii Hook. f.	*109*
Monnieria trifolia L.	*110*
Pitavia punctata Molina	*111*
Ptelea trifoliata L.	*112*
Ruta angustifolia Pers.	*113*
R. chalepensis L.	*113*
R. graveolens L.	*114*
R. montana Dill.	*113*
Skimmia foremanii Hort.	*115*
S. japonica Thunb.	*37, 116*
Zanthoxylum arnottianum Maxim	*117*
Z. belizense Ludell	*65*
Z. decaryi	*118*
Z. parviflorum Benth.	*41*
Pteleine (6-Methoxydictamnine)	
Helietta longifoliata	*108*
Medicosma cunninghamii	*109*
Platydesma campanulata Mann (Volume IX)	*119*
Ptelea trifoliata	*120*
Evolitrine (7-Methoxydictamnine)	
Evodia belahe	*101*
γ-Fagarine (8-Methoxydictamnine)	
Chloroxylon swietenia DC	*121*
Dictamnus albus L.	*122*
D. caucasicus	*100, 123*
Fagara mayu	*102*
Geigeria salicifolia Schott.	*40*
Glycosmis pentaphylla	*104*

TABLE VI (*Continued*)

Plant source	Ref.
Haplophyllum bucharicum	*105*
H. kowalenskyi	*124*
H. schelkovnikovii	*124*
H. villosum	*124*
Pitavia punctata	*111*
Ruta chalepensis	*113*
Skimmia japonica	*3*
Thamnosma montana Torr. and Frem.	*125*
Zanthoxylum tsihanimposa H. Pett.	*126*
Z. piperitum DC	*127*
Kokusaginine (6,7-Dimethoxydictamnine)	
Araliopsis soyauxii Engl.	*44*
Choisya arizonica Standl.	*128*
C. mollis Standl.	*128*
Evodia alata F. Muell.	*129*
E. belahe	*101*
Halfordia kendack	*22*
Haplophyllum suaveolens	*70*
Helietta longifoliata	*108*
H. parvifolia Benth.	*130*
Lunasia amara (Blanco)	*131*
Melicope confusa (Merr.) Liu	*132*
M. perspicuinervia Merr & Perry	*42*
Pelea barbigera	*43*
Ptelea aptera Parry	*32*
Ruta chalepensis	*113*
R. montana	*113*
Teclea unifoliata	*133*
Vepris ampody H. Perr	*134*
Zanthoxylum pluviatile Hartley	*135*
Maculosidine (6,8-Dimethoxydictamnine)	
Esenbeckia hartmanii H.B.K.	*128*
Philoteca hasseli F. Muell	*136*
Ptelea trifoliata	*120*
Skimmianine (7,8-Dimethoxydictamnine)	
Araliopsis soyauxii	*44*
A. tabouensis Aubrev et Pellegr.	*66*
Choisya arizonica	*128*
C. mollis	*128*
C. ternata H.B. et K. (Volume IX)	*63, 137*
Decatropis bicolor	*98*
Dictamnus angustifolius	*99*
Diphasia klaineana	*66*
Eriostemon difformis A. Cunn.	*136*
Esenbeckia febrifuga Juss.	*138*

(*Continued*)

TABLE VI (*Continued*)

Plant source	Ref.
E. hartmanii	*128*
Evodia alata	*129*
E. elleryana F. Muell.	*139*
Fagara capensis Thunb.	*140*
F. chalybea Engl.	*141, 142*
F. leprieurii Engl.	*26, 141, 143*
F. macrophylla (Oliv.) Engl.	*141*
F. mantsurica Honda	*144*
F. mayu	*102*
F. okinawensis Nakai	*145*
F. rubescens Engl.	*146*
Flindersia pimenteliana	*103*
Geijera salicifolia	*40*
Gleznowia verrucosa Turz.	*147*
Haplophyllum acutifolium (DC.) G. Don	*148*
H. bungei	*21*
H. dubium Eug. Kov.	*149*
H. obtusifolium Ldb.	*105*
H. kowalenskyi	*124*
H. latifolium	*150*
H. pedicellatum Bge.	*151*
H. popovii Eug. Kov.	*152*
H. ramossisium	*21, 106*
H. schelkovnikovii	*124*
H. suaveolens	*107*
H. villosum	*124*
H. tenue	*124*
Helietta longifoliata	*108*
H. parvifolia	*130*
Melicope confusa	*132*
M. perspicuinervia	*42*
Monnieria trifolia	*110, 153*
Philoteca hasseli F. Muell	*136*
Ptelea aptera	*32*
P. crenulata Greene	*32*
Ruta chalepensis	*113*
R. montana	*113*
Teclea unifoliata	*133*
Thamnosma montana	*125*
Toddalia aculeata Pers.	*154*
Vinca herbaceae	*155*
Zanthoxylum alatum Roxb.	*156*
Z. belizense	*65*
Z. decaryi	*118*
Z. dinklagei Waterm.	*157*
Z. parviflorum	*41*

TABLE VI (*Continued*)

Plant source	Ref.
Z. piperitum	*127*
Z. pluviatile	*135*
Z. tsihanimposa	*126, 158*
Maculine (6,7-Methylenedioxydictamnine)	
Araliopsis soyauxii	*44*
Esenbeckia febrifuga	*138*
Helietta longifoliata	*159*
Ptelea aptera	*32*
Teclea unifoliata	*133*
Vepris bilocularis Engl.	*160*
Flindersiamine (8-Methoxy-6,7-methylenedioxydictamnine)	
Araliopsis soyauxii	*44*
Esenbeckia febrifuga	*138*
Helietta longifoliata	*108*
H. parvifolia	*134*
Robustine (8-Hydroxydictamnine)	
Dictamnus caucasicus	*100, 123*
Haplophyllum bucharicum	*105*
H. dubium	*15*
H. pedicellatum	*105*
H. robustum (Volume IX)	*161*
Zanthoxylum arnottianum	*117*
Haplopine (7-Hydroxy-8-methoxydictamnine)	
Aegle marmelos Correa	*162*
Haplophyllum bucharicum	*105, 163*
H. dubium	*15*
H. foliosum	*163, 164*
H. pedicellatum	*105*
H. perforatum Kar. et Kar. (Volume IX)	*165*
Zanthoxylum arnottianum	*117*
Evoxine (= haploperine) [7-(2,3-Dihydroxy-3-methylbutyl)oxy-8-methoxydictamnine]	
Choisya ternata	*63, 137*
Evodia alata	*129*
E. xanthoxyloides F. Muell. (Volume VI, p. 98)	*166*
Haplophyllum dubium	*15*
H. hispanicum	*167*
H. latifolium	*150*
H. obtusifolium	*105*
H. perforatum (Volume IX)	*168*
H. popovi	*152*
H. ramossisium	*21*
H. suaveolens	*70, 169*

(*Continued*)

TABLE VI (*Continued*)

Plant source	Ref.
Choisyine (**203**)	
Choisya arizonica	*128*
C. mollis	*128*
C. ternata (Volume IX)	*63, 137*
Maculosine [4-(2,3-Dihydroxy-3-methylbutyl)oxy-6,7-methylenedioxydictamnine]	
Flindersia maculosa F. Muell. (Volume IX)	*170*
Evodine [7-(2-Hydroxy-3-methylbut-3-enyl)oxy-8-methoxydictamnine]	
Evodia xanthoxyloides (Volume IX)	*166*

[a] Supplementary to Volume IX, Chapter 6, Table I.

A. Further Structural Studies and New Alkaloids

The occurrence of new furoquinoline alkaloids and their physical properties are indicated in Table VII. *N*-Methyl-4-quinolones, the so-called "isoalkaloids" are well known; one example is isodictamnine (**164**), obtained by heating dictamnine with methyl iodide. Isodictamnine, isopteleine, isomaculosidine, and isoflindersiamine have now been recognized as constituents of rutaceous species; the *N*-isopentyl alkaloids of *Acronychia* are also in this category. The alkaloids discussed in this section are grouped in accord with the pattern of oxygenation in the aromatic ring.

1. Confusameline, Evellerine, and Epoxide **167**

7-Hydroxydictamnine, described as 7-*O*-demethylevolitrine, was isolated first from *Evodia elleryana* F. Muell (*139*); it was later obtained from *Melicope confusa* (Mesr.) Liu and named confusameline (*132*). The structure was established by conversion with diazomethane into 7-methoxydictamnine (evolitrine).

Evodia elleryana also contains evellerine (**165**), a derivative of 7-hydroxydictamnine. Evellerine was shown to be a 4-methoxyfuroquinoline by conversion into the isoalkaloid isoevellerine (**166**) by heating with methyl iodide. Isoevellerine furnished a monoacetyl derivative, and yielded 7-methoxyisodictamnine when heated with sodium hydroxide and dimethylsulfate. The structure of the side chain in evellerine was established by reaction of isoevellerine with periodate and the identification of acetone (*139*).

Dreyer (*172*) reinvestigated the alkaloids of *Evodia xanthoxyloides* F. Muell and isolated a new furoquinoline alkaloid, $C_{17}H_{17}NO_4$, with a UV spectrum similar to that of 7-methoxydictamnine. Structure **167** was established by the NMR spectrum, which showed a one-proton triplet at 6.88 coupled to a two-proton doublet at 5.80τ. Confirmation of the epoxide structure was obtained by hydrolysis with aqueous oxalic acid to the glycol evellerine (**165**).

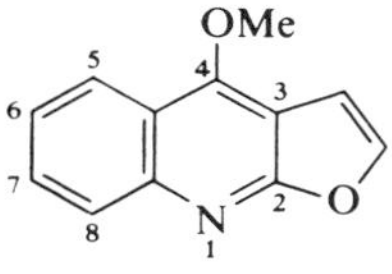

163 Dictamnine

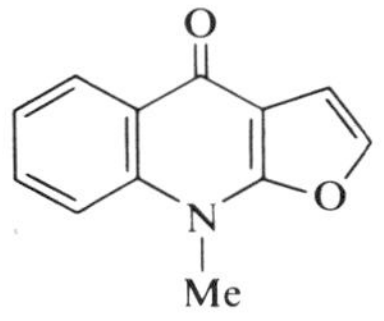

164 Isodictamnine

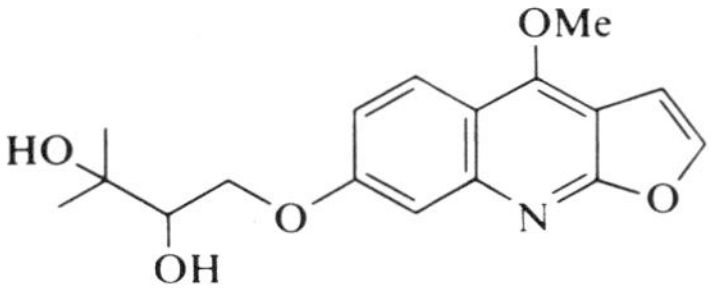

165 Evellerine

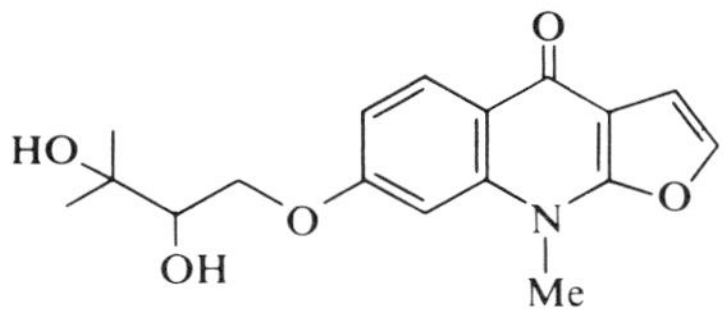

166 Isoevellerine

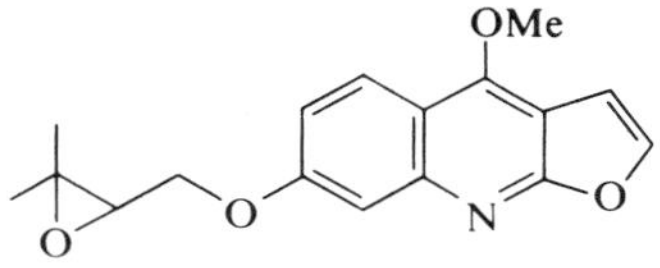

167

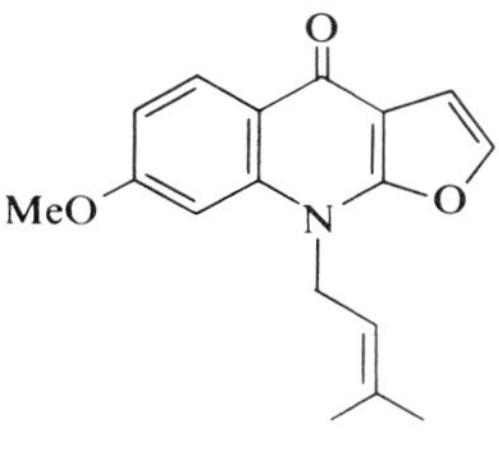

168 Acrophylline

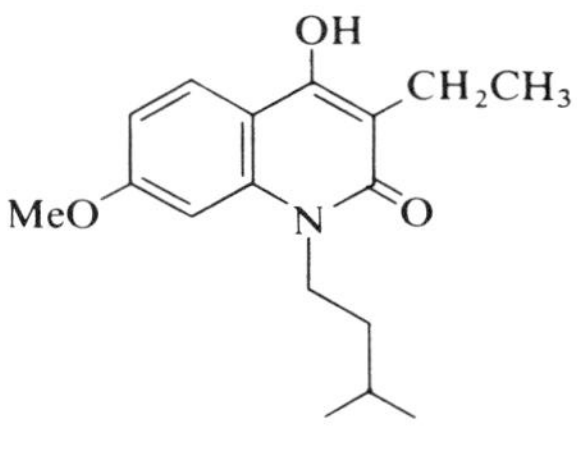

169

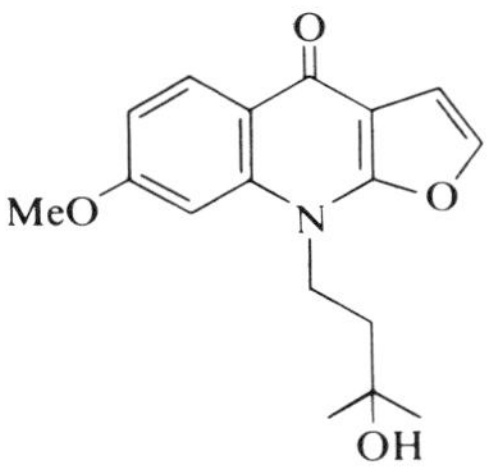

170 Acrophyllidine

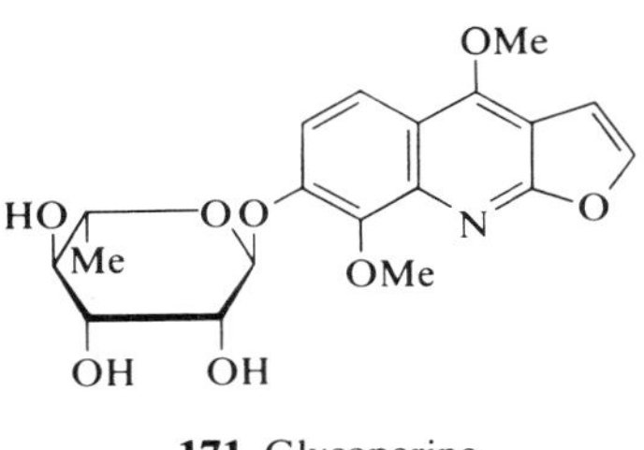

171 Glycoperine

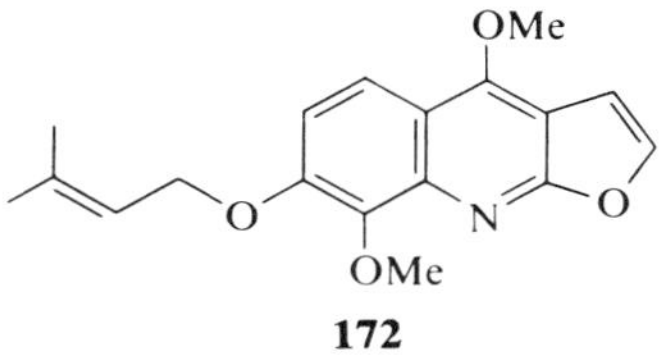

172

TABLE VII
OCCURRENCE OF FUROQUINOLINE ALKALOIDS FIRST ISOLATED 1966–1976

Name	Structure number	Source	Melting point (°C)	$[\alpha]_D$ (Solvent)	Ref.
Acrophyllidine	**170**	*Acronychia haplophylla* (F. Muell) Engl.	176–177		*171*
Acrophylline	**168**	*A. haplophylla*	119–120		*171*
Confusameline	**228**	*Evodia elleryana*	240–242		*139*
		Melicope confusa	239–240		*132*
Epoxide (unnamed)	**167**	*Evodia zanthoxyloides*	145–147	+50° ($CHCl_3$)	*172*
Epoxide (unnamed)	**174**	*E. zanthoxyloides*	141–142	+13° ($CHCl_3$)	*172*
Evellerine	**165**	*E. elleryana*	166–168	+21° ($CHCl_3$)	*139*
Folifinine		*Haplophyllum foliosum* Vved.	181–182		*173*
Foliminine	**190**	*H. foliosum*			*174*
Glycoperine	**171**	*Haplophyllum perforatum* Kar. and Kir.			*175*
Halfordinine	**208**	*Araliopsis tabouensis*	144		*45*
		Halfordia schleroxyla F. Muell	150–151		*22*
		Melicope perspicuinervia	150–152		*42*

Haplatine	**176**	*Haplophyllum latifolium*	139–140		*176*
Haplophydine	**189**	*H. perforatum*	111–112		*177*
Haplophyllidine	**181**	*H. perforatum* (Volume IX)	110–111	−16.2° (Me_2CO)	*178, 179*
Heliparvifoline	**187**	*Helietta parvifolia*	245–247		*180*
Isodictamnine	**164**	*Dictamnus albus*			*181*
		D. caucasicus			*100*
		Helietta longifoliata			*108*
Isoflindersiamine		*Helietta parvifolia*	212–213		*180*
Isomaculosidine		*Dictamnus albus*	170–172		*23, 181*
		D. caucasicus			*100*
		Ptelea trifoliata	169–170		*57*
7-Isopentenyloxy-γ-fagarine	**172**	*Ptelea aptera*	100–103		*32*
		Choisya ternata	100–103		*3*
		H. perforatum	105–106		*182*
Isopteleine	**214**	*Dictamnus caucasicus*			*100*
Methylevoxine	**177**	*Haplophyllum perforatum*	55–56		*183*
Perfamine	**184**	*H. perforatum*	142–144	−38°	*184*
Perforine	**179**	*H. perforatum*	182–183	+14.56° (CH_3OH)	*178*
				+14.32° ($CHCl_3$)	*179*

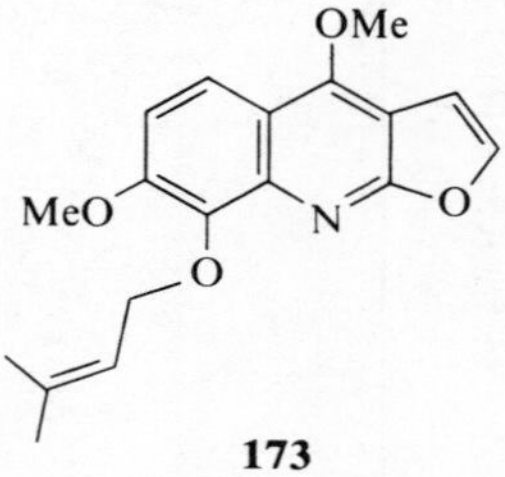

173

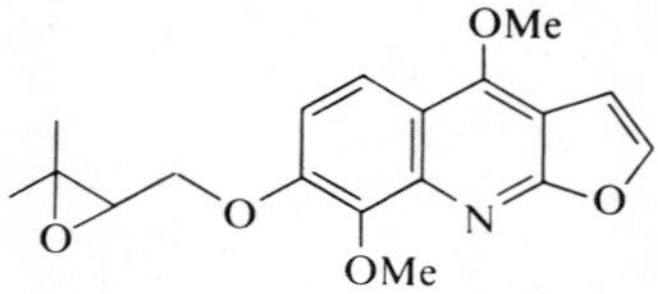

174

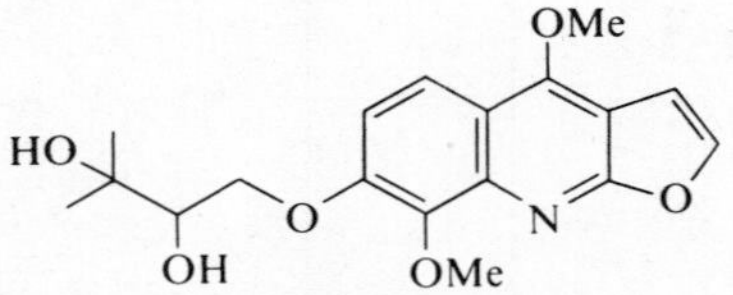

175 Evoxine

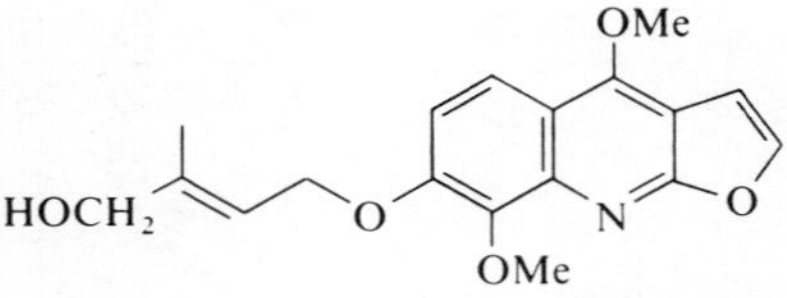

176 Haplatine

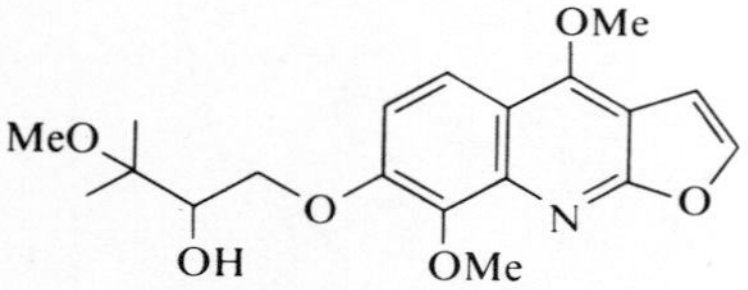

177 Methylevoxine

2. Acrophylline and Acrophyllidine

Lahey and his collaborators (*171, 185*) investigated the constituents of the leaves of *Acronychia haplophylla* (F. Muell.) Engl. and isolated the alkaloids acrophylline and acrophyllidine.

Acrophylline was shown by NMR and UV spectroscopy to be a methoxyisodictamnine containing a dimethylallyl substituent. The Australian group undertook an intensive study of the mass spectra of acrophylline and other isofuroquinolone alkaloids and suggested that the base peak at M − 68 in the mass spectrum of acrophylline arose from a McLafferty rearrangement and indicated that the prenyl group was attached to the electronegative nitrogen atom rather than to carbon. Furthermore, the abundance of the M − 85 mass peak pointed to the presence of a 7-methoxyl substituent in acrophylline. Structure **168** was proposed for the alkaloid, and this was confirmed by a study of the hexahydro derivative **169**. Hexahydroacrophylline prepared by catalytic reduction of the alkaloid was clearly a 3-ethyl-*N*-isopentyl-2-quinolone, and it was synthesized by reaction of *N*-isopentyl-*m*-methoxyaniline with diethyl ethylmalonate.

Acrophyllidine, $C_{17}H_{19}NO_4$, appeared from its NMR spectrum to be the tertiary alcohol **170**, and this was confirmed by conversion of acrophylline into acrophyllidine through acid-catalyzed hydration. Because aqueous acid was used in the isolation procedure, acrophyllidine may be an artifact.

3. Glycoperine, 7-Isopentenyloxy-γ-fagarine, Epoxide **174**, Haplatine, and Methylevoxine

7-Hydroxy-8-methoxydictamnine (haplopine) has now been found in a number of rutaceous plants (Table I). The alkaloid glycoperine, isolated from *Haplophyllum perforatum* (M.B) Kar and Kir (*175*), was shown to be the rhamnoside of haplopine by hydrolysis to L-rhamnose and haplopine.

A new furoquinoline alkaloid, $C_{18}H_{19}NO_4$, was obtained by Dreyer (*32*) from the fruit of *Ptelea aptera* Parry. The NMR spectrum indicated the presence of an isopentenyl group, and because the alkaloid had a UV spectrum identical to that of skimmianine, it was formulated as an isopentenyloxymethoxydictamnine (**172** or **173**). The resonance of the methoxy group did not shift upfield in benzene solution relative to chloroform, indicating that the group was flanked on each side by other groups. Thus, structure **172** for the alkaloid was preferred. The compound has since been obtained from *Choisya ternata* (3) and from *Haplophyllum perforatum* (*175*). Yunusov and his co-workers (*182*) prepared the alkaloid from 7-hydroxy-8-methoxydictamnine and 1-chloro-3-methylbut-2-ene, thereby finally establishing the structure of the alkaloid as 7-isopentenyloxy-γ-fagarine **172**.

Dreyer (*172*) obtained another new alkaloid, the epoxide **174**, from *Evodia xanthoxyloides*; its structure was determined by NMR spectroscopy and by acid hydrolysis to evoxine (**175**). The isolation of epoxides **167** and **174** from *E. xanthoxyloides* and the failure to detect evoxine, previously obtained from this species, are of considerable interest; because of the fact that acid was employed in the earlier isolation work, it was suggested that evoxine may be an artifact derived from the corresponding epoxide.

Haplatine from *Haplophyllum latifolium* was shown by spectroscopic studies and by hydrolysis to haplopine (**230**) to be the allylic alcohol **176** (*176*). Methylevoxine (**177**) was isolated from *H. perforatum* (*183*).

4. Haplophyllidine, Perforine, and Perfamine

Haplophyllidine and perforine are two related alkaloids of unusual structure obtained by Yunusov *et al.* from *Haplophyllum perforatum* (Volume IX, pp. 252, *178*, *179*).

Perforine, $C_{18}H_{25}NO_5$, is optically active and contains two hydroxyl groups and two methoxyl substituents but no *N*-methyl group. Reaction with chromic acid afforded a ketone, apparently by oxidation of a secondary hydroxyl function. The NMR spectrum showed coupled doublets at $\tau 2.57$ and 3.18 indicative of the presence of a furan ring, three-proton singlets at 5.87 and 6.93 (2 OMe), but no aromatic proton resonances. Consequently, perforine was assigned structure **179**. In accord with this structure, perforine with zinc and hydrochloric acid gave the anhydro derivative **180** and with acid alone afforded the cyclic compound **178**, shown by NMR to contain no olefinic protons. Both cyclic derivatives **178** and **180** gave iso compounds when heated, with methyl iodide (*179*).

The tentative structure **183** was earlier suggested for haplophyllidine (see Volume IX of this treatise), but later spectroscopic studies indicated that this alkaloid was also a furoquinoline containing a reduced aromatic ring. The new structure **181** was proposed as a result of correlation with perforine and its derivatives (*178*). Thus, compound **182**, obtained from perforine, was converted by an elimination reaction into haplophyllidine acetate. Further, the cyclic compound **178** was also obtained from haplophyllidine by treatment of the alkaloid with acid.

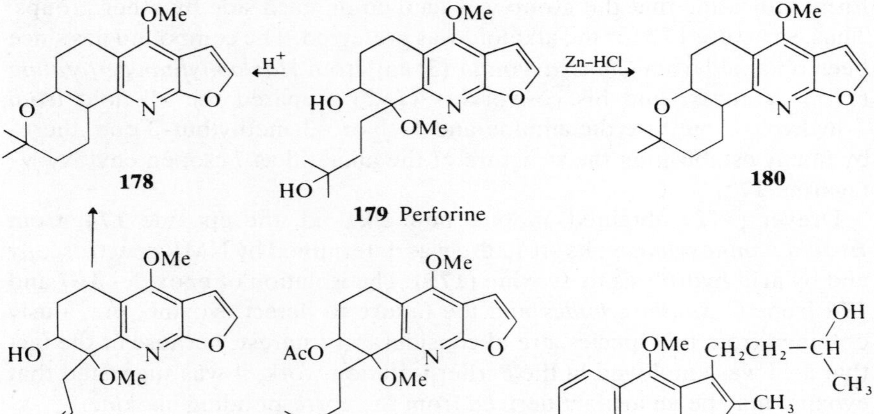

OMe H$^+$ OMe Zn–HCl OMe

HO N O OMe HO N O

178 **179** Perforine **180**

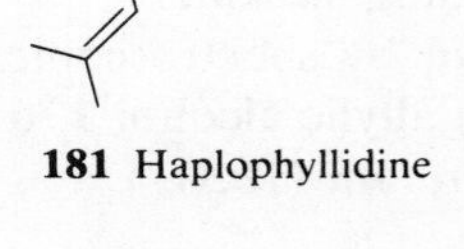

181 Haplophyllidine  **182**  **183**

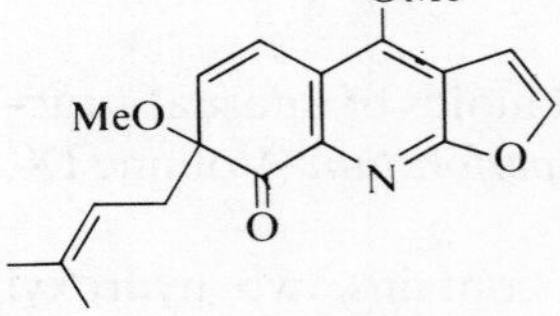

184 Perfamine 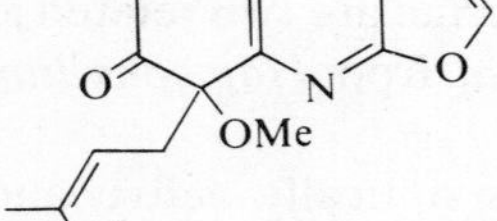 **185** 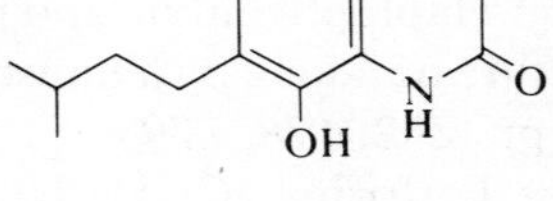 **186**

Perfamine is a furoquinoline alkaloid isolated more recently from *H. perforatum* (*184*). It appears to be related to haplophyllidine. Thus, the NMR spectrum indicates that perfamine is a 4-methoxyfuroquinoline containing as additional substituents a methoxy group, a C-prenyl group, and an oxygen function. Infrared absorption at 1670 cm^{-1} and the UV spectrum show that a conjugated ketonic carbonyl group is present. Resonances in the NMR spectrum at τ6.96 (OMe at saturated carbon) and at 2.10 and 3.88 (–CH=CH–) and the optical activity of the alkaloid are in accord with the proposed structure, **184**, although the isomeric structure **185** apparently was not excluded. Reaction of perfamine with concentrated sulfuric acid and treatment of the product with diazomethane furnished 7,8-dimethoxydictamnine. A recent report (*186*) indicates that reaction of perfamine with acid yields 8-hydroxy-7-methoxydictamnine and that, on hydrogenation, the alkaloid is converted into the 2-quinolone **186**; although full details are not available, the new evidence clearly supports structure **184** for perfamine.

A plausible biosynthetic pathway to perfamine and thence to more reduced alkaloids of the haplophyllidine type involves C-prenylation of a hydroxymethoxydictamnine.

The spectroscopic evidence for the structures of haplophyllidine, perforine, and perfamine seems convincing, but confirmation is desirable. It would be of particular interest to see whether dehydrogenation of the dihydrofuroquinoline **178** yields an isomer of the alkaloid, foliminine (**190**).

5. Heliparvifoline

The biological activity of extracts of *Helietta parvifolia* (A. Gray) Benth. prompted an investigation of the constituents by Farnsworth *et al.* (*180*) and led to the isolation of a new phenolic furoquinoline alkaloid, heliparvifoline. Since reaction with diazomethane gave 6,7-dimethoxydictamnine (kokusaginine), heliparvifoline is either 7-hydroxy-6-methoxydictamnine or 6-hydroxy-7-methoxydictamnine. The structural problem was resolved by NMR spectroscopy; compared with data for kokusaginine (**188**), the chemical shifts of aromatic protons of heliparvifoline in deuterated DMSO (τ2.57

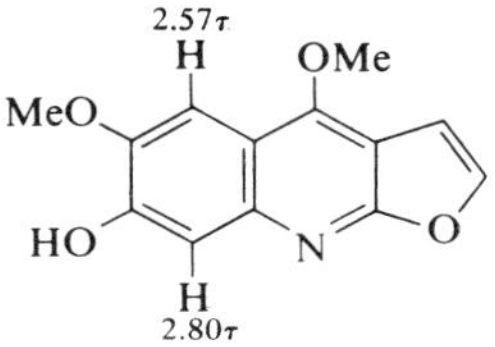

187 Heliparvifoline

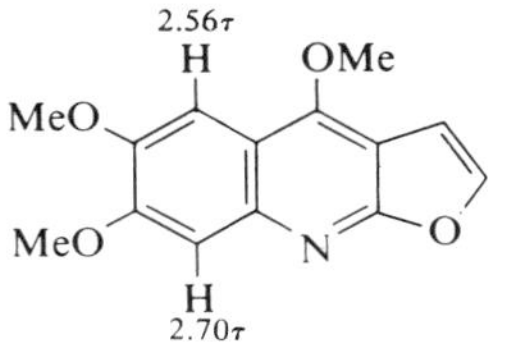

188 Kokusaginine

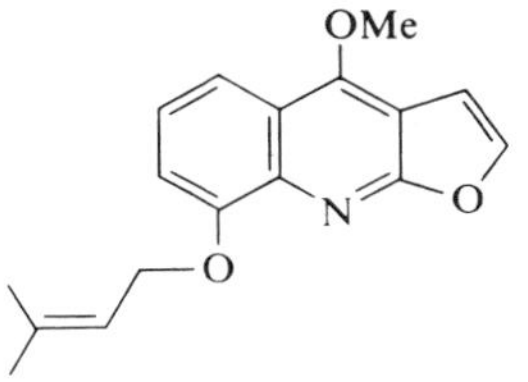

189 Haplophydine

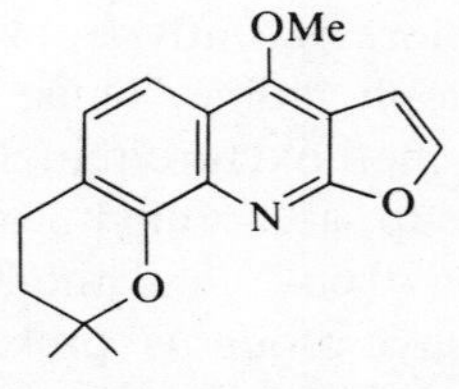

190 Foliminine 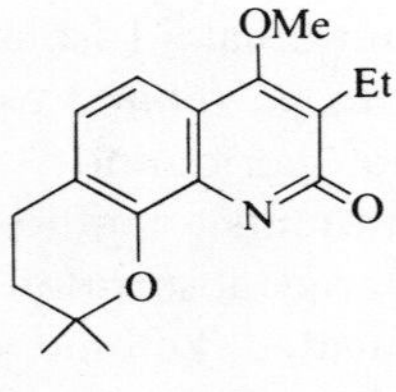 **191** 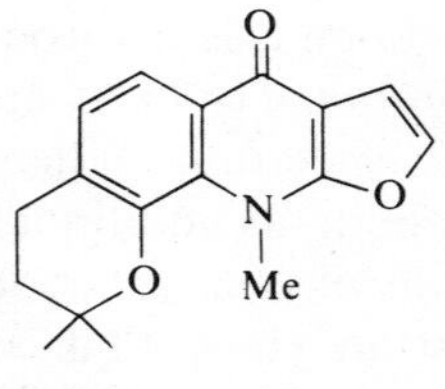 **192**

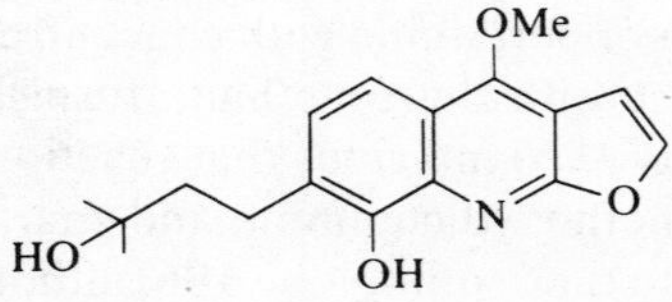

193 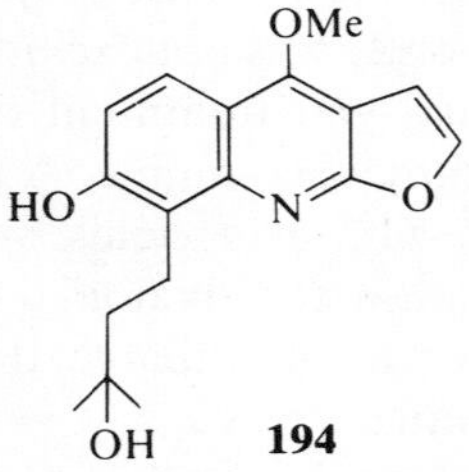 **194**

and 2.80) and in the presence of D_2O–NaOD (τ2.63 and 2.97) showed that the higher-field resonance was due to a C-8 proton adjacent to a phenolic group, as in structure **187**.

6. Haplophydine, Foliminine, and Folifinine

Robustine (8-hydroxydictamnine) is a constituent of several *Haplophyllum* spp. studied intensively by Yunusov and his co-workers, and several alkaloids that apparently are robustine derivatives have also been isolated.

Haplophydine obtained from *H. perforatum* (*177*) was shown by spectroscopy to be 8-(3,3-dimethylallyloxy)dictamnine (**189**), but confirmation, for example by correlation with robustine, is clearly desirable.

Foliminine (**190**) is a constituent of *H. foliosum* and was shown to be a 4-methoxyfuroquinoline by catalytic reduction to a 3-ethyl-2-quinolone, **191**, and by conversion by methyl iodide into isofoliminine (**192**). The structures of the alkaloid and transformation products were established by spectroscopic means (*174*).

In 1968, the Russian group described the isolation from *H. foliosum* of an optically inactive alkaloid, $C_{17}H_{19}O_4N$, named folifinine (*173*). Folifinine was soluble in aqueous base, gave an oily *O*-methyl ether with diazomethane, and furnished a diacetate showing IR absorption at 1735 and 1765 cm^{-1}. On this basis, folifinine was assigned alternative structures **193** or **194**. Although no further work has been reported, the subsequent isolation of foliminine, the notional cyclization product of folifinine, from the same species would seem to favor the 8-hydroxy structure; it is possible that foliminine **190** is an artifact derived from folifinine.

7. Acronidine, Medicosmine, and Choisyine

An earlier investigation indicated that the furoquinoline alkaloids acronidine and medicosmine contained dimethylpyrano rings, but it was not possible to differentiate between the angular structures **195** and **197** for acronidine or between **196** and **198** for medicosmine. This problem has now been resolved in two ways. Johns, Lamberton, and Sioumis (*63*) compared the NMR spectrum of acronidine with that of the 4-quinolone, isoacronidine. The resonance at 2.45τ (C-4′–H) was shifted downfield to 1.58τ in isoacronidine because of the proximity of the carbonyl group in **199**; this effect would not be expected in the alternative formula based on structure **197** for acronidine. An upfield shift of 0.77 ppm in the C-8–H resonance of isoacronidine also accords with structures **195** and **199** for the two compounds. Similarly, medicosmine was shown to possess structure **196**.

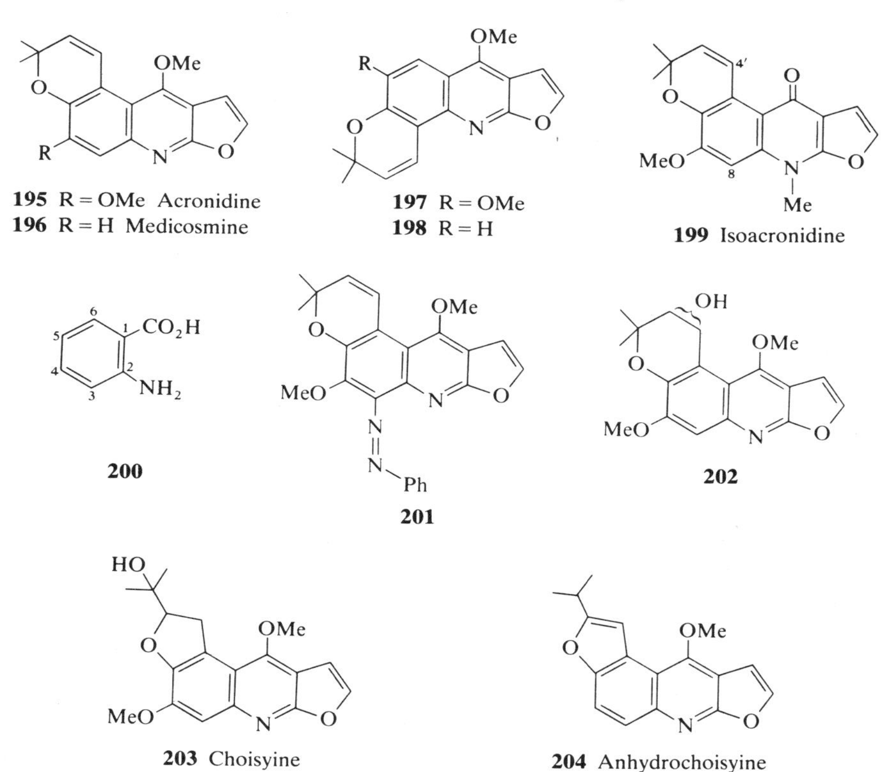

195 R = OMe Acronidine
196 R = H Medicosmine

197 R = OMe
198 R = H

199 Isoacronidine

200

201

202

203 Choisyine

204 Anhydrochoisyine

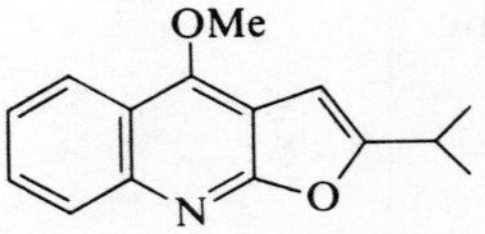

205 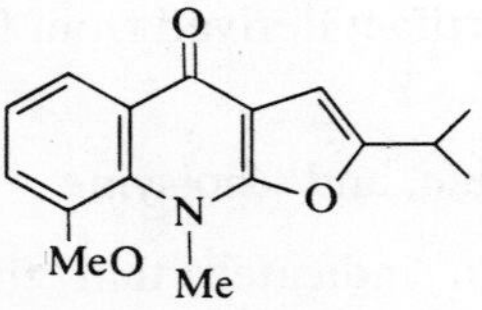 206 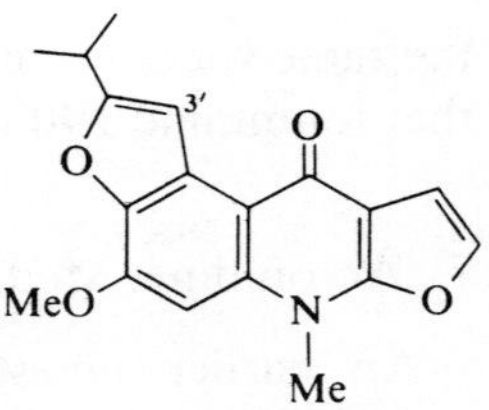 207

Prager and Skurray (*187*) confirmed the structure of acronidine by an ingenious application of tracer feeding methods. Anthranilic acid was shown to be a precursor of acronidine in *Acronychia baueri* Schott., but when [6-^{3}H]anthranilic acid (**200**) was fed no incorporation of tritium into acronidine was observed; this result is consistent with structure **195** but rules out the alternative angular arrangement **197**. As expected, application of [3,6-$^3H_2$]anthranilic acid gave acronidine containing tritium at C-8, which was lost on conversion of the alkaloid into the azo derivative **201**.

The furoquinoline alkaloid choisyine, first isolated from *Choisya ternata* (Volume IX, p. 233), has now been shown to be a constituent of both *C. arizonica* Standl. (*128*) and *C. mollis* Standl. (*128*). Choisyine was assigned partial structure **202** because reaction with concentrated sulfuric acid gave anhydrochoisyine, which was thought to be identical with acronidine (**195**) on the basis of a correspondence of melting point. Johns, Lamberton, and Sioumis (*63*) showed that the two compounds were not identical. NMR and mass spectrometry indicated that choisyine was the hydroxyisopropyldihydrofuran derivative **203**. Anhydrochoisyine is then the isopropylfuran **204**, analogous to the dehydration products **205** and **206** derived from platydesmine and from balfourodine, respectively. As in the case of acronidine, the downfield shift of the 3′-H in the NMR spectra of isoanhydrochoisyine (**207**) compared to anhydrochoisine (**204**) confirmed the arrangement of rings in choisyine and its derivatives.

8. Halfordinine

Halfordinine, $C_{15}H_{15}NO_5$, a minor component of the bark of *Halfordia scleroxyla*, was isolated by Crow and Hodgkin (*22*). There was insufficient material for a full structural investigation, but the alkaloid was shown to be a trimethoxydictamnine on the basis of the NMR spectrum. Halfordinine (mp 150–151°) clearly was not identical with 5,7,8-trimethoxydictamnine (acronycidine) (mp 136–137°). Halfordinine has been obtained recently from the leaves of *Melicope perspicuinervia* Mer. and Perry. Murphy, Ritchie, and Taylor (*42*) noted that the NMR signal at $\tau 2.69$ indicated that the lone aromatic proton was not flanked by methoxyl groups and hence was at C-5 or at C-8. Hydrogenation of halfordinine followed by acid hydrolysis gave the 4-hydroxy-2-quinolone derivative **209**, identified by synthesis from 2,3,4-trimethoxyaniline and diethyl ethylmalonate. Halfordinine is therefore 6,7,8-trimethoxydictamnine (**208**). Halfordinine and kokusaginine did not show the expected nuclear Overhauser effects in the NMR spectra, irradiation of the C-4–OMe signals resulting in enhancement of the C-3–H resonances rather than the C-5–H signals.

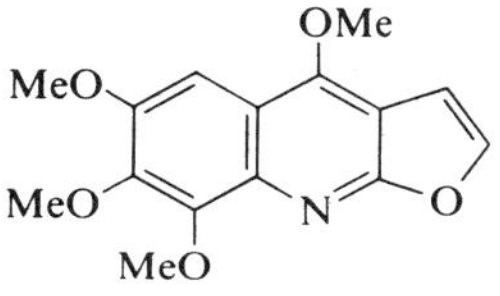

208 Halfordinine

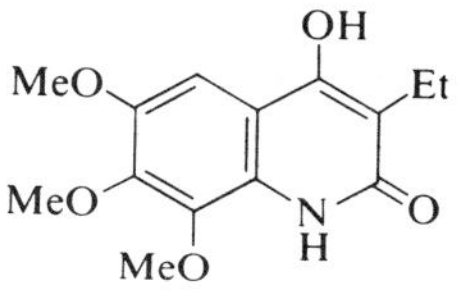

209

B. Synthesis

The most important advances in synthesis of furoquinoline alkaloids during the last decade involve modification of the Tuppy and Böhm synthesis (Volume VII, p. 237) and the use of lithiated 2,4-dimethoxyquinoline derivatives.

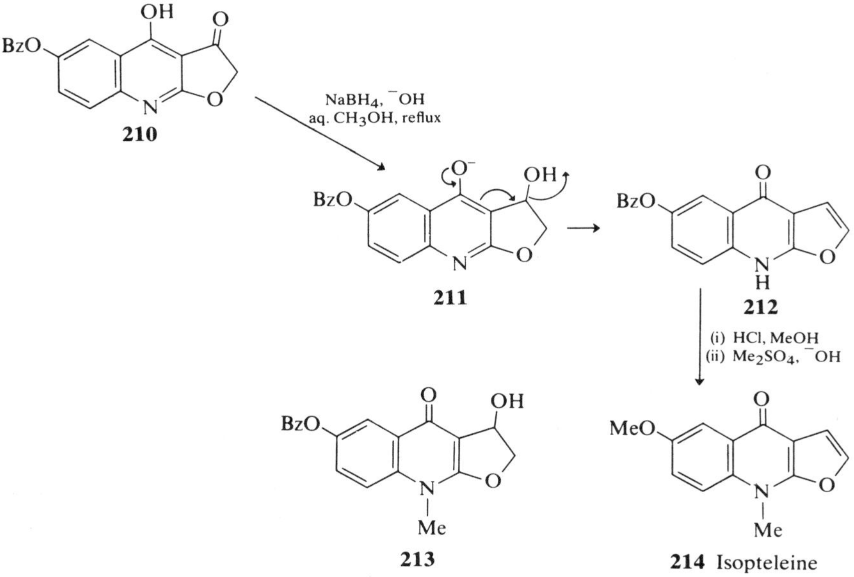

210

$NaBH_4$, ^{-}OH
aq. CH_3OH, reflux

211

212

(i) HCl, MeOH
(ii) Me_2SO_4, ^{-}OH

213

214 Isopteleine

SCHEME 15. *Synthesis of isopteleine.*

1. Modified Tuppy–Böhm Procedure

During a projected synthesis of medicosmine (**221**), Govindachari *et al.* (*188*) prepared ketone **210** by the Tuppy–Böhm procedure, but conversion with diazomethane into the corresponding 4-methoxyketone proceeded in low yield. Unexpectedly, reaction of ketone **210** with an excess of sodium borohydride in refluxing aqueous sodium hydroxide gave the furoquinolone **212**, probably via alcohol **211**. The anion **211** probably participates in the elimination reaction as shown, for the *N*-methylquinolone alcohol **213** is stable to base. Debenzylation with methanolic hydrochloric acid then gave 6-hydroxynordictamnine, which was converted into 6-methoxyisodictamnine (**214**) on methylation with dimethyl sulfate (Scheme 15). 6-Methoxyisodictamnine (isopteleine) was identical with a degradation product of normedicosmine and has since been identified as a natural product (see Table VII).

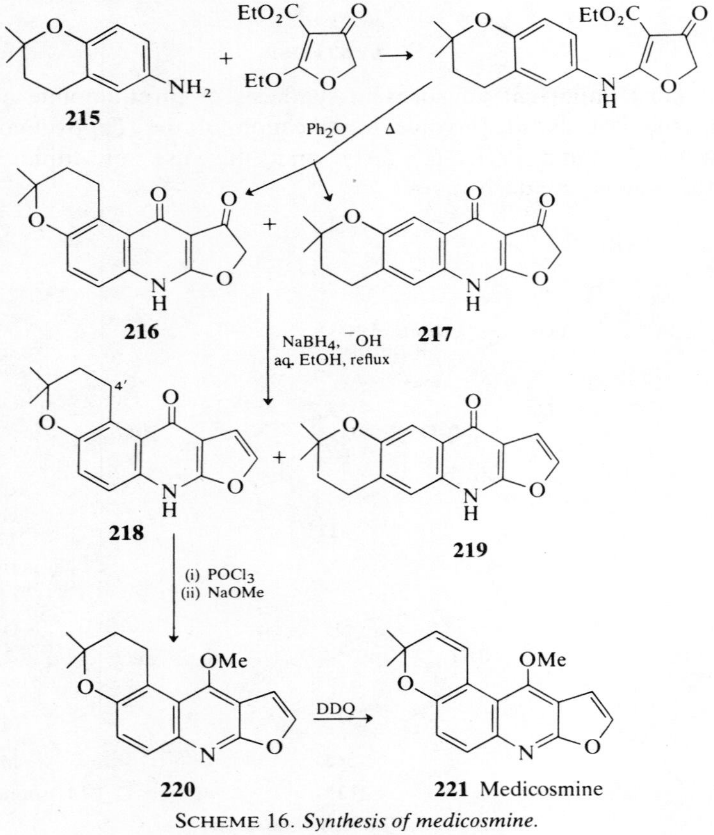

EtO_2C O NH$_2$ EtO O 215 + → EtO_2C O N H O Ph$_2$O Δ

216 + 217 NaBH$_4$, $^-$OH aq. EtOH, reflux 4′ 218 + 219 (i) POCl$_3$ (ii) NaOMe OMe 220 DDQ OMe 221 Medicosmine

SCHEME 16. *Synthesis of medicosmine.*

The borohydride reaction was applied to the inseparable mixture of ketones **216** and **217** derived from the aminochroman **215**, and the result was the isolation of the angular and linear compounds **218** and **219**, respectively. The two products were distinguished by NMR spectroscopy, the angular compound showing ortho-coupled aromatic protons at $\tau 2.5$ and 2.9 ($J = 9$ Hz). The methylene group at C-4′ in angular compound **218** appears as a two-proton triplet at $\tau 6.38$, deshielded by 0.62 ppm compared with linear compound **219** because of the adjacent 4-quinolone carbonyl group. The quinolone **218** was converted into dihydromedicosmine (**220**) by standard procedures; dehydrogenation with DDQ then furnished the alkaloid, medicosmine (**221**) (*188*) (Scheme 16).

The modified Tuppy–Böhm synthesis was also applied to the synthesis of the 4-quinolones **222–226**. The 7-methoxy derivative **223** reacted with isopentenyl bromide to give the alkaloid acrophylline **227** (*189*). In the first synthesis of a phenolic furoquinoline alkaloid, 7-benzyloxydictamnine, obtained from the 4-quinolone **224** in the usual way, was converted with ethanolic hydrochloric acid into confusameline (**228**) (*189*). Two other phenolic alkaloids, robustine (**229**) and haplopine (**230**), were similarly synthesized (*190*). Acronycidine (**231**) was prepared from ketone **232** (*191*), but in this case borohydride reduction of the ketone gave two products, 5,7,8-trimethoxynordictamnine (**233**) and its dihydro derivative **234**; the latter was converted into the *N*-methylquinolone **235** that had been obtained previously from dihydroacronycidine. 6-Methoxydictamnine (pteleine) was synthesized by treatment of 6-methoxynordictamnine (**222**) with diazomethane and separation by chromatography from 6-methoxyisodictamnine (**214**) (*188*).

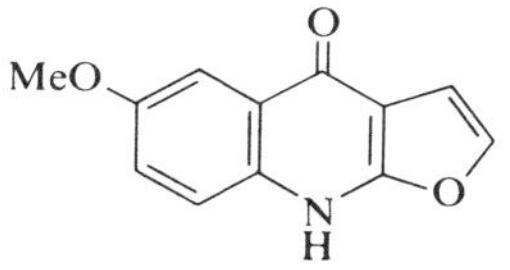

222

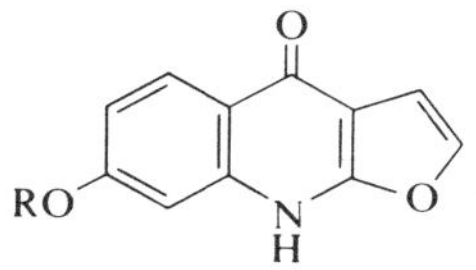

223 R = Me
224 R = CH_2Ph

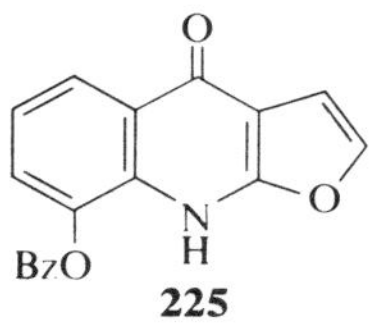

225

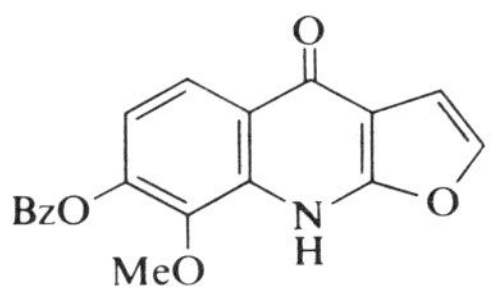

226

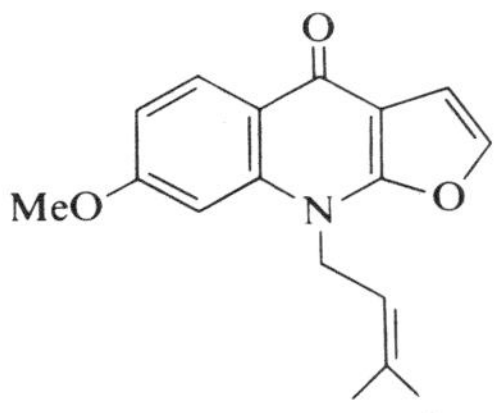

227 Acrophylline

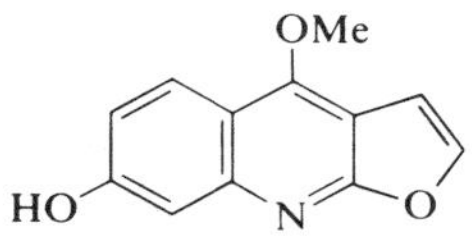

228 Confusameline

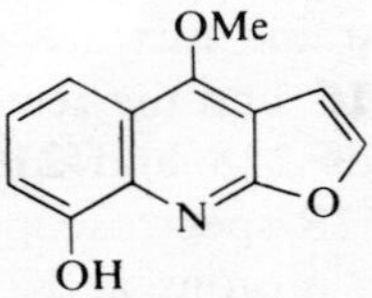

229 Robustine

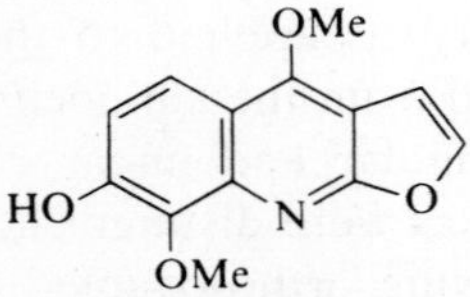

230 Haplopine

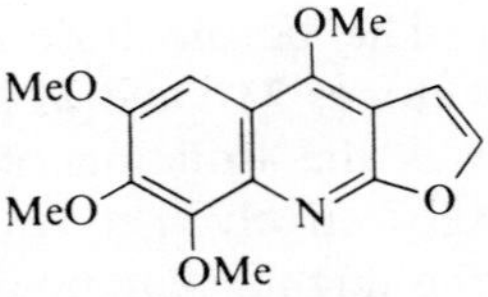

231 Acronycidine

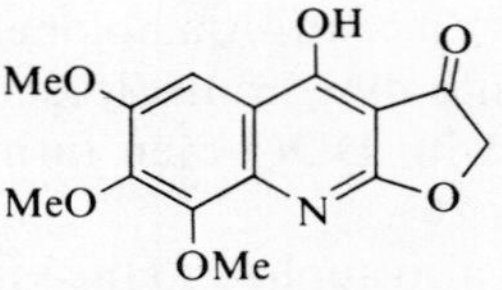

232

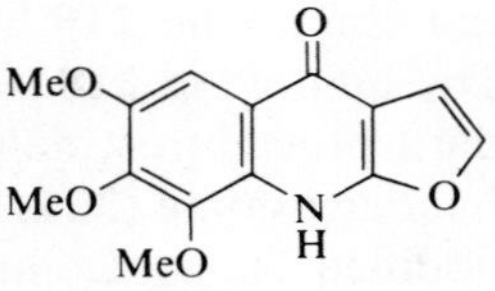

233

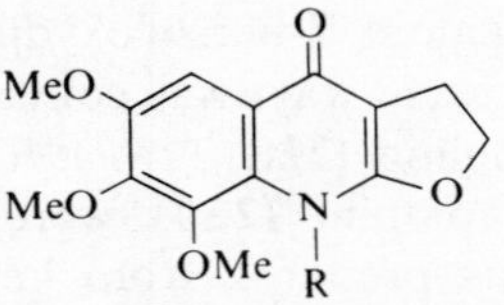

234 R = H
235 R = Me

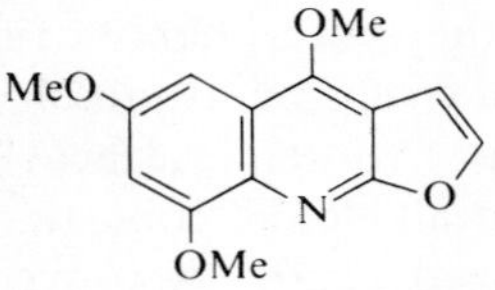

236

It should be noted that one advantage of the new method (and those to be described under Sections IV,B,3 and IV,B,4) is that they avoid the need to dehydrogenate a dihydrofuroquinoline. This process is accomplished in certain cases by bromination–dehydrobromination but sometimes fails (Volume IX, p. 232). The method of choice is the reaction of the dihydro derivatives with DDQ. Piozzi, Venturella, and Bellino (*192*) first used this reagent in the quinoline series to convert dihydrodictamnine and its angular derivative into the corresponding furoquinolines; maculosidine (**236**), evolitrine, γ-fagarine, and pteleine were prepared similarly from their dihydro derivatives (*193*).

2. Cyclization of Hydroxyethyl Derivatives

Narasimhan *et al.* (*194*) instituted a new approach to the synthesis of furoquinoline alkaloids by introducing a C_2 side chain into a preformed quinoline nucleus. Lithiation of 2,4-dimethoxyquinoline, 2,4,6-trimethoxyquinoline, and 2,4,8-trimethoxyquinoline with *n*-butyllithium in ether and subsequent reaction with ethylene oxide furnished hydroxyethyl derivatives **237**, which with 20% hydrochloric acid were converted into dihydrodictamnine (**239**), dihydropteleine (**240**), and dihydro-γ-fagarine

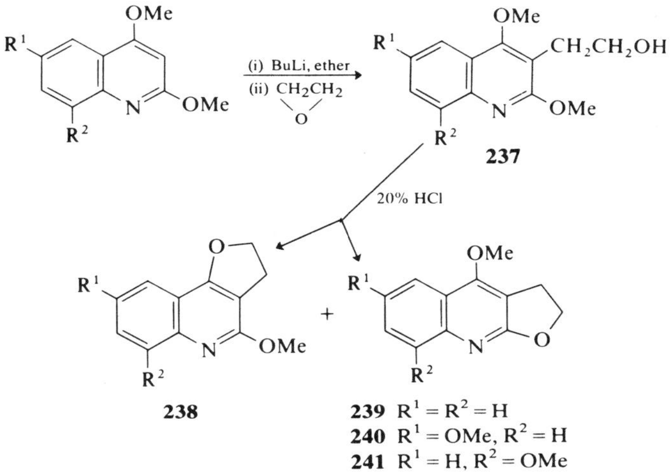

(i) BuLi, ether
(ii) CH_2CH_2 O

237

20% HCl

238

239 $R^1 = R^2 = H$
240 $R^1 = OMe$, $R^2 = H$
241 $R^1 = H$, $R^2 = OMe$

SCHEME 17. *Synthesis of dihydrofuroquinolines by alkylation with ethylene oxide.*

(**241**), respectively (Scheme 17). The overall yields were not high in the synthesis, principally because the angular compounds **238** were the major products of cyclization.

3. Synthesis from 3-Prenylquinolines and Rearrangement of Furoquinolines

Furoquinoline alkaloids specifically labeled with ^{14}C in the furan ring were required for biosynthetic studies (see Section VII of this chapter), and since existing routes produced low yields, Grundon and co-workers (*195, 196*) developed a more efficient synthesis from 4-methoxy-3-prenyl-2-quinolones **244–246**. These compounds had previously been prepared from aromatic amines and substituted malonates, but direct allylation is more suitable for the preparation of labeled compounds, employing, for example, [^{14}C]-3,3-dimethylallyl bromide. It was found that reaction of 2,4-dimethoxyquinolines with *n*-butyllithium and the allylic bromide gave the prenyl derivatives **242** and **243** in high yield; these were converted readily into the required 2-quinolones with dry hydrogen chloride. Oxidative cleavage of prenylquinolones **244** and **246** was carried out by reaction with osmium tetroxide–periodate to give aldehydes **247** and **248**, cyclized by heating with polyphosphoric acid to dictamnine and skimmianine, respectively (Scheme 18). The overall yield of dictamnine from 2,4-dimethoxyquinoline was 50%, a clear improvement on previous methods.

Oxidation of the dimethoxyquinolone **245** was unsatisfactory, but ozonolysis of the trimethoxyquinolone **243** and cyclization of the resultant aldehyde **249** gave γ-fagarine.

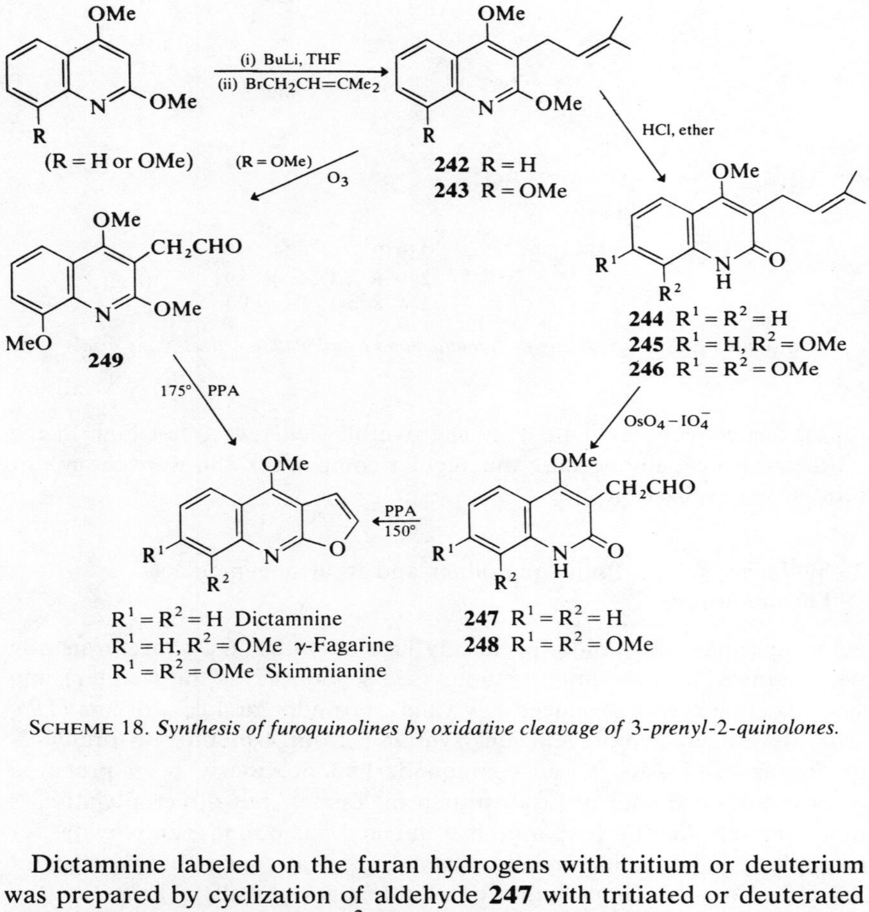

(R = H or OMe)

(i) BuLi, THF
(ii) $BrCH_2CH{=}CMe_2$

242 R = H
243 R = OMe

HCl, ether

244 $R^1 = R^2 = H$
245 $R^1 = H, R^2 = OMe$
246 $R^1 = R^2 = OMe$

(R = OMe) O_3

249

175° PPA

$OsO_4-IO_4^-$

PPA 150°

$R^1 = R^2 = H$ Dictamnine
$R^1 = H, R^2 = OMe$ γ-Fagarine
$R^1 = R^2 = OMe$ Skimmianine

247 $R^1 = R^2 = H$
248 $R^1 = R^2 = OMe$

SCHEME 18. *Synthesis of furoquinolines by oxidative cleavage of* 3-*prenyl*-2-*quinolones.*

Dictamnine labeled on the furan hydrogens with tritium or deuterium was prepared by cyclization of aldehyde **247** with tritiated or deuterated PPA (*3*). Treatment of [2,3-$^2H_2$]dictamnine with aqueous acid did not result in exchange of deuterium atoms. In the formation of [2,3-$^2H_2$]dictamnine, exchange of deuterium between the 2- and 3-positions of the furan ring may occur by a 1,2-shift, (**251**) $\rightleftarrows$ (**252**) (Scheme 19), although an alternative mechanism for exchange of the aldehyde proton involves addition of D^+ to the enol form of aldehyde **250** assisted by the 4-methoxy group.

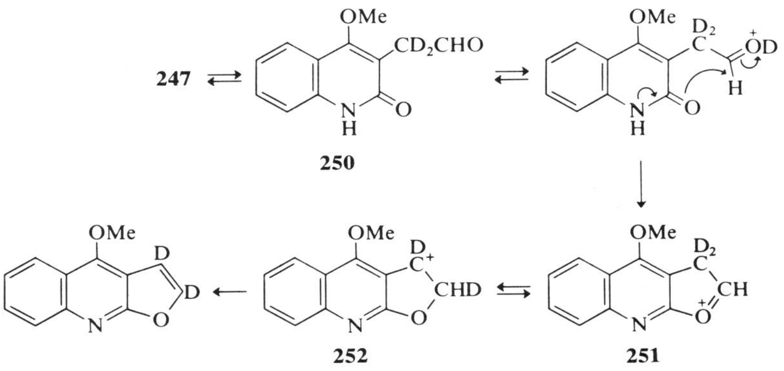

SCHEME 19. *Synthesis of* [2,3-$^2H_2$]*dictamnine.*

The angular isomer **254** which was obtained previously (Volume VII, pp. 235, 237) was a minor product (5%) of the cyclization of aldehyde **247**. Heating dictamnine with PPA at 180° resulted in partial conversion (20%) to the angular 2-quinoline, indicating that to obtain maximum yields of furoquinoline alkaloids in the aldehyde cyclization reaction temperature and reaction time must be carefully controlled. The mechanism of the dictamnine rearrangement may involve initial cleavage of the protonated vinyl ether **253** followed by formation of aldehyde **255** and cyclization to the more stable angular derivative (Scheme 20) (*196*).

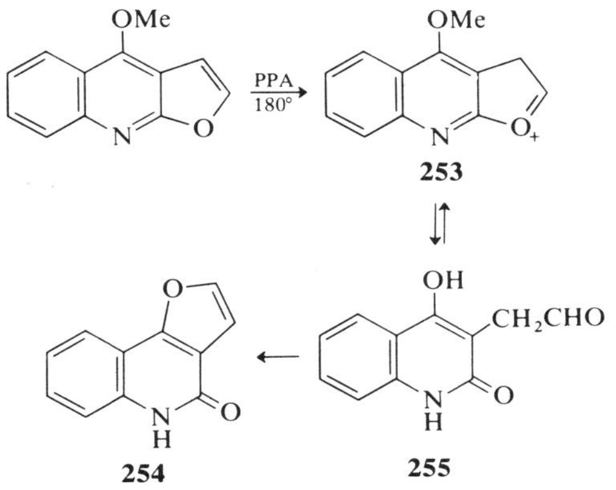

SCHEME 20. *Rearrangement of dictamnine.*

Cyclization of the *N*-methylaldehyde **257**, obtained by periodate oxidation of edulinine (**256**), gave isodictamnine (**258**) (47%) and its angular isomer (**259**) (29%); isodictamnine also undergoes acid-catalyzed rearrangement to the angular 2-quinolone (*196*).

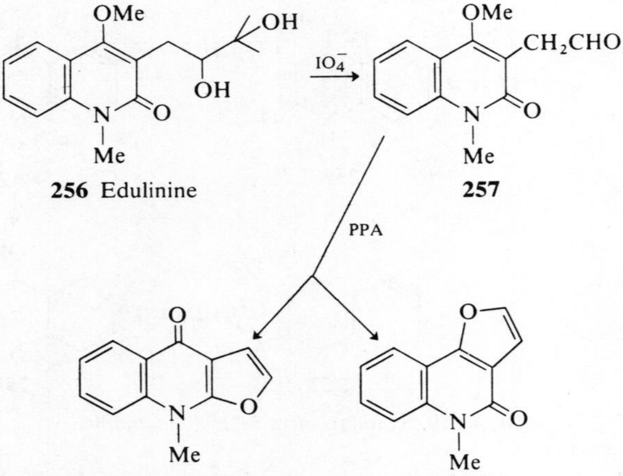

OMe OH OH N O Me **256** Edulinine $\xrightarrow{IO_4^-}$ OMe CH_2CHO N O Me **257**

PPA

O N O Me **258** Isodictamnine   O N O Me **259**

4. Synthesis from 3-Formylquinolines

Another high-yield synthesis of furoquinoline alkaloids, devised by Narasimhan and Mali (*197*), also involves the preparation of aldehydes **247** and their cyclization with polyphosphoric acid, and is illustrated in Scheme 21 for dictamnine. In this route, treatment of lithiated 2,4-dimethoxyquinolines with *N*-methylformanilide gave 3-formylquinolines **260**. Wittig reactions with methoxymethylene triphenylphosphonium chloride and *tert*-butoxide led to a mixture of *cis*- and *trans*-enol ethers, which were hydrolyzed to the 2-quinolone aldehydes. Dictamnine, evolitrine, pteleine, and γ-fagarine were synthesized by this method.

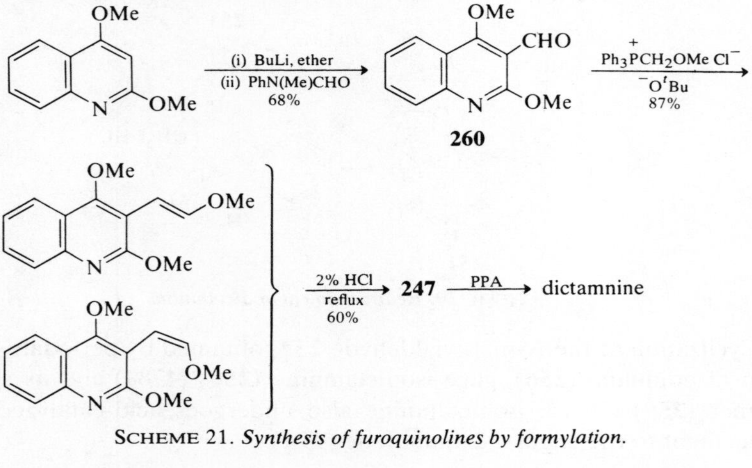

OMe N OMe $\xrightarrow[\text{(ii) PhN(Me)CHO, 68\%}]{\text{(i) BuLi, ether}}$ OMe CHO N OMe **260** $\xrightarrow[{}^-O^tBu,\ 87\%]{Ph_3\overset{+}{P}CH_2OMe\ Cl^-}$

OMe OMe N OMe; OMe N OMe OMe $\xrightarrow[\text{reflux, 60\%}]{2\%\ HCl}$ **247** $\xrightarrow{PPA}$ dictamnine

SCHEME 21. *Synthesis of furoquinolines by formylation.*

V. 1,1- and 1,2-Dimethylallyl Alkaloids

A. Introduction

Quinoline alkaloids containing a C_5 side chain at the 3-position of the quinoline ring typically have a potential 3,3-dimethylallyl structure, but in 1963 a dihydrofuroquinoline, ifflaiamine, was isolated from *Flindersia ifflaiana* F. Muell which proved to be the first example of a quinoline alkaloid containing the alternative 1,1-dimethylallyl arrangement (Volume IX). Chamberlain and Grundon synthesized ifflaiamine and reexamined the alkaloids of *F. ifflaiana*, while Bose and Paul and Talapatra and his colleagues studied the constituents of *Ravenia spectabilis* and identified quinoline alkaloids related to ifflaiamine. In this account the preparative work will be described first, as three of the alkaloids of the group were synthesized before they were isolated.

B. Synthesis of Ifflaiamine, Ravenine, Ravenoline, and Lemobiline

Chamberlain and Grundon (*198*) planned to synthesize ifflaiamine (**264**) by Claisen rearrangement of the prenyl ether **261** followed by cyclization of the 1,1-dimethylallyl derivative **263**, but they encountered unexpected difficulties (Scheme 22). The ether **261** (ravenine), prepared from 4-hydroxy-1-methyl-2-quinolone and 3,3-dimethylallyl bromide, rearranged readily when heated at 130–140° or when refluxed in *N*-methylpiperidine, but "normal" Claisen rearrangement products were not obtained; the three isomeric compounds that were isolated were the 1,2-dimethylallyl derivatives, **265**, **267**, and **268**. Apparently, the products were formed by an "abnormal" Claisen rearrangement to olefin **265** followed by alternative cyclization in which the 2- or 4-oxygen substituents participated. The acidic 4-hydroxyl group is necessary for rearrangement of normal to abnormal rearrangement product, **263** → **265**, and in the presence of sodium carbonate this rearrangement was prevented, but the angular isomer of ifflaiamine (**266**) was the product obtained. Ifflaiamine was successfully synthesized by trapping the normal Claisen product as acetate **262**, which was cyclized to ifflaiamine with hydrogen bromide. Compound **265** (ravenoline) is readily prepared in 80% yield by reaction of ravenine (**261**) in boiling acetic anhydride and subsequent hydrolysis of ravenoline acetate (*84*).

In the synthetic reactions, isomers were separated by means of their different properties; for example, the 4-hydroxy derivative **265** is acidic, the 4-quinolone **267** is basic, and the 2-quinolone **268** is neutral. NMR

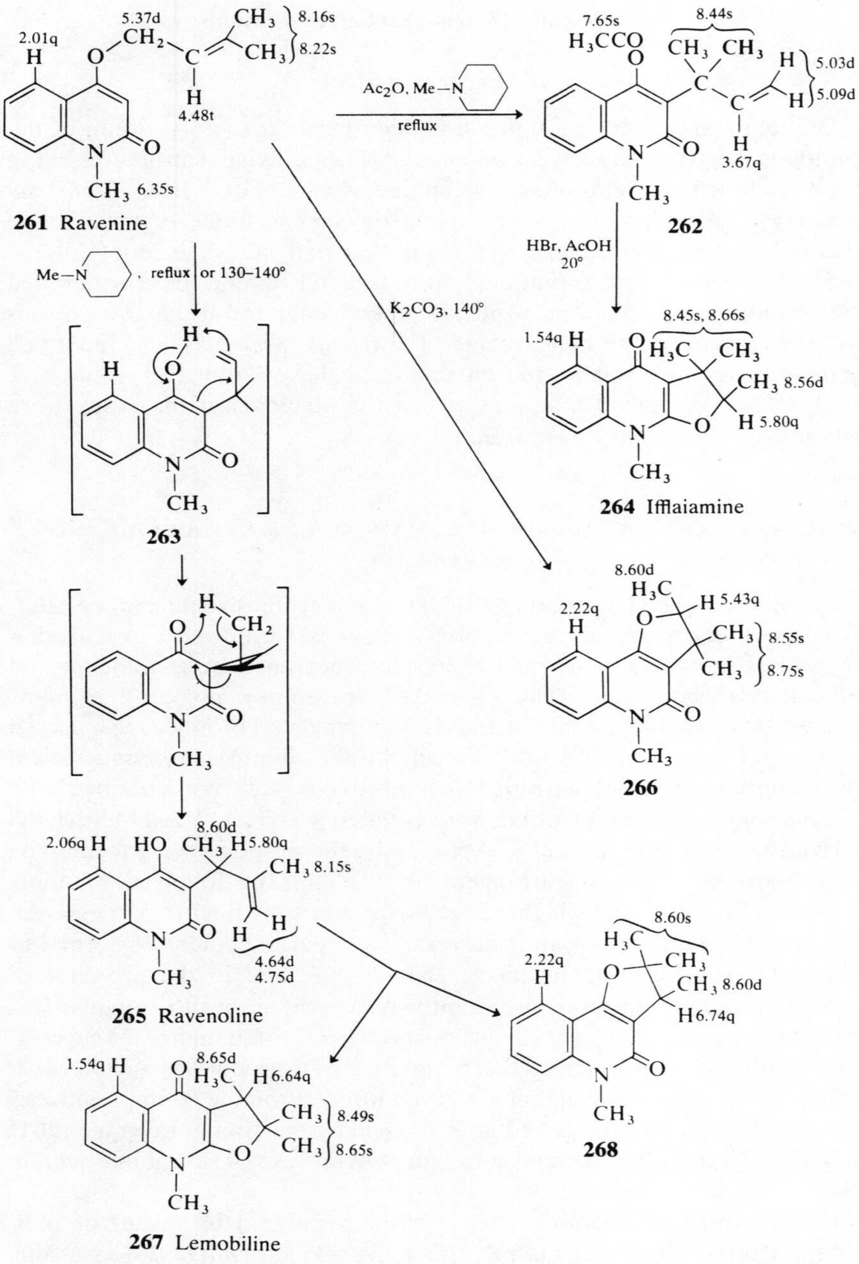

261 Ravenine

262

263

264 Ifflaiamine

265 Ravenoline

266

267 Lemobiline

268

SCHEME 22. *Synthesis of ravenine, ravenoline, ifflaiamine, and lemobiline.*

spectroscopy was particularly useful in establishing the structures of the compounds, and relevant chemical shifts are included in Scheme 22; 4-quinolones are distinguished by low-field resonances of C-5 protons, and the downfield shift of C–H quartets adjacent to heterocyclic oxygen in "normal" Claisen products **264** and **266** compared to C–H signals in "abnormal" compounds **267** and **268** is also noteworthy.

C. ALKALOIDS OF *FLINDERSIA IFFLAIANA* F. MUELL AND *RAVENIA SPECTABILIS* ENGL.

1. Structural Studies

In order to compare synthetic ifflaiamine with the natural alkaloid, Chamberlain and Grundon (*198, 199*) reexamined a crude alkaloid extract derived from the wood of *Flindersia ifflaiana* and obtained two isomeric alkaloids. Ifflaiamine (**264**) {mp 122–125°; $[\alpha]_D -6.2°$ in methanol and $-9.15°$ in chloroform; picrate, mp 189–192°} had properties different from those reported earlier (Volume IX). The second alkaloid (mp 53–54°; picrate, mp 188–191°) was shown to be the 1,2-dimethylallyl derivative **267** by comparison with the synthetic racemate (mp 106–108°; picrate, mp 186–190°). The alkaloid was levorotatory, but too little was available to obtain an accurate value of the specific rotation.

Paul and Bose (*28*) examined the alkaloids of *Ravenia spectabilis* (*Lemonia spectabilis* Lindl.), and by NMR spectroscopy they identified the 4-prenyloxyquinolone **261** (ravenine) and the 1,2-dimethylallyl derivative **265** {mp 144°; $[\alpha]_D +6°$ ($CHCl_3$)}, to which the name ravenoline was given.

Talapatra and co-workers (*200*) also studied the *Ravenia* alkaloids and recorded the isolation of (−)-ravenoline {mp 98–99°; $[\alpha]_D -5 \pm 1°$ ($CHCl_3$)} rather than the dextrorotatory enantiomer obtained by Paul and Bose. This discrepancy has not yet been resolved. Another alkaloid isomeric with ravenoline was first called spectabiline, but to avoid confusion with another compound of the same name this was later changed to lemobiline. (−)-Lemobiline (**267**) (mp 198°; picrate, mp 198°) is identical with the alkaloid isolated from *F. ifflaiana*. Lemobiline and ifflaiamine retain water of crystallization tenaciously, and this probably accounts for the widely different melting points recorded for the alkaloids.

2. Reactions of Ravenoline and Lemobiline

Ravenoline (**265**) is converted into lemobiline (**267**) almost quantitatively when heated with 6 *N* hydrochloric acid at 95° for 1 hr. The reverse process occurs on prolonged reaction with base, (−)-lemobiline giving a mixture of (−)-ravenoline and the tertiary alcohol **269**; because the products are unaffected by base it was suggested that the hydrolysis of lemobiline occurs by means of competitive reactions, as illustrated (Scheme 23) (*200*).

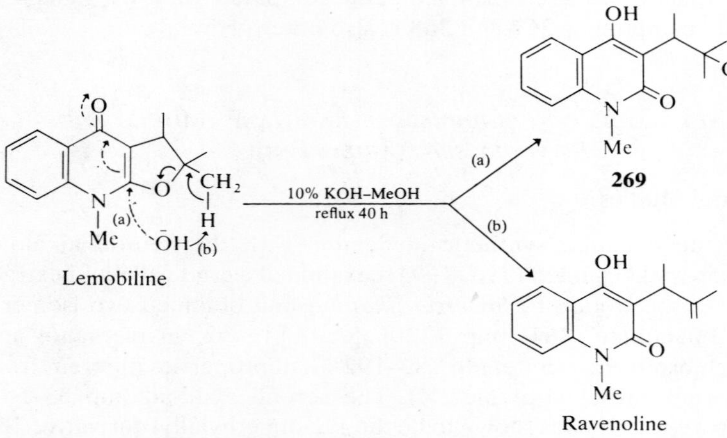

SCHEME 23. *Reaction of lemobiline with base.*

The 4-methoxyquinolone **270**, prepared from ravenoline with diazomethane, was ozonized to give the ketone **271** (*84*). Reaction of the ketone with PPA afforded the dimethylfuroquinolone **272** presumably via the hemiketal **273**.

Grundon and James (*95*) showed that reaction of (±)-lemobiline with *N*-bromosuccinimide gave 6-bromolemobiline (**274**) and the two vinyl

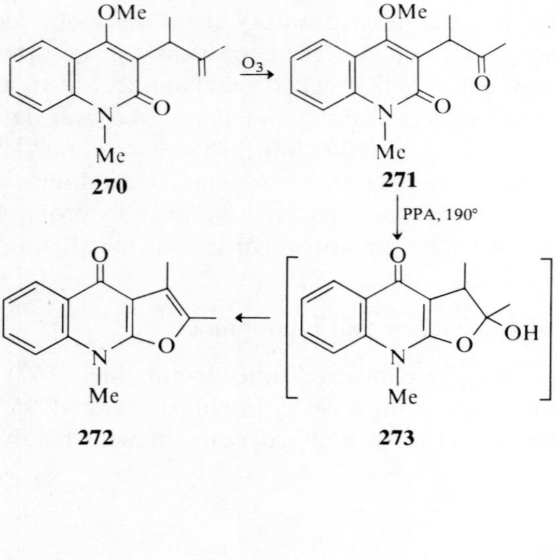

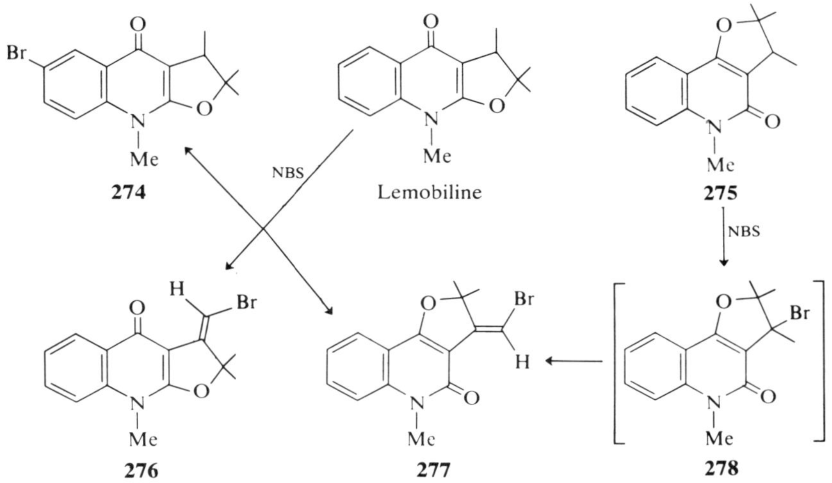

274 Lemobiline 275 NBS NBS 276 277 278

bromides **276** and **277**. The angular isomer **275** of lemobiline, named ψ-lemobiline by analogy with the angular isomers of balfourodine and isobalfourodine (Section III,F), is readily prepared by pyrolysis of ravenine (**261**) and with *N*-bromosuccinimide gave a single product **277**. This compound was formulated as the *Z*-stereoisomer, because the NMR signal at τ2.59 indicates that the olefinic proton is deshielded by the adjacent carbonyl group. The bromination of lemobiline and of ψ-lemobiline thus follows a similar course to that of dihydroflindersine (Section III,G,3) and an analogous mechanism can be written. It appears from an inspection of molecular models that the bromide **278** initially formed is likely, as a result of molecular overcrowding, to undergo rapid elimination.

D. Bucharaine, Bucharidine, and Bucharamine

These unique alkaloids were obtained by Yunusov and co-workers from *Haplophyllum bucharicum* and appear to be derived from geranyl pyrophosphate. Although all structures have not been established, the alkaloids seem to be geranyl analogs of the *Ravenia* bases and are therefore considered in this section.

Bucharaine ($C_{19}H_{25}NO_4$; mp 151–152°) gave a mono-*O*-acetate, an *N*-methyl derivative, and a dibromide, and with platinum and hydrogen it furnished 4-hydroxy-2-quinolone and an oil, $C_{10}H_{22}O_2$. On this basis bucharaine was assigned partial structure **279**, which was also in accord with the IR and UV spectra; the alkaloid was believed to contain a double bond and a secondary and a tertiary hydroxyl function (*201*). The alkaloid

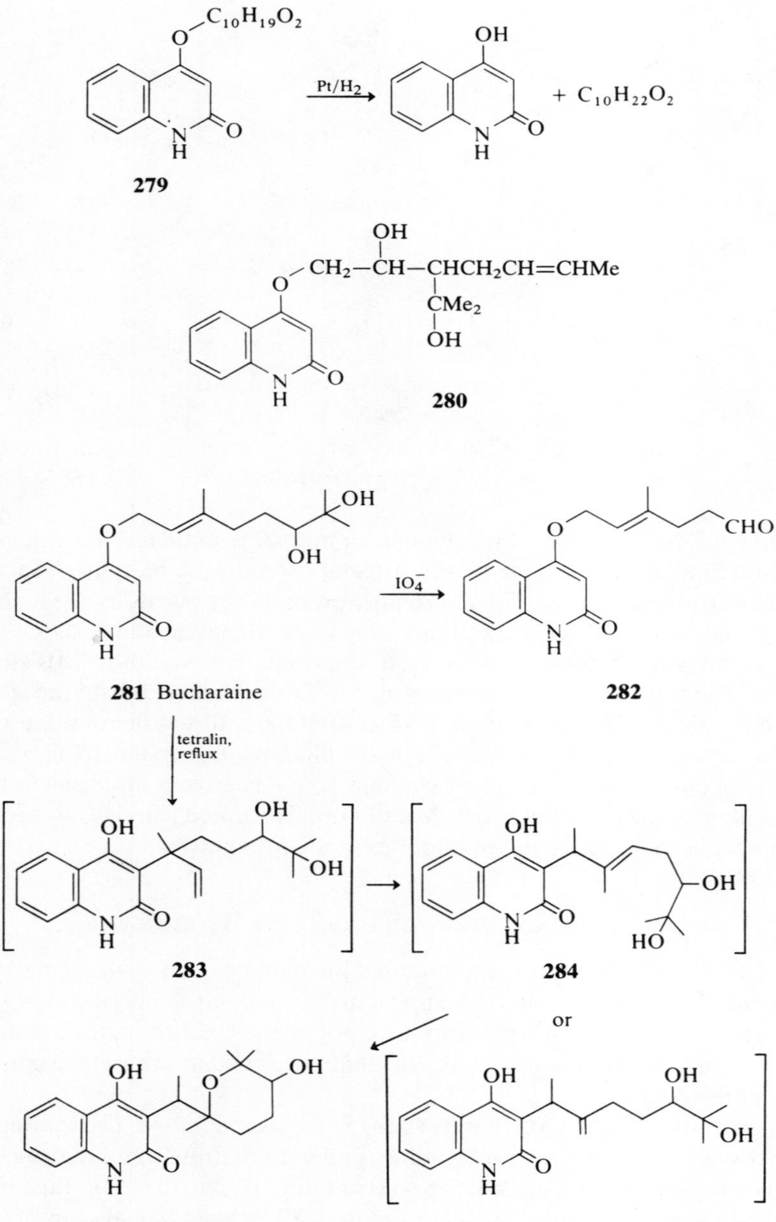

$C_{10}H_{19}O_2$ — Pt/H_2 → OH + $C_{10}H_{22}O_2$

279

OH

CH_2—CH—CHCH$_2$CH=CHMe

CMe$_2$

OH

280

281 Bucharaine — IO_4^- → 282 (CHO)

tetralin, reflux

283 → 284

or

285 Bucharidine

SCHEME 24. *Structures of bucharaine and bucharidine.*

was later given structure **280** (*202*), because ozonolysis was reported to yield bucharainic acid ($C_{17}H_{21}NO_6$; mp 288–289°) and acetaldehyde; in the NMR spectrum, a doublet at τ5.4 and a quartet at 6.41 were attributed, respectively, to the methylene and methine protons of a Ar–O–CH_2–CH(OH) group. The revised structure **281** was then proposed, mainly as a result of mass spectral studies and reaction with periodate to give bucharainal (**282**) (*203*). The new structure is certainly more acceptable in biosynthetic terms and allows interpretation of catalytic reduction as proceeding through allylic hydrogenolysis and subsequent reduction of the C_{10} moiety (see Scheme 24). The reported formation of acetaldehyde in ozonolysis of bucharaine has yet to be explained.

Bucharaine is reported to undergo rearrangement in the mass spectrometer, giving the Claisen product **283** at 60° and 40 eV and the abnormal Claisen rearrangement product **284** at higher temperature (*204*). On the other hand, the acetonide of bucharaine is believed to be transformed in the mass spectrometer into the abnormal product (acetonide of **284**) and its linear and angular cyclization products at 90–110° and into the cyclization products of the normal Claisen rearrangement product at higher temperatures (*205*).

The structure of bucharaine has now been confirmed by synthesis (Scheme 25). The geranyl ether **286** was prepared from 4-hydroxy-2-quinolone and geranyl chloride. Selective epoxidation with *m*-chloroperbenzoic acid yielded monoepoxide **287**. Reaction of the epoxide with formic acid gave the formate ester **288**, which on treatment with base was converted into bucharaine (*206*).

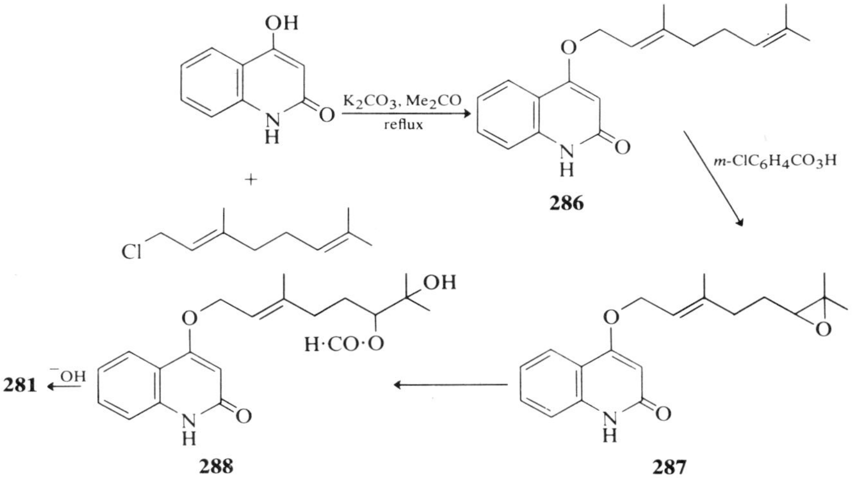

OH
K_2CO_3, Me_2CO
reflux
m-$ClC_6H_4CO_3H$
286
Cl
OH
H·CO·O
281 $\xleftarrow{^{-}OH}$
288
287

SCHEME 25. *Synthesis of bucharaine.*

Bucharidine ($C_{19}H_{25}NO_2$; mp 251–252°) contains no methoxyl or *N*-methyl groups and is resistant to catalytic hydrogenation; it was shown by its solubility in base, by IR absorption at 1645 cm^{-1} and by its UV spectrum to be a 4-hydroxy-2-quinolone (*207*). Structure **285** was proposed for bucharidine on the basis of the NMR spectrum, which showed a one-proton quartet at τ2.12 (C-5–H), a two-proton multiplet at 6.10 (>C*H*–OH and >C*H*–Me), a four-proton multiplet at 8.08 (–CH_2CH_2–), a three-proton singlet at 8.70 (Me–C–O–), a three-proton doublet at 8.77 (>CH–Me), and a six-proton singlet at 8.84 (>CMe_2) (*207*). Bucharidine appears to be a product of abnormal Claisen rearrangement of bucharaine (Scheme 24). Although this conversion was carried out *in vitro* by heating bucharaine in tetralin (*203*), the structure of the alkaloid is not yet certain; the NMR resonance at 8.84τ, for example, is at abnormally high field for a *gem*-dimethyl group in structure **285**.

The third new alkaloid of *H. bucharicum* was named bucharamine, and was an optically inactive base ($C_{22}H_{29}NO_4$; mp 223°) containing no methoxyl or methylamino groups (*105*). It was shown by UV and IR spectroscopy to be a 4-quinolone. In the NMR spectrum there were one-proton multiplets at τ5.4 and at 6.4 attributed to the presence of –CH–O groups and a 22-proton signal at 7.65–9.2 incorporating resonances for six methyl groups. On this basis bucharamine was formulated as the acetonide of a rearrangement product of bucharaine with structure **289** or **290**. The fragmentation peak at *m/e* 313 (M–C_3H_6O) in the mass spectrum of bucharamine supported the presence of an acetonide group. Acid hydrolysis of bucharamine furnished acetone and a new base, $C_{19}H_{23}NO_3$, which was assigned structure **291**. The IR spectrum showed the absence of a hydroxyl group but the presence of a carbonyl group producing absorption at 1710 cm^{-1}. Clearly, hydrolysis of bucharamine

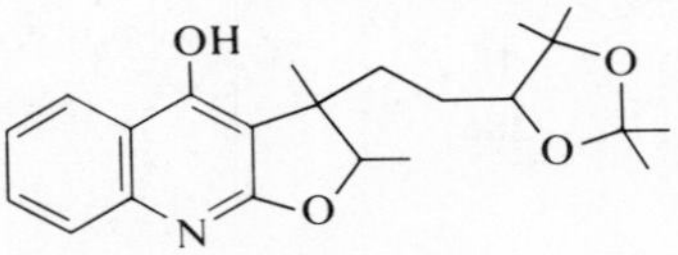

289 Bucharamine

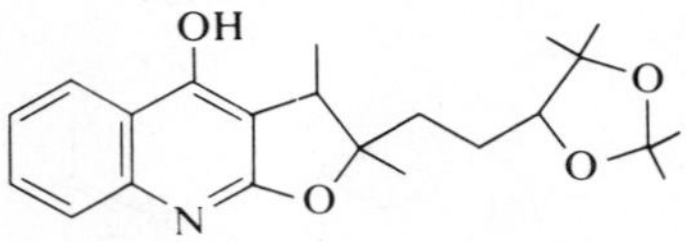

H_3O^+

290

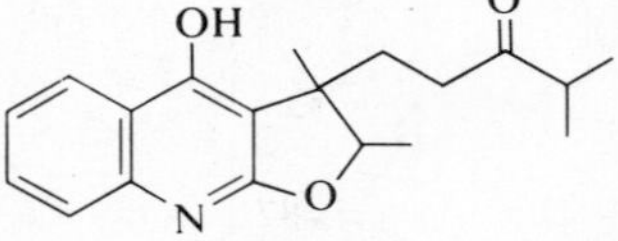

291

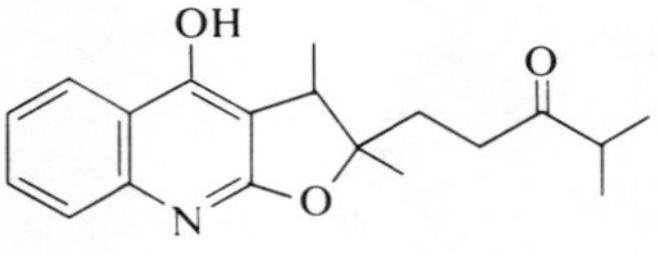

292

resulted in subsequent elimination to give a ketone. The resonance at $\tau 5.46$ ($\gtrdot$*CH*–Me) in the NMR spectrum of the hydrolysis product supports structure **291** (cf. ifflaiamine **264**), rather than the "abnormal" Claisen product **292** (cf. lemobiline **267**). Hence, the alkaloid bucharamine probably has structure **289**.

Because bucharamine was crystallized from acetone, it is possible that the acetonide group was introduced in this way and that the extract of *H. bucharicum* contains the corresponding diol. Although it contains a chiral center, bucharamine is optically inactive, and the diol, therefore, may in any case be an artifact arising through Claisen rearrangement of bucharaine followed by cyclization.

VI. 2-Alkyl- and 2-Aryl-4-quinolone Alkaloids

Data for 2-alkyl- and 2-aryl-4-quinolone alkaloids are given in Table VIII (*3, 69, 134, 208–217*).

A. 2-ALKYL-4-QUINOLONE ALKALOIDS

4-Quinolones containing a long unbranched alkyl chain in the 2-position (pseudans) were first obtained from microorganisms, but during the last decade pseudans with C_7–C_{15} alkyl groups have been isolated from a number of rutaceous species.

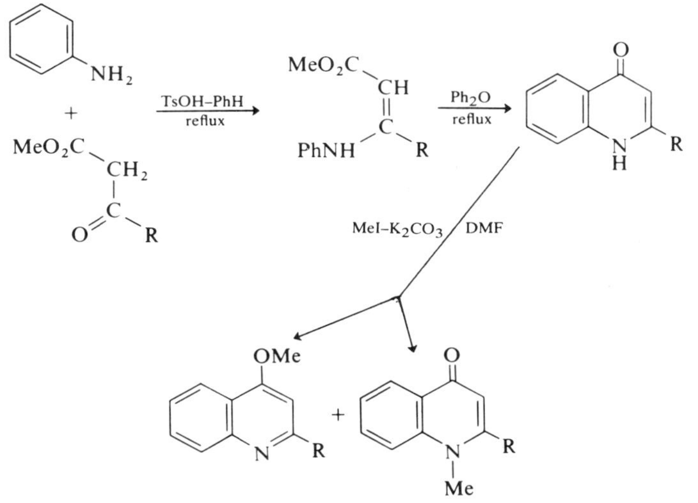

NH2 + MeO2C–CH2–C(=O)–R → TsOH–PhH reflux → MeO2C–CH=C(R)–PhNH → Ph2O reflux → 2-R-4-quinolone (NH) → MeI–K2CO3 / DMF → 4-OMe-2-R-quinoline + 2-R-N-Me-4-quinolone

SCHEME 26. *Synthesis of 2-alkyl-N-methyl-4-quinolones.*

TABLE VIII

2-ALKYL AND 2-ARYL-4-QUINOLONE ALKALOIDS[a]

Compound	Structure number	Molecular formula	Melting point (°C)	Source	Ref.
5-Hydroxy-1-methyl-2-phenyl-4-quinolone	**312**	$C_{16}H_{13}NO_2$	174–175	*Lunasia quercifolia* Lauterb and K. Schum	*69*
2-(Hept-1-enyl)-4-quinolone	**305**	$C_{16}H_{18}NO$		*Pseudomonas aeruginosa*	*3, 208*
Acutine	**304**	$C_{16}H_{19}NO$	122–123	*Haplophyllum acutifolium*	*209*
Folimidine	**313**	$C_{17}H_{15}NO_3$		*Haplophyllum foliosum*	*210*
Japonine	**317**	$C_{18}H_{17}NO_3$	143	*Orixa japonica* Thunb.	*211*
2-(Nona-2,6-dienyl)-4-quinolone	**300**	$C_{18}H_{21}NO$	103	*Vepris ampody*	*134*
2-(9-Hydroxynonyl)-4-quinolone	**301**	$C_{18}H_{25}NO_2$	<50	*V. ampody*	*134*
1-Methyl-2-nonyl-4-quinolone	**296**	$C_{19}H_{27}NO$	71–75	*Ruta graveolens*	*212*
Rutavarin	**306**	$C_{20}H_{20}NO_3$	224	*R. graveolens*	*213*
2-(10-Oxoundecyl)-4-quinolone	**302**	$C_{20}H_{27}NO_2$	126	*V. ampody*	*134*
2-Undecyl-4-quinolone	**303**	$C_{20}H_{29}NO$	130–132	*Ptelea trifoliata*	*214*
1-Methyl-2-undecyl-4-quinolone	**294**	$C_{21}H_{31}NO$	68.5–70	*Evodia rutaecarpa* Hook f. & Thoms	*215*
Melochinone	**308**	$C_{22}H_{21}NO_2$	316–318	*Melochia tomentosa* L.	*216*
Evocarpine	**297**	$C_{23}H_{33}NO$	34–38	*E. rutaecarpa*	*217*
1-Methyl-2-tridecyl-4-quinolone	**293**	$C_{23}H_{35}NO$	74–75	*E. rutaecarpa*	*215*
1-Methyl-2-pentadecyl-4-quinolone	**295**	$C_{25}H_{39}NO$	80	*E. rutaecarpa*	*215*

[a] Supplementary to Volume IX, Chapter 6, Table II.

Evocarpine, $C_{23}H_{33}NO$, was isolated from the fruits of *Evodia rutaecarpa* Hook. f. and Thoms by Tschesche and Werner (*217*) and shown to be a 4-quinolone by its UV spectrum and by IR absorption at 1623 cm^{-1}. The alkaloid was converted by catalytic hydrogenation into a dihydro derivative, identified by synthesis (Scheme 26) as 1-methyl-2-tricodecyl-4-quinolone (**293**). The presence in evocarpine of a nonterminal double bond was indicated by NMR resonances at τ4.71 (2H, triplet, –CH=CH–) and at 9.1 (3H, –Me), but the position of the unsaturated function was not readily determined. Ozonolysis of the alkaloid and treatment of the products with 2,4-dinitrophenylhydrazine resulted in the formation of a hydrazone mixture consisting mainly of valeraldehyde hydrazone but containing some C_6 and C_4 aldehyde derivatives. It was concluded that the alkaloid was the $\Delta^{8'}$-olefin **297** but was mixed with some $\Delta^{7'}$ and some $\Delta^{9'}$ isomers. The mass spectrum of evocarpine also was studied; in addition to a substantial molecular ion peak at *m/e* 339, fragmentation ions at 186 (C-2′–C-3′ cleavage) and at 173 (C-1′–C-2′ cleavage) were observed; the latter (**298**) was the base peak and apparently arises by rearrangement of the molecular ion as illustrated.

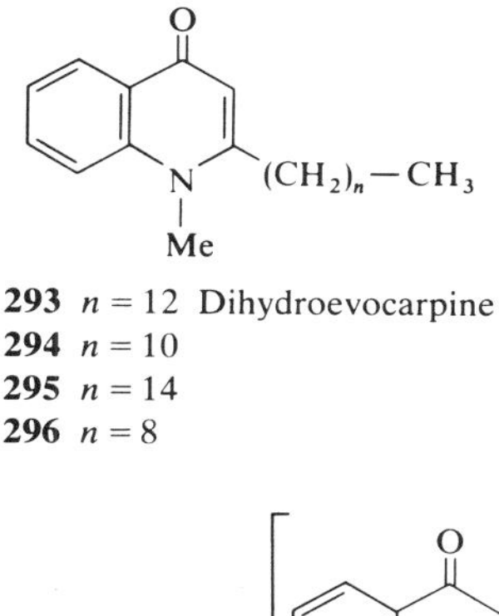

$(CH_2)_n-CH_3$

293 $n = 12$ Dihydroevocarpine
294 $n = 10$
295 $n = 14$
296 $n = 8$

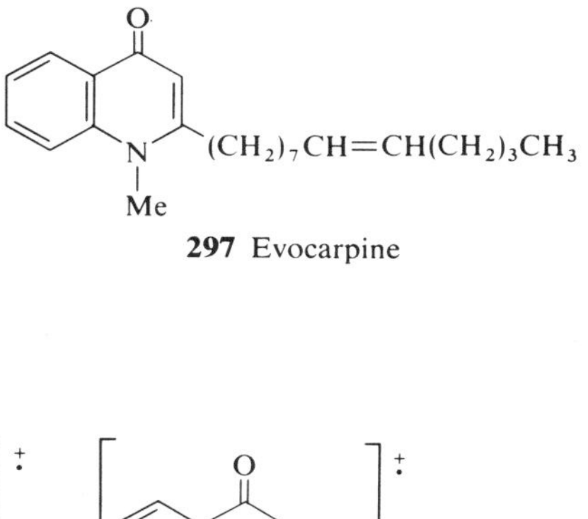

$(CH_2)_7CH=CH(CH_2)_3CH_3$

297 Evocarpine

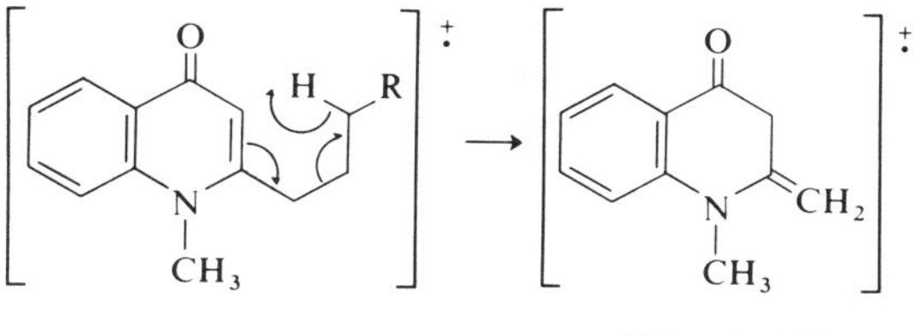

298 *m/e* 173

A later study (*215*) of the leaves and fruit of *E. rutaecarpa* led to the isolation of three new alkaloids, shown by spectroscopy to be dihydroevocarpine (**293**) and the analogs **294** and **295** containing, respectively, C_{11} and C_{15} saturated side chains. The mass spectra of these compounds were particularly informative; in each case the base peaks occurred at *m/e*

173, and a succession of fragmentation peaks at mass intervals of 14 indicated the length of the 2-alkyl group. The structures of alkaloids **294** and **295** were confirmed by synthesis, using the procedure developed for dihydroevocarpine (Scheme 26). Another member of this series containing a C_9 side chain (**296**) has been isolated from the aerial parts of *Ruta graveolens* and its structure was established by NMR and mass spectrometry (*212*).

The constituents of *Vepris ampody* H. Perr. were studied by Potier and co-workers (*134*), and three new alkaloids were identified mainly by spectroscopic means as pseudans with C_9 or C_{11} side chains. One alkaloid, $C_{18}H_{21}NO$, was clearly a 2-nonadienyl-4-quinolone, for catalytic reduction furnished 2-nonyl-4-quinolone. In the NMR spectrum, a two-proton multiplet at τ8.05 and a three-proton triplet at 9.15 indicated that the group $-CH{=}CH-CH_2-CH_3$ was present, while a six-proton multiplet at 7.3 was attributed to three methylene groups adjacent either to double bonds or to the quinolone nucleus. Hence, the alkaloid had structure **299** or **300**. Structure **299** was preferred on the basis of the mass spectrum,

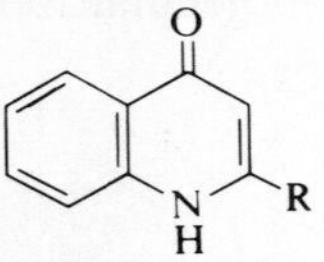

299 $R = CH_2CH_2CH{=}CHCH_2CH{=}CHCH_2CH_3$ (positions 1′–9′)
300 $R = CH_2CH{=}CHCH_2CH_2CH{=}CHCH_2CH_3$
301 $R = (CH_2)_8CH_2OH$
302 $R = (CH_2)_9COCH_3$
303 $R = (CH_2)_{10}CH_3$
304 $R = (CH_2)_3CH{=}CHCH_2CH_3$ Acutine
305 $R = CH{=}CH(CH_2)_4Me$
306 $R = (CH_2)_4$–(3,4-methylenedioxyphenyl) Rutavarin
307 $R = CH{=}CH-CH{=}CH$–(3,4-methylenedioxyphenyl)

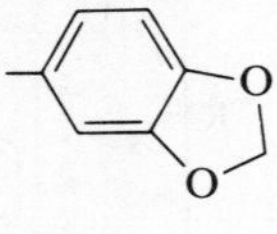

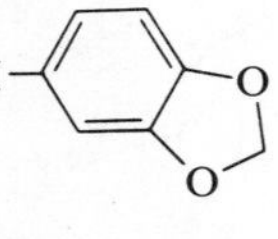

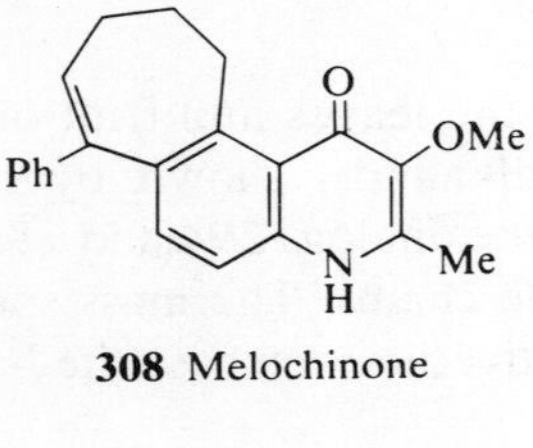

308 Melochinone

which showed a fragmentation peak *m/e* 172 resulting from cleavage of the C-2′–C-3′ bond.

A second alkaloid from *V. ampody* was shown to be the primary alcohol **301** (*134*). The NMR spectrum showed a resonance at $\tau 6.35$ (2H, triplet, $-CH_2OH$), appearing at lower field after acetylation. The third alkaloid was the 4-quinolone **302** with a C_{11} side chain. The presence of a ketonic function was indicated by IR absorption at 1705 cm^{-1}, and the structure was confirmed by NMR and mass spectrometry.

Reisch and co-workers (*214*) recently identified 2-undecyl-4-quinolone (**303**) as a constituent of leaves of *Ptelea trifoliata*, and they isolated from the roots of *Ruta graveolens* an unresolved mixture of 4-quinolones in which the 2-alkyl groups contain 11, 12, 13, or 14 carbon atoms.

Structure **304** for acutine, an alkaloid of *Haplophyllum acutifolium* (DC) G. Don, has been established by spectroscopy and by synthesis of dihydroacutine (*209*).

A microbial metabolite isolated from *Pseudomonas aeruginosa* was shown to be the $\Delta^{1'}$-heptenylquinolone **305** by a study of the NMR and mass spectra and by its synthesis from aniline and methyl-3-keto-non-4-enyloate (*208*).

Rutaverin, an alkaloid of *R. graveolens*, was shown (*213*) to be the 2-alkyl-4-quinolone **306** related to the well-known quinoline alkaloids of *Galipea officinalis* Hancock. The butadiene derivative **307**, prepared from 2-methyl-4-quinolone and 3,4-methylenedioxycinnamaldehyde, was converted into rutaverin by catalytic reduction.

An alkaloid, melochinone ($C_{22}H_{21}NO_2$), was obtained from *Melochia tomentosa* L. (Sterculiaceae) and the unusual structure **308** was established by X-ray analysis (*216*). The constitution of melochinone was not apparent from spectroscopic data, partly because the presence of a 4-quinolone system was not recognized from IR absorption at 1620 cm^{-1}. The NMR spectrum indicated that the alkaloid contained one phenyl and two methyl groups; a resonance at 6.44τ initially suggested that the terminal methylene group in the sequence $>C{=}CHCH_2CH_2CH_2-$ was attached to an electronegative group, but the X-ray determination showed that the signal was at unexpectedly low-field because of the adjacent carbonyl group.

B. 2-Aryl-1-methyl-4-quinolone Alkaloids

Alkaloids of this type have been described previously, and during the last ten years new sources of known alkaloids have been recorded. Thus, the simplest member of the group **309**, previously obtained only from *Balfourodendron riedelianum*, has now been isolated from *Haplophyllum foliosum* (*164*). Graveoline (rutamine) (**310**), a characteristic alkaloid of

Ruta species, is a constituent of *R. angustifolia* Peers., *R. chalepensis* L. (*113*), and *R. bracteosa* D.C. (*218*) as well as *R. graveolens* (see Volume IX of this treatise). Eduline (**311**) has been found in the leaves of *Skimmia japonica* Thunb. (*37*).

New alkaloids of the group include the 5-hydroxy-4-quinolone **312** isolated from *Lunasia quercifolia* (*69*) and later from an *S. japonica* variety (*3*). The structure of the alkaloid was established by spectroscopy (*69*), the NMR resonance at τ−4.0 being due to a hydroxyl group at C-5 linked by intramolecular hydrogen bonding to the 4-quinoline carbonyl group.

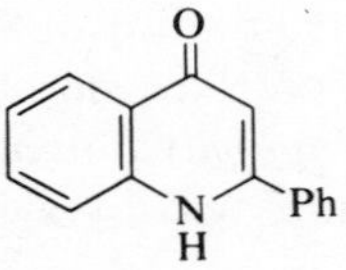

309

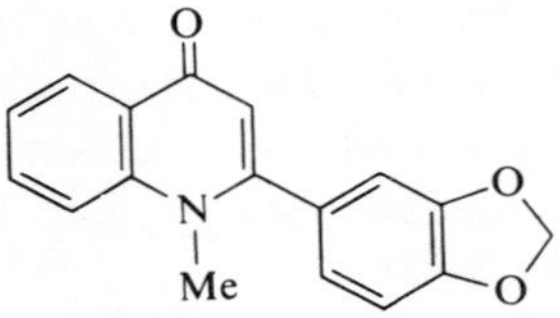

310

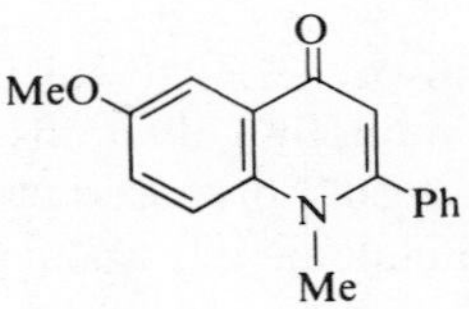

311 Eduline

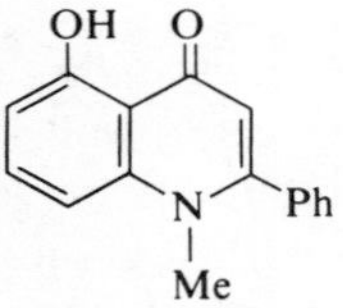

312

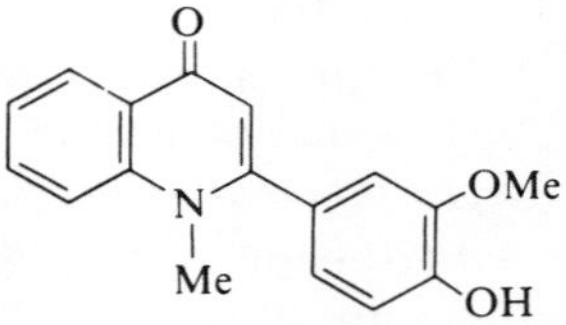

313 Folimidine

Folimidine was obtained from *Haplophyllum foliosum* and is assumed to have structure **313** on the basis of spectroscopic studies on the alkaloid and its *O*-methyl derivative (*210*).

A number of alkaloids were isolated from the root bark of *Orixa japonica* (Volume IX, p. 250), and a study of the aerial parts of the plant has now yielded a new alkaloid, japonine ($C_{18}H_{17}NO_3$), shown by spectroscopic methods to be 3-methoxyeduline (**317**) (*211*). The additional methoxyl group was assigned to the 3-position because of the presence of eight aromatic proton resonances in the NMR spectrum of japonine and the absence of a resonance attributable to a proton at C-3.

Piozzi *et al.* (*219*) have confirmed **317** as the structure of japonine by synthesis (Scheme 27). Application of Darzen's reaction to 2-nitro-5-methoxybenzaldehyde furnished epoxide **314**. This was converted with hydrogen chloride and hydroquinone into **315**, which was reduced to the 3-hydroxy-4-quinolone **316**; methylation then afforded japonine as the major product.

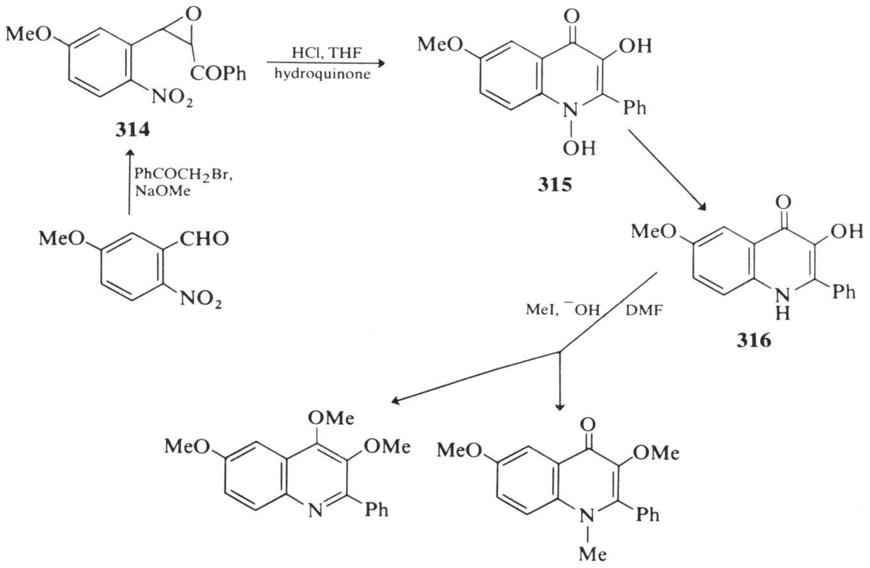

MeO, O, COPh, NO_2, **314**; PhCOCH$_2$Br, NaOMe; MeO, CHO, NO_2; HCl, THF, hydroquinone; MeO, O, OH, N, Ph, OH, **315**; MeO, O, OH, N, H, Ph, **316**; MeI, $^{-}$OH, DMF; OMe, MeO, OMe, N, Ph; MeO, O, OMe, N, Ph, Me, **317** Japonine

SCHEME 27. *Synthesis of japonine.*

VII. Biosynthesis

Earlier theories of the biosynthesis of quinoline alkaloids found in the Rutaceae have been discussed in Volume IX of this treatise, and the first experimental results using radioactive tracers were also reviewed. Spenser and his collaborators showed that anthranilic acid is incorporated intact into dictamnine in *Dictamnus albus* and that C-10 and C-11 (cf. **328**) are derived from C-1 and C-2 of acetate, respectively (*220*). Matsuo *et al.* (*221*) demonstrated almost simultaneously that the quinoline ring of skimmianine in *Skimmia japonica* was formed similarly. Since then, the biosynthesis of furo- and isopropyldihydrofuroquinoline alkaloids has been studied intensively, with particular interest in the origin of the C_2 and C_5 side chains present at the 3-position of the quinoline ring of these alkaloids.

The existence of furoquinoline alkaloids containing C_5 isoprenyl groups elsewhere in the molecule [for example, evoxine (**331**) and choisyine (**332**)] and the cooccurrence of furoquinoline alkaloids such as dictamnine with isopropyldihydrofuroquinolines like *N*-methylplatydesminium salt (**326**) provided convincing taxonomic evidence for the suggestion that the C_2

furan ring was formed by loss of a 3-carbon fragment from a mevalonate-derived isoprenyl group. Earlier attempts to obtain experimental evidence for this theory were unsuccessful, and mevalonic acid was not incorporated into dictamnine (**328**) in *D. albus* (*220*) or in *S. japonica* (*222*), although Colonna and Gros (*223*) later showed that C-4 and C-5 of mevalonic acid are incorporated specifically into C-2 and C-3, respectively, of skimmianine (**330**) in *Fagara coco* Engl.

Confirmation of the isoprenoid nature of the furan rings of dictamnine (**328**) and of *N*-methylplatydesminium salt (**326**) in *S. japonica* was provided indirectly by Grundon and his co-workers (*224*), who showed that the 3-prenylquinolones **321** and **322** labeled at C-1′ with ^{14}C were good precursors of both alkaloids (3.6–4.8% incorporation). Specific incorporation of the precursors into dictamnine was indicated by oxidative degradation by a method similar to that used previously. *N*-Methylplatydesminium salt was counted as its base-cleavage product edulinine (**256**), which was converted into isodictamnine (Section IV,B,3); oxidation via 3-carboxy-4-hydroxy-1-methyl-2-quinolone gave ^{14}C-labeled carbon dioxide.

The biosynthetic pathway to furoquinoline alkaloids apparently involves formation of the quinoline ring before introduction of the side chains, for 4-hydroxy-2-quinolone (**319**) is a specific precursor of skimmianine (**330**) in *R. graveolens* (*225*) and 4-hydroxy-2-quinolone (**319**) and 4-methoxy-2-quinolone (**320**) are incorporated into dictamnine (**328**) and *N*-methylplatydesminium salt (**326**) in *S. japonica* (*224*). Labeled 2,4-dihydroxyquinoline is incorporated into kokusaginine (6,7-dimethoxydictamnine) in *Ruta graveolens*, but degradation of the alkaloid indicated that randomization of the label had occurred, suggesting that 2,4-dihydroxyquinoline is not a direct precursor (*226*).

Although the work described apparently established the general biosynthetic route to furoquinoline alkaloids, i.e., **318** → **319** → **321** → **322** → **328** or → **326**, many details of the pathway were unresolved; in particular, speculation about the fragmentation of the isoprenyl side chain leading to an unsubstituted furan ring stimulated further research with radioactive tracers. Seshadri and his collaborators (*227*) had proposed that oxidative cleavage of a prenyl group at the double bond could lead to an aldehyde which by cyclization might lead to a benzofuran; this scheme led to *in vitro* syntheses of furoquinolines of the dictamnine group (Section IV,B,3). The biosynthetic theories of Birch and Smith (*228*) applied to furoquinoline alkaloids by Diment, Ritchie, and Taylor (*229*) implied that platydesmine (**325**) was an intermediate in the pathway from prenylquinolones **322** to dictamnine. This process was realized *in vitro* by the isolation of dictamnine from the reaction of platydesmine with lead tetraacetate and iodine (*229*).

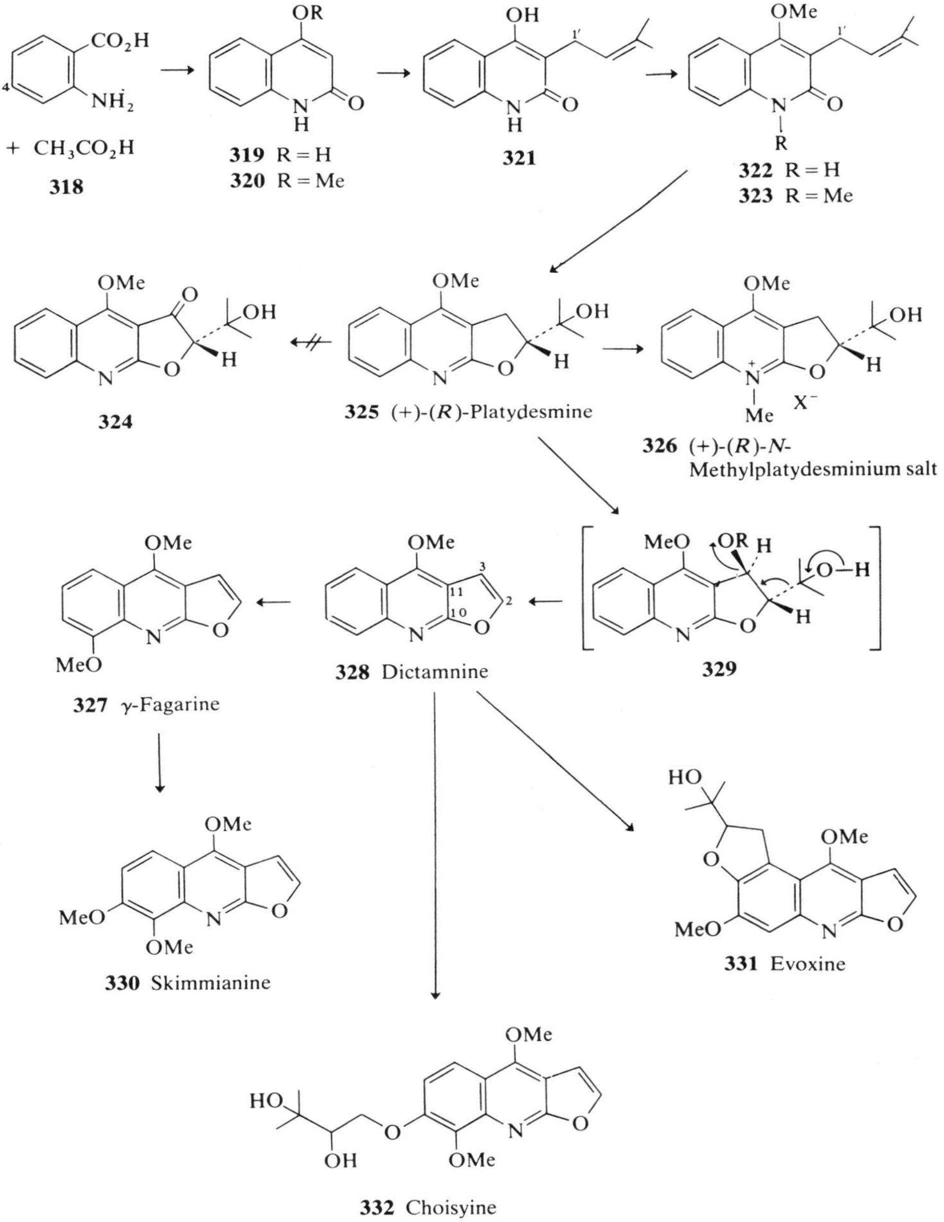

CO_2H
NH_2
+ CH_3CO_2H
318

OR
319 R = H
320 R = Me

OH
321

OMe
322 R = H
323 R = Me

OMe
324

OMe
325 (+)-(*R*)-Platydesmine

OMe
Me X^-
326 (+)-(*R*)-*N*-Methylplatydesminium salt

MeO OR H
O—H
329

OMe
328 Dictamnine

OMe
MeO
327 γ-Fagarine

OMe
MeO
OMe
330 Skimmianine

HO
OMe
MeO
331 Evoxine

OMe
HO
OH
OMe
332 Choisyine

SCHEME 28. *Biosynthesis of furoquinoline and hydroxyisopropyldihydrofuroquinoline alkaloids.*

The Birch–Ritchie proposals were supported by further feeding experiments with *S. japonica* in which (±)-[^{14}C]platydesmine (**325**) was shown to be an excellent specific precursor (18.8% incorporation) of dictamnine (*224*). Although platydesmine was not detected in *S. japonica*, even by gas–liquid chromatographic analysis of the alkaloid extract, the alkaloid was shown to be an intermediate in the biosynthetic pathway, as well as an efficient precursor, by *in vivo* isotope-trapping; [^{14}C]prenylquinolone **322** fed with unlabeled (±)-platydesmine gave radioactive platydesmine and dictamnine.

Studies of the biomechanism of the transformation of platydesmine into furoquinoline alkaloids were carried out with *Choisya ternata* (*230*); skimmianine is the principal furoquinoline alkaloid of this species (*3*), but as it arises by hydroxylation of dictamnine (see below), results obtained are applicable to both alkaloids. The 4-methoxyquinolone **322**, in which both hydrogen atoms attached to C-1′ were labeled with tritium, was mixed with the [1′-^{14}C]quinolone and administered to shoots of *Choisya ternata*. Doubly labeled skimmianine (**330**) was isolated, the ^{3}H : ^{14}C ratio indicating that approximately half the tritium label was retained. This result eliminates the possibility that the furan rings of dictamnine and skimmianine arise from platydesmine via a ketone **324**, and is consistent with a pathway involving stereospecific oxidation to an alcohol derivative, e.g. **329**, or direct hydrogen ion abstraction, followed in either case by loss of the isopropyl group. Myrtopsine (**329**, R = H unknown stereochemistry) has recently been isolated from *Myrtopsis sellingii* (*238*); a study of its reactions and of its role as a biosynthetic precursor should further clarify the transformation of platydesmine into dictamnine *in vivo*.

Subsequent work was designed to indicate at what point in the pathway methylation of the 4-oxygen function occurs. It was shown by feeding quinolone **322** doubly labeled in the 4-methoxy group and in the side chain that the methoxy group remains intact in the formation of skimmianine in *C. ternata* (*230*). 4-Methoxy[3-^{14}C]-2-quinolone is incorporated efficiently into dictamnine in *S. japonica* (*224*), but may be an unnatural precursor because the quinolone labeled on the 4-methoxy group is not incorporated into skimmianine in *C. ternata* and apparently undergoes demethylation to 4-hydroxy-2-quinolone (*67*). The incorporation of the *N*-methyl-3-prenyl[3-^{14}C]-2-quinolone **323** into dictamnine indicates that *N*-demethylation also occurs in *S. japonica* (*231*).

Several groups have studied the hydroxylation of furoquinoline alkaloids. Anthranilic acid is a specific precursor of acronidine (**195**, Section IV,A,7) in *Acronychia baueri* (*187*). Hall and Prager (*232*) fed [4-^{3}H]anthranilic acid to the plant and estimated that, during hydroxylation to acronidine and skimmianine, migration of tritium occurred to the extent of

19% and 10%, respectively. It appears, therefore, that partly specific N.I.H. shifts are involved, although this work did not show at what stage of the pathway hydroxylation occurred.

Grundon, Harrison, and Spyropoulos (*3*) showed that dictamnine is a specific precursor of skimmianine in *C. ternata* and in a variety of *S. japonica* (1.3–6.6% incorporation); it was a specific although less efficient precursor of choisyine (**332**) and evoxine (**331**) in *C. ternata*. Labeled dictamnine was incorporated into γ-fagarine (**327**), skimmianine (**330**), and kokusaginine in cell cultures of *Ruta graveolens*, and γ-fagarine was found to be a precursor of skimmianine, but in this work specific labeling in the products was not established (*2*). It appears that a general route to furoquinoline alkaloids oxygenated in the aromatic ring involves hydroxylation of dictamine, although the results do not exclude additional pathways, such as the formation of skimmianine from a 7,8-dimethoxy-2-quinolone, as already suggested (*31*).

Although the general biosynthetic pathway to 3-prenylquinoline alkaloids and their relatives is apparent from the feeding experiments with labeled precursors discussed above (Scheme 28), many details remain to be resolved. Theories are based on chemotaxonomic and stereochemical evidence and have been reviewed for isoprenoid quinoline alkaloids and coumarins (*38*); some aspects will be discussed here.

The proposal that epoxides are intermediates in the formation of hydroxyisopropyldihydrofuroquinoline alkaloids from 3-prenylquinolones is based on *in vitro* analogy and on the occurrence of epoxides in rutaceous species. Reference to Scheme 5 (Section III,E,1) indicates that if an (*S*)-epoxide in *Balfourodendron riedelianum*, for example, cyclized with inversion of configuration to balfourodine (**100**) and to the pyrano derivative isobalfourodine without affecting the chiral center, the two alkaloids would be expected to have (*R*) and (*S*) configurations, respectively. The fact that both alkaloids have the (*R*) configuration is consistent with a pathway in which the furo derivative originates directly from an epoxide and the pyrano isomer is formed by subsequent rearrangement involving retention of configuration; there is *in vitro* analogy for this rearrangement (Section III,F).

The biosynthesis of isopropylfuroquinolines such as lunasine (**338**) and lunacrine (**335**) has been discussed (cf. *80*), but not yet established; their cooccurrence with hydroxy derivatives (cf. **333**) in *Lunasia amara* and in *Ptelea trifoliata* suggests that the two groups of alkaloids are interrelated biosynthetically, for example by a pathway involving successive dehydration and reduction via terminal olefins (Scheme 29). This scheme is supported by the presence of terminal olefins in *P. trifoliata* but not by the stereochemistry of the alkaloids (−)-(*R*)-lunacrine (**335**) and

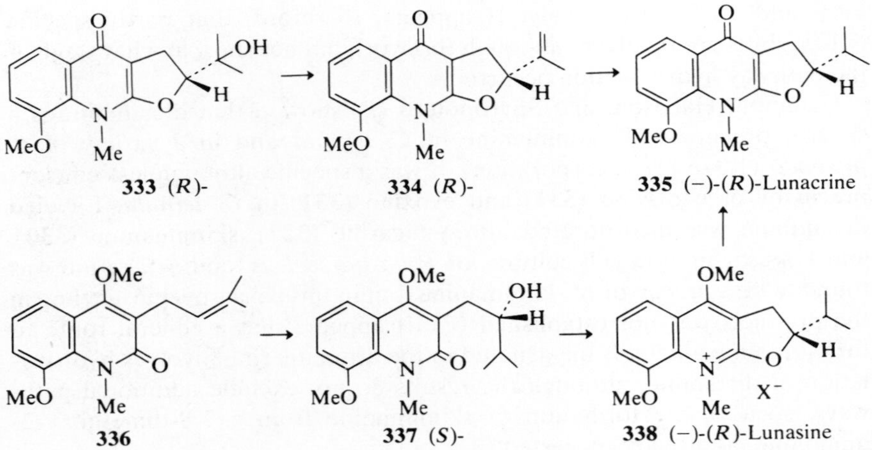

333 (*R*)- → **334** (*R*)- → **335** (−)-(*R*)-Lunacrine

336 → **337** (*S*)- → **338** (−)-(*R*)-Lunasine

SCHEME 29. *Biosynthesis of isopropyldihydrofuroquinoline alkaloids.*

(−)-(*S*)-hydroxylunacrine (cf. **333**) found in *L. amara*. The biosynthesis of (−)-(*R*)-lunasine is more likely to occur by anti-Markovnikov hydration of a prenylquinolone to an (*S*)-hydroxy derivative followed by cyclization to lunasine with inversion at the chiral center and thence conversion into lunacrine (**335**) (Scheme 29); direct cyclization of a prenylquinolone is less probable because at least *in vitro* acid-catalyzed reactions lead exclusively to pyranoquinolines. The (*R*)-hydroxy derivative lunacridine (cf. **337**) has been obtained from *L. amara* but is probably an artifact arising from hydrolysis of (−)-(*R*)-lunasine during isolation, rather than a biosynthetic intermediate.

The occurrence of (1,1-dimethylallyl)- and (1,2-dimethylallyl)quinoline derivatives in *Flindersia* and *Ravenia* species and the isolation of the prenyl ether **261** from *Ravenia* (Section V,B) suggested that the biosynthesis of this group of alkaloids might proceed via Claisen and abnormal Claisen rearrangements as in the synthetic route (Scheme 22, Section V, B).

Support for this proposition was provided by preliminary ^{14}C-feeding experiments with *R. spectabilis* whereby it was shown that the prenyl ether **261** (ravenine) is a precursor of ravenoline **265** (0.75% incorporation) (*199*).

The biosynthesis of 2-alkyl- and 2-arylquinoline alkaloids was investigated by Luckner and co-workers. For example, anthranilic acid was incorporated intact into 2-heptyl-4-hydroxyquinoline **339** in the microorganism *Pseudomonas aeruginosa*. Acetate and malonate are also precursors, and degradation experiments showed that the side chain was

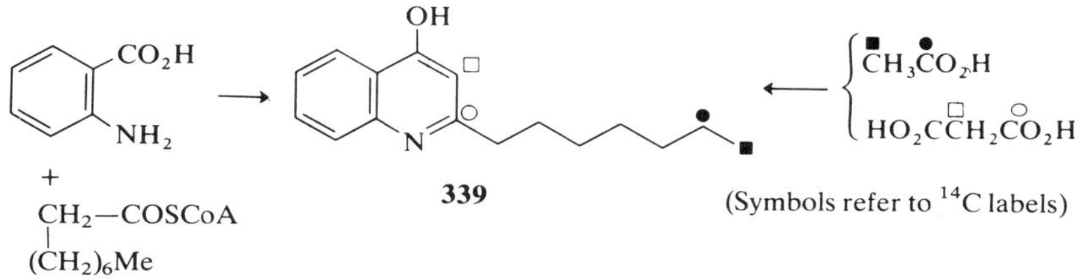

CO2H
NH2
+
CH2—COSCoA
(CH2)6Me
OH
N
339
CH3CO2H
HO2CCH2CO2H
(Symbols refer to 14C labels)

SCHEME 30. *Biosynthesis of a pseudan.*

evolved by the acetate–polymalonate route and that C-2 and C-3 were derived from the carbonyl and methylene groups, respectively, of malonate; the pathway shown in Scheme 30 was proposed (*233*).

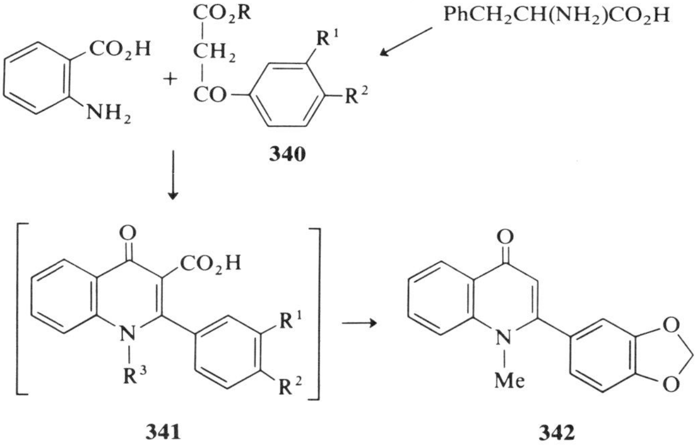

CO2R
CH2
CO
CO2H
NH2
+
R1
R2
PhCH2CH(NH2)CO2H
340
O
CO2H
N
R3
O
N
Me
O
O
341
342

SCHEME 31. *Biosynthesis of graveoline.*

Anthranilic acid is also a precursor of the 2-aryl-4-quinolone alkaloid graveoline (**342**) of *Ruta angustifolia* Pers. (*234*), but the labeled carboxyl group is mainly lost, apparently because of interconversion with tryptophan before formation of the quinoline nucleus. Other feeding experiments indicated that the side chain is derived from phenylalanine, with hydroxylation of the aromatic ring preceding cyclization to the quinoline system. A β-keto ester **340** and a 3-carboxyquinoline **341** are probable intermediates in the biosynthesis of graveoline (Scheme 31), as in the case of 2-alkylquinoline alkaloids. In contrast to the results obtained with graveoline, preliminary work on the biosynthesis of eduline (**311**) in *Skimmia japonica* using doubly labeled anthranilic acid indicated that the carboxyl group of the precursor was retained in the formation of the alkaloid (*235*).

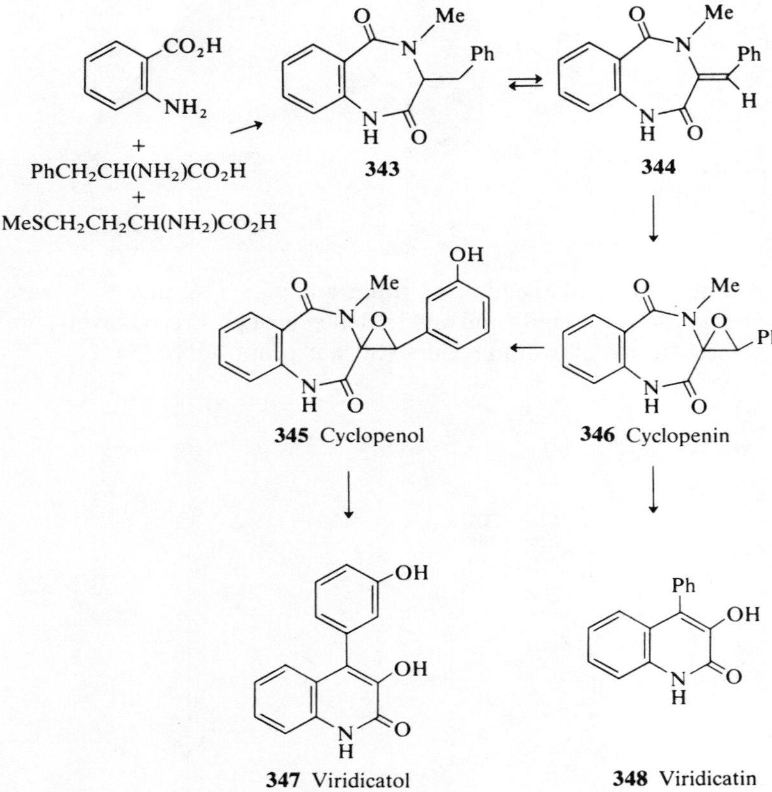

$C_6H_4(CO_2H)(NH_2)$
+
$PhCH_2CH(NH_2)CO_2H$
+
$MeSCH_2CH_2CH(NH_2)CO_2H$

343 ⇄ **344**

345 Cyclopenol ← **346** Cyclopenin

347 Viridicatol **348** Viridicatin

SCHEME 32. *Biosynthesis of viridicatin and viridicatol.*

An intensive study of the biosynthesis of the fungal alkaloid viridicatin (**348**) in *Penicillium viridicatum* has been carried out by Luckner and Mothes and co-workers (*236*). Tracer feeding experiments showed that phenylalanine was an efficient precursor of viridicatin, but that the carboxyl group of anthranilic acid was not incorporated. The benzodiazepine, cyclopenin (**346**), itself a constituent of *P. viridicatum*, is converted *in vitro* and *in vivo* into viridicatin, and the enzyme responsible for this transformation has been identified. It is now known that the cyclic peptide **343** found in the fungus is derived from phenylalanine and anthranilic acid and is converted through the dehydro derivative **344** into cyclopenin (*237*). The pathway to viridicatin established by this work is shown in Scheme 32. The biosynthesis of the related metabolite, viridicatol (**347**), occurs via the benzodiazepine epoxide cyclopenol (**345**).

Acknowledgments

I am grateful to Dr. H. T. Openshaw for providing references for Table VI, and to colleagues and research students who have helped to maintain my interest in quinoline alkaloids during the last 25 years.

References

1. Zs. Rózsa, K. Szendrei, I. Novák, E. Minker, M. Koltai, and J. Reisch, *J. Chromatogr.* **100**, 218 (1974).
2. D. Boulanger, B. K. Bailey, and W. Steck, *Phytochemistry* **12**, 2399 (1973).
3. M. F. Grundon, D. M. Harrison, and C. G. Spyropoulos, *J. Chem. Soc., Perkin Trans. 1* 2181 (1974).
4. M. Vlassa, *Stud. Cercet. Chim.* **19**, 515 (1971).
5. K. L. Seitanidi and M. R. Yagudaev, *Khim. Prir. Soedin.* 755 (1974).
6. J. Reisch, K. Szendrei, I. Novák, and E. Minker, *Acta Pharm. Hung.* **44**, 107 (1974).
7. P. J. Scheuer, *in* "Chemistry of the Alkaloids" (S. W. Pelletier, ed.), p. 355. Van Nostrand-Reinhold, Princeton, New Jersey, 1970.
8. M. F. Grundon, *in* "Enciclopedia della Chimica," Vol. III, p. 344. Uses Edizioni Scientifiche, Firenze, 1974.
9. V. A. Snieckus, *Alkaloids (London)* **1**, 96 (1971); **2**, 86 (1972); **3**, 104 (1973); **4**, 117 (1974); **5**, 103 (1975); M. F. Grundon, *ibid.* **6**, 103 (1976).
10. I. Mester, *Fitoterapia* **44**, 123 (1973).
11. L. A. Mitscher, T. Suzuki, and G. Clark, *Heterocycles* **5**, 565 (1976).
12. M. N. S. Nayar, C. V. Sutar, and M. K. Bhan, *Phytochemistry* **10**, 2843 (1971).
13. H. Ishii, H. Ohida, and J. Haginiwa, *Yakugaku Zasshi* **92**, 118 (1972); *CA* **77**, 16530.
14. D. M. Razzakova, I. A. Bessonova, and S. Yu. Yunusov, *Khim. Prir. Soedin.* **8**, 133 (1972); *CA* **77**, 72563.
15. S. A. Sultanov and S. Yu. Yunusov, *Khim. Prir. Soedin.* 131 (1969).
16. Z. Ch. Faizutdinova, I. A. Bessonova, and S. Yu. Yunusov, *Khim. Prir. Soedin.* 257 (1967).
17. T. P. Taube, J. W. Murphy, and A. D. Cross, *Tetrahedron* **23**, 2061 (1967).
18. P. Venturella, A. Bellino, and F. Piozzi, *Chim. Ind. (Milan)* **50**, 451 (1968).
19. T. Kappe, H. Schmidt, and E. Ziegler, *Z. Naturforsch., Teil B* **25**, 328 (1970).
20. N. S. Narasimhan, M. V. Pradkar, and R. H. Alurkar, *Tetrahedron* **27**, 1351 (1971).
21. D. Kurbanov and S. Yu. Yunusov, *Khim. Prir. Soedin.* 289 (1967).
22. W. D. Crow and J. H. Hodgkin, *Aust. J. Chem.* **21**, 3075 (1968).
23. R. Storer and D. W. Young, *Tet. Lett.* 1555 (1972); *Tetrahedron* **29**, 1215 (1973).
24. P. Venturella, A. Bellino, and F. Piozzi, *Chem. Ind. (London)* 887 (1972); *Gazz. Chim. Ital.* **104**, 297 (1974).
25. R. Storer, D. W. Young, D. R. Taylor, and J. M. Warner, *Tetrahedron* **29**, 1721 (1973).
26. I. T. Eshiett and D. A. H. Taylor, *Chem. Commun.* 467 (1966); *J. Chem. Soc. C* 481 (1968).
27. R. M. Bowman and M. F. Grundon, *Chem. Commun.* 334 (1965); *J. Chem. Soc. C* 1504 (1966).
28. B. D. Paul and P. K. Bose, *J. Indian Chem. Soc.* **45**, 552 (1968); *Indian J. Chem.* **7**, 678 (1969).
29. F. Bohlmann and V. S. Bhaskar Rao, *Ber.* **102**, 1774 (1969).
30. L. A. Mitscher, G. W. Clark, T. Suzuki, and M. S. Bathala, *Heterocycles* **3**, 913 (1975).

31. R. Storer and D. W. Young, *Tet. Lett.* 2199 (1972); *Tetrahedron* **29**, 1217 (1973).
32. D. L. Dreyer, *Phytochemistry* **8**, 1013 (1969).
33. J. Reisch, K. Szendrei, I. Novák, E. Minker, J. Körösi, and K. Csedo, *Tet. Lett.* 449 (1972).
34. J. Reisch, J. Körösi, K. Szendrei, I. Novák, and E. Minker, *Phytochemistry* **14**, 1678 (1975).
35. J. L. Gaston and M. F. Grundon, *Tet. Lett.* 2629 (1978).
36. D. Lavie, N. Danieli, R. Weitman, and E. Glotter, *Tetrahedron* **24**, 3011 (1968).
37. D. R. Boyd and M. F. Grundon, *Tet. Lett.* 2637 (1967); *J. Chem. Soc. C* 556 (1970).
38. M. F. Grundon and I. S. McColl, *Phytochemistry* **14**, 143 (1975).
39. J. Reisch, K. Szendrei, E. Minker, and I. Novák, *Pharmazie* **24**, 699 (1969).
40. S. R. Johns and J. A. Lamberton, *Aust. J. Chem.* **19**, 1991 (1966).
41. J. A. Diment, E. Ritchie, and W. C. Taylor, *Aust. J. Chem.* **20**, 565 (1967).
42. S. T. Murphy, E. Ritchie, and W. C. Taylor, *Aust. J. Chem.* **27**, 187 (1974).
43. T. Higa and P. J. Scheuer, *Phytochemistry* **13**, 1269 (1974).
44. J. Vaquette, M. S. Hifnawy, J. L. Pousset, A. Fournet, A. Bouquet, and A. Cavé, *Phytochemistry* **15**, 743 (1976).
45. F. Fish, I. A. Meshel, and P. G. Waterman, *Planta Med.* **29**, 310 (1976).
46. R. A. Corral and O. O. Orazi, *Tet. Lett.* 583 (1967); R. A. Corral, O. O. Orazi, and I. A. Benages, *Tetrahedron* **29**, 205 (1973).
47. Dr. L. Tökés, personal communication.
48. S. R. Johns, J. A. Lamberton, and A. A. Sioumis, *Aust. J. Chem.* **23**, 419 (1970).
49. W. Steck, B. K. Bailey, J. P. Skyluk, and O. L. Gamborg, *Phytochemistry* **10**, 191 (1971).
50. E. V. Lassak and J. T. Pinkey, *Aust. J. Chem.* **22**, 2175 (1969).
51. E. F. Nesmelova, I. A. Bessonova, and S. Yu. Yunusov, *Khim. Prir. Soedin.* 815 (1975); *CA* **84**, 150808 (1976).
52. R. A. Corral and O. O. Orazi, *Tetrahedron* **22**, 1153 (1966).
53. K. Szendrei, J. Reisch, I. Novák, L. Simon, Zs. Rózsa, E. Minker, and M. Koltai, *Herba Hung.* **10**, 131 (1971); *CA* **79**, 15853 (1973).
54. L. A. Mitscher, M. S. Bathala, and J. L. Beal, *Chem. Commun.* 1040 (1971); L. A. Mitscher, M. S. Bathala, G. W. Clark, and J. L. Beal, *Lloydia* **38**, 109 (1975).
55. J. Reisch, K. Szendrei, I. Novák, E. Minker, and V. Pápay, *Tet. Lett.* 3803 (1969).
56. J. Reisch, J. Körösi, K. Szendrei, I. Novák, and E. Minker, *Phytochemistry* **14**, 2722 (1975).
57. L. A. Mitscher, M. S. Bathala, G. W. Clark, and J. L. Beal, *Lloydia* **38**, 117 (1975).
58. K. Szendrei, M. Petz, I. Novák, J. Reisch, H. E. Bailey, and V. L. Bailey, *Herba Hung.* **13**, 49 (1974); *CA* **83**, 40169 (1975).
59. W. J. Donnelly and M. F. Grundon, *J. Chem. Soc., Perkin Trans. 1* 2116 (1972).
60. N. K. Hart and J. R. Price, *Aust. J. Chem.* **19**, 2185 (1966).
61. H. Rapoport and K. G. Holden, *J. Am. Chem. Soc.* **82**, 4395 (1960).
62. R. A. Corral and O. O. Orazi, *Tetrahedron* **21**, 909 (1965).
63. S. R. Johns, J. A. Lamberton, and A. A. Sioumis, *Aust. J. Chem.* **20**, 1975 (1967).
64. R. Garestier and M. Rideau, *C. R. Acad. Sci., Ser. D* **274**, 3541 (1972).
65. S. Najjar, G. A. Cordell, and N. R. Farnsworth, *Phytochemistry* **14**, 2309 (1975).
66. P. G. Waterman, *Biochem. Syst.* **1**, 153 (1973).
67. M. F. Grundon and S. A. Surgenor, unpublished work.
68. J. Körösi, K. Szendrei, I. Novák, J. Reisch, G. Blazso, E. Minker, and M. Koltai, *Herba Hung.* **15**, 9 (1976); *CA* **85**, 30608 (1976).

69. N. K. Hart, S. R. Johns, J. A. Lamberton, and J. R. Price, *Aust. J. Chem.* **21**, 1389 (1968).
70. M. Ionescu, I. Mester, and M. Vlassa, *Rev. Roum. Chim.* **13**, 1641 (1968).
71. J. Reisch, K. Szendrei, V. Pápay, E. Minker, and I. Novák, *Tet. Lett.* 1945 (1970).
72. J. Reisch, K. Szendrei, V. Pápay, I. Novák, and E. Minker, *Tet. Lett.* 3365 (1970).
73. J. Reisch, Y. W. Mirhom, J. Körösi, K. Szendrei, and I. Novák, *Phytochemistry* **12**, 2552 (1973); J. Körösi, K. Szendrei, I. Novák, J. Reisch, and E. Minker, *Herba Hung.* **13**, 17 (1974).
74. I. A. Bessonova and S. Yu. Yunusov, *Khim. Prir. Soedin.* 29 (1969); *CA* **71**, 3523 (1969).
75. I. A. Bessonova, Z. Sh. Faizutdinova, Ya. V. Raskes, and S. Yu. Yunusov, *Khim. Prir. Soedin.* 446 (1970); *CA* **73**, 135145 (1970).
76. I. A. Bessonova and S. Yu. Yunusov, *Khim. Prir. Soedin.* 629 (1971).
77. M. F. Grundon and K. J. James, *Tet. Lett.* 4727 (1971).
78. R. M. Bowman, J. F. Collins, and M. F. Grundon, *Chem. Commun.* 1131 (1967); *J. Chem. Soc., Perkin Trans. 1* 626 (1973).
79. R. M. Bowman and M. F. Grundon, *J. Chem. Soc. C* 2368 (1967).
80. R. M. Bowman, G. A. Gray, and M. F. Grundon, *J. Chem. Soc., Perkin Trans. 1* 1051 (1973).
81. J. F. Collins and M. F. Grundon, *Chem. Commun.* 1078 (1969); *J. Chem. Soc., Perkin Trans. 1* 161 (1973).
82. J. Lemmich and B. E. Nielson, *Tet. Lett.* 3 (1969).
83. M. F. Grundon and K. J. James, *Chem. Commun.* 337 (1970).
84. M. F. Grundon and K. J. James, unpublished work.
85. D. L. Dreyer and A. Lee, *Phytochemistry* **11**, 763 (1972).
86. C. D. Adams, D. R. Taylor, and J. M. Warner, *Phytochemistry* **12**, 1359 (1973).
87. V. I. Akhmedzhariova, I. A. Bessonova, and S. Yu. Yunusov, *Khim. Prir. Soedin.* 109 (1974); *CA* **80**, 121153 (1974).
88. M. O. Abe and D. A. H. Taylor, *Phytochemistry* **10**, 1167 (1971).
89. F. Piozzi, P. Venturella, and A. Bellino, *Gazz. Chim. Ital.* **99**, 711 (1969).
90. P. Venturella, A. Bellino, and F. Piozzi, *Heterocycles* **3**, 367 (1975).
91. L. Maat, A. W. Buijen Van Weelderen, and H. C. Beyerman, *Recl. Trav. Chim. Pays-Bas* **92**, 1399 (1973).
92. R. M. Bowman, M. F. Grundon, and K. J. James, *Chem. Commun.* 666 (1970); *J. Chem. Soc., Perkin Trans. 1* 1055 (1973).
93. J. W. Huffman and T. M. Hsu, *Tet. Lett.* 141 (1972).
94. A. De Groot and B. J. M. Jansen, *Tet. Lett.* 3407 (1975).
95. M. F. Grundon and K. J. James, *Chem. Commun.* 1427 (1970), and unpublished work.
96. J. A. Mears, *Phytochemistry* **12**, 2265 (1973).
97. J. R. Cannon and C. D. Shilkin, *Aust. J. Chem.* **24**, 2181 (1971).
98. X. A. Dominguez, D. Butruille, and J. Wapinsky, *Phytochemistry* **10**, 2554 (1971).
99. S. A. Sultanov and S. Yu. Yunusov, *Khim. Prir. Soedin.* 195 (1969); *CA* **71**, 98965 (1969).
100. I. M. Kikvidze, I. A. Bessonova, K. S. Mudjiri, and S. Yu. Yunusov, *Khim. Prir. Soedin.* 675 (1971).
101. J. Rondest, B. C. Das, M. N. Ricroch, C. Kan-Fan, P. Potier, and J. Polonsky, *Phytochemistry* **7**, 1019 (1968).
102. I. A. Benages, M. E. A. De Juarez, S. M. Albonico, A. Urzua, and B. K. Cassels, *Phytochemistry* **13**, 2891 (1974).

103. B. F. Bowden, L. Cleaver, P. K. Ndalut, E. Ritchie, and W. C. Taylor, *Aust. J. Chem.* **28**, 1393 (1975).
104. D. P. Chakraborty and B. K. Barman, *Trans. Bose. Res. Inst., Calcutta* **24**, 121 (1961); *CA* **56**, 14394 (1962).
105. K. Ubaidulaev, I. A. Bessonova, and S. Yu. Yunusov, *Khim. Prir. Soedin.* 343 (1972); *CA* **78**, 2010 (1973).
106. D. Kurbanov, G. P. Sidyakin, and S. Yu. Yunusov, *Khim. Prir. Soedin.* 67 (1967).
107. M. Ionescu and I. Mester, *Phytochemistry* **9**, 1137 (1970).
108. C. A. Mammarella and J. Comin, *An. Asoc. Quim. Argent.* **59**, 239 (1971).
109. E. Bianchi, C. C. J. Culvenor, and J. A. Lamberton, *Aust. J. Chem.* **21**, 2357 (1968).
110. I. Fouraste, J. Gleye, and E. Stanislas, *Plant. Med. Phytother.* **7**, 216 (1973).
111. H. N. Millan and O. M. Silva, *Rev. R. Acad. Cienc. Exactas, Fis. Nat. Madrid* **63**, 637 (1969); *CA* **72**, 107836 (1970).
112. M. Kowalska and B. Borkowski, *Acta Pol. Pharm.* **23**, 295 (1966); *CA* **66**, 17029 (1967).
113. T. N. Vasudevan and M. Luckner, *Pharmazie* **23**, 520 (1968).
114. G. Schneider, *Planta Med.* **13**, 425 (1965).
115. B. Weinstein and A. R. Craig, *Phytochemistry* **10**, 2556 (1971).
116. Y. Asahina, T. Ohta, and M. Inubuse, *Ber.* **63**, 2045 (1930).
117. I. Ishii, K. Hosoya, T. Ishikawa, E. Ueda, and J. Haginiwa, *J. Pharm. Soc. Jpn.* **94**, 322 (1974); *CA* **81**, 132753 (1974).
118. J. Vaquette, J. L. Pousset, R. R. Paris, and A. Cavé, *Phytochemistry* **13**, 1257 (1974).
119. F. Werny and P. J. Scheuer, *Tetrahedron* **19**, 1293 (1963).
120. V. I. Frolova, A. D. Kuzovkov, and P. N. Kibal'chich, *Zh. Obshch. Khim.* **34**, 3499 (1964).
121. J. Vrkoc and P. Sedmera, *Phytochemistry* **11**, 2647 (1972).
122. Ha-Huy-Ke and M. Luckner, *Pharmazie* **21**, 771 (1966).
123. V. S. Asatiani, I. M. Kikvidze, I. A. Bessonova, K. S. Mukzipi, and S. Yu. Yunusov, *Svobschch. Akad. Nauk Graz. SSR* **64**, 85 (1971); *CA* **76**, 56582 (1972).
124. I. Ya. Isaev and I. A. Bessonova, *Khim. Prir. Soedin.* 815 (1974); *CA* **82**, 121677 (1975).
125. D. L. Dreyer, *Tetrahedron* **22**, 2923 (1966).
126. N. Decaudain, N. Kunesch, and J. Poisson, *Phytochemistry* **13**, 505 (1974).
127. F. Abe, S. Yahara, K. Kubo, G. Nonaka, H. Okabe, and I. Nishioka, *Chem. Pharm. Bull.* **22**, 2650 (1974); F. Abe, M. Furukawa, G. Nonaka, H. Okabe, and I. Nishioka, *Yakagaku Zasshi* **93**, 624 (1973); *CA* **79**, 50741 (1973).
128. D. L. Dreyer, M. V. Pickering, and P. Cohan, *Phytochemistry* **11**, 705 (1972).
129. S. R. Johns and J. A. Lamberton, *Aust. J. Chem.* **19**, 895 (1966).
130. X. A. Dominguez, A. Canales, J. A. Garza, E. Gomez, and L. Garza, *Phytochemistry* **10**, 1966 (1971).
131. S. Goodwin, A. F. Smith, A. A. Velasquez, and E. C. Horning, *J. Am. Chem. Soc.* **81**, 6209 (1959).
132. T.-H. Yang, S.-T. Lu, S.-J. Wang, T.-W. Wang, J.-H. Lin and I.-S. Chen, *Yakugaku Zasshi* **91**, 782 (1971).
133. J. Vaquette, J. L. Pousset, and A. Cavé, *Plant. Med. Phytother.* **8**, 72 (1974).
134. C. Kan-Fan, B. C. Das, P. Boiteau, and P. Potier, *Phytochemistry* **9**, 1283 (1970).
135. J. E. T. Corrie, G. H. Green, E. Ritchie, and W. C. Taylor, *Aust. J. Chem.* **23**, 133 (1970).
136. A. M. Duffield, P. R. Jefferies, and P. H. Lucich, *Aust. J. Chem.* **15**, 812 (1962).
137. V. I. Frolova, A. I. Bankovski, and M. B. Volinskaya, *Med. Ind. SSSR* **12**, 35 (1958).

138. J. C. Vitagliano and J. Comin, *An. Asoc. Quim. Argent.* **58**, 59 (1970).
139. S. R. Johns, J. A. Lamberton, and A. A. Sioumis, *Aust. J. Chem.* **21**, 1897 (1968).
140. J. M. Calderwood, N. Finkelstein, and F. Fish, *Phytochemistry* **9**, 675 (1970).
141. J. M. Calderwood and F. Fish, *J. Pharm. Pharmacol.* **18**, Suppl., 1195 (1966).
142. F. Fish and P. G. Waterman, *Phytochemistry* **11**, 1866 (1972).
143. F. Fish and P. G. Waterman, *Phytochemistry* **10**, 3322 (1971).
144. R. Goto, *Yakugaku Zasshi* **61**, 91 (1941); *CA* **35**, 7971 (1941).
145. N. Morita, M. Arisawa, and T. Takegaki, *Yakugaku Zasshi* **87**, 1017 (1967); *CA* **68**, 195892 (1968).
146. F. Fish and P. G. Waterman, *J. Pharm. Pharmacol.* **23**, Suppl., 132S (1971).
147. P. F. Jefferies, cited by J. R. Price, *in* "Chemical Plant Taxonomy" (T. Swain, ed.), p. 429. Academic Press, New York, 1963.
148. D. M. Gulyamova, I. A. Bessonova, and S. Yu. Yunusov, *Khim. Prir. Soedin.* 850 (1971).
149. S. A. Sultanov, V. I. Pastuklova, and S. Yu. Yunusov, *Khim. Prir. Soedin.* 355 (1967).
150. E. F. Nesmelova and G. P. Sidyakin, *Khim. Prir. Soedin.* 550 (1973); *CA* **80**, 105845 (1973).
151. S. Yu. Yunusov and G. P. Sidyakin, *Zh. Obshch. Khim.* **22**, 1055 (1952).
152. Z. Sh. Fayzutdinova, G. P. Sidyakin, and S. Yu. Yunusov, *Dokl. Akad. Nauk Uzb. SSR* No. 1, 35 (1966).
153. R. Rouffiac, I. Fovraste, and E. Stanislas, *Planta Med.* **17**, 361 (1969).
154. T. R. Govindachari and N. Wiswanathan, *Indian J. Chem.* **5**, 280 (1967).
155. V. Yu. Vachnadze, V. M. Malikov, K. S. Mudzhiri, and S. Yu. Yunusov, *Khim. Prir. Soedin.* 676 (1971).
156. H. Ishii and K. Harada, *Yakugaku Zasshi* **81**, 238 (1961); *CA* **55**, 14495 (1961); M. Tomita and H. Ishii, *ibid.* **79**, 1228 (1959); *CA* **55**, 1677 (1961).
157. F. Fish, I. A. Meshal, and P. G. Waterman, *Phytochemistry* **14**, 2094 (1975).
158. N. Weber, *Ber.* **106**, 3769 (1973).
159. D. F. Theumann and J. H. Comin, *An. Asoc. Quim. Argent.* **55**, 253 (1967).
160. A. K. Ganguly, T. R. Govindachari, A. Manmade, and P. A. Mohamed, *Indian J. Chem.* **4**, 334 (1966).
161. I. M. Fakhrutdinova, S. P. Sidyakin, and S. Yu. Yunusov, *Khim. Prir. Soedin.* 107 (1965).
162. D. Basu and R. Sen, *Phytochemistry* **13**, 2329 (1974).
163. S. M. Sharafutdinova and S. Yu. Yunusov, *Khim. Prir. Soedin.* 198 (1968).
164. D. Kurbanov, I. A. Bessonova, and S. Yu. Yunusov, *Khim. Prir. Soedin.* 58 (1968).
165. S. P. Sidyakin and S. Yu. Yunusov, *Dokl. Akad. Nauk Uzb. SSR* No. 4, 39 (1962).
166. G. K. Hughes, K. G. Neill, and E. Ritchie, *Aust. J. Sci. Res., Ser. A* **5**, 401 (1952).
167. A. G. Gonzáles, R. Moreno Ordonez, and F. Rodriquez Luis, *An. Quim.* **68**, 1133 (1972).
168. T. Shakirov, G. P. Sidyakin, and S. Yu. Yunusov, *Dokl. Akad. Nauk Uzb. SSR* **6**, 28 (1959).
169. M. Ionescu, M. Vlassa, I. Mester, and E. C. Vicol, *Rev. Roum. Biochem.* **8**, 123 (1971); *CA* **75**, 115908.
170. R. F. C. Brown, P. T. Gilham, G. K. Hughes, and E. Ritchie, *Aust. J. Chem.* **7**, 181 (1954); R. H. Prager, E. Ritchie, and W. C. Taylor, *ibid.* **13**, 380 (1960).
171. F. N. Lahey and M. McCamish, *Tet. Lett.* 1525 (1968); F. N. Lahey, M. McCamish, and T. McEwan, *Aust. J. Chem.* **22**, 447 (1969).
172. D. L. Dreyer, *J. Org. Chem.* **35**, 2420 (1970).
173. D. Kurbanov, I. A. Bessonova, and S. Yu. Yunusov, *Khim. Prir. Soedin.* 373 (1968).

174. I. A. Bessonova and S. Yu. Yunusov, *Khim. Prir. Soedin.* 52 (1974); *CA* **80**, 121152m (1974).
175. V. I. Akhmedzhanova, I. A. Bessonova, and S. Yu. Yunusov, *Khim. Prir. Soedin.* 680 (1974); *CA* **82**, 73261 (1975).
176. E. F. Nesmelova, I. A. Bessonova, and S. Yu. Yunusov, *Khim. Prir. Soedin.* 666 (1975); *CA* **84**, 105863 (1976).
177. Kh. A. Abdullaeva, I. A. Bessonova, and S. Yu. Yunusov, *Khim. Prir. Soedin.* 684 (1974); *CA* **82**, 73260 (1975).
178. Z. Sh. Fayzutdinova, I. A. Bessonova, and S. Yu. Yunusov, *Khim. Prir. Soedin.* 356 (1967); *CA* **68**, 69160 (1968).
179. Z. Sh. Fayzutdinova, I. A. Bessonova, and S. Yu. Yunusov, *Khim. Prir. Soedin.* 360 (1968); *CA* **71**, 3517 (1969).
180. P. T. O. Chang, G. H. Aynilian, G. A. Cordell, M. Tin-Wa, H. H. S. Fong, R. E. Perdue, and N. Farnsworth, *J. Pharm. Sci.* **64**, 561 (1976).
181. M. Gellert, I. Novak, K. Szendrei, J. Reisch, and E. Minker, *Herba Hung.* **10**, 123 (1971); *CA* **79**, 2768 (1973).
182. I. A. Bessonova, V. I. Akhmedzhanova, and S. Yu. Yunusov, *Khim. Prir. Soedin.* 677 (1974); *CA* **82**, 86462 (1975).
183. V. I. Akhmedzhanova, I. A. Bessonova, and S. Yu. Yunusov, *Khim. Prir. Soedin.* 272 (1975); *CA* **83**, 97662 (1975).
184. D. M. Razakova, I. A. Bessonova, and S. Yu. Yunusov, *Khim. Prir. Soedin.* 812 (1975); *CA* **84**, 180443 (1976).
185. F. N. Lahey, I. Lauder, and M. McCamish, *Aust. J. Chem.* **22**, 431 (1969).
186. D. M. Razakova, I. A. Bessonova, and S. Yu. Yunusov, *Khim. Prir. Soedin.* 791 (1976); *CA* **86**, 140310 (1977).
187. R. A. Prager and G. R. Skurray, *Aust. J. Chem.* **28**, 1037 (1968).
188. T. R. Govindachari, S. Prabhakar, V. N. Ramachandran, and B. R. Pai, *Indian J. Chem.* **9**, 1031 (1971).
189. S. Prabhakar, B. R. Pai, and V. N. Ramachandran, *Indian J. Chem.* **8**, 857 (1970).
190. V. N. Ramachandran, B. R. Pai, N. Somasundaran, and C. S. Swaninathan, *Indian J. Chem.* **11**, 1088 (1973).
191. S. Prabhakar, B. R. Pai, and V. N. Ramachandran, *Indian J. Chem.* **9**, 191 (1971).
192. F. Piozzi, P. Venturella, and A. Bellino, *Org. Prep. Proced. Int.* **3**, 223 (1971).
193. T. Sekiba, *Bull. Chem. Soc. Jpn.* **46**, 577 (1973).
194. N. S. Narasimhan and M. V. Paradkar, *Chem. Ind.* (*London*) 831 (1967); N. S. Narasimhan and R. H. Alurkar, *ibid.* 515 (1968).
195. J. F. Collins, G. A. Gray, M. F. Grundon, D. M. Harrison, and C. G. Spyropoulos, *Chem. Commun.* 1029 (1972).
196. J. F. Collins, G. A. Gray, M. F. Grundon, D. M. Harrison, and C. G. Spyropoulos, *J. Chem. Soc., Perkin Trans. 1* 94 (1973).
197. N. S. Narasimhan and R. S. Mali, *Tet. Lett.* 843 (1973); *Tetrahedron* **30**, 4153 (1974).
198. T. R. Chamberlain and M. F. Grundon, *Tet. Lett.* 3547 (1967); *J. Chem. Soc. C* 910 (1971).
199. T. R. Chamberlain, J. F. Collins, and M. F. Grundon, *Chem. Commun.* 1269 (1969).
200. S. B. Talapatra, B. C. Maiti, B. Talapatra, and B. C. Das, *Tet. Lett.* 4789 (1969); S. B. Talapatra, B. C. Maiti, and B. Talapatra, *ibid.* 2683 (1971).
201. S. M. Sharafutdinova and S. Yu. Yunusov, *Khim. Prir. Soedin.* 198 (1968); *CA* **69**, 87260 (1968).
202. S. M. Sharafutdinova and S. Yu. Yunusov, *Khim. Prir. Soedin.* 394 (1969); *CA* **72**, 67152 (1970).

203. Z. Sh. Fayzutdinova, I. A. Bessonova, Ya. V. Rashkes, and S. Yu. Yunusov, *Khim. Prir. Soedin.* 239 (1970); *CA* **73**, 131179 (1970).
204. Ya. V. Rashkes, Z. Sh. Faizutdinova, I. A. Bessonova, and S. Yu. Yunusov, *Khim. Prir. Soedin.* 577 (1970); *CA* **74**, 54044 (1971).
205. Ya. V. Rashkes, I. A. Bessonova, and S. Yu. Yunusov, *Khim. Prir. Soedin.* 336 (1972); *CA* **77**, 152403 (1972) and *Khim. Prir. Soedin.* 364 (1974); *CA* **81**, 152456 (1974).
206. M. E. Donnelly and M. F. Grundon, unpublished results.
207. Z. Sh. Fayzutdinova, I. A. Bessonova, and S. Yu. Yunusov, *Khim. Prir. Soedin.* 455 (1969); *CA* **72**, 79267 (1970).
208. A. G. Kozlovskii, M. U. Arinbasarov, G. I. Yakovlev, A. M. Zyakan, and V. M. Adanin, *Izv. Akad. Nauk SSSR, Ser. Khim.* 1146 (1976); *CA* **86**, 29964 (1977).
209. D. M. Razzakova, I. A. Bessonova, and S. Yu. Yunusov, *Khim. Prir. Soedin.* 206 (1973); *CA* **79**, 32147 (1973).
210. D. M. Razzakova, I. A. Bessonova, and S. Yu. Yunusov, *Khim. Prir. Soedin.* 755 (1972); *CA* **78**, 84605 (1973).
211. Ha-Huy-Ke, M. Luckner, and J. Reisch, *Phytochemistry* **9**, 2199 (1970).
212. M. F. Grundon and H. M. Okely, unpublished results.
213. J. Reisch, I. Novák, K. Szendrei, and E. Minker, *Naturwissenschaften* **54**, 517 (1967).
214. J. Reisch, Zs. Rózsa, K. Szendrei, and J. Körösi, *Phytochemistry* **14**, 840 (1975).
215. T. Kamikado, C.-F. Chang, S. Murakoshi, and A. Sakurai, *Agric. Biol. Chem.* **40**, 605 (1976).
216. G. J. Kapdia, B. D. Paul, J. V. Silverton, M. H. Fales, and E. A. Sokoloski, *J. Am. Chem. Soc.* **97**, 6814 (1975).
217. R. Tschesche and W. Werner, *Tetrahedron* **23**, 1873 (1967).
218. A. G. Gonzalez, R. Estevez Reyes, and E. Diaz Chicho, *An. Quim.* **70**, 281 (1974).
219. P. Venturella, A. Bellino, F. Piozzi, and M. L. Marino, *Heterocycles* **4**, 1089 (1976).
220. I. Monković and I. D. Spenser, *Chem. Commun.* 204 (1966); I. Monković, I. D. Spenser, and A. D. Plunkett, *Can. J. Chem.* **45**, 1935 (1967).
221. M. Matsuo, M. Yamazaki, and Y. Kashida, *Biochem. Biophys. Res. Commun.* **23**, 679 (1966); M. Matsuo and Y. Kashida, *Chem. Pharm. Bull.* **14**, 1108 (1966).
222. E. Atkinson, D. R. Boyd, and M. F. Grundon, unpublished work.
223. A. O. Colonna and E. G. Gros, *Chem. Commun.* 674 (1970); *Phytochemistry* **10**, 1515 (1971).
224. J. F. Collins and M. F. Grundon, *Chem. Commun.* 622 (1969); M. F. Grundon and K. J. James, *ibid.* 1311 (1971); J. F. Collins, W. J. Donnelly, M. F. Grundon, and K. J. James, *J. Chem. Soc., Perkin Trans. 1* 2177 (1974).
225. M. Cobet and M. Luckner, *Eur. J. Biochem.* **4**, 76 (1968).
226. M. Cobet and M. Luckner, *Phytochemistry* **10**, 1031 (1971).
227. R. Aneja, S. K. Mukerjee, and T. R. Seshadri, *Tetrahedron* **4**, 256 (1958).
228. A. J. Birch and H. Smith, *Chem. Soc., Spec. Publ.* **12**, 1 (1958); A. J. Birch, M. Muang, and A. Pelter, *Aust. J. Chem.* **22**, 1923 (1969).
229. J. A. Diment, E. Ritchie, and W. C. Taylor, *Aust. J. Chem.* **22**, 1797 (1969).
230. M. F. Grundon, D. M. Harrison, and C. G. Spyropoulos, *J. Chem. Soc., Chem. Commun.* 51 (1974); *J. Chem. Soc., Perkin Trans. 1* 302 (1975).
231. J. F. Collins and M. F. Grundon, unpublished work.
232. C. R. Hall and R. A. Prager, *Aust. J. Chem.* **22**, 2437 (1969).
233. C. Ritler and M. Luckner, *Eur. J. Biochem.* **18**, 391 (1971).
234. M. Blasche-Cobet and M. Luckner, *Phytochemistry* **12**, 2393 (1973).
235. M. F. Grundon, D. M. Harrison, and C. G. Spyropoulos, unpublished work.

236. M. Luckner and K. Mothes, *Tet. Lett.* 1035 (1962); M. Luckner, *Eur. J. Biochem.* **2**, 74 (1967); M. Luckner, K. Winter, and J. Reisch, *ibid.* **7**, 380 (1969); L. Nover and M. Luckner, *ibid.* **10**, 268 (1969).
237. J. Framm, L. Nover, A. El Azzouny, H. Richter, K. Winter, S. Werner, and M. Luckner, *Eur. J. Biochem.* **37**, 78 (1973); E. A. Aboutabl and M. Luckner, *Phytochemistry* **14**, 2573 (1975).
238. M. S. Hifnawy, J. Vaquette, T. Sévenet, J-L. Pousset, and A. Cavé, *Planta Med.* **29**, 346 (1976).

—CHAPTER 3—

THE *ASPIDOSPERMA* ALKALOIDS

GEOFFREY A. CORDELL

College of Pharmacy
University of Illinois
Chicago, Illinois

ISBN 0-12-469517-5

I. Introduction

The range of work conducted on the alkaloids derived from *Aspidosperma* and related species of plants has broadened considerably since this area was last discussed in Volume XI of this series. Because of this, several revisions have been necessary in order to adequately present this material. A major compromise has been the omission of essentially all data on the dimeric alkaloids which, it is hoped, will be treated in a subsequent volume in this series.

The organization of this chapter is therefore along the following lines. First, the isolation of new monomeric *Aspidosperma* alkaloids and those of *Melodinus, Tabernaemontana*, and *Ochrosia* species are discussed. This is followed by successive discussions of the chemical and extensive synthetic reactions carried out on these alkaloids. Finally, the recent use of X-ray crystallography and ^{13}C NMR techniques in the structure elucidation of these alkaloids is summarized.

Several reviews are pertinent to the discussion. The Swiss group has discussed a different approach to the classification of indole alkaloids along biogenetic grounds (*1*) and has described the biogenetic routes in fascinating detail. In a complementary review, this author has discussed comprehensively the biosynthetic experimentation on monoterpenoid indole alkaloids (*2*). In addition, Kapil and Brown have reviewed the closely related area of nitrogenous glycosides in this volume (Chapter 5).

Gabetta (*3*) has summarized the indole alkaloids isolated between 1968 and mid-1972, and Aliev and Babaev (*4*) have discussed the physical properties of the many *Aspidosperma*-type alkaloids isolated from *Vinca* species.

The numbering system used throughout this review is that of Le Men and Taylor (*5*).

II. Isolation and Structure Elucidation of the *Aspidosperma* Alkaloids

A. β-Anilinoacrylate Alkaloids

A substantial number of the new *Aspidosperma* alkaloids isolated during the period of this review are of the β-anilinoacrylic ester type. That is, they contain the unit **1**, and depending on additional aromatic substitution, give rise to a quite characteristic UV λ_{max} at 328 nm. Of major import in the structure elucidation of these alkaloids is the mass spectrum. The retro-Diels–Alder fragmentation pathway occurs in classic fashion in ring C of these alkaloids, and after cleavage through the tryptamine bridge two ions are apparent. One of these contains the indole nucleus and carbons 6, 16, and 17 (ion A). The second ion contains carbon 5, the piperidine ring, and the ethyl side chain (ion B). An alternative major pathway involves loss of the two carbon side chain (radical C) at C-20 to afford the pentacyclic system (ion D).

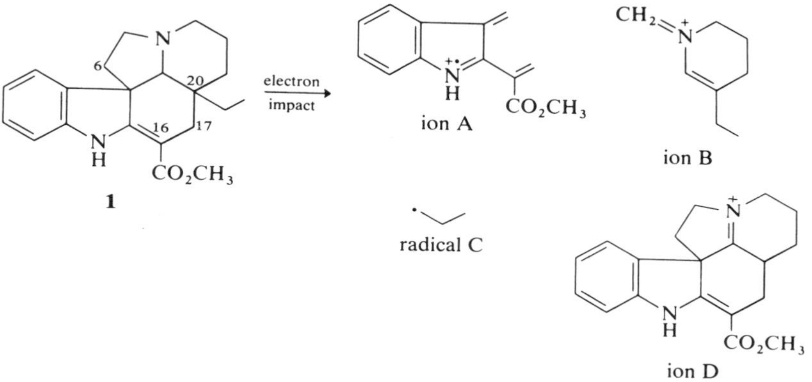

These ions make available a considerable amount of structure information. Substitution in the indole nucleus leads to an increase in the mass of ion A, whereas additional functionalization of the piperidine rings increases the mass of species B and D. If substitution is limited to the two-carbon side chain, then the masses of B and C increase but not the mass of ion D. Examination of these four fragments therefore reveals the location of additional functional groups. For the simplest anilinoacrylate derivative, vincadifformine (**2**), the mass values of these ions are: ion A, 214; ion B, 124; ion C, 29; and ion D, 309.

1. 11-Methoxyvincadifformine (**4**)

Although numerous alkaloids have been isolated from *Vinca minor* L., Dopke and Meisel have isolated a new amorphous β-anilinoacrylate derivative from the leaves (*6*).

Typical UV (λ_{max} 245 and 327 nm) and IR (ν_{max} 1685 and 1620 cm^{-1}) spectral data confirmed the presence of the anilinoacrylate unit, but in addition the IR spectrum indicated the presence of a 1,2,4-trisubstituted aromatic nucleus. The mass spectrum with a base peak at m/e 244 (ion A) indicated that the substituent was probably a methoxy group. This group was placed at the 11-position, even though no NMR studies or chemical correlations with 11-methoxytabersonine (**3**) were carried out. The compound is therefore 11-methoxyvincadifformine (**4**) (*6*).

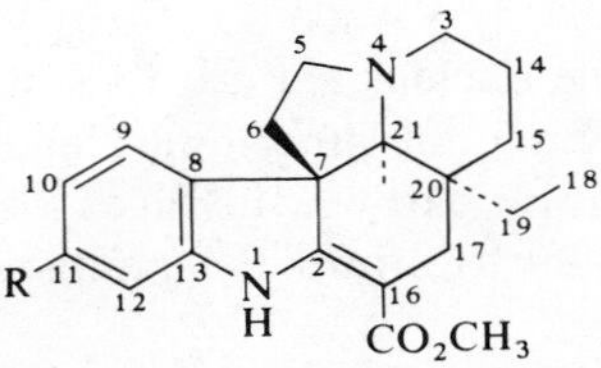

2 R = H
4 R = OCH_3
6 R = OCH_3, C-21Hβ

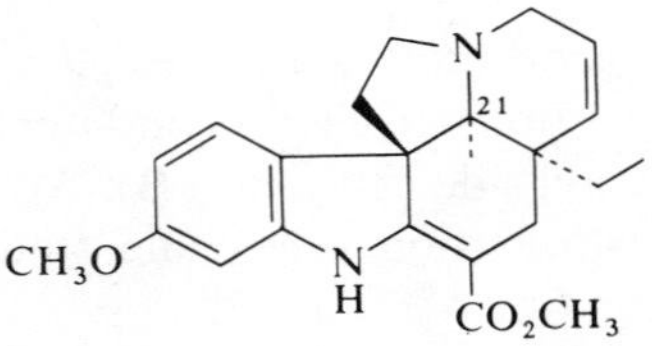

3
7 C-21Hβ

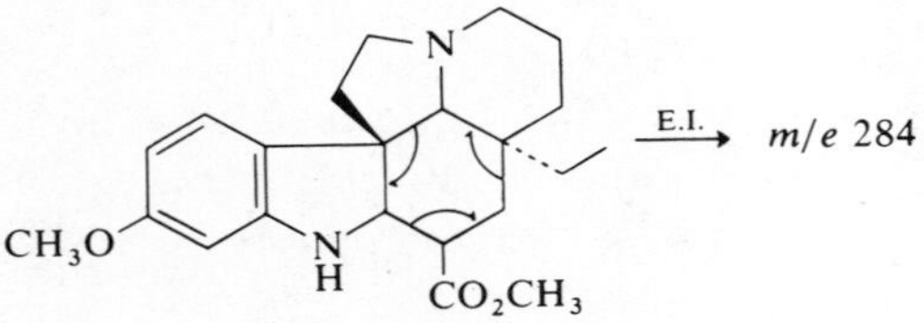

5 $\xrightarrow{\text{E.I.}}$ *m/e* 284

2. Ervinceine (**6**) and Ervamicine (**7**)

The Tashkent group has, in the recent past, provided a number of interesting alkaloids from the apocynaceous plant *Vinca erecta* Rgl. et Schmalh. The petroleum ether-soluble alkaloid fraction gave a base, ervinceine {mp 99–100°; $[\alpha]_D -488°$}, exhibiting λ_{max} 248 and 328 nm and carbonyl absorption at 1672 cm^{-1} (*7*). The mass spectrum confirmed the β-anilinoacrylate nature of the compound, as a base peak was observed at *m/e* 124 (ion B). In the PMR spectrum of ervinceine three aromatic protons (6.14–6.91 ppm), a three-proton singlet (3.69 ppm) for a carbomethoxy group, and a second three-proton singlet (3.66 ppm) for an aromatic methoxyl group were observed. The methyl triplet of the ethyl side chain appeared at 0.53 ppm, strongly shielded by the aromatic nucleus. The coupling constants of the aromatic nucleus (*8*) clearly indicated a 1,2,4-trisubstituted aromatic nucleus.

Reduction of ervinceine with zinc in methanolic sulfuric acid gave a dihydroderivative showing λ_{max} 249, 305 nm, typical of an indoline. The mass spectrum showed a molecular ion at M^+ 370, a base peak at *m/e* 124, and an important fragment ion at *m/e* 284 (*7*). The latter is derived by retro-Diels–Alder reaction in ring C to give a loss of C-16 and C-17. Comparison of the UV spectrum of this dihydro derivative with 7-methoxytetrahydrocarbazole indicated that the methoxy group in dihydroervinceine (**5**) was located at C-11. Ervinceine therefore has the structure **6** (*7, 8*) and is isomeric with 11-methoxyvincadifformine (**4**).

Also obtained from *V. erecta* was an amorphous base, ervamicine (*8*) {$[\alpha]_D -264.4°$}. Ervamicine exhibited a number of structural similarities to ervinceine (**6**); in particular, three aromatic protons were observed at 6.94, 6.29, and 6.18 ppm showing coupling constants in agreement with a 1,2,4-substitution pattern. Carbomethoxyl (3.63 ppm) and aromatic methoxyl (3.66 ppm) were found, and the methyl of the ethyl group was located at 0.58 ppm. The mass spectrum of ervamicine indicated a molecular ion at 366 mu, two mass units less than ervinceine. The additional degree of unsaturation was traced to the 14,15-position in the piperidine ring, for two olefinic protons were observed at 5.61 and 5.53 ppm showing a coupling constant of 10 Hz, and a two-proton, complex doublet of doublets at 3.30 ppm for the N–CH_2–C=C group.

Catalytic hydrogenation of ervamicine gave a 14,15-dihydro derivative identical with ervinceine (**6**). Ervamicine therefore has the structure **7** (*8*) and is a stereoisomer of 11-methoxytabersonine (**3**) (*9*). The melting points of the hydrochlorides of these two compounds are very different, 213–214° for ervamicine hydrochloride and 184–186° for 11-methoxytabersonine hydrochloride (*9*).

3. 11-Methoxyminovincinine (**10**)

Continuing their work on the isolation of alkaloids of *Vinca minor* Döpke *et al.* (*10*) obtained a new amorphous β-anilinoacrylate alkaloid ($C_{22}H_{28}N_2O_4$).

The mass spectrum exhibited losses of 15, 18, and 45 mu from the parent ion and a base peak at m/e 140 (ion B), thereby establishing a hydroxy group in the side chain. Lithium aluminum hydride reduction gave a diol **8** identical with that obtained from 11-methoxyminovincine (**9**), and Oppenauer oxidation gave 11-methoxyminovincine. No PMR data were obtained, but on this evidence the structure of 11-methoxyminovincinine (**10**) was proposed (*10*).

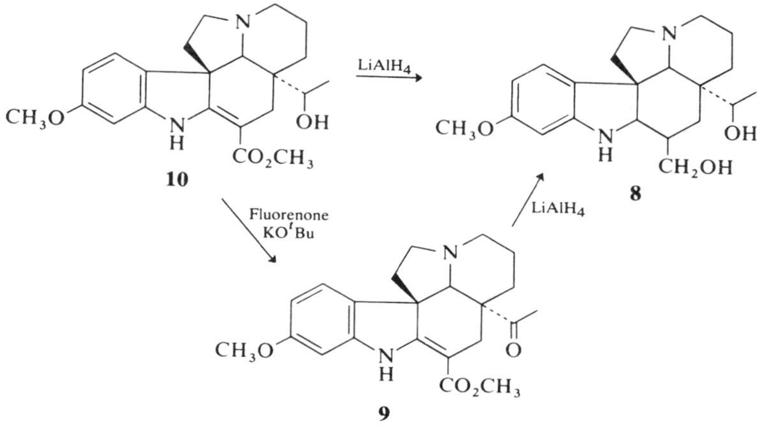

CH_3O N H OH CO_2CH_3 **10**

$LiAlH_4$

CH_3O N H OH CH_2OH **8**

Fluorenone KO^tBu

$LiAlH_4$

CH_3O N H O CO_2CH_3 **9**

The stereochemistry of C-21 in 11-methoxyminovincine and 11-methoxyminovincinine was determined by a method developed for the minovincine/minovincinine series (*11*).

Reduction of minovincine (**11**) with sodium borohydride gave a mixture of the 19-epimers of minovincinine, (**12**) and (**13**). When each of these compounds was treated with zinc in methanolic sulfuric acid, 2,16-reduction and lactonization occurred to afford **14** and **15**, respectively. From molecular models, once the stereochemistry of the tryptamine bridge and of C-20 are described, only the stereochemistry in which the C/D ring junction is *cis* will permit lactonization. Further work indicated that in **14** C-19 had the *R* configuration whereas in **15** this center was *S*. Interestingly, there was quite a substantial difference in the $[\alpha]_D$ of these two compounds (*11*).

Application of this sequence in the 11-methoxyminovincine series (*12*) indicated the same stereochemical relationship for the C-21 proton, i.e., *cis* to the C-20 substituent.

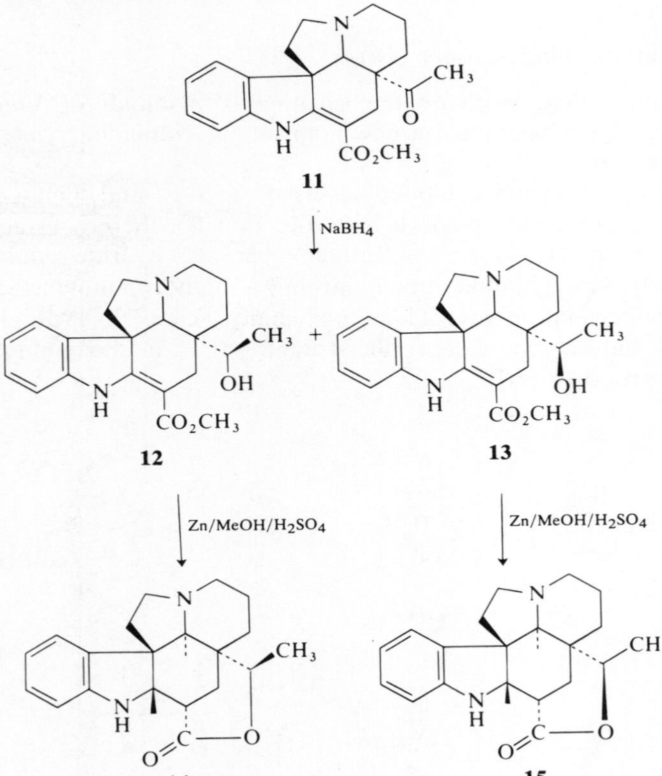

11

$NaBH_4$

12 + **13**

$Zn/MeOH/H_2SO_4$ $Zn/MeOH/H_2SO_4$

14 **15**

4. 19*R*-Hydroxytabersonine (**18**)

The roots of *Catharanthus lanceus* Baj ex A. DC. have afforded another new β-anilinoacrylate alkaloid (*13*). From the UV and IR spectra and the high negative optical rotation, the alkaloid was determined to be a member of the β-anilinoacrylate series. The NMR spectrum confirmed the NH and carbomethoxyl groups and also showed four aromatic protons. Two olefinic protons were observed at 5.9 ppm, confirming the unsaturation at C-14–C-15. The C-18 methyl group appeared as a doublet ($J = 7.0$ Hz) at 0.90 ppm, and because an alcohol function had been indicated from the IR spectrum, this group could be placed at C-19. The mass spectrum, showing M^+ at *m/e* 352, although generally similar to that of related alkaloids, showed *m/e* 151 (**16**) rather than *m/e* 138 (**17**) as the base peak. Catalytic hydrogenation of the alkaloid afforded 19-epiminovincinine (**13**) (*11*), and the parent compound could therefore be described as 19*R*-hydroxytabersonine (**18**) (*13*).

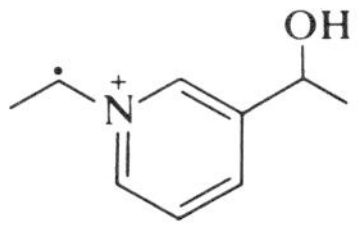

OH N $\overset{+}{N}$ H CO_2CH_3 CH_2

m/e 151

16

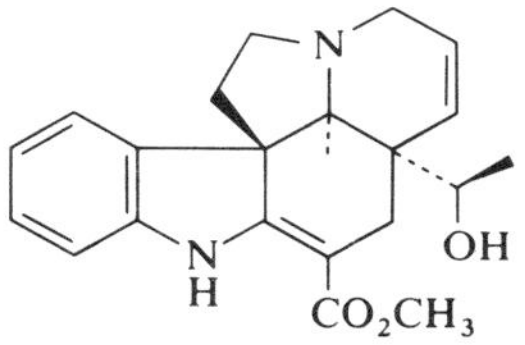

18

HO

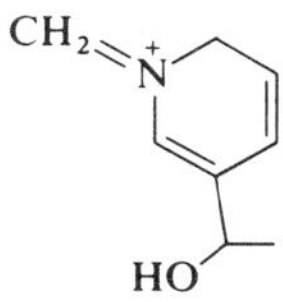

m/e 138

17

5. (+)-3-Oxominovincine (**19**), (+)-Minovincine (**20**), and (−)-5-Oxominovincine (**23**)

A recent development in the isolation of the β-anilinoacrylate alkaloids is the discovery of lactam alkaloids which are only very weakly basic. The first of these alkaloids to be obtained was (+)-3-oxominovincine (**19**) from *Tabernaemontana riedelii* Muell. Arg. (*14*).

The UV spectrum (λ_{max} 298 and 330 nm) and the IR spectrum (ν_{max} 3390, 1680, 1620 cm^{-1}) demonstrated the presence of a β-anilinoacrylate derivative, but an additional carbonyl absorption was also observed at 1660 cm^{-1} The molecular weight from mass spectrometry was 366, and the base peak in the mass spectrum was at *m/e* 214 (ion A) with a major fragment ion at *m/e* 154.

The PMR spectrum indicated a methyl ketone side chain (three-proton singlet at 1.93 ppm) and a carbomethoxyl group at 3.75 ppm. Four aromatic protons confirmed the location of the additional oxygenation to be in the D or E rings.

Before discussing the location of the lactam carbonyl, it should be mentioned that the major alkaloid from the aboveground parts of *T. riedelii* was also a β-anilinoacrylate derivative (*14*) showing a carbomethoxyl signal (3.71 ppm) and a methyl ketone (1.82 ppm) in the PMR spectrum. In the mass spectrum major ions were observed at *m/e* 309 (ion D), *m/e* 214 (ion A), and *m/e* 138 (ion B) and comparison with (−)-minovincine (**11**) confirmed the close similarity. An important difference was noted, however, in the optical rotation of this compound: Instead of exhibiting the usual negative rotation, an $[\alpha]_D$ of +340° was observed. This compound is therefore (+)-minovincine (**20**) (*14*).

The lactam alkaloid showed $[\alpha]_D + 269°$ and was chemically correlated with (+)-minovincine (**20**). Permanganate oxidation of **20** gave, in very low yield, a mixture of two lactams one of which was identical with the natural product. The second (major) lactum showed an IR band at 1670 cm^{-1} which was attributed to a γ-lactam having the structure **21**. The natural lactam must therefore be (+)-3-oxominovincine (**19**) with the absolute stereochemistry shown.

A second lactam alkaloid obtained from natural sources turned out to be quite closely related to the second synthetic lactam obtained by Cava *et al.* (*14*).

From *Vinca minor* Meisel and Döpke (*15*) obtained a small quantity of an anilinoacrylate alkaloid having a molecular weight of 366 and major ions at *m/e* 214 and 154. A broad band at 1696–1684 cm^{-1} was interpreted as being due to a γ-lactam functionality, and this was supported by the mass spectrum which showed an ion at *m/e* 241, thought to be due to **22**. Potassium permanganate oxidation of (−)-minovincine (**11**) gave a mixture of two lactams, one of which was apparently identical with the natural product, which therefore has the structure (−)-5-oxominovincine (**23**). It does not appear that the samples from the two oxidation reactions were ever directly compared.

6. Ervinidinine (**25**)

Ervinidinine {mp 265–266°; $[\alpha]_D - 160.6°$} isolated from *V. erecta* Regl. et Schmalh. (*16*) gave typical UV (298 and 332 nm) and IR (1680, 1600 cm^{-1}) spectral data for a β-anilinoacrylate derivative. The molecular ion appeared at *m/e* 352 and the base peak was found at *m/e* 214 (ion A), indicating a lack of substitution in the indole nucleus. The additional

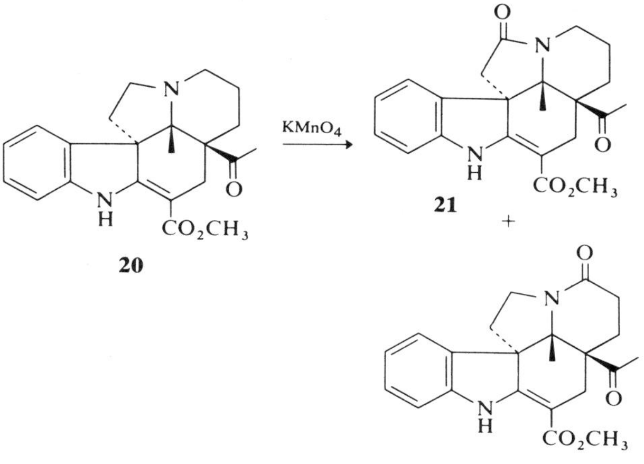

20 $\xrightarrow{KMnO_4}$ **21** + **19**

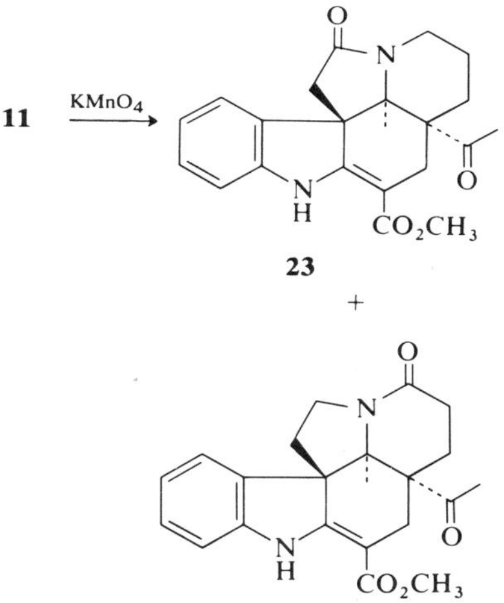

11 $\xrightarrow{KMnO_4}$ **23** +

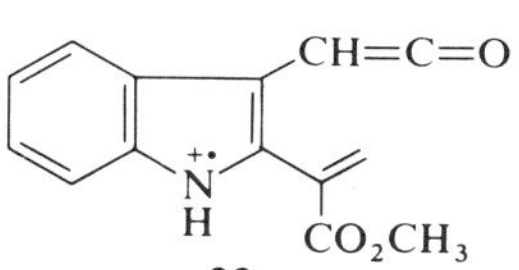

22

oxygen could therefore be traced to the side chain or the piperidine ring. The PMR spectrum indicated the methyl of the ethyl group to be at 0.62 ppm. Together, these data were taken to indicate a structure **24** for ervinidine, isomeric with lochnericine (mp 190–193°) (*17*).

Subsequently, Malikov and Yunusov (*18*) reinvestigated the structure of ervinidinine. Reduction with zinc in methanolic sulfuric acid gave the 2,16-dihydro derivative. The IR spectrum showed a nonconjugated carbonyl group at 1725 cm^{-1} attributed to the carboxymethyl group, and a second carbonyl at 1675 cm^{-1} which was attributed to a lactam carbonyl group in a five-membered ring. Ervinidinine therefore has the revised structure **25**, and the stereochemistry of C-21 remains unknown (*18*).

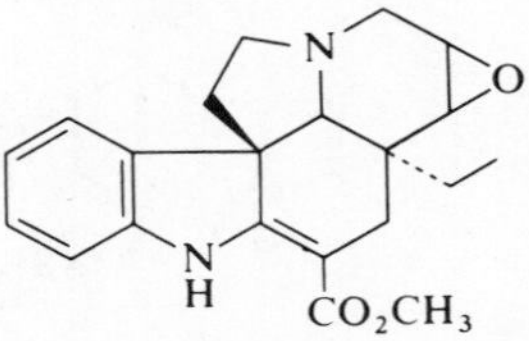

24 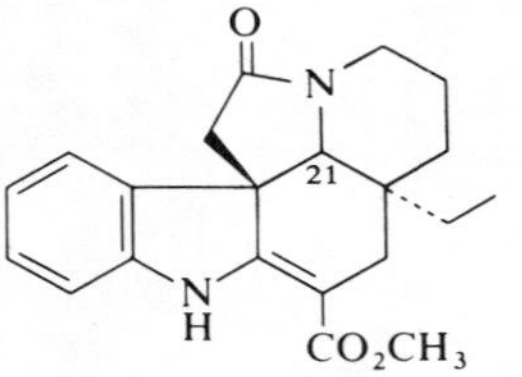 **25**

7. Baloxine (**26**)

From the leaves of *Melodinus balansae* Baill. Mehri and co-workers (*19*) obtained a new anilinoacrylate derivative, baloxine (**26**). The molecular ion at *m/e* 368 analyzed for $C_{21}H_{24}N_2O_4$ and the remainder of the mass spectrum showed typical fragments at *m/e* 214, 144, and 130 for the indole nucleus and a base peak at *m/e* 154 (ion B). The two additional oxygens were therefore in the piperidine nucleus and/or in the side chain. The IR spectrum indicated a hydroxyl group (3590 cm^{-1}) and a nonconjugated ketone (1735 cm^{-1}). A loss of 45 mu (ion D) from the molecular ion showed the hydroxy group to be present in the side chain although its position could not be conclusively proved.

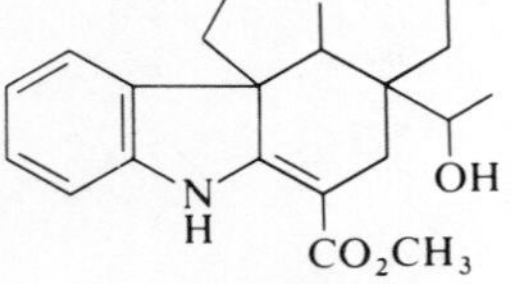

26

The absence of lactam carbonyl indicated the carbonyl group to be at either C-14 or C-15, and the former position was preferred on the evaluation of a dilute IR spectrum which showed no hydrogen-bonded hydroxyl. Baloxine was therefore suggested to be 14-oxominovincinine (**26**).

8. Vandrikidine (**27**) and Vandrikine (**29**)

Few details are available for these alkaloids, although the carbon magnetic resonance (CMR) data substantiate the structural assignments (*20*).

Vandrikidine (**27**) showed some similarity to tabersonine (**28**), but from other data an aromatic methoxy group and a secondary hydroxyl needed to be placed. From the chemical shift of the C-11 carbon at 159.9 ppm, this was clearly the site of the aromatic methoxy group. The hydroxyl group was located at C-19 from the chemical shift of C-19 at 67.9 ppm and the smaller downfield shifts of both C-18 and C-21. Vandrikidine therefore was assigned structure **27** (*20*). The relationship with 11-methoxymino-vincine (**9**) was not established.

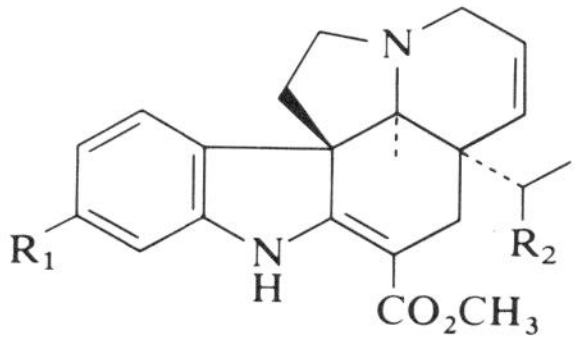

27 $R_1 = OCH_3$, $R_2 = OH$
28 $R_1, R_2 = H$

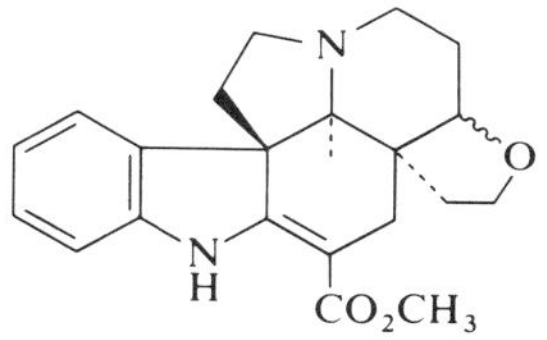

29

Vandrikine (**29**), isomeric with vandrikidine, gave an identical aromatic region of the CMR with **27** but was closer to vincadifformine in the frequencies of the ring D carbons. Other data indicated vandrikine to contain an ether linkage, and from the chemical data the linkage was placed between C-15 (79.8 ppm) and C-18 (64.7 ppm) and vandrikine was given structure **29** (*20*).

9. Echitovenaldine (**32**)

Echitovenaldine was obtained from the leaves of *Alstonia venenata* R. Br. by Majumder and co-workers (*21*). The UV spectrum (λ_{max} 328 nm) and the IR spectrum (ν_{max} 3300, 1675, and 1610 cm^{-1}) indicated an anilinoacrylate derivative, and three proton singlets at 3.77 ($-CO_2CH_3$), 3.80 ($Ar-OCH_3$), and 1.53 ppm ($-OCOCH_3$) demonstrated the presence of the indicated groups. The C-19 proton appeared as a quartet at 4.70 ppm coupled to a three-proton doublet at 0.97 ppm. Location of the acetoxy group is therefore in the side chain and the methoxyl group in the aromatic ring.

The above deductions were confirmed by the mass spectrum, which showed a molecular ion at *m/e* 426 and base peak at *m/e* 182 (ion B) (**30**).

The aromatic protons showed coupling constants characteristic of a 1,2,4-trisubstituted benzene so that the aromatic methoxyl group could potentially be located at C-10 or C-11. A distinction between these two possibilities was made on the basis of chemical correlation.

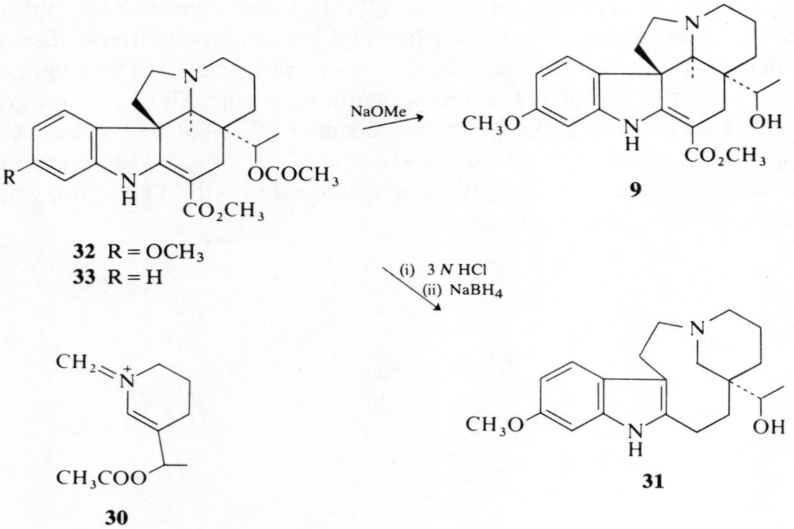

NaOMe

CH_3O N H OH CO_2CH_3

9

R N H $OCOCH_3$ CO_2CH_3

32 $R=OCH_3$
33 $R=H$

(i) 3 *N* HCl
(ii) $NaBH_4$

$CH_2=\overset{+}{N}$

CH_3COO

30

CH_3O N H OH

31

Methanolysis of echitovenaldine gave a deacetyl compound (M^+ 384) having spectral properties similar to 11-methoxyminovincinine (**9**). Acid-catalyzed hydrolysis gave an indolenine (*22*), which was reduced by sodium borohydride to a quebrachamine derivative **31** having a UV spectrum (λ_{max} 230, 270, and 300 nm) typical of a 6-methoxy-2,3-disubstituted indole (*23*). Echitovenaldine therefore has the structure **32** and is the 11-methoxy analog of echitovenine (**33**).

10. Echitoserpine (34) and Echitoserpidine (35)

Continuing with the isolation of alkaloids from the fruits of *Alstonia venenata*, Majumder *et al.* have obtained two unusual β-anilinoacrylate bases, echitoserpine (**34**) (*24*) and echitoserpidine (**35**) (*25*). Because these alkaloids are so closely related they will be discussed together, although the main emphasis will be on echitoserpidine. The UV and IR spectra, and the characteristic chromogenic reaction with ceric ammonium sulfate, indicated the presence of the anilinoacrylate chromophore, but the IR spectrum also indicated the presence of an ester carbonyl (1720 cm^{-1}). Six aromatic protons were discerned, but the pattern was too complex for interpretation in terms of substitution. A two-proton singlet at 6.01 ppm was assigned to a methylenedioxy group; three-proton singlets at 3.90 and 3.45 ppm were assigned to aromatic methoxyl and carbomethoxyl groups, respectively.

The two-carbon side chain was found to be substituted at C-19, because the C-19 proton appeared as a quartet ($J = 6.5$ Hz) at 4.93 ppm and the methyl group as a doublet at 1.04 ppm.

The location of the various functionalities was determined from the mass spectrum. Ion B was observed at *m/e* 214 (no substitution in the indole nucleus) and the molecular ion at 532 mu. Because ion B (*m/e* 318, **36**) fragmented to give ions at *m/e* 123 and 122, all the functionality is located at C-19 in the form of an aromatic acid containing a methylenedioxy and a methoxy group. A fragment ion at *m/e* 179 was interpreted as resulting from the standard ester cleavage to give an ion radical of structure **37**.

Acid-catalyzed hydrolysis of echitoserpidine gave an indolenine (*22*) and an aromatic acid identical with myristicinic acid (**38**). Reduction of the indolenine with sodium borohydride gave an indole derivative identified as **39**. Methanolysis of echitoserpidine gave an anilinoacrylate derivative showing a molecular ion at 354 mu and a base peak at *m/e* 140. This compound was identical to (−)-minovincinine (**13**). Echitoserpidine therefore has structure **35** (*25*).

Echitoserpine showed both a molecular ion at *m/e* 562, 30 mu higher than echitoserpidine, and all the characteristic spectral characteristics of an anilinoacrylate alkaloid. Typical of these were IR absorptions at 3400, 1685, and 1620 cm^{-1}, a UV λ_{max} at 325 nm, and a three-proton singlet at 3.50 ppm for the carbomethoxyl group. Two other methoxyl singlets were also observed; one of these, like that in echitoserpidine, was deshielded to 3.93 ppm, typical of a methoxy group adjacent to a methylenedioxy group. The other methoxyl singlet was observed at 3.73 ppm, characteristic of an aromatic methoxyl group. From the observation of a quartet at 4.97 ppm the benzoic acid substituent could again be placed at C-19 in the side chain.

The mass spectrum of echitoserpine showed a base peak at *m/e* 318 (**36**), indicating that the additional methoxyl group was in the indolic nucleus. Acid hydrolysis gave myristicinic acid (**38**), and methanolysis gave both **38** and 11-methoxyminovincinine (**10**). Echitoserpidine therefore has the structure **34** and is the 11-methoxy analog of echitoserpidine (*24*).

11. Cathaphylline (**42**)

Catharanthus trichophyllus (Baker) Pichon, a plant native to the Malagasy Republic, had been found by the Illinois group to exhibit both cytotoxic and antitumor activity (*26, 27*). At least part of the cytotoxic

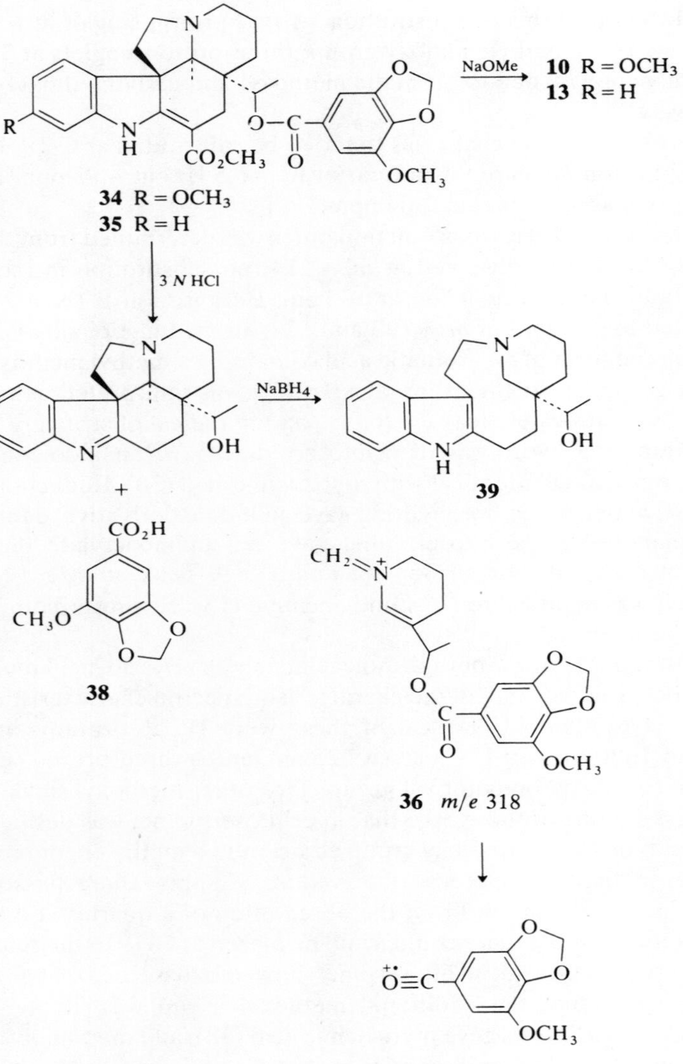

NaOMe → **10** R = OCH_3; **13** R = H

34 R = OCH_3
35 R = H

3 *N* HCl

$NaBH_4$

39

38

36 *m/e* 318

37 *m/e* 179

activity in the roots of this plant was traced to the two known alkaloids lochnericine (**40**) and hörhammericine (**41**), and the structure elucidation of this latter compound will be discussed subsequently. A new inactive β-anilinoacrylate derivative, cathaphylline, was isolated by Cordell and Farnsworth (*28*).

Cathaphylline exhibited a UV spectrum typical of a β-anilinoacrylate with absorptions at 299 and 328 nm; this was supported by the two characteristic bands at 1675 and 1615 cm^{-1}. In addition, however, cathaphylline showed IR absorptions typical of both hydroxyl (3450 cm^{-1}) and ketone (1710 cm^{-1}) functionalities. The PMR spectrum indicated four aromatic protons (7.40–6.70 ppm), a carbomethoxyl group (3.75 ppm), and a two-proton "doublet" (4.13 ppm). As with many β-anilinoacrylate alkaloids, it was the mass spectrum which gave the clues to the structure of cathaphylline. The molecular ion at *m/e* 368 analyzed for $C_{21}H_{24}N_2O_4$, and major fragments were observed at *m/e* 337 ($C_{20}H_{21}N_2O_3$), *m/e* 309 ($C_{19}H_{21}N_2O_2$), *m/e* 214 ($C_{13}H_{12}NO_2$), and *m/e* 154 ($C_8H_{12}NO_2$). As expected, the latter ion was the base peak. Acetylcathaphylline, formed by reaction with acetic anhydride/pyridine, was a monoacetyl derivative, but only two ions in the mass spectrum were shifted, the molecular ion and the base peak, each of which increased in mass by 42 mu. From previous discussions, the ion at *m/e* 309 (ion D) indicates that all the additional oxygenation (two oxygen atoms) is located in the side chain, which therefore has the molecular formula $C_2H_3O_2$ and contains both the ketone and hydroxyl functions. The two possibilities for this arrangement are a hydroxyaldehyde or a hydroxyketone. Several pieces of evidence favored the latter interpretation. No aldehyde proton was observed in the PMR spectrum. The mass spectra of both acetylcathaphylline and cathaphylline showed a fragment ion at *m/e* 337, corresponding to initial losses of CH_3O and $C_3H_5O_2$, respectively. The acetylatable (hydroxy) function should therefore be at C-18. Subsequently in each case, this ion (*m/e* 337) lost CO to give the ion *m/e* 309.

Following acetylation, the two-proton multiplet at 4.13 ppm was shifted to become two doublets ($J = 16.5$ Hz) at 4.84 and 4.34 ppm, thereby establishing C-18 to be a primary hydroxyl group. Cathaphylline therefore has the structure **42** with a novel α-hydroxyketone side chain (*28*).

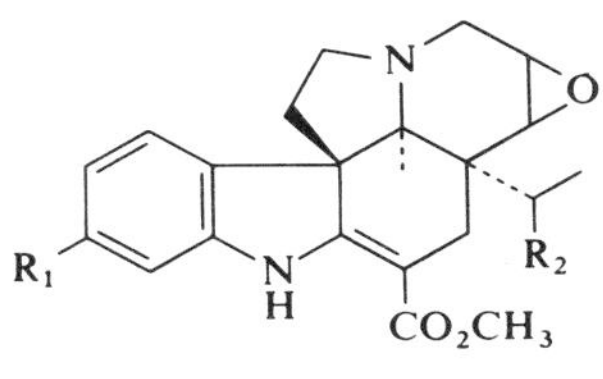

40 $R_1 = H, R_2 = H$
41 $R_1 = H, R_2 = OH$
43 $R_1 = OCH_3, R_2 - OH$
45 $R_1 = OCH_3, R_2 = H$

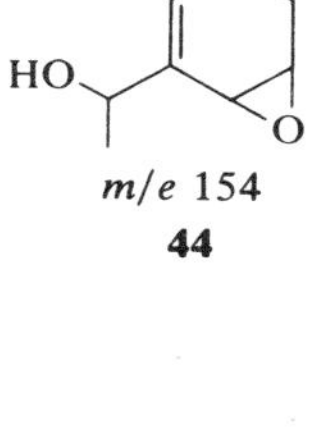

m/e 154
44

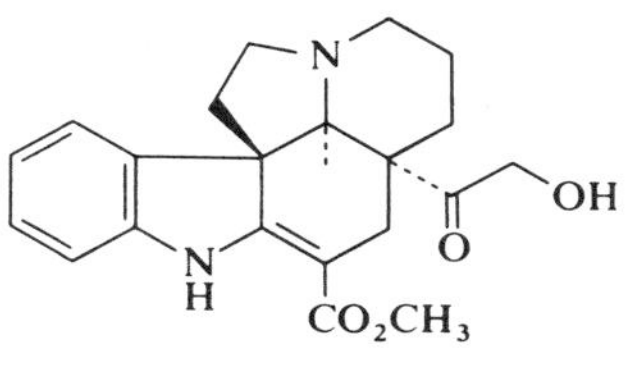

42

12. Hörhammericine (**41**) and Hörhammerinine (**43**)

Hörhammericine (**41**) was isolated by Abraham and co-workers from the roots of *Catharanthus lanceus* (*29*) and hörhammerinine (**43**) from the leaves of *C. lanceus* (*30*).

Hörhammericine {$[\alpha]_D - 403°$} showed the typical IR and UV spectral data of a β-anilinoacrylate. Similar observations were made for hörhammerinine, but on the basis of subtle differences in the UV spectra hörhammericine was deduced to be the demethoxy analog of hörhammerinine (*29*). The structures of these alkaloids were determined mainly on the basis of their mass spectral fragmentations (*31*).

Hörhammericine and hörhammerinine each showed a base peak at *m/e* 154 (**44**) (ion B) and a loss of 45 mu from the molecular ions. Each compound showed a doublet at 1.00 ppm for the methyl of the ethyl side chain, which must therefore contain a hydroxy group at C-19. The remaining functionality is therefore in the piperidine ring and is an epoxide at C-14–C-15 (no ketonic carbonyl). Ion B, containing the piperidine ring and the side chain, has the structure **44**. The ion A appeared in mass spectra of hörhammericine and hörhammerinine at *m/e* 214 and *m/e* 244, respectively. In the case of hörhammerinine, the aromatic methoxy group was placed at C-11 on the basis of similarity to the aromatic region of lochnerinine (**45**). Hörhammericine therefore has the structure **41**, and hörhammerinine is **43** (*31*). No stereochemical information about the substituent groups was deduced. Hörhammericine has also been isolated from the roots of *C. trichophyllus* (*28*).

13. Ervincinine (**46**) and Hazuntine (**47**)

From the epigeal part of *Vinca erecta* Yunusov and co-workers isolated a further β-anilinoacrylate alkaloid (*32*). Ervincinine (mp 247–248°), showing a molecular ion at *m/e* 382, gave a surprisingly low $[\alpha]_D$ of −80.5°. The UV and IR spectra again indicated the presence of the β-anilinoacrylate unit, and a band at 840 cm^{-1} was assigned to a 1,2,4-trisubstituted benzene. Reduction with zinc in methanolic sulfuric acid gave a dihydro derivative having an indoline UV spectrum. The PMR spectrum of ervincinine exhibited the methyl of an ethyl group at 0.61 ppm, two methoxyl singlets at 3.71 ppm, and three aromatic protons in the region 6.34–8.01 ppm. No Bohlmann bands were observed, and therefore, the nitrogen lone pair and the C-21 proton were assumed to be *cis*. Structure **46** was proposed for ervincinine on the basis of the UV spectrum of the dihydro derivative. The stereochemistry of the epoxy group was not determined.

Hazuntine {mp 152°; $[\alpha]_D -45°$} was obtained by Potier and co-workers (*33*) from the leaves of *Hazunta velutina* Pichon, and from a molecular ion at *m/e* 382 a molecular formula $C_{22}H_{26}N_2O_4$ was deduced. The UV and IR spectra indicated the presence of a β-anilinoacrylate, and in the mass spectrum the base peak was at *m/e* 138 (ion B) with an important fragment ion at *m/e* 244 (ion A). The PMR spectrum showed an aromatic methoxyl and a carbomethoxyl group as an overlapping singlet at 3.78 ppm. In the aromatic region three protons were observed in a pattern characteristic of 1,2,4-trisubstitution. The ethyl group was observed in the region of 0.7 ppm, and consequently hazuntine (**47**) has the gross structure of lochnerinine (**45**) but is an isomer, probably at the epoxy group.

At this point an important PMR spectral observation was made. Comparison of the PMR spectrum in the 0–3.7 ppm region of hazuntine and pachysiphine (*34*) indicated identity. Pachysiphine (**48**) is therefore related to lochnericine (**40**) in the same way that hazuntine (**47**) is related to lochnerinine (**45**).

Although the stereochemistry of the epoxy group in lochnericine has not yet been determined, it seems clear that lochnericine and lochnerinine are in one stereochemical series while pachysiphine, hazuntine, and hazuntinine are in the opposite stereochemical series as far as the epoxy group is concerned.

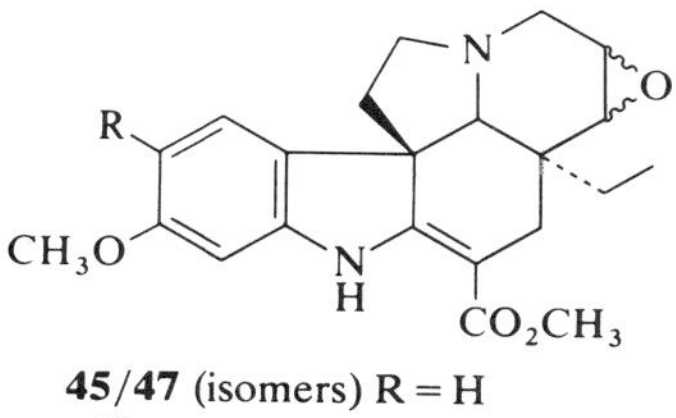

45/47 (isomers) R = H
49 R = OCH_3

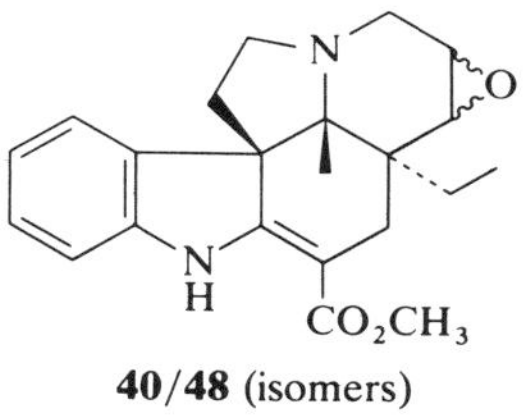

40/48 (isomers)

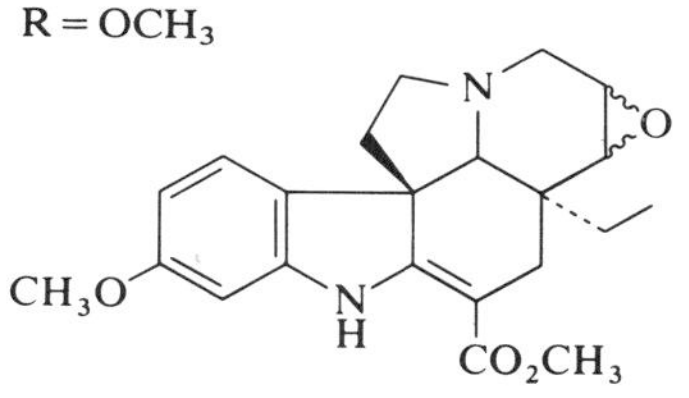

46

14. Hazuntinine (**49**)

A second new β-anilinoacrylate alkaloid was also obtained from *Hazunta velutina* by Potier and co-workers (*33*). Hazuntinine {$[\alpha]_D -482°$} showed a molecular ion at 412 mu for a probable molecular formula of $C_{23}H_{28}N_2O_5$. The mass spectrum showed a base peak at *m/e* 138 and a major fragment ion at *m/e* 274. This ion corresponds to ion *m/e* 244 (ion A) in hazuntine. Because hazuntinine showed only *para*-coupled aromatic protons and an additional methoxy group, hazuntinine has the structure **49**. The aliphatic region of the PMR spectrum of hazuntinine was very close to that of both hazuntine (**47**) and pachysiphine (**48**).

Confirmation of the structure of hazuntinine (**49**) came from the CMR spectrum, which showed close similarity with the spectrum of vincadifformine (**2**) except for C-14 and C-15, and C-10 and C-11 (*20*). The former two carbons at 51.8 and 57.0 ppm were clearly attached to oxygen from their chemical shift and are therefore the site of the epoxide linkage. The chemical shift differences of C-10 (+23.0 ppm) and C-11 (+22 ppm)* in comparison with vincadifformine (**2**) demonstrate the *ortho* relationship of the two aromatic methoxyl groups.

15. Hedrantherine (**50**) and 12-Hydroxyhedrantherine (**51**)

The leaves of *Hedranthera barteri* (Hook. f.) Pichon have afforded a number of interesting alkaloids, including two new representatives of the β-anilinoacrylate type, hedrantherine (**50**) and its 12-hydroxy derivative (**51**) (*36*).

Hedrantherine was an amorphous alkaloid $\{[\alpha]_D -458.5°\}$ showing a typical anilinoacrylate chromophore and a molecular ion at *m/e* 368. Two principal fragment ions were observed at *m/e* 214 (ion A) and *m/e* 154 (ion B). Substitution could therefore be limited to two additional oxygens in the piperidine ring and side chain, with the loss of two hydrogens. No loss of the side chain occurred and there was no terminal methyl group observed in the PMR spectrum. On this basis it could be reasonably assumed that there was substitution in both the piperidine ring and the terminal carbon of the ethyl side chain and that the latter was in some way linked to the piperidine ring. The only possible group reasonably satisfying these requirements is an acetal linkage and, although several structures are possible, the acetal linkage between C-18 and C-15 was found to be correct on the basis of the following evidence.

In hedrantherine (**50**), a doublet of doublets was observed at 5.34 ppm ($J = 4, 6$ Hz) which was assigned to the acetal proton. A triplet ($J =$ 2.5 Hz) at 4.16 ppm defined not only the position of the ether oxygen on the piperidine ring but also the stereochemistry. The ether is therefore

* In Wenkert *et al.* (*20*) C-10 and C-11 of hazuntinine are assigned the chemical shifts 149.3 and 143.5 ppm, respectively. In the opinion of the reviewer these assignments should be reversed in accord with an expected matching chemical shift difference in comparison with vincadifformine and the assigned frequencies (*35*) for the corresponding carbons in deacetylgeissovelline.

attached at C-15 in an α-orientation, with the β-proton at C-15 bisecting those on C-14.

Acetylation with acetic anhydride–pyridine gave a monoacetyl derivative **52** showing a doublet of doublets at 6.05 ppm ($J = 4, 6$ Hz). The methylene protons on C-19 each appeared as a doublet of doublets, one at 1.73 ppm ($J = 14, 6$ Hz) and the other at 1.48 ppm ($J = 14, 4$ Hz). The stereochemical integrity at C-18 is therefore confirmed.

A number of other reactions confirmed the structural features. On treatment with acetic anhydride in dimethyl sulfoxide, hedrantherine gave a monoacetyl derivative **52** and **53**, as a mixture of C-18 epimers. The low $[\alpha]_D$ of $-294°$ for this mixture is an interesting phenomenon. With BF_3–etherate in methanol at room temperature, the methyl ether **54** was produced. Sodium borohydride reduction of **50** gave a diol (M^+ 370) (**55**) which with acetic anhydride–pyridine gave a diacetate (**56**). Reaction of hedrantherine with zinc in methanolic sulfuric acid gave a mixture of 2,16-dihydrohedrantherine methyl ethers **57** and **58**, as expected.

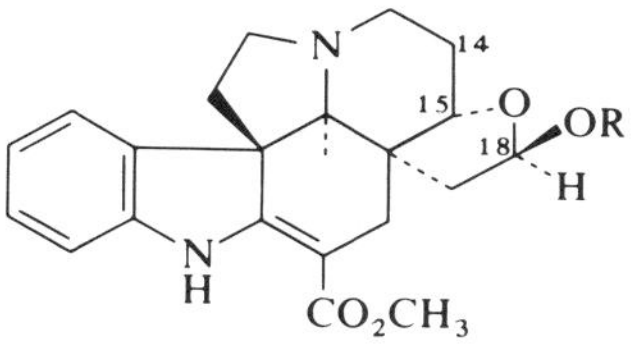

50 R = H
52 R = Ac
53 R = Ac, 18-epi
54 R = CH_3

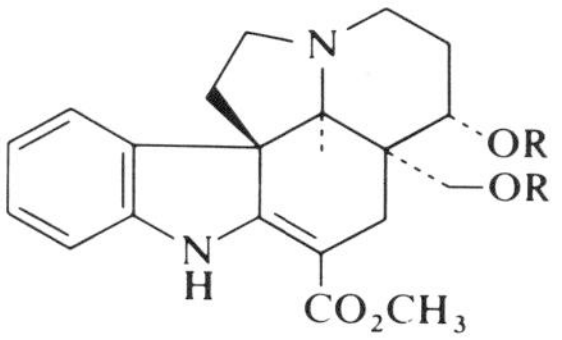

55 R = H
56 R = Ac

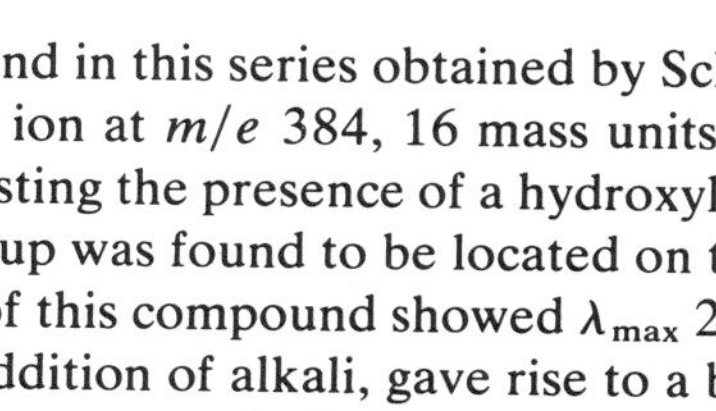

57/**58** (isomers)

The second compound in this series obtained by Schmid and co-workers (*36*) gave a molecular ion at m/e 384, 16 mass units higher than hedrantherine (**50**) and suggesting the presence of a hydroxyl group somewhere in the molecule. This group was found to be located on the aromatic nucleus, for the UV spectrum of this compound showed λ_{max} 234, 290, and 336 nm which, following the addition of alkali, gave rise to a bathochromic shift to 246, 294, and 366 nm, respectively. Treatment with diazomethane in (DMF) gave a monomethyl ether (**59**). Three aromatic protons and two methoxyl groups (3.83 and 3.73 ppm) were observed in the PMR spectrum together with a doublet of doublets ($J = 6$ and 4 Hz) for the C-18 proton at 5.31 ppm and a triplet ($J = 2$ Hz) for the C-15 proton at 4.13 ppm.

The position of oxygenation in the aromatic ring was determined by reduction of the β-anilinoacrylate unit. Treatment of the compound with zinc in methanolic sulfuric acid gave a mixture of dimethyl ethers ($M^+ = 414$) (**60/61**). The UV spectra of these ethers corresponded closely to that of 9-methoxy-1,2,3,4-tetrahydrocarbazole (**62**); consequently, the original material is 12-hydroxyhedrantherine (**51**).

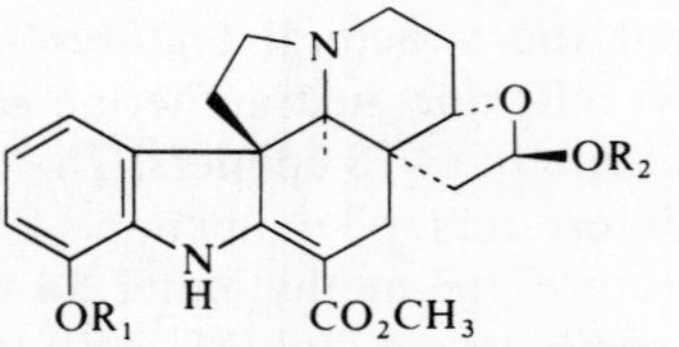

51 $R_1, R_2 = H$
59 $R_1 = CH_3, R_2 = H$

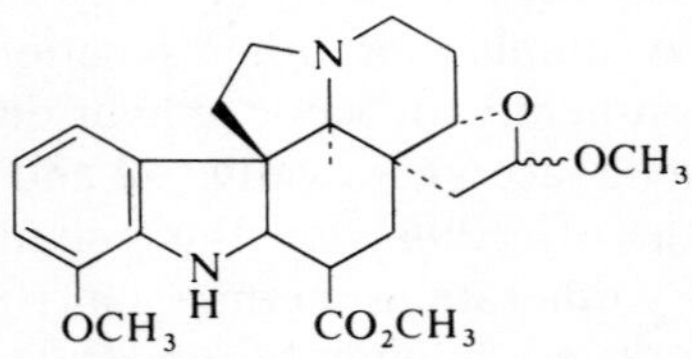

60/61 (isomers)

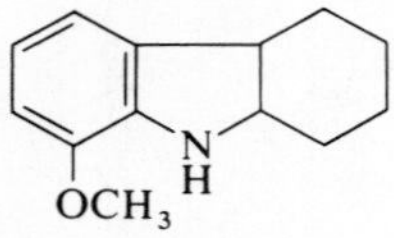

62

16. Apodine (**63**) and Deoxoapodine (**66**)

Iglesias and Diatta (*37, 38*) have isolated two new alkaloids from *Tabernaemontana armeniaca* very closely related to hedrantherine (**50**). Apodine was assigned structure **63** on the basis of spectral and chemical evidence. Zinc and acid gave the 2,16-dihydro derivative **64**, and hydrolysis with *p*-toluenesulfonic acid followed by reductive decarboxylation gave **65** (*38*).

Deoxoapodine (**66**) differed in the absence of a lactone carbonyl group, as shown by the spectral data. Decarboxylation with *p*-toluenesulfonic acid gave the 1,2-dehydro derivative **67** (*37*).

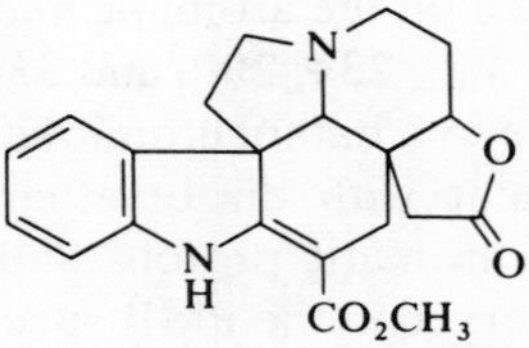

63

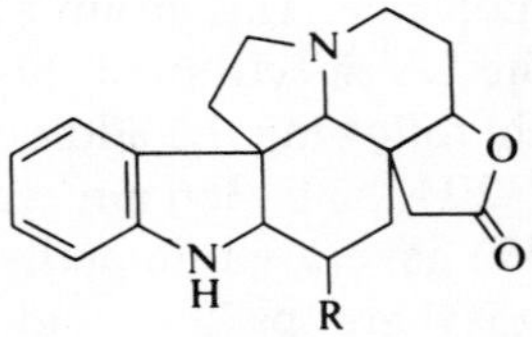

64 $R = CO_2CH_3$
65 $R = H$

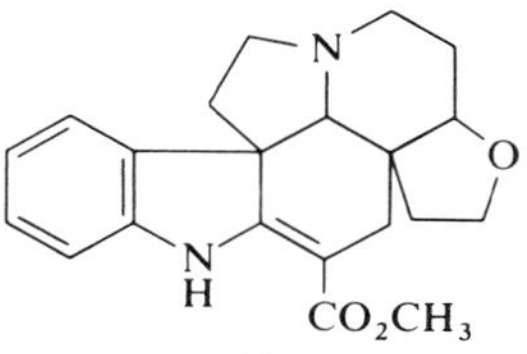

66 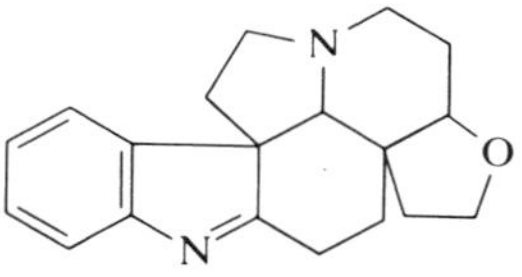 **67**

17. Ervinidine (**68**)

Also obtained from the epigeal part of *Vinca erecta* by Malikov and Yunusov (*18*) is the interesting alkaloid ervinidine. Only fragmentary (no pun intended) evidence is available for the assignment of the structure, and this leads to a certain amount of ambiguity.

The molecular formula $C_{21}H_{24}N_2O_4$ was deduced from the molecular ion at *m/e* 368 which showed important fragment ions at *m/e* 340, 228, 214 (ion A), 168, and 154 (ion B). Three carbonyl functions were observed, at 1720, 1690, and 1660 cm^{-1}, and the UV spectrum was said to indicate a β-anilinoacrylate. Four aromatic protons were observed and the methoxy of an ester group at 3.85 ppm. The methyl of an ethyl group was found at 1.13 ppm.

The structure of ervinidine was determined as **68** on the basis of the above information and the rationale that follows. The two carbonyls of the lactam and the ketone were placed adjacent to each other in order to account for the facile loss of 28 mu, as shown in Scheme 1, and from the same intermediate the important fragment at *m/e* 168 is produced. An alternative initial fragmentation (Scheme 1) leads to the ions at 228, 214, and 154 mu. Lithium aluminum hydride afforded the aspidospermine derivative **69** by cyclization. One of the problems associated with this particular structure is the possibility that it would not exist as described but rather as the 3-acylindole derivative **70**. In this event, the carbonyl at 1720 cm^{-1} would be assigned to the ester group and that at 1690 cm^{-1} to the 3-acylindole function. The C-20 stereochemistry was not determined.

18. Eburine (**76**), Eburcine (**81**), and Eburenine (**73**)

Le Men and co-workers have examined the seeds of *Hunteria eburnea* Pichon and, as well as several known alkaloids, three new alkaloids were isolated (*39*) and their gross structures defined. In a subsequent publication (*40*), the absolute stereochemistries of these alkaloids were deduced.

Eburine {$[\alpha]_D - 18°$} showed an indoline UV spectrum and IR spectrum, but more particularly, the mass spectrum was compatible with formulation as a dihydrovincadifformine derivative. The mass spectrum

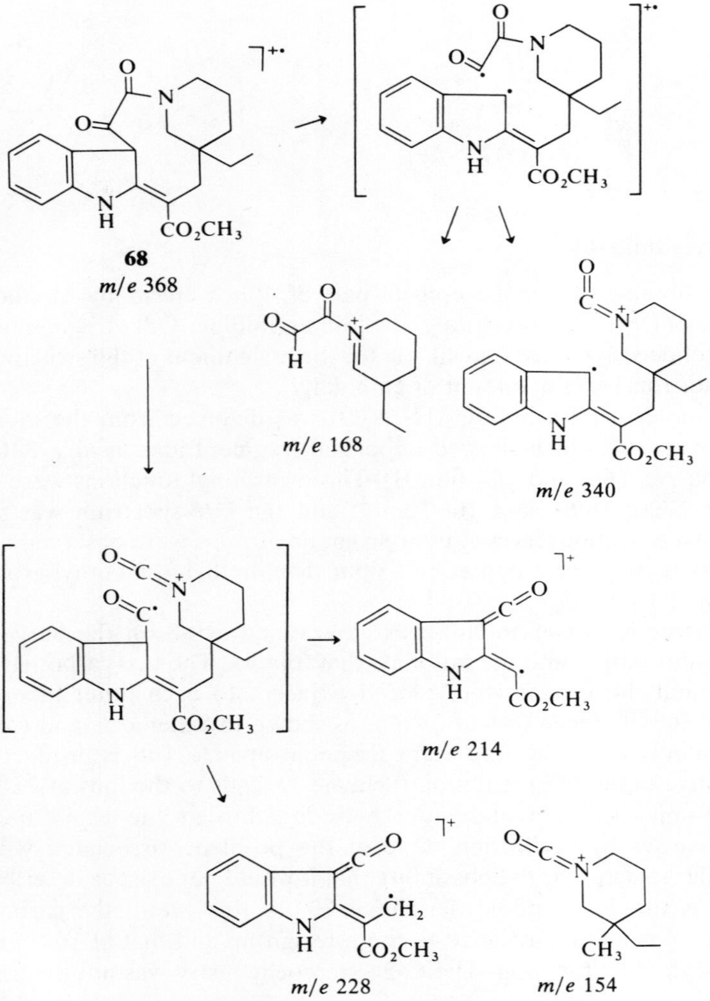

68

m/e 368

m/e 168

m/e 340

m/e 214

m/e 228

m/e 154

SCHEME 1

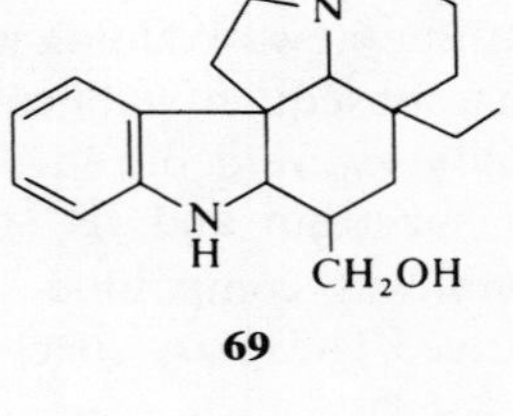

69

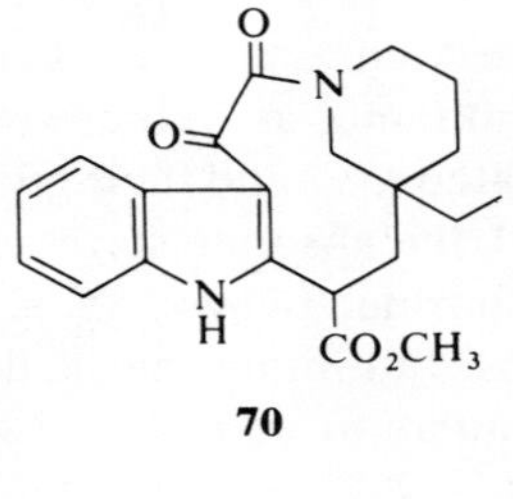

70

exhibited a molecular ion at *m/e* 340 ($C_{21}H_{28}N_2O_2$) with characteristic ions at *m/e* 254 (loss of methyl acrylate), 130, and 124 (base peak). Eburine was concluded to have structure **71** with no stereochemistry defined.

Eburcine was suspected to be an *N*-carboxyindoline on the basis of its UV spectrum (λ_{max} 242, 285, 290 nm). Two carbonyl bands were observed at 1753 cm^{-1} (nonconjugated ester) and 1710 cm^{-1} (urethane) confirming the indications of the UV spectrum. The molecular weight was 398 from the mass spectrum, and a major fragment ion at *m/e* 312, loss of methyl acrylate, indicated that eburcine was *N*-carbomethoxy-2,16-dihydrovincadifformine (**72**) of undefined stereochemistry.

Eburenine (M^+ 280) was an indolenine (λ_{max} 225 and 262 nm) having a molecular formula $C_{19}H_{24}N_2$. Structure **73** was suggested for eburenine, identical with (+)-1,2-dehydroaspidospermidine, but with an undefined stereochemistry at C-21.

The stereochemistry of eburine was established by correlation with (−)-vincadifformine (**2**) (*40*). Reduction of (−)-vincadifformine (**2**) with zinc in acetic acid gave the 2,16-dihydro derivative **74**, identical with eburine except for optical rotation. Epimerization of eburine with sodium methoxide gave 16-epieburine, which could be oxidized with lead tetraacetate to (+)-vincadifformine (**75**). This reaction could not be carried out on eburine itself; consequently, eburine has the structure **76** in which the carbomethoxyl group is β (*40*).

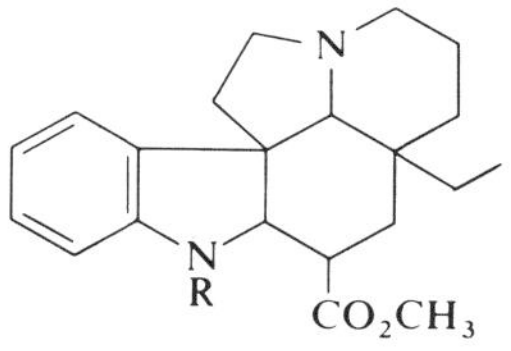

71 R = H
72 R = CO_2CH_3

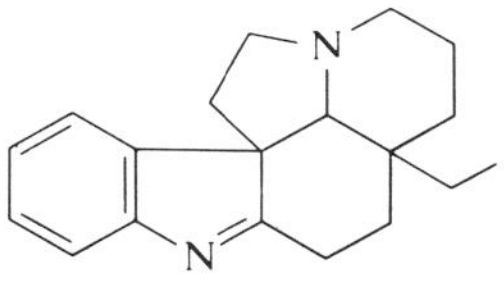

73

Reduction of eburine (**76**) and 16-epieburine with lithium aluminum hydride gave two alcohols, **77** and **78**, respectively. Neither of these was the optical antipode of the lithium aluminum hydride reduction product of (−)-vincadifformine (**2**). This is in agreement with previous work in the (−)-akuammicine series, where lithium aluminum hydride introduces a 2α-proton and catalytic reduction a 2β-proton. By analogy, eburine would therefore have the C-2 proton α (opposite absolute stereochemistry), as deduced previously.

Lithium aluminum hydride reduction of eburcine gave three derivatives. The most polar was eburinol (**77**), identical with the product obtained from eburine (**76**). The other products were *N*-methyleburinol (**79**) and methyleneeburinol (**80**). Eburcine is therefore *N*-carbomethoxyeburine (**81**) (*40*).

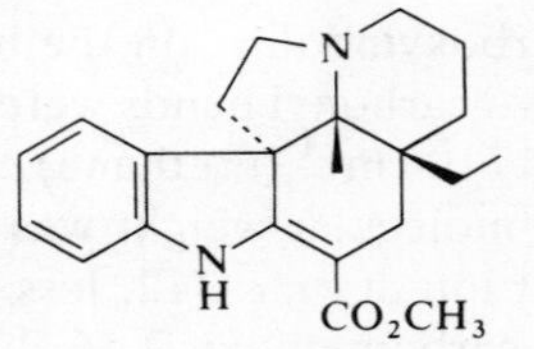

75

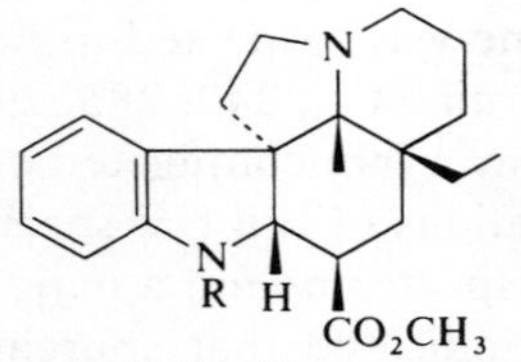

76 R = H
81 R = CO_2CH_3

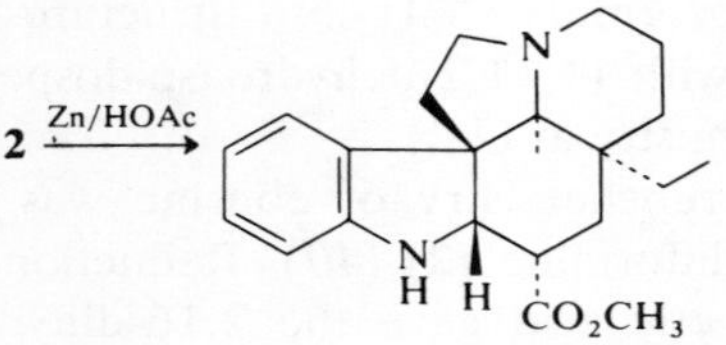

2 $\xrightarrow{Zn/HOAc}$

74

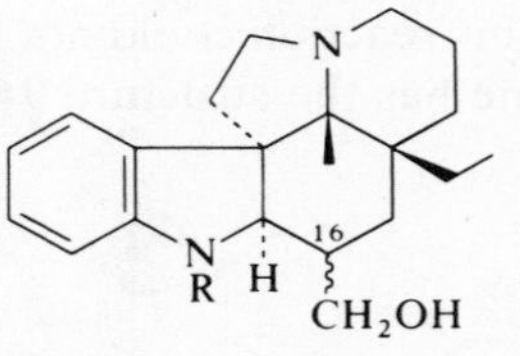

77 16α-H, R = H
78 16β-H, R = H
79 16α-H, R = CH_3

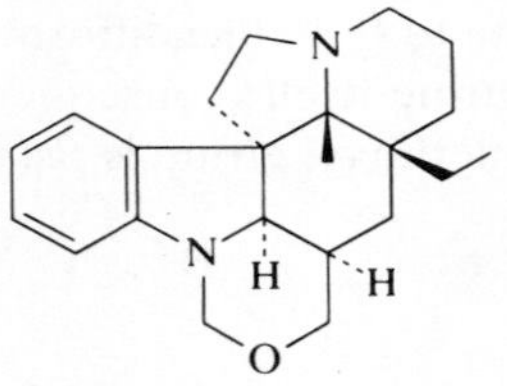

80

B. Aspidospermidine-Type Alkaloids

1. Fendlispermine (**83**)

From the root bark of *Aspidosperma fendleri* Woods. Medina and co-workers (*41*) isolated a new crystalline alkaloid, fendlispermine. The UV spectrum indicated the dihydroindole moiety, and the mass spectrum, which showed a molecular ion at *m/e* 298, suggested a molecular formula of $C_{19}H_{26}N_2O$. Reduction of fendlispermine with lithium aluminum hydride gave quebrachamine (**82**).

The 220 MHz PMR spectrum of fendlispermine confirmed the presence of four aromatic protons, an NH, and an OH group. The methyl of the ethyl group was found at 0.85 ppm but the typical signals for the C-2, C-3, and C-5 protons were all shifted to higher field.

The location of the hydroxyl group was deduced mainly from the mass spectrum, which showed a base peak at $M^+ - 17$, *m/e* 281. Because no $M^+ - 18$ peak was observed, it was thought the hydroxy group must be located at a position which not only allows reduction by lithium aluminum hydride but has no neighboring carbons bearing hydrogen. The only position which satisfies these requirements is position 21. Fendlispermine therefore has the structure **83**.

In order to account for the quebrachamine-like mass fragmentation of fendlispermine (**83**), it was proposed that, after loss of the hydroxy group, a C-2 to C-21 hydrogen migration occurred.

2. De-*O*-methylaspidocarpine (**84**)

The aerial bark of the Venezuelan shrub *Aspidosperma cuspa* Blake ex Pittier afforded the alkaloid de-*O*-methylaspidocarpine (**84**) (*42*), previously obtained as a degradation product (*43*).

The IR spectrum showed an *N*-acetylcarbonyl at 1635 cm^{-1} and a nonhydrogen-bonded hydroxyl group at 3360 cm^{-1}. From the molecular ion at *m/e* 356 a molecular formula of $C_{21}H_{28}N_2O_3$ could be deduced. The remainder of the mass spectrum showed a loss of ethylene, as well as loss of an acetyl residue, but the principal indole containing ions, at *m/e* 190, 176, and 162, indicated the aromatic portion to contain two hydroxyl groups. The aromatic protons, however, appeared as a two-proton singlet at 6.36 ppm, so that the substitution on the aromatic ring was deduced from the PMR spectrum of the dimethyl ether derivative, where two doublets ($J = 8$ Hz) at 6.49 and 6.68 ppm were observed. The compound therefore has the structure **84** (*42*).

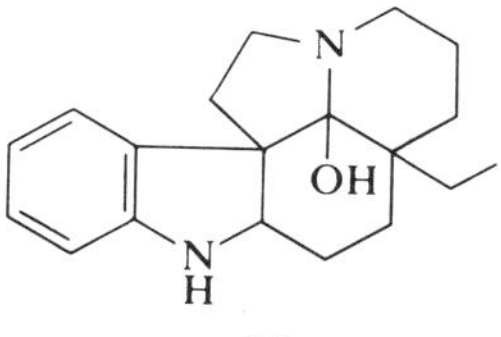

N OH N H

83

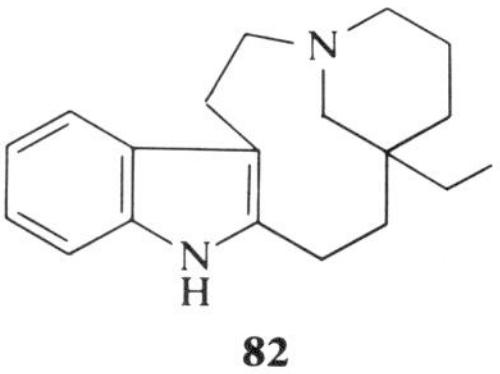

N N H

82

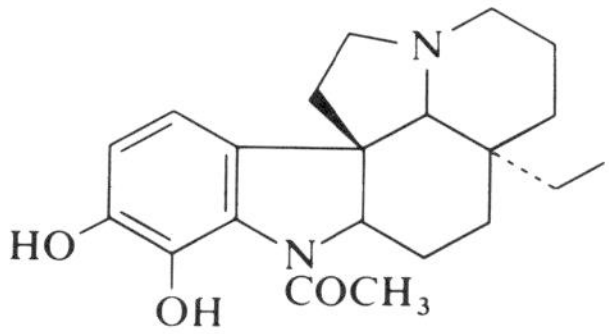

N HO N OH $COCH_3$

84

3. Vincoline (**85**) and 19-Epivincoline (**86**)

After considerable confusion it appears that the structures of vincoline (**85**) and 19-epivincoline (**86**) have been established. Each of these compounds was initially assigned an incorrect structure, but a chemical correlation and CMR study have established the correct structure of vincoline to be **85**.

Vincoline was first isolated by Svoboda *et al.* from *Catharanthus roseus* (L.) G. Doh (*44*), then by Farnsworth *et al.* from *C. lanceus* (*45*), and by Aynilian *et al.* from *Vinca libanotica* Zucc. (*46*).

Reisolation from *C. roseus* (*47*) gave a quantity of material adequate for structure elucidation studies. The UV spectrum indicated a dihydroindole (λ_{max} 245, 299 nm), and from the IR spectrum, hydroxyl (3480 cm^{-1}), secondary amine (3350 cm^{-1}), ester (1715 cm^{-1}), and 1,2-disubstituted benzene (740 cm^{-1}) functionalities were apparent.

A considerable amount of information could be deduced from the PMR spectrum. Four aromatic protons were evident together with a carbomethoxyl group (3.82 ppm), an indoline NH (5.56 ppm), and two adjacent olefinic protons (5.89–5.37 ppm). A methyl group was observed at 0.60 ppm as a doublet ($J = 6.6$ Hz), and a singlet at 3.81 ppm was assigned to the C-2 proton in **91**.

The mass spectrum of vincoline gave a molecular ion at *m/e* 368, analyzing for $C_{21}H_{24}N_2O_4$, and major fragments at *m/e* 267 (base peak) (**87**, $C_{17}H_{19}N_2O$), 146 (**88**, C_9H_8NO), 121 (**89**, $C_8H_{11}N$), and 93 (C_6H_7N). These fragments were typical for an aspidospermine type of alkaloid for which a retro-Diels–Alder reaction of ring C would be expected. The ion at *m/e* 267 (**87**) was apparently derived by such a process indicating three of the four oxygen atoms to be associated in some way with either C-16 or C-17.

Acetylation of vincoline gave a monoacetate in which the base peak was at *m/e* 309; several other ions, including those originally at *m/e* 160 and *m/e* 146, were also shifted by 42 mu. Sodium borohydride had no effect on vincoline, and on this basis the possibility of a carbinolamine at C-2, C-3, C-5, or C-21 was ruled out. The only position remaining was therefore C-6; the *m/e* 146 (**88**) ion arises as shown in Scheme 2. Vincoline was consequently assigned structure **90**, in which an oxygen atom remained to be placed at either C-16 or C-17.

The fourth oxygen must be in the form of an ether, and one of the points of attachment was determined from the irradiation of the methyl doublet. A methine proton previously evident as a quartet at 3.89 ppm ($J = 6.6$ Hz) was now sharpened to a singlet. The low-field nature of this proton indicated attachment to oxygen. Because no other methine proton was observed, the other bond from oxygen must be attached to C-16. Vincoline was deduced to have the structure **91** (*47*).

Melobaline was obtained from the leaves of *Melodinus balansae* (*19, 48*), and was found to have the molecular formula $C_{21}H_{24}N_2O_4$. The UV spectrum was dihydroindolic and the IR spectrum indicated both hydroxyl and ester functionalities. A secondary methyl group was observed at 1.03 ppm together with four aromatic protons, a carbomethoxyl singlet (3.90 ppm), and two olefinic protons. On deuterium exchange the signals of two protons disappeared, one in the region of 2.5 to 3.0 ppm and one between 5 and 6 ppm. Melobaline was therefore suggested to be a dihydroxyvindolinine, and analysis of the mass fragments indicated one of the hydroxyl groups to be associated with C-6 or the indolic nucleus because the ions at *m/e* 130, 144, and 250 in vindolinine were shifted by 16 mu in melobaline. Since the ion *m/e* 121 (**89**) was still observed, it was reasoned that the second hydroxy group must be at C-16 or C-17, and because it was acetylatable with acetic anhydride–pyridine, C-17 was chosen. Melobaline could therefore be described by the structure **92** in which the stereochemistry at C-17 remained unknown.

Andriamialisoa *et al.* isolated from *Catharanthus ovalis* Mgf (*49*) two alkaloids closely related to vincoline. The main peaks of the mass spectrum, however, were shifted 14 mu, suggesting the presence of a methoxy rather than a hydroxy group. Indeed, the PMR spectrum showed an additional three-proton singlet at 3.18 ppm. The remainder of the spectrum was very similar to that of vincoline; in one of the alkaloids a methyl doublet was observed at 0.55 ppm, a carbomethoxy singlet was at 3.75 ppm and a disubstituted double bond and an exchangeable NH proton were also observed. In the other compound the only major difference was the chemical shift of the methyl doublet which now appeared at 1.02 ppm. The second compound is therefore merely the C-19 epimer of the first and each would be a methyl ether of the general vincoline skeleton and have the structure **93**.

Structures **91** and **93** were not in agreement with the coupling constants as derived by examination of the 240 MHz spectrum of vincoline,

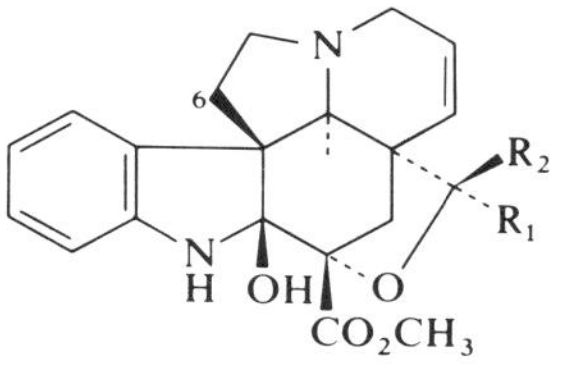

6 N R_2 R_1 N H OH O CO_2CH_3

85 $R_1 = CH_3$, $R_2 = H$
86 $R_1 = H$, $R_2 = CH_3$

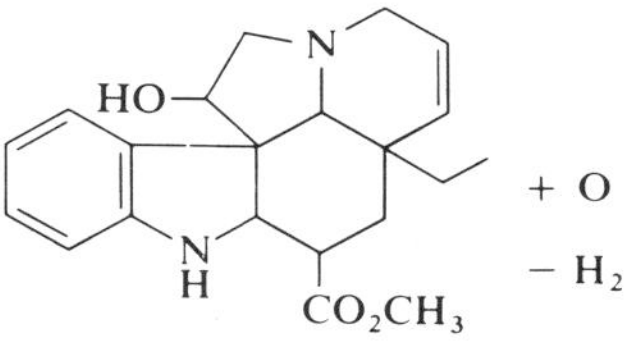

HO N N H CO_2CH_3 + O − H_2

90

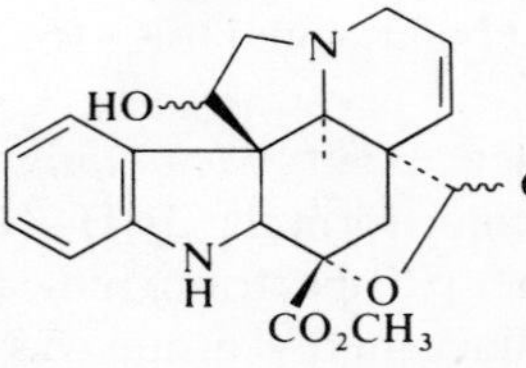

91

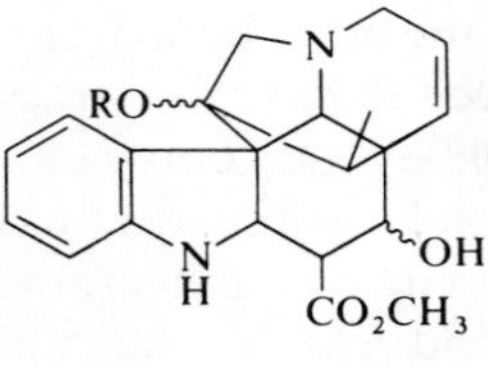

92 R = H
93 R = CH_3

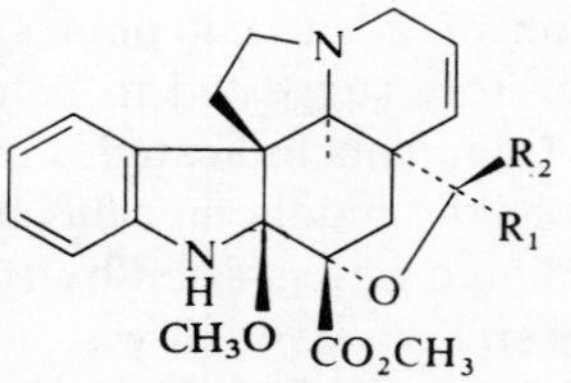

94 $R_1 = CH_3$, $R_2 = H$
95 $R_1 = H$, $R_2 = CH_3$

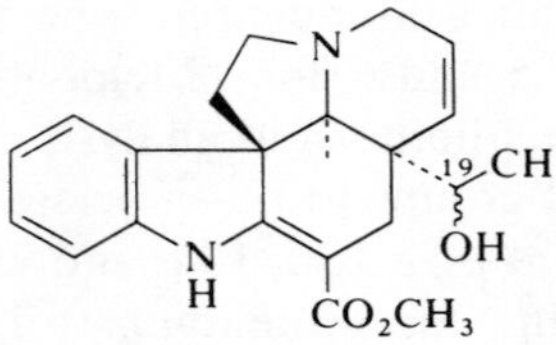

18 19*R*
96 19*S*

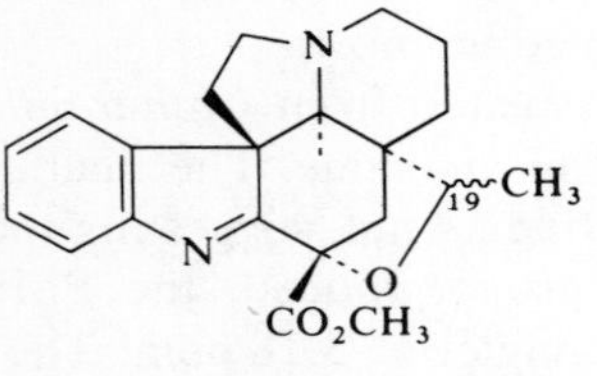

97 19*S*
98 19*R*

O-methylvincoline, and *O*-methyl-19-epivincoline (*49*). The PMR spectrum of vincoline confirmed all the previous structure assignments except for the C-2 proton, which was absent, and the expected C-6 methine proton, which was also absent. Instead, *two* C-6 protons were observed, one at 1.55 ppm and the other strongly deshielded to 3.0 ppm. The two C-3 protons were assigned to resonances at 3.48 ppm ($J = 16.5$ and 4 Hz) and 2.87 ppm. The hydroxy group in vincoline should therefore be at C-2 and not at C-6, as shown in **85**. *O*-Methylvincoline and *O*-methyl-19-epivincoline have structures **94** and **95**, respectively.

These structures were confirmed by partial synthesis from the two 19-hydroxytabersonines **96** and **18**. Reaction of each of these compounds with lead tetraacetate gave unstable indolenines **97** and **98** which, by reaction with sodium methoxide, gave **94** and **95** in approximately 85% yield. It was demonstrated that compound **94** from 19*S*-hydroxytabersonine (**96**) was identical with that obtained from natural sources, thereby establishing the stereochemistry at C-19.

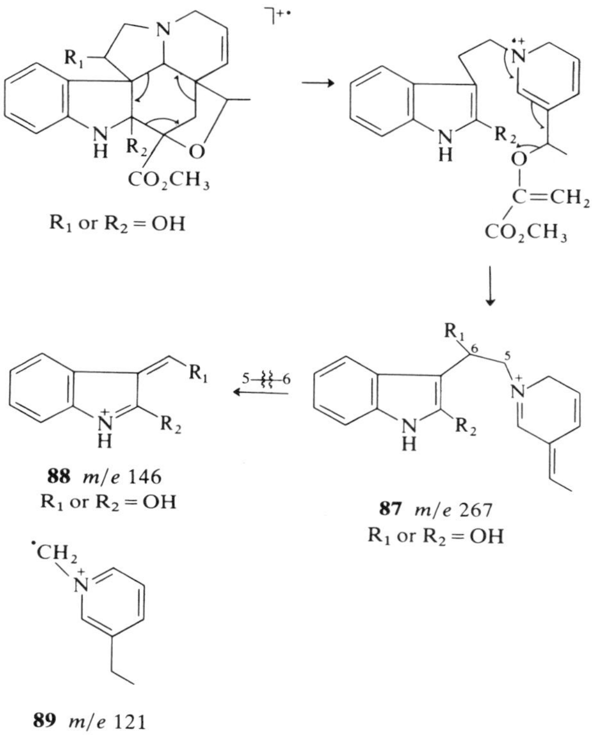

R_1 or $R_2 = OH$

88 *m/e* 146
R_1 or $R_2 = OH$

87 *m/e* 267
R_1 or $R_2 = OH$

89 *m/e* 121

SCHEME 2. *Mass spectral fragmentation of vincoline.*

When the indolenine **98** was treated with silica in the presence of water, it gave a stable dihydroindole identical with vincoline (**85**) (*49*).

The inability of the carbinolamine at C-2 to be reduced by sodium borohydride was explained on the basis of steric constraints prohibiting loss of water and the formation of an intermediate iminium species (*49*). However, this same imine is an intermediate in the partial synthesis of vincoline.

With the structure of vindolinine revised (*50*), Potier and co-workers turned their attention to melobaline, which had previously (*19*) been assigned a structure in the vindolinine series. A reevaluation of the PMR spectrum and determination of the CMR spectrum resulted in a revision of the structure of melobaline (*51*).

Deuterium exchange removed the signals of two protons at 5.50 ppm (NH) and 3.40 ppm (OH), and irradiation of the methyl doublet at 1.06 ppm collapsed a quartet at 3.56 ppm to a singlet. The C-19 carbon, rather than being attached to C-6, should be attached to oxygen. Attempts to acetylate melobaline failed, indicating that the hydroxyl was probably tertiary and the remaining oxygen an ether.

Comparison of spectral data of melobaline* with those of vincoline indicated the very close relationship of the compounds; indeed, the mass spectra were essentially identical. The PMR spectrum firmly established that carbons 5 and 6 were unsubsituted, although, as with *O*-methylvincoline (*49*), the C-6 protons had quite dissimilar chemical shifts (2.48 and 1.45 ppm).

It was the ^{13}C NMR spectrum, however, which indicated that melobaline was actually 19-epivincoline (**86**) (*51*). In particular, carbon 6 was observed as a triplet at 35.7 ppm and carbon 2 as a deshielded singlet at 95.0 ppm in the single frequency off-resonance decoupled spectrum.

4. Cathovaline (**100**)

Two groups independently and simultaneously isolated an alkaloid having the structure **100** from different species of *Catharanthus* (*52, 53*). Aynilian and co-workers (*53*) isolated cathovaline (as cathanneine) from the leaves of *C. lanceus*, whereas Langlois and Potier (*52*) obtained cathovaline from *C. ovalis*.

Cathovaline (mp 88–90°) showed a molecular ion at *m/e* 426, analyzing for $C_{24}H_{30}N_2O_5$. The UV spectrum was typical of a dihydroindole (λ_{max} 255, 308 nm), and the PMR spectrum showed a number of close similarities with that of vindorosine (**99**). In particular, four aromatic protons (6.3–7.2 ppm), an *N*-methyl group (2.78 ppm), a carbomethoxyl group (3.76 ppm), and an acetyl group (1.93 ppm) were observed. A singlet at 5.28 ppm was attributed to the acetyl methine proton at C-17. No olefinic protons were observed, so that this degree of unsaturation must involve the C-16 oxygen in the form of an ether linkage. One new peak was observed in the PMR spectrum at 4.05 ppm and assigned to an ether bridge proton. The assignment of the ether bridge linkage to either C-14 or C-15 was made after double-irradiation studies indicated coupling of this ether bridge proton to only two other protons. The ether linkage is therefore between C-15 and C-16, and cathovaline has the structure **100** (*52, 53*). The mass spectral fragmentation of cathovaline (**100**), like that of vindoline (**101**) and vindorosine (**99**), involves an initial retro-Diels–Alder reaction in the C-ring. In this instance, C-16 and C-17 are not lost as a fragment immediately, but instead give rise to the three fragments at *m/e* 267 (**102**)

* Potier and co-workers (*19, 51*) indicated that their "melobaline" was actually a mixture of two isomers at C-19 in the approximate ratio 9 : 1 based on the methyl doublets at 1.06 and 0.61 ppm, respectively.

(base peak), *m/e* 282, and *m/e* 144. Sodium borohydride reduction of cathovaline gave a diol **103** in which the ions at *m/e* 267 and *m/e* 144 remained, but the ion at *m/e* 282 was now shifted to *m/e* 212 (*52*). Similar conclusions about the mass spectral fragmentation of cathovaline (**100**) were made by Aynilian *et al.* (*53*), although they were unable to reduce cathovaline with sodium borohydride. Reduction with lithium aluminum hydride (*54*) did produce the diol **103** which was compared with a product available from vindorosine.

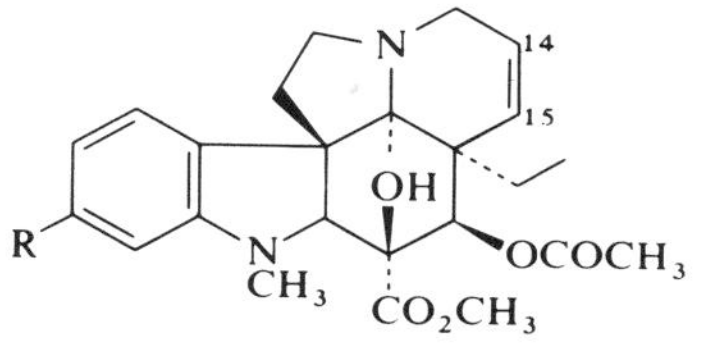

99 R = H
101 R = OCH_3

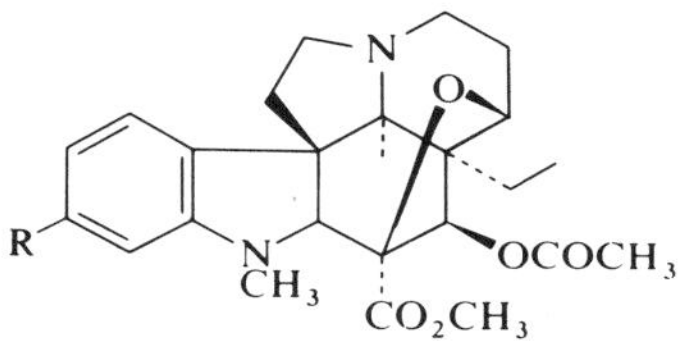

100 R = H
104 R = OCH_3

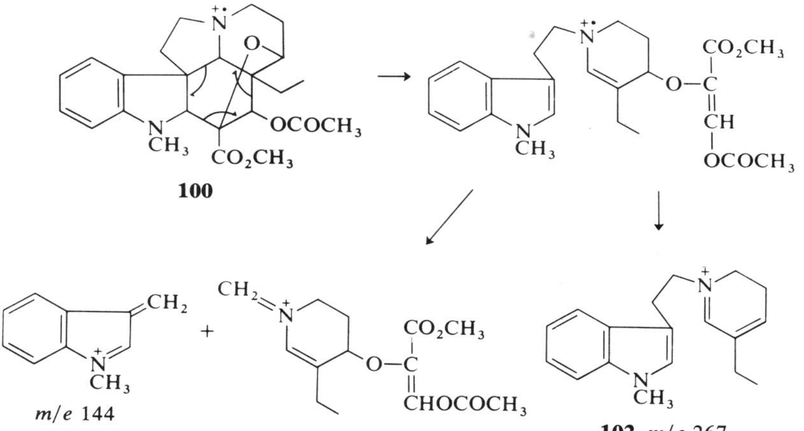

100

m/e 144

m/e 282

102 *m/e* 267

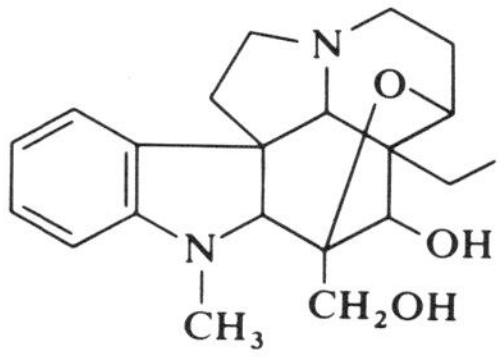

103

Battersby and Gibson (*55*) had found that oxidation of vindoline (**101**) with chromic acid in pyridine gave an ether of structure **104** as one of the products. Oxidation of vindorosine (**99**) (*54*) under similar conditions gave a mixture of two products which were reduced with lithium aluminum hydride. One of the products from this reduction was identical with the diol **103** produced from cathovaline (**100**).

Potier and co-workers (*56*) developed their structure proof along more discriminating chemical lines. Hydrolysis of vindorosine (**99**) afforded a deacetyl derivative which readily gave the N_b-oxide **105** on treatment with *p*-nitroperbenzoic acid. Using a modification of the Polonovskii reaction, the N_b-oxide was reacted with trifluoroacetic anhydride in methylene chloride at 0°, and the crude product reduced with sodium borohydride to afford deacetylcathovaline (**106**) in 39% yield. Other examples of the work by Potier and co-workers on their recent uses of the modified Polonovskii reaction are discussed later in this chapter. In this particular reaction elimination of the adjacent hydrogen gives an intermediate iminium species **107** which undergoes internal attack by the C-16 hydroxy group to give an enamine ether **108**. Hydride reduction of the enamine affords **106**.

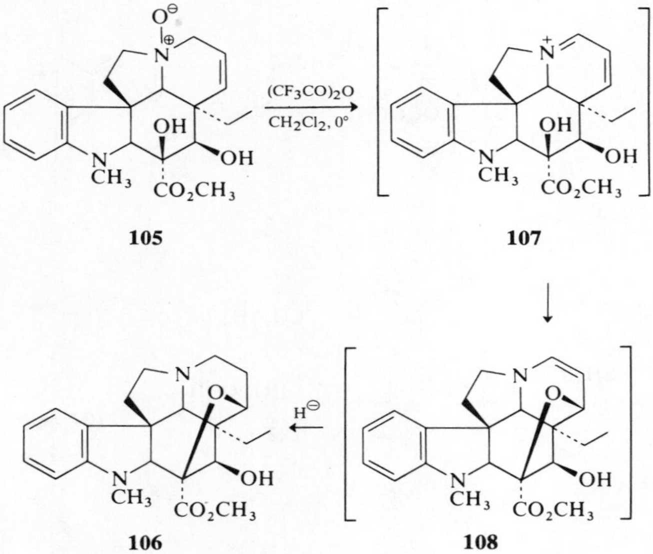

$(CF_3CO)_2O$, CH_2Cl_2, 0°

105 **107**

$H^{\ominus}$

106 **108**

5. 14-Hydroxycathovaline (**109**)

An alkaloid closely related to cathovaline (**100**) has also been isolated by Langlois and Potier from *C. ovalis* (*57*). The alkaloid was determined to be a dihydroindole from its characteristic red coloration with the ceric reagent, and this was confirmed from the UV spectrum. The mass spectrum differed from that of cathovaline (**100**) in that the molecular ion was now increased by 16 mu to m/e 442 and the base peak was observed at m/e 283 (m/e 267 + 16). The additional oxygen function was therefore at some point in the piperidine ring, and its precise location was readily determined from the PMR spectrum. The methyl of the ethyl group was found as a triplet at 0.78 ppm, thereby excluding side chain substitution. The remainder of the spectrum was quite similar to that of cathovaline (**100**), except that instead of a multiplet at 4.05 ppm, this proton now appeared as a doublet ($J = 3.8$ Hz) at 4.04 ppm, together with a multiplet at 4.21 ppm. Acetylation gave a monoacetyl derivative in which two acetate signals were observed at 1.94 and 2.08 ppm. The multiplet had now been shifted downfield to 5.32 ppm. Irradiation of this multiplet collapsed a doublet ($J = 3.5$ Hz) at 3.97 ppm to a singlet. The new hydroxyl group was therefore located at C-14 as in **109**, although the stereochemistry could not be assigned with certainty.

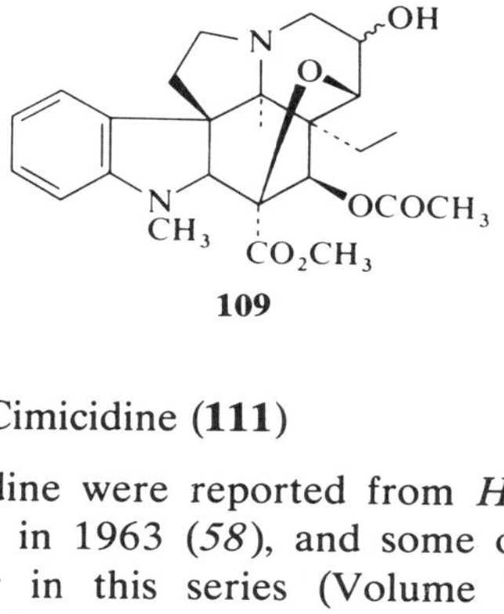

109

6. Cimicine (**110**) and Cimicidine (**111**)

Cimicine and cimicidine were reported from *Haplophyton cimicidum* A. DC. by Cava *et al.* in 1963 (*58*), and some of their chemistry has been described earlier in this series (Volume VIII, p. 451). More recently, a paper (*59*) has appeared on these alkaloids including data leading to the stereoformulas **110** and **111** for cimicine and cimicidine, respectively.

Cimicine (**110**) {$[\alpha]_D + 113°$} showed neither methoxyl nor *N*-methyl groups, but a strongly hydrogen-bonded hydroxyl at 10.70 ppm and three aromatic protons were observed in the PMR spectrum. Also noted were a three-proton triplet at 1.32 ppm and a two-proton quartet at 2.64 ppm, suggesting a *N*-propionyl residue. In neutral solution cimicine showed λ_{max} 221, 257, and 295 nm, shifting in base to 235 and 311 nm. These data were very similar to the corresponding data for haplocine (**112**). Cimicine, however, showed an additional carbonyl band at 1750 cm^{-1}. Haplocine (**112**) has the molecular formula $C_{22}H_{28}N_2O_3$; cimicine has $C_{22}H_{26}N_2O_4$.

Cimicine would therefore be the lactonic analog of haplocine, and this was confirmed by chromium trioxide–pyridine oxidation of (+)-haplocine (**112**), which gave (+)-cimicine (**110**) in low yield. The CMR spectrum of cimicine indicated a carbonyl at 175.4 ppm for the lactone and at 172.2 ppm for the amide carbonyl, demonstrating the correct structure to be **110**.

Cimicidine was found to be very closely related to cimicine, showing similar IR and PMR spectral properties. Two areas in the PMR spectrum were different, however: an additional three-proton singlet was observed at 3.87 ppm, and only two aromatic protons ($J = 8$ Hz) could be found. Cimicidine (**111**) is therefore the 11-methoxy analog of cimicine.

Oxidation of cimicine with alkaline hydrogen peroxide gave a neutral compound, 5-oxocimicine (**113**), which showed carbonyl absorptions in the IR spectrum at 1800, 1720, and 1630 cm^{-1} and in the CMR spectrum at 172.8, 172.5, and 172.1 ppm for the γ-lactone, γ-lactam, and *N*-acyl-carbonyls, respectively. The phenolic group survives this oxidation.

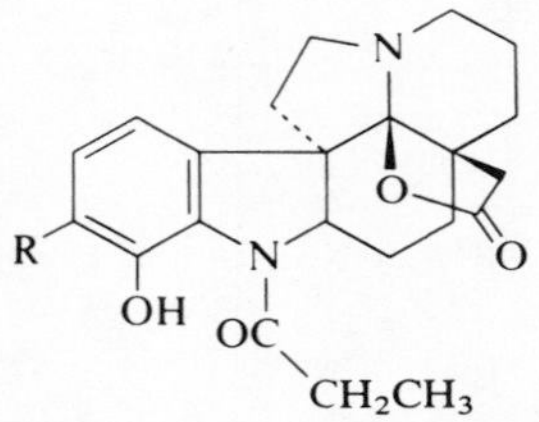

110 R = H
111 R = OCH_3

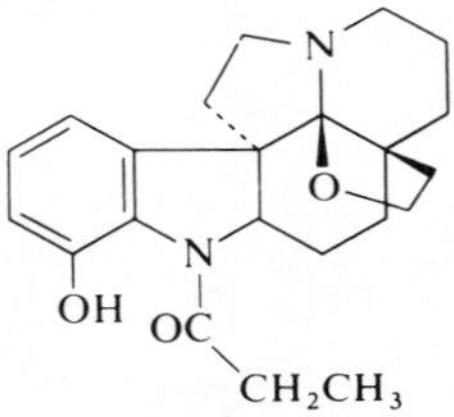

112

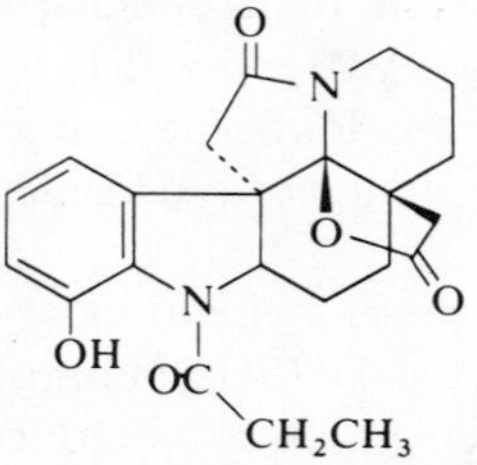

113

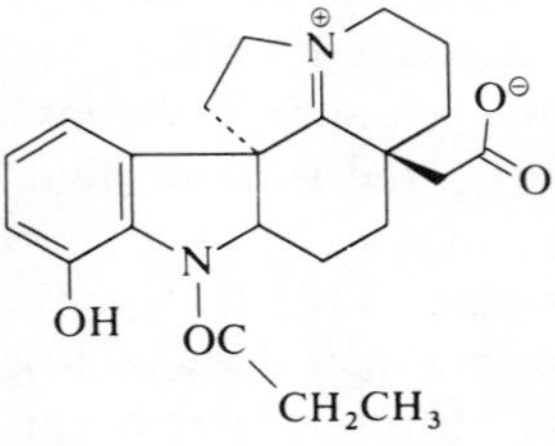

114

Cava *et al.* have concluded (*59*) that the dramatic difference between the lactone carbonyl absorptions of cimicine (1750 cm^{-1}) and 5-oxocimicine (1800 cm^{-1}) is due to a major contribution in the former case of the ring-opened zwitterionic form **114**.

C. Cylindrocarine Alkaloids

1. *N*-Propionylcylindrocarine (**118**) and 12-Demethoxy-*N*-acetylcylindrocarine (**116**)

The basic fraction of the root bark of *Tabernaemontana amygdalifolia* Sieber ex A. DC. gave two dihydroindole alkaloids (*60*), the structures of which were deduced mainly on the basis of mass spectrometry.

N-Propionylcylindrocarine $\{[\alpha]_D -82°\}$ showed a molecular ion at *m/e* 412 and important fragment ions at *m/e* 384 ($M^+ - C_2H_4$), *m/e* 338 [$M^+ - CH_2{=}C(OH)OCH_3$], and *m/e* 168 (**115**) (base peak). The latter ion is derived from *m/e* 384 by cleavage at the 5,6-position in the tryptamine bridge. In the PMR spectrum three aromatic protons were observed as well as aromatic methoxyl (3.71 ppm) and carbomethoxyl (3.39 ppm) groups.

12-Demethoxycylindrocarpidine $\{[\alpha]_D -49°\}$ gave a molecular ion at *m/e* 368 and fragment ions at *m/e* 340, 294 [$M^+ - CH_2{=}C(OH)OCH_3$], and 168. By analogy with *N*-propionylcylindrocarine and other compounds in the series, this alkaloid contained a carbomethoxy methyl group at C-20. The PMR spectrum of this compound showed four aromatic protons and a singlet at 2.10 ppm attributed to a *N*-acetyl residue. The structure **116** could be assigned to 12-demethoxy-*N*-acetylcylindrocarine on this basis (*60*).

Each of the alkaloids showed a carbonyl band at 1660 cm^{-1} for an acyl indole, and acid hydrolysis of *N*-propionylcylindrocarine followed by treatment with diazomethane gave deacetylcylindrocarpidine (**117**). *N*-Propionylcylindrocarine therefore has the structure **118**. Hydrolysis of the other alkaloid and reaction with diazomethane gave 12-demethoxycylindrocarine (**119**) (*60*).

From the facile McLafferty rearrangement, which accounts for the loss of the $CH_2{=}C(OH)OCH_3$ side chain, it would appear that the C-21 proton is *cis* to the two-carbon side chain, as shown.

2. Alkaloids of *Aspidosperma cylindrocarpon* Muell. Arg.

In 1961, Djerassi and co-workers discussed the isolation and structure elucidation of cylindrocarpine (**120**) and cylindrocarpidine (**121**) from *A. cylindrocarpon* (*61*). Subsequently, Milborrow and Djerassi (*62*) reported the chromatographic isolation of 12 new alkaloids closely related to those isolated previously.

Because so many of these compounds were very close in structure, a new name was introduced for the parent compound (**117**): cylindrocarine. Cylindrocarpine therefore has a revised name of *N*-cinnamoylcylindrocarine (**120**), and cylindrocarpidine is *N*-acetylcylindrocarine (**121**).

Some of the alkaloids obtained by Milborrow and Djerassi were simple modifications of the previously isolated alkaloids; others were more elaborate. For the most part, they were derived from either cylindrocarine (**117**) or 19-hydroxycylindrocarine (**122**).

N-Benzoylcylindrocarine (**123**) (M^+460) shows three significant peaks at *m/e* 386, 168, and 105 in its mass spectrum, as well as a loss of 28 mass units (ethylene) from the molecular ion. The ion at *m/e* 168 is the piperdine unit (**115**), analogous to the *m/e* 124 commonly found in the mass spectra of *Aspidosperma* alkaloids. Ethylene arising from the parent ion signifies the loss of carbon atoms 16 and 17, and the strong peak at *m/e* 105 is the benzoyl fragment (**124**).

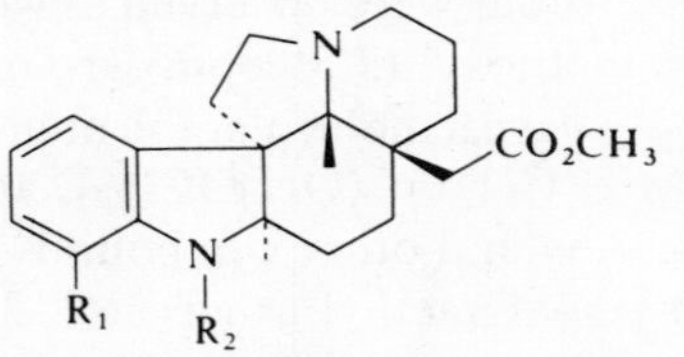

116 $R_1 = H, R_2 = COCH_3$
117 $R_1 = OCH_3, R_2 = H$
118 $R_1 = OCH_3, R_2 = COC_2H_5$
119 $R_1 = H, R_2 = H$
120 $R_1 = OCH_3, R_2 = COCH{=}CHC_6H_5$
121 $R_1 = OCH_3, R_2 = COCH_3$
123 $R_1 = OCH_3, R_2 = COC_6H_5$
125 $R_1 = OCH_3, R_2 = CHO$
126 $R_1 = OCH_3, R_2 = CH_3$

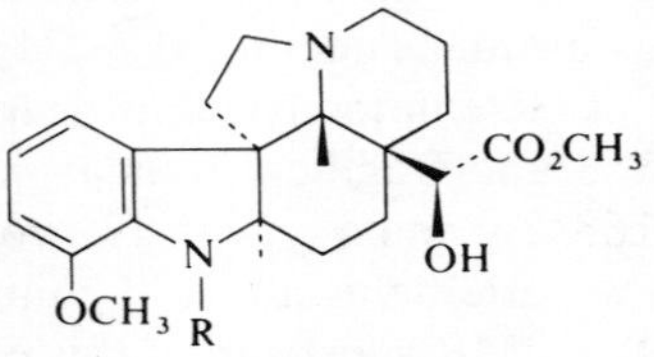

122 $R = H$
131 $R = COC_6H_5$
133 $R = COCH{=}CHC_6H_5$
134 $R = CHO$
135 $R = COCH_3$
136 $R = COCH_2CH_2C_6H_5$

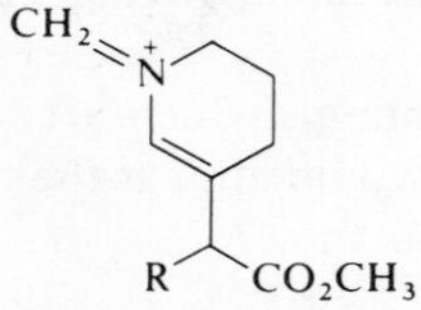

115 $R = H$ *m/e* 168
132 $R = OH$ *m/e* 184

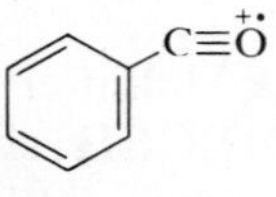

124 *m/e* 105

The PMR spectrum showed two three-proton singlets at 3.37 ppm and 3.56 ppm, the former signal being sharp, the latter quite broad. This was interpreted as being the influence of the *N*-benzoyl group on the adjacent 12-methoxy group, because this phenomenon was only observed in this compound. The C-21 proton was observed as a singlet at 2.50 ppm, with the C-17 protons appearing as a doublet of doublets at 4.34 ppm.

An ion at *m/e* 386, a loss of 74 mu, arises as the result of the McLafferty rearrangement involving the C-21 proton and the C-18–C-19 side chain discussed previously.

The most abundant alkaloid of the extract was *N*-acetylcylindrocarine (**121**, previously cylindrocarpidine), and since its structure had been deduced previously, it was used as a reference for the identification of the parent compound for the series, cylindrocarine (**117**). Only a small quantity of **117** was available, but the UV spectrum was typical of a dihydroindole and the IR spectrum indicated NH and ester functionalities. The molecular ion was found at *m/e* 356 together with important fragment ions at *m/e* 282 (M^+ − side chain) and *m/e* 168 (base peak). Confirmation of the structure came by acetylation to give *N*-acetylcylindrocarine (**121**), identical with the natural material.

Treatment of **117** with formic–acetic anhydride afforded *N*-formylcylindrocarine (**125**), and this too was obtained as a natural product. The molecular ion of *m/e* 384 and a base peak at *m/e* 168 confirmed the additional substitution to be in the aromatic nucleus. Three aromatic protons were observed in the PMR spectrum together with aromatic and carbomethoxyl groups. No NH protons were found, but a singlet at 9.3 ppm could be ascribed to the *N*-formyl group.

The most minor of the alkaloids obtained (only 0.5 mg) was *N*-methylcylindrocarine (**126**), which showed no NH or OH in the IR spectrum and only one carbonyl band (1730 cm^{-1}). Three-proton singlets were observed in the PMR spectrum, at 3.53 ppm for the aromatic methoxy, at 3.75 ppm for the carbomethoxy group, and at 3.04 ppm for the *N*-methyl group. The base peak in the mass spectrum remained at *m/e* 168.

Another of the major alkaloids obtained was 12-demethoxy-*N*-acetylcylindrocarine (**116**), previously isolated by Achenbach (*60*). Similar spectral properties were found, the only exception being that according to Milborrow and Djerassi (*62*) the C-12 proton was shifted downfield to 8.1 ppm, whereas Achenbach observed this proton together with the other aromatic protons at 6.6–7.0 ppm (*63*).

Two other alkaloids were isolated in this general series. *N*-Formylcylindrocarpinol (**127**) showed a singlet at 9.30 ppm for the formyl proton and an aromatic methoxy group. No carbomethoxy group was observed and only one carbonyl band (1625 cm^{-1}) could be located. The base peak at *m/e* 140 was rationalized in terms of a hydroxyethyl side chain and was assigned structure **128**. Milborrow and Djerassi (*62*) suggested the name cylindrocarpinol for the parent compound **129** in this series. A second compound was characterized as the *N*-acetyl derivative **130**, previously obtained in the same laboratories from *A. dispermum* (*63*).

N-Benzoyl-19-hydroxycylindrocarine (**131**) gave a molecular ion at *m/e* 460 and losses of *m/e* 28 and 74 mu. The ion at *m/e* 168 was now shifted to *m/e* 184 (**132**), indicating the additional hydroxy group to be in the side chain. In confirmation, a singlet methine proton (C-19 proton) was observed at 4.1 ppm and shifted after acetylation to 4.97 ppm. Eight aromatic protons were observed in groups of five and three. The C-2 proton was observed as a doublet of doublets at 4.32 ppm and the C-1 proton as a singlet at 3.02 ppm. Again, the ester methoxy appeared as a sharp singlet (3.89 ppm) and the aromatic methoxy as a broad singlet (3.35 ppm). Consideration of these data led to assignment of structure **131** for *N*-benzoyl-19-hydroxycylindrocarine.

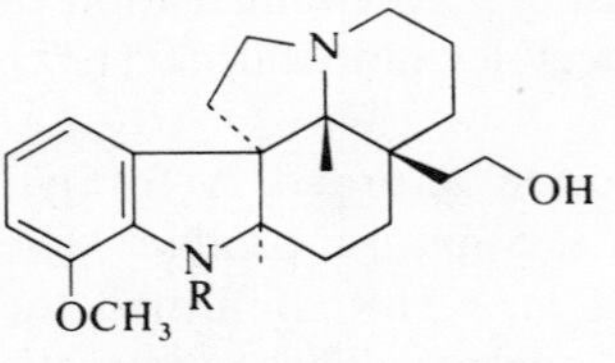

127 R = CHO
129 R = H
130 R = $COCH_3$

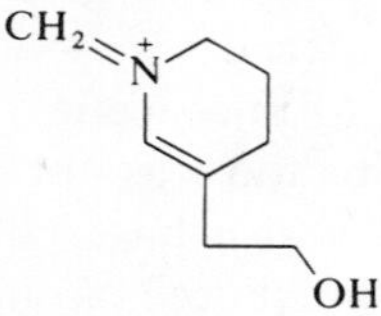

128

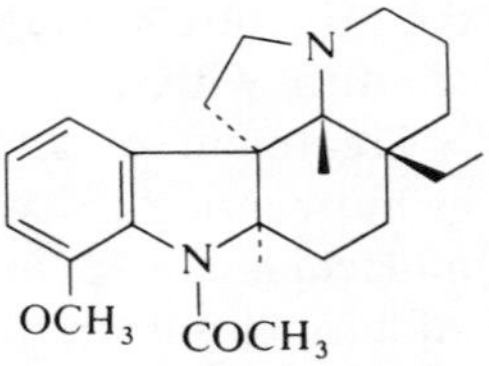

137

The structure was confirmed by the direct N-benzoylation of 19-hydroxycylindrocarine (**122**). Attempts to remove the hydroxy group by oxidation and reduction of a thioketal failed.

The simplest compound isolated in the 19-hydroxy series was 19-hydroxycylindrocarine (**122**) itself. A molecular ion was observed at *m/e* 372 together with important fragment ions at *m/e* 283 [$M^+ - CH(OH)$-CO_2CH_3] and *m/e* 184.

N-Acylation of 19-hydroxycylindrocarine (**122**) with cinnamoyl chloride gave *N*-cinnamoyl-19-hydroxycylindrocarine (**133**) which was also obtained as a natural product. Two olefinic protons at 6.74 and 7.70 ppm were observed as doublets ($J = 16$ Hz) and attributed to the cinnamoyl protons. An aromatic methoxy group at 3.79 ppm was equally as intense as the ester carbomethoxyl at 3.89 ppm. The C-21 proton was a sharp singlet at 3.00 ppm and the C-19 proton a singlet at 4.06 ppm. In the mass spectrum a molecular ion was observed at *m/e* 502 with a fragment ion at *m/e* 413 corresponding to loss of the hydroxy ester side chain. A base peak was discerned at *m/e* 184, as expected.

Reaction of 19-hydroxycylindrocarpine (**122**) with formic acid–acetic anhydride gave *N*-formyl-19-hydroxycylindrocarine (**134**), identical with a natural alkaloid. The structure was again evident from the spectral data. Two almost equivalent base peaks were observed at *m/e* 311 (M^+ − side chain) and *m/e* 184 (**132**). The aromatic ether and ester methoxyl groups appeared at 3.86 and 3.89 ppm, and the C-21 and C-19 protons at 2.99 and 4.19 ppm. The *N*-formyl proton was evident as a singlet at 9.30 ppm.

Acetylation of **122** with acetic anhydride in benzene gave a product which was the most abundant of the natural 19-hydroxy compounds, namely, *N*-acetyl-19-hydroxycylindrocarpine (**135**). The molecular ion at *m/e* 414 was followed by losses of ethylene and the side chain (89 mu), and as expected, the base peak was at *m/e* 184. A three-proton singlet at 2.18 ppm was ascribed to the *N*-acetyl group. Two other three-proton singlets were observed at 3.83 and 3.87 ppm together with characteristic one-proton singlets at 2.96 and 3.99 ppm for the C-21 and C-19 protons.

The final 19-hydroxy derivative isolated was originally observed as an impurity in the mass spectrum of **131**. The 19-hydroxy side chain could be deduced from a base peak at *m/e* 184, loss of 89 mu from the molecular ion at *m/e* 504, and from the C-19 proton at 3.98 ppm. Eight aromatic protons were observed but no olefinic protons. The compound was concluded to be *N*-dihydrocinnamoyl-19-hydroxycylindrocarine (**136**), and catalytic reduction of **133** confirmed this structure assignment.

From the $[\alpha]_D$ and ORD of these alkaloids it was concluded that they all belonged in absolute stereochemical series of (−)-aspidospermine (**137**). The stereochemistry of the 19-hydroxy group was determined by examining the rotatory dispersion of the methylxanthate derivative of **135**, which indicated the *R*-stereochemistry for this center, as shown (*62*).

D. Kopsane-Type Alkaloids

1. Hydroxykopsinines of *Melodinus australis* (F. Mueller) Pierre

Previously in this series (Volume XI, p. 250), there was some preliminary discussion of the structure of two isomeric hydroxykopsinines. The data for these compounds have since been presented in more detail (*64*). Hydroxykopsinine-1 was assigned structure **138**, in which the hydroxyl group was placed at either C-14 or C-15, and hydroxykopsinine-2 was afforded structure **139** based on mass spectral considerations.

Acetylation of hydroxykopsinine-1 gave a single monoacetyl derivative in which the acetyl methine proton appeared as a broadened triplet at 4.88 ppm. From the band width at half-band height it was deduced that the proton was only coupled to two other protons and was also equatorial. Hydroxykopsinine-1 is therefore 15α-hydroxykopsinine (**140**) (*64*).

Base-catalyzed epimerization of hydroxykopsinine-2 gave an isomer. In the mass spectrum of this isomer, only a very small fragment ion was observed at *m/e* 140, the base peak being at *m/e* 109. In this isomer both the carbonyl and hydroxyl groups were shifted to shorter wavelengths as a result of hydrogen bonding which could only be explained if the C-19 hydroxyl group were beta. Hydroxykopsinine-2 is therefore 19β-hydroxykopsinine (**141**) (*64*).

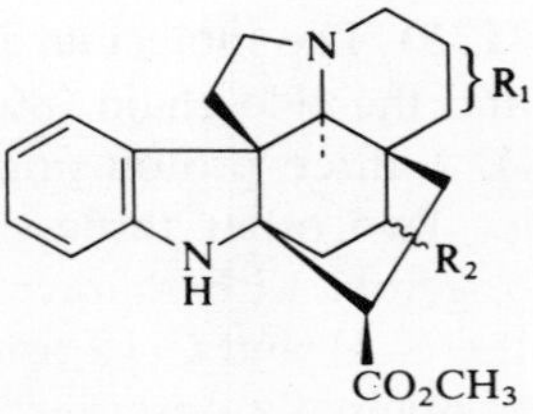

138 $R_1 = OH, R_2 = H$
139 $R_1 = H, R_2 = OH$

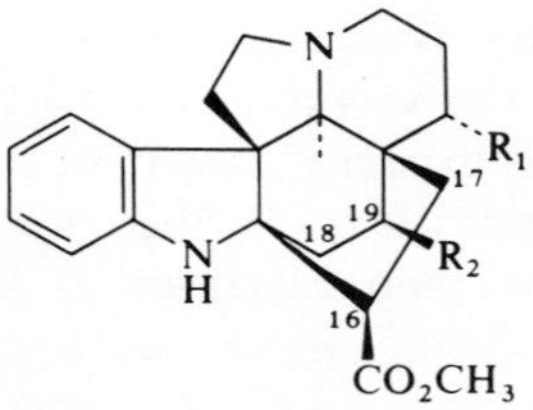

140 $R_1 = OH, R_2 = H$
141 $R_1 = H, R_2 = OH$

2. Kopsinine *N*-Oxide (**142**)

Yunusov and co-workers have isolated several alkaloids from the epigeal part of *Vinca erecta* (*65*), and one of these was the *N*-oxide of kopsinine (**142**). The alkaloid showed typical data for a dihydroindole, and the mass spectrum indicated a molecular formula $C_{21}H_{26}N_2O_3$. Two of these oxygens could be assigned to a carbomethoxyl group (3.72 ppm). The mass spectrum was quite similar to that of kopsinine (**143**), and from the water solubility of the isolated alkaloid it was deduced that the compound was kopsinine *N*-oxide (**142**). No chemical conversion was carried out.

3. *N*-Formylkopsanol (**144**)

From *Aspidosperma verbascifolium* Müll.-Arg. Braekman *et al.* (*66*) isolated four known kopsane alkaloids and a new derivative, *N*-formylkopsanol (**144**).

The compound crystallized from methanol and showed a typical *N*-acylindoline UV spectrum. The IR spectrum indicated both hydroxyl (3610 cm^{-1}) and tertiary amide (1670 cm^{-1}) functionalities, and the PMR spectrum showed a hydroxy methine proton at 4.45 ppm and two signals for the formyl proton at 8.41 and 8.87 ppm, totaling one proton.

Hydrolysis under acidic conditions afforded kopsanol (**145**), and formylation of the latter compound with formic–acetic anhydride gave *N*-formylkopsanol (**144**) identical with the natural product (*66*).

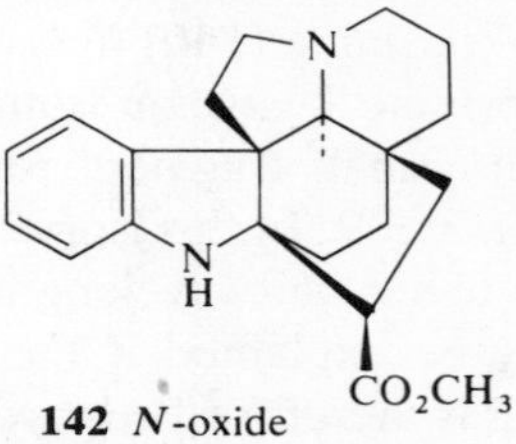

142 *N*-oxide
143

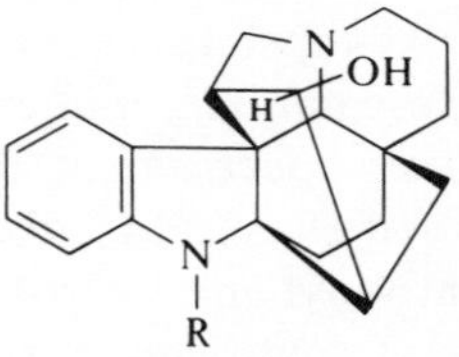

144 R = CHO
145 R = H

4. Pyrifoline and Refractidine

Linde has commented on the stereochemistry of pyrifoline and refractidine (*64*), which are closely related to the hydroxykopsinines just discussed. Each of these compounds showed "quartets" (really doublets of doublets) for the C-6 proton having line widths of 10–15 Hz.

Linde assigned the C-15 methoxy group the configuration *cis* to the C-16–C-17 unit. The methoxy group should therefore be placed in a β-configuration and pyrifoline and refractidine have the structures **146** and **147**, respectively.

5. Vellosine (Refractine)

In 1877, Hesse isolated from the Brazilian febrifuge *Geissospermum laeve* a crystalline alkaloid, geissospermine ($C_{19}H_{24}N_2O_2$) (*67*). Subsequently, Freund and Fauret isolated the same alkaloid, to which they assigned the name vellosine and a molecular formula $C_{23}H_{28}N_2O_4$ (*68*). Raymond-Hamet (*69*) concluded that one part of the original crystalline alkaloid was of this latter molecular formula, but that the main component of the base was aspidospermine and not geissospermine.

From *Aspidosperma refractum* Martius Djerassi *et al.* (*70*) obtained an alkaloid ($C_{23}H_{28}N_2O_4$) which they named refractine and to which structure **148** was assigned.

Raymond-Hamet *et al.* (*71*) compared the spectral data, particularly the PMR spectra, of refractine and vellosine, and found them to be identical. Vellosine therefore has the structure **148** and is the preferred name for this compound.

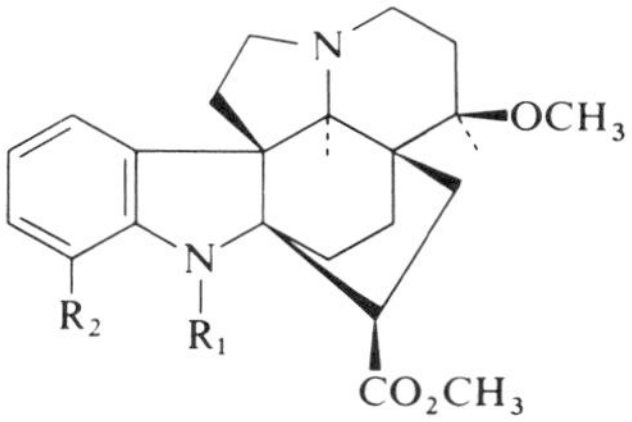

N, OCH_3, N, R_2, R_1, CO_2CH_3

146 $R_1 = COCH_3$, $R_2 = OCH_3$
147 $R_1 = CHO$, $R_2 = H$

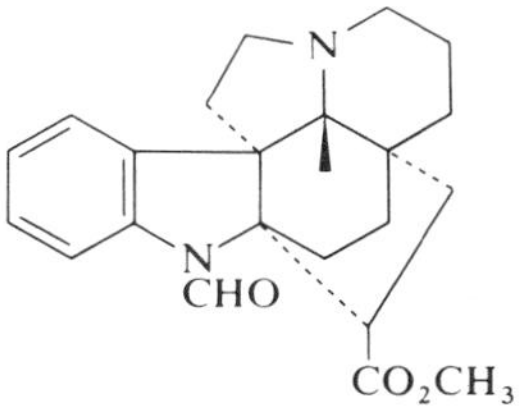

N, N, CHO, CO_2CH_3

148

E. VINDOLININE-TYPE ALKALOIDS

1. Vindolinine (**150**)

In the previous report on the structures of these alkaloids (Volume XI, p. 237), the evidence leading to the structure **149** for vindolinine was presented in detail. The data, particularly the mass spectra of a number of derivatives, seemed to point to this novel skeleton being correct—in particular, the doublet methyl group, the carbomethoxyl singlet, and two olefinic protons were required. Moreover, a doublet ($J = 4.5$ Hz) at 3.85 ppm was attributed to the C-2 proton in an α-configuration coupling with the C-16 proton in a β-configuration. These interpretations have now to be revised, because three independent studies (*50, 72–74*) have concluded that the original structure proposed for vindolinine was incorrect.

The first indications that the structure of vindolinine would need to be altered came from a ^{13}C NMR analysis (*50*). Most of these carbon resonance assignments are discussed in Section V below, and it will be sufficient at this point to mention the key absorptions. In particular, there was a nonaromatic methine signal (C-2) missing, but an additional resonance for a nonprotonated nonaromatic carbon was observed. This resonance, at 81.4 ppm, would be appropriate for an oxygenated or aminated quaternary carbon in a highly strained system. N-Methylation of vindolinine shifted this resonance to 84.4 ppm and the C-16 resonance from 39.2 ppm to 37.0 ppm. The structure of vindolinine was consequently revised to **150**. No mention was made at this time of the needed new assignment to the doublet at 3.85 ppm.

The alkaloid pseudokopsinine was reported in 1967 (*75*), and structure **151** assigned. The relationship to vindolinine was not established, although it was suggested by the previous reviewer (Volume XI, p. 241) that pseudokopsinine was the C-2 epimer of dihydrovindolinine.

Confirmation of the structural revision of vindolinine was obtained by single-crystal X-ray crystallography of the hemihydrochloride hemiperchlorate which crystallized in the C2 monoclinic space group (*73*).

The Tashkent group has examined the X-ray structure of the hydrate of (−)-pseudokopsinine hydrobromide (*72, 74*). The data firmly established that pseudokopsinine has the structure and absolute configuration shown in **152**. It has the same structure as dihydrovindolinine, which is the preferred name.

2. 19-Epivindolinine (**153**)

Melodinus balansae has afforded several new alkaloids, as discussed elsewhere in this chapter (Section II,I). The French group (*48*) also obtained a number of compounds related to vindolinine (**150**). Since the structure of vindolinine has been revised it is the revised structures which will be used for these compounds.

19-Epivindolinine (**153**) was obtained as an amorphous gum having a mass spectrum identical with that of vindolinine. The only important difference in the PMR spectrum was that the *C*-methyl group now appeared as a three-proton doublet at 0.51 ppm. This is somewhat shielded compared to the corresponding absorption in vindolinine, which appears in the 0.95–1.00 ppm region. On this basis, structure **153** was assigned.

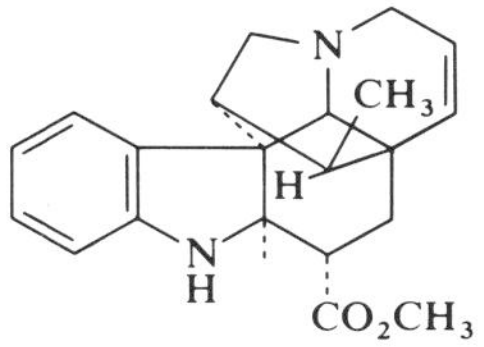

149

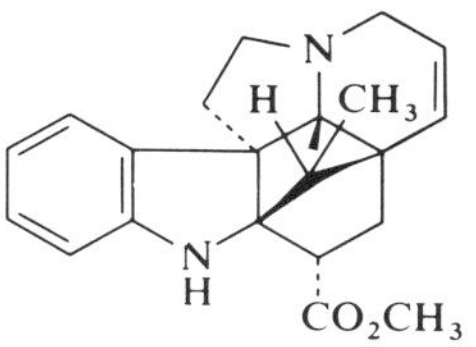

150
154 *N*-oxide

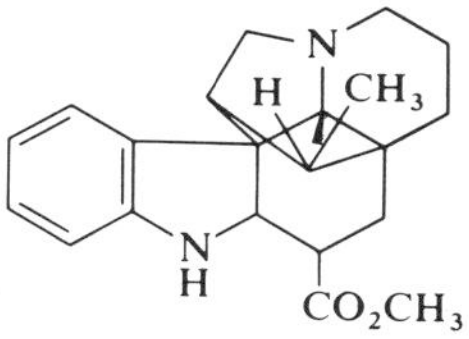

151

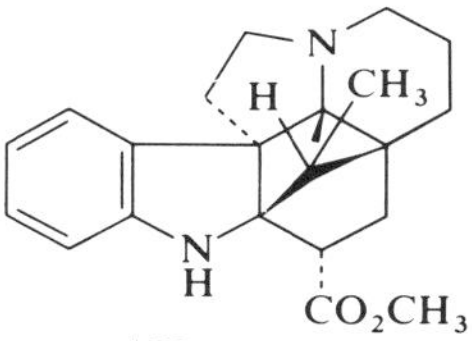

152
156 *N*-oxide

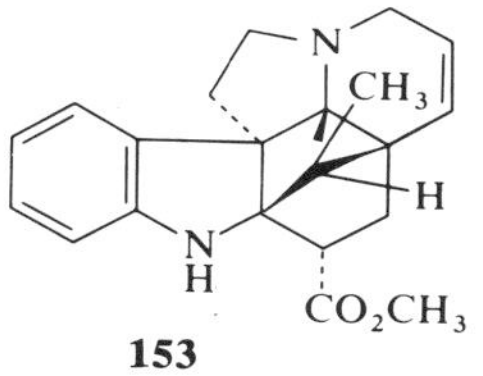

153
155 *N*-oxide

3. Vindolinine *N*-Oxide (**154**) and 19-Epivindolinine *N*-Oxide (**155**)

Also isolated from *M. balansae* were two closely related alkaloids, vindolinine *N*-oxide (**154**) and 19-epivindolinine *N*-oxide (**155**) (*48*).

The mass spectrum in each instance showed a molecular ion at *m/e* 352, 16 mu above that of vindolinine (**150**), and a base peak at *m/e* 336. The fragmentation pattern was analogous to that of **150**. The only major difference in the PMR spectra of these compounds was in the chemical shift of the methyl doublet, which in the case of **154** appeared at 0.95 ppm and in the spectrum of **155** at 0.57 ppm. Each compound was reduced with ferrous ion at room temperature to afford the parent alkaloid identical with the natural material.

4. Dihydrovindolinine *N*-Oxide (**156**)

The epigeal part of *Vinca erecta* continues to yield new alkaloids, and one of these was an alkaloid analyzing for $C_{21}H_{26}N_2O_3$ (*65*). The base was soluble in chloroform and water and the UV spectrum indicated a dihydroindole (λ_{max} 247, 301 nm). The mass spectrum showed a molecular ion at *m/e* 338, indicating a facile loss of oxygen from an expected molecular ion at *m/e* 354. The remainder of the mass spectrum was essentially identical with that of dihydrovindolinine (**152**), and reduction of the parent compound with zinc and hydrochloric acid gave **152**. The parent compound is therefore dihydrovindolinine *N*-oxide (**156**). One interesting feature of the PMR spectrum was the observation of a doublet ($J = 8$ ppm) for the C-9 proton deshielded to 7.67 ppm (*65*).

F. Miscellaneous Aspidospermine-Type Alkaloids

1. Aspidodispermine (157) and Deoxyaspidodispermine (158)

Two of the structurally more novel *Aspidosperma* alkaloids obtained in the recent past are aspidodispermine (**157**) and deoxyaspidodispermine (**158**), from *A. dispermum* (*76*). In each of these alkaloids the side chain has been removed and replaced by a hydroxy group.

Deoxyaspidodispermine gave a molecular ion at *m/e* 312, analyzing for $C_{19}H_{24}N_2O_2$, and its UV spectrum indicated an *N*-acyldihydroindole. The PMR spectrum confirmed this, showing a three-proton singlet at 2.26 ppm. A similar signal (2.30 ppm) was observed in the PMR spectrum of aspidodispermine, but the latter compound showed only three aromatic protons in contrast to deoxyaspidodispermine which showed four. The molecular ion in aspidodispermine was at *m/e* 328 ($C_{19}H_{25}N_2O_3$), and analysis of the aromatic-containing fragments indicated this oxygen to be located in the aromatic nucleus. The phenolic nature of this group was demonstrated by a 12 nm bathochromic shift in alkali. Acetylation of aspidodispermine gave a phenolic monoacetate (**159**) (ν_{max} 1760 cm^{-1}) which still showed hydroxyl absorption (3610 cm^{-1}), and methylation gave monomethyl ether **160** (three-proton singlet at 3.84 ppm). The location of the original hydroxyl group was deduced from the PMR spectrum, which showed a hydrogen-bonded hydroxyl group at 10.82 ppm. With the aromatic moiety firmly established, attention was turned to the aliphatic part of the molecule and the location of the second hydroxyl group.

In the alkaloid and its derivatives an intense ion was observed at *m/e* 112, analyzing for $C_6H_{10}NO$ which could be assigned structure **161**. No signal for an ethyl side chain was observed in the PMR spectrum of any of these compounds. The location of the hydroxyl group on the piperidine ring was deduced after rigorous acetylation of aspidodispermine, which gave a diacetyl derivative showing no acetyl methine protons. Hydrolysis of the acetates regenerated the parent compounds.

Reduction of deoxyaspidodispermine with lithium aluminum hydride in tetrahydrofuran (THF) gave the *N*-ethyl derivative (M^+ 298) still showing a base peak at *m/e* 112, and this eliminated the possibility of a carbinolamine at C-3 or C-21.

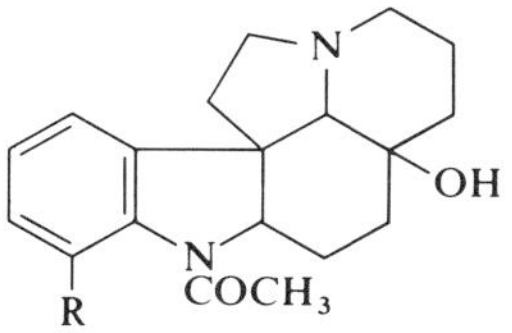

157 R = OH
158 R = H
159 R = OAc
160 R = OCH_3

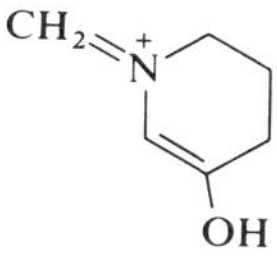

m/e 112
161

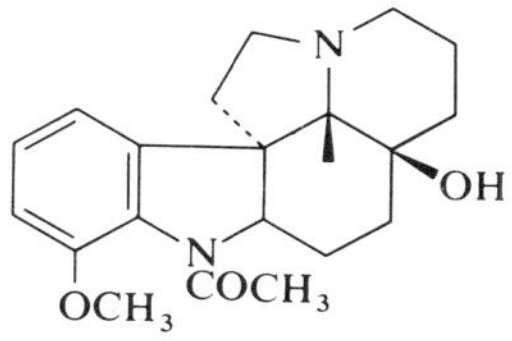

162

Deoxyaspidodispermine and aspidodispermine were formulated as **158** and **157** on the basis of the above evidence, with no stereochemical implications for the C-21 proton or the C-20 hydroxy group (*76*).

Because of the novel nature of the structure of aspidodispermine, confirmation of structure was sought from X-ray analysis. The derivative chosen was the hydrobromide of methyl ether **160**. A three-dimensional bromine-phased difference Fourier map was calculated, and after several least-squares refinements the structure indicated previously was confirmed together with the stereochemistry for the ring junctions as shown in **162**. Aspidodispermine showed a positive optical rotation {$[\alpha]_D + 119°$}, so the stereochemistry shown is also the absolute stereochemistry.

2. Vincatine (**163**)

Döpke *et al.* (*77*) obtained an interesting *seco* ring C *Aspidosperma* alkaloid from *Vinca minor* L. to which they gave the name vincatine (**163**).

Vincatine (mp 111–112°) gave a molecular ion at *m/e* 370, in agreement with a molecular formula of $C_{22}H_{30}N_2O_3$. The IR spectrum showed carbonyl absorptions at 1705 and 1735 cm^{-1} and the UV spectrum indicated the presence of a 2-oxindole. Four aromatic protons were observed in the PMR spectrum together with *N*-methyl (3.20 ppm) and carbomethoxyl (3.62 ppm) groups. Lithium aluminum hydride reduction gave a diol **164** but acetylation of this compound gave only a monoacetyl derivative. The structure of vincatine (**163**) was deduced on the basis of a rationalization of the mass spectra of the parent compound and its various derivatives, particularly the ions at *m/e* 297, 211, 182, and 124 in the parent compound (Scheme 3) and the base peak at *m/e* 184 (**165**) in the mass spectrum of the diol.

3. Rhazinilam (**172**)

One of the structurally most interesting alkaloids in the *Aspidosperma* series to be obtained in the recent past is rhazinilam. The compound was

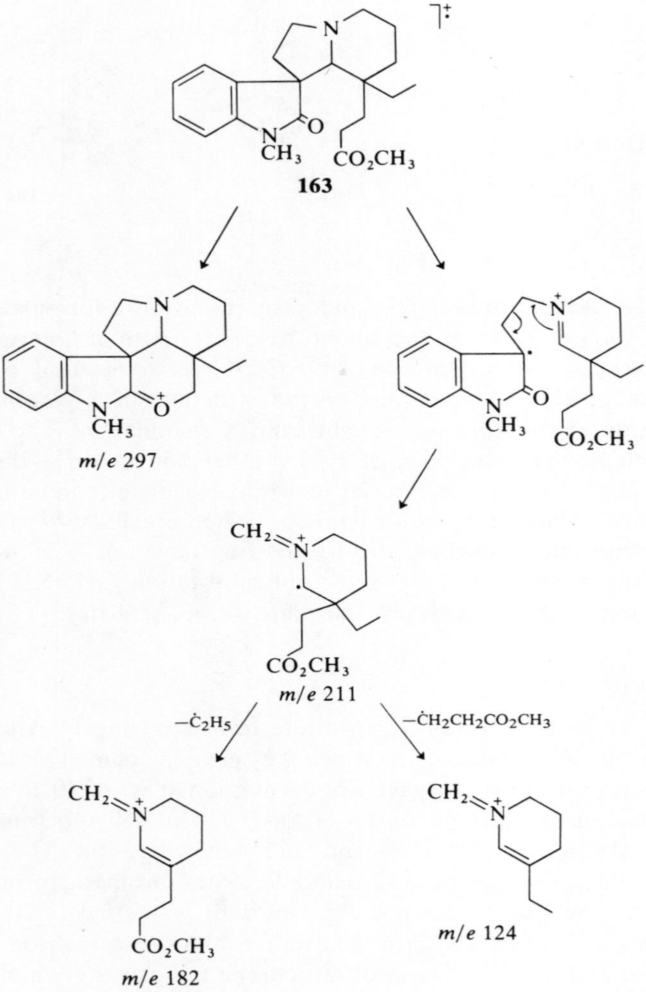

N

$]^{+\cdot}$

N O

CH_3 CO_2CH_3

163

N

N $\overset{+}{O}$

CH_3

m/e 297

$\overset{+}{N}$

N O

CH_3

CO_2CH_3

$CH_2=\overset{+}{N}$

CO_2CH_3

m/e 211

$-\dot{C}_2H_5$

$-\dot{C}H_2CH_2CO_2CH_3$

$CH_2=\overset{+}{N}$

CO_2CH_3

m/e 182

$CH_2=\overset{+}{N}$

m/e 124

SCHEME 3. *Mass spectral fragmentation of vincatine.*

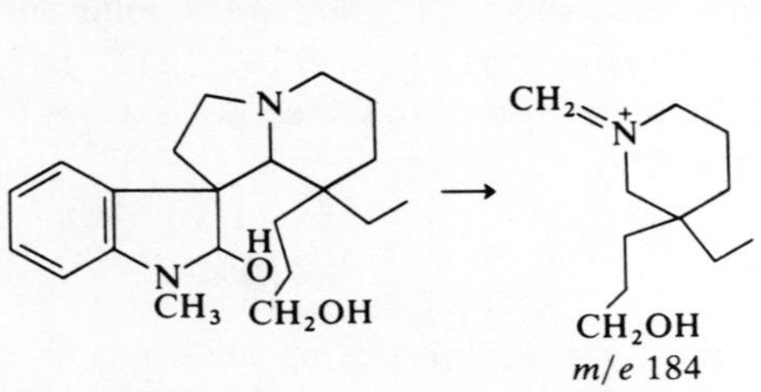

N

H

N O

CH_3 CH_2OH

164

→

$CH_2=\overset{+}{N}$

CH_2OH

m/e 184

165

first obtained by Linde as Ld 82 from *Melodinus australis* (*78*). Chatterjee and co-workers (*79*) subsequently isolated from *Rhazya stricta* Decaisne a compound which they named rhazinilam. They deduced that rhazinilam contained a secondary amide and a tertiary ethyl group.

The same compound was isolated by the author from basic fractions of *R. stricta* root material (*80*), and subsequent work (*81*) permitted elucidation of the structure. Quite independently, rhazinilam was obtained from *A. quebracho-blanco* Schlecht. (*82, 83*), and in collaboration with Abraham and Rosenstein the structure was obtained by X-ray crystallography (*82*).

The Manchester group was intrigued by the fact that rhazinilam accumulated *in vitro* in the basic fractions of *R. stricta* and also by the dramatic irreversible shift in the UV spectrum to long wavelengths under strongly acidic conditions. This UV spectrum in strong acid was found to be quite similar to that of 3*H*-3,4-dimethylpyrrolo[2,3-*c*]quinoline (**166**). Working on a larger scale, the UV spectrum was found to be due to a mixture of bases of molecular formula $C_{19}H_{20}N_2$, and the PMR spectrum of the aromatic region of these bases was also quite similar to that of **166**. Because no C-4 proton was observed in the PMR spectrum of the bases, it was deduced that the bases from rhazinilam were also disubstituted in this manner. Rhazinilam is a neutral alkaloid, and as mentioned previously, contains an amide which was placed by the Manchester group at the 2-position of the original indole nucleus. The partial structure **167** then followed, as two sharp doublets were observed at 6.46 and 5.73 ppm ($J = 1.5$ Hz) quite characteristic of α- and β-protons, respectively, on a pyrrole ring. Formylation with acetic–formic anhydride gave a compound ($C_{20}H_{22}N_2O_2$) the UV spectrum of which (λ_{max} 303 nm, ϵ 16200) corresponded to a 2-formylpyrrole. The PMR spectrum of rhazinilam indicated only one terminal methyl group as a tertiary ethyl side chain (triplet at 0.7 ppm, $J = 7.0$ Hz, and quartet at 1.3 ppm), and a complex multiplet centered at 3.85 ppm was assigned to a methylene attached to a pyrrolic nitrogen. The base peak in the mass spectrum of rhazinilam occurs at $M^+ - C_2H_5$, and it was reasoned that this ion was stabilized by resonance with pyrrolic nitrogen thereby leading to a partial structure **168** in which $x + y = 5$. Mild alkaline hydrolysis of rhazinilam gave an internal amino acid, acetylation of which afforded an amide carboxylic acid $C_{21}H_{26}N_2O_3$.

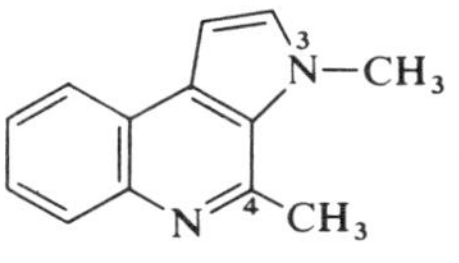

3 N—CH_3 N 4 CH_3

166

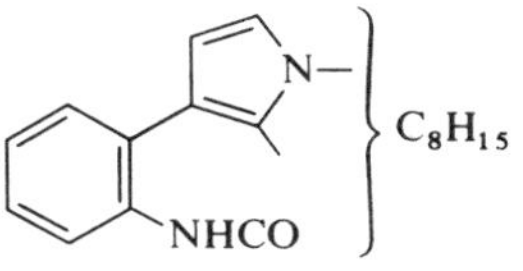

N— NHCO C_8H_{15}

167

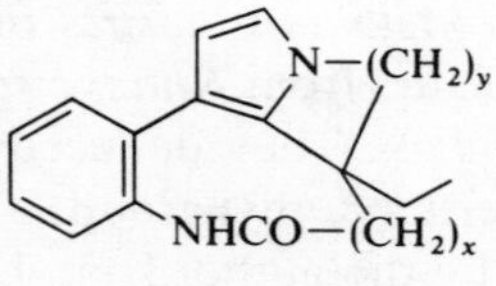

168

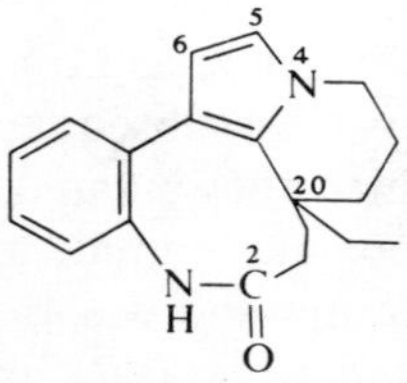

169

In the mass spectrum of this compound the base peak was observed at *m/e* 281 and analyzed for $C_{17}H_{21}N_2O$. This corresponds to a loss of $CH_2CH_2CO_2H$ from the molecular ion and establishes *x* to be 2 and rhazinilam to have the gross structure **169**.

In the X-ray structure analysis of Abraham and co-workers (*82*), which confirmed the correctness of the Manchester groups structure, one of the interesting features to emerge was that the pyrrole nucleus was displaced by almost 90° away from the plane of the benzene ring. Hence the two chromophores do not interact and the absence of π-orbital overlap accounts for the quite low λ_{max} [224, 264 (sh) nm] for rhazinilam. The Manchester group proposed a fascinating mechanism for the *in vitro* formation of rhazinilam involving the fragmentation of 2,21-dihydroxy-5,6-dehydroaspidospermidine (**170**; Scheme 4).

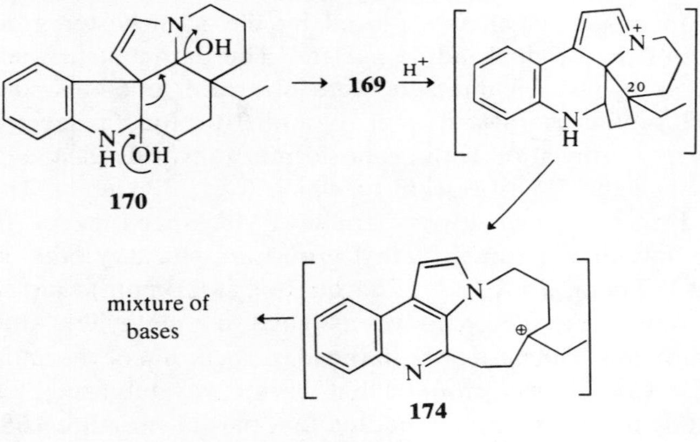

170

174

SCHEME 4

This derivation of rhazinilam from an *Aspidosperma* precursor received at least circumstantial support from a facile and moderately efficient partial synthesis (*84*). Treatment of (+)-1,2-dehydroaspidospermidine (**171**) with *m*-chloroperbenzoic acid and aqueous iron (II) sulfate gave (−)-rhazinilam in about 30% yield. This chemical correlation establishes the absolute configuration of (−)-rhazinilam to be as shown in **172**. The steps involved in this transformation are without doubt very complex. Smith and co-workers suggested Scheme 5, which proposes a 5,6-dihydrorhazinilam (**173**) as a key intermediate.

In addition, the same group postulated the formation of the mixture of bases in acid to involve transannular nucleophilic attack of N-4 on the protonated amide with subsequent dehydration and cleavage of the C-20–C-21 bond. The resulting cation **174** has several possible fates, depending on four alternative proton losses and stabilization of the carbonium ion by addition of water (see Scheme 4). The structure of rhazinilam was also confirmed by total synthesis, and this will be described in Section IV,A,27 of this chapter.

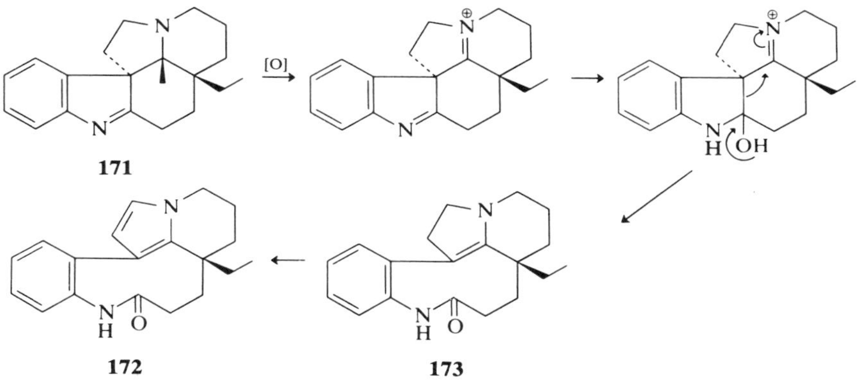

SCHEME 5. *Partial synthesis of* (−)-*rhazinilam* (**172**).

G. QUEBRACHAMINE-TYPE ALKALOIDS

1. 14,15-Dihydroxyquebrachamine (**179**) and Hydroxyindolenine (**176**)

Numerous alkaloids have been isolated from *Voacanga africana* Stapf., and Poisson and co-workers (*85*) have reported on two new compounds having the quebrachamine skeleton.

One of these alkaloids exhibited an indolic UV spectrum and was characterized in the form of its diacetyl derivative. The IR spectrum of the parent compound showed both NH and OH functions and the diacetate derivative exhibited NH and ester groups. Analysis of the diacetate indicated a molecular formula $C_{23}H_{30}N_2O_4$.

Two three-proton singlets were observed at 2.05 and 1.95 ppm, with the corresponding methine protons at 4.0 ppm. A triplet at 0.98 ppm indicated the presence of an ethyl group. The parent compound was unaffected by lithium aluminum hydride; consequently, the structure 14,15-dihydroxyquebrachamine (**175**) was assigned to this compound.

The second compound was identified as an indolenine (λ_{max} 265, 285 nm), which crystallized as the hydrochloride ($C_{19}H_{24}N_2O_2 \cdot HCl$). The IR spectrum of the parent compound indicated both OH and NH functionalities. Poisson assigned structure **176** to this compound (*85*).

The isolation of 14,15-dihydroxyquebrachamine (**175**) was just described. However, the stereochemistry at C-14–C-15 in both this compound and voaphylline had not been determined. Poisson and coworkers established the stereochemistry and absolute configuration of voaphylline to be as shown in **177** by X-ray crystallography (*86*). Previous chemical work had established the relationship between voaphylline, its 14,15-diol, and (+)-quebrachamine (**178**) (*87, 88*). 14,15-Dihydroxyquebrachamine therefore has the structure **179**, in which the two hydroxy groups are *trans*.

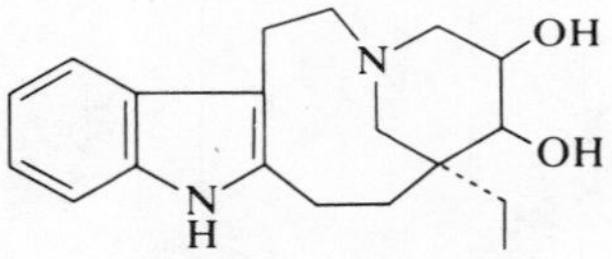

175 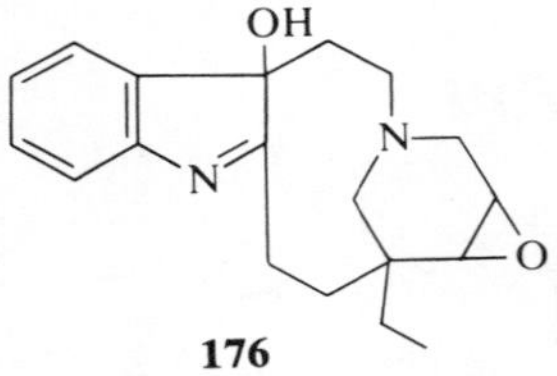 **176**

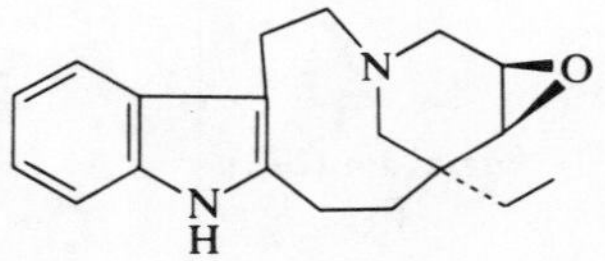

177 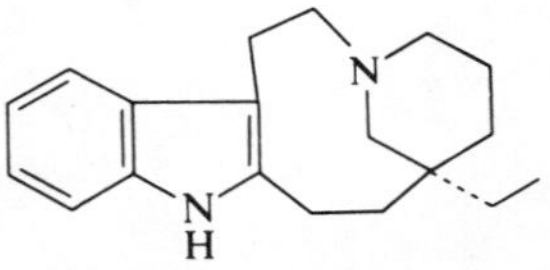 **178**

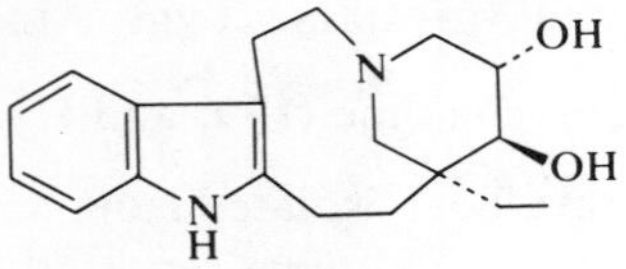

179

2. Vincadine Alkaloids of *Amsonia tabernaemontana* Walt.

The leaves of *A. tabernaemontana* afforded two closely related alkaloids: one showing $[\alpha]_D -92°$, the other being optically inactive (*89*). The UV, IR, and mass spectra of these alkaloids were identical, and high-resolution data indicated a molecular formula $C_{21}H_{28}N_2O_2$.

Heating either compound in 1 *N* hydrochloric acid gave (−)-quebrachamine (**180**) and (+)-quebrachamine (**82**), respectively. The isolated compounds are therefore carbomethoxy derivatives of quebrachamine. The location of the carbomethoxy group was deduced from the mass spectrum; in particular, the ions at *m/e* 210 (**181**) and 138 (**182**) indicated it to be at C-16. (−)-Vincadine was therefore assigned structure **183**, in which the stereochemistry at C-16 was undefined (*89*). (+)-Vincadine (**184**) had been obtained previously (*90*).

In a full paper concerning these compounds (*91*), it was made clear that (−)-vincadine was indeed the optical antipode of (+)-vincadine and therefore had the structure **185**. Reduction of (+)-(**75**) and (−)-vincadifformine (**2**) with formic acid–formamide (*91*) gave (−)-vincadine (**185**) and (+)-vincadine (**184**), respectively. Methylation of (−)-vincadine (**185**) gave (−)-vincaminoreine (**186**).

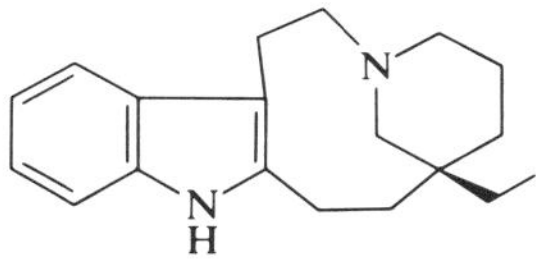

180

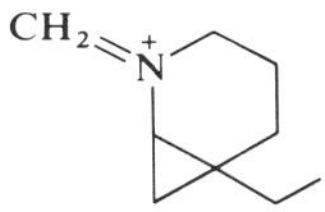

m/e 138
182

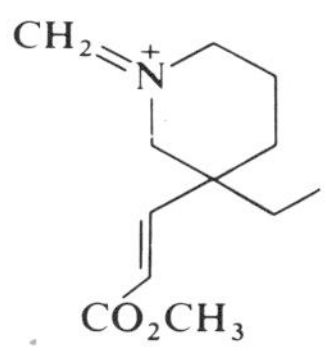

CO_2CH_3
m/e 210
181

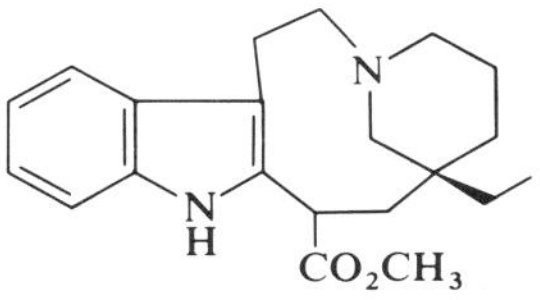

183

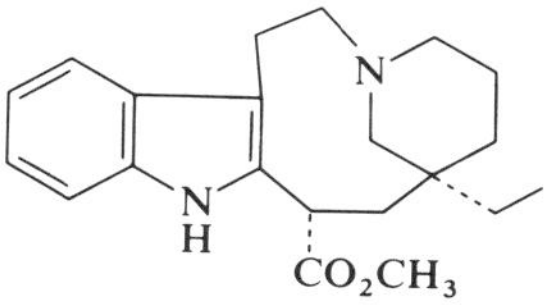

184

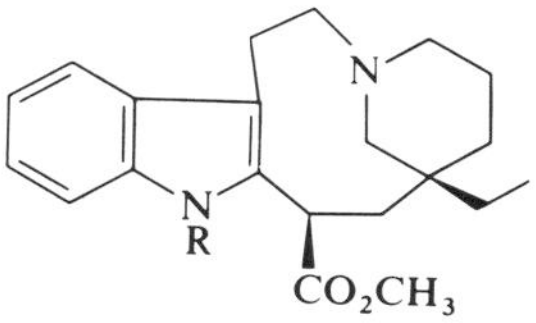

185 R = H
186 R = CH_3

Three new alkaloids in the vincadine series have also been reported by the Hungarian group (*92*). The first compound obtained was (+)-14,15-dehydrovincadine (**187**) {$[\alpha]_D + 65°$}, whose structure was proved by catalytic hydrogenation to (+)-vincadine (**184**) and hydrolysis/decarboxylation to (+)-14,15-dehydroquebrachamine (**188**). The two other compounds were the C-16 epimers of vincadine (**184**) and **187**. 16-Epivincadine (**189**) was racemic, and hydrolysis gave (+)-quebrachamine (**82**). The reason for the racemic nature of 16-epivincadine (**189**) was not evaluated. The second compound, 14,15-dehydro-16-epivincadine (**190**), was optically active {$[\alpha]_D + 85°$}, and catalytic reduction gave (+)-16-epivincadine (**189**). Hydrolysis of **190** gave (+)-14,15-dehydroquebrachamine (**188**).

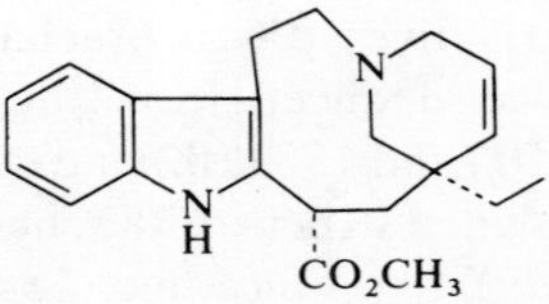

187 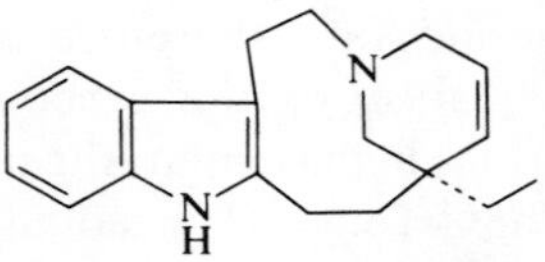 **188**

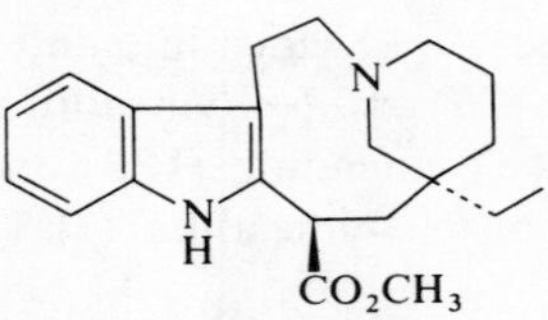

189 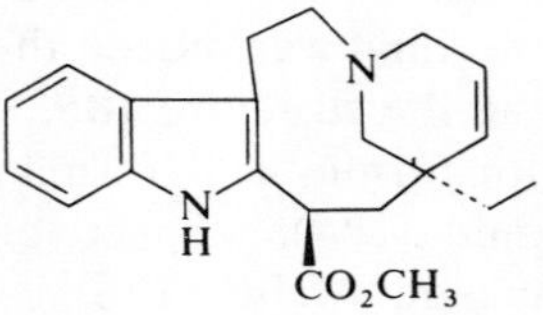 **190**

The mass spectra of **187** and **190** differed in the region of the ions *m/e* 215 and 208. In **187**, *m/e* 208 is more abundant than *m/e* 215 whereas in **190** the reverse is true.

3. Vincaminoridine (**191**)

Mokry and Kompis reported in 1963 the isolation of the minor alkaloid, vincaminoridine, from *Vinca minor* (*93*), and subsequently described the structure (*94*).

Vincaminoridine (**191**) showed M^+ 384, analyzing for $C_{23}H_{32}N_2O_3$. The PMR spectrum indicated the presence of aromatic methoxy (3.82 ppm) and carbomethoxy (3.61 ppm) groups and the latter was confirmed from the IR spectrum (ν_{max} 1730 cm^{-1}). Also observed was a singlet at 3.51 ppm for an indolic *N*-methyl group. The UV spectrum showed λ_{max} 232 and 300 nm, typical of an *N*-methyl substituted indole. In the mass spectrum vincaminoridine gave ions typical of a quebrachamine derivative, especially *m/e* 138 (**182**) and 210 (**181**). These peaks indicate a quebrachamine skeleton with a 16-carbomethoxy group, leaving only the aromatic methoxy group to be placed. The aromatic region of the PMR spectrum showed a doublet ($J = 10$ Hz) at 7.36 ppm and a two-proton multiplet at 6.78 ppm; the methoxy group can therefore be assigned to C-11.

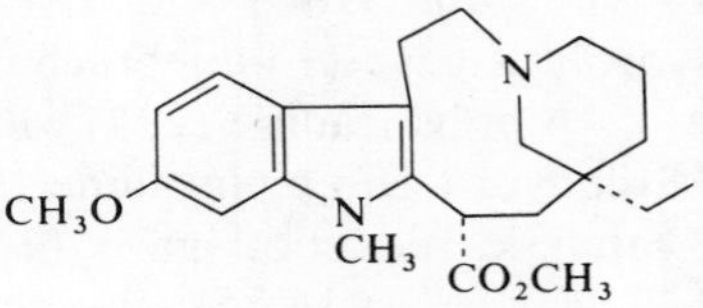

191 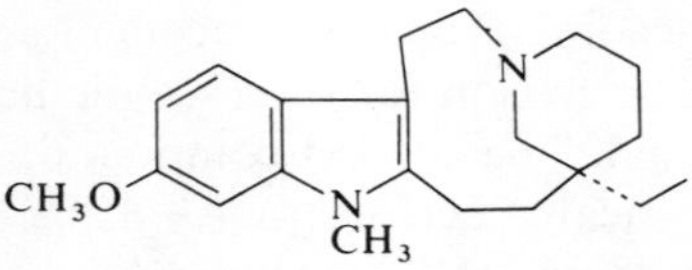 **192**

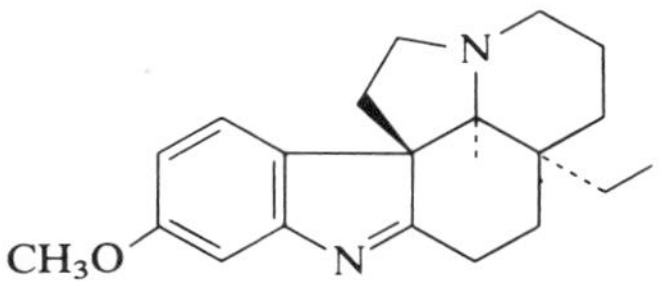

193

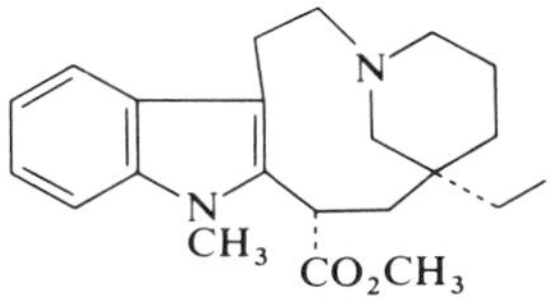

194

The stereochemistries at C-16 and C-20 were determined by chemical correlation. Acid hydrolysis of vincaminoridine gave (+)-11-methoxy-*N*-methylquebrachamine (**192**) identical with a product obtained by sodium borohydride reduction and N-methylation of (−)-11-methoxy-1,2-dehydroaspidospermidine (**193**). The ORD curve also established the C-20 configuration to be α. The stereochemistry at C-16 was determined from the chemical shift of the C-16 proton. As in vincaminorine (**194**), this proton was shifted downfield to 6.10 ppm (*90*). On this basis, vincaminoridine could be ascribed the structure **191**.

4. 12-Methoxyvoaphylline (**195**)

The leaves of *Crioceras dipladeniiflorum* K. Schum. afforded a new alkaloid, identified as 12-methoxyvoaphylline (**195**) (*95*).

The alkaloid $\{[\alpha]_D + 61°\}$ gave a molecular ion at *m/e* 326 which analyzed for $C_{20}H_{26}N_2O_2$ and showed a typical indolic UV spectrum. The PMR spectrum showed only three aromatic protons together with a three-proton singlet at 3.95 ppm for an aromatic methoxy group and a triplet at 0.71 ppm for the methyl of an ethyl group.

The mass spectrum gave ions at *m/e* 140 and 122 for the piperidine ring and at *m/e* 186, 174, and 173 for the indole nucleus. Consequently, there is an ether linkage in the piperidine ring and a methoxy group in the aromatic ring.

Location of the methoxy group at C-12 was determined from the PMR spectrum in the presence of a europium shift reagent.

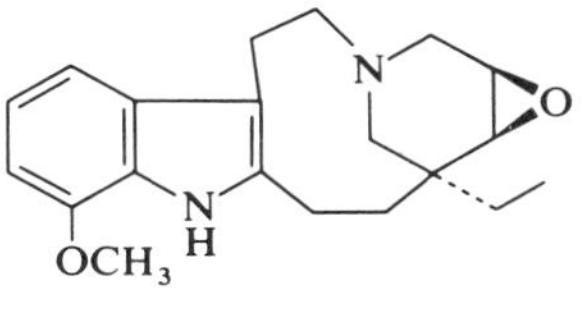

195

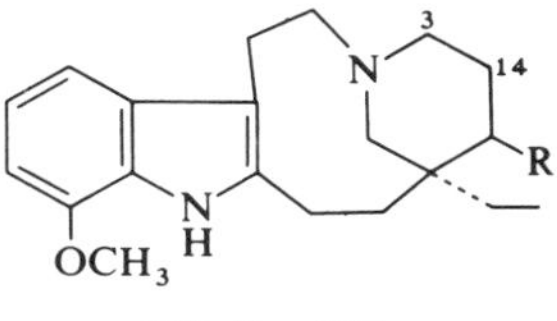

196 R = OH
197 R = O
198 R = O, $\Delta^{3,14}$
199 R = H

Reduction with lithium aluminum hydride gave alcohol **196**, oxidized under Oppenauer conditions to a mixture of ketones (**197**, **198**). Wolff–Kishner reduction of ketone **197** gave a methoxyquebrachamine. The mass spectrum of this product was identical with that of 12-methoxyquebrachamine (**199**), obtained previously by Biemann (*96*). The optical rotation, however, was exactly opposite. Final confirmation of the position of the methoxy group came from the CMR spectrum.

5. Rhazidigenine *N*-Oxide (200)

From the bark of *Aspidosperma quebracho-blancho* Tunmann and Wolf (*97*) have isolated a new hydroxyindolenine, rhazidigenine *N*-oxide (**200**).

In the UV spectrum rhazidigenine *N*-oxide gave absorptions at 237 and 293 nm for an indolenine, shifting in alkali to 308 nm. Losses of 17 and 29 mu from the molecular ion (M^+ 314) indicated the presence of hydroxy and ethyl groups, and a loss of 16 from $M^+ - 17$ suggested an *N*-oxide. Ions at *m/e* 144, 143, and 130 indicated an unsubstituted indole nucleus. Rhazidigenine *N*-oxide was assigned structure **200** on this basis.

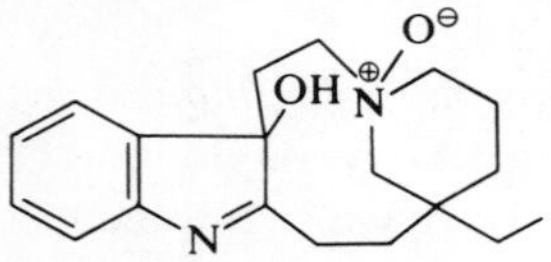

$O^{\ominus}$ OH $\overset{\oplus}{N}$ N

200

H. Pseudovincadifformine-Type Alkaloids

Previous chemical work on the chemistry of catharanthine (**201**) (*98*) had established alkaloids of the pseudovincadifformine type as viable structure possibilities. For many years, however, this structure type remained only an *in vitro* product, and it was not until 1974 that the first natural alkaloid was obtained having this skeleton.

1. Pandoline (202)

From *Pandaca calcarea* and *P. debrayi* Le Men and co-workers (*99, 100*) isolated an alkaloid having $[\alpha]_D + 417°$ which they named pandoline; the structure was determined subsequently (*100*). The alkaloid had a molecular ion at 354 mu, in agreement with a molecular formula $C_{21}H_{26}N_2O_3$. The UV and IR spectral properties confirmed a β-anilinoacrylate derivative. Pandoline gave a monoacetyl derivative (ν_{max} 1710 and 1240 cm^{-1}), indicating the third oxygen function to be an alcohol. The general location of this alcohol was determined from the mass spectrum. In the parent compound the base peak was at *m/e* 140; in the monoacetyl derivative, at *m/e* 182. The hydroxy function could therefore be placed in the piperidine ring or the ethyl side chain. The PMR spectrum, which showed a methyl triplet at 0.95 ppm, excluded the possibility of hydroxyl being at C-18 or C-19, and positions 3 and 21 could be excluded on the basis of nonreduction with lithium aluminum hydride. The possibilities for the location of the hydroxyl groups are therefore C-14, C-15, and C-20.

The tertiary nature of the hydroxyl groups was deduced from the CMR spectrum of pandoline in which a quaternary carbon at 71.0 ppm was shifted to 82.0 ppm on acetylation. In addition, three methylene carbons were shifted by this reaction. These data are only in agreement with a structure such as **202** (*100*).

The *cis* nature of the C/D ring junction was determined from the 250 MHz PMR spectrum, which showed a broad singlet for H-3 at 2.86 ppm, excluding a *trans* stereochemistry between C-3 and C-14. Pandoline was therefore suggested to have structure **202** including absolute stereochemistry, in which the stereochemistry of the hydroxyl group was not defined.

Pandoline (**202**) underwent a number of reactions typical of the β-anilinoacrylate chromophore. Treatment with sodium cyanoborohydride gave 2,16-dihydropandoline (**203**) with a carbomethoxyl group axial and β, and reduction of **203** with lithium aluminum hydride gave 2,16-dihydropandolinol (**204**). Hydrolysis and decarboxylation of pandoline afforded an indolenine (**205**) which was reduced to the indoline (**206**) by lithium aluminum hydride.

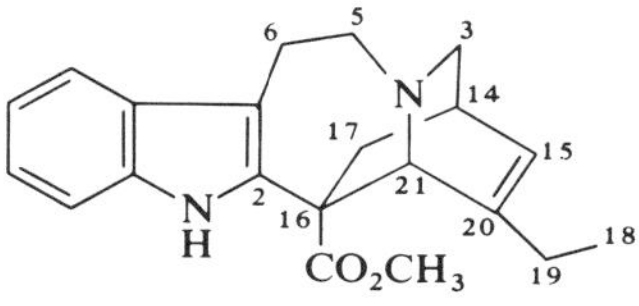

201

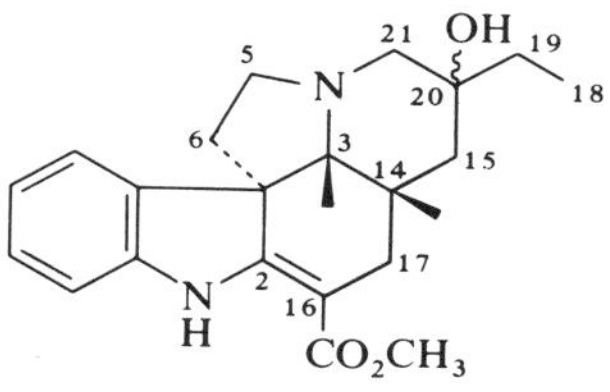

202

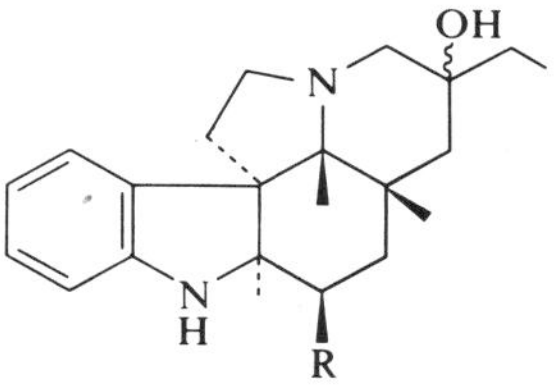

203 R = CO_2CH_3
204 R = CH_2OH
206 R = H

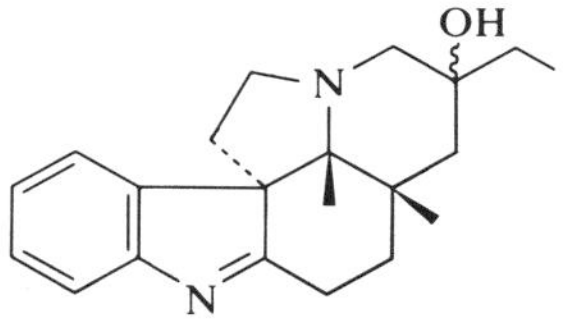

205

Additional evidence for the structure of pandoline (**202**) was derived from two separate chemical correlations with (+)-catharanthine (**201**) (*101*). Dehydration of pandoline (**202**) with concentrated sulfuric acid gave three isomeric dervatives showing M^+ 336, namely **207** in 39% yield and the ethylidine isomers **208** and **209** in 34% overall yield. The mass spectra of these compounds, although showing similar fragment ions, were quite different for the m/e 135 fragment, which in the case of **207** was the base peak but for the ethylidene isomers was only 5%. In **207** the methyl of the ethyl group appeared as a triplet at 1.02 ppm and the C-15 olefinic proton as doublet ($J = 8$ Hz) at 5.47 ppm.

Oxidation of 16α-carbomethoxycleavamine (**210**) with mercuric acetate in acetic acid gave a mixture of products, one of which, (−)-pseudotabersonine (**211**), was identical with **207** except for its optical rotation.

The second correlation involved reactions in the decarbomethoxy series. As described previously, pandoline (**202**) was hydrolyzed and decarboxylated to give the indolenine **205** which with lithium aluminum hydride gave **206**. Dehydration with sulfuric acid gave the ethylidene derivative **212** in 35% yield {$[\alpha]_D + 48°$}. This compound could also be produced from (−)-pseudotabersonine (**211**). Hydrolysis and decarboxylation of **211** followed by lithium aluminum hydride reduction gave **213**, having $[\alpha]_D -$ 50°, but otherwise identical with the product **212** from pandoline (**202**).

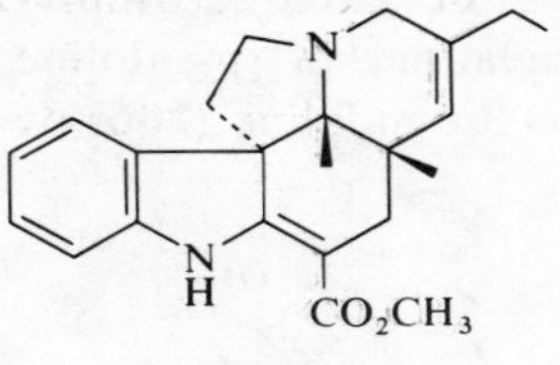

207

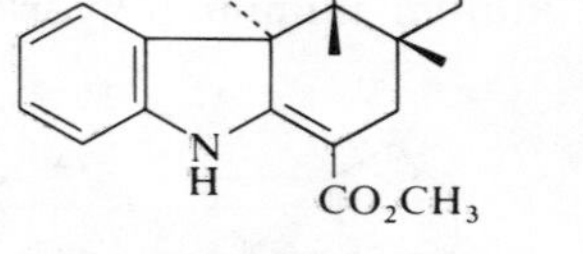

208/209 (isomers)

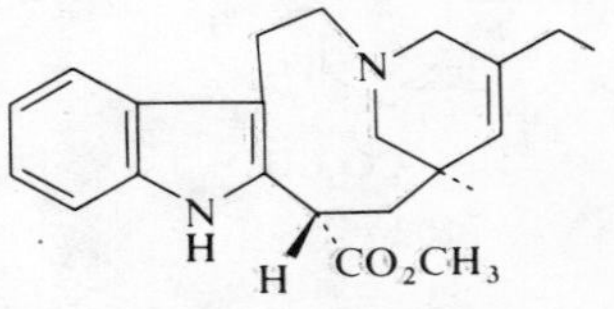

210

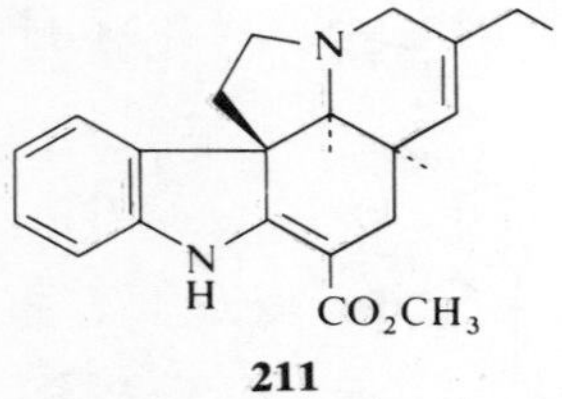

211

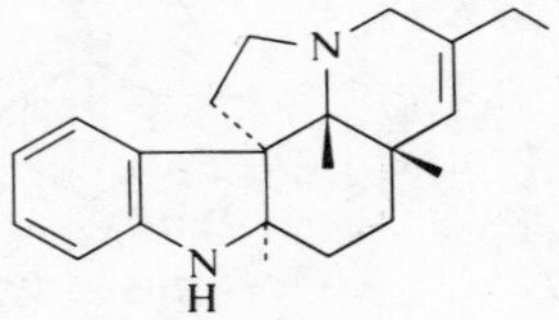

212

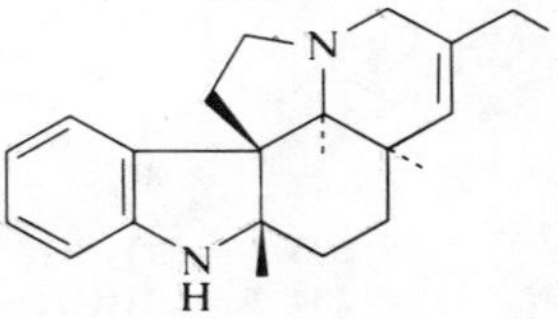

213

A number of interesting observations can be made as a result of these experiments. Perhaps the most important one is that the acid-catalyzed cyclization takes place stereospecifically, so the absolute configuration of pandoline is firmly established to be that represented in **202**.

2. Pandine (**215**)

The structure elucidation of pandoline (**202**) opened up a new area, and several closely related alkaloids have been subsequently characterized.

The first of these was a novel hexacyclic alkaloid, pandine, obtained from three *Pandaca* species, *P. calcarea*, *P. debrayi* (*99*), and *P. caducifolia* Mgt. (*99, 102*). Pandine had IR and UV spectra typical of a β-anilinoacrylate but an atypical mass spectrum showing no products arising from a simple retro-Diels–Alder reaction and 5,6-fission sequence.

The molecular formula of pandine ($C_{21}H_{24}N_2O_3$) indicated one additional degree of unsaturation compared with pandoline (**202**) and, because no additional chromophores or IR absorptions were observed, this unsaturation must be in the form of a new ring.

That pandine contained a tertiary alcohol was evident from the shift of a quaternary carbon from 80.3 to 90.4 ppm upon acetylation and from the dehydration product which showed an ethylidene group. The location of the additional C–C bond was determined by analysis of the PMR and CMR spectra.

Compared with pandoline, two carbon atoms in pandine were shifted downfield by about 15 ppm in the CMR spectrum, in agreement with a new carbon–carbon bond being formed between these two carbon atoms. Thus in pandoline (**202**) C-21 is observed at 61.3 ppm and C-17 at 25.7 ppm. In pandine these carbons resonate at 77.0 and 40.8 ppm, respectively. The new C–C bond is therefore between C-21 and C-17.

Reduction of pandine with sodium cyanoborohydride gave the 2,16-dihydro derivative (**214**) which could not be epimerized with alkali (because of steric factors). The 2 and 16 protons were observed in the form of two doublets ($J = 6$ Hz) at 3.80 and 2.62 ppm, respectively. In the PMR spectrum of this compound, three quite narrow singlets were observed at 2.55, 3.10, and 2.35 ppm which could be attributed to the H-3, H-17, and H-21 protons, respectively, thereby establishing the stereochemistry of H-3, H-14, H-17, and H-21. Pandine was assigned structure **215** on the basis of this evidence and additional chemical reactions as follows. Alkaline hydrolysis of pandine (**215**) and brief heating in acid gave an indolenine which could be reduced to indoline **216** with sodium cyanoborohydride. Reduction of **215** with lithium aluminum hydride did not give

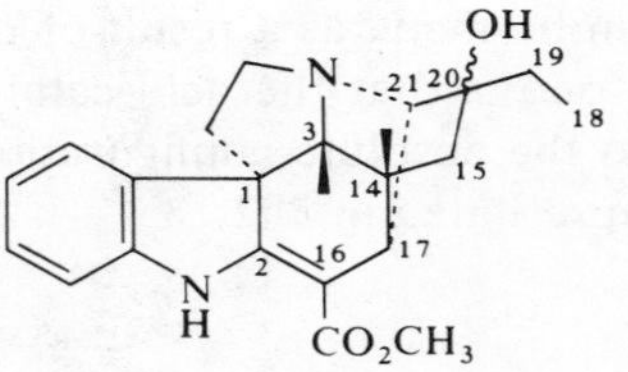

215

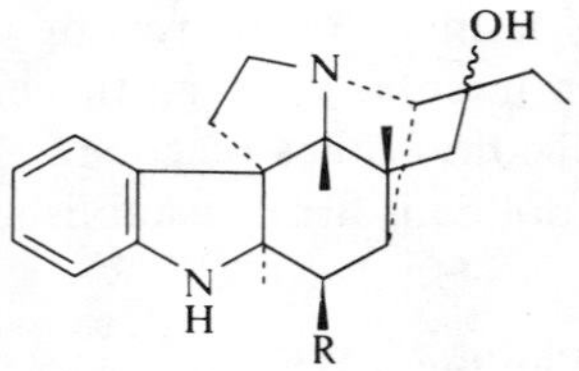

214 R = CO_2CH_3
216 R = H
217 R = CH_2OH

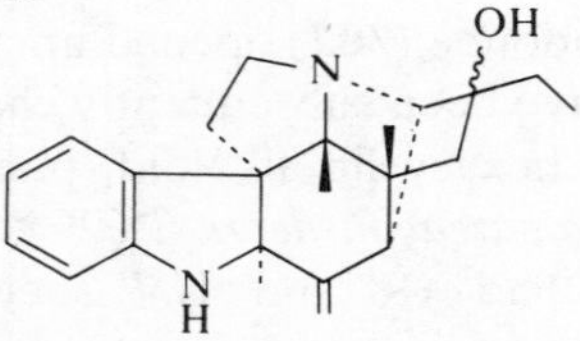

218

the expected carbinol **217**, but rather the exomethylene derivative **218** in a manner similar to that of akuammicine (*102*).

3. (+)-Pseudotabersonine (**207**) and (+)-(20*R*)-Pseudovincadifformine (**219**)

Examination of the leaves of *Pandaca caducifolia* gave two pseudo *Aspidosperma* alkaloids isolated previously, (+)-pandoline (**202**) and (+)-pandine (**215**), and three new alkaloids in this series (*103*).

One of these alkaloids was identified as the amorphous base (+)-pseudotabersonine (**207**) {$[\alpha]_D + 320°$}, identical except for optical rotation with the compound derived by acid rearrangement or other chemical transformations from (+)-catharanthine (**201**) (*102–106*). The compound proved to be identical with one of the products obtained by the dehydration of pandoline (**202**) (*100*). The UV and IR spectra were typical for a β-anilinoacrylate alkaloid, and the PMR spectrum showed a triplet for three protons at 1.02 ppm, an olefinic signal as a doublet ($J = 8$ and 2 Hz) at 5.47 ppm, and a three-proton singlet at 3.74 ppm. The mass spectrum showed a molecular ion at *m/e* 336, a base peak at *m/e* 135, and characteristic ions at *m/e* 122, 121, and 107.

The second new alkaloid was readily identified as (+)-(20*R*)-pseudovincadifformine (**219**) {$[\alpha]_D + 430°$}. This alkaloid, showing a molecular ion at *m/e* 388, exhibited a typical base peak at *m/e* 124, a characteristic ion at *m/e* 214 (ion A), but no *m/e* 309 (ion D). The PMR spectrum showed a triplet at 0.95 ppm for the methyl of the ethyl group, and no olefinic proton was observed. The compound was identical with (−)-(20*S*)-pseudovincadifformine prepared from (+)-catharanthine (**201**) except for optical rotation.

4. (+)-20-Epipandoline (**221**)

The third new monomeric alkaloid obtained from *P. caducifolia* (*103*) was very similar in all spectral properties to pandoline (**202**). In particular, the UV and IR spectral data verified the nature of the aromatic nucleus, and from the PMR spectrum an ethyl side chain attached at a hydroxy substituent was evident. This could also be demonstrated by the mass spectrum, where the base peak was found at *m/e* 140. On this basis, the alkaloid was speculated to be the 20-isomer of pandoline. Because pandoline was more polar than the epipandoline, it was reasoned that pandoline (**220**) had the 20-hydroxy group in the α-position (axial) and that in 20-epipandoline (**221**) the hydroxy group was equatorial and β.

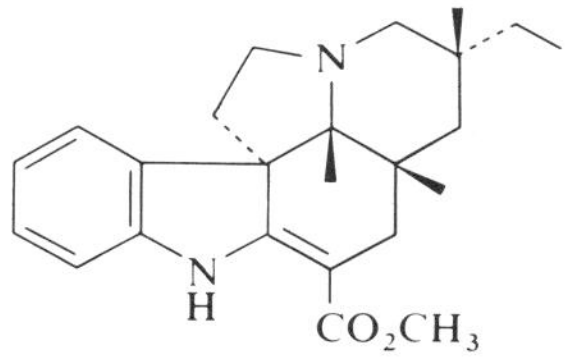

N H CO$_2$CH$_3$

219

OH N H CO$_2$CH$_3$

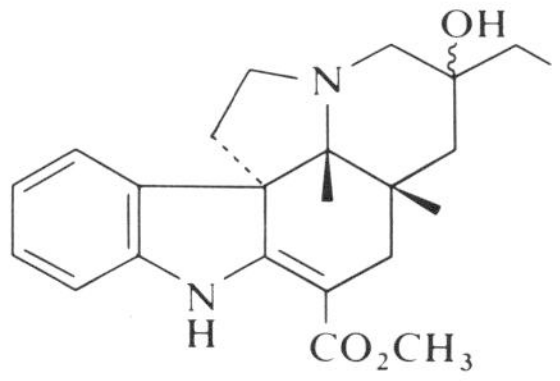

220 α-OH
221 β-OH

5. Capuronidine (**222**) and Capuronine (**224**)

The Gif sur-yvette group has examined the Madagascan plant *Capuronetta elegans* Mgf. (Apocynaceae) and obtained a number of interesting alkaloids, two of which are biogenetically very closely related (*107*).

Capuronidine (**222**), an amorphous alkaloid {$[\alpha]_D + 220°$}, was obtained from the leaves of *C. elegans* and exhibited a typical indolenine chromophore (λ_{max} 220 and 260 nm, λ_{max} H^+, 275 and 341 nm) and a molecular ion at *m/e* 296. Reduction with sodium borohydride gave an indoline (**223**), and in the presence of base decyclization occurred to give the cleavamine derivative **224**. This compound was identical with another alkaloid, capuronine, isolated from the same plant.

Capuronine showed a molecular ion at *m/e* 298, in agreement with a molecular formula $C_{19}H_{26}N_2O$, and exhibited an indolic UV spectrum. The mass spectrum of capuronine gave a number of fragments characteristic of a velbanamine derivative, in particular a base peak at *m/e* 154 containing the ethyl piperidine unit. The single oxygen function was traced to a secondary alcohol by its facile acetylation and oxidation to a nonconjugated ketone (**225**). Capuronine did not react with sodium borohydride, and consideration of all these data indicated that the only possible positions for substitutions are C-15 and C-17. Substitution at C-17 was eliminated by base-catalyzed deuterium exchange of the ketone **225** which gave only a dideuterio derivative. Capuronine therefore has the structure **224**. The question of the stereochemistry of the hydroxy and ethyl groups was determined by chemical correlation.

Hydroboration of (+)-cleavamine (**226**) gave a mixture of two epimeric alcohols in each of which the hydroxyl and ethyl groups are *cis*; capuronine was identical with one of these compounds. X-Ray analysis of capuronine acetate (**227**) distinguished between these possibilities and indicated structure **224** to be correct. Dehydration of (+)-capuronine with sulfuric acid gave (−)-cleavamine (**228**), thereby establishing the absolute configuration of (+)-capuronine to be as shown in **224** (*107*).

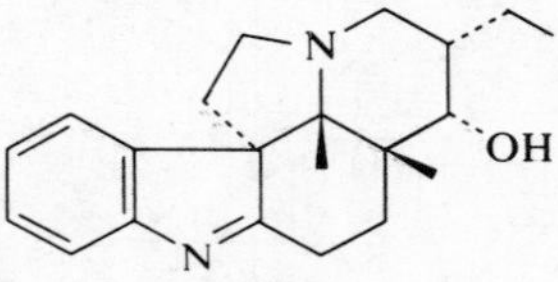

222

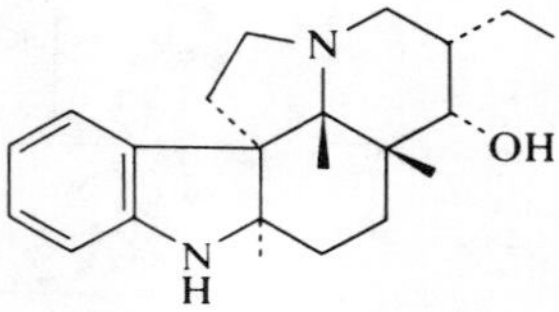

223

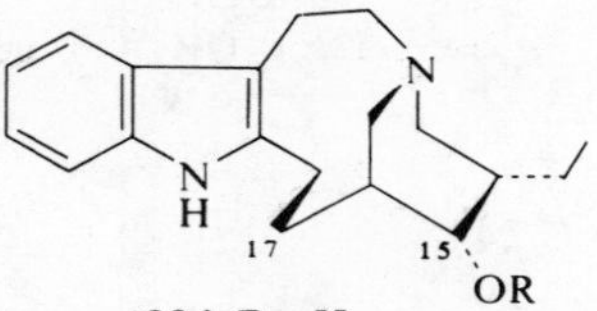

224 R = H
227 R = Ac

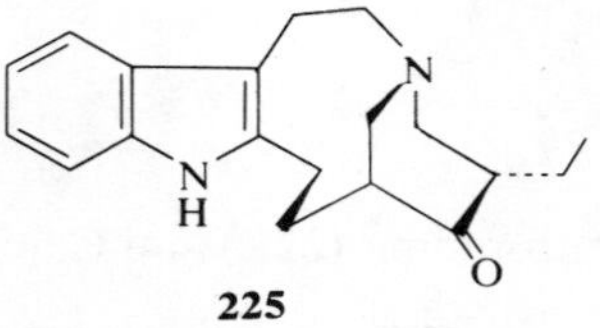

225

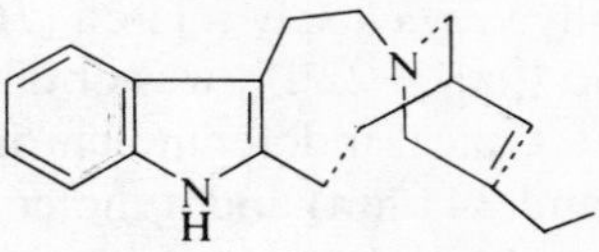

226

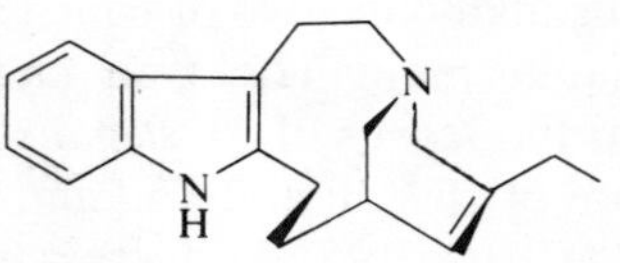

228

6. (20*S*)-1,2-Dehydropseudoaspidospermidine (**229**)

Le Men and co-workers have also examined *Pandaca ensepala* Mgf. and, in addition to a number of known alkaloids, they obtained a new pseudo *Aspidosperma* alkaloid (*108*).

(*20S*)-1,2-Dehydroseudoaspidospermidine (**229**) was isolated as an unstable oil {$[\alpha]_D + 153°$} having an indolenine UV chromophore and a base peak in the mass spectrum at *m/e* 137. Reduction with lithium aluminum hydride gave a dihydro derivative {**230**; $[\alpha]_D + 60°$}. The mass spectrum now showed a base peak at *m/e* 124 typical of an aspidospermidine-type alkaloid. This compound was identical, except for optical rotation, with (–)-(20*R*)-pseudoaspidospermidine (**231**) formed from catharanthine (**201**) (*108*).

Two other alkaloids also isolated from the same plant were of considerable biogenetic interest (*108*): these alkaloids were (+)-(20*R*)-dihydrocleavamine (**232**) and (–)-(20*S*)-dihydrocleavamine (**233**).

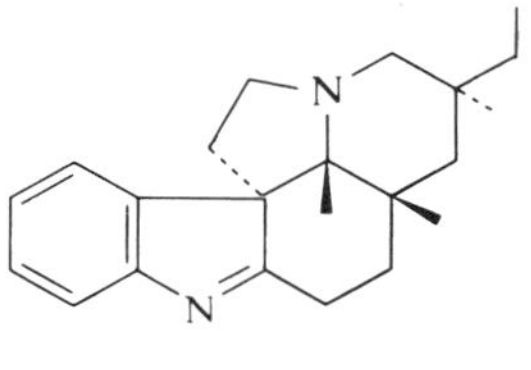

229

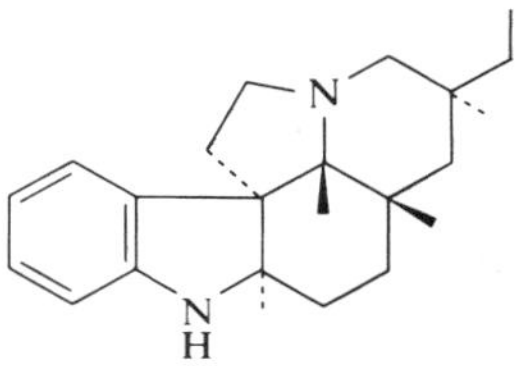

230

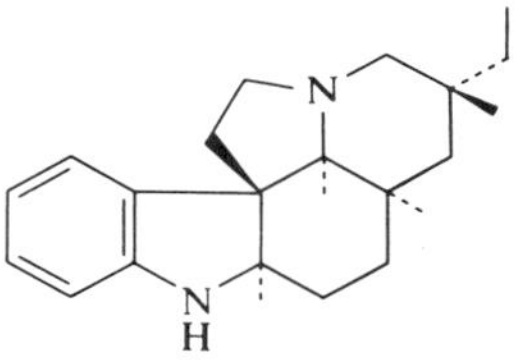

231

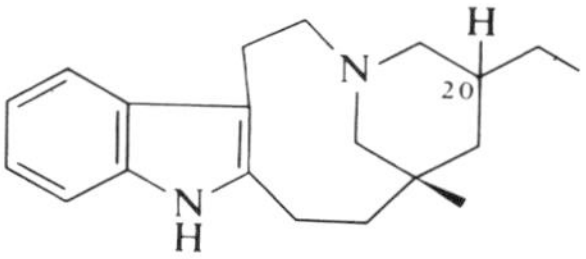

232 C-20Hβ
233 C-20Hα

7. Ibophyllidine (**235**) and Iboxyphylline (**237**)

Two new indole alkaloid skeleta have been obtained from *Tabernanthe iboga* Baillon and *T. subsessilis* Stapf in the form of the alkaloids ibophyllidine and iboxyphylline (*109*).

Ibophyllidine showed a molecular ion at *m/e* 324 ($C_{20}H_{24}N_2O_2$) and a β-anilinoacrylate chromophore (λ_{max} 302 and 333 nm). The PMR spectrum indicated the presence of an ethyl group, and the mass spectrum a base peak at *m/e* 110 (**234**), 14 mass units less than for an aspidospermine-type skeleton. The nature of the skeleton was deduced from the CMR spectrum. Except for the aromatic ring carbons only one quaternary carbon was observed (C-7 at 55.8 ppm), indicating that the ethyl group was not at a ring junction. Three carbons bearing only a single hydrogen were observed: C-14 at 37.8 ppm, C-3 at 65.7 ppm, and C-20 at 75.6 ppm. The ethyl group is therefore on the carbon adjacent to nitrogen. The stereochemistry was deduced from the PMR spectrum where H-3 was observed as a doublet ($J = 3$ Hz), indicating a *cis* relationship to H-14 and H-20 as a multiplet. The C-19 protons were nonequivalent, indicating some type of restricted rotation about the C-19–C-20 bond which occurs (according to models) when the ethyl group is α between the C-5 and C-15 methylenes. The structure of ibophyllidine is therefore **235**.

Iboxyphylline was crystalline, had a molecular formula of $C_{21}H_{26}N_2O_3$, and exhibited a β-anilinoacrylate chromophore. The third oxygen was in the form of a secondary alcohol, showing a one-proton multiplet at 3.95 ppm shifted downfield on acetylation to 5.05 ppm. The base peak at *m/e* 140 (**236**) contained this functionality. A methyl group appeared as a doublet at 0.93 ppm but was affected only slightly when the ketone of iboxyphylline was reduced. Again the CMR spectrum was very important in determining the structure. Four secondary carbons were observed, at 70.3 ppm for C-3 and at 41.1 ppm for C-19, but more importantly, at 37.6 ppm for C-14 and at 73.1 ppm for C-20. The D ring is therefore confirmed to contain both a secondary methyl and a secondary alcohol. Confirmation of the structure and demonstration of the stereochemistry came from a single-crystal X-ray analysis of the free base, which indicated **237** to be correct (*109*).

These two compounds, ibophyllidine (**235**) and iboxyphylline (**237**), were suggested to have a quite novel biogenesis in which pandoline (**220**) is the key intermediate (Scheme 6) (*109*).

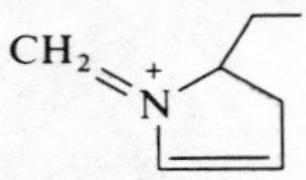

$CH_2=N^+$ *m/e* 110 **234**

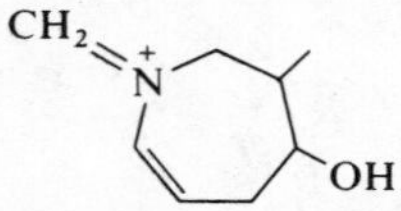

$CH_2=N^+$ OH *m/e* 140 **236**

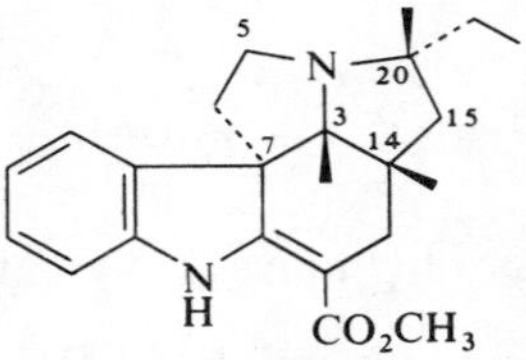

5 N 20 3 15 7 14 N H CO_2CH_3 **235**

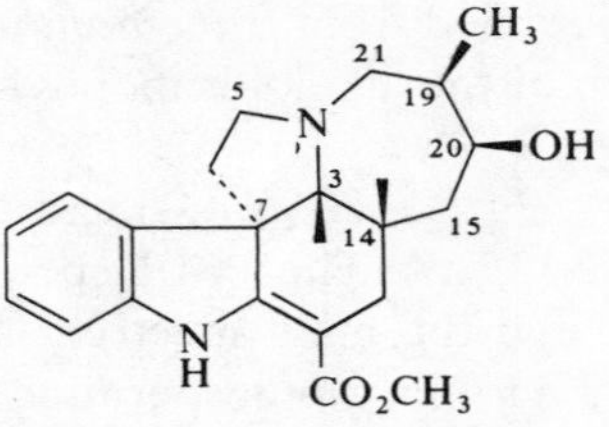

CH_3 21 5 19 N 20 OH 3 7 14 15 N H CO_2CH_3 **237**

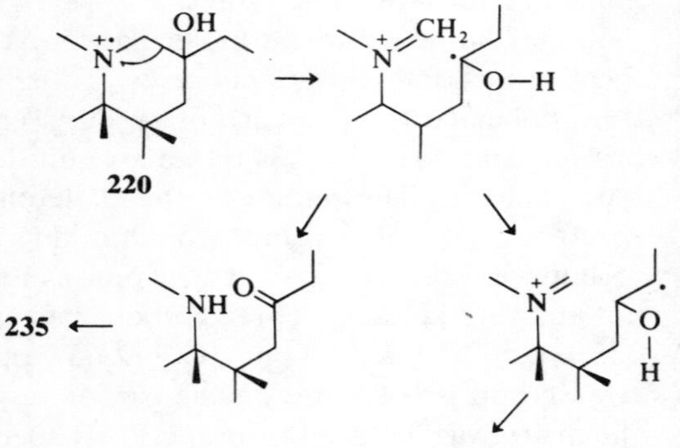

OH $N^{+\bullet}$ **220** → N^+ CH_2 O–H

235 ← NH O

N^+ O H → **237**

SCHEME 6

I. *Melodinus* Alkaloids

In a previous discussion of the genus *Melodinus* (Volume XI, p. 242), the isolation and structure elucidation of meloscine (**238**) and 16-epimeloscine (**239**) were discussed. The Hoffmann–La Roche group has subsequently published a full presentation of the structure elucidation of these novel alkaloids (*110*).

A third alkaloid, isolated with the others from *M. scandens* Forst. (*111*), on acid hydrolysis underwent decarboxylation to give meloscine (**238**), and was ascribed the structure **240**. In the more recent publication (*110*), the alkaloid is assigned the name scandine and its structure is conclusively demonstrated.

Scandine exhibited a molecular ion at *m/e* 350 analyzing for $C_{21}H_{22}N_2O_3$ and an important fragment ion at $M^+ - C_2H_3O_2$ for loss of a carbomethoxy group. The IR spectrum showed two carbonyl frequencies at 1671 cm^{-1} for the quinolone carbonyl and at 1748 cm^{-1} for the ester group. Two major differences were observed in the PMR spectrum of scandine compared to that of meloscine: (a) the C-21 proton which appeared as a singlet at 3.53 ppm in meloscine (**238**) was shifted upfield to 3.02 ppm in scandine, and (b) a three-roton singlet was now evident at 3.54 ppm. The remainder of the spectrum was very similar except for a simplification in the region of the C-17 protons (AB system at 3.07 and 2.85 ppm in scandine methiodide).

The diamagnetic anisotropic effect of the carbomethoxyl group indicated it to be α and the B/C ring junction *trans*. Because (+)-scandine gave rise to (+)-meloscine (**238**), it was assigned the structure and absolute stereochemistry shown in **241** (*110*).

Two new alkaloids of *M. scandens* have been isolated by Plat and co-workers (*112*). One of these showed a molecular ion at *m/e* 308 with a fragment ion at $M^+ - 16$. This suggested loss of oxygen from an *N*-oxide was confirmed by reduction with ferrous ion at room temperature to give 16-epimeloscine (**239**). Oxidation of 3-epimeloscine (**239**) with oxygen in ethanol gave the *N*-oxide of 16-epimelosine (**242**), identical with the natural product.

The second new alkaloid, meloscandonine (*112, 113*); gave a molecular ion at *m/e* 320 analyzing for $C_{20}H_{20}N_2O_2$, a UV spectrum similar to meloscine (λ_{max} 216, 254 nm), and an IR spectrum having carbonyl bands at 1750 and 1685 cm^{-1}. Four aromatic protons were seen, but unlike the other *Melodinus* alkaloids only two olefinic protons were observed, at 5.88 and 6.01 ppm. These olefinic protons could be assigned to those in the piperideine ring, indicating that the vinyl group present in all the other meloscine alkaloids was involved in some other functionality.

The clues to the nature of this functionality were evident from the PMR spectrum, which showed a doublet ($J = 7.6$ Hz) methyl group at 1.12 ppm and the 1750 cm^{-1} carbonyl absorption characteristic of a cyclopentanone. At this point structure **243** was proposed. Reduction of meloscandonine with sodium borohydride gave two isomeric alcohols [**244** and **245** (M^+ 322)] each of which was readily acetylated (to **246** and **247**, respectively). Evaluation of the PMR spectra of these compounds defined the stereochemistry at C-20.

In alcohol **244**, a methine doublet ($J = 2.9$ Hz) was observed at 3.81 ppm which shifted to 4.90 ppm following acetylation. On the other hand, in acetate **247** from the epimeric alcohol **245**, the acetate methine proton (H-22) appeared as a doublet of doublets ($J = 7.4$, 2.0 Hz) at 5.00 ppm. From Dreiding models the long-range coupling could be ascribed to a "W" arrangement between the β proton and H-22 in an "endo" configuration. The 7.4 Hz coupling between H-22 and H-19 indicated these protons to be *cis*. Supporting evidence for these stereochemical assignments came from an evaluation of the acetate singlets. In acetate **247** this singlet appeared in a "normal" position at 1.97 ppm. In acetate **246**, however, this singlet was shielded to 1.23 ppm by being held over the aromatic nucleus. A Nuclear Overhauser Effect (NOE) effect was observed between H-22 and the C-19 methyl group in **246**.

The only proton remaining to be defined stereochemically is H-21. In acetate **247**, the 17α proton appears as a doublet of doublets ($J = 11.2$, 1.2 Hz) at 2.23 ppm. The long-range coupling of H-4α is with H-19 (by double irradiation) and establishes the 21 proton to be α. The circular dichroism of meloscandinone was almost identical with that of the previously isolated meloscine alkaloids; therefore, **248** also represents the absolute stereochemistry of meloscandinone (*112, 113*).

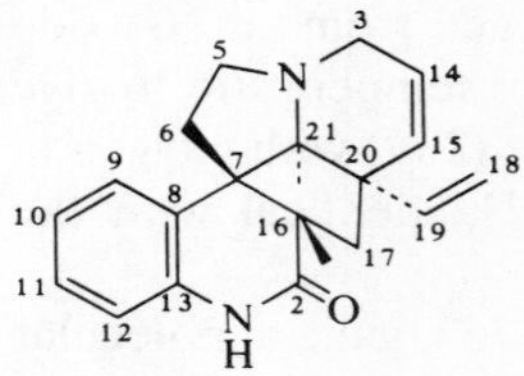

238 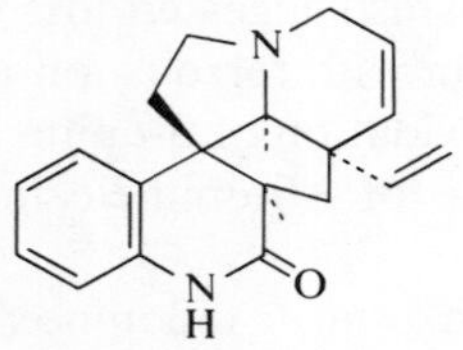 **239** 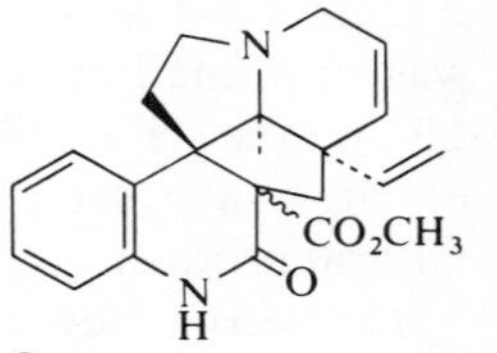 **240**

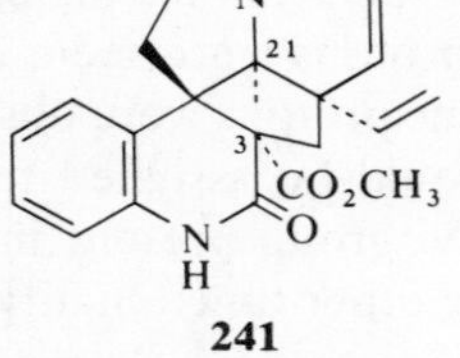

241 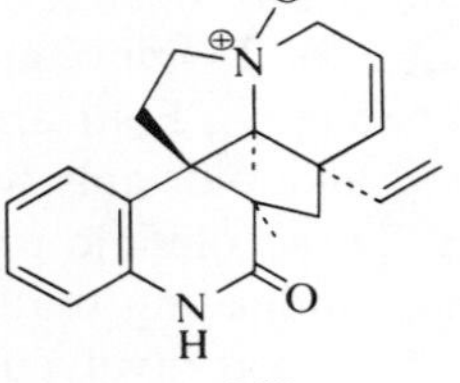 **242**

243

244 $R_1 = H, R_2 = OH$
245 $R_1 = OH, R_2 = H$
246 $R_1 = H, R_2 = OCOCH_3$
247 $R_1 = OCOCH_3, R_2 = H$

248

J. Miscellaneous Alkaloids

1. Alkaloids of *Aspidosperma neblinae*

Biemann and co-workers (*114*) have investigated the alkaloids of *A. neblinae* by combination gas chromatography–mass spectrometry. The alkaloids detected were aspidospermidine (**249**), 1,2-dehydroaspidospermidine (**171**), deacetylpyrifolidine (**250**), 1,2-dehydrodeacetylpyrifolidine (**251**), *N*-acetylaspidospermidine (**252**), aspidospermine (**253**), demethylaspidospermine (**254**), pyrifolidine (**255**), aspidocarpine (**256**), eburnamonine (**257**), and neblinine (**258**). Both the power and the limitations of the technique are evident here. The technique is powerful in the sense that identification of microgram quantities of an alkaloid is easily made with computerized comparison of mass spectra. It has limitations, however, because compounds which are not stable to 300° (injector temperature) either will not be detected or will give alternate compounds, and that isolation must be made if the taxonomically important optical rotations of the alkaloids are to be determined.

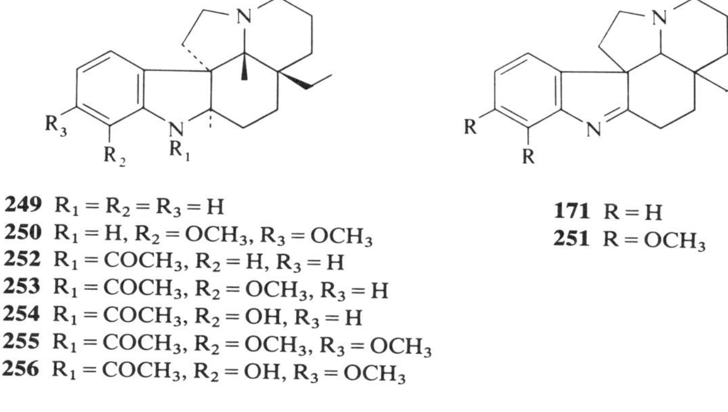

249 $R_1 = R_2 = R_3 = H$
250 $R_1 = H, R_2 = OCH_3, R_3 = OCH_3$
252 $R_1 = COCH_3, R_2 = H, R_3 = H$
253 $R_1 = COCH_3, R_2 = OCH_3, R_3 = H$
254 $R_1 = COCH_3, R_2 = OH, R_3 = H$
255 $R_1 = COCH_3, R_2 = OCH_3, R_3 = OCH_3$
256 $R_1 = COCH_3, R_2 = OH, R_3 = OCH_3$

171 $R = H$
251 $R = OCH_3$

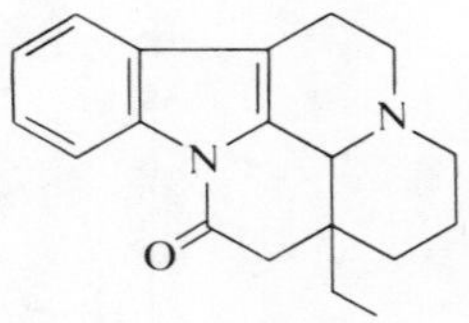

257 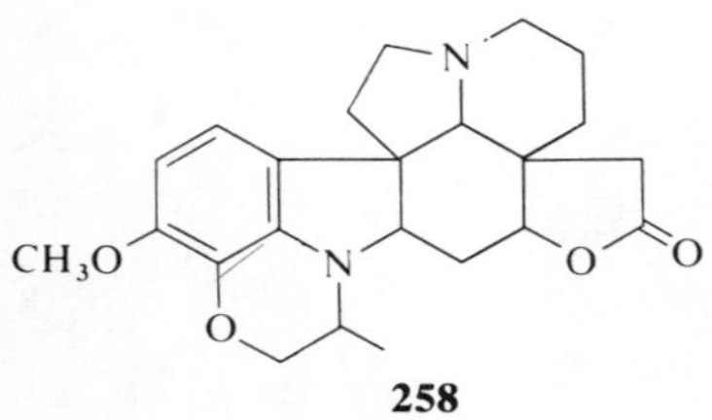 **258**

2. 16-Decarbomethoxy-14,15,16,17-tetrahydrosecodine (**260**)

In numerous reviews of the biosynthesis of indole alkaloids (*2*), an intermediate of the type **259** has been postulated. This skeleton remained merely a topic for chemical comment until a compound having this type of skeleton was actually isolated (*115*).

The first monomeric compound of this skeletal type to be isolated was **260**, obtained by Crooks *et al.* from the leaves of *Tabernaemontana cumminsii* (*116*). High-resolution peak matching gave a molecular formula $C_{19}H_{28}N_2$ (M^+ 284) showing an extremely intense base peak at *m/e* 126. This ion had previously been observed only in the spectra of the tetrahydrosecamines (*115*), where structure **261** had been demonstrated for ions of this mass.

Because only 1.9 mg of material were obtained, the only other physical data obtainable was a UV spectrum, which indicated the compound to be indolic. With this in mind, and ions at *m/e* 158 and 143 prominent in the mass spectrum, structure **260** was proposed.

This postulate was confirmed by synthesis. Reaction of 2-ethylindole with oxalyl chloride gave **262**, which reacted with 3-ethylpiperidine to afford the amide **263**. Lithium aluminum hydride reduction gave the amine **260**, identical with the natural material. The correct name for this compound, based on secodine nomenclature (*117*), is 16-decarbomethoxy-14,15,16,17-tetrahydrosecodine.

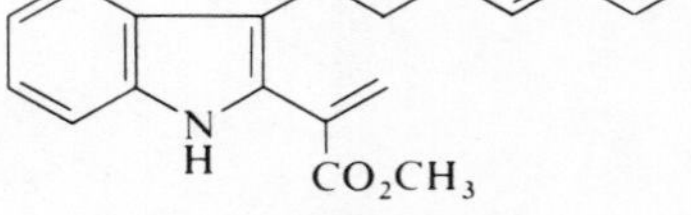

259 **260**

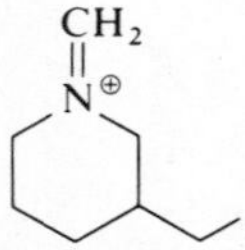

m/e 126
261

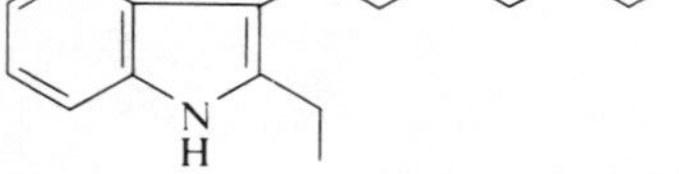

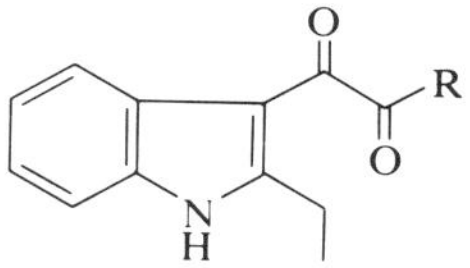

262 R = Cl

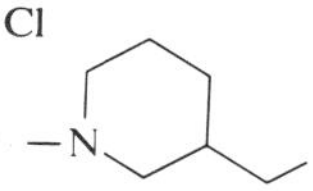

263 R = —N

3. New Alkaloids from *Ochrosia* Species

Preaux *et al.* have obtained two new alkaloids, ochromianine (**264**) and ochromianoxine (**265**), from the bark of *Ochrosia miana* H. Bn. ex Guill. (*118, 119*). The stereochemistry of C-20 in each of these alkaloids remains to be determined.

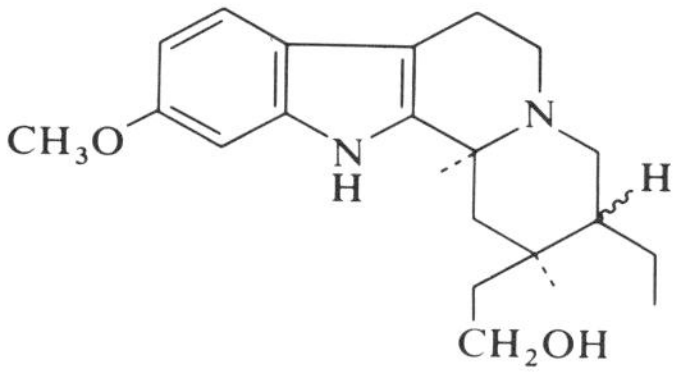

CH_3O N H H CH_2OH

264

CH_3O N H O N H OH

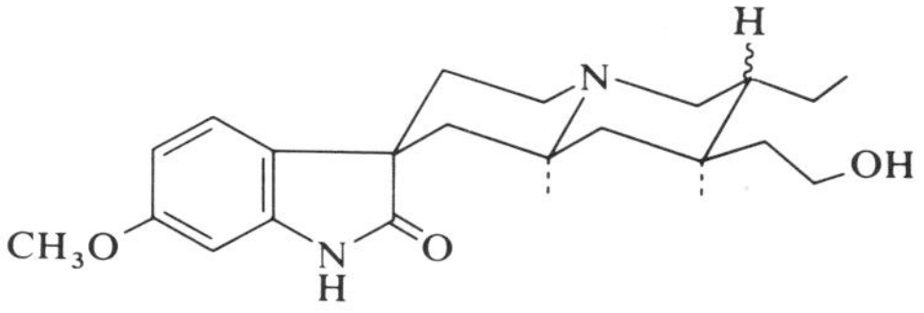

265

The bark of *O. viellardii* Guill. afforded seven known alkaloids and an eighth new alkaloid, ellipticine *N*-oxide (**266**) (*120*). In the mass spectrum the molecular ion was observed at *m/e* 262, indicating the presence of an additional oxygen compared to ellipticine (**267**). The oxygen was not in the form of a carbonyl (IR) or phenolic (UV) function. Reduction with zinc and hydrochloric acid gave ellipticine (**267**), identical with the authentic alkaloid. Attempts to produce the *N*-oxide with classical reagents failed.

4. Harman Carboxylic Acid (**268**)

Sanchez and Brown have obtained a new harman derivative from *Aspidosperma exalatum* Monachino (*121*). The alkaloid was obtained from a butanol extract, and in the IR spectrum it showed principal absorptions at 1925 cm^{-1} for an immonium ion, at 1610 cm^{-1} for a carboxylate anion, and at 748 cm^{-1} for a disubstituted benzene. In the mass spectrum a molecular ion was observed at *m/e* 226 and a base peak at *m/e* 182 ($M^+ - 44$). An aromatic methyl group and five aromatic protons were observed in the PMR spectrum. Consideration of this data together with the UV spectrum led to deduction of the structure of the alkaloid as **268**, harman carboxylic acid. The ethyl ester (**269**) was also isolated but it was probably an artifact.

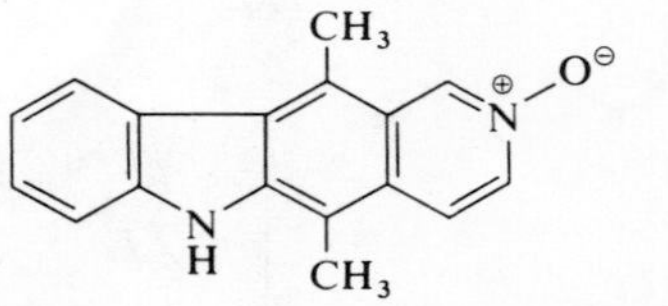

266

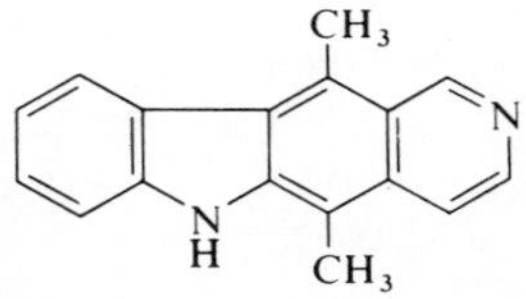

267

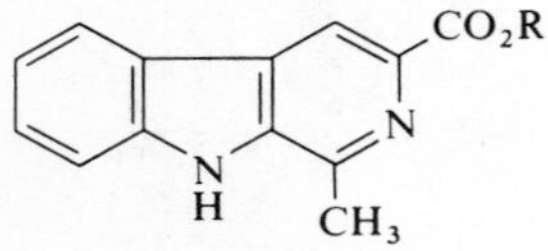

268 R = H
269 $R = C_2H_5$

5. Subincanine (**271**)

Gaskell and Joule (*122*) obtained a novel alkaloid from *A. subincanum* Mart. ex A. DC. which they named subincanine. The UV spectrum indicated subincanine to be a carbazole derivative lacking any phenolic group. A single NH was observed (ν_{max} 3460 cm^{-1}) which was exchanged with D_2O but which was not acetylated with Ac_2O/pyridine. Five aromatic protons were in two groups, four adjacent and one as a singlet. The PMR spectrum also indicated a methyl ketone (3-proton singlet at 1.95 ppm), an aromatic ethyl group, and an *N*-methyl group. As the molecular formula ($C_{22}H_{24}N_2O_2$) indicated two oxygens, the second oxygen must be in the form of an ether ring.

Reduction of subincanine with sodium borohydride gave a tetrahydro derivative which afforded a diacetate on acetylation. The base peak in the mass spectrum of the diacetate appeared at *m/e* 263 ($C_{18}H_{19}N_2$), a loss of $C_8H_{13}O_4$. The fragment being acetylated is therefore $C_4H_9O_2$, containing two hydroxyl groups and a terminal methyl group. The mass spectrum of dihydrosubincanine (from $LiAlH_4$, low temperature) gave an important ion at *m/e* 221 ($C_{16}H_{16}N$) which was rationalized as being the result of a retro-Diels–Alder reaction in the D ring. This ion was not observed in the mass spectrum of the diacetate, indicating that the side chain must be attached to a basic unit (**270**). The only unit capable of being reduced by borohydride in the appropriate way is the N–C–O–C unit which has the acetyl group attached at a position capable of carrying a positive charge.

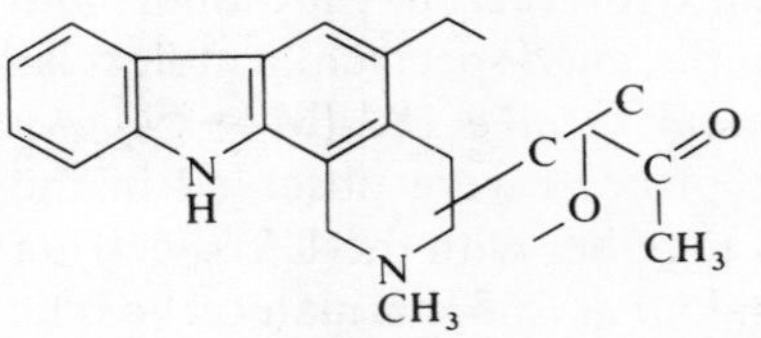

270

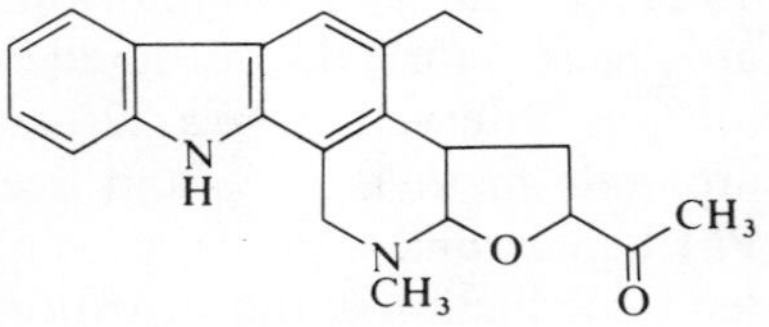

271

On this basis, structure **271** was proposed for subincanine, although a number of other possibilities could not be absolutely excluded (*122*).

6. Tubotaiwine *N*-Oxide (273)

From the root bark of *Conopharyngia johnstonii* Stapf (Apocynaceae), Schmid and co-workers (*123*) isolated tubotaiwine (**272**) and a new alkaloid which was shown to be tubotaiwine *N*-oxide (**273**).

Tubotaiwine *N*-oxide (**273**) showed a strong positive rotation $\{[\alpha]_D + 588°\}$ and an ORD spectrum entirely analogous to that of **272**. The UV spectrum indicated a β-anilanoacrylate (λ_{max} 293, 326 nm), and the mass spectrum confirmed a mass 16 mu greater than **272** but an identical fragmentation pattern. Reduction with sulfur dioxide afforded **272**, and oxidation of **272** with *m*-chloroperbenzoic acid gave **273**.

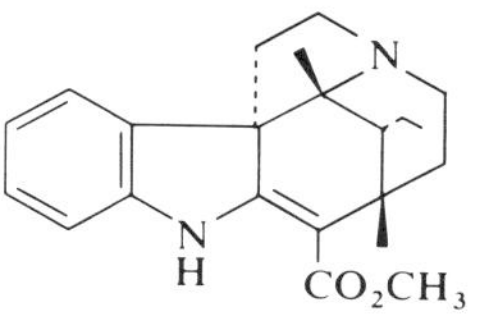

N H CO_2CH_3 **272**

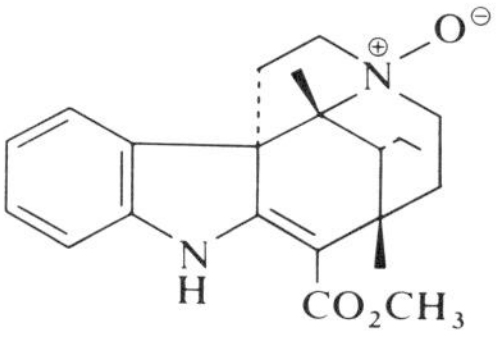

$\overset{\oplus}{N}$–$O^{\ominus}$ N H CO_2CH_3 **273**

Table I (*124–126*, etc.) summarizes the isolation of new *Aspidosperma* alkaloids and Table II (*127–161*, etc.) the isolation of established alkaloids.

III. Chemistry of the *Aspidosperma* Alkaloids

A. Rearrangements of the *Aspidosperma* Skeleton

Considerable efforts in the past have been devoted to the skeletal interconversion of alkaloids, if possible along biogenetic lines. Scott and co-workers have carried out numerous experiments in this area, and this work has been reviewed (*162*). The early work by Scott on the rearrangements of certain alkaloids, the subject of so much controversy, has been discussed elsewhere (*2*).

Le Men and co-workers have investigated a number of reactions of *Aspidosperma* alkaloids, and some of these are of considerable significance.

Treatment of tabersonine hydrochloride (**28**) with zinc and copper sulfate at 100° in glacial acetic acid afforded three products (*163*). One of these, **274** (1% yield), is the result of simple reduction; the major indole

TABLE I

NEW *ASPIDOSPERMA* AND RELATED ALKALOIDS

Compound	Source and ref.	mp (°C)	$[\alpha]_D$	Spectral data ref.	M^+	Molecular formula
A. β-Anilinoacrylate-type alkaloids						
Vincadifformine, 11-methoxy- (**4**)	*Vinca minor* (*6*)	—	−368°	(*6*)	368	$C_{22}H_{28}N_2O_3$
Ervinceine (**6**)	*V. erecta* (*7, 8*)	99–100	−448°	(*7, 8*)	368	$C_{22}H_{28}N_2O_3$
Ervamicine (**7**)	*V. erecta* (*8*)	213–214	−264°	(*8*)	366	$C_{22}H_{26}N_2O_3$
Minovincine, 11-methoxy (**10**)	*V. minor* (*100*)	—	−395°	(*11*)	384	$C_{22}H_{28}N_2O_4$
19*R*-Hydroxytabersonine (**18**)	*Catharanthus lanceus* (*13*)	163	−337°	(*13*)	352	$C_{21}H_{24}N_2O_3$
Minovincine, 3-oxo- (**19**)	*Tabernaemontana riedelii* (*14*)	252–253	+268°	(*14*)	366	$C_{21}H_{22}N_2O_4$
Minovincine (**20**)	*T. riedelii* (*14*)	—	+340°	(*14*)	352	$C_{21}H_{24}N_2O_3$
5-Oxo- (**23**)	*V. minor* (*15*)	—	−270°	(*15*)	366	$C_{21}H_{22}N_2O_4$
Ervinidinine (**25**)	*V. erecta* (*18*)	265–266	−160.6°	(*16, 18*)	352	$C_{21}H_{24}N_2O_3$
Baloxine (**26**)	*Melodinus balansae* (*19, 48*)	—		(*19, 48*)	368	$C_{21}H_{24}N_2O_4$
Vandrikidine (**27**)				(*20*)	382	$C_{22}H_{26}N_2O_4$
Vandrikine (**28**)				(*20*)	382	$C_{22}H_{26}N_2O_4$
Echitovenaldine (**32**)	*Alstonia venenata* R. Br. (*21*)	148	−485°	(*21*)	426	$C_{24}H_{30}N_2O_5$
Echitoserpine (**34**)	*A. venenata* (*24*)	154	−444°	(*24*)	562	$C_{31}H_{34}N_2O_8$
Echitoserpidine (**35**)	*A. venenata* (*25*)	110	−427°	(*25*)	532	$C_{30}H_{32}N_2O_7$
Cathaphylline (**42**)	*C. trichophyllus* (*28*)	—	−438°	(*28*)	368	$C_{21}H_{24}N_2O_4$
Horhammericine (**41**)	*C. lanceus* (*29*) *C. trichophyllus* (*28*)	140–144	−403°	(*28, 29, 31*)	368	$C_{21}H_{24}N_2O_4$
Horhammerinine (**43**)	*C. lanceus* (*30*)	209.5–211	−381°	(*30, 31*)	398	$C_{22}H_{26}N_2O_5$
Ervincinine (**46**)	*V. erecta* (*8, 32*)	247–248	−90.5°	(*8, 32*)	382	$C_{22}H_{26}N_2O_4$
Hazuntine (**47**)	*Hazunta velutina* (*33*)	152	−450°	(*33*)	382	$C_{22}H_{26}N_2O_4$
Hazuntinine (**49**)	*H. velutina* (*33*)	132	−482°	(*33*)	412	$C_{23}H_{28}N_2O_5$
Hedrantherine (**50**)	*Hedranthera barteri* (*36*)	—	−459°	(*36*)	368	$C_{21}H_{24}N_2O_4$
12-Hydroxy- (**51**)	*H. barteri* (*36*)	245	—	(*36*)	384	$C_{21}H_{24}N_2O_5$
Apodine (**63**)	*T. armeniaca* (*38*)		−520°	(*38*)	366	$C_{21}H_{22}N_2O_4$

Deoxoapodine (**66**)	*T. armeniaca (37)*		−432°	*(37)*	352	$C_{21}H_{24}N_2O_3$
Ervinidine (**68**)	*V. erecta (18)*	282–283	+17°	*(18)*	368	$C_{21}H_{24}N_2O_4$
Eburine (**76**)	*Hunteria eburnea* Pichon *(39)*	—	−18°	*(39, 40)*	340	$C_{21}H_{28}N_2O_2$
Eburcine (**81**)	*H. eburnea (39)*	—	−61°	*(39, 40)*	414	$C_{23}H_{30}N_2O_4$
B. Aspidospermidine-type alkaloids						
Eburenine (**73**)	*H. eburnea (39)*	—	+143°	*(39, 40)*	280	$C_{19}H_{24}N_2$
Fendlispermine (**83**)	*Aspidosperma fendleri (41)*	284–286		*(41)*	298	$C_{19}H_{26}N_2O$
Aspidocarpine, de-*O*-methyl- (**84**)	*A. cuspa (42)*			*(42)*	356	$C_{21}H_{28}N_2O_3$
Vincoline (**85**)	*C. roseus (44, 47)*	228–232	−242°	*(19, 46–49)*	368	$C_{21}H_{24}N_2O_4$
	V. libanotica (46)					
	M. balansae (19, 46)					
19-Epivincoline (**86**)	*M. balansae (19, 48)*	200–204	−205°	*(19. 48, 49)*	368	$C_{21}H_{24}N_2O_4$
Cathovaline (**100**)	*C. ovalis (52, 57)*	88–90	−73°	*(52–54, 57)*	426	$C_{24}H_{30}N_2O_5$
	C. lanceus (53, 54)					
14-Hydroxy- (**109**)	*C. ovalis (57)*	133–134	−73°	*(57)*	442	$C_{24}H_{30}N_2O_6$
Cimicine (**110**)	*Haplophyton cimicidum (58)*	229–231	+113°	*(59)*	382	$C_{22}H_{26}N_2O_4$
Cimicidine (**111**)	*H. cimicidum (58)*	266–268		*(59)*	412	$C_{23}H_{28}N_2O_5$
C. Cylindrocarine-type alkaloids						
Cylindrocarine (**117**)	*Aspidosperma cylindrocarpon* Muell. Arg. *(62)*	204–205	−280°	*(62)*	356	$C_{21}H_{28}N_2O_3$
N-Methyl- (**126**)	*A. cylindrocarpon (62)*	—	−110°	*(62)*	370	$C_{22}H_{30}N_2O_3$
N-Formyl- (**125**)	*A. cylindrocarpon (62)*	161–162	−140°	*(62)*	384	$C_{22}H_{28}N_2O_4$
N-Propionyl- (**118**)	*T. amygdalifolia (60)*	—	−82°	*(60)*	412	$C_{24}H_{32}N_2O_4$
N-Benzoyl- (**123**)	*A. cylindrocarpon (62)*	—	−131°	*(62)*	460	$C_{28}H_{32}N_2O_4$
12-Demethoxy-*N*-acetyl- (**116**)	*A. cylindrocarpon (62)*	160–162	0°	*(60, 62)*	368	$C_{22}H_{28}N_2O_3$
	T. amygdalifolia (60)	—	−49°			
19-Hydroxy- (**122**)	*A. cylindrocarpon (62)*	—	−300°	*(62)*	372	$C_{21}H_{28}N_2O_4$
N-Formyl-19-hydroxy- (**134**)	*A. cylindrocarpon (62)*	—	−250°	*(62)*	400	$C_{22}H_{28}N_2O_5$
N-Acetyl-19-hydroxy- (**135**)	*A. cylindrocarpon (62)*	207–210	−400°	*(62)*	414	$C_{23}H_{30}N_2O_5$
N-Benzoyl-19-hydroxy- (**131**)	*A. cylindrocarpon (62)*	215–216	−150°	*(62)*	476	$C_{28}H_{32}N_2O_5$

(*Continued*)

TABLE I (*Continued*)

Compound	Source and ref.	mp (°C)	$[\alpha]_D$	Spectral data ref.	M^+	Molecular formula
N-Cinnamoyl-19-hydroxy- (**133**)	*A. cylindrocarpon* (*62*)	123–124	−384°	(*62*)	502	$C_{30}H_{34}N_2O_5$
N-Dihydrocinnamoyl-19-hydroxy (**136**)	*A. cylindrocarpon* (*62*)	—	−278°	(*62*)	504	$C_{30}H_{36}N_2O_5$
Cylindrocarpinol *N*-Formyl- (**127**)	*A. cylindrocarpon* (*62*)	101–105	−150°	(*62*)	358	$C_{21}H_{30}N_2O_3$
N-Acetyl- (**130**)	*A. cylindrocarpon* (*62*)	210–215		(*62*)	370	$C_{22}H_{30}N_2O_3$
D. Kopsane-type alkaloids						
Kopsinine						
15α-Hydroxy- (**140**)	*M. australis* (*64*)			(*64, 78*)	354	$C_{21}H_{26}N_2O_3$
19β-Hydroxy- (**141**)	*M. australis* (*64*)			(*64, 78*)	354	$C_{21}H_{26}N_2O_3$
Kopsinine *N*-oxide (**142**)	*V. erecta* (*65*)	159–161		(*65*)	354	$C_{21}H_{26}N_2O_3$
Kopsanol, *N*-formyl (**144**)	*Aspidosperma verbascifolium* Muell. Arg. (*66*)			(*66*)	366	$C_{22}H_{26}N_2O_3$
E. Vindolinine-type alkaloids						
19-Epivindolinine (**153**)	*M. balansae* (*48*)			(*48*)	336	$C_{21}H_{24}N_2O_2$
19-Epivindolinine *N*-oxide (**154**)	*M. balansae* (*48*)		+8°	(*48*)	352	$C_{21}H_{24}N_2O_3$
Vindolinine *N*-oxide (**155**)	*M. balansae* *V. erecta* (*65*)	186–187	+7°	(*48*)	352	$C_{21}H_{24}N_2O_3$
Dihydrovindolinine *N*-oxide (**153**)	*V. erecta* (*65*)			(*65*)	354	$C_{21}H_{26}N_2O_3$
F. Miscellaneous aspidospermine-type alkaloids						
Aspidodispermine (**154**)	*Aspidosperma dispermum* Muell. Arg. (*76*)	—	+119°	(*76*)	328	$C_{19}H_{24}N_2O_3$
Deoxyaspidodispermine (**158**)	*A. dispermum* (*76*)	—	−20°	(*76*)	312	$C_{19}H_{24}N_2O_2$

Vincatine (**163**)	*V. minor (77)*	111–112	−13.5°	*(77)*	346	$C_{20}H_{30}N_2O_3$
Rhazinilam (**172**)	*Rhazya stricta (79–81)* *A. quebracho-blanco (83)* *M. australis (78)*	214–215	−432°	*(81–83)*	326	$C_{19}H_{22}N_2O$
G. Quebrachamine-type alkaloids						
Quebrachamine, 14,15-dihydroxy- (**179**)	*Voacanga africana (85)*	—	+132°	*(85)*	314	$C_{19}H_{26}N_2O_2$
(−)-Vincadine (**185**)	*Amsonia tabernaemontana (89, 91–92a)*	75	−70° (91) −92° (89)	*(89, 91)*	340	$C_{21}H_{28}N_2O_2$
(±)-Vincadine	*A. tabernaemontana (89, 91, 92)*	126	0°	*(89, 91)*	340	$C_{21}H_{28}N_2O_2$
16-Epivincadine (**189**)	*A. tabernaemontana (92, 92a)*	157	0°	*(92)*	340	$C_{21}H_{28}N_2O_2$
Vincadine, 14,15-dehydro- (**187**)	*A. tabernaemontana (92, 92a)*	—	+65°	*(92)*	340	$C_{21}H_{28}N_2O_2$
16-Epivincadine, 14,15-dehydro- (**190**)	*A. tabernaemontana (92, 92a)*	—	+85°	*(92)*	338	$C_{21}H_{26}N_2O_2$
Vincaminoridine (**191**)	*Vinca minor (94)*	99–100	+57.7°	*(94)*	384	$C_{23}H_{32}N_2O_3$
Voaphylline, 12-methoxy (**195**)	*Crioceras dipladeniiflorum (95)*	149	+61°	*(95)*	326	$C_{20}H_{26}N_2O_2$
Rhazidigenine *N*-oxide (**200**)	*Aspidosperma quebracho-blanco (97)*			*(97)*	314	$C_{19}H_{26}N_2O_2$
H. Pseudoaspidosperma-type alkaloids						
Pandoline (**220**)	*Pandaca calcarea (99)* *P. caducifolia (103)* *P. debrayi (99)*	—	+417°	*(99)*	354	$C_{21}H_{26}N_2O_3$
Pandine (**215**)	*P. calcarea (99)* *P. caducifolia (103)* *P. debrayi (99)*	108–113	+273°	*(99)*	352	$C_{21}H_{24}N_2O_3$
(+)-Pseudotabersonine (**207**)	*P. caducifolia (103)*	—	+320°	*(103)*	336	$C_{21}H_{24}N_2O_2$
(+)-Pseudovincadifformine (**219**)	*P. caducifolia (103)*	—	+430°	*(103)*	338	$C_{21}H_{26}N_2O_2$
20-Epipandoline (**221**)	*P. caducifolia (103)*	—	+462°	*(103)*	354	$C_{21}H_{26}N_2O_3$
Capuronidine (**222**)	*Capuronetta elegans (107)*	—	+220°	*(107)*	296	$C_{19}H_{24}N_2O$

(Continued)

TABLE I (*Continued*)

Compound	Source and ref.	mp (°C)	$[\alpha]_D$	Spectral data ref.	M^+	Molecular formula
(+)-1,2-Dehydropseudoaspidospermidine (**229**)	*P. eusepala* (*108*)	—	+153°	(*108*)	282	$C_{19}H_{26}N_2$
Ibophyllidine (**235**)	*Tabernanthe iboga* Baill. (*109*) *T. subsessilis* (*109*)	—	+134°	(*109*)	324	$C_{20}H_{24}N_2O_2$
Iboxyphylline (**237**)	*T. iboga* (*109*) *T. subsessilis* Benth. (*109*)	245	+444°	(*109*)	354	$C_{21}H_{26}N_2O_3$
I. *Melodinus* alkaloids						
Scandine (**241**)	*M. scandens* (*110*)	180–182	+245°	(*110*)	350	$C_{21}H_{22}N_2O_3$
Epimeloscine *N*-oxide (**242**)	*M. scandens* (*112*)	203–207	+310°	(*112*)	308	$C_{19}H_{24}N_2O_2$
Meloscandonine (**248**)	*M. scandens* (*112*)	318–320	+72°	(*112, 113*)	320	$C_{20}H_{20}N_2O_2$
J. Miscellaneous alkaloids						
14,15,16,17-Tetrahydrosecodine, 16-decarbomethoxy- (**260**)	*Tabernaemontana cuminsii* (*116*)	—	—	(*116*)	284	$C_{19}H_{28}N_2$
Ellipticine *N*-oxide (**266**)	*Ochrosia viellardii* (*120*)	—	—	(*120*)	262	$C_{17}H_{14}N_2O$
Harman-3-carboxylic acid (**268**)	*Aspidosperma exalatum* (*121*)	300	—	(*121*)	226	$C_{13}H_{10}N_2O_2$
Subincanine (**271**)	*A. subincanum* (*124*)	160–163		(*122*)	348	$C_{22}H_{24}N_2O_2$
Tubotaiwine *N*-oxide (**273**)	*Conopharyngia johnstonii* (*123*)	224	+588°	(*123*)	340	$C_{20}H_{24}N_2O_3$
Andranginine (**378**)	*Craspidospermum verticillatum* var. *petiolare* (*126a*)	132	−42°	(*125, 126*)	334	$C_{21}H_{22}N_2O_2$

TABLE II
FURTHER ISOLATION OF *Aspidosperma* ALKALOIDS

Compound	Plant source	Plant part[a]	Ref.
(+)-Aspidospermidine	*Amsonia tabernaemontana*	LF	*128*
1,2-Dehydro-	*A. tabernaemontana*	WP	*91*
		SD	*127*
		LF	*128, 129*
		RT	*129*
		FR	*92a*
	Aspidosperma neblinae		*114*
N-Formyl-	*Amsonia tabernaemontana*	LF	*128*
Aspidospermine	*Aspidosperma neblinae*		*114*
	A. quebracho-blanco	LF	*83*
Demethoxy-	*A. neblinae*		*114*
Demethyl-	*A. neblinae*		*114*
Aspidocarpine	*A. formosanum*	BK	*130*
	A. neblinae		*114*
Pyrifolidine	*A. neblinae*		*114*
Deacetyl-	*A. neblinae*		*114*
Deacetyl-1,2-dehydro	*A. neblinae*		*114*
Beninine	*Voacanga chalotiana*	RB	*131*
Neblinine	*A. neblinae*		*114*
Minovincine	*Cabucala erythrocarpa* var. *erythrocarpa*	WP	*132*
	Catharanthus trichophyllus	RT	*28*
Minovincinine	*C. trichophyllus*	RT	*28*
(−)-Tabersonine	*Amsonia angustifolia*	LF	*83*
	A. tabernaemontana	RT	*129*
		WP	*91*
		FR	*92a*
	Melodinus balansae	LF	*48*
	M. celastrodies	LF	*133*
	M. scandens	LF	*111, 112*
	Pandaca retusa	RT	*134*
	V. africana	SD	*85*
	V. grandifolia	FR	*135*
(+)-Vincadifformine	*Amsonia angustifolia*	LF	*83*
	A. tabernaemontana	RT	*129*
		WP	*91*
		LF	*128, 129, 136*
		FR	*92a*
	M. scandens	LF-FL-FR	*112*
	P. minutiflora	LF	*137*
	Tabernaemontana riedelii	LF-TW	*14*
Tabersonine			
11-Hydroxy-	*M. balansae*	LF-FR	*48*
11-Methoxy-	*Craspidospermum verticillatum* var. *petiolare*	LF	*126a*

(*Continued*)

TABLE II (*Continued*)

Compound	Plant source	Plant part[a]	Ref.
Lochnericine	*A. angustifolia*	LF	*83*
	A. tabernaemontana	LF	*84, 129*
	Catharanthus trichophyllus	RT	*28*
	T. divaricata	LF	*138*
Lochnerinine	*C. pusillus*	LF	*139*
Echitovenine	*C. trichophyllus*	RT	*28*
Vindorosine	*C. longifolius*	AP	*140*
	C. pusillus	LF	*139, 141*
	C. trichophyllus	RT	*26, 28*
Vindoline	*C. longifolius*		*140*
	C. pusillus		*141*
Aspidolimidine	*Aspidosperma* sp.		*142*
Cylindrocarpidine	*A. cylindrocarpon*		*62*
Cylindrocarpine	*A. cylindrocarpon*		*62*
Aspidoalbine	*A. desmanthum*	BK	*130*
Vindolinine	*C. longifolius*	AP	*130*
	C. trichophyllus	RT	*139*
	M. balansae	LF	*48*
Kopsanone	*A. cuspa*		*143*
	A. verbascifolium		*66*
Kopsanol	*A. cuspa*		*143*
	A. verbascifolium		*66*
Epikopsanol	*A. cuspa*		*143*
	A. verbascifolium		*66*
10-Oxo-	*A. verbascifolium*		*66*
Venalstonine	*C. ovalis*		*144*
	Craspidospermum verticillatum var. *petiolare*	LF	*126a*
	M. balansae	LF	*48*
	M. scandens	LF-FL-FR	*112*
Venalstonidine	*Catharanthus ovalis*		*144*
	M. balansae	LF	*48*
	M. scandens	LF-FL-FR	*112*
Pyrifolidine	*A. quebracho-blanco*	LF	*83*
Voaphylline	*V. africana*	LF	*85*
	P. retusa	RT	*134*
	T. divaricata	LF	*138*
(−)-Quebrachamine	*Amsonia tabernaemontana*	WP	*91*
		SD	*127*
		LF	*128, 129*
(+)-Condylocarpine	*P. minutiflora*		*144*
19,20-Dihydro-	*A. tabernaemontana*		*128*
	P. eusepala	SB, LF	*108*
	P. ochrascens		*145*
Tubotaiwine	*Conopharyngia johnstonii*	RB	*123*
	P. minutiflora	LF	*144*

TABLE II (*Continued*)

Compound	Plant source	Plant part[a]	Ref.
Pericalline	*Catharanthus longifolius*	AP	*140*
	C. trichophyllus	RT	*27*
	Ervatamia orientalis (R. Br) Domin	LF	*146*
	Muntafaria sessilifolia (Baker) Pichon	LF	*147*
	Ochrosia silvatica Däniker	BK	*148*
	P. calcarea	WP	*99*
	P. debrayi	WP	*99*
	P. eusepala	SB, LF	*108*
	P. ochrascens	LF, RT, BK	*145*
	T. cuminsii	LF	*149*
Uleine	*Aspidosperma formosanum*	BK	*130*
3-Epiuleine	*A. formosanum*	BK	*130*
Uleine, 1,13-dihydro-13-hydroxy-	*A. formosanum*		*130*
Ellipticine	*Bleekeria vitiensis*	BK, RT	*150*
	O. balansae (Guillaumin) Baill.	LF	*151*
		BK	*152*
	O. confusa Pichon	LF	*153*
	O. vieillardii	BK	*120*
		LF	*154*
1,2-Dihydro-	*O. balansae*	LF	*151*
	O. vieillardii	BK	*120*
1,2,3,4-Tetrahydro-	*O. balansae*	BK	*151*
	O. vieillardii	BK	*120*
1,2,3,4-Tetrahydro-2-methyl-	*A. vargasii* A. DC.	BK	*155*
9-Methoxy-	*B. vitiensis*	LF, BK, RT	*150, 156*
	O. balansae	BK	*151*
9-Methoxy-1,2-dihydro-	*O. balansae*	BK	*151*
Olivacine	*A. campus-belus*	BK	*130*
	A. vargasii	BK	*155*
	Peschiera affinis (Muell. Arg.)	RT, BK	*157*
	P. lundii Miers		*158*
Pleiocarpamine	*Alstonia glabriflora* Margf.		*159*
	A. muelleriana Damin		*160*
	A. spectabilis		*159*
	A. vitiensis Seem. var. *nova ebudica Monachino*		*161*
2,7-Dihydro-	*Alstonia muelleriana*		*160*

[a] LF, Leaf; WP, whole plant; SD, seed; RT, root; BK, bark; AP, aerial parts.

isolated (15% yield) had the structure **275**, and the major indoline, vincamsonine, (53% yield), was assigned structure **276**. Because Kuehne (*164*) has already converted dihydro-**275** to vincamine (**277**), the reaction constitutes a conversion of tabersonine (**28**) to vincamine.

Oppenauer oxidation of either **275** or **276** gave the same compound which, on the basis of spectral properties, was assigned the structure **278**. In particular, the UV spectrum was close to that of eburnamonine.

The mechanism postulated to account for the formation of **275** and **276** is shown in Scheme 7.

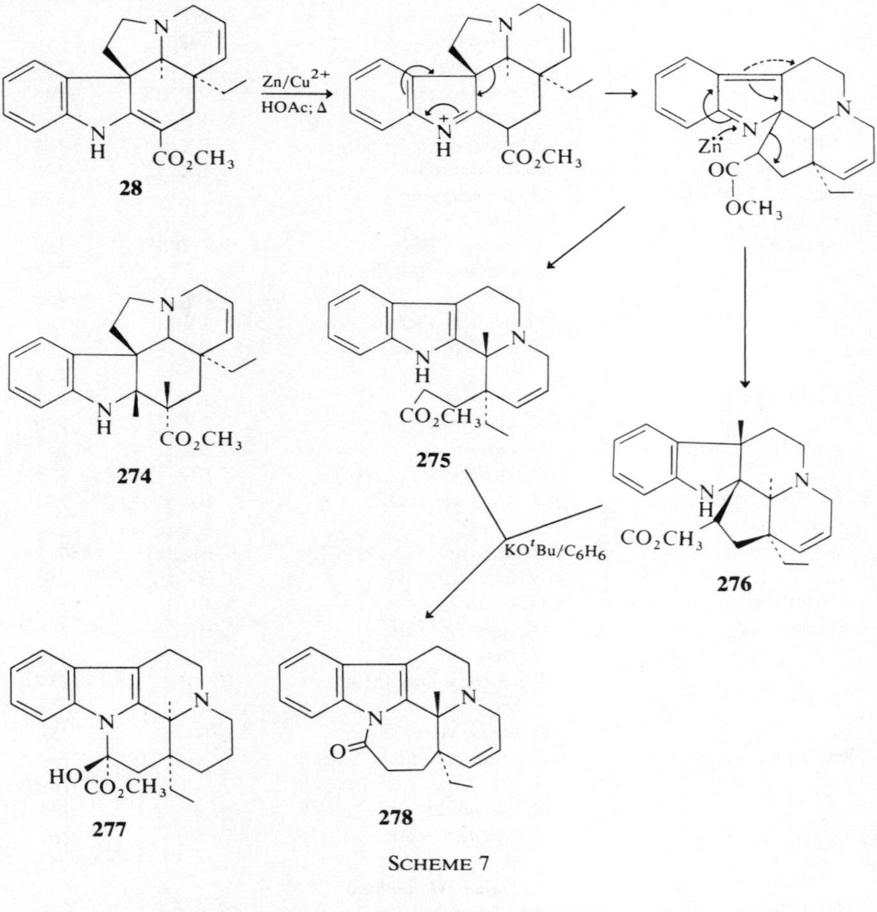

SCHEME 7

Reaction of 1,2-dehydroaspidospermidine (**171**) with zinc and copper sulfate in glacial acetic acid at 100° gave aspidospermidine (**249**) as a minor product and a major product, **279** (24% yield) (*165*). The major product was formylated with formic–acetic anhydride and reduced first with lithium aluminum hydride and with Adams catalyst to give vallesamidine (**280**), identical with the natural alkaloid (except for optical rotation).

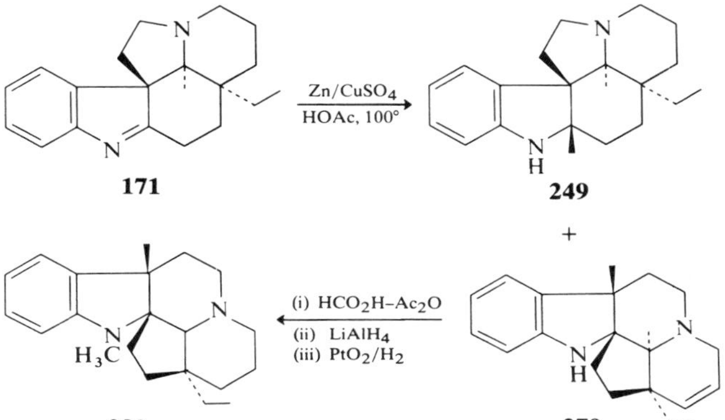

$Zn/CuSO_4$, HOAc, 100°; (i) $HCO_2H–Ac_2O$ (ii) $LiAlH_4$ (iii) PtO_2/H_2

171 **249** + **280** **279**

(−)-1,2-Dehydroaspidospermidine (**171**) was treated with *p*-nitroperbenzoic acid at 0° to give the *N*-oxide **281**. This compound was rearranged with triphenylphosphine in aqueous acetic acid at 0° for two days to give a mixture in which the major product (42% yield) was (+)-eburnamine (**282**) (*166*). The hydroxyindolenine **283** is a probable intermediate.

Two other reactions were also studied at this time. One of these was on 16-oxoaspidospermidine (**284**). When **284** was heated with acetic acid at 90°, eburnamenine (**285**) was produced in 40% yield (*166*).

The oxime of **284** gave a mixture of two nitriles (**286** and **287**) on standing with phosphous oxychloride at room temperature, presumably *via* **288**. Acid hydrolysis of the nitriles gave (+)-vincamone (**289**) and (+)-21-epivincamone (**290**), respectively (*166*).

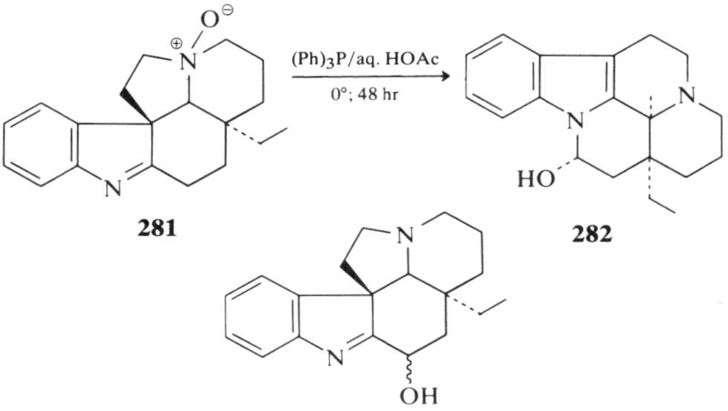

$(Ph)_3P$/aq. HOAc, 0°; 48 hr

281 **282** **283**

HOAc, 90°

284

285

(i) NH_2OH
(ii) $POCl_3$, RT

288

286 C-21Hα
287 C-21Hβ

289 C-21Hα
290 C-21Hβ

As well as eburnamine (**282**), the *N*-oxide of 1,2-dehydroaspidospermidine (**281**) gave a new product on reaction with triphenylphosphine and acetic acid (*167*). The compound showed a molecular ion at *m/e* 296 analyzing for $C_{19}H_{24}N_2O$. The UV spectrum was that of an *N*-acylindoline and the IR spectrum indicated a lactam (ν_{max} 1680 cm^{-1}). The product was unaffected by catalytic hydrogenation or by sodium cyanoborohydride reduction. In aqueous hydrochloric acid for 2 hr at 90°, the product was transformed into an *N*-acylindole which was assigned structure **291**. The parent compound for these transformations was therefore **292**. Lithium aluminum hydride reduction of **292** gave the diamine **293**, which afforded the indole **294** on treatment with acid.

A structurally similar product was observed when vincadifformine (**2**) was treated with 4 equivalents of *m*-chloroperbenzoic acid in benzene under reflux (*168*). The product was assigned the novel structure **295** on the basis of the following spectral evidence. The molecular formula was shown to be $C_{21}H_{26}N_2O_4$ and the UV spectrum indicated an *N*-acylindoline. A carboxylic ester was evident from the IR spectrum (ν_{max}

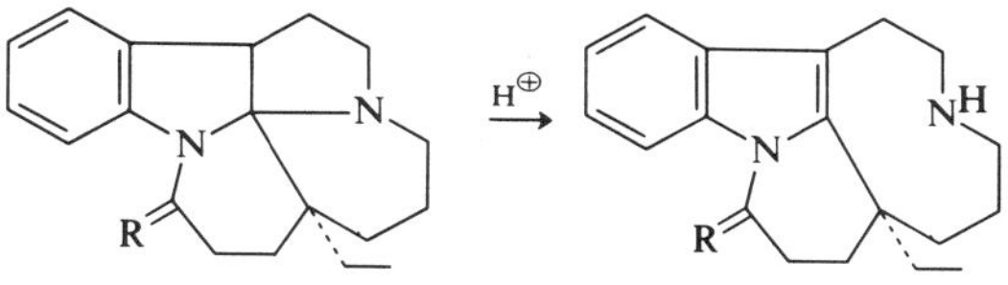
H⊕

292 R = O
293 R = H_2

291 R = O
294 R = H_2

1745 cm^{-1}) and PMR spectrum (singlet 3.90 ppm). The fourth oxygen was traced to a hydroxy group. The skeleton was deduced by analysis of the CMR spectrum. A singlet at 98.0 ppm was attributed to a quaternary carbon (C-21) located between two nitrogens, and a singlet at 78.0 ppm to a carbon (C-16) attached to oxygen and further deshielded. A lactam carbonyl carbon (C-2) was observed at 170 ppm.

The general scheme proposed (*166*) to account for these rearrangements is shown in Scheme 8. The close similarity of intermediate **296** to rhazinilam (**172**) should be noted.

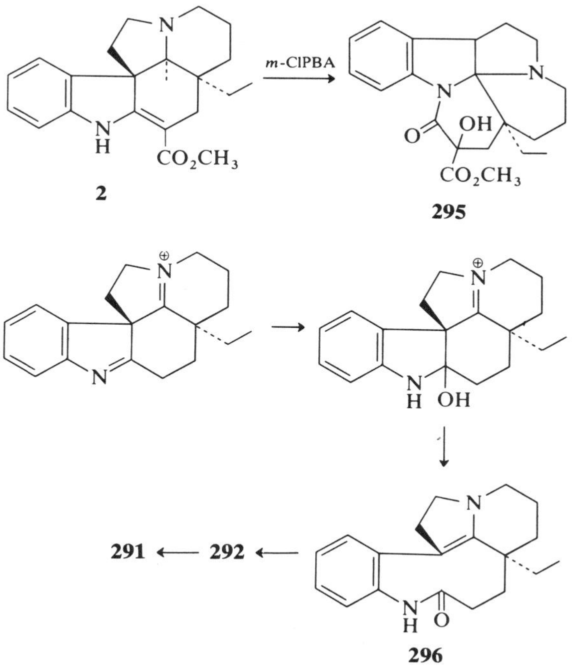
m-ClPBA
CO_2CH_3
OH
CO_2CH_3
H OH
2
295
291 ← **292** ←
296

SCHEME 8

Treatment of tabersonine (**28**) or vincadifformine (**2**) with formic acid and formamide (*169*) gave products arising from the reduction of the intermediate iminium species (**297**), itself derived by the familiar Smith cleavage reaction. The products from vincadifformine were established to be (+)-vincadine (**184**) (57%) and quebrachamine (**178**) (5%). A trace product obtained was 16-epivincadine (**189**), and analogous products in the 14,15-dehydro series were obtained from tabersonine (*169*). 16-Epivincadine (**189**) was converted almost quantitatively to the more stable epimer (**184**) on formic acid/formamide treatment.

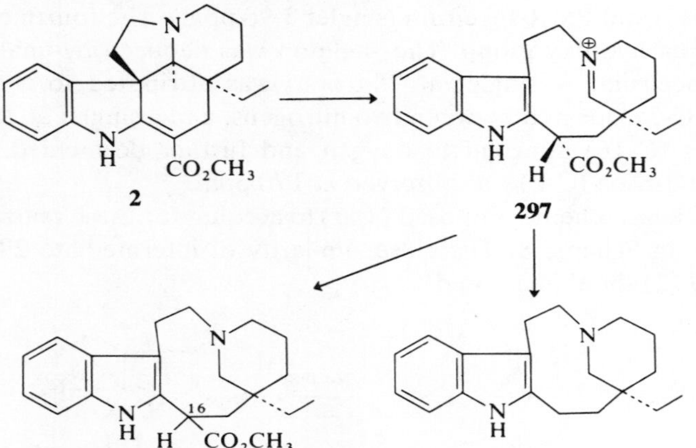

2 → **297** → **184**, **178**

184 16Hα
189 16Hβ

The rearrangement of tabersonine (**28**) to vincamsonine (**276**) was described previously. Le Men and Lévy *et al.* (*170–174*), however, reasoned that there was a more direct route to vincamine (**277**) along biogenetic lines (*2*).

Reaction of (−)-vincadifformine (**2**) with lead tetraacetate in benzene gave the acetoxy indolenine **298** in about 50% yield. This compound, when treated with trifluoroacetic acid in chloroform at 0° followed by reaction with sodium acetate in aqueous acetic acid, gave a mixture of (+)-vincamine (**277**), (−)-16-epivincamine (**299**), and (+)-apovincamine (**300**) (*170, 172, 174*). Under the most favorable conditions (+)-vincamine (**277**) could be produced from (−)-vincadifformine (2) in 36% overall yield.

The 16-hydroxyindolenine **301**, which is probably an intermediate in the first reaction, could also be produced by reaction of (−)-vincadifformine with *p*-nitroperbenzoic acid to give **302**. Treatment with triphenylphosphine in aqueous acetic acid gave the hydroxyindolenine **301** as an intermediate which subsequently underwent rearrangement to (+)-vincamine (**277**) in 66% yield and to (−)-16-epivincamine (**299**) in 21% (*170, 171, 174*).

The same sequence of reactions was also conducted on (+)-vincadifformine (**75**) to give (−)-vincamine (**303**), (+)-16-epivincamine (**304**), and (−)-apovincamine (**305**) (*175*).

Catalytic reduction of (−)-11-methoxytabersonine (**3**) and oxidation with peracid gave 1,2-dehydro-16-carbomethoxy-16-hydroxy-11-methoxyaspidospermidine *N*-oxide (**306**), which was rearranged by triphenylphosphine in acid to a mixture of vincine (**307**), 16-epivincine (**308**), and apovincine (**309**) (*173*).

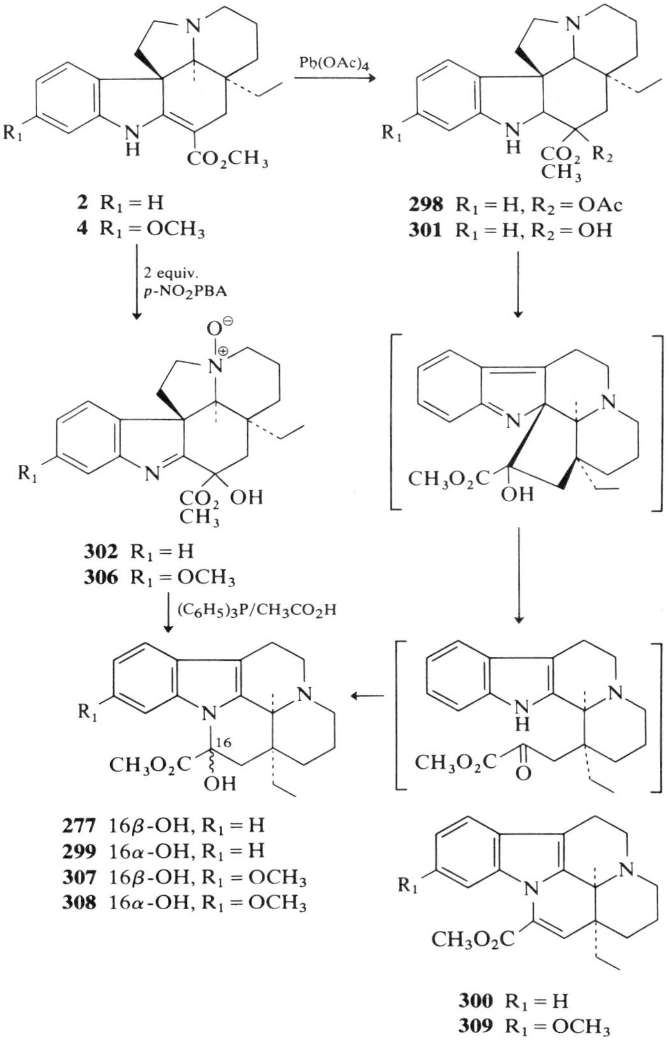

$Pb(OAc)_4$

2 $R_1 = H$
4 $R_1 = OCH_3$

298 $R_1 = H, R_2 = OAc$
301 $R_1 = H, R_2 = OH$

2 equiv.
p-NO_2PBA

302 $R_1 = H$
306 $R_1 = OCH_3$

$(C_6H_5)_3P/CH_3CO_2H$

277 16β-OH, $R_1 = H$
299 16α-OH, $R_1 = H$
307 16β-OH, $R_1 = OCH_3$
308 16α-OH, $R_1 = OCH_3$

300 $R_1 = H$
309 $R_1 = OCH_3$

With (−)-tabersonine (**28**) as the starting material and *m*-chloroperbenzoic acid as the oxidizing agent followed by reaction with triphenylphosphine in acid, the corresponding 14,15-dehydro compounds, 14,15-dehydrovincamine (**310**), 16-epi-14,15-dehydrovincamine (**311**) and 14,15-dehydroapovincamine (**312**) (*176*), were obtained. Analogous reactions on (−)-11-methoxytabersonine (**3**) gave the corresponding compounds (**313**, **314**, and **315**) in the vincine series (*176*).

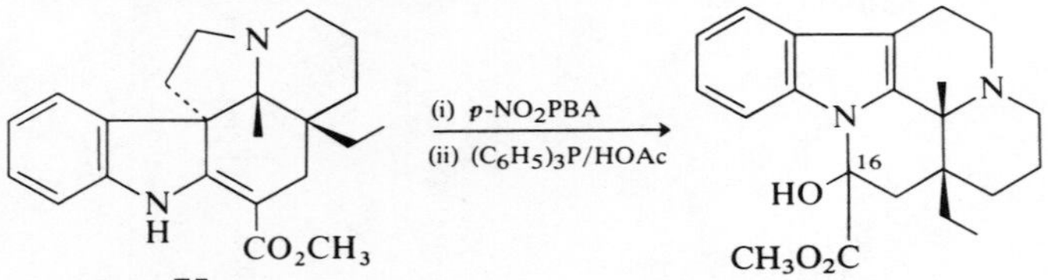

N H CO_2CH_3

75

(i) *p*-NO_2PBA

(ii) $(C_6H_5)_3P/HOAc$

N N 16 HO CH_3O_2C

303 16α-OH
304 16β-OH

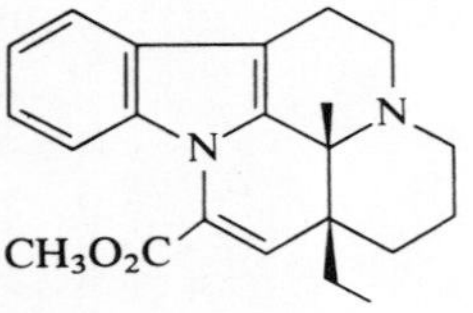

N N CH_3O_2C

305

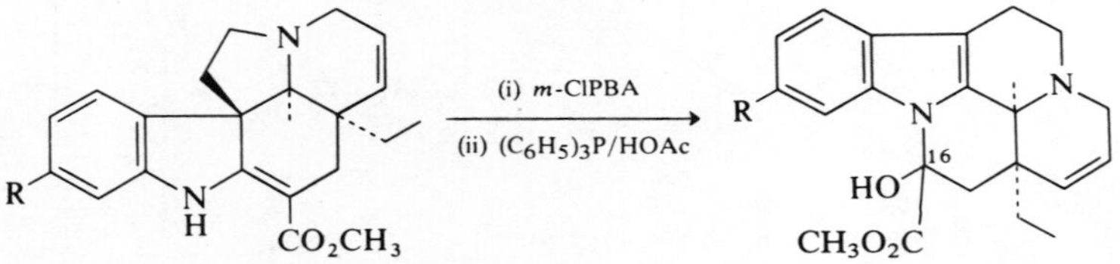

R N H CO_2CH_3

3 R = OCH_3
28 R = H

(i) *m*-ClPBA

(ii) $(C_6H_5)_3P/HOAc$

R N N 16 HO CH_3O_2C

310 16β-OH, R = H
311 16α-OH, R = H
313 16β-OH, R = OCH_3
314 16α-OH, R = OCH_3

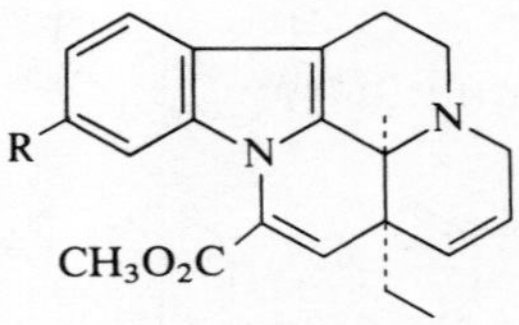

R N N CH_3O_2C

312 R = H
315 R = OCH_3

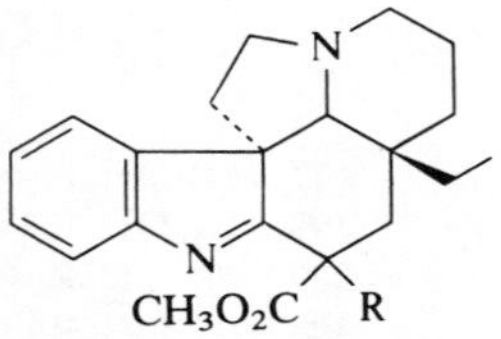

N N CH_3O_2C R

316 R = OH, O alkyl, OAc, Cl

Lévy has also reported briefly (*175*) that when compounds of the type **316** in which R = OH, Oalkyl, OAc, or Cl are treated with trifluoroacetic acid in methylene chloride **303**, **304**, and **305** are produced. The compound in which R = OH was produced by PtO_2/O_2 oxidation, and that where R = Oalkyl by treatment of **75** with an alkyl peroxide.

The availability of both (+)-(**219**) and (−)-pseudovincadifformines (**317**) allowed the rearrangement to be carried out in these series of compounds as well (*177*). Thus pseudovincamine (**318**) and 16-epipseudovincamine (**319**) were prepared by rearrangement of the *N*-oxide of pseudovincadifformine in each antipodal series.

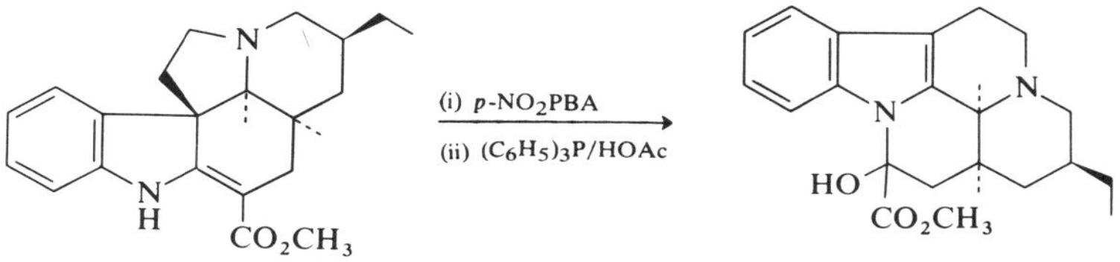

(i) *p*-NO_2PBA
(ii) $(C_6H_5)_3P$/HOAc

317

318 16β-OH
319 16α-OH

B. Tabersonine Chloroindolenine (320)

Le Men and co-workers have performed several studies of the chemistry of the chloroindolenine of tabersonine (**320**) (*174*, *178*). This compound is produced in high yield when tabersonine (**28**) is treated with a slight excess of *tert*-butyl hypochlorite in anhydrous methylene chloride at −16° in the presence of triethylamine (*178*). In order to demonstrate that no skeletal changes had occurred, they carried out reconversion to tabersonine with potassium *tert*-butoxide in benzene under reflux. Similarly prepared was the chloroindolenine of vincadifformine. Subsequent reactions demonstrated that these chloroindolenines were highly reactive.

Refluxing **320** in methanol for 5 min gave a mixture of four basic compounds, each of which had an indolic UV spectrum. The major component (38%) was deduced to have the structure **321**, and the very minor component (3%) was the C-16 epimer. A closely related compound (16% yield) was assigned the α-hydroxy ester structure **322**. The fourth product (15%) was the result of a more extensive rearrangement, for in the PMR spectrum it exhibited three methoxyl singlets and an acetal proton at 4.40 ppm. It was assigned the structure **323**. The stereochemistries of C-16 and C-20 were deduced by lithium aluminum hydride reduction which afforded the cyclic acetal **324** (*178*). The reaction sequence was also carried out in ethanol and the corresponding products obtained. The proposed mechanism of formation of these products is shown in Scheme 9 (*178*).

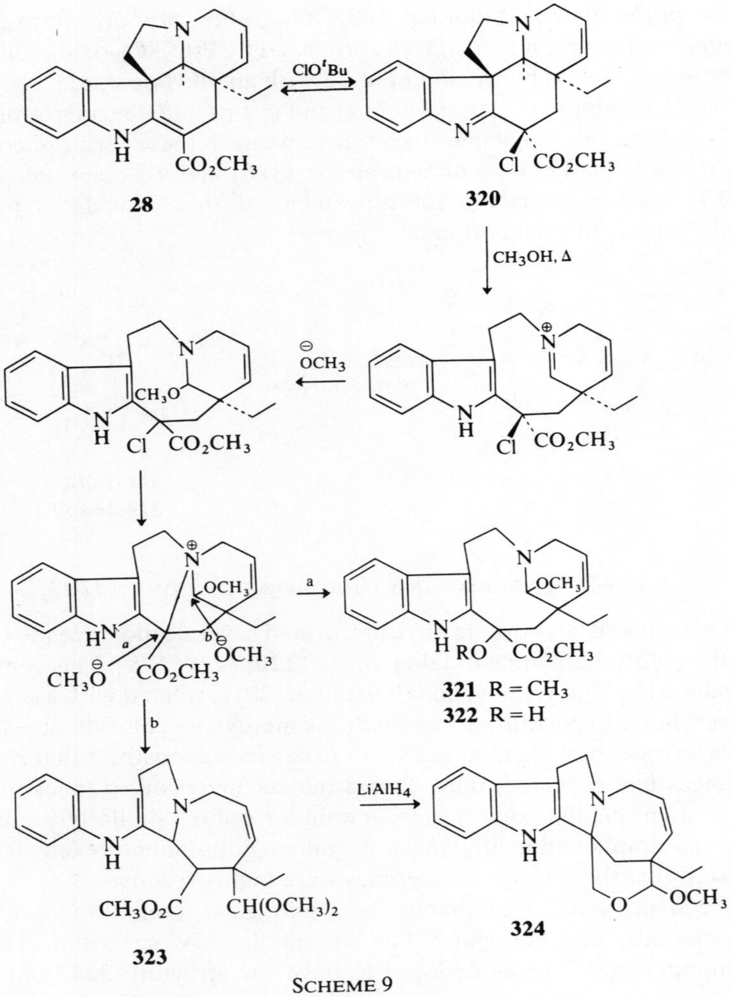

ClO^tBu

28

320

CH_3OH, Δ

$\ominus OCH_3$

a

b

321 $R = CH_3$
322 $R = H$

$LiAlH_4$

323

324

SCHEME 9

C. DEGRADATION OF (−)-KOPSINE (325)

Schmid and co-workers (*124*) have described a new degradation of (−)-kopsine (**325**) to (−)-aspidofractinine (**326**). Sodium periodate cleavage of the hydroxyketone under acidic conditions gave a keto acid which was esterified. Oxidation with chromium trioxide in pyridine at room temperature for 48 hr then gave the 5,6-dehydro compound **327**. Hydrolysis of **327** under acidic conditions gave the unstable enamine **328** which was reduced successively with zinc and sulfuric acid to an amine and under Clemmensen conditions to eliminate the ketone. Alkaline hydrolysis afforded a product identical with (−)-aspidofractinine (**326**) in optical rotation and ORD curve.

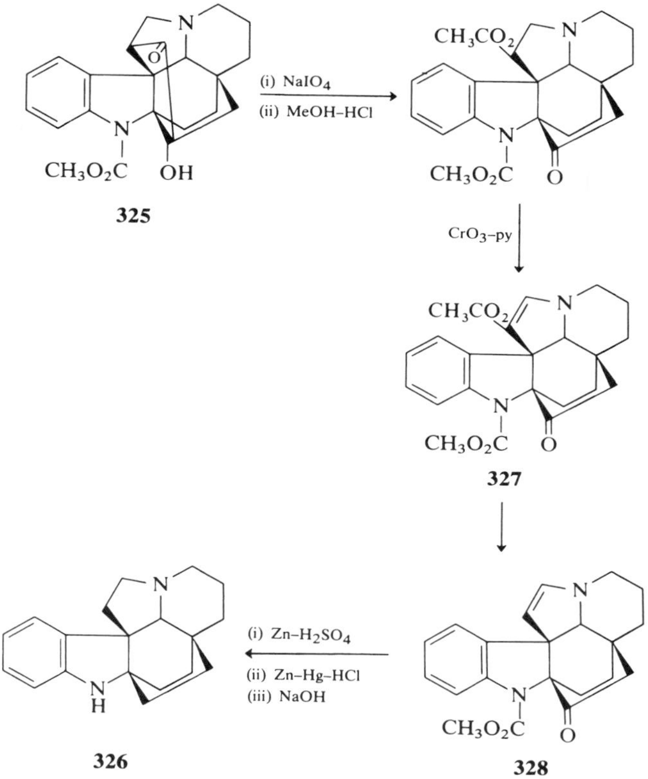

CH_3O_2C OH

325

(i) $NaIO_4$
(ii) MeOH–HCl

CH_3CO_2 N CH_3O_2C O

CrO_3–py

327

326

(i) Zn–H_2SO_4
(ii) Zn–Hg–HCl
(iii) NaOH

328

D. Correlation of (−)-Minovincine (11) with the (−)-Kopsane Alkaloids

Confirmation of the absolute configuration of (−)-kopsine (**325**) came from a correlation of (−)-aspidofractinine (**326**) with (−)-minovincine (**11**) (*124*), a compound of known absolute configuration (*179*).

When (−)-minovincine (**11**) was heated at 105° in the presence of 3 *N* hydrochloric acid, hydrolysis, decarboxylation, enolization, and cyclization occurred to afford in high yield (−)-19-oxoaspidofractinine (**329**). This compound could be reduced under Clemmensen conditions to (−)-**326**. (−)-Aspidofractinine (**326**) therefore is an important relay compound in determining the absolute stereochemistry of the kopsane alkaloids, all of which appear so far to belong in the (−)-stereochemical series.

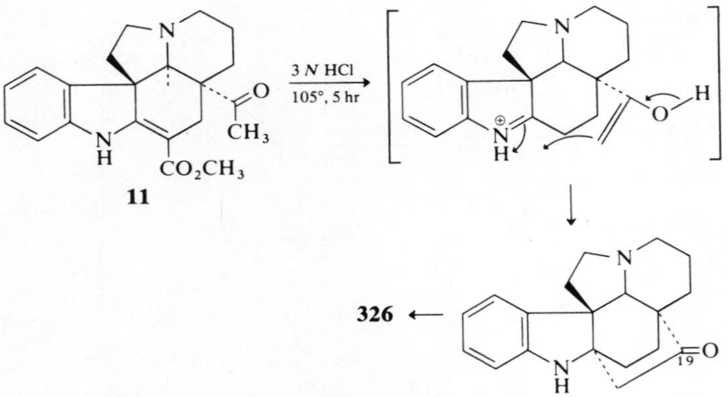

3 *N* HCl
105°, 5 hr

11

326 ←

329

E. Oxidative Alkoxylation of 12-Alkoxyaspidospermidines

A novel alkoxylation reaction has been observed by Schmid and coworkers (*180*) on compounds in the 12-methoxyaspidospermidine series. Treatment of *N*-deacetylaspidospermine (**330**) with sodium hypoiodite in methanol gave, in moderate yield, a product showing λ_{max} 327 nm. The mass spectrum gave a molecular ion at 342 and a base peak at *m/e* 124. That oxidation had occurred in the aromatic ring was evident from the PMR spectrum, which showed three nonaromatic vinyl protons in the region 5.2–6.5 ppm and two methoxy singlets at 3.33 and 3.43 ppm. Reduction of the product with sodium borohydride afforded **330**, thereby indicating that no skeletal changes had occurred. The compound was assigned structure **331**.

As expected, oxidative attack on a compound such as **250** also occurred regiospecifically at the C-12 position to afford **332** (*180*).

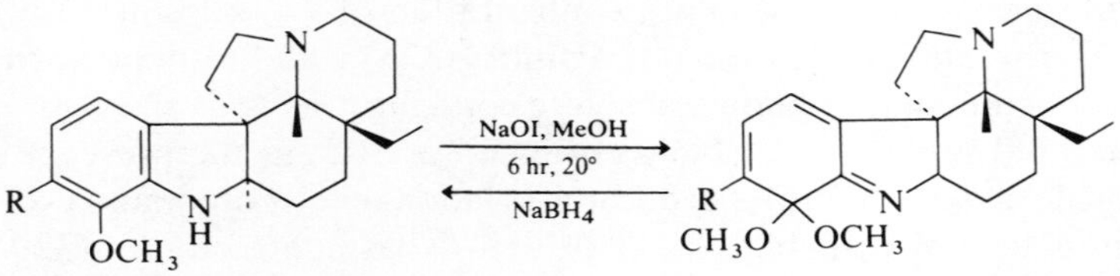

NaOI, MeOH
6 hr, 20°
$NaBH_4$

330 R = H
250 R = OCH_3

331 R = H
332 R = OCH_3

F. Conversion of the *Aspidosperma* to the *Melodinus* Skeleton

The biosynthesis of the *Melodinus* skeleton from the *Aspidosperma* has been suggested (Volume XI, p. 242) to occur by rearrangement of a diol such as **333**, as shown in Scheme 10. The synthesis of such a diol poses a number of problems, but it was thought that an adequate leaving group could be provided at C-16 by forming a chloroindolenine which could then be attacked at C-2 with subsequent rearrangement.

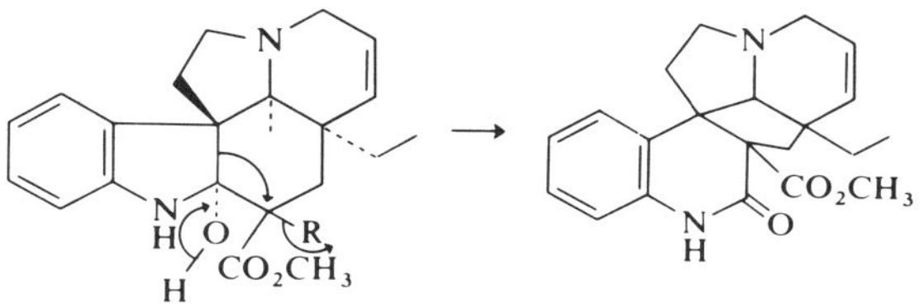

333 R = OH

Scheme 10

Two groups (*181, 182*) have investigated the rearrangement of tabersonine chloroindolenine (**320**). The French group (*181*) investigated the reaction of **320** under reflux with 1 : 1 aqueous THF. The major product (78% yield) gave a molecular ion (322 mu) in agreement with a molecular formula $C_{20}H_{22}N_2O_2$, and exhibited an indolic UV spectrum. The Swiss group (*182*) treated **320** in a 2 : 1 mixture of acetone : water with silver perchlorate. The product (87% yield) also showed an indolic UV spectrum and M^+ 322. Completely independently, these two products were determined to have the structure **334**, the result of a quite extensive rearrangement.

The IR spectrum confirmed an indolic NH and a carbomethoxy group. In the 270 MHz (*182*) NMR spectrum all 20 protons were observed and many of the key aspects of the structure were obvious; in particular, the indolic nucleus, a part structure **335** for the piperideine ring and ethyl side chain, and the unit **336** for the C-17 methylene were evident. These elements in the structure were substantiated by the CMR spectrum, which indicated two nonaromatic, noncarbonyl quaternary carbons as singlets at 73.9 and 47.6 ppm. The spectral evidence was interpreted in terms of structure **334**. This structure was confirmed by single-crystal X-ray crystallographic analysis (*182*).

Each group proposed a slightly different mechanism for this rearrangement. The most probable mechanism would appear to be that of the French group (*181*), which does not involve cleavage of the tryptamine bridge but does involve loss of the C-21 carbon (Scheme 11).

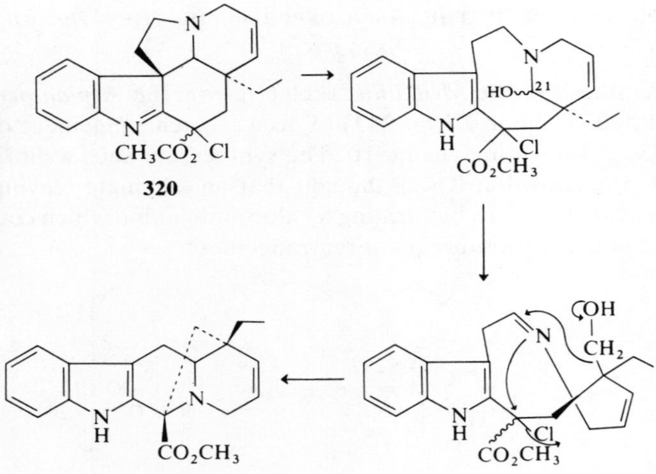

N
N
CH_3CO_2 Cl
320

N
HO 21
N
H
Cl
CO_2CH_3

OH
N
CH_2
N
H
Cl
CO_2CH_3

N
N
H
CO_2CH_3
334

SCHEME 11

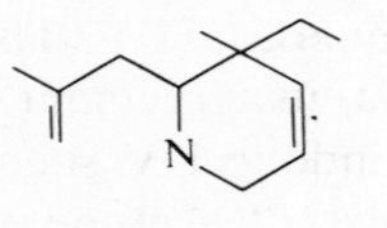

N
335

$-\overset{|}{\underset{|}{C}}-CH_2-\overset{|}{\underset{|}{C}}-$

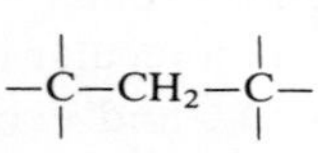

336

In a further attempt to produce the *Melodinus* alkaloid skeleton, **320** was treated with 2 *N* sulfuric acid at 90–100° for 30 min (*183*). The product, obtained in 92% yield, was a quinolone, not of the *Melodinus* type but rather a 3-quinolone having the structure **337**. Instead of the 7-carbon migrating, the indole nitrogen had migrated and displaced the leaving group. This structure type has not yet been found naturally but must be considered a likely candidate, in view of the relatively facile conversion process.

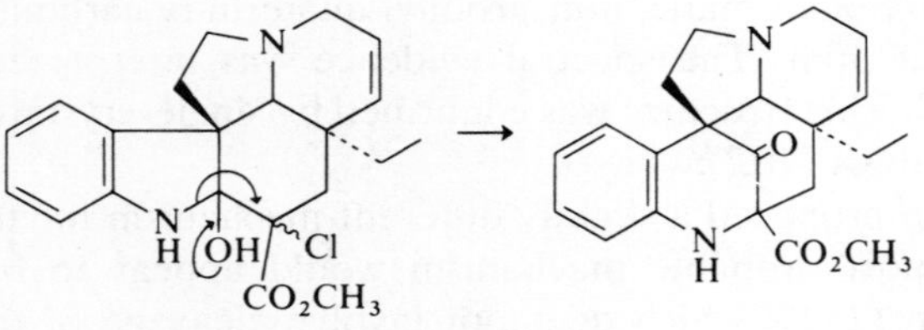

N
N
H OH Cl
CO_2CH_3

N
O
N
H
CO_2CH_3
337

G. Conversion of 18,19-Dehydrotabersonine (**340**) to Andranginine (**338**)

Andranginine was obtained from *Craspidospermum verticillatum* Boj. var. *petiolare* (*125*), but the mass spectral and CMR spectral data did not allow deduction of its carbon skeleton. Using the INDOR technique the structure was deduced to be **338** (*126*).

Potier and co-workers (*184*) reasoned that andranginine was produced biosynthetically from a didehydrosecodine such as **339**. In order to accomplish this process *in vitro* 18,19-dehydrotabersonine (**340**) was needed. This compound was produced by elimination of HI from 19-iodotabersonine (**341**). When **340** was heated in degassed methanol at 145° for 40 hr, three products were formed. One of these was identical with andranginine (**338**, 16% yield), and a second indolic product (1% yield) gave a mass spectrum identical with that of **338** and from the chemical shift of the C-21 proton (4.67 ppm) was shown to be 21-epiandranginine (**342**). A further indolic product gave a molecular ion at *m/e* **366** (addition of methanol), and the PMR spectrum indicated two methoxy singlets at 3.75 and 3.35 ppm. A methoxy methine proton (3.25 ppm) limited the position of substitution, but no olefinic protons were observed save for a vinyl group. A key observation was that the product was optically active, indicating that the C-17–C-20 bond was not broken in the rearrangement. Taken together, these data indicated the structure to be a 15-methoxy-18,19-dehydroallocatharanthine (**343**).

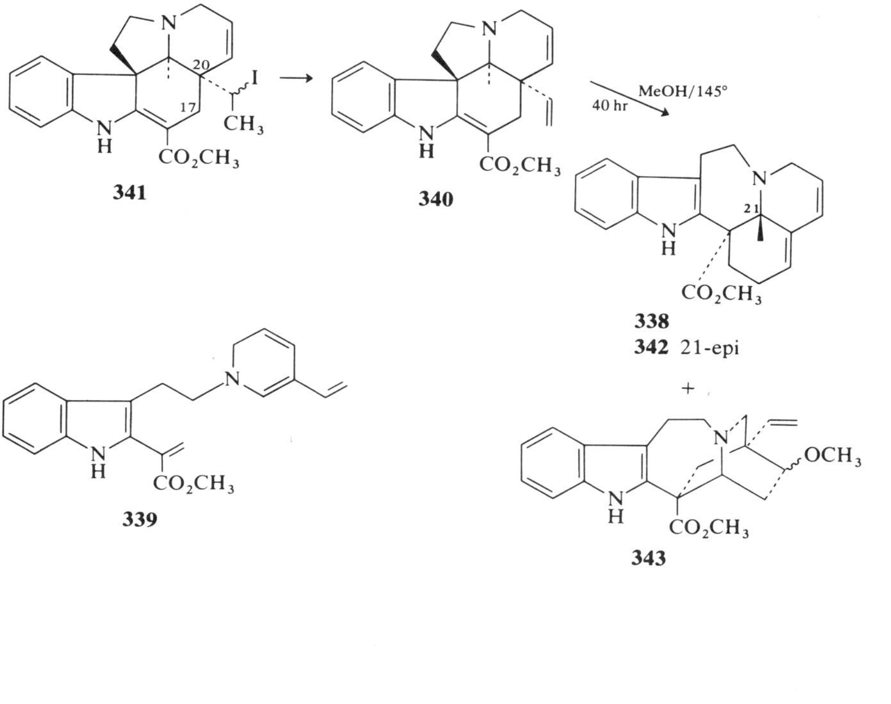

20 17 I CH_3 N N H CO_2CH_3

341

N N H CO_2CH_3

340

MeOH/145° 40 hr

N N H 21 CO_2CH_3

338

342 21-epi

+

N N H CO_2CH_3

339

N N H OCH_3 CO_2CH_3

343

The first partial synthesis of andranginine (**338**) was reported by the French group in collaboration with Scott and Wei (*125*). Thermolysis of precondylocarpine acetate (**344**) at 100° in ethyl acetate afforded andranginine (**338**) in 28% yield, again *via* the didehydrosecodine (**339**). The other products of the reaction were not discussed at that time.

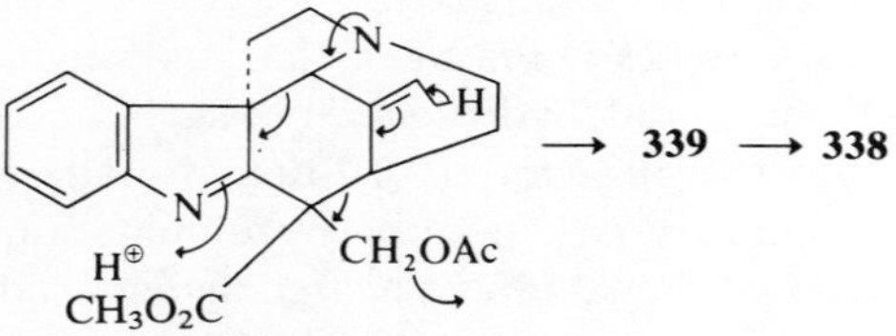

$H^{\oplus}$ CH_3O_2C CH_2OAc $\longrightarrow$ **339** $\longrightarrow$ **338**

344

H. Reduction of Tabersonine (28) and Vincadifformine (2)

Zsadon and Horvath-Otta have investigated some of the simple reactions of tabersonine (**28**) and vincadifformine (**2**) (*185, 186*). Zinc and hydrochloric acid in methanol gave the 2,16-dihydro derivatives **274** and **74**, which could be hydrolyzed to the corresponding carboxylic acids. Reduction of the dihydro esters with lithium aluminum hydride gave the carbinols **345** and **78**.

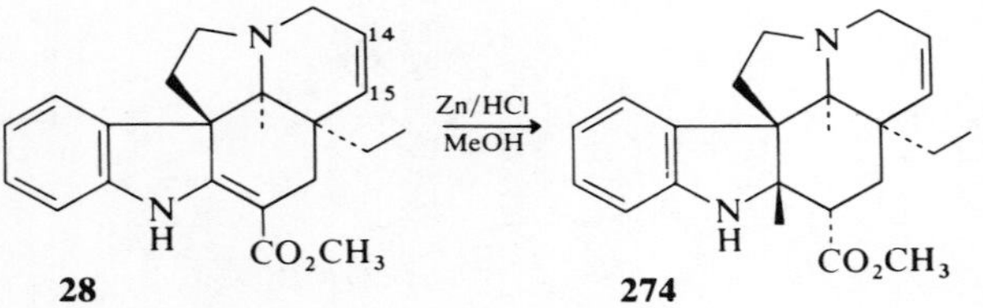

CO_2CH_3

28
2 14,15-dihydro

Zn/HCl, MeOH

CO_2CH_3

274
74 14,15-dihydro

$LiAlH_4$

CH_2OH

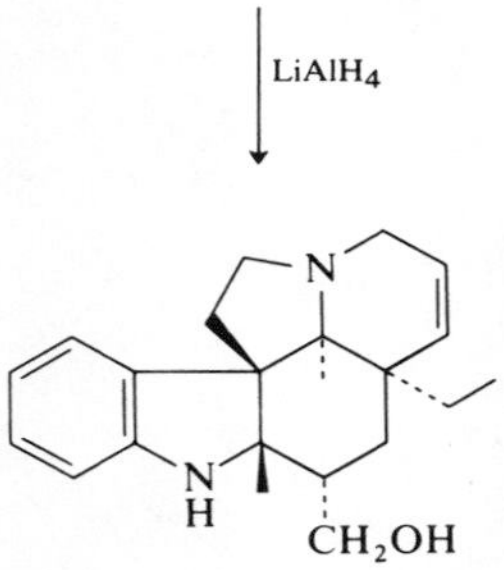

345
78 14,15-dihydro

I. Removal of the Angular Ethyl Group of Aspidospermine

The isolation by Ikeda and Djerassi (*76*) of alkaloids lacking the angular ethyl group prompted Djerassi to attempt to oxidatively remove this group (*187*).

Oxidation of 19-oxoaspidospermine (**346**) with *m*-chloroperbenzoic acid gave two products, the N_b-oxide (**347**) and a second product, analyzing for $C_{22}H_{26}N_2O_5$, which exhibited a methyl singlet at 2.10 ppm, a doublet of doublets at 4.4 ppm for the C-2 proton, and an exchangeable singlet at 6.28 ppm. The UV spectrum was similar to that of the starting material, but the mass spectrum showed a gain of 30 mass units. These data were interpreted in terms of structure **348** (*187*).

When **348** was treated with acetic anhydride in the presence of a trace of sulfuric acid, a product ($C_{20}H_{22}N_2O_3$) was formed. The PMR spectrum lacked both the hydroxyl absorption and the methyl singlet. The loss of 60 mass units compared to **348** indicated the enamide structure **349** as the most likely. Attempts to reduce **349** failed, but alkaline hydrolysis and diborane reduction afforded a mixture, the major component of which was identified as *N*-deacetyl-20-deethylaspidospermine (**350**). The stereochemistry at C-21 and C-20 were not determined, but the three minor products were isomers of the major product showing a base peak in the mass spectrum at *m/e* 96 (**351**).

J. Microbiological Conversions of Vindoline

One of the major stumbling blocks in the clinical use of leurocristine (VCR) is the substantial difference in availability of this compound compared to vincaleukoblastine (VLB), itself a rare alkaloid. Clearly one of the ways to improve the yield of VCR is to oxidize the *N*-methyl group to an *N*-formyl group either chemically or microbiologically. Because the unit requiring refunctionalization is the vindoline portion, it was reasoned that initial work would be most effectively carried out with vindoline (**101**) as the model compound.

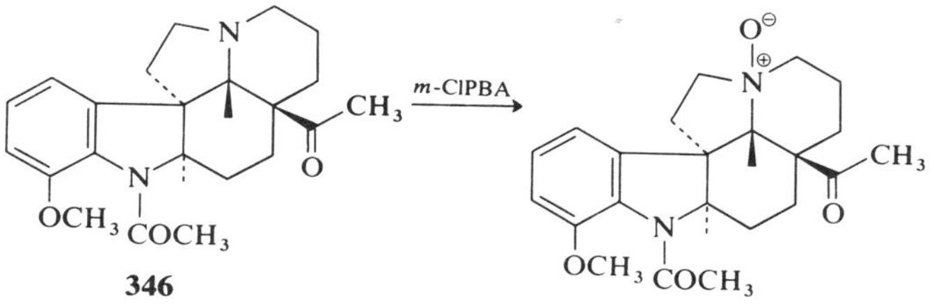

346

347

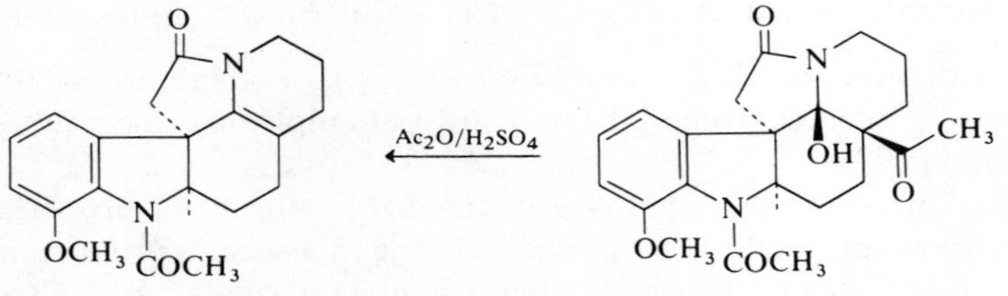

349

Ac2O/H2SO4

348

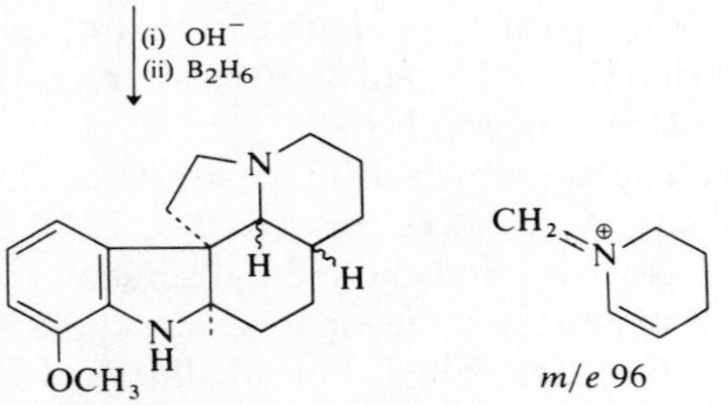

(i) OH⁻
(ii) B2H6

350

m/e 96

351

Many *Actinomycetes* and *Streptomycetes* demonstrated the ability to remove the *O*-acetyl group, but very few carried out other transformations (*188*). A *Streptomyces* sp. A17000, when incubated with vindoline (**101**) for five days afforded several compounds of which two were characterized. One of these, termed dihydrovindoline ether or more correctly 11-methoxycathovaline (**104**), was identified by comparison of the spectral data, particularly the PMR and mass spectra with those of cathovaline (**100**). A minor product proved to have a novel skeleton.

The UV spectrum was of a dihydroindole but the IR spectrum exhibited a five-membered lactam carbonyl at 1710 cm^{-1} and the ester and acetate carbonyls at 1750 cm^{-1}. In the mass spectrum a molecular ion at *m/e* 456 analyzed for $C_{24}H_{48}N_2O_7$, indicating a loss of a carbon and four hydrogens and a gain of oxygen in comparison with vindoline. The C-5 and C-6 methylene groups were clearly evident in the 220 MHz PMR spectrum but the major clue to the structure was the observation of a *singlet* for the C-15 proton at 4.22 ppm, demonstrating the lack of hydrogen at C-14. Structure **352** was therefore proposed for this product, in agreement with all the PMR and mass spectral data.

When incubated with *Streptomyces albogriseolus*, vindoline (**101**) afforded yet another new product whose mass spectrum indicated an additional 56 mu (C_3H_4O) in comparison with 11-methoxycathovaline (**104**). This mass was found in the piperidine ring, which now appeared at *m/e* 178 (corresponding to *m/e* 122 in vindoline). The C-15 proton was observed as a multiplet at 4.06 ppm confirming a C-14 methylene.

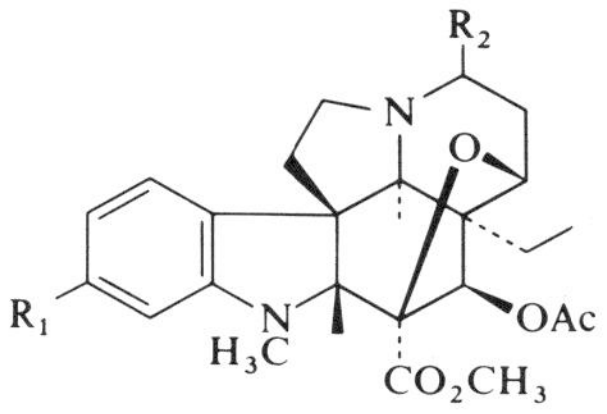

R_2 N O R_1 N H_3C OAc CO_2CH_3

CH_3O N H_3C N O O OAc CO_2CH_3

352

100 $R_1 = H, R_2 = H$
104 $R_1 = OCH_3, R_2 = H$
353 $R_1 = OCH_3, R_2 = CH_2COCH_3$

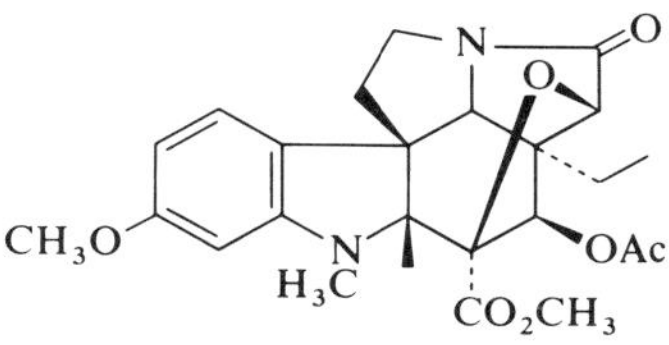

In addition, however, a new methyl singlet was observed at 2.17 ppm, characteristic of a methyl ketone. The additional 56 mu was therefore an acetonyl group located at the C-3 position, and the metabolite was assigned the structure **353** (*188*).

K. Correlation of Corynanthe and Pleiocarpamine-Type Alkaloids

Although pleiocarpamine (**354**) and related alkaloids are well known and quite commonly isolated, until recently the absolute configuration of these alkaloids had not been determined.

Sakai and Shinma were interested in the chemical correlation of the Corynanthe alkaloids with those of the C-mavacurine type and have carried out a number of successful transformations in this area (*189, 190*).

Treatment of hirsutine (**355**) with acetone/HCl at 0° gave the aldehyde–ester **356** in 61% yield (*189*). Cleavage of the C/D ring junction was carried out with cyanogen bromide in ethanolic chloroform to give a mixture of the 3-(*R*)- and (*S*)-ethoxy isomers (**357**). Oxidation with *tert*-butyl hypochlorite gave an epimeric mixture of the C_{16}-deformylchloro compound **358**. The molecule is now set up for formation for the N–C-16 bond by reaction of the chloro compound with sodium hydride in dimethyl sulfoxide at 75°. The UV spectrum of the product **359** indicated the formation of the new bond. Treatment with aqueous ammonium acetate/acetic acid gave 19,20-dihydro-16-epipleiocarpamine (**360**). In addition, an indoline derivative (**361**) was obtained (Scheme 12).

To try to obviate some of the problems involved in dealing with diastereoisomers, the same sequence was carried out with dihydrocorynantheine (**362**). Treatment of demethyldihydrocorynantheine (**363**) with cyanogen bromide in ethanol–chloroform gave only the 3-(*R*)-ethoxy derivative **364**, albeit in only 25% yield. Upon reaction with *tert*-butyl hypochlorite the 3-(*R*)-16-deformyl-16-chlorinated compound **365** was

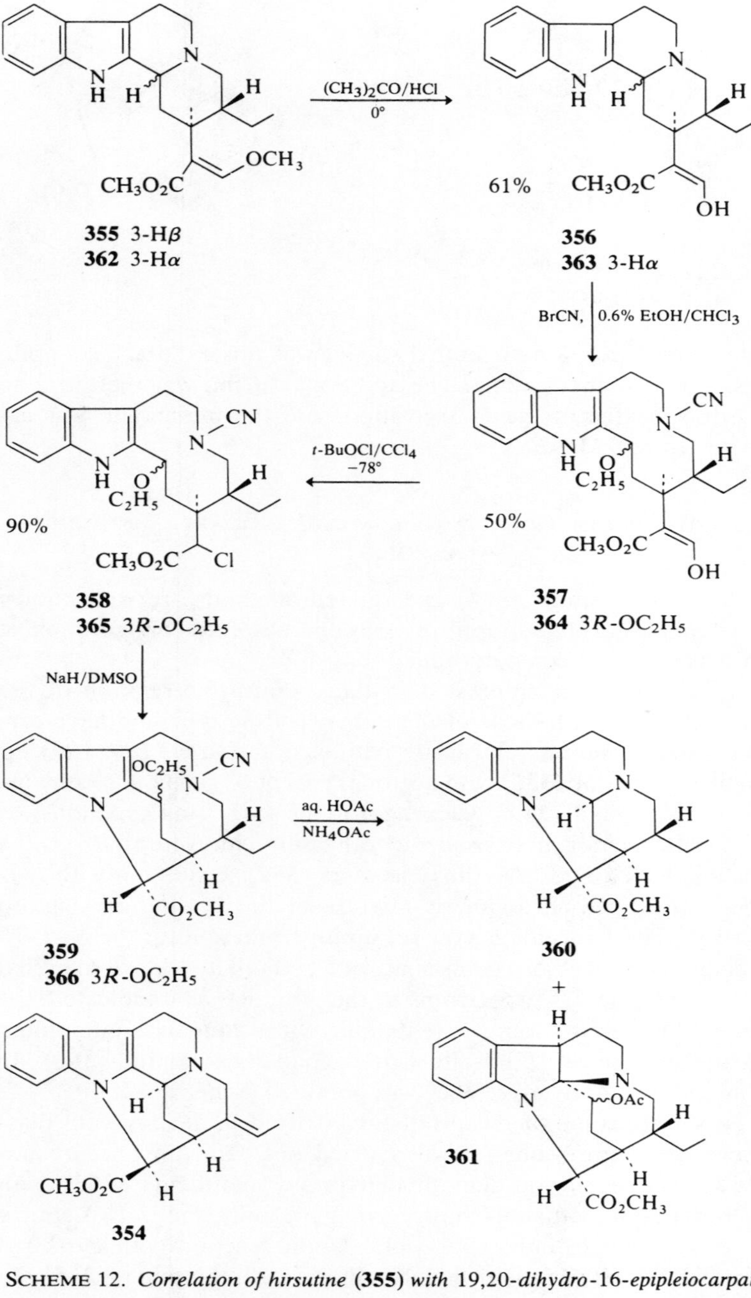

355 3-Hβ
362 3-Hα

$(CH_3)_2CO/HCl$, 0°

61%

356
363 3-Hα

BrCN, 0.6% $EtOH/CHCl_3$

50%

357
364 3*R*-OC_2H_5

t-$BuOCl/CCl_4$, −78°

90%

358
365 3*R*-OC_2H_5

NaH/DMSO

359
366 3*R*-OC_2H_5

aq. HOAc, NH_4OAc

360
+
361

354

SCHEME 12. *Correlation of hirsutine* (**355**) *with* 19,20-*dihydro*-16-*epipleiocarpamine* (**360**).

formed in 56% yield. Cyclization was carried out in the same way as before, and the product **366** on treatment with aqueous buffered acetate solution gave the desired compound **360**. Again, however, the 3-acetoxy-*seco*-C/D-derivative **361** was also produced.

More recently, the same group (*190*) has described a partial synthesis of 16-epipleiocarpamine (**367**) from geissoschizine methyl ether (**368**) in a quite analogous series of reactions (Scheme 13). Treatment of geissoschizine methyl ether (**368**) with acetone/dry HCl gave an aldehyde, and reaction with ethyl chlorocarbonate afforded the carbonate derivative **369**

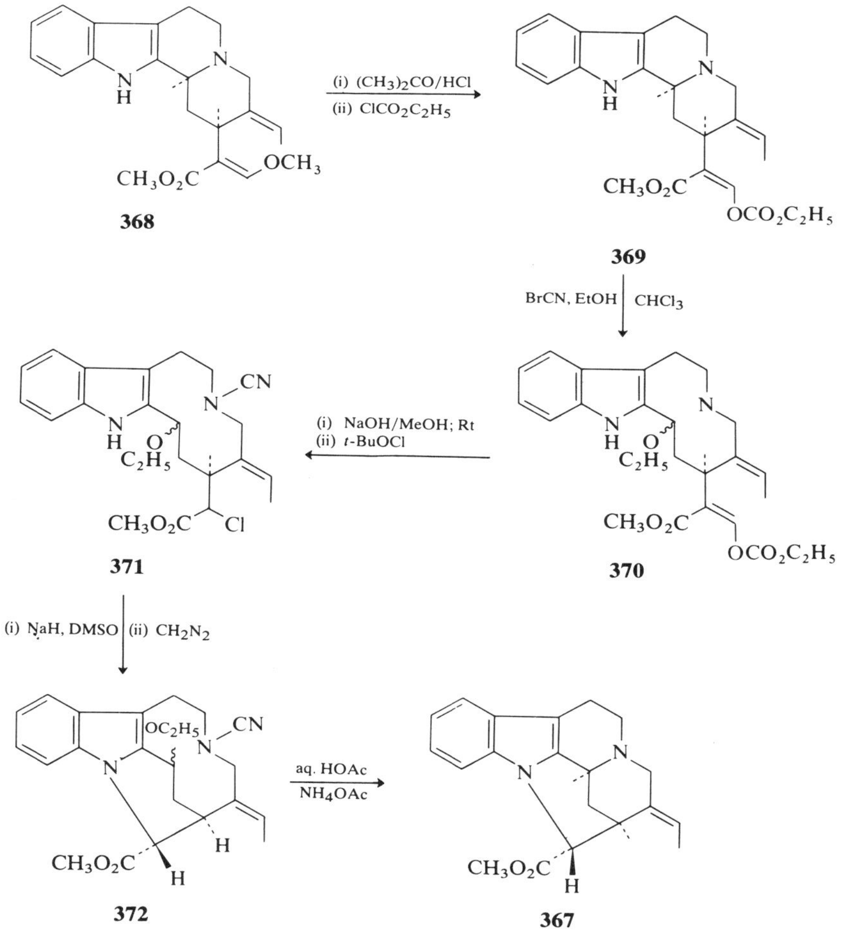

N H N CH_3O_2C OCH_3

368

(i) $(CH_3)_2CO/HCl$
(ii) $ClCO_2C_2H_5$

N H N CH_3O_2C $OCO_2C_2H_5$

369

BrCN, EtOH | $CHCl_3$

N H O C_2H_5 N CH_3O_2C $OCO_2C_2H_5$

370

(i) NaOH/MeOH; Rt
(ii) *t*-BuOCl

N H O C_2H_5 N CN CH_3O_2C Cl

371

(i) NaH, DMSO | (ii) CH_2N_2

N OC_2H_5 N CN CH_3O_2C H H

372

aq. HOAc
NH_4OAc

N N CH_3O_2C H

367

SCHEME 13. *Correlation of geissoschizine methyl ether* (**368**) *with* 16-*epipleiocarpamine* (**367**).

which was subjected to the ring cleavage reaction as before. The 3-(*R*)-ethoxy derivative (**370**) obtained was hydrolyzed to the aldehyde under mild basic conditions and then submitted to chlorination with *tert*-butyl hypochlorite to give the 16-deformyl-16-chloro compound **371** in 80% yield. Ring closure of the indole nitrogen and C-16 was accomplished with sodium hydride in dimethyl sulfoxide, and the product (**372**) was treated with diazomethane to convert partially hydrolyzed material back to the methyl ester. The C-18 proton of **372** appeared as a doublet at 0.1 ppm, being highly shielded by proximity to the indole ring. The C/D ring was reclosed with aqueous ammonium acetate/acetic acid as before to afford 16-epipleiocarpamine (**367**), identical with material obtained from pleiocarpamine (**354**) itself. This is the first correlation of pleiocarpamine with any alkaloids of established absolute configuration.

IV. Synthesis of *Aspidosperma* and Related Alkaloids

A. Alkaloids with the *Aspidosperma* Skeleton

A number of synthetic approaches to the *Aspidosperma* skeleton have been developed in recent years. This section attempts to highlight some of the features of these routes.

Several reviews of the synthesis of *Aspidosperma* alkaloids are available (*191–195*). The review by Winterfeldt (*194*) deals with the stereoselective aspects of synthesis, and the review by Kutney (*193*) is the most detailed as far as *Aspidosperma* alkaloids are concerned.

1. Ban Tricyclic Amino Ketone Approach

The details of the initial efforts of Ban and co-workers were discussed in a previous review in this series (Volume XI, p. 222). Since then, Ban and co-workers have extended these synthetic efforts to complete syntheses of several *Aspidosperma* alkaloids or their stereoisomers.

One of the crucial points which arose during the syntheses of the *Aspidosperma* skeleton by the groups of Ban and Stork was a controversy over the stereochemistry of the tricyclic amino ketone intermediate. Thus Stork assigned structure **373** to this compound and Ban assigned structure **374** to his intermediate prepared by a different route.

In order to establish the stereochemical relationships with greater certainty, Ban and co-workers (*196*) synthesized the *trans*-7-ketodecahydroquinoline **375**, which was converted to the keto lactam **376** via chloroketone **377**. The PMR spectrum of this compound was similar to that of compound **374**. There are two possibilities for the stereochemistry of **376**: namely, **378** and **379**.

Reduction of **376** with sodium borohydride gave the lactam–alcohol **380**, which was reduced to amine **381** with lithium aluminum hydride. No Bohlmann bands were observed in the IR spectrum, and the alcohol was oxidized to ketone **382**.

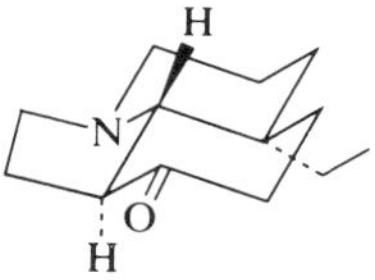

373 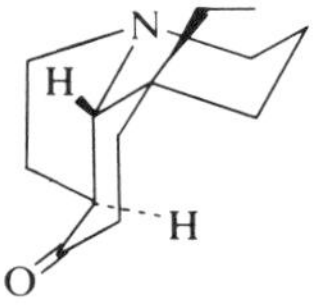 **374** 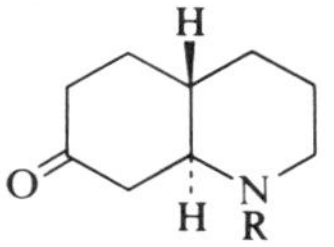 **375** R = H **377** R = $COCH_2Cl$ **376**

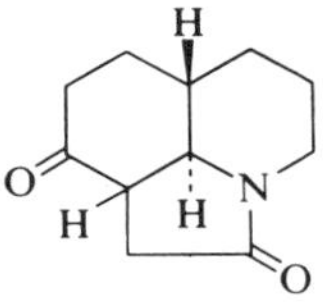

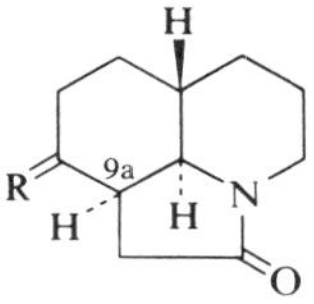

378 R = O **384** R = H_2 **379** R = O **385** R = H_2

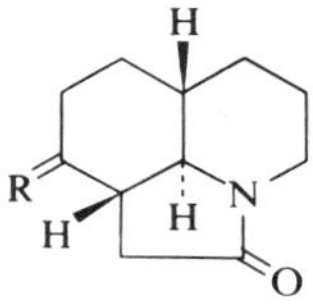

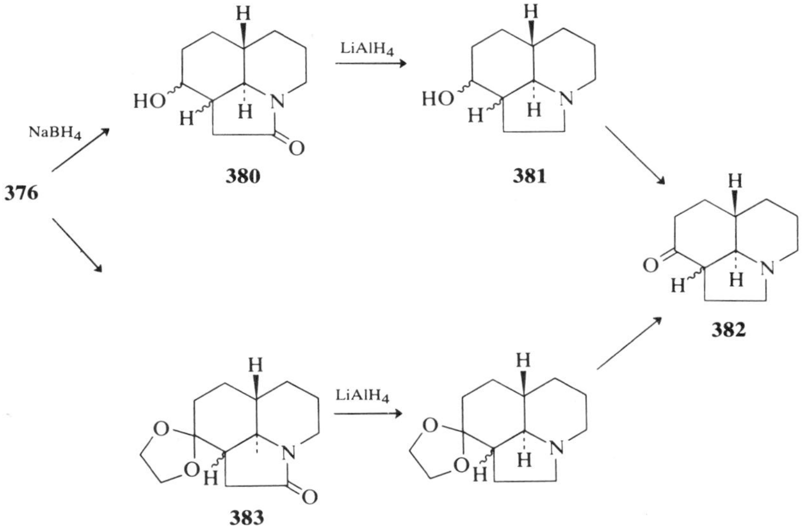

$NaBH_4$ $LiAlH_4$ **376** **380** **381** **382** **383** $LiAlH_4$

The same ketone was also produced by protection of ketone **376** as the ketalamine **383**, which could then be reduced with lithium aluminum hydride and hydrolyzed to ketone **382**. In this ketone Bohlmann bands were observed, and it was rationalized that the correct configuration of the keto lactam **376** was **378** and that epimerization of C-9a must have occurred in the hydrolysis step to give an equilibrum mixture of stereoisomeric lactams **384** and **385**.

In an effort to establish the epimerization at C-9a, ketone **382** was reduced to an alcohol with sodium borohydride and the *four* isomeric alcohols separated. One of the pairs of alcohols (**386** and **387**) exhibited Bohlmann bands; the other pair (**388** and **389**) did not. The stereochemistries indicated are thus established for these isomers, and the easy equilibration of **382** is verified. The stereochemistry for intermediate **378** is therefore demonstrated for the keto lactam in Ban's synthetic route in which the A/B ring juncture can now be assigned a *trans* stereochemistry.

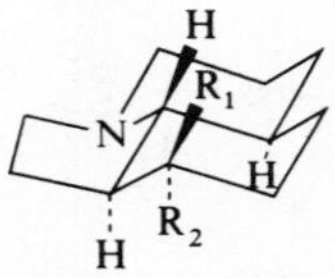

386 $R_1 = OH, R_2 = H$
387 $R_1 = H, R_2 = OH$

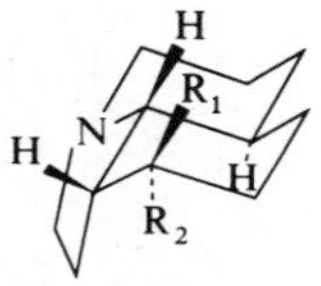

388 $R_1 = OH, R_2 = H$
389 $R_1 = H, R_2 = OH$

With this information in hand, it became necessary to return to the system of interest containing an angular ethyl group, for the previous stereochemical assignments would now appear to be suspect. A simple series of reactions highlights the problem (*197*).

Reduction of the ketolactam **390** (having the newly assigned A/B *trans* stereochemistry) with sodium borohydride followed by reduction of the amide with lithium aluminum hydride gave an amino alcohol. Oxidation of this amino alcohol with chromic acid followed by rereduction with lithium aluminum hydride gave an amino alcohol different from that obtained previously. In the oxidation to the ketone **373**, therefore, epimerization at C-9a must have occurred, and the two amine alcohols must have the configurations **391** and **392**. Since in both compounds Bohlmann bonds were observed in the IR spectrum, the A/B ring juncture is *trans*.

The compounds used by Ban in the synthesis of aspidospermine therefore have the structures **390** and **373** rather than the A/B *cis*, A/C *trans* structures previously assigned. In addition, the corresponding intermediate in the Stork synthesis which had previously been assigned the stereochemistry **373** must now be revised to the all-*cis* compound **393**.

(i) $NaBH_4$ (ii) $LiAlH_4$ — CrO_3 — $LiAlH_4$

390 **391** **373** **393** **392**

A key observation as a result of these revised stereochemical assignments is that the stereochemistry of the Ban amino ketone (**373**) is now different from that of the natural series of compounds in which the C/E ring junction (A/B in **373**) is *cis*. Stork's amino ketone (**393**) therefore has the correct stereochemistry, and the elaborate mechanism proposed to account for the total synthesis of natural aspidospermidine (**249**) (which originally involved isomerization at C-21) is unnecessary. What became important, however, was the identity of the product from a Fischer indole synthesis of **373**. If Stork was correct and epimerization at C-21 is possible, then aspidospermidine should still be the product. If epimerization is not possible a new stereoisomer of aspidospermidine should result in which the C-21 proton and the C-20 ethyl group are *trans*.

Treatment of the *o*-methoxyphenylhydrazone of **373** with formic acid under reflux (*198*) followed by hydrolysis and N-acetylation gave a compound having the following key NMR properties. The C-2 proton was observed as a triplet at 4.92 ppm, and the methyl of the ethyl group was found as a triplet at 0.94 ppm. These chemical shifts should be contrasted with those of aspidospermine, which are 4.50 ppm (doublet of doublets) and 0.67 ppm. In the new stereoisomer, therefore, it would appear that the ethyl group is placed away from the shielding environment of the benzene nucleus and has the structure **394**. This structure was confirmed by X-ray analysis of the hydroiodide salt (*199*).

When the *o*-methoxyphenylhydrazone of **373** was heated for 8 hr in acetic acid two products were obtained (*198*), the *N*-deacetyl compound **395** and *N*-deacetylaspidospermine (**330**). No proportions were given but the process must involve a different initial attack of the benzene ring than in the synthesis of **395**.

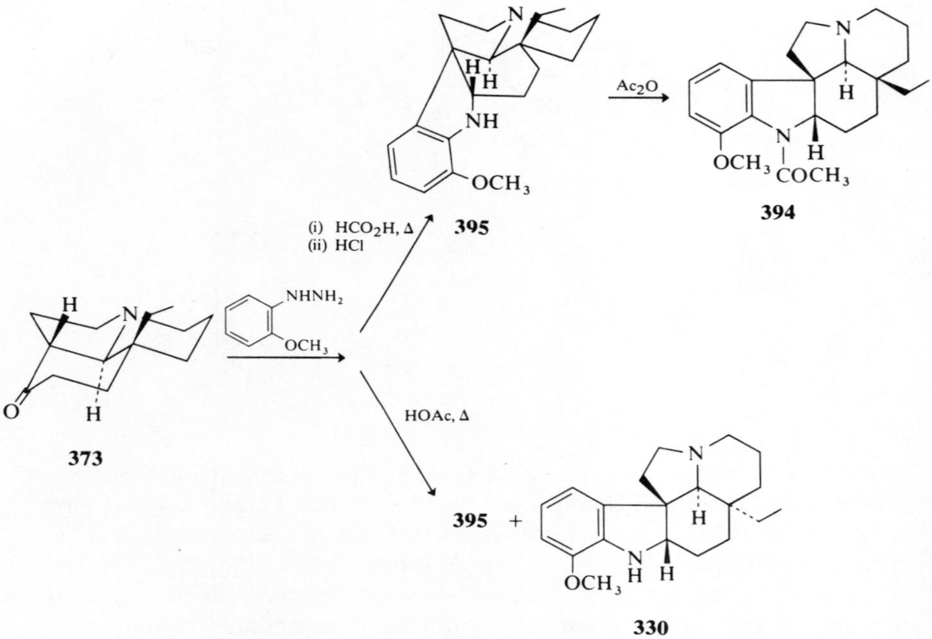

373 → (o-methoxyphenylhydrazine; (i) HCO_2H, Δ (ii) HCl) → 395 → (Ac_2O) → 394

373 → (o-methoxyphenylhydrazine; HOAc, Δ) → 395 + 330

2. Attempted Synthesis of Limaspermine (**396**)

Ban and co-workers next turned their attention to more complex alkaloids. The first system to receive attention (*197*) was the alkaloid limaspermine (**396**) obtained previously by Schmid and co-workers (*200*).

Condensation of 5-phenoxypentan-2-one with acrylonitrile gave the lactam ketone **397** which upon hydrogenation gave a mixture of two isomeric alcohols (**398** and **399**) differing in configuration at the ring juncture.

Reduction of the lactam with lithium dimethoxyaluminum hydride (to prevent hydrogenolysis of the phenoxy group) gave the corresponding amino alcohol, which was converted by the previously outlined methods to the tricyclic keto lactam. By carrying out this series of reactions with each lactam alcohol, they were able to obtain the two tricyclic keto lactams **400** and **401**. The configurations of these compounds were established by comparison with the compounds in the 6a-ethyl and 6a-H series.

In carrying out the further steps to the *Aspidosperma* skeleton, only the compound in the major series—the tricyclic keto lactam **400**—was used (*201*). Ketalization, reduction with lithium dimethoxyaluminum hydride, and mild acid hydrolysis gave the keto amine **404**, which exhibited strong Bohlmann bands. Fischer condensation using formic acid on the *o*-

methoxyphenylhydrazone of **402** gave the *N*-formyldihydroindole **403** in 16% yield. This compound showed a triplet for the C-2 proton at 4.9 ppm and consequently was assigned the stereochemistry shown in which the C/E ring is *trans*. This compound, having the unnatural C-20 stereochemistry, is 18-phenoxyisopalosine (**403**; Scheme 14) (*201*).

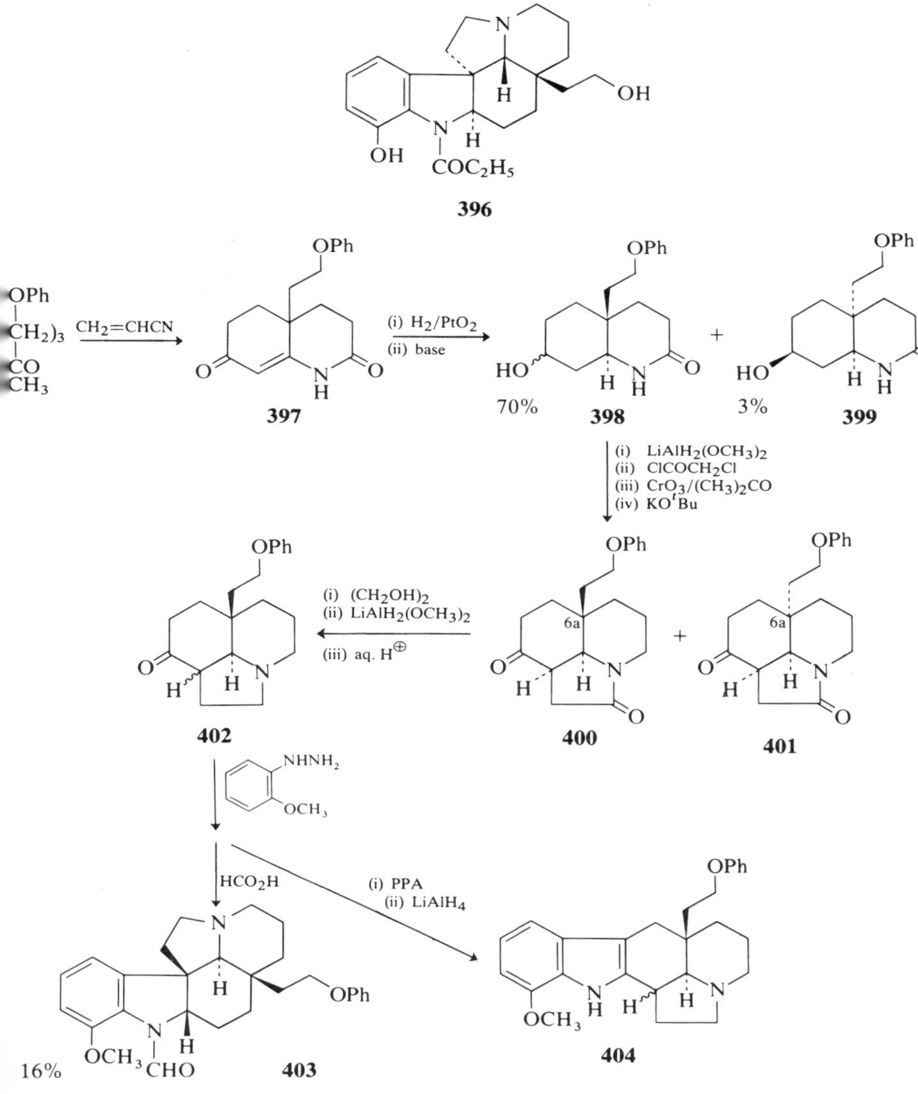

SCHEME 14. *Attempted synthesis of limaspermine* (**396**).

It is worth pointing out that reaction of the phenylhydrazone of **402** in polyphosphoric acid did not give a compound in the *Aspidosperma* series but one derived by Fischer cyclization at the other α-position. Lithium aluminum hydride reduction then gave the indole derivative **404**.

3. Ban Oxindole Approach

With the failure of the previous approach involving a Fischer cyclization, Ban and co-workers turned their attention to a general approach used by Harley-Mason (*202*) and co-workers (Volume XI, p. 225). This approach centers on the formation of an oxindole such as **405** produced by condensation of 2-hydroxytryptamine hydrochloride (**406**) with 3-oxobutanal ethylene ketal (**407**), which can then be elaborated by two separate routes (*203*).

The compound **408** as a mixture of stereoisomers was acylated with β-chloropropionyl chloride and hydrolyzed with acid to give the two diastereoisomers of **405**. Treatment of each isomer with sodium hydroxide in ethanol–methylene chloride gave the corresponding acryloyl derivatives **409** in high yield. One of these isomers, when reacted with the Meerwein reagent, gave the tetracyclic imino ether **410** as a mixture of two isomers. The isomeric acryloyl derivatives gave yet another isomeric imino ether. The stereochemistry of these products was not established at this point; rather, a mixture of isomers **410** was heated with sodium hydride in dimethyl sulfoxide for 60 hr to give a mixture of pentacyclic products in which one isomer (**411**) predominated (~30–50%). Reduction of this compound with lithium aluminum hydride in THF, acetylation, and then hydrogenation gave *N*-acetyl-20-deethylaspidospermidine (**412**; Scheme 15). This compound was identical with a product derived from the Fischer indole cyclization of **413**. These results do *not* establish the stereochemistry of the vinylogous keto amide, **411** (*203*).

The success of the oxindole to vinylogous amide approach just discussed has been used subsequently by Ban and co-workers in a number of synthetic efforts.

The first compound having an ethyl side chain to be synthesized by this method was a compound in the pseudo-*Aspidosperma* series (*204*). The key compound in this sequence is the vinylogous amide **414**, which was prepared from 2-hydroxytryptamine hydrochloride (**406**) and 3-oxobutanal ethylene ketal (**407**), followed by hydrolysis, cyclization with the Meerwein reagent, and acetylation (*205*).

Reduction of **414** with sodium borohydride in isopropanol gave a mixture of three products. Two of these were the stereoisomeric alcohols **415** and **416**, as shown by their facile oxidation with DCC–DMSO to the same

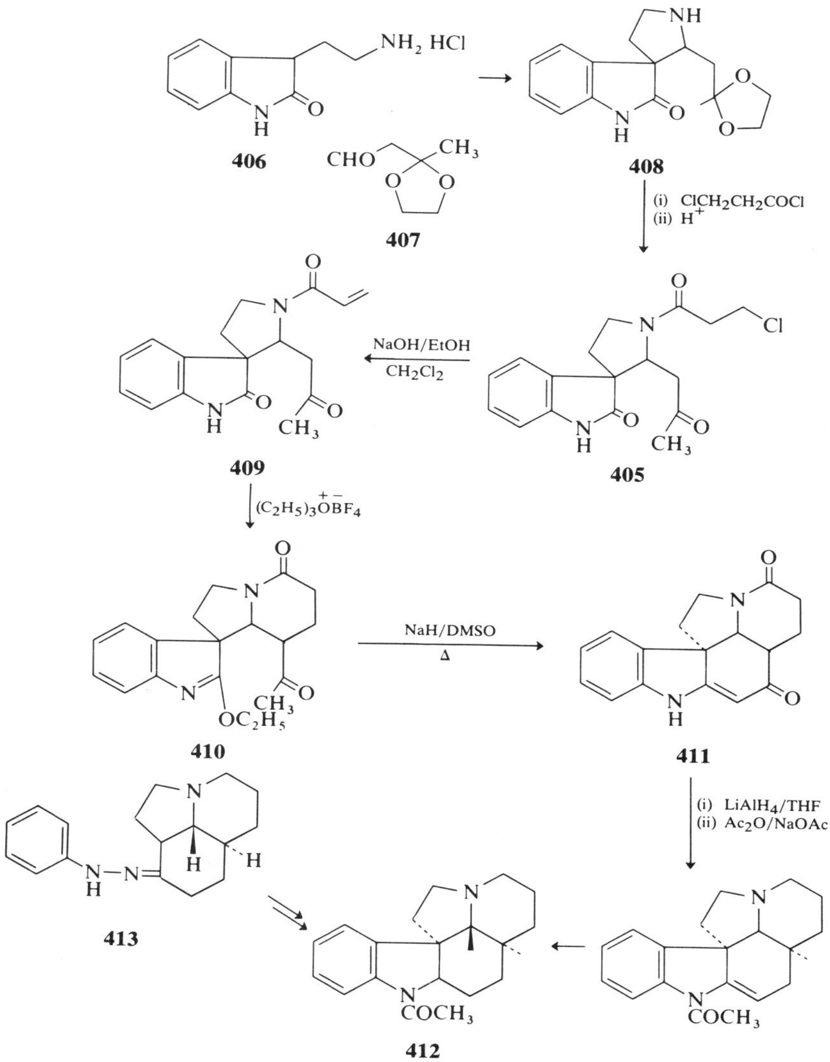

SCHEME 15. *Synthesis of* N-*acetyl*-20-*deethylaspidospermidine* (**412**).

ketoamide, **417**. The major alcohol (**416**) was selectively tosylated at the N-position with tosyl chloride in methylene chloride and 10% aqueous sodium hydroxide solution, and the amide group was removed with 5% aqueous ethanolic potassium hydroxide to give the tosyl amino alcohol **418**.

Two series of reactions were carried out on this compound. In one of these, the tosyl amino alcohol **418** was reacted with β-chloropropionyl chloride followed by oxidation with Jones reagent to give the tosyl keto amide **419**. Elimination of HCl with potassium carbonate in methylene chloride and cyclization of the acrylamide with Meerwein's reagent gave the *N*-tosyl keto lactam **420**. This compound was also prepared by the previously discussed route (*203*).

The next sequence of reactions was designed to examine the feasibility of synthesizing compounds in a natural series. The tosyl amino alcohol **418** was condensed with α-chloromethylbutyryl chloride. The product **421** was treated with base to afford the acrylamide **422**, which after oxidation of the alcohol function was cyclized in low yield with Meerwein's reagent to a mixture of keto lactams **423** and **424**. The major keto lactam exhibited a PMR spectrum similar to that of **420**, indicating a *cis* C/D ring juncture as required in the natural series.

The ketone at C-17 was removed by thioketalization and Raney nickel reduction, and the resulting tosyl lactam **425** was reduced with lithium aluminum hydride and acetylated to give *N*-acetyl-14β-ethyl-20-deethyl-aspidospermidine (**426**; Scheme 16) (*204*).

4. Synthesis of (±)-17-Hydroxyaspidofractinine (**432**)

In 1964, Schnoes and Biemann (*206*) reported that minovincine (**11**) could be successfully transformed into 19-hydroxyaspidofractinine (**427**), thereby establishing a relationship between the pentacylic *Aspidosperma* alkaloids and those which are hexacyclic.

As the next step in the synthetic program, Ban envisaged a synthesis of the *Aspidosperma* skeleton *via* a double Michael cyclization in which attack occurs successively at the α- and β-positions of an α,β-unsaturated ketone. The success of this novel cyclization reaction was first achieved on compound **428** to give a compound having the aspidofractinine-type skeleton (*207*).

The vinylogous amide lactam **411** prepared by the route described previously (*203*) was N-tosylated in 44% yield by reaction with sodium hydride in boiling monoglyme followed by treatment with tosyl chloride. On treating the tosyl derivative **428** with excess acrylonitrile in *tert*-butanol–dimethyl sulfoxide in the presence of potassium *tert*-butoxide, two products were formed: the pentacylic derivative **429** and the hexacyclic compound **430**. Further base treatment of the former compound gave **430** in quantitative yield, thereby bringing the overall reaction yield to a respectable 50%.

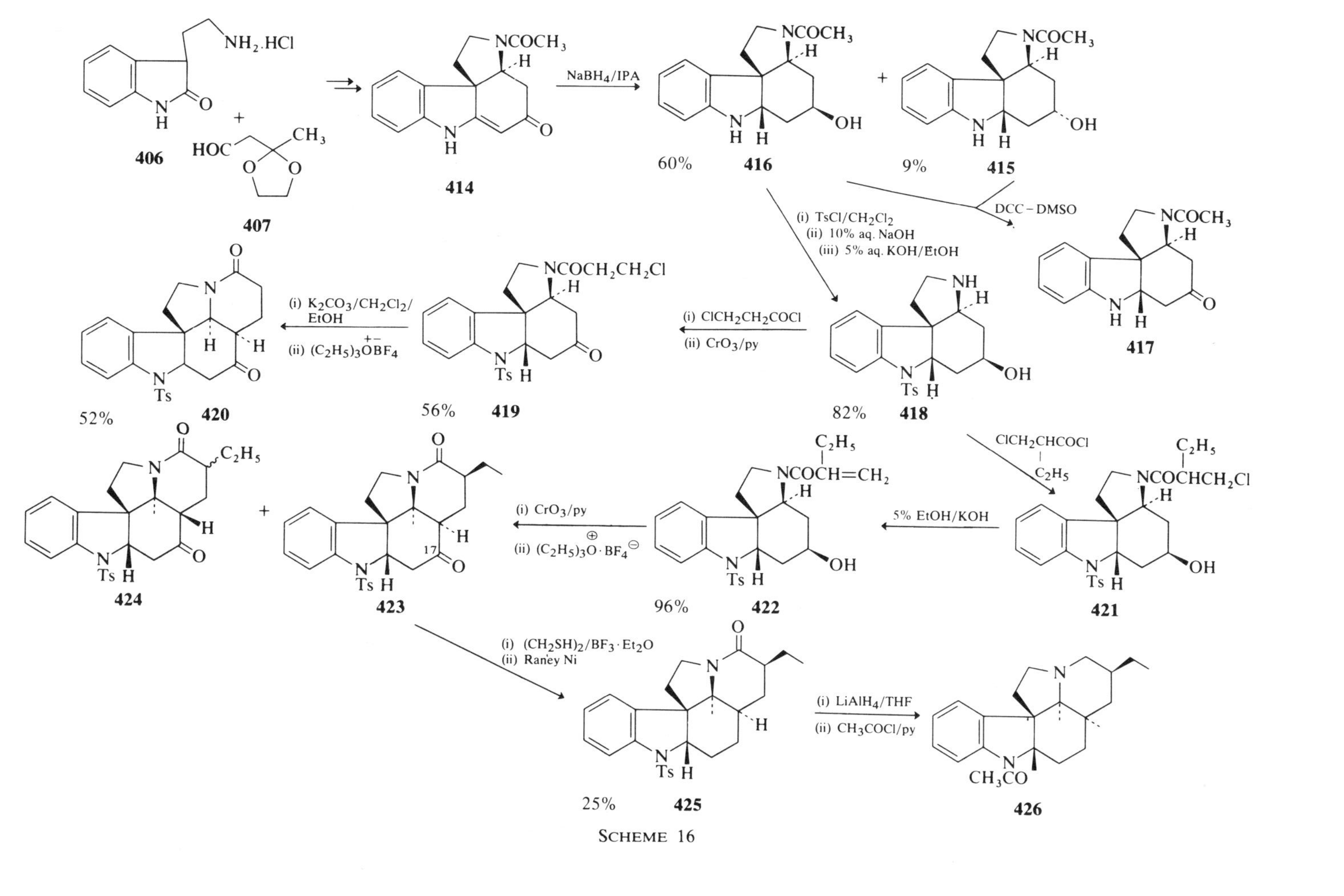
406
407
414
NaBH4/IPA
60%
416
9%
415
DCC – DMSO
417
(i) TsCl/CH2Cl2
(ii) 10% aq. NaOH
(iii) 5% aq. KOH/EtOH
82%
418
(i) ClCH2CH2COCl
(ii) CrO3/py
56%
419
(i) K2CO3/CH2Cl2/EtOH
(ii) (C2H5)3OBF4
52%
420
ClCH2CHCOCl C2H5
421
5% EtOH/KOH
96%
422
(i) CrO3/py
(ii) (C2H5)3O·BF4
423
424
(i) (CH2SH)2/BF3·Et2O
(ii) Raney Ni
25%
425
(i) LiAlH4/THF
(ii) CH3COCl/py
426

SCHEME 16

Quite analogously, reaction of **428** with methyl vinyl sulfone gave the hexacyclic product **431** as a mixture of C-18 stereoisomers. Reduction of **431** with di(2-methoxyethoxy)sodium aluminum hydride in dioxane under reflux gave a hydroxythiomethylamine, which was desulfurized with Raney nickel to afford a racemic hydroxyaspidofractinine.

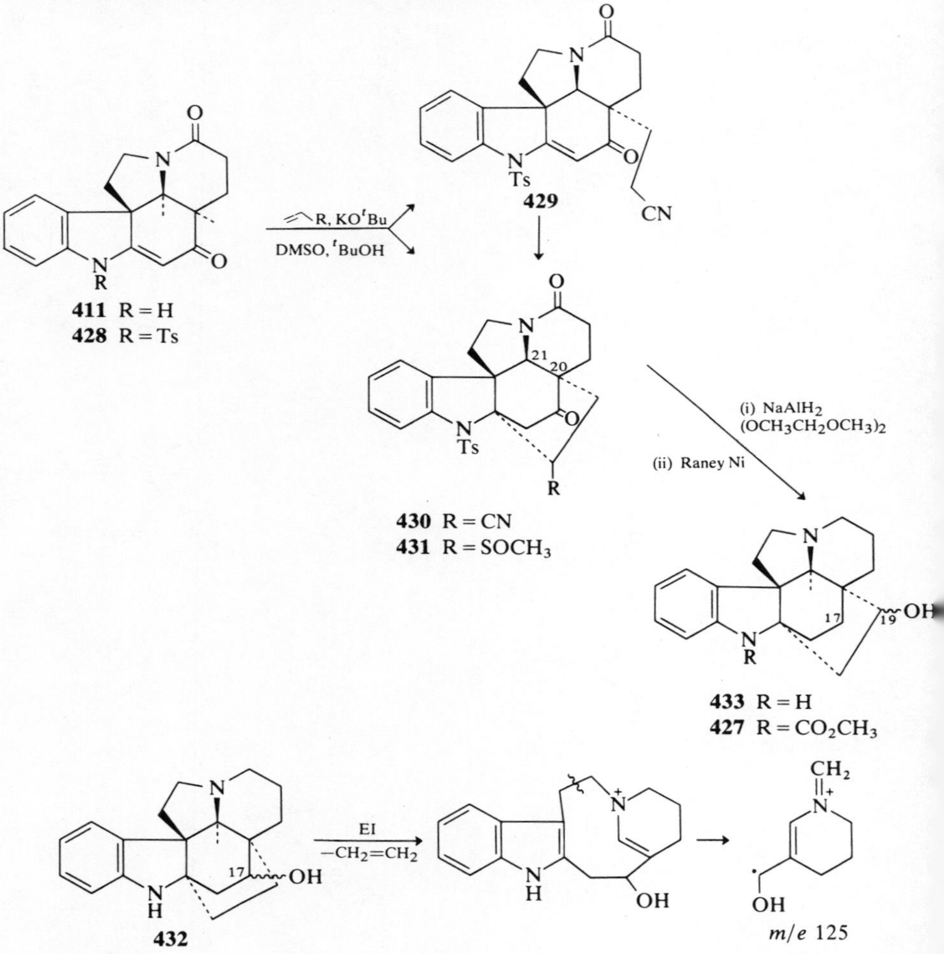

411 R = H
428 R = Ts

=\R, KO^{t}Bu
DMSO, ^{t}BuOH

429

430 R = CN
431 R = $SOCH_3$

(i) $NaAlH_2(OCH_3CH_2OCH_3)_2$
(ii) Raney Ni

433 R = H
427 R = CO_2CH_3

432

EI
$-CH_2{=}CH_2$

m/e 125
434

SCHEME 17

The question of the C-20 stereochemistry of this product then arose, because theoretically one could postulate two alternative products, **432** and **433**, depending upon the stereoselectivity of the initial Michael addition. The stereochemistry could easily be deduced from the mass spectrum.

In this series of compounds, it is established that retro-Diels–Alder reaction occurs in the bond *cis* to the C-21 proton. Thus, if structure **433** were correct, the initial loss expected is of 44 mu whereas if the 17-hydroxy isomer **432** is correct, the initial loss should be of 28 mu with a subsequent ion at *m/e* 125 (**434**). In fact, the latter situation was observed; the stereochemistry of the product at C-20 is therefore established together with the location (C-17) of the hydroxy group (Scheme 17) (*207*).

The stereochemistry of the initial Michael attack can also be explained on the basis of attack from the least hindered side, as shown in Scheme 18.

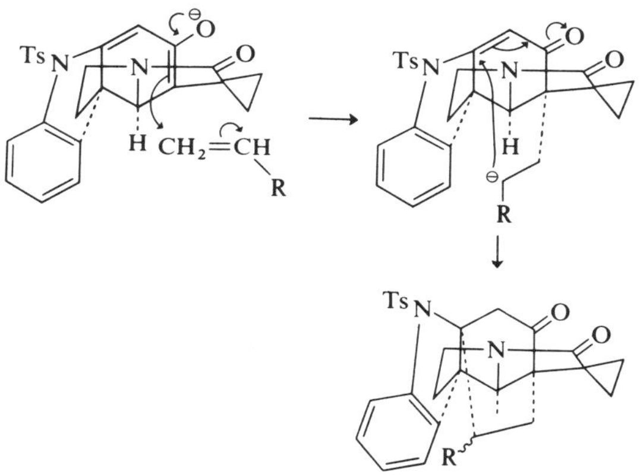

430/431 (isomers)

SCHEME 18

5. Synthesis of (±)-Deoxyaspidodispermine (158)

As discussed in Section II,F,1, one of the more novel *Aspidosperma* alkaloid types isolated in the recent past is that having the aspidodispermine skeleton, in which the ethyl side chain is replaced by a hydroxy group. One of the alkaloids obtained was (−)-deoxyaspidodispermine (**158**) (*76*), and Ban and co-workers have recently reported a synthesis of this compound (*208*). Oxidation of **428** at the 20-position by reaction with oxygen at −78° gave a mixture of two ketols, the major one (57%) of which had a *cis* C/D ring junction with a hydroxyl group at C-20 β, as shown in **435**. Lithium aluminum hydride reduction gave three products, each in about 30% yield. The two undesired products were alcohols **436** and **437**, and the desired product was the enamine, **438**. Acetylation of **438** with acetyl chloride in 5% sodium hydroxide followed by catalytic hydrogenation under pressure gave (±)-deoxyaspidodispermine (**158**), identical with the natural product (*208*).

O_2, NaH
DMF, −78°

428

435

$LiAlH_4$

(i) CH_3COCl/5% NaOH
(ii) PtO_2/H_2/56 psi/H^+

158

438

+

436/437 (isomers)

6. Synthesis of *N*-Acetylaspidospermidine (**252**)

In the synthesis of the aspidofractinine nucleus, Ban and co-workers had developed a route involving a double Michael reaction to construct the additional ring. This process was both regioselective and stereoselective to give a compound in which attack occurred initially at C-20 and from the same side as the C-21 proton to give compounds in the natural C/D *cis* series. Because most *Aspidosperma* alkaloids have an ethyl side chain, it became extremely important to modify the synthetic procedure to produce compounds in the 20-ethyl series. The first compound to be thus synthesized was (+)-*N*-acetylaspidospermidine (**252**) (*209*). Again beginning with the preformed pentacylic intermediate **428**, reaction with methyl vinyl sulfone using lithium diisopropylamide as the base at reduced temperatures gave the Michael addition product **439** as colorless needles in 82% yield. Desulfurization with Raney nickel gave the residual ethyl side chain and also some removal of the tosyl group and reduction of the vinylogous amide. Thus, two products were formed: the *N*-tosyl vinylogous amide **440** in 75% yield and the keto lactam **441** in 10% yield.

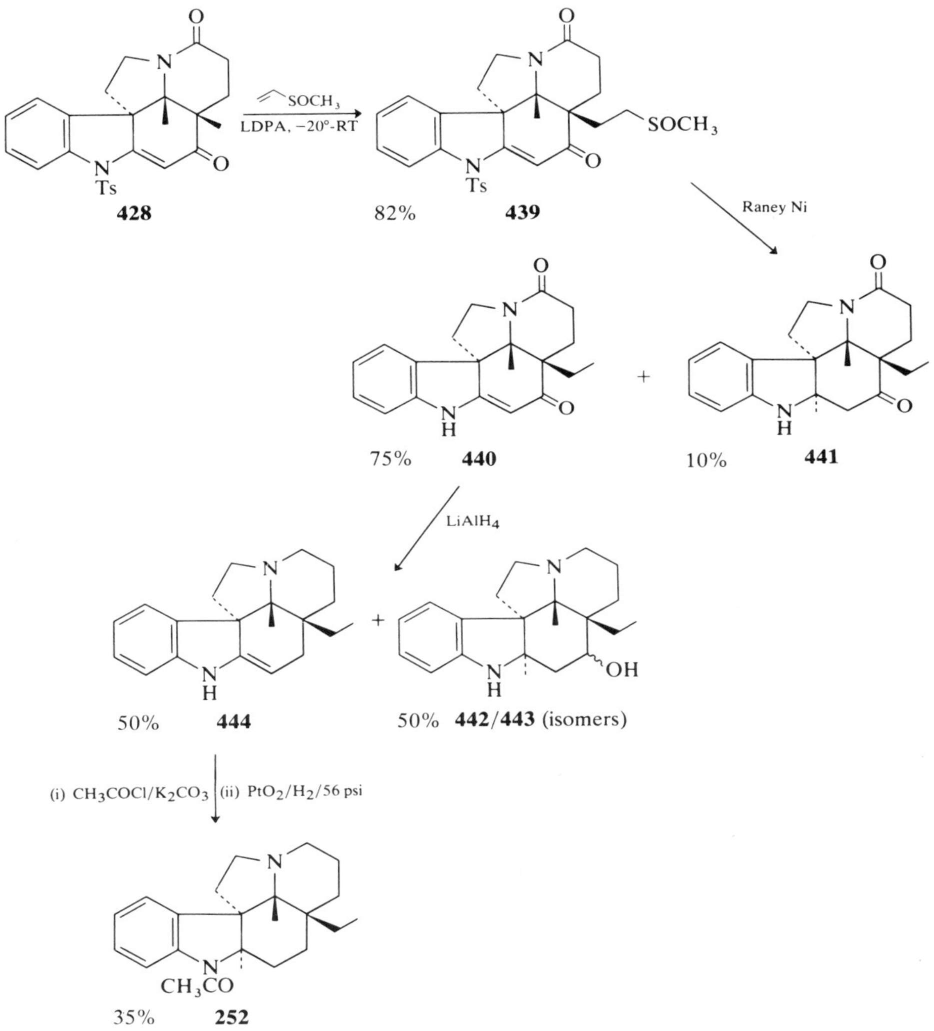

SCHEME 19. *Synthesis of* N-*acetylaspidospermidine* (**252**).

Lithium aluminum hydride reduction gave the hydroxy diastereoisomers **442** and **443** together with the enamine **444**, each in about 50% yield. Acetylation of **444** and high-pressure catalytic hydrogenation gave *N*-acetylaspidospermidine (**252**) in 35% overall yield (Scheme 19). The efficacy of this route in stereospecifically introducing the ethyl side chain is clearly established.

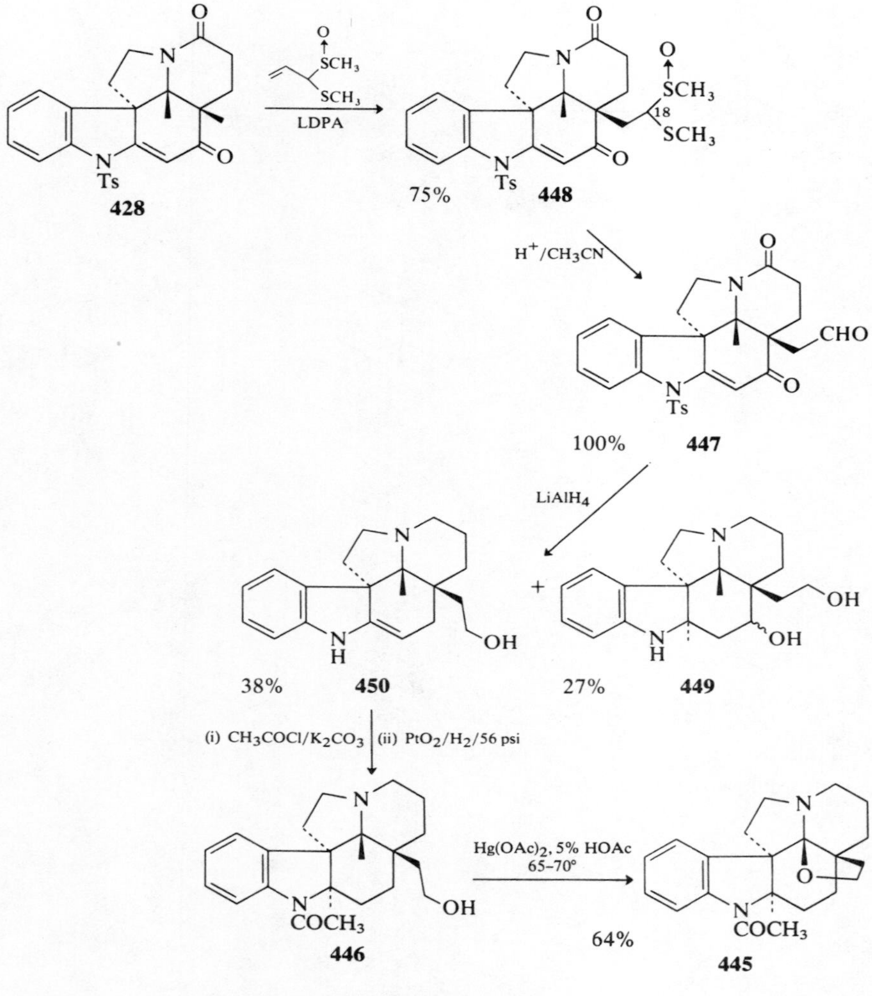

SCHEME 20. *Synthesis of* N-*acetylaspidoalbidine* (**445**).

7. Synthesis of *N*-Acetylaspidoalbidine (**445**)

Yet another variation of the *Aspidosperma* skeleton is that of fendleridine (aspidoalbidine) (*210*), in which the C-18 carbon is attached to C-21 via an ether bridge. By a route quite analogous to that used for *N*-acetylaspidospermidine (**252**), Ban has completed a synthesis of *N*-acetylaspidoalbidine (**445**) using deoxylimapodine (**446**) as the key intermediate (*211*).

The terminally oxygenated two-carbon side chain was inserted by Michael addition of ketene thioacetal monoxide to the now familiar vinylogous amide **428** followed by deacetalization to give the aldehyde **447** in 75% yield. In fact, the intermediate **448** existed as two separable isomers, but because each was converted to **447**, they were assumed to be isomers at C-18.

Reduction with excess lithium aluminum hydride gave the diol **449** and the enamine alcohol **450**. Reaction of the latter with acetyl chloride and aqueous sodium hydroxide solution gave the 1-acetyl derivative, which was reduced catalytically under pressure to deoxylimapodine (**446**) in almost quantitative yield. When this compound was heated with mercuric acetate in 5% acetic acid at 65–70°, *N*-acetylaspidoalbidine (**445**) was isolated in 64% yield (Scheme 20). The product was identical with an authentic sample (*212*).

8. Synthesis of (±)-Aspidofractinine (326)

The synthesis of 17-hydroxyaspidofractinine (**432**) was discussed previously, but attempts to transform this compound into the parent compound aspidofractinine (**326**) failed. Consequently, an alternative procedure (*213*) was needed to introduce the two-carbon unit to form the new ring.

The key reaction is the Diels–Alder addition of nitroethylene to the diene **451**, which, as expected, gave a product derived by closure of the most stable radicals.

Reduction of the vinylogous amide **428** with sodium borohydride gave an alcohol which could be dehydrated with anhydrous pyridine and phosphorus tribromide in 64% overall yield to give the desired diene, **451**. Treatment with nitroethylene at room temperature overnight afforded **452** in 80% yield as the result of regio- and stereoselective addition.

Catalytic reduction under pressure gave amine **453**, which was diazotized to afford a mixture of the alcohols **454** and **455**, and the olefin **456**. Lithium aluminum hydride reduction and catalytic hydrogenation completed the synthesis of (±)-aspidofractinine (**326**; Scheme 21) (*213*).

9. Synthesis of (±)-14,15-Dehydroaspidospermine (465)

Klioze and Darmory (*214*) have recently reported a synthesis of the tricyclic ketone **457** which differs from that obtained previously by Stork in that it is functionalized with a double bond at the prospective 14,15-position. Because it was considered likely that the third ring of the tricyclic

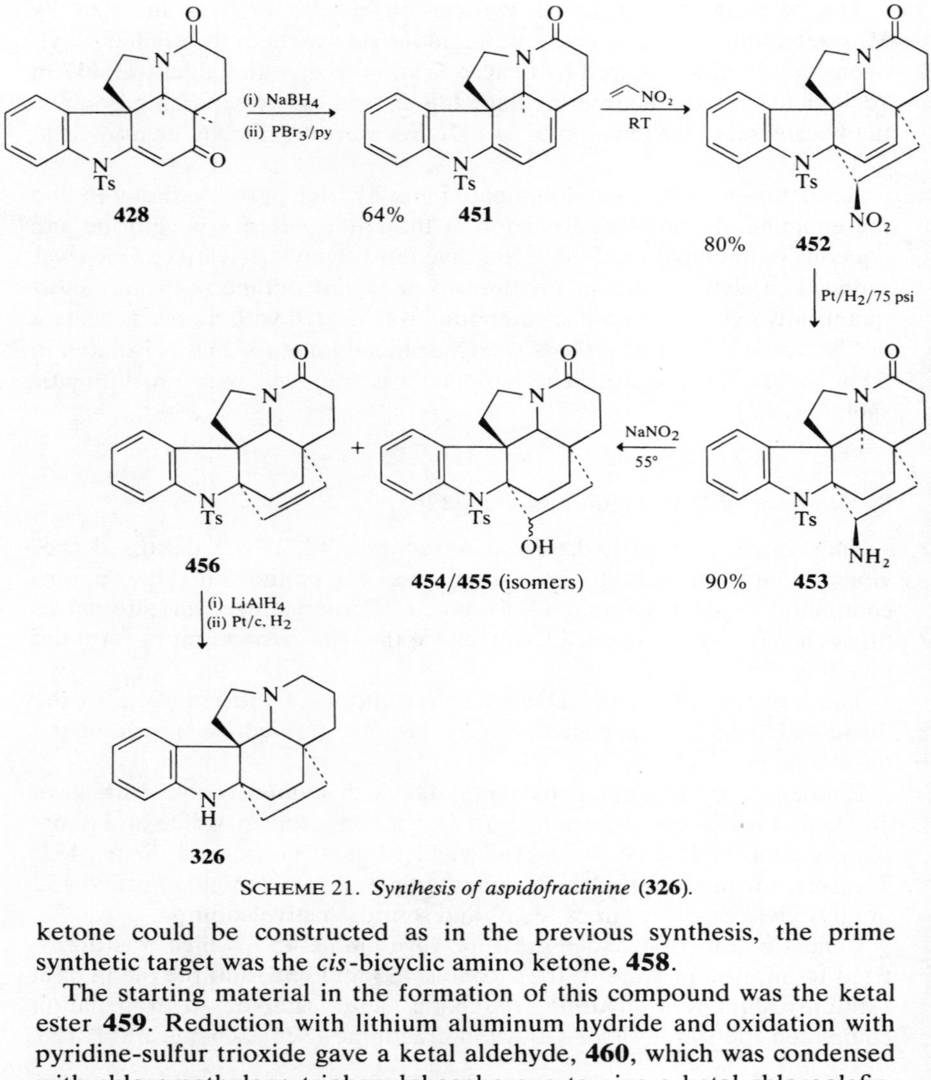

SCHEME 21. *Synthesis of aspidofractinine* (**326**).

ketone could be constructed as in the previous synthesis, the prime synthetic target was the *cis*-bicyclic amino ketone, **458**.

The starting material in the formation of this compound was the ketal ester **459**. Reduction with lithium aluminum hydride and oxidation with pyridine-sulfur trioxide gave a ketal aldehyde, **460**, which was condensed with chloromethylene triphenylphosphorane to give a ketal chloroolefin. Treatment with potassium *tert*-butoxide in 1:1 hexamethylphosphoramide:glyme at room temperature gave acetylene **461**. The lithium salt of the acetylene with excess formaldehyde gave an acetylenic alcohol which could be reduced catalytically to a 3:1 mixture of *cis: trans* allylic alcohols **462**. The hydroxy group was converted to an azide via the mesylate and sodium azide and this was followed by aluminum amalgam reduction to the amine. Hydrolysis of the ketal led to spontaneous cyclization of the *cis* isomer and formation of the required *cis*-bicyclic amino ketone, **458**. Following these synthetic procedures, **458** was treated with chloroacetyl chloride in the presence of triethylamine, and the chloroacetylamide was cyclized with potassium *tert*-butoxide to **463**. The stereochemistry of this intermediate was established by reduction of the double bond and comparison with the Stork–Dolfini keto lactam whose all-*cis* configuration had been confirmed by Ban (*215*). The amide functionality was reduced with lithium aluminum hydride after ketalization. Deprotection by acid hydrolysis gave the amino ketone **457**. Fischer cyclization of the *o*-methoxyphenyl hydrazone of **457** in refluxing acetic acid gave an indolenine (**464**) which was reduced with lithium aluminum hydride and acetylated to give 14,15-dehydroaspidospermine (**465**; Scheme 22).

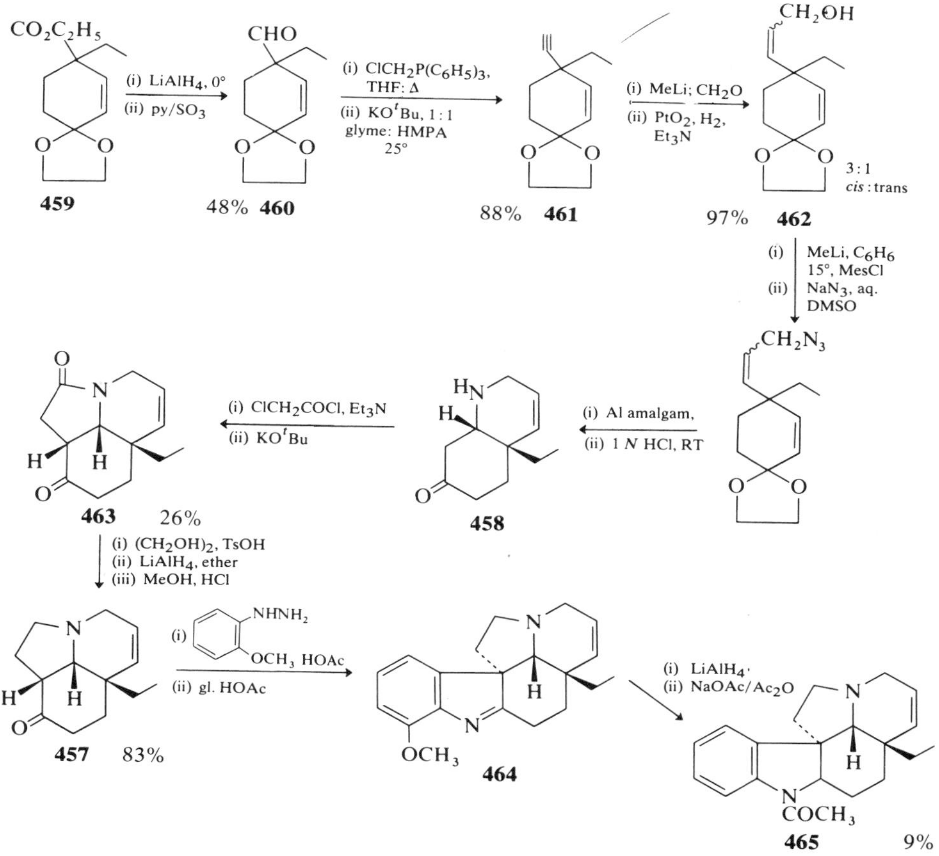

SCHEME 22. *Synthesis of* 14,15-*dehydroaspidospermine* (**465**).

10. Ziegler Synthesis of the Quebrachamine Skeleton

The simplest of all *Aspidosperma* alkaloids is quebrachamine (**82**), having only one stereochemical center and no carbomethoxy group. A number of syntheses of quebrachamine have been discussed previously in these volumes (see, e.g., Volume XI, p. 277).

Ziegler and co-workers (*216, 217*) have described a new approach to the quebrachamine skeleton involving internal cyclization of an acetic acid side chain to a 2-unsubstituted indole. The key intermediate is the disubstituted piperidine **466**, which is treated with indolyl-3-acetyl chloride in aqueous sodium carbonate–methylene chloride to afford a lactam ester, **467**.

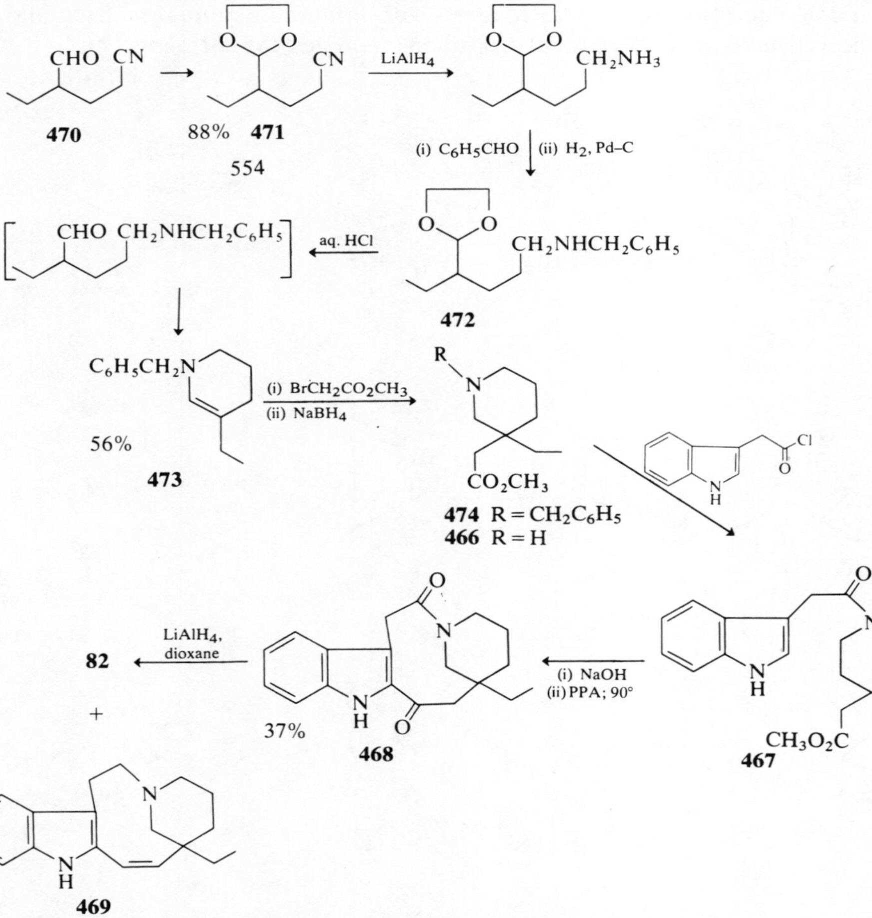

SCHEME 23. *Ziegler synthesis of quebrachamine* (**82**).

Hydrolysis with alkali and heating the resultant acid with polyphosphoric acid gave the keto lactam (**468**) having the quebrachamine skeleton.

Reduction of the lactam ketone **468** with lithium aluminum hydride in dioxane gave a low (6%) yield of quebrachamine (**82**) and 16,17-dehydroquebrachamine (**469**) in 57% yield (Scheme 23).

The latter compound displayed a molecular ion at *m/e* 280, two mass units less than quebrachamine, but the UV spectrum indicated only a dialkylindole. Therefore, the double bond must be considerably out of the plane of the indole nucleus (*217*).

The piperidine was prepared from 2-(2-cyanoethyl)-butryraldehyde (**470**) via ketal **471**, which was reduced with lithium aluminum hydride and reductively alkylated with benzaldehyde over 10% palladium on charcoal to **472**. Acid hydrolysis led to spontaneous cyclization and formation of enamine **473**. Treatment of this enamine with methyl bromoacetate and reduction of the iminium species with sodium borohydride gave the 1-benzyl-3,3-disubstituted piperidine **474** which was debenzylated with palladium charcoal under acidic conditions to give the desired piperidine, **466** (*216*).

11. Synthesis of 14,15-Dehydroquebrachamine (475)

Prior to their synthesis of tabersonine (**28**) (*218, 219*) (see next section) Ziegler and Bennett had used all the early stages in a synthesis of 14, 15-dehydroquebrachamine (**475**) (*220*). The only difference was a change in the solvent used for the reduction of the keto lactam **476**. With lithium aluminum hydride in refluxing dioxane, elimination of water occurred to give 14,15,16,17-didehydroquebrachamine (**477**) in addition to **475**. The relative proportions of these materials from the reductive process were not disclosed.

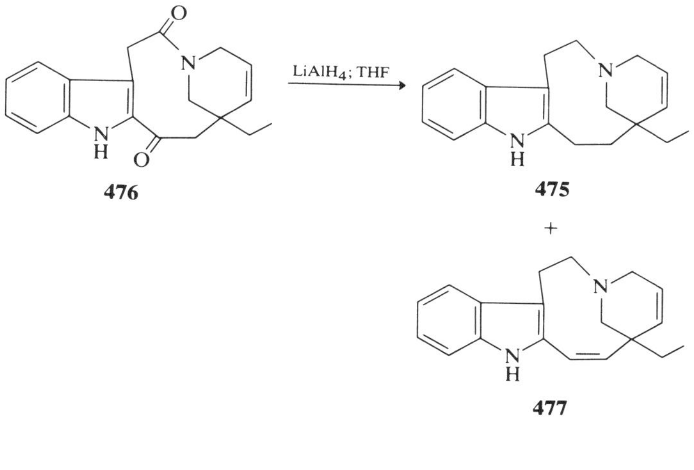

$LiAlH_4$; THF

476 → **475** + **477**

12. Ziegler–Bennett Synthesis of Tabersonine (**28**)

Tabersonine (**28**), another β-anilinoacrylate derivative, has been postulated to be a key biosynthetic precursor of all the *Aspidosperma* alkaloids and possibly also the iboga alkaloids. It is therefore a key compound for synthetic endeavors, but any pctential synthesis of tabersonine poses an interesting problem because of the 14,15-double bond. Thus a biogenetic-type synthesis would involve the acrylic ester **478**, the very ester postulated as the *biosynthetic* precursor of the *Aspidosperma* and iboga alkaloids. To date, efforts to produce this highly reactive ester have failed.

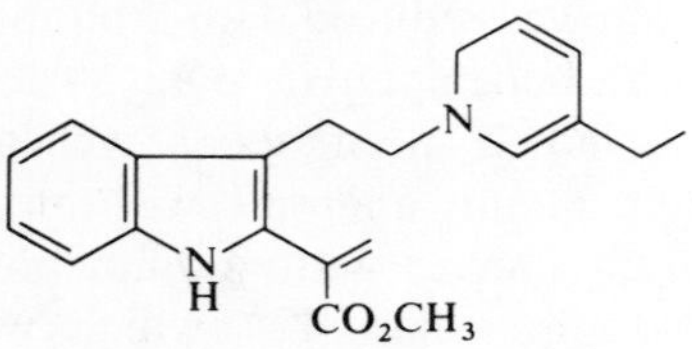

478

The key steps in the Ziegler–Bennett synthesis (*218, 219*) of tabersonine (**28**) are the Claisen rearrangement of the allylic alcohol **479** to the amino ester and transannular cyclization of the amino alcohol **481** to the quaternary mesylate **482** followed by cyanolysis.

The amino alcohol **479** was prepared by a somewhat circuitous route. Treatment of this allylic alcohol with ethyl orthoacetate in the presence of pivalic acid at 140° gave the Claisen rearrangement product **480** in 74% yield. Debenzylation was achieved with ethyl chloroformate in refluxing benzene to give the carbonate ester **483**. Hydrolysis and decarboxylation, followed by reesterification and N-acylation with indole-3-acetyl chloride gave the amide ester **484**. Hydrolysis of **484** and condensation with polyphosphoric acid at 85° gave the unsaturated tetracyclic keto lactam **476**. Reduction with lithium aluminum hydride in THF at reflux gave, as expected, the mixture of alcohols **481** and 14,15-dehydroquebrachamine (**475**). The structure of the latter compound was proved by reduction to quebrachamine (**82**). The mixture of C-16 alcohols (**481**) underwent transannular cyclization with methanesulfonyl chloride/pyridine to give the quaternary mesylate **482**. Cyanide cleavage of compounds of this type is extremely complex and rarely gives clean products. The major product from the cyanolysis of the quaternary mesylate **482** was a mixture of several nitriles, two of which were the diastereomeric nitriles at C-16 (**485**). The nitrile in which the C-16 proton was beta was hydrolyzed and esterified, and the resulting ester was cyclized with platinum–oxygen to give racemic tabersonine (**28**; Scheme 24) (*218, 219*).

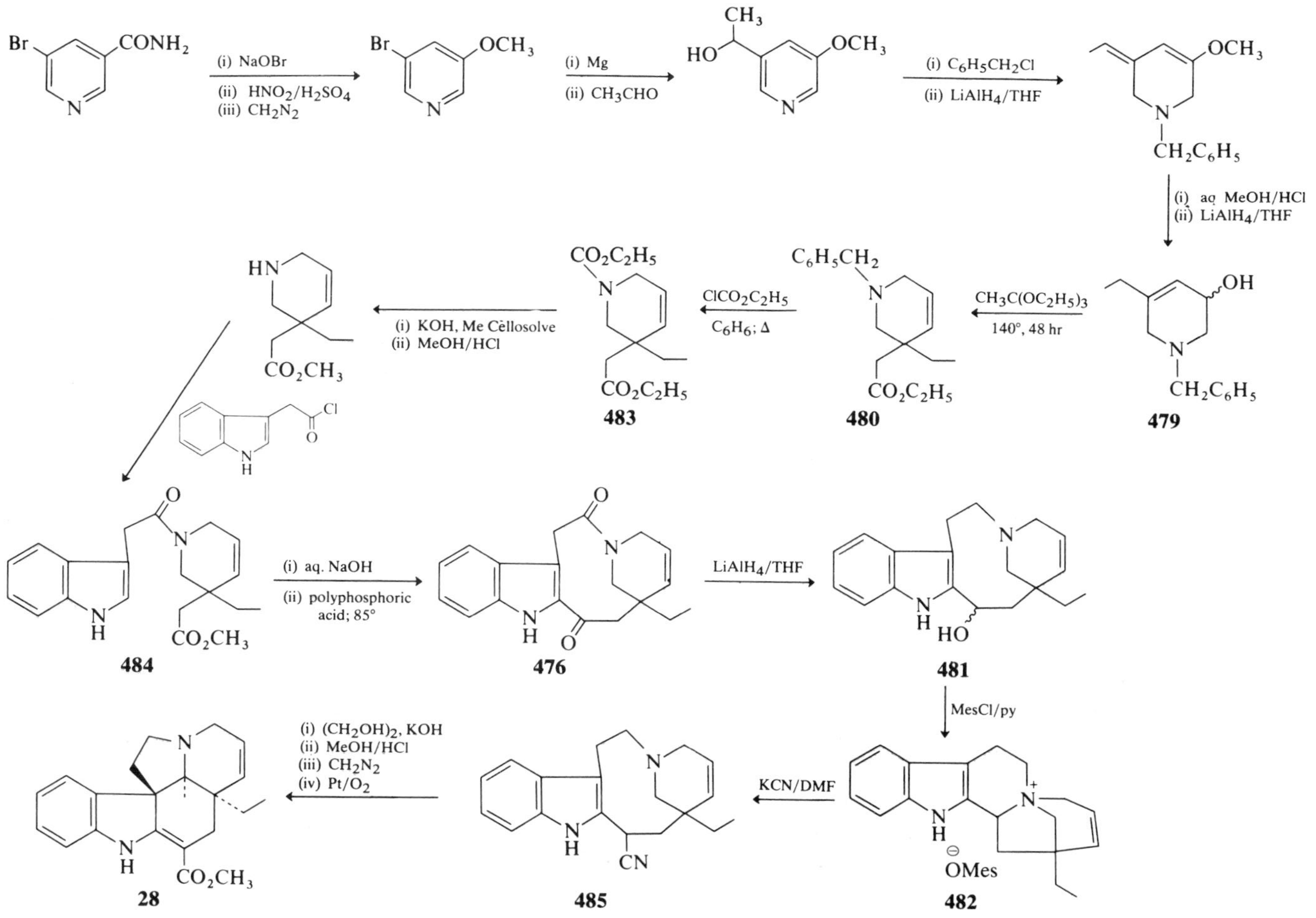

(i) NaOBr
(ii) HNO_2/H_2SO_4
(iii) CH_2N_2

(i) Mg
(ii) CH_3CHO

(i) $C_6H_5CH_2Cl$
(ii) $LiAlH_4/THF$

(i) aq MeOH/HCl
(ii) $LiAlH_4/THF$

479

$CH_3C(OC_2H_5)_3$
140°, 48 hr

480

$ClCO_2C_2H_5$
C_6H_6; Δ

483

(i) KOH, Me Cellosolve
(ii) MeOH/HCl

484

(i) aq. NaOH
(ii) polyphosphoric acid; 85°

476

$LiAlH_4/THF$

481

MesCl/py

482

KCN/DMF

485

(i) $(CH_2OH)_2$, KOH
(ii) MeOH/HCl
(iii) CH_2N_2
(iv) Pt/O_2

28

SCHEME 24. *Synthesis of tabersonine* (**28**).

13. Takano Synthesis of Quebrachamine (**82**) and Tabersonine (**28**)

Recently, Takano and co-workers (*221*) have reported an improved synthesis of quebrachamine (**82**) and of 16-cyano-14,15-dehydroquebrachamine (**485**), an intermediate in the synthesis of tabersonine (**28**) (*218, 219*). The synthesis was aimed at improving the formation of the two quaternary mesylate salts **486** and **482**. The synthesis of mesylate **486** is the most direct procedure and is described first.

The ethylene ketal of 4-carbethoxycylohexanone (**487**) was converted to the keto ester **488** by standard methods and then transformed to the α-diketone monothioketal **489** by Woodward's procedure (*222*). Cleavage of **489** with sodium hydride (*223*) and DCC condensation of tryptamine with the resultant dithianyl half-ester gave the dithianyl amide **490** in 60% yield from **489**. Hydrolysis of the protecting group allowed spontaneous cyclization to give a mixture of stereoisomeric lactams (**491**). All the subsequent transformations to the mesylate **486** were carried out on the separated isomer. Reduction of **491** in refluxing THF with lithium aluminum hydride gave an amino alcohol which upon mesylation and heating in chloroform afforded the quaternary mesylate salt **486**. Reduction with sodium in liquid ammonia gave (±)-quebrachamine (**82**) in 20% overall yield from **487** (*221*).

The 14,15-double bond present in tabersonine was introduced into the β-ethyl isomer of lactam **491** by treatment with LDA–diphenyl disulfide followed by oxidation and elimination to give the α,β-unsaturated lactam **492**. Murphy's law operated at this point, for upon lithium aluminum hydride reduction, **492** gave only a 5% yield of the amino alcohol **493**. An alternative procedure was therefore developed. Hydrolysis gave a carboxylic acid, and treatment with ethyl chloroformate and triethylamine followed by sodium borohydride in aqueous THF gave a lactam alcohol **494**, which after silylation was reduced with lithium aluminum hydride to amino alcohol **493** in 58% overall yield from **492**. Mesylation and elimination of HCl in refluxing chloroform gave the unsaturated mesylate salt **482**. The same salt was prepared previously by Ziegler and Bennett (*218, 219*). Cyanolysis of **482** gave 16-cyano-14,15-didehydroquebrachamine (**485**) in 27% yield (*221*), which can be readily converted to tabersonine (**28**; Scheme 25).

14. Second Takano Synthesis of Quebrachamine (**82**)

Takano and co-workers (*224*) have also reported a new synthetic approach to the cleavamine and quebrachamine skeleta via a thio-Claisen rearrangement. The synthesis begins with the previously described lactam

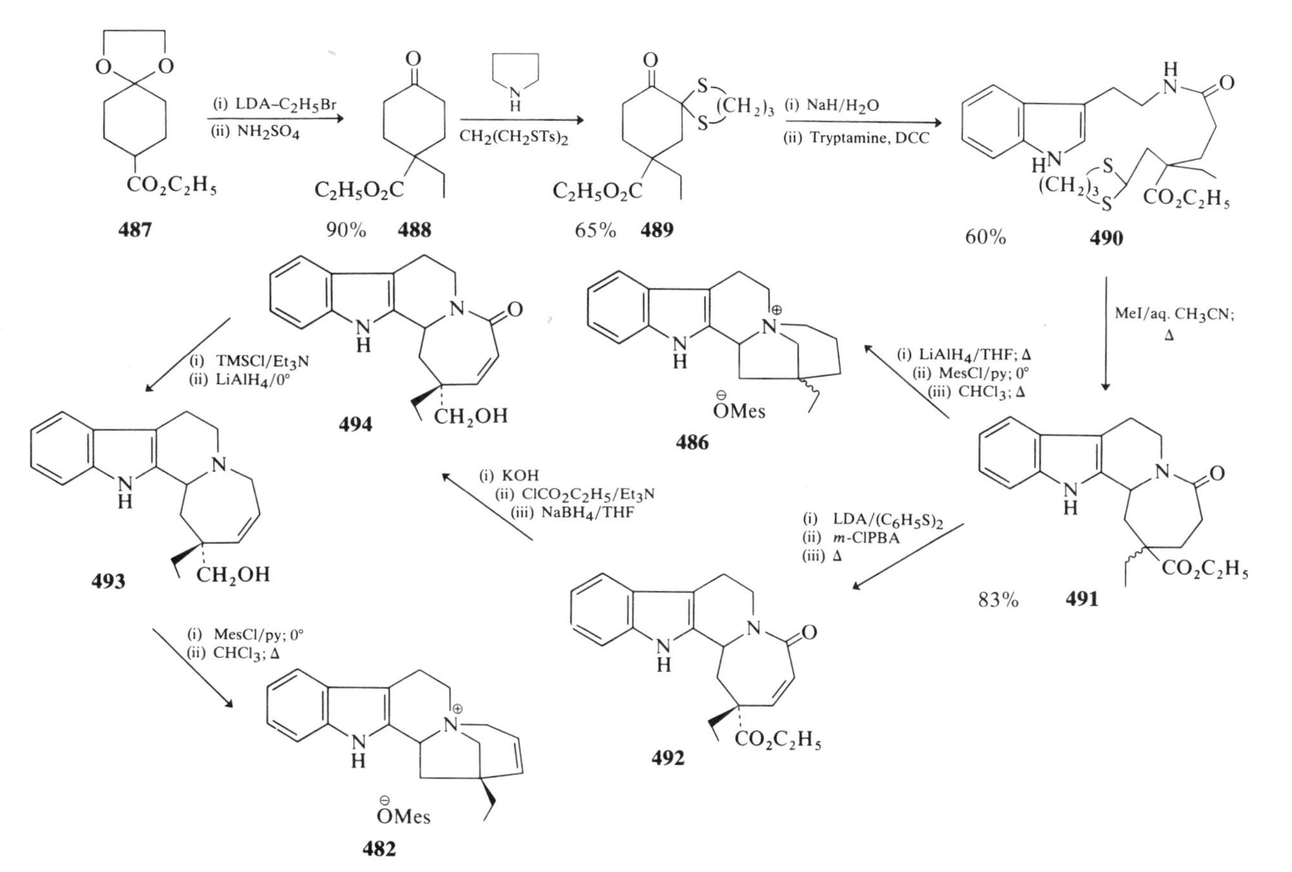

SCHEME 25. *Takano synthesis of quebrachamine* (**82**) *and tabersonine* (**28**).

495, which is converted to the thiolactam **496** and then to the sulfonium bromide **497** by treatment with 2-bromomethylbut-1-ene. Reaction of the salt **497** with potassium *tert*-butoxide gave a mixture of ethyl isomers **498**. Hydroboration of each isomer in THF under reflux permitted reductive desulfurization, and oxidative hydrolysis gave alcohols **499** and **500**, respectively. The stereochemical integrity of the hydroboration reaction should be noted. By means of the standard procedure developed by Kutney and co-workers (*225*), the alcohol **500** was converted to the dihydrocleavamine **233**.

Reaction of the ethyl thiolactam **501** with allyl bromide gave the salt **502** which on base-catalyzed rearrangement afforded **503** and a mixture of isomers. Hydroboration and alkaline oxidation of each isomer gave a mixture of alcohols (**504** and **505**) which were converted to quebrachamine (**82**) in the standard manner (Scheme 26).

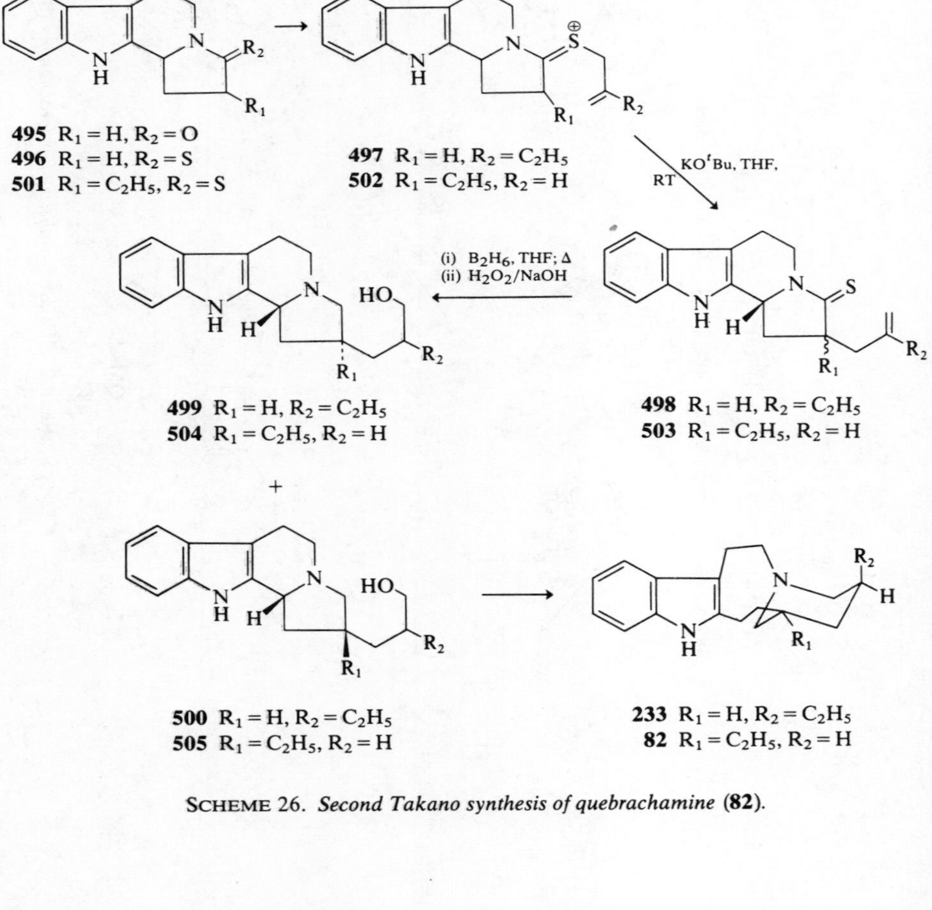

495 $R_1 = H, R_2 = O$
496 $R_1 = H, R_2 = S$
501 $R_1 = C_2H_5, R_2 = S$

497 $R_1 = H, R_2 = C_2H_5$
502 $R_1 = C_2H_5, R_2 = H$

KO^tBu, THF, RT

498 $R_1 = H, R_2 = C_2H_5$
503 $R_1 = C_2H_5, R_2 = H$

(i) B_2H_6, THF; Δ
(ii) H_2O_2/NaOH

499 $R_1 = H, R_2 = C_2H_5$
504 $R_1 = C_2H_5, R_2 = H$

+

500 $R_1 = H, R_2 = C_2H_5$
505 $R_1 = C_2H_5, R_2 = H$

233 $R_1 = H, R_2 = C_2H_5$
82 $R_1 = C_2H_5, R_2 = H$

SCHEME 26. *Second Takano synthesis of quebrachamine* (**82**).

15. Synthesis of (±)-Minovine (**506**)

The intermediacy of a C_7-C_3 acrylic ester in the formation of the *Aspido-sperma* alkaloids, although not established *in vivo* with certainty, has a number of biogenetic attractions. Not surprisingly, therefore, attempts have been made to mimic this Diels–Alder reaction *in vitro*. One of the syntheses (*226, 227*) to have been developed along these lines is that of minovine (**506**), which may be regarded as *N*-methylvincadifformine.

The two critical components required for the Diels–Alder reaction were considered to be the tetrahydropyridine **507** and the acrylic ester **508**. The acrylic ester was prepared by an unexceptional route from *N*-methylindole (**509**). Addition of lithiated *N*-methylindole to ethyl oxalate in ether and base hydrolysis gave a glyoxylic acid which was converted to its methyl ester (**510**) with diazomethane. Treatment of this ester with methylene triphenylphosphonium bromide and butyllithium under reflux gave the acrylate **508** in 40% overall yield.

The tetrahydropyridine **507** and the acrylate **508** were heated in methanol to afford a mixture of two tetracylic bases (**511**) which were separated by fractional crystallization. By deuterium exchange it was deduced that these bases were epimers, differing only in the stereochemis-try of the ester group. The stereochemistry of the ring junction was established spectroscopically. The α-ester proton in the least ther-modynamically stable isomer appeared as a doublet of doublets ($J = 7$ and 10 Hz). In the hydrochloride of this ester, the α-ester proton appeared as a triplet ($J = 9$ Hz), indicating a change in conformation of ring C from a half-chair to a half-boat. The latter conformation was preferred by the hydrogen bonding between the ester carbonyl and the protonated nitrogen. These data necessitate a *cis*-fused C-D ring junction which can also be rationalized by the possible transition states leading to the tetracyclic products.

The two amines were efficiently debenzylated with Pd/C in acidic methanol at room temperature, and the remaining tryptamine bridge carbons inserted by double alkylation at the indole β-position and the secondary amine. Deprotonation of the resulting indolenine gives the β-anilinoacrylate, minovine (**506**; Scheme 27) (*226, 227*).

16. Büchi Syntheses of Vindorosine (**99**) and Vindoline (**101**)

The ultimate goal of the syntheses of compounds with the *Aspidosperma* nucleus is vindoline (**101**). One component of many of the dimeric indole alkaloids of *Catharanthus roseus*, vindoline (**101**) is a structural component of vincaleukoblastine, the powerful oncolytic agent, and *N*-demethyl-*N*-

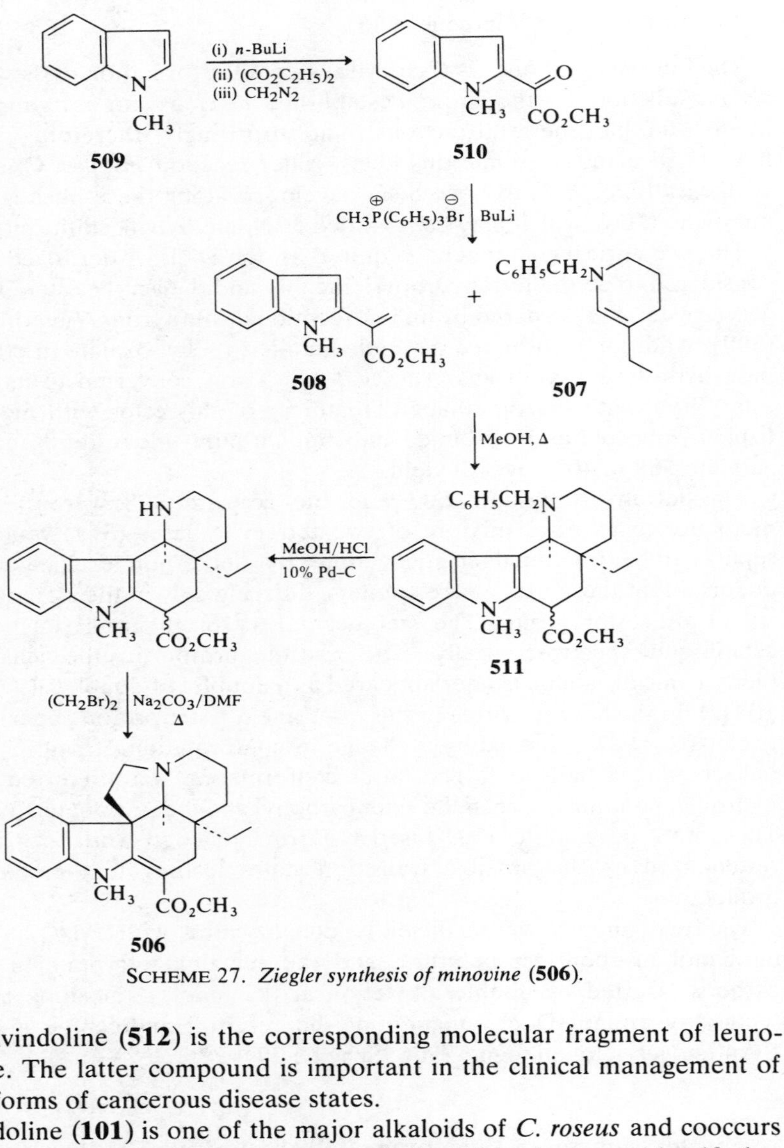

SCHEME 27. *Ziegler synthesis of minovine* (**506**).

formylvindoline (**512**) is the corresponding molecular fragment of leurocristine. The latter compound is important in the clinical management of many forms of cancerous disease states.

Vindoline (**101**) is one of the major alkaloids of *C. roseus* and cooccurs with its demethoxy analog, vindorosine (**99**). Although not yet identified as a component of any of the dimeric species, vindorosine (**99**) was considered for synthesis as a model for the slightly more elaborate vindoline (**101**). As the syntheses developed, unexpected complications arose when the route developed for vindorosine (**99**) was attempted for vindoline (**101**). These

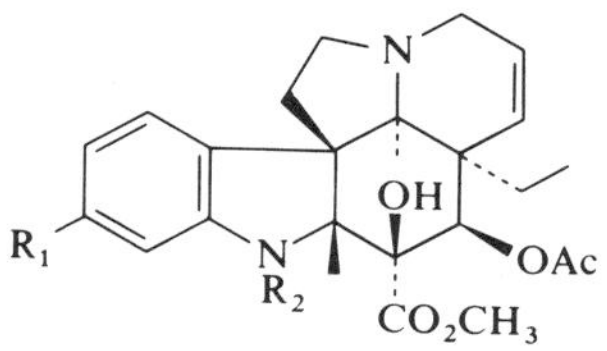

99 $R_1 = H, R_2 = CH_3$
101 $R_1 = OCH_3, R_2 = CH_3$
512 $R_1 = OCH_3, R_2 = CHO$

complications will be discussed at a later point. The first synthesis described will be that of vindorosine (*228*).

The basic strategy of the synthesis involves sequential building of the A, B, C, D, and E rings. In the crucial intermediate the D and E rings are functionalized at two points, a ketone at C-17 and a 15,20-double bond, permitting elaboration of this nucleus with the remaining functional groups with excellent regio- and stereospecificity.

Treatment of *N*-methyltryptamine (**513**) with 1-chloro-3-ketobut-1-ene (**514**) under basic conditions gave an enamino ketone (**515**) in 92% yield which existed in the *cis* configuration. This compound was resistant to cyclization attempts, but the *N*-acetyl-*trans* derivative **516** was cyclized in low yield (38%) to a tetracyclic dihydroindole which was hydrolyzed under acidic conditions to the corresponding amine **517** in 92% yield. This compound is a β,β'-diamino ketone and therefore has two positions potentially susceptible to epimerization. On the basis of molecular models, it was concluded that the most stable diastereoisomer was that shown. Treatment of **517** with acrolein under basic conditions gave the crucial pentacyclic ketone **518** in 40% yield, now beautifully functionalized for the remaining steps. The α,β-unsaturated ketone was stereospecifically alkylated at the α-position (C-20) with ethyl iodide under strongly basic conditions to give the β,γ-unsaturated ketone **519** in 44% yield. In the PMR spectrum of both vindoline and vindorosine, the methyl of the ethyl group is observed at 0.4 ppm because of shielding due to the benzene ring. The ketone **519** also showed the methyl triplet at 0.4 ppm, in agreement with an α-stereochemistry for the ethyl group.

The subsequent reactions are concerned solely with elaboration of the C-ring. The ketone **519** was carbomethoxylated at C-16 with dimethyl carbonate and sodium hydride to give the keto ester **520**. Oxidation with oxygen/hydrogen peroxide in the presence of potassium *tert*-butoxide at −35° gave stereo-specifically, in 31% yield, the α-hydroxy β-keto ester **521**. The key to the assignment of the stereochemistry of the hydroxy group was the observation of hydrogen bonding analogous to that found in the IR spectrum of vindoline (**101**). This reaction is assumed to proceed by peroxide anion attack at C-17 and subsequent nucleophilic attack by a carbanion at C-16 on the peroxy group to give stereospecifically a hydroxy epoxide intermediate (**522**) which, under basic conditions, is isomerized to the hydroxy ketone **523**. Partial reduction with lithium aluminum hydride

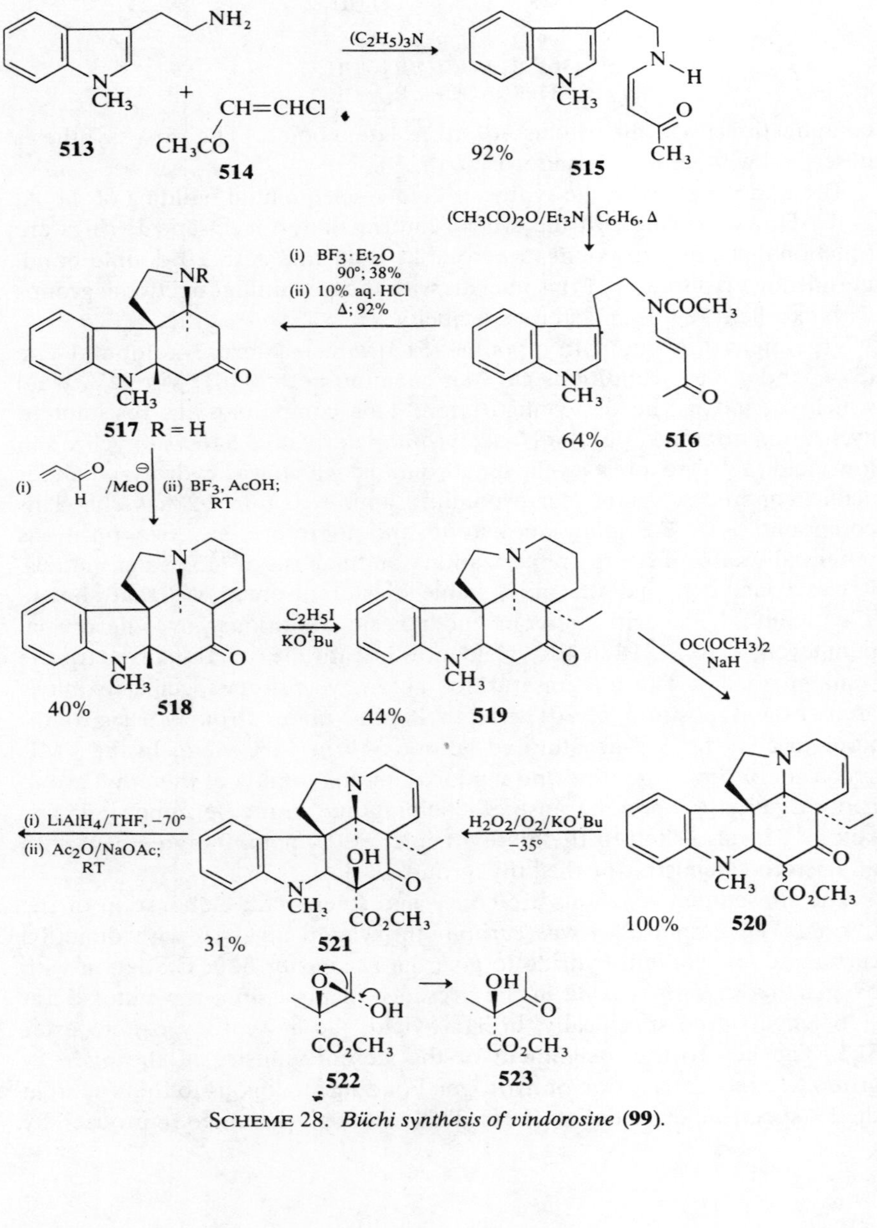

SCHEME 28. *Büchi synthesis of vindorosine* (**99**).

in THF at $-70°$ gave a mixture of diols. Acetylation of the more polar of these afforded a monoacetate derivative identical with vindorosine (**99**; Scheme 28) (*228*).

Büchi and co-workers (*229*) next attempted to carry out an analogous route in order to prepare vindoline (**101**) itself. However, the reaction sequence failed at an early stage in the 6-methoxytryptamine series when it was found that the vinylogous amide **524** cyclized in only 9% yield to the desired tetracyclic dihydroindole **525**. Several groups were examined in place of methoxy, but the most efficacious was the tosyloxy group. Cyclization of the *trans*-acetamide **526** gave the desired dihydroindole **527** in 89% yield. The protecting groups were removed sequentially. Base hydrolysis afforded a phenol which was methylated with dimethyl sulfate/potassium carbonate in acetone. Removal of the *N*-acetyl group was accomplished in 82% yield with triethyloxoniumfluoroborate/sodium bicarbonate in methylene chloride to give diamine **528**. Subsequent steps in the synthesis were analogous to those in the vindorosine synthesis up to the point of reducing the ketone at C-17. Using sodium bis(2-methoxyethoxy)aluminum hydride at $-20°$ with prior addition of aluminum chloride gave stereospecifically the C-17 β-hydroxy isomer in 56% yield. Complexation of the C-16 hydroxy group and N-4 with aluminum chloride explains the failure of hydride to attack from the β-face of the molecule. Acetylation of this alcohol produced vindoline (**101**; Scheme 29) (*229*).

17. Kutney Approach

A previous chapter in this series on this topic discussed the details of the synthetic procedures used by Kutney and co-workers for the synthesis of the *Aspidosperma* and iboga series of compounds (Volume XI., p. 205). These approaches have subsequently been published in full detail (*105, 225, 230–232*).

The general techniques used were also applied to the synthesis of natural alkaloids in the quebrachamine and *Aspidosperma* series (*233*). The oxidative cyclization step in the previous step proceeded in only 30–40% yield. In order to improve this process, a more reactive intermediate was used. Condensation of tryptamine with the highly reactive aldehyde–ester **529** gave the lactam **530** as a mixture of C-3 isomers in 90% yield. Lithium aluminum hydride reduction, catalytic reduction, and mesylation gave the familiar mesylate salt **486**. Nucleophilic attack by cyanide ion in dimethylformamide produced the nine-membered ring intermediates **531** and **532**. Hydrolysis and methylation with diazomethane gave vincadine (**185**) and 16-epivincadine (**189**).

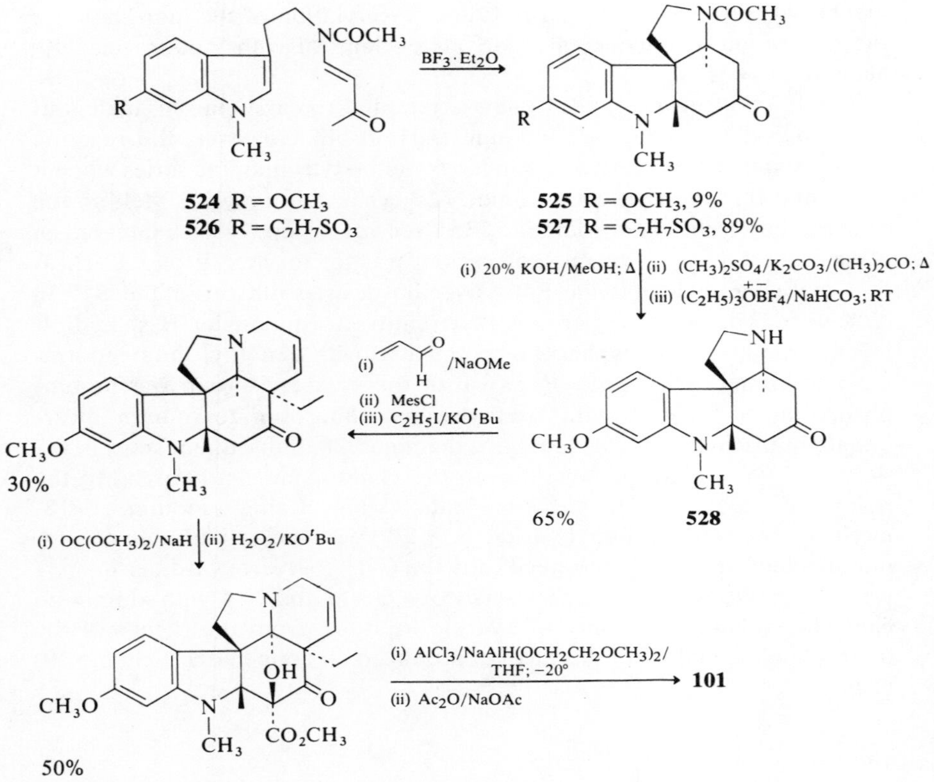

$BF_3 \cdot Et_2O$

524 R = OCH_3
526 R = $C_7H_7SO_3$

525 R = OCH_3, 9%
527 R = $C_7H_7SO_3$, 89%

(i) 20% KOH/MeOH; Δ (ii) $(CH_3)_2SO_4/K_2CO_3/(CH_3)_2CO$; Δ
(iii) $(C_2H_5)_3\overset{+}{O}\overset{-}{B}F_4/NaHCO_3$; RT

65% **528**

(i) acrolein/NaOMe
(ii) MesCl
(iii) C_2H_5I/KO^tBu

30%

(i) $OC(OCH_3)_2/NaH$ (ii) H_2O_2/KO^tBu

50%

(i) $AlCl_3/NaAlH(OCH_2CH_2OCH_3)_2/$THF; −20°
(ii) $Ac_2O/NaOAc$ → **101**

SCHEME 29. *Büchi synthesis of vindoline* (**101**).

Cyclization to the pentacyclic *Aspidosperma* series could be accomplished by two alternative routes: reaction with mercuric acetate or oxygen in the presence of a platinum catalyst; in this way, vincadifformine (**2**) and minovine (**506**) were produced (Scheme 30).

The reaction scheme was extended to the methoxytryptamine series to afford racemic 11-methoxyvincadifformine (**4**) and 11-methoxyminovine (**533**). At the time, neither of these alkaloids was known naturally. It is important to note the stereospecificity in the transannular cyclization process, which apparently only gave the natural C/D *cis* stereochemistry.

18. Kutney Synthesis of Vindoline (**101**)

A second synthesis of vindoline (**101**) was recently completed by Kutney and co-workers (*234*). The early steps in the synthesis, those leading to

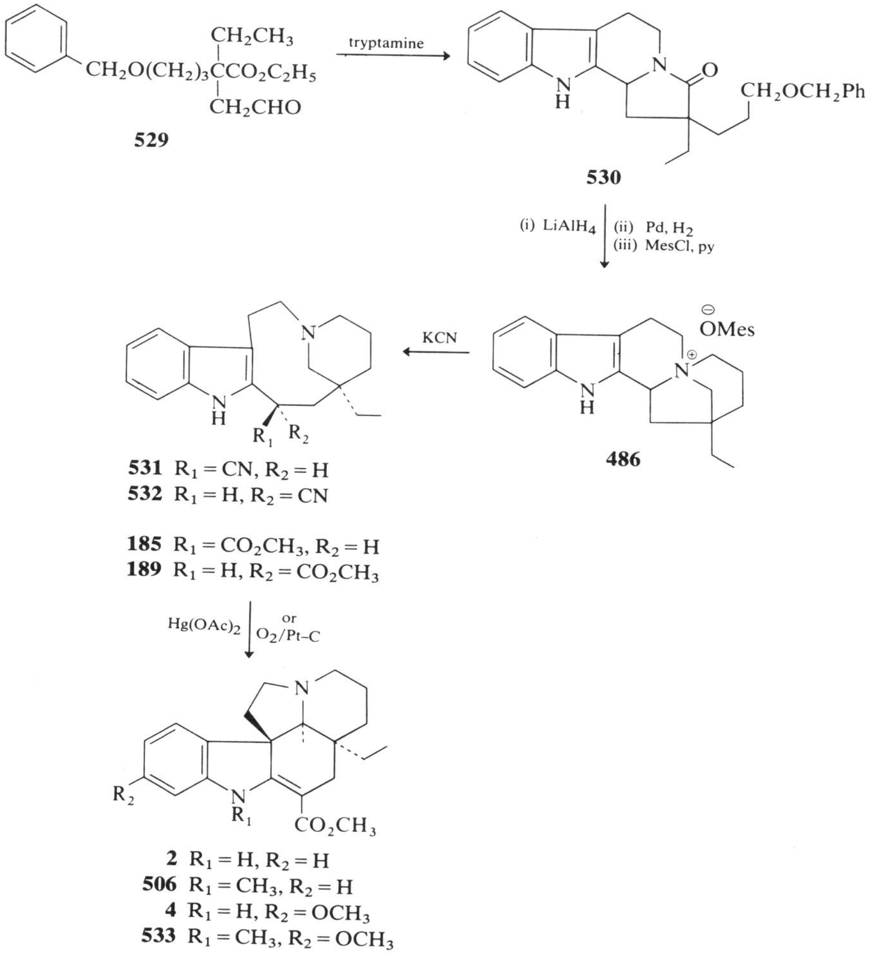
CH₂CH₃
CH₂O(CH₂)₃CCO₂C₂H₅
CH₂CHO
529
tryptamine
CH₂OCH₂Ph
530
(i) LiAlH₄
(ii) Pd, H₂
(iii) MesCl, py
OMes
486
KCN
531 R₁ = CN, R₂ = H
532 R₁ = H, R₂ = CN
185 R₁ = CO₂CH₃, R₂ = H
189 R₁ = H, R₂ = CO₂CH₃
Hg(OAc)₂
or
O₂/Pt–C
CO₂CH₃
2 R₁ = H, R₂ = H
506 R₁ = CH₃, R₂ = H
4 R₁ = H, R₂ = OCH₃
533 R₁ = CH₃, R₂ = OCH₃

SCHEME 30. *Kutney synthesis of vincadifformine* (**2**) *and minovine* (**506**).

533, were described in Section IV,A,17, and it remains only to discuss the conversion of this compound to **101**. Amazingly, the key intermediate in the synthesis is dihydrovindoline (**534**), and the final steps involve the introduction of the 14,15-double bond.

The β-keto ester **535** available from **533** is α-hydroxylated under basic conditions using the method of Büchi *et al.* (*228*). Low-temperature reduction gave the secondary alcohol which was then acetylated to afford dihydrovindoline (**534**).

Reaction of dihydrovindoline (**534**) with mercuric acetate in refluxing dioxane gave in one step the ether lactam **536**. The intermediates in this interesting reaction are the six-membered lactam and the α,β-unsaturated lactam, which then undergoes Michael attack at the β-position to afford **536**. The ether bridge was opened by removal of a proton alpha to the lactam carbonyl with the anion of triphenylmethane, and the tertiary hydroxyl group was acetylated to afford **537**. The lactam carbonyl was removed by conversion to the imino ether followed by borohydride reduction. Treatment of the product with moist silica gel gave vindoline (**101**; Scheme 31).

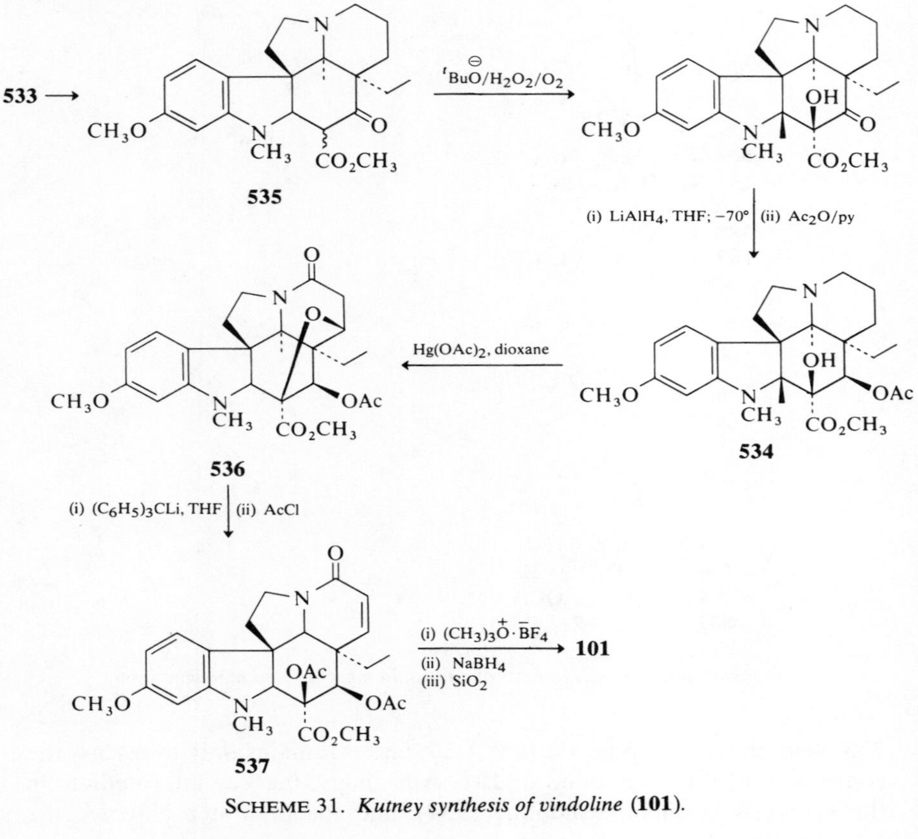

533 →

535

$^tBuO^{\ominus}/H_2O_2/O_2$

(i) $LiAlH_4$, THF; −70° (ii) Ac_2O/py

534

$Hg(OAc)_2$, dioxane

536

(i) $(C_6H_5)_3CLi$, THF (ii) AcCl

537

(i) $(CH_3)_3\overset{+}{O}\cdot\overset{-}{B}F_4$ (ii) $NaBH_4$ (iii) SiO_2 → **101**

SCHEME 31. *Kutney synthesis of vindoline* (**101**).

19. Synthesis of the Reduction Product of Vincatine (163)

The structure elucidation of vincatine (**163**) was described in Section II,F,2. Harley-Mason and co-workers (*202*) have reported a synthesis of a reduction product of vincatine, the diol **538**. The hydrochloride of 2-hydroxy-1-methyltryptamine (**539**), on treatment with dimethyl 4-ethyl-4-formylpimelate (**540**) (*164*) in buffered aqueous ethanol under reflux, gave the diamide 3-oxovincatine (**541**) as a mixture of C-20 diastereoisomers. Reduction with lithium aluminum hydride gave a mixture of diols (**538**), one of which was identical with a diol obtained by lithium aluminum hydride reduction of vincatine (**163**).

NH_2 CO_2CH_3 CHO CH_3 **539** CO_2CH_3 **540** NaOAc aq. EtOH; Δ O CO_2CH_3 CH_3 **541** $LiAlH_4$, ether OH CH_3 CH_2OH **538**

20. Le Men Synthesis of Aspidospermidine (249) and Vincadifformine (2)

The previous syntheses of aspidospermidine (**249**) and vincadifformine (**2**) had been quite long and perhaps unduly concerned with carefully creating the C/D/E ring stereochemistry. This led to an increase in the number of overall steps.

Utilizing a route similar to that used by Harley-Mason in the synthesis of the reduced product from vincatine, the French group has successfully extended this work to a more select group of alkaloids (*235, 236*).

Treatment of 2-hydroxytryptamine (**406**) with dimethyl 4-ethyl-4-formylpimelate (**540**) gave two isomeric oxindoles (**542** and **543**) differing in the stereochemistry of the ethyl group. The most polar product (**542**) was heated for 2 hr under nitrogen in the presence of polyphosphoric acid to afford 3-oxo-1,2-dehydroaspidospermidine (**504**) in 68% yield (Scheme 32). The stereochemistry of this polar product must therefore be that

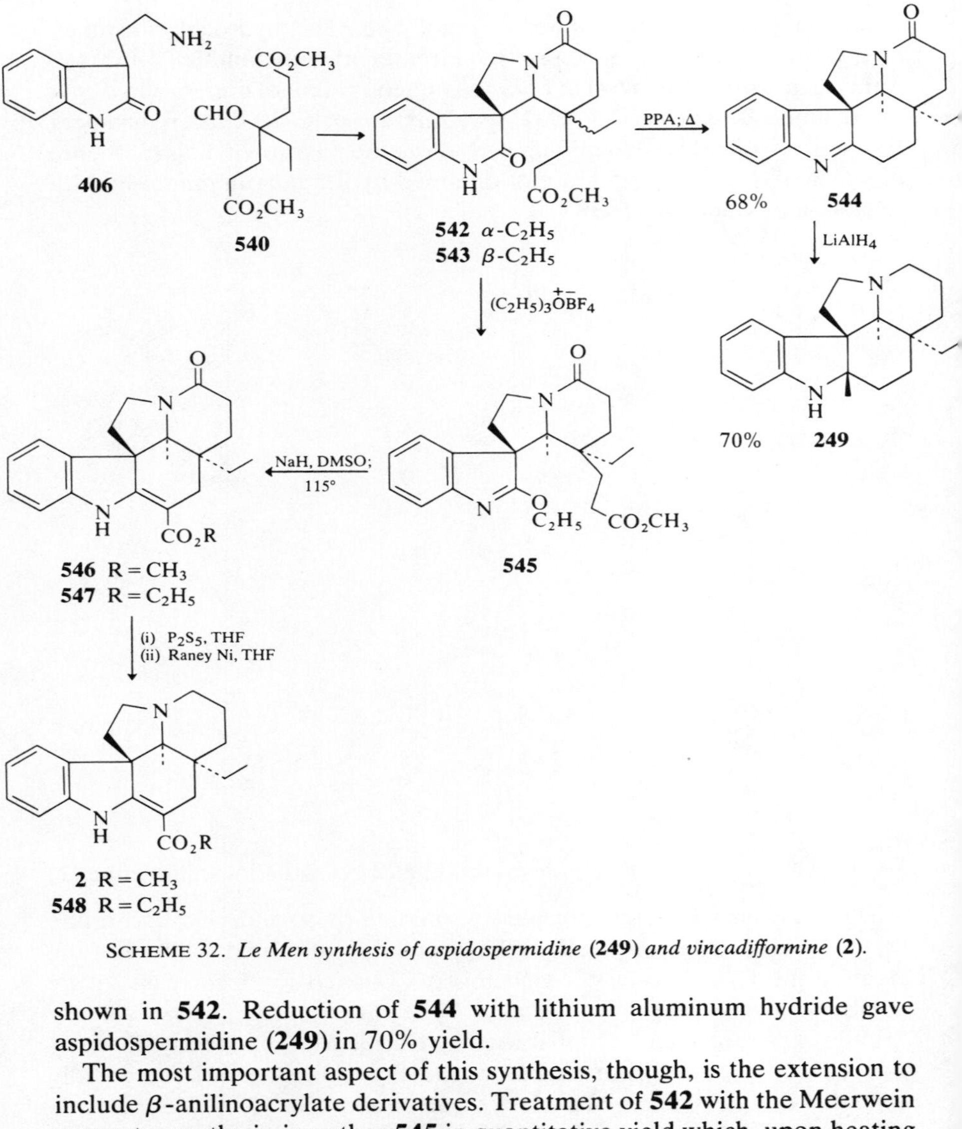

NH_2 CO_2CH_3 CHO CO_2CH_3

406 **540**

CO_2CH_3

542 α-C_2H_5
543 β-C_2H_5

PPA; Δ

68% **544**

$LiAlH_4$

70% **249**

$(C_2H_5)_3\overset{+}{O}\overset{-}{B}F_4$

C_2H_5 CO_2CH_3

545

NaH, DMSO; 115°

CO_2R

546 $R = CH_3$
547 $R = C_2H_5$

(i) P_2S_5, THF
(ii) Raney Ni, THF

CO_2R

2 $R = CH_3$
548 $R = C_2H_5$

SCHEME 32. *Le Men synthesis of aspidospermidine* (**249**) *and vincadifformine* (**2**).

shown in **542**. Reduction of **544** with lithium aluminum hydride gave aspidospermidine (**249**) in 70% yield.

The most important aspect of this synthesis, though, is the extension to include β-anilinoacrylate derivatives. Treatment of **542** with the Meerwein reagent gave the imino ether **545** in quantitative yield which, upon heating with sodium hydride in dimethyl sulfoxide at 115° for $2\frac{1}{2}$ hr, gave a 2:1 mixture of the 3-oxo derivatives **546** and **547**, which were not separated at this point.

A key problem then became to eliminate the oxygen from the lactam without reducing the β-anilinoacrylate. This was successfully achieved by exchange of the lactam oxygen by sulfur with phosphorus pentasulfide in THF, and reduction with Raney nickel in THF at reflux to give a mixture of two esters. The more polar product was identical with vincadifformine (**2**); the less polar product was identified as the corresponding ethyl ester (**548**) (*235, 236*).

21. Stevens Approach

The Stork–Dolfini synthesis of the amino ketone **458** requires some 11 steps; clearly, this is an area for improvement. One method used has been that of methyl vinyl ketone annelation (*237*). This procedure had been used previously in a synthesis of mesembrine (*238*) involving Δ^2-pyrrolines. Stevens hoped to extend this technique to endocyclic enamines to give compounds in the octahydroquinolone series. 3-Ethylpiperid-2-one (**549**) was prepared from the γ-lactone **550** according to standard procedures (*239*). N-Benzylation followed by reduction with diisobutylaluminum hydride (DIBAL) gave enamine **473**. Addition of the enamine to methyl vinyl ketone (**551**) gave an almost quantitative yield of a hydroquinoline **552**, which could be readily debenzylated to the known amino ketone **458**. As with the Stork method of synthesis, this route to the *Aspidosperma* nucleus is efficient for the parent compounds but appears to lack the potential for diversification required to synthesize the more complex alkaloids such as vindoline (**101**) or even tabersonine (**28**).

CH_2CH_2CN

550 → **549** —(i) $C_6H_5CH_2Cl$, (ii) DiBAL→ **473** (PhCH$_2$) —Δ→ **458** R = H; **552** R = $CH_2C_6H_5$

551

Stevens and co-workers have also attempted to develop a fundamentally different approach to the tricyclic amino ketone **463** which is used in the Fischer cyclization approach to the *Aspidosperma* skeleton (*240*). Condensation of the aldehyde–ester **553** with protected keto amine **554** gave an imine (**555**) which, upon heating with ammonium chloride at 160°, afforded the 2-pyrroline ester **556** in 70% yield. Treatment with dry hydrochloric acid gas in ether was followed by acid hydrolysis of the ketal, and base-catalyzed cyclization produced a mixture of two enol ethers (**557** and **558**) the latter predominating. The major isomer was reduced with lithium aluminum hydride and the hydroxy enol ether dehydrated in hot

CHO
$CO_2C_2H_5$
553
+
H_2N
554
$-H_2O$
$CO_2C_2H_5$
555
NH_4Cl
160°
$CO_2C_2H_5$
556
(i) H^+
(ii) K_2CO_3
$CO_2C_2H_5$
NaOMe/MeOH
557
OCH_3
558
CH_3O
(i) $LiAlH_4$
(ii) H^+
559
H_2/Pd–C
373

acid to the α,β-unsaturated ketone **559**. Catalytic reduction gave the tricyclic ketone **373**, whose stereochemistry was established by comparison with authentic samples. This compound is identical with Bans' compound and therefore will not produce (±)-aspidospermine (**253**) on cyclization but rather the unnatural 20-ethyl isomer, as discussed previously.

22. Harley-Mason Synthesis of Aspidospermidine

The Harley-Mason approach to the *Aspidosperma* skeleton was discussed in Volume XI (p. 225), and very brief mention was made of the successful synthesis of aspidospermidine (**249**) using this route (*241*). In view of the complication involving *cis/trans* C/D ring stereochemistry involved in a number of other approaches, it is amazing that the Harley-Mason approach should be stereoselective. The process in question is typically reaction of the hydroxyester **560** with tryptamine to give a compound (**561**) having a *seco*-eburnane skeleton. When this compound is treated with 40% sulfuric acid or boron trifluoride etherate at 100°, the indolenine lactam **562** is produced. Lithium aluminum hydride reduction then gives racemic aspidospermidine (**249**; Scheme 33).

The stereospecificity of the reaction may well be set in the first intermediate where molecular models appear to indicate that attack of the indole 2-position on C-16 cannot occur if the C-3 proton and the ethyl side chain are *trans* to each other. Except for the possibility of hydrogen bonding between the 16-hydroxy and lactam carbonyl groups in **561**, there

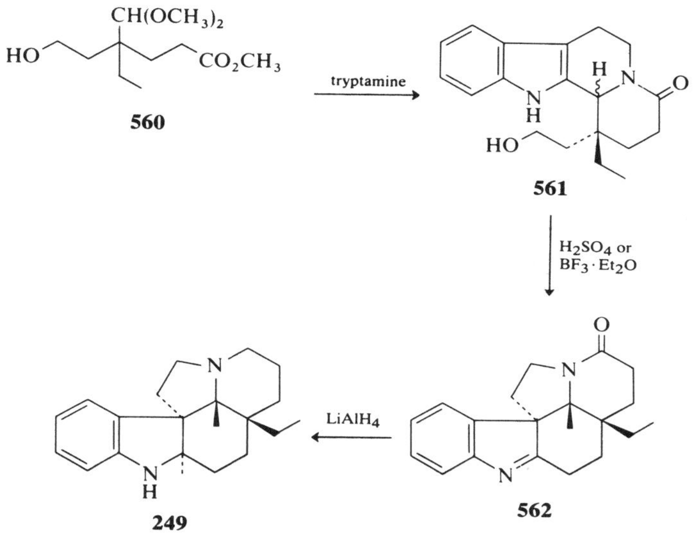

$CH(OCH_3)_2$ HO CO_2CH_3

560

tryptamine

561

H_2SO_4 or $BF_3 \cdot Et_2O$

LiAlH₄

249 **562**

SCHEME 33. *Harley-Mason synthesis of aspidospermidine* (**249**).

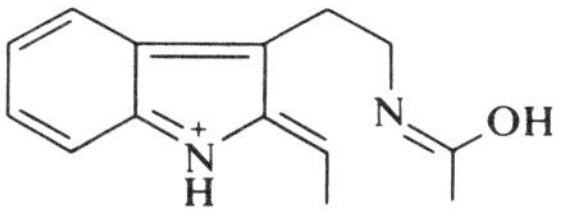

563

would appear to be no incentive for stereospecific formation of **562**. If **561** is a mixture of isomers, then C-3 epimerization probably occurs (as in **563**) in an attempt to maintain the equilibrium with the ethyl C-3 *cis* isomer being removed by rearrangement.

23. Wenkert Approach

Yet another approach to the *Aspidosperma* nucleus has been reported by Wenkert and co-workers (*242*). This approach was modeled along biosynthetic lines, allowing an iminium species in a piperidine ring to be attacked by the nucleophilic indole β-position. This concept met with some success when it was demonstrated that treatment of the vinylogous urethane **564** with HBr gave, after borohydride reduction, the spiro derivative **565**. Extension of this reaction to the indoleacetic acid series was also successful. The chloroester **566**, on treatment with methyl nicotinate, gave a quaternary salt **567** which was reduced to the tetrahydropyridine **568**. Treatment of the latter with methanolic hydrochloric acid gave the tetracycle **569** in 80% yield. No stereochemical assignments were established rigorously although it was suggested that the carbomethoxyl group was close to the π-system. This approach to the *Aspidosperma* skeleton does not appear to have been extended further.

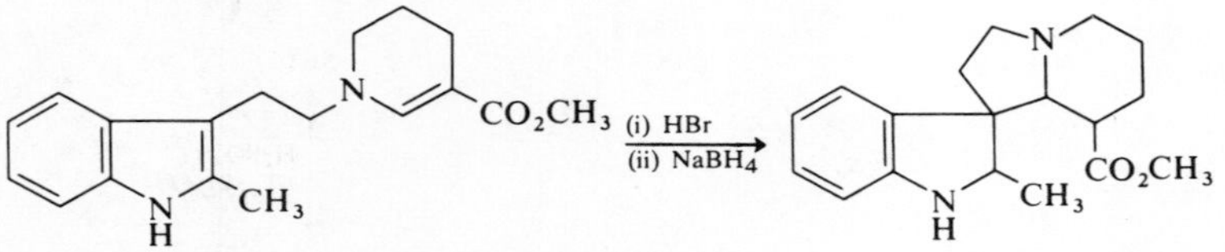

CO_2CH_3 (i) HBr (ii) $NaBH_4$ CO_2CH_3 CH_3 CH_3

564 **565**

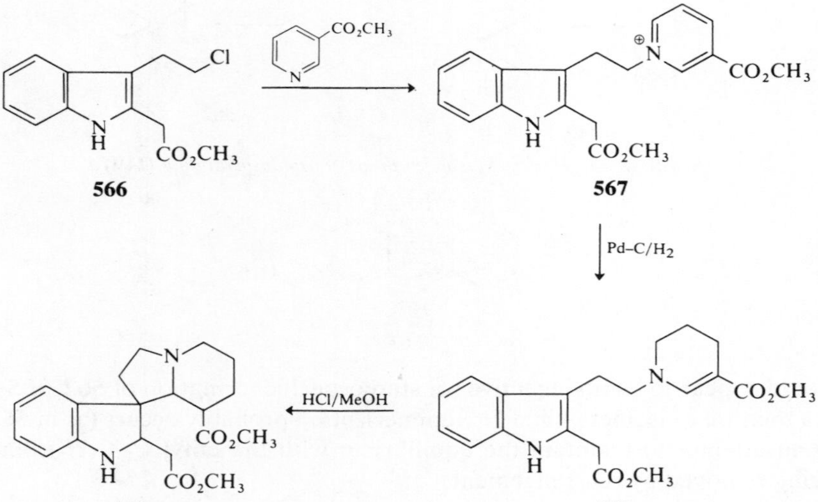

CO_2CH_3 Cl CO_2CH_3 CO_2CH_3 CO_2CH_3

566 **567**

Pd–C/H_2

HCl/MeOH CO_2CH_3 CO_2CH_3 CO_2CH_3 CO_2CH_3

569 **568**

A more successful approach has been reported briefly by Wenkert (*243*), although, as will be observed, it is not without complication.

The technique of cyclization of a 1,4,5,6-tetrahydropyridine to an indole nucleus is already familiar (*191*), and this method was extended to compound **570**. Treatment of **570** with boron trifluoride–etherate gave the keto lactam **571** in 62% yield. Thioketalization and Raney nickel reduction gave a lactam (**572**) which, upon lithium aluminum hydride reduction, gave 20-iso-20-deethylaspermidine (**573**). However, although the initial keto lactam **571** possessed the required C/D *cis* stereochemistry, under the thioketalization conditions C-20 isomerization occurred to give the more stable C/D *trans* ring juncture.

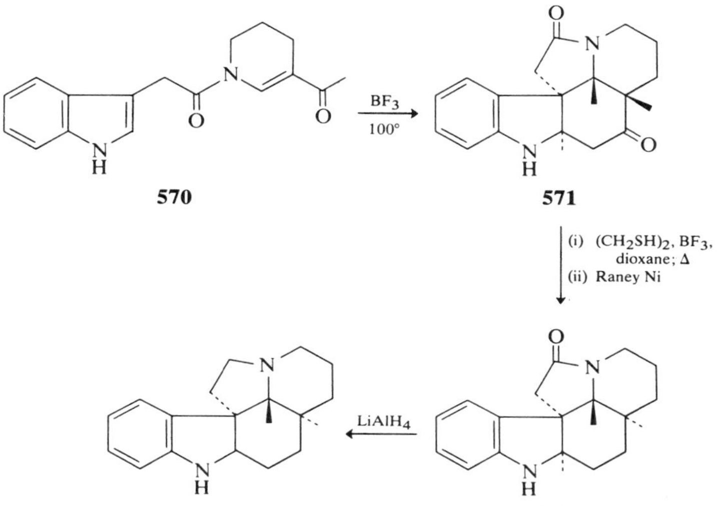

570 $\xrightarrow[100°]{BF_3}$ **571**

(i) $(CH_2SH)_2$, BF_3, dioxane; Δ
(ii) Raney Ni

572 $\xrightarrow{LiAlH_4}$ **573**

24. Potier–Wenkert Approach

The interim aim of this synthetic approach (*244*) was the synthesis of the nitrile **574** which could potentially be doubly alkylated to form the tryptamine bridge and the ethyl side chain.

Condensation of α-indolylacetonitrile with nicotinaldehyde gave **575**, which was oxidatively photocyclized to **576** in 30% yield. Irradiation in ethanol, however, led to **577** in 30% yield, probably by decyclization of **578** (*244*).

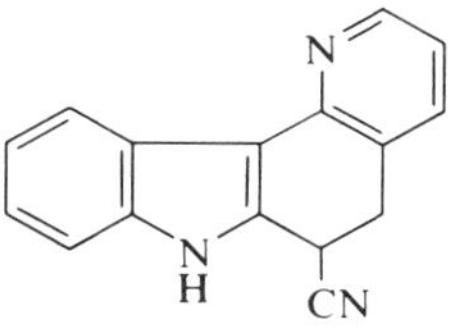

574 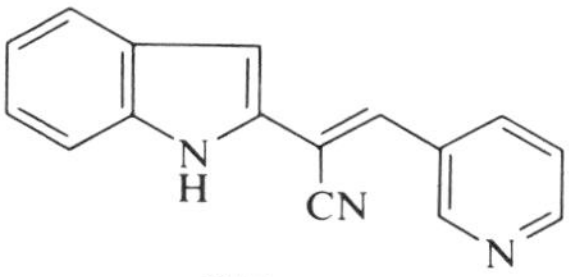 **575** 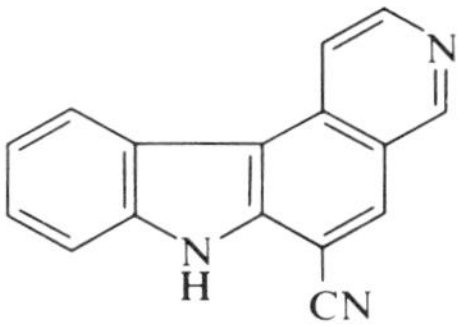 **576**

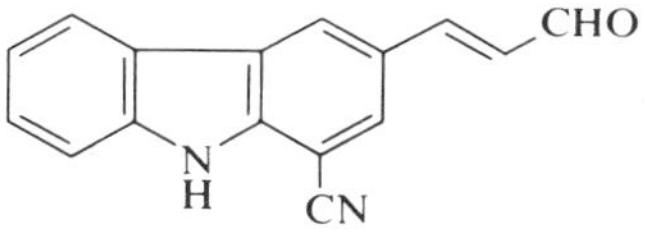

577 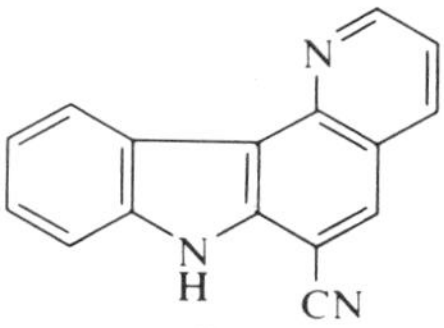 **578**

25. Potier Approach

A further example of synthetic approaches to the *Aspidosperma* skeleton is provided by Potier's synthesis of 5-oxo-20-deethyl vincadifformamide (**579**) (*245*). This approach differs fundamentally from the available syntheses in that it begins with a 2-substituted indole followed by cyclization in a single step to the A,B,C,D system lacking the tryptamine bridge carbons.

The simple reaction sequence begins with the pyridylindolylacrylonitrile **575**, which is reduced with sodium borohydride to a nitrile (**580**) and then hydrolyzed to the amide **581**. Catalytic reduction under acidic conditions gave the tetracyclic amide **582** in 45% yield. This cyclization apparently gave the *trans* C/D ring junction stereochemistry, for insertion of the tryptamine bridge proceeded in poor yield. No further synthetic elaboration of this scheme has been reported.

A number of miscellaneous attempts at the synthesis of the *Aspido-sperma* skeleton have been reported in preliminary forms (*246–248*).

N H CN N **575** $\xrightarrow{NaBH_4}$ N H R N
580 R = CN
581 R = $CONH_2$
$PtO_2/H_2/H^+$, MeOH
HN N H $CONH_2$ **582**
(i) $BrCH_2COBr$
(ii) $NaHCO_3$/DMF
N N H $CONH_2$ **579**

26. Saxton Synthesis of Cylindrocarine-Type Alkaloids

Saxton and co-workers have reported their route for the synthesis of *Aspidosperma* alkaloids of the cylindrocarine type (*249*). Aware of problems involving *cis* and *trans* C/D ring junction ions, they reasoned that having the least bulky substituent at C-20 would help limit the proportion of undesired *trans* isomer. Thus, the target compound became the tricyclic amino ketone **583** having a potential C/D *cis* stereochemistry and an angular allyl group.

Successive alkylation of the pyrrolidine enamine of pent-4-enal with methyl acrylate and methyl vinyl ketone gave **584** after acid-catalyzed cyclization. Treatment with ammonia gave a mixture of amidoketones which could be protected and reduced to the protected amino ketones **585**. Chloroacetylation and acid hydrolysis followed by *tert*-butoxide-initiated cyclization gave the *cis* and *trans* amino ketones **583** and **586**. By means of the method developed by Stork, the *cis* isomer **583** was converted to the *Aspidosperma* base **587**. The terminal carbon of the allyl side chain was easily removed with osmium tetroxide–sodium periodate to give the aldehyde **588**. Sodium borohydride reduction gave racemic *N*-acetylcylindrocarpinol (**130**), and acetylation gave *N,O*-diacetylcylindrocarpinol (**589**), each identical with the natural product.

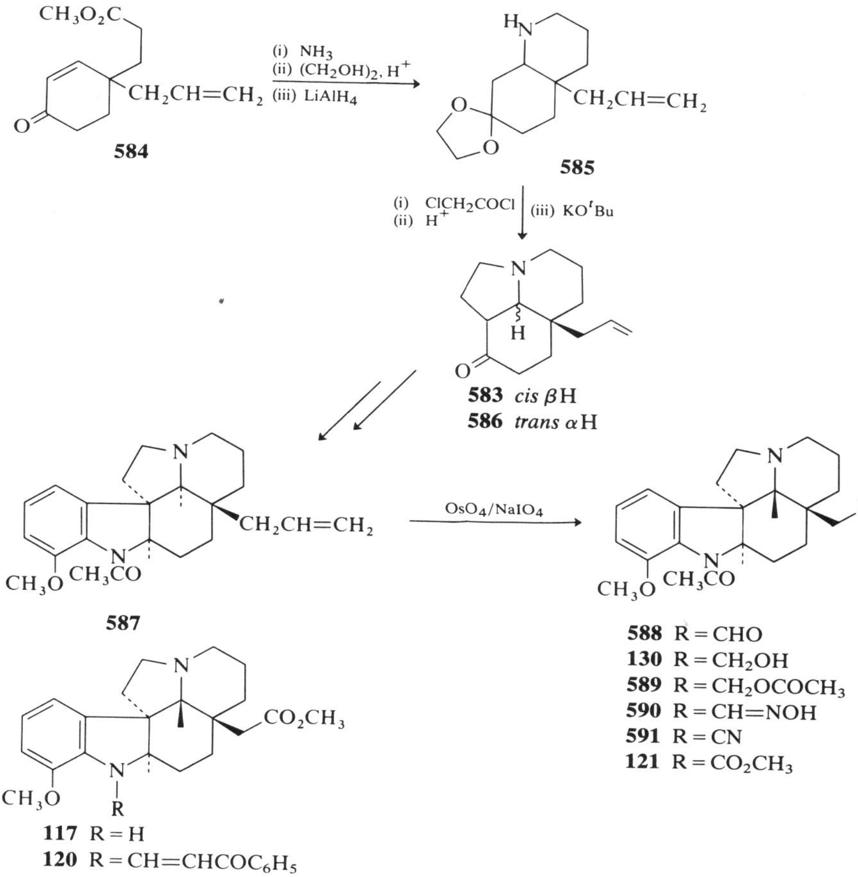

SCHEME 34. *Saxton synthesis of cylindrocarpine alkaloids.*

The oxime **590** derived from the aldehyde **588** was dehydrated with acetic anhydride, and the resulting nitrile (**591**) methanolized to (±)-cylindrocarine (**117**). As expected, acetylation of **117** gave (±)-cylindrocarpidine (**121**), and cinnamoylation gave (±)-cylindrocarpine (**120**; Scheme 34) (*249*).

27. Smith Synthesis of Rhazinilam (172)

The structure elucidation of rhazinilam (**172**) by spectral and chemical means (*81*) and by X-ray crystallography (*82*) was discussed in Section II,F,3. In addition, a partial synthesis from (+)-1,2-dehydroasidospermidine (**171**) was described (*84*). Smith and co-workers have also described an elegant total synthesis of rhazinilam (**172**) (*84*).

The key steps were the N-alkylation of 2-methoxycarbonyl-4-(2′-nitrophenyl)pyrrole (**592**) with 4-ethyl-4-hydroxy-7-tosyloxyheptanoic acid γ-lactone (**593**) and subsequent cyclization to give the phenyl tetrahydroimidazoline, **594**. Reduction of the nitro group and lactamization with dicyclohexylcarbodiimide in THF gave 5-methoxycarbonylrhazinilam (**595**) in 82% yield from **594**. Hydrolysis and decarboxylation gave (±)-rhazinilam (**172**) in 88% yield (Scheme 35).

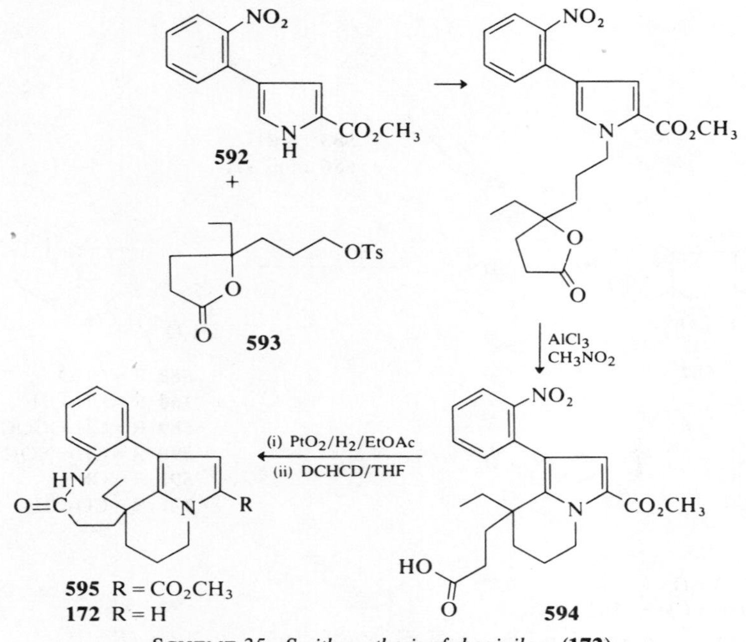

NO_2 CO_2CH_3 **592** + OTs **593** $AlCl_3$ CH_3NO_2 **594** (i) $PtO_2/H_2/EtOAc$ (ii) DCHCD/THF **595** $R = CO_2CH_3$ **172** R = H

SCHEME 35. *Smith synthesis of rhazinilam* (**172**).

B. Alkaloids with the Fluorocurine–Mavacurine Skeleton

The fluorocurine–mavacurine series of alkaloids has been the subject of only very limited synthetic interest. Boekelheide and co-workers (*250*) have reported their work on the synthesis of 19,20-dihydronormavacurine (**596**) and 19,20-dihydronorfluorocurarine (**597**), and this appears to be the limit of investigations in this area.

The initial synthetic target was the tricyclic ketone **598**. Reaction of 3,4-dihydro-β-carboline (**599**) with ethyl α-ethylacetoacetate (**600**) and formaldehyde gave a β-keto ester in 60% yield which was readily hydrolyzed and decarboxylated to the tricyclic ketone **598**. Condensation of **598** with methoxymethylenetriphenylphosphorane gave an enol ether which on acid hydrolysis afforded the aldehyde **601**. Treatment of this aldehyde with dimethyloxosulfonium methylide afforded the epoxide **602**. When this epoxide was reacted with sodium hydride in dimethyl sulfoxide under nitrogen, an *N*-alkylindole derivative was produced. The UV spectrum of this product ruled out the formation of the pleiocarpamine skeleton, and the compound was assigned structure **603** in the apogeissoschizine series. In contrast, reaction of **602** with sodium hydride and dimethyl sulfoxide under oxygen gave a yellow crystalline product in 62% yield which from its UV spectrum was an *N*-alkylpseudoindoxyl. The spectral properties were in agreement with those of 19,20-dihydronorfluorocurarine (**597**), and this structure assignment was confirmed by conversion to 19,20-dihydronormavacurine (**596**; Scheme 36).

C. Alkaloids with the Uleine Skeleton

Four groups have reported their synthetic efforts in this area of indole alkaloids which apparently lack one carbon of the tryptamine bridge. The two series of compounds differ on the basis of the presence or absence of a C-16 exomethylene group.

1. Joule Approach

The key intermediate in the Joule syntheses (*251*) in this series is the pyridyl keto indole **604**, which was prepared by Scheme 37. Standard reactions gave the γ-amino enone, **605**. Reaction of **605** with the salt of dimethyl sulfoxide and acetic acid-catalyzed rearrangement of the tetracyclic intermediate **606** gave dasycarpidone (**607**) and 3-epidasycarpidone (**608**) via the iminium **609**. Dasycarpidone was converted to uleine (**610**) by the Wittig reaction, but 3-epidasycarpidone required somewhat different conditions for conversion to 3-epiuleine (**611**; Scheme 37).

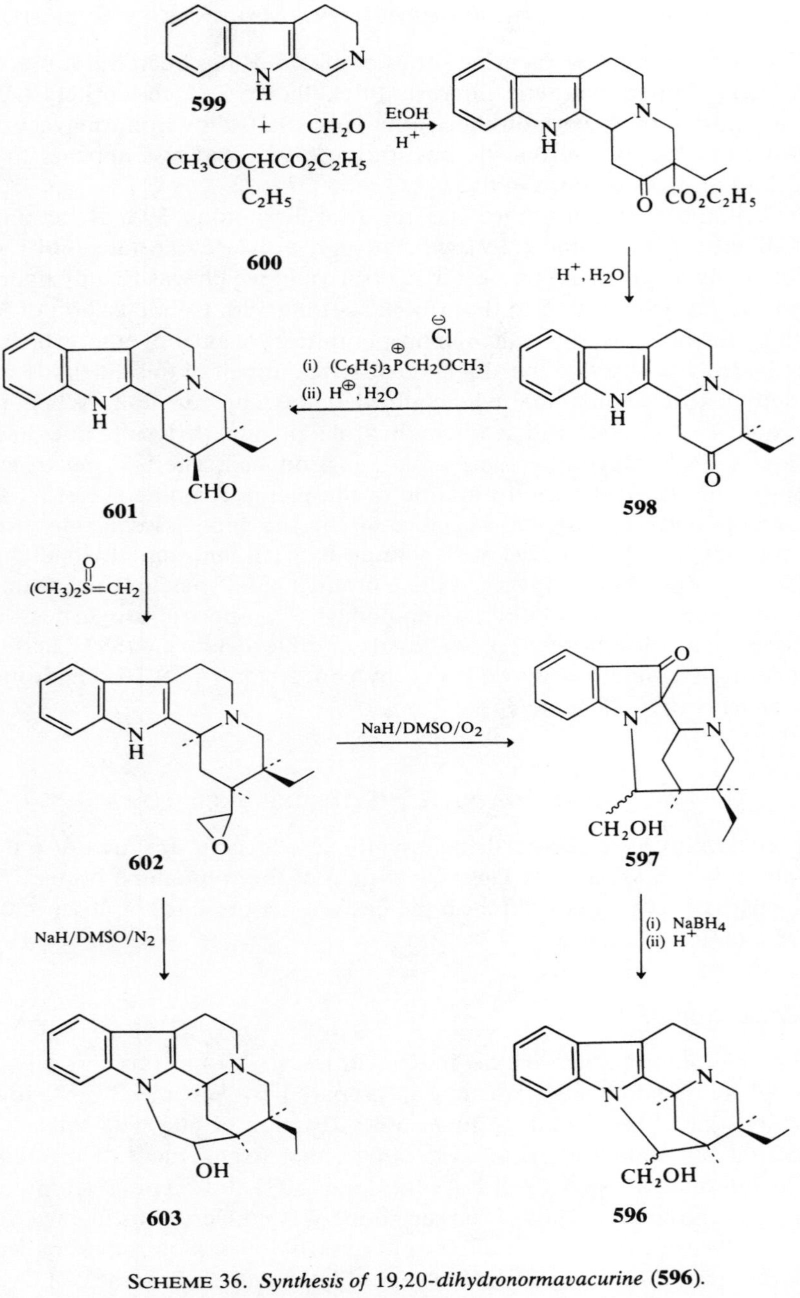

599

+ CH_2O

$CH_3COCHCO_2C_2H_5$

C_2H_5

600

EtOH

H^+

$CO_2C_2H_5$

H^+, H_2O

(i) $(C_6H_5)_3\overset{\oplus}{P}CH_2OCH_3\ \overset{\ominus}{Cl}$

(ii) $H^{\oplus}, H_2O$

CHO

601

598

$(CH_3)_2\overset{O}{\overset{\|}{S}}{=}CH_2$

NaH/DMSO/O_2

CH_2OH

602

597

NaH/DMSO/N_2

(i) $NaBH_4$

(ii) H^+

OH

CH_2OH

603

596

SCHEME 36. *Synthesis of* 19,20-*dihydronormavacurine* (**596**).

SeO_2

(i) $SOCl_2$
(ii) MeOH
(iii) $C_2H_5CO_2CH_3$/NaH

CO_2H

CO_2CH_3

(i) H_2SO_4, HOAc
(ii) pyrrolidine
(iii) $\overset{\oplus}{N}_2$

H_3PO_4

C_6H_5NH

604

(i) MeI
(ii) $NaBH_4$
(iii) MnO_2

NCH_3

605

DMSO
Na

CH_3N

606

HOAc

$\overset{\oplus}{N}CH_3$

609

607

$(C_6H_5)_3P{=}CH_2$

610

608

Mg(Hg)
CH_2I_2

611

SCHEME 37. *Joule synthesis of dasycarpidone* (**607**), 3-*epidasycarpidone* (**608**), *uleine* (**610**), *and* 3-*epiuleine* (**611**).

2. Dolby Approach

Dolby and Biere (*252, 253*) adopted a quite different tactic to produce these systems. The key step is the trapping of a Vilsmeier salt with sodium borohydride. All the functionality required to effect the condensation reactions is built into a 2-pyridone derivative **612**. This compound is produced by the route outlined in Scheme 38. Treatment of **612** with indole in the presence of phosphorus oxychloride gives a Vilsmeier salt, which is reduced with sodium borohydride to the piperidylindole **613**. Hydrolysis of the ester group and cyclization with polyphosphoric acid gave a mixture consisting mainly of epidasycarpidone (**608**) (54% yield) together with some **607** (15% yield). The **608** was converted to the carbinol **614** with methyllithium and dehydrated to **611** with alumina.

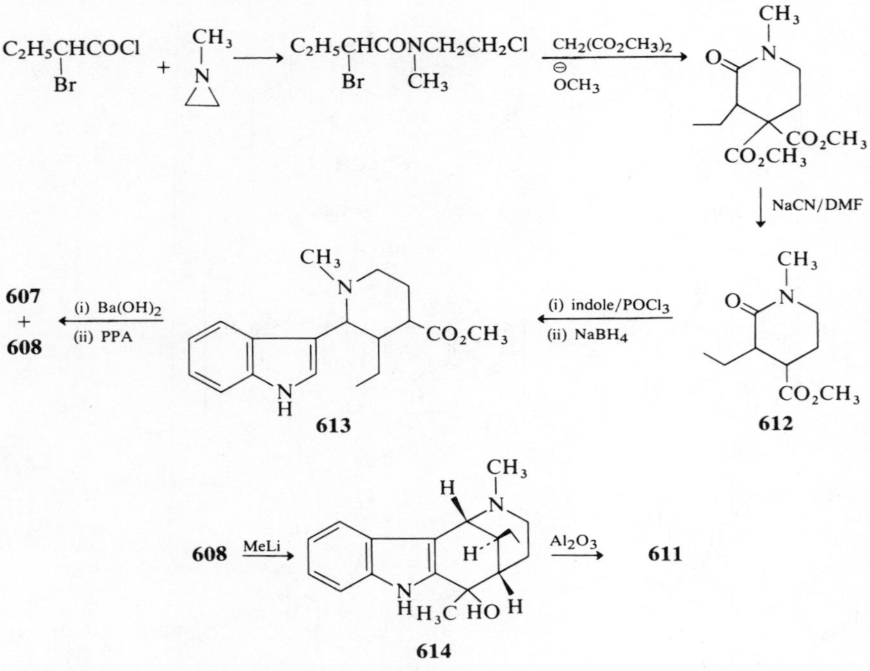

$C_2H_5CH(Br)COCl$ + N-methylaziridine → $C_2H_5CH(Br)CON(CH_3)CH_2CH_2Cl$ —($CH_2(CO_2CH_3)_2$, $^{\ominus}OCH_3$)→ (dimethyl ester pyridone) —NaCN/DMF→ **612** —(i) indole/$POCl_3$, (ii) $NaBH_4$→ **613** —(i) $Ba(OH)_2$, (ii) PPA→ **607** + **608**

608 —MeLi→ **614** —Al_2O_3→ **611**

SCHEME 38

3. Kametani Approach

Kametani and co-workers (*254–256*) adopted yet another approach to this skeleton. Grignard condensation of indolyl magnesium bromide with methyl 3-ethyl isonicotinate 1-oxide (**615**) in the presence of benzoyl chloride gave a mixture of structural isomers **616** and **617**. N-Methylation and catalytic reduction of **616** gave a mixture of amino esters **613**. Saponification and cyclization with polyphosphoric acid then afforded a 2:1 mixture of **607** and **608** (*255*). The overall yield was extremely low (Scheme 39).

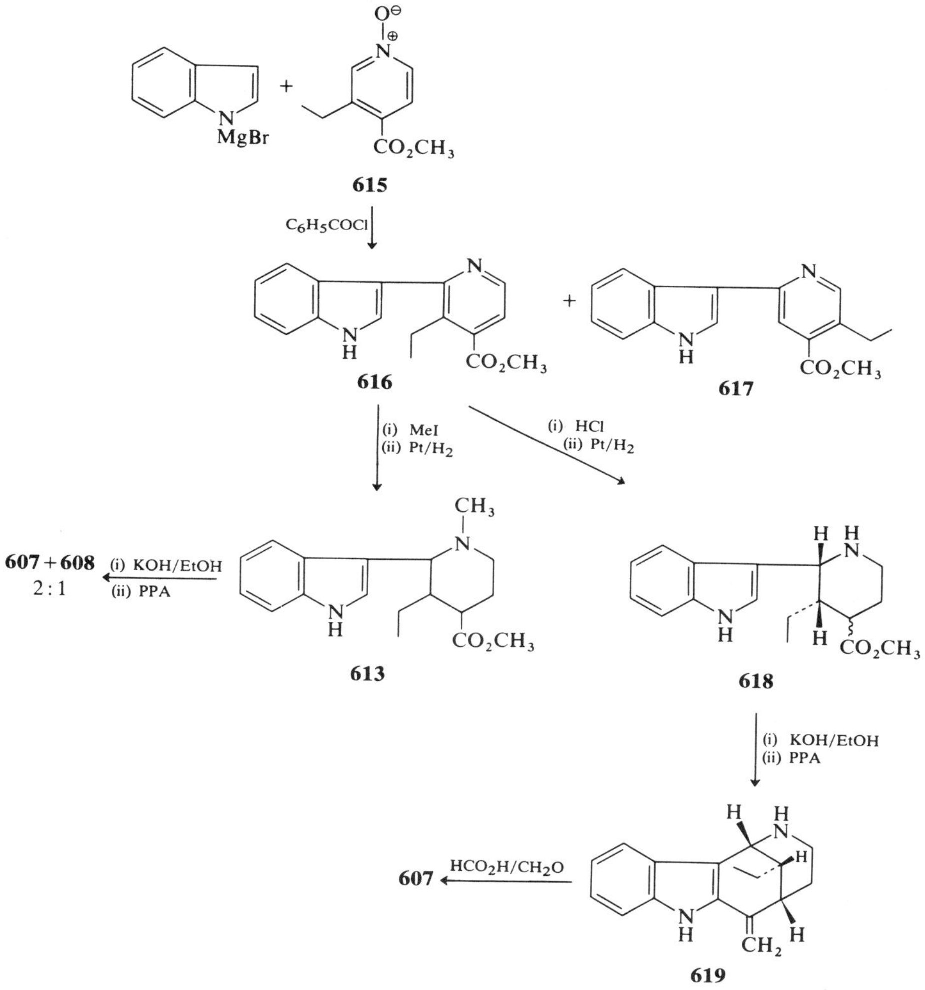

MgBr + CO_2CH_3 **615**

C_6H_5COCl

616 + **617**

(i) MeI (ii) Pt/H_2 (i) HCl (ii) Pt/H_2

607 + 608 2:1 ← (i) KOH/EtOH (ii) PPA — **613**   **618**

(i) KOH/EtOH (ii) PPA

607 ← HCO_2H/CH_2O — **619**

SCHEME 39. *Kametani synthesis of uleine* (**607**) *and* 3-*epiuleine* (**608**).

In a slightly different approach, some stereochemical control was achieved. Catalytic reduction of the hydrochloride of **616** gave the *cis*-2,3 product **618**, in which no stereochemistry was assigned to the carbomethoxy group. Saponification and cyclization afforded only *N*-nordasycarpidone (**619**), which could be converted to **607** under Eschweiler–Clarke conditions (*256*).

4. Büchi Approach

In the early syntheses no attention was paid to the separation of stereoisomers at any point; consequently, epimeric mixtures resulted in the condensation processes. Büchi and co-workers (*257*) were the first to report a stereospecific synthesis of **610** and **611**.

The synthesis begins with the condensation of 3-formylindole (**620**) with 1-aminohexan-3-one (**621**) to give the *trans*-diequatorial ketone **622**. The compound was quite unstable and was converted to the formamide **623**. Reaction with potassium acetylide in *tert*-butyl alcohol–tetrahydrofuran gave, stereospecifically, the ethynyl carbinol **624**, which was rearranged with mercuric acetate to the *O*-acetoxyketone **625**. The acetoxy group was removed by treatment with lithium metal in ammonia to give the methyl ketone **626** in which all three substituents are *trans* and equatorial. Facile cyclization and dehydration occurred when **626** was treated with boron trifluoride etherate to afford **627**, and lithium aluminum hydride reduction gave epiuleine (**611**).

The synthesis of uleine (**610**) required that the stereochemistry of the ethyl side chain be inverted and this was accomplished by thermal (*cis*) elimination of acetic acid from the acetoxy ketone **625**. One of the products was the tetracyclic olefin **628**, and reduction with palladium catalyst and a trace of pyridine gave almost exclusively the required *cis,cis*-ketone **629**. This compound was cyclized and reduced as before to give uleine (**610**; Scheme 40).

D. Alkaloids with the Ellipticine–Olivacine Skeleton

Earlier synthetic approaches to ellipticine (**267**) were described in Volume XI (p. 279) of this series. In the intervening years, several new approaches to this important alkaloid have been developed.

1. Le Goffic *et al.* Approaches

Le Goffic *et al.* (*258, 259*) used an approach which built almost all the ring carbons in one step by reaction of the piperidone enamine of *N*-benzyl-4-piperidone (**630**) and α-methylgramine (**631**) in refluxing dioxane. Reaction of **632** with sodium acetylide in liquid ammonia gave the

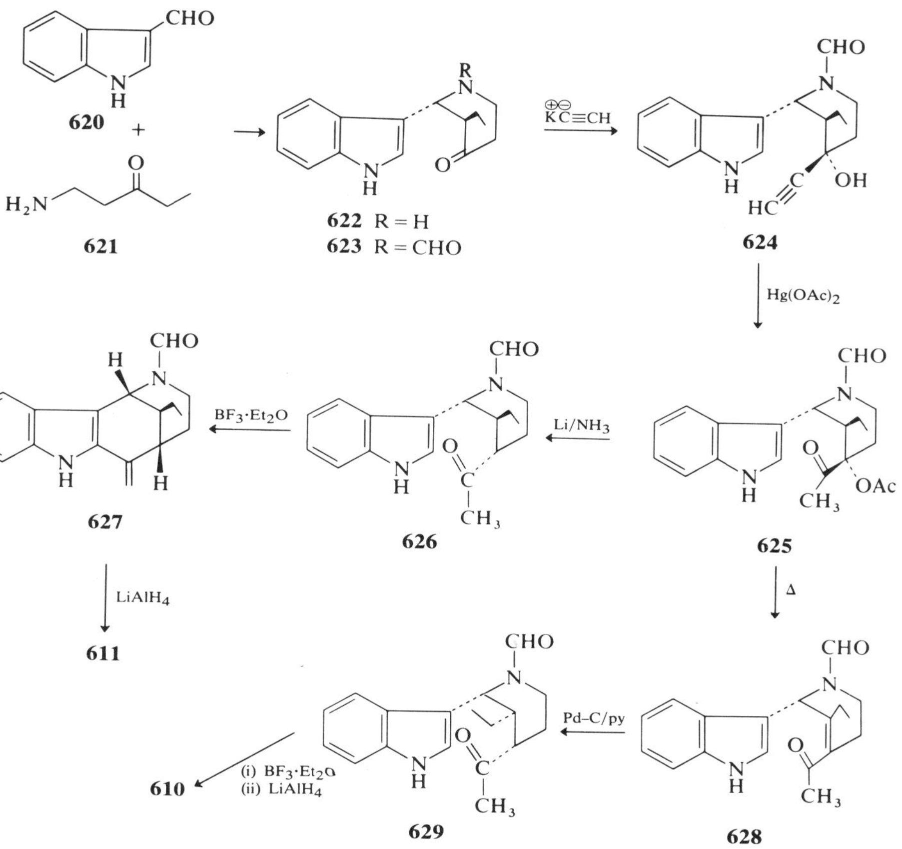

SCHEME 40. *Büchi synthesis of uleine* (**610**).

propargyl alcohol **633** as a mixture of epimers which in formic acid gave in 93% yield the tetrahydroellipticine **634**. Palladium–charcoal dehydrogenation/debenzylation gave ellipticine (**267**) in high yield (Scheme 41).

A second route to the 6*H*-pyrido[4,3-*b*]carbazole nucleus of ellipticine involves as a key step the internal condensation of a 2-alkylskatolylpiperidone. Le Goffic *et al.* (*259*) developed this synthesis in the 5,11-bisdemethyl series. When enamine **630** was condensed with 2-methylgramine (**635**), hydrolysis afforded the skatolylpiperidone (**636**). Cyclization was carried out with glacial acetic acid under reflux, debenzylation with sodium/liquid ammonia, and dehydrogenation with palladium–charcoal to give **637**.

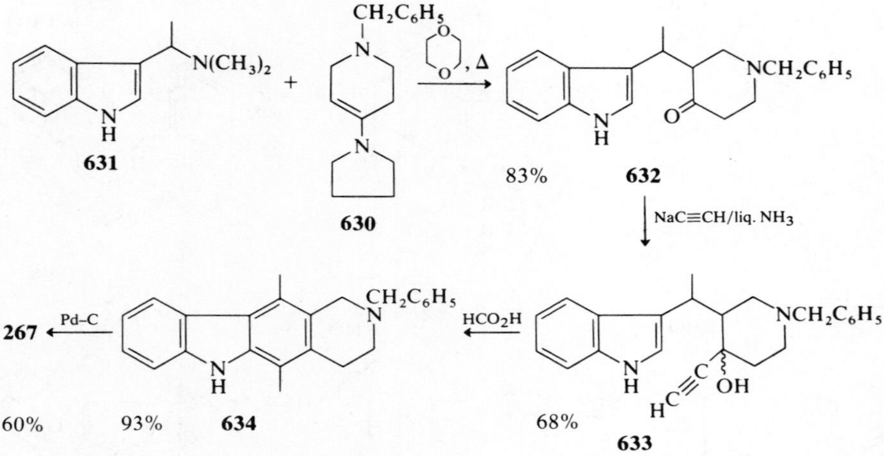

$CH_2C_6H_5$

$N(CH_3)_2$ + , Δ $NCH_2C_6H_5$

631 630 83% 632

$NaC{\equiv}CH$/liq. NH_3

$CH_2C_6H_5$ HCO_2H $NCH_2C_6H_5$

267 ← Pd–C

OH

60% 93% 634 68% 633

SCHEME 41. *Le Goffic* et al. *synthesis of ellipticine* (**267**).

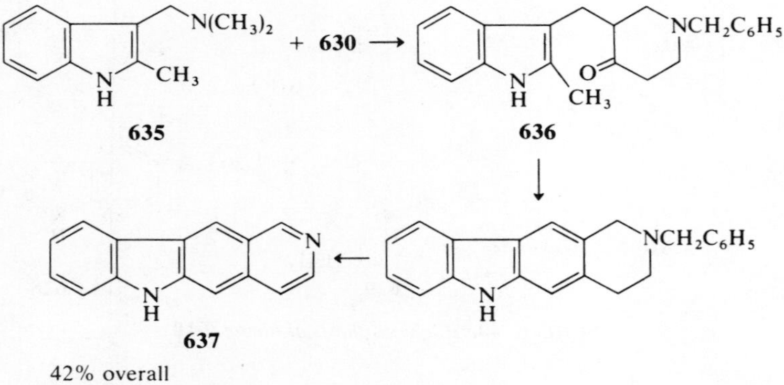

$N(CH_3)_2$ + **630** → $NCH_2C_6H_5$

CH_3 CH_3

635 636

$NCH_2C_6H_5$

637

42% overall

2. Kilminster–Sainsbury Approach

The initial idea of Kilminster and Sainsbury (*260*) was to improve the Woodward route (*261*) to ellipticine. The successful route involved the base-catalyzed condensation of the 1,3-diacetyl derivative of indolin-3-one (**638**) with 4-acetyl-3-(1-methoxyethyl)pyridine (**639**) to give the unsaturated ketone **640** as a mixture of isomers. Reduction with sodium borohydride and dehydration with acid gave indole **641**. When this material was heated with hydrogen bromide under reflux and then adsorbed onto silica, extraction of the silica with chloroform gave **267** in 40% yield (Scheme 42).

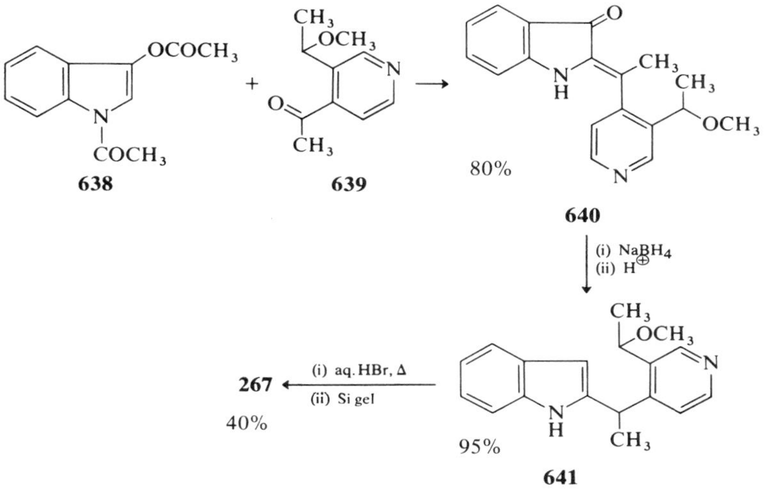

SCHEME 42. *Kilminster–Sainsbury synthesis of ellipticine* (**267**).

3. Potier Approaches

Potier and co-workers have described two synthetic routes to ellipticine (*262, 263*). In the first of these (*263*) the lithium salt of 1-sulfobenzoylindole is condensed with 4-acetylpyridine to give **642**. Hydrolysis and sodium borohydride reduction in the presence of a large excess of KCN gave the cyanopiperideine **643**. Condensation of **643** with the Mannich reagent from acetaldehyde and dimethylamine followed by sodium borohydride reduction of the intermediate iminium species **644** gave *N*-methyltetrahydroellipticine (**645**). Palladium–charcoal in boiling decalin with **645** afforded **267** (Scheme 43). The overall yield for the sequence was 8.4%.

The second synthesis (*262*), although a little longer, is interesting for some of the chemistry involved, particularly the use of the Polonovskii reaction to complete cyclization of ring C.

Condensation of the allylic alcohol **646** with methyl orthopropionate in butyric acid gave ester **647** which, after hydrolysis, was treated with methyllithium to give ketone **648**. Fischer indole cyclization with polyphosphoric acid of the phenylhydrazone of **648** gave the indole **649**. The *N*-oxide of **649** in trifluoroacetic anhydride at 0° gave in 92% yield the hexahydroellipticine **650**, convertible with palladium–charcoal to **267** in 35% yield. The overall yield from **646** is on the order of 19% (Scheme 43).

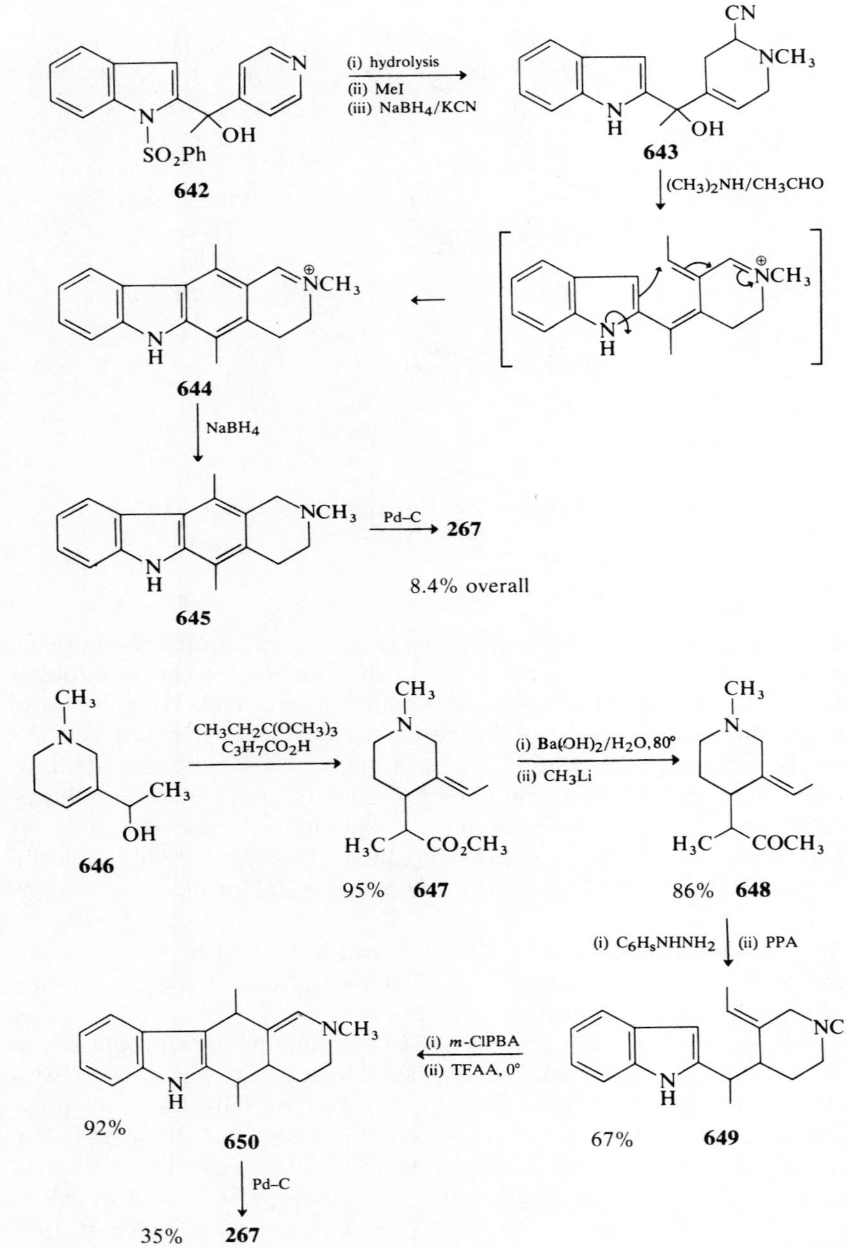

(i) hydrolysis
(ii) MeI
(iii) $NaBH_4/KCN$

642 → 643

$(CH_3)_2NH/CH_3CHO$

644

$NaBH_4$

645

Pd–C → 267

8.4% overall

646

$CH_3CH_2C(OCH_3)_3$
$C_3H_7CO_2H$

95% 647

(i) $Ba(OH)_2/H_2O, 80°$
(ii) CH_3Li

86% 648

(i) $C_6H_5NHNH_2$ (ii) PPA

67% 649

(i) m-ClPBA
(ii) TFAA, 0°

92% 650

Pd–C

35% 267

SCHEME 43. *Potier syntheses of ellipticine* (**267**).

4. Synthesis of 9-Aminoellipticine (**658**)

Attempts to use the previous synthesis (*260*) of ellipticines to the synthesis of 9-nitroellipticine (**651**) failed when the mixture of isomers **652** on reduction with sodium borohydride gave mainly resinous material (*264*). However, condensation of the amide **653** with ketone **654** gave a mixture of isomers **655** in 35% yield. Reduction with sodium borohydride then gave the alcohol **656** which, with methanolic hydrogen chloride, gave indole **657**. When **657** was hydrolyzed with aqueous hydrogen bromide, cyclization, dehydrogenation, and hydrolysis occurred to afford 9-amino-ellipticine (**658**) directly (Scheme 44) (*264*).

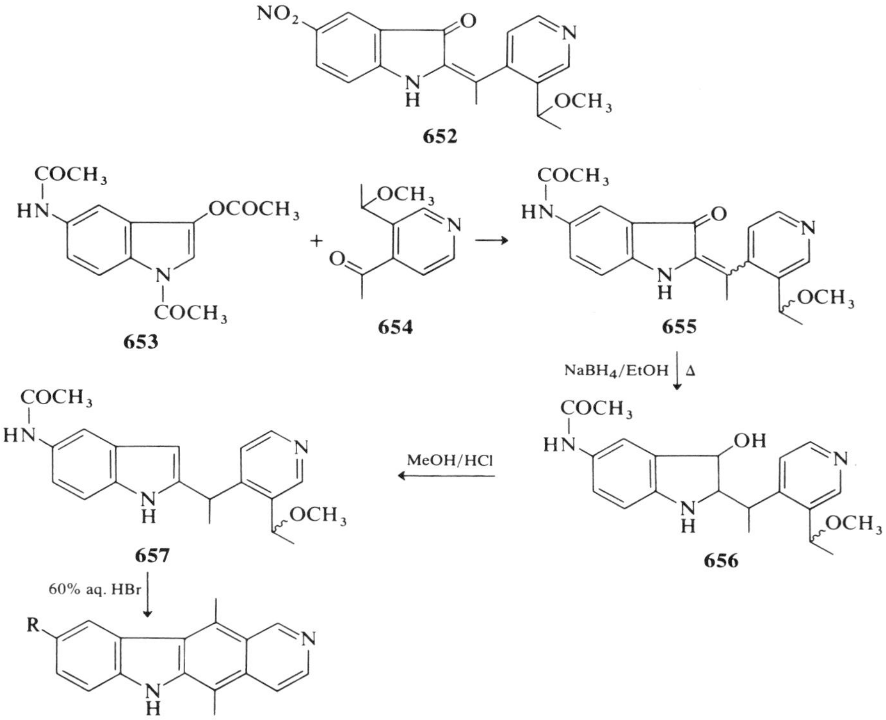

652

653 + **654** → **655**

$NaBH_4/EtOH$, Δ → **656**

MeOH/HCl → **657**

60% aq. HBr →

651 R = NO_2
658 R = NH_2

SCHEME 44. *Synthesis of 9-aminoellipticine* (**658**).

5. Roche Approach

The Roche group (*265*) has also been interested in the synthesis of ellipticine derivatives. In particular, 8,9-dimethoxy- and 8,9-methyl-enedioxyellipticines (**659**) and **660** were their synthetic targets.

Condensation of 5,6-methylenedioxyindole (**661**) with hexane-2,5-dione gave the carbazole **662** which was formylated and condensed with amino acetal. The resulting Schiff base **663** was reduced with sodium borohydride in methanol to give an amine. Tosylation followed by reaction with 6 *N* hydrochloric acid in dioxane at room temperature gave 8,9-methylenedioxyellipticine (**660**; Scheme 45). The overall yield was low.

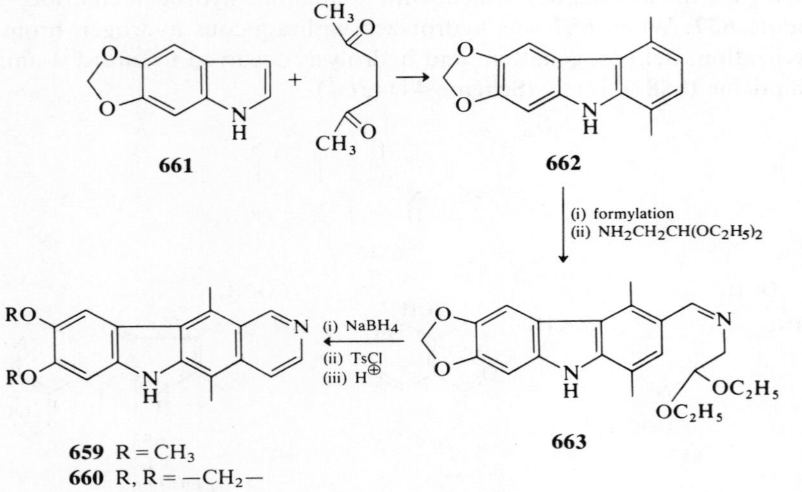

661 **662**

(i) formylation
(ii) $NH_2CH_2CH(OC_2H_5)_2$

(i) $NaBH_4$
(ii) TsCl
(iii) $H^{\oplus}$

663

659 $R = CH_3$
660 R, R = $-CH_2-$

SCHEME 45

6. Sainsbury–Schinazi Approach

Still fascinated by the simplicity of Woodward's synthesis (*261*), Sainsbury and Schinazi (*266*) have continued to improve this, the most direct route.

The aim was to produce a 1:1 adduct of indole and an appropriately substituted pyridine which can be ring closed to give the C-ring.

Reaction of indolemagnesium bromide with the chloroethylpyridine **664** gave the 3-ethylindole derivative (**665**) in 50% yield. Following the procedure of Suzue (*267*), **665** was converted into the salt **666** by successive reaction with *o*-mesitylsulfonylhydroxylamine, acetic anhydride, and methyl iodide, and then treated with potassium cyanide in the presence of ammonium chloride to afford the nitrile **667**. Deacetylation by passage over alumina was followed by reaction with methyllithium, and the intermediate imine **668** was hydrolyzed and cyclized to ellipticine (**267**; Scheme 46). Overall yield for the latter steps is on the order of 25–30%. A similar procedure was used to prepare 8,9-methylenedioxyellipticine (**660**).

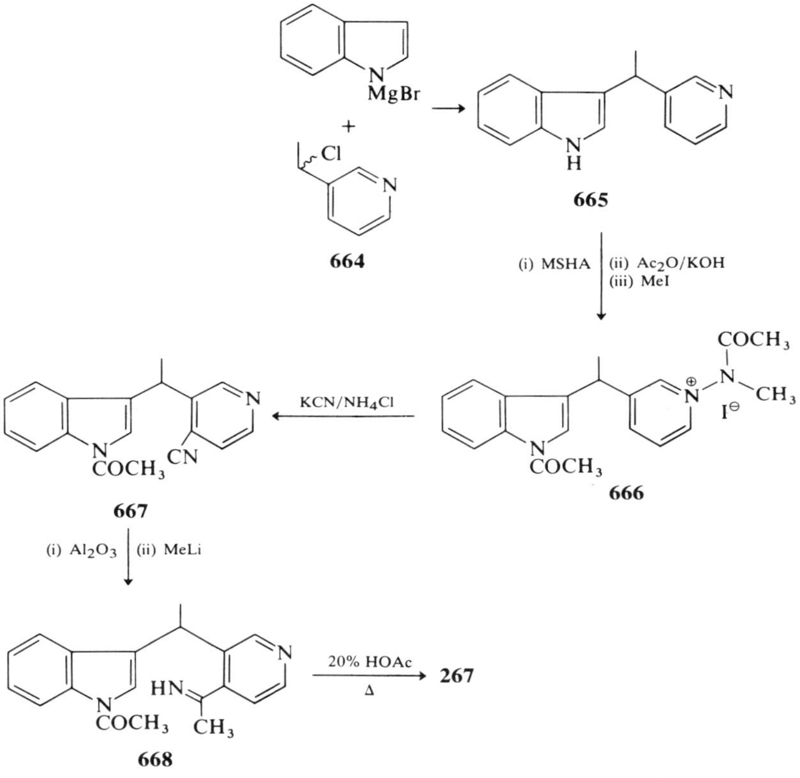

664 **665**

(i) MSHA (ii) Ac_2O/KOH (iii) MeI

KCN/NH_4Cl

667 **666**

(i) Al_2O_3 (ii) MeLi

20% HOAc, Δ → **267**

668

SCHEME 46. *Sainsbury–Schinazi synthesis of ellipticine* (**267**).

7. Kutney–Grierson Approach to Olivacine (**669**)

Like ellipticine (**267**), the antileukemic properties of olivacine (**669**) have attracted the attention of synthetic chemists, although the area has been dormant until quite recently.

The first of the new syntheses was that of Kutney and Grierson (*268*), who investigated a route involving the sequential building of the C and D rings. Condensation of tryptophyl bromide with the sodium salt of methyl acetoacetate at 100° gave the alkylation product **670** in 80% yield. Cyclization with 2% hydrochloric acid in methanol at 0° gave a mixture of **671** and **672** in quantitative yield. The former was converted into the latter by dehydrogenation with chloranil in xylene. The cyclization reaction gave only **672** when carried out in the presence of chloranil. Reduction with lithium aluminum hydride and oxidation with chromium trioxide in pyridine gave the aldehyde **673** in 70% yield. The synthesis of this aldehyde in only three high-yield steps is a great improvement over previous routes (*269*). The remaining synthetic steps were identical with those of a previous procedure (Scheme 47) (*268*).

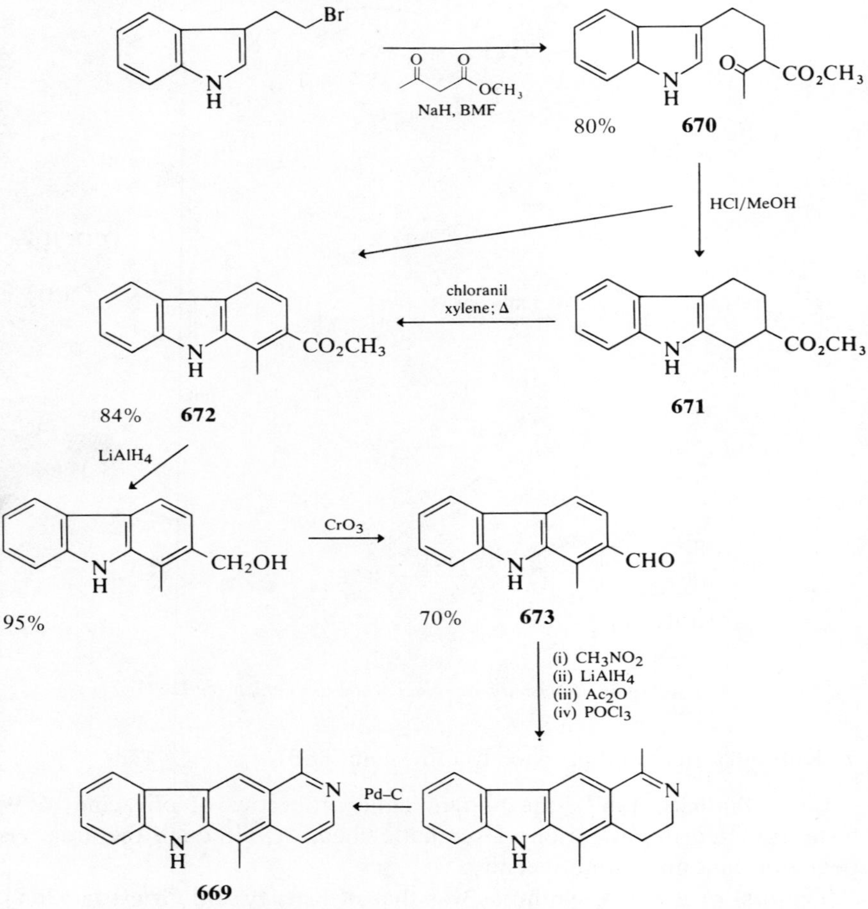

Br

NaH, BMF

OCH_3

CO_2CH_3

80% **670**

HCl/MeOH

chloranil
xylene; Δ

CO_2CH_3

CO_2CH_3

84% **672**

671

$LiAlH_4$

CrO_3

CH_2OH

CHO

95%

70% **673**

(i) CH_3NO_2
(ii) $LiAlH_4$
(iii) Ac_2O
(iv) $POCl_3$

Pd–C

669

SCHEME 47. *Kutney–Grierson synthesis of olivacine* (**669**).

8. Kametani Approach

Kametani and co-workers (*270, 271*) have made some progress in reducing the length of the synthesis.

Cyanation of 3-methoxymethyl-2-methylpyridine *N*-oxide (**674**) gave the 4-cyano derivative **675** in 20% yield which, with methylmagnesium bromide, yielded the 4-acetylpyridine **676**. Reduction with sodium borohydride gave **677**, and subsequent treatment with aqueous hydrogen bromide gave a dibromide **678** which condensed with indole under heating to give olivacine (**669**) in 30% yield (Scheme 48) (*270, 271*).

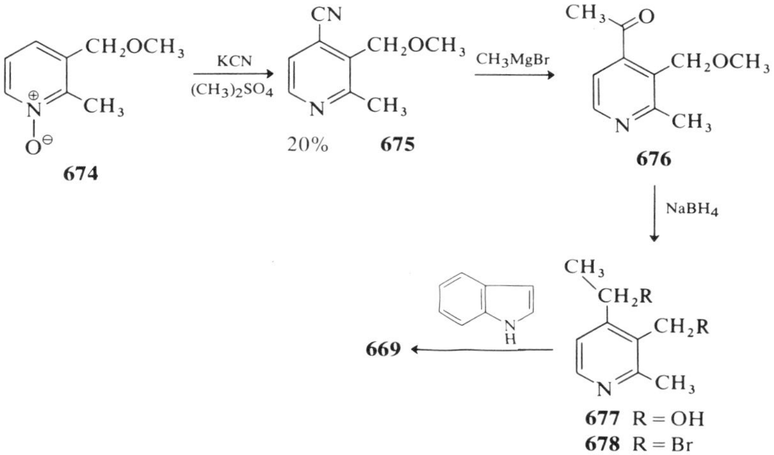

SCHEME 48. *Kametani synthesis of olivacine* (**669**).

9. Besseliëvre–Husson Approach

The most dramatic synthesis, however, is probably that of Besseliëvre and Husson (*272*). Heating indole and the piperideine **679** in 50% aqueous acetic acid for 56 hr afforded **680** in 74% yield. Acetylation and Bischler–Napieraliski cyclization with phosphorus oxychloride and dehydrogenation of the product with palladium–charcoal gave **669** in 45% yield. The interesting condensation reaction is thought to occur via the intermediate immonium species **681**, as shown in Scheme 49.

10. Martinez–Joule Approach

Martinez and Joule (*273*) have described a very general route to the 6*H*-pyrido[4,3-*b*]carbazole system involving γ-aminoenone intermediates. Ketalization of **682** followed by indole N-methylation, pyridine N-methylation, and sodium borohydride reduction afforded **683**. Mannich condensation gave a product which was hydrolyzed to the γ-aminoenone **684**. When **684** was refluxed in 50% aqueous acetic acid the hydroxy tetrahydro 6*H*-pyrido[4,3-*b*]carbazole **685** was produced in 25% yield.

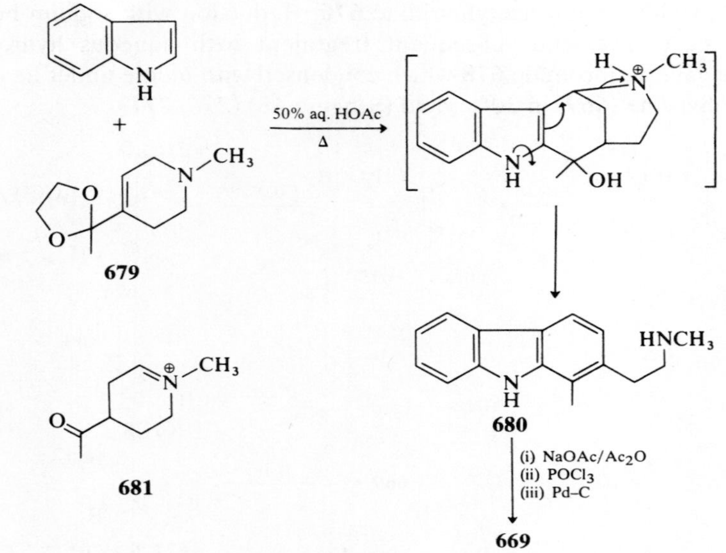
50% aq. HOAc
Δ
OH
679
680
HNCH₃
681
(i) NaOAc/Ac₂O
(ii) POCl₃
(iii) Pd–C
669

SCHEME 49

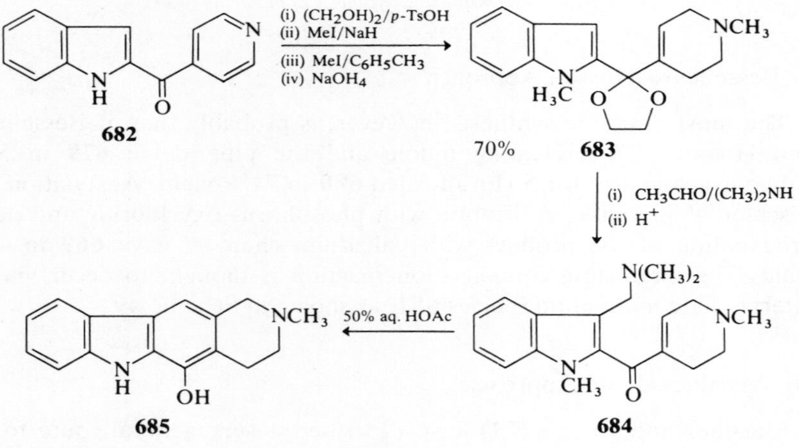
(i) (CH₂OH)₂/p-TsOH
(ii) MeI/NaH
(iii) MeI/C₆H₅CH₃
(iv) NaOH₄
682
70%
683
(i) CH₃CHO/(CH₃)₂NH
(ii) H⁺
N(CH₃)₂
50% aq. HOAc
685
684

V. New Physical Methods in the Structure Elucidation of *Aspidosperma* Alkaloids

Two techniques have revolutionized the structure elucidation of natural products in the past eight years in much the same way that mass spectrometry and proton nuclear magnetic resonance (NMR) spectroscopy did in the early 1960's. These techniques are single-crystal X-ray crystallography and carbon magnetic resonance (CMR) spectroscopy.

It is not the intent of this section to discuss these techniques, but rather to indicate their use in the structure determination of *Aspidosperma* alkaloids. Some of these determinations have been mentioned elsewhere in this chapter.

The Barton–Harley-Mason synthesis (*274*) of *N*-acetyl-16-methylaspidospermidine methiodide (**686**) proceeded in a remarkably stereospecific manner, as indicated elsewhere.

In order to confirm the stereochemical assignments for this compound, an X-ray crystallographic analysis was carried out (*275*). The structure was solved by the heavy-atom method using Patterson and Fourier syntheses. Least-squares refinement reduced the *R* value to 13.6%. Two molecules were observed in the asymmetric unit and were found to correspond to the optical antipodes having the same relative stereochemistry. Thus, the stereochemistry is firmly established to be as shown in **686** (*275*).

A number of alkaloids having the kopsane skeleton have been isolated and their relative stereochemistry deduced (Volume XI, p. 244). To date, all the alkaloids having this skeleton appear to belong to the same stereochemical series, but the absolute stereochemistry has not been determined. The first demonstration of the stereochemistry came by analogy with (−)-aspidospermine (*276*). This assignment was reversed when a chemical correlation with minovincine was carried out (*277*).

In order to firmly establish the absolute stereochemistry, an X-ray crystallographic analysis of (−)-kopsanone methiodide was made (*278*). These data confirmed the assignment of the structure **687** to (−)-kopsanone and determined the D ring to be in a chair conformation.

The X-ray analysis of (−)-aspidospermine methiodide established the absolute configuration **253** for this compound (*279*).

In 1965 Walser and Djerassi isolated vallesamidine from *Vallesia dichotoma* Ruiz *et* Pav. (*272*). Its novel structure yet apparent close relationship to the *Aspidosperma* alkaloids indicated a need for establishment of the absolute stereochemistry.

Vallesamidine methiodide crystallized as monoclinic prisms, and 6375 diffraction intensities were collected. Evaluation of these data not only confirmed the structure but gave the absolute stereochemistry shown in **688** for vallesamidine (*280*).

In a continuation of their efforts to firmly establish the nature of their isolates Djerassi and Ling have examined the X-ray crystal structure of one of the novel alkaloids isolated from *Aspidosperma dispermum* (*76*). These alkaloids lack the characteristic ethyl side chain and are therefore of considerable biosynthetic interest. The compound chosen was the methyl

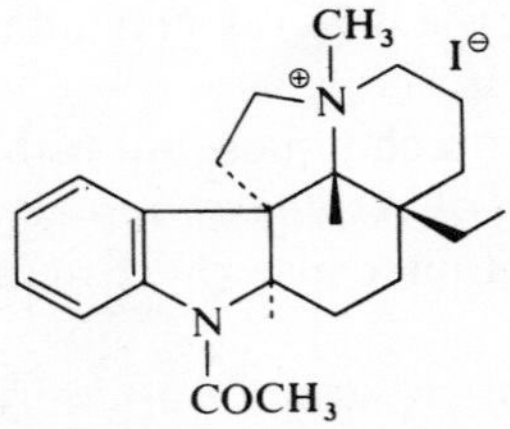

686 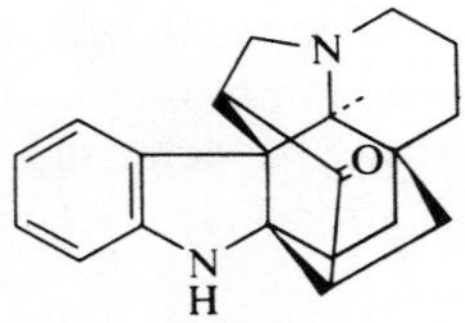 **687** 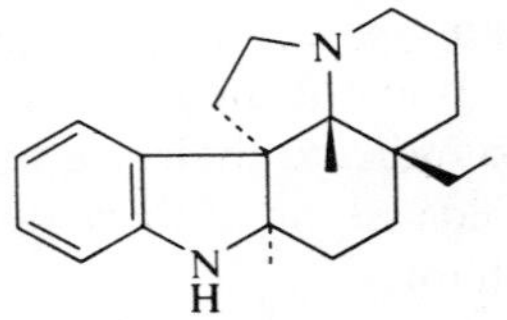 **253**

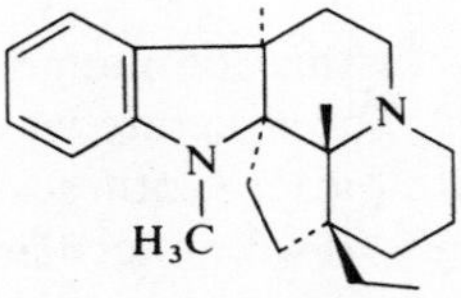

688 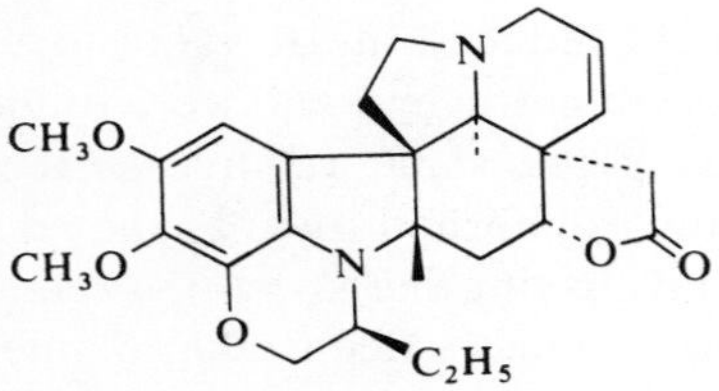 **689** 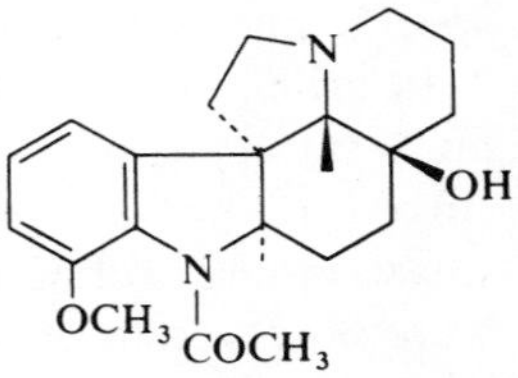 **162**

ether of aspidodispermine as the hydrobromide. The compound has a negative rotation. The structure was solved by determination of the position of bromine followed by further refinement by a full-matrix least-squares calculation. The calculated Fourier map revealed all the atoms of the alkaloid and indicated the structure to be **162** (*281*). That this was also the absolute configuration was confirmed by comparison of the ORD curve with that of (−)-aspidospermine (**253**).

Brown and Djerassi have described the isolation of several novel *Aspidosperma* alkaloids from *A. obscurinervium* Azembuja (*282*). Typical of these alkaloids is obscurinervine (**689**). The unique nature of the heptacyclic system was confirmed by an X-ray crystallographic analysis of the hydrobromide of obscurinervine (*283*).

The structure determined from this analysis indicated that the original structure (**689**) proposed for obscurinervine was correct including all aspects of the relative stereochemistry.

The structure of haplophytine was determined by crystallographic analysis of the dihydrobromide to be **690** (*284*). However, it was subsequently established that in the formation of the dihydrobromide a rearrangement must have taken place. Haplophytine was then suggested to have structure **691**. Clearly, a rearrangement such as this required more careful evaluation, and one approach was to examine the X-ray crystallographic structure of haplophytine itself (*285*).

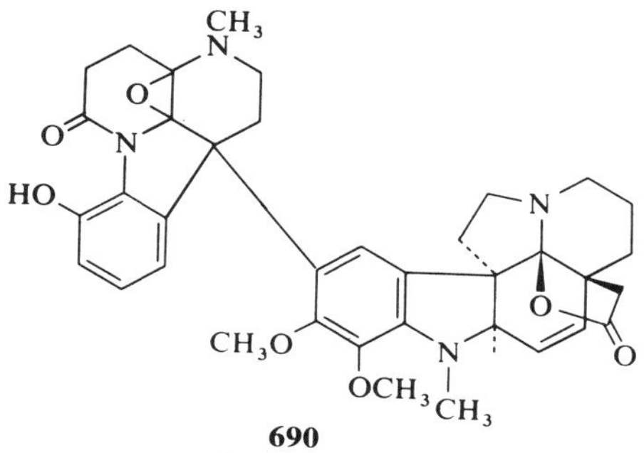

690

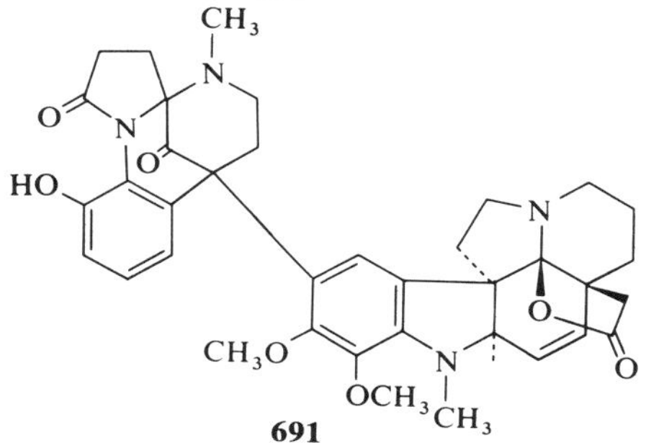

691

Haplophytine crystallized as the methanolate in the orthorhombic space group. Refinement by full-matrix anisotropic least-squares three times gave a difference map containing all C, N, and O atoms. The resulting structure is in agreement with **691**, including all the stereochemical assignments (*285*).

Oberhänsli (*286*) has examined the crystal structure of the methobromide of meloscine and confirmed the structure and stereochemistry deduced by chemical and spectroscopic methods.

The results of the X-ray structure determinations of voaphylline (**177**) (*86*), capuronine acetate (**227**) (*107*), and dihydrovindolinine (**152**) (*72*) have been discussed elsewhere.

Without doubt, however, the most important technique to come to the fore in the last ten years is that of ^{13}C NMR aided by Fourier transform instrumentation. The ^{13}C resonances for many carbons in indole alkaloids have characteristic values, and consequently considerable structural information can be gained from the spectrum. In Table III the ^{13}C NMR shifts of the carbon atoms of a number of alkaloids in the *Aspidosperma* series are given (*20, 50, 51, 59, 100, 102, 109, 287–290*).

Clearly, a detailed discussion of either the technique or individual resonances is beyond the scope of this review. A number of generalizations may be made, however, based mainly on the work of Wenkert *et al.* (*20*).

TABL

^{13}C NMR OF *ASPIDOSPERMA*

	Carbo											
	2	3	5	6	7	8	9	10	11	12	13	14
Vincadifformine	167.4	50.4	51.6	45.1	55.4	137.7	121.0	120.5	127.3	109.2	143.7	21.9
	167.8	51.7	50.7	45.3	55.5	138.0	121.0	120.5	127.4	109.3	143.4	22.2
N-Methyl-2,16-dihydro	75.5*	52.2	53.5	42.9	52.4	135.5	122.3	117.6	127.3	106.9	152.4	21.1
N-Methyl-14β-hydroxy-2,16-dihydro-	77.0*	59.0	54.0	42.5*	52.9	136.0	122.7	118.3	128.0	107.5	153.0	66.7
N-Methyl-18α-hydroxy-2,16-dihydro-	77.0	45.4	53.0	43.6	52.2	136.0	123.0	118.3	128.0	107.5	153.0	28.0*
Tabersonine	166.7	50.3*	50.8*	44.6	55.0	137.8	121.4	120.5	127.6	109.2	143.1	124.8
2,16-Dihydro-	66.3	51.3	52.2	43.0	52.9	133.5	122.8	117.5	127.3	107.9	149.9	122.8
N-Methyl-2,16-dihydro-	75.7	51.4	52.8	43.3	52.8	134.3	122.1	117.4	127.5	106.7	152.3	122.1
Hazuntinine	166.0	49.2	51.4	43.6	54.8	128.7	103.5	149.3	143.5	95.3	137.0	51.8
Vindoline	83.2	50.9	51.9	43.9	52.6	124.9	122.4	104.5	161.1	95.6	153.6	123.9
Dihydro-	83.5	52.4	51.4	43.4	52.4	125.2	122.7	104.1	160.6	95.6	154.0	22.4
Vincoline												
19-Epi-	95.0	53.0	54.1	35.7	55.9	135.7	121.9	118.8	126.0	108.2	147.4	128.5*
Cimicine	67.0	42.2*	48.1	33.6*	58.5	137.8	118.2	128.3	115.0	146.9	127.2	24.6*
Cimicidine	67.7	42.3*	48.3	33.7*	58.0	137.2	114.5	110.3	149.9	129.6	128.0	25.6*
Aspidophytine	71.8	43.3	47.3	35.4	57.3	133.8	120.2	102.3	153.9	143.6	149.4	21.5
Vandrikidine	167.4	49.9	50.8	44.2	54.8	130.4	122.0	105.2	159.9	96.6	144.0	127.6
Vandrikine	167.4	45.7	51.2	45.1	54.2	130.5	121.5	104.8	159.8	96.5	144.1	27.4*
Vindolinine	81.4	58.0	50.3	36.3	59.8	139.8	123.6	121.0	127.2	112.0	149.4	128.5
N-Methyl-	84.4	58.0	50.0	36.0	58.8	135.8	123.0	117.8	127.7	105.6	150.2	127.7
16-Epi-	80.5	57.4	50.1	35.0	60.7	135.7	123.1	118.9	126.9	109.0	148.7	128.2
14,15-Dihydro-	80.6	55.0	48.1	37.3	60.3	140.1	123.6	121.1	127.2	112.7	149.5	20.7
Venalstonine	66.5	49.0	50.0	36.4	56.1	139.5	121.1	119.0	126.8	110.9	149.0	126.5
Pandoline	165.9	68.6	51.1	45.2	55.6	137.2	121.3	120.5	127.9	109.4	143.7	36.0
2,16-Dihydro-	65.7	73.5	54.6	39.4	52.2	134.6	122.5	118.3	127.5	108.7	150.2	35.2
Pandine	164.8	68.9	50.3	41.4	60.2	130.5	121.2	120.9	127.6	108.6	144.3	43.0
2,16-Dihydro-	63.0	74.6	52.6	37.7	54.8	131.7	121.6	118.6	127.5	109.5	149.7	43.6
Ibophyllidine	165.1	65.7*	47.6	41.3	55.8	138.6	123.2	121.3	127.6	108.7	143.1	37.8
Iboxyphylline	164.1	70.3	53.1	41.4	57.8	137.1	122.2	119.9	127.3	108.7	142.8	37.6
Vincamine	131.4	44.5	50.9	16.9	105.9	128.9	118.4	121.5	120.1	110.2	134.1	20.8
Vincine												
14,15-Dehydro-	130.2	43.3	49.3	16.4	106.0	123.4	118.6	109.2	156.0	95.2	134.8	125.3
14,15-Dehydro-16-epi-	131.5	43.4	49.3	16.4	106.0	123.4	118.6	109.2	156.0	95.2	134.8	125.3
Meloscine	171.9	45.6	52.4	43.2	56.8	126.5	127.2	123.6	172.2	115.4	134.8	126.4
Epimeloscine	173.0	45.7	51.7	35.4	55.3	135.8*	122.3*	123.2*	126.7	116.2	136.5*	120.8
Scandine	170.2	47.6	53.2	39.8	57.7	128.5	126.7*	123.4	127.2*	115.5	134.1	122.7
Melascandonine	169.0	47.2	54.8	38.1	54.8	130.5	123.5*	123.4*	127.6	116.3	136.5	124.0

[a] Asterisk indicates some ambiguity regarding the assignment.

III
ALKALOIDS[a]

15	16	17	18	19	20	21	N–CH_3	Ester CO	OCH_3	Ar –OCH_3	CO	Acetyl CH_3	Ref.
32.3	92.4	25.6	7.0	29.3	38.1	72.4		168.9	50.8				*109*
32.9	92.8	25.6	7.3	29.3	38.2	72.7		169.2	50.9				*20*
36.4	37.9	27.5	7.8	34.2	33.4	76.1*	35.2	175.5	51.1				*287*
42.1*	37.6	28.3	7.8	35.7	33.2	75.7*	35.4	175.8	51.6				*287*
72.6	38.0	25.8	7.3	29.0*	38.0	70.2	35.6	176.2	51.4				*287*
132.9	92.2	26.7*	7.3	28.4*	41.2	69.9		168.8	50.8				*20*
134.5	39.2	31.8	8.2	33.7	37.3	69.1		175.2	51.3				*287*
135.0	37.7	31.0	8.3	33.7	37.0	70.0	35.8	174.9	51.1				*287*
57.0	90.7	23.3	7.3	26.3	36.8	70.8		168.8	50.7	55.9			*20*
130.2	79.5	76.2	7.5	30.6	42.8	67.0	38.0	170.4	51.9	55.1	171.7	20.8	*20*
33.0	78.3	75.6	7.8	29.9	40.0	72.3	37.7	170.0	52.0	55.0	172.3	20.7	*20*
127.5*	86.4	34.7	18.9	81.8	48.0	76.2		172.6	52.1				*51*
34.3*	24.6*	20.0	175.4	43.3*	40.2	106.1							*59*
34.8*	25.6*	20.0	175.6	43.5*	40.3	106.3				56.2			*59*
34.6	125.6*	130.5*	175.7	47.8	41.4	107.2	35.4			55.8			*288*
129.6	90.8	28.1	17.7	67.9	46.0	66.3		168.3	50.8	55.3			*20*
79.8	93.9	26.6*	64.7	34.6	46.4	68.7		168.5	50.8	55.2			*20*
130.7	39.2	29.1	7.4	48.4	46.2	78.0		174.2	51.8				*50*
130.8	37.0	28.0	9.0	47.0	45.6	77.0	30.0	174.0	52.0				*50*
130.6	39.4	31.9	7.8	44.8	47.8	76.4		174.5	51.7				*50*
31.2	40.2	29.0	7.5	51.0	44.5	78.8		175.0	52.0				*50*
132.5	43.4	29.6	31.6	34.0	35.0	66.8		173.7	51.6				*50*
39.3	97.4	25.7	7.4	32.6	71.0	61.3		168.6	51.1				*100, 109*
38.8*	37.9	23.6	7.1	33.1	71.0	63.6		175.7	51.5				*100*
44.2	96.3	40.8	8.2	30.2	82.0	77.0		168.1	51.1				*102*
44.0	39.2	36.6	8.3	29.8	80.3	75.5		173.2	51.8				*102*
34.9	92.2	25.6	12.4	31.8	75.6*			168.9	50.8				*109*
36.5	94.2	26.0	16.5	41.1	73.1	59.3		167.9	50.7				*109*
25.2	81.9	44.5	7.6	28.8	35.1	59.1		174.3	54.1				*289*
127.9	82.0	43.1	8.1	34.5	36.6	57.3		172.9	53.7	55.4			*289*
127.9	84.0	45.6	8.1	34.9	38.0	56.9		171.8	52.2	55.4			*289*
134.2	50.0	40.8	112.2	142.4	47.3	81.1							*290*
130.9	47.9	34.0	112.1	144.3	44.9	71.5							*290*
131.2	63.6	44.0	144.4	142.0	46.5	83.5		169.3	52.4				*290*
127.4	67.7	36.0	11.0	50.7	44.3	69.9		210.0					*290*

As with most ^{13}C NMR spectra, those of *Aspidosperma* alkaloids can be divided into two regions: above 90 ppm, where there is the region of sp^2 carbon atoms (olefinic, aromatic, carbonyl); and below 90 ppm, where there is the complex region of sp^3 carbons attached to hydrogen, oxygen, or nitrogen. The indole or indoline nucleus shows quite characteristic carbon resonances when there is no substitution on the benzene ring and the order is typically C-13, C-8, C-10, C-11, C-9, C-12, and C-7. In the β-anilinoacrylate series this order is C-13, C-8, C-11, C-9, C-10, and C-12.

Substitution by methoxy in the aromatic nucleus is readily apparent. The typical shift for the carbon bearing the methoxy group is 30 ppm down field, for an *ortho* carbon 15 ppm and for a *para* carbon 7 ppm. The point of attachment is therefore simply determined.

In general the next group of carbon frequencies is those associated with heteroatoms, e.g., C-21 in *Aspidosperma* alkaloids, where the heteroatom is oxygen, or C-20 in pandoline (**220**), where the atom is oxygen. As expected, attachment to both oxygen and an electron-withdrawing group has a substantial effect, e.g., C-16 in vindoline.

A third more diffuse group of resonances is those carbons bearing electron-withdrawing groups or which are allylic, e.g., C-16 in venalstonine and C-20 in tabersonine. The remaining carbons are those associated with aliphatic carbons, and in many cases close chemical shift approximations can be accomplished by application of established equations or, preferably, good skeletal models.

One of the important observations to develop the work on *Aspidosperma* alkaloids was an "endocyclic homoallylic effect" (*20*) in alkaloids having a double bond in the piperidine ring. For example, C-21 in tabersonine (**28**) is observed at 69.9 ppm, yet is at 72.7 ppm in vincadifformine (**2**).

The 300 MHz spectrum of vindolinine (**150**) was carried out by Durham *et al.* (*291*) to confirm the structure revision suggested by the C-13 NMR spectrum (*50*) and confirmed by the X-ray analysis (*73*).

As these workers pointed out, a structure such as the one now proved for vindolinine was ruled out by the assignment of a "doublet" at δ3.83 ppm in the 60 MHz spectrum to the C-2 proton. This assignment has been used subsequently by numerous workers in structural assignments.

The 300 MHz spectrum of vindolinine (**150**) still contained some overlapping signals but was essentially first order. Because of this, not only did the proton assignments become greatly simplified, but most of the coupling constants could be determined. The key question to be answered initially was, "Which proton gave rise to the signal at 3.97 ppm which was now observed as a pair of doublets ($J = 18$, 5 Hz)?" By measurement of coupling constants and appropriate spin decoupling (at 100 MHz), it was shown that this proton was the β-proton at C-3 being strongly deshielded by the nitrogen and the double bond. The C-3α proton appeared at 3.29 ppm. Because these assignments and the coupling constants are so important in the structure determination of *Aspidosperma* alkaloids, they are reproduced in Table IV.

TABLE IV
PROTON ASSIGNMENTS AND COUPLING CONSTANTS OF VINDOLININE[a]

	δ(ppm)	J(Hz)		δ(ppm)	J(Hz)
H-3α	3.45	3.3, 2, 18	H-15	6.16	10, 3.3
H-3β	3.97	5, 18	H-16α	3.04	6, 12
H-5α	3.29	9, 10.5, 7	H-17α	2.47	6, 14.5
H-5β	3.40	9, 9	H-17β	1.80	12, 14.5
H-6α	2.17	7, 16	H-19α	2.08	7.0
H-6β	1.85	10.5, 16, 9	H-21	3.46	
H-14	5.78	10, 5, 2			

[a] From Bruneton and Cavé (*153*).

The use of INDOR experiments in the Fourier transform (FT) mode has recently been applied to the structure elucidation of andranginine (**378**) and brings yet another PMR technique to the structure elucidation of *Aspidosperma* alkaloids.

The FT technique permits the storage of the original free induction decay (FID) signals. The proton under study is subjected to the RF irradiating field, and the same number of FID signals are accumulated. The two are then subtracted and the difference Fourier transformed. Signals which are not coupled to the proton under study disappear after subtraction, although those of close chemical shift may be observed at low intensity and slightly shifted. The proton under study reappears after subtraction, as expected. The coupling constants can be obtained from the difference spectrum. By judiciously irradiating critical protons, most of the protons could be assigned and their stereochemistries deduced (*126*).

The diagnostic potential of proton nuclear relaxation times is an underdeveloped NMR technique which has only recently been applied to complex molecules. The first determination on vindoline (**101**) used the audio-frequency pulse technique (*292*), but with the advent of FT techniques the longitudinal relaxation times (T_1) of all the protons of vindoline could be determined (*293*) (see **692**). The T_1 values vary from 0.45 to 2.7 sec, the smaller times being quite characteristic of methylene protons. Also important was the finding that peripheral groups showed longer relaxation times than groups in the center of the molecule.

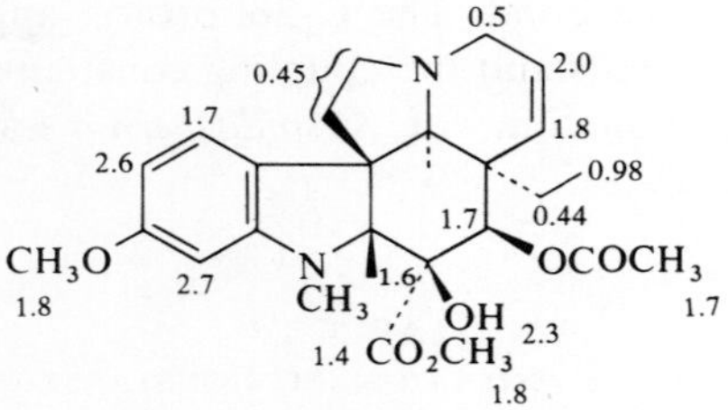

692

Appendix

In the time that has elapsed since this review was first submitted for publication there have been several papers published in the area of *Aspidosperma* alkaloid chemistry. In addition, a small number of papers omitted from the prior discussion have been brought to the attention of the reviewer. An appendix was therefore deemed necessary. It is organized along the same lines as the main body of the review, and appropriate section designations are given for the new material.

II.A.19. 14α,15α-Epoxy-19*S*-hydroxytabersonine (Cathovalinine) (**693**)

Continuing studies on *Catharanthus ovalis* Mgf. have afforded another new β-anilinoacrylate alkaloid, cathovalinine {mp 223°, $[\alpha]_D -492°$} (*294*). The molecular formula, $C_{21}H_{24}O_4$, suggested the addition of two oxygen atoms to a tabersonine skeleton, and from the mass spectrum which showed important fragment ions at *m/e* 214 and 154, these should be located in the piperidine ring or the ethyl side chain. An ion at $M^+ - 45$ (loss of radical B) indicated that one of these oxygens was a hydroxy group in the ethyl side chain. The NMR spectrum located this group at C-19 from a three-proton doublet at 1.13 ppm for the C-18 methyl group, and a quartet at 3.30 ppm for the C-19 methine proton. Absorptions in the IR spectrum indicated the presence of hydroxyl and β-anilinoacrylate units, but gave no indication of additional carbonyl absorption. The oxygen in the piperidine ring must therefore be present as an ether group. The possible presence of a C-3 lactam was eliminated when cathovalinine gave, with sodium cyanoborohydride–acetic acid, a 2,16-dihydro derivative showing only saturated ester absorption at 1740 cm^{-1}. A minor product of this reaction was an internal lactone which showed a C-19 proton at 4.45 ppm. Cathovalinine was therefore assigned the gross structure **694** identical to that of hörhammericine (**41**). Direct comparison with hörhammericine, however, indicated that these compounds were not identical. The structure of cathovalinine (**693**) was solved by single-crystal X-ray analysis which indicated a 14α,15α-epoxy and a 19*S* hydroxy group (*294*).

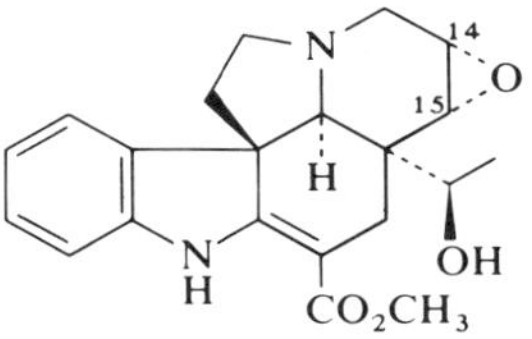

693

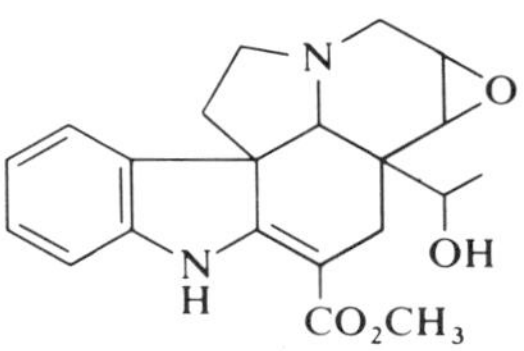

694

II.B.7. Alkaloids of *Melodinus celastroides*

Rabaron and Plat (*295*) have reported the isolation and structure elucidation of four new alkaloids from the leaves and twigs of *Melodinus celastroides* Baillon.

20,21-Epi-(+)-aspidospermidine (**695**) {$[\alpha]_D + 82°$} gave a molecular ion at *m/e* 282 ($C_{19}H_{26}N_2$) and a CD curve indicating the chirality at C-2 and C-7 to be the same as that in (+)-aspidospermidine (**249**). The major differences, however, were observed between the mass spectra of **695** and **249**. In particular, the molecular ion of **695** is the base peak and the $M^+ - 1$ ion is quite intense, and whereas *m/e* 124 is the base peak in the spectrum of **249**, it is only a weak signal in the spectrum of **695**. These data suggest that the C-21 proton in **695** has the α configuration.

The NMR spectrum of **695** shows a triplet for the C-18 protons at 0.64 ppm. In **249** this signal appears at 0.88 ppm, indicative of a C-20 α-ethyl group. On this basis, the alkaloid was assigned the structure 20,21-epi-(+)-aspidospermidine (**695**). Also isolated was the *N*-oxide **696** (mp 215°), which was reduced with ferrous ion to give **695**.

The two other alkaloids isolated were found to be structural isomers. Meloceline exhibited a molecular ion at *m/e* 296 ($C_{19}H_{24}N_2O$) and shows an indoline UV spectrum. The IR spectrum displayed a carbonyl absorption at 1660 cm^{-1} characteristic of a γ-lactam. Two protons in the NMR spectrum at 2.19 and 2.72 ppm were assigned to a methylene adjacent to a carbonyl group and a quaternary carbon. The mass spectrum of meloceline had a number of interesting features, namely, a low mass region having ions at *m/e* 130, 143, 144, and 156, and an $M^+ - 42$ ion suggesting a loss of ketene. The structure **697** for meloceline is in agreement with these findings.

Melocelinine (mp 188°), although quite similar to meloceline overall, gave a carbonyl absorption at 1635 cm^{-1} typical of a δ-lactam and a base peak at *m/e* 140 in the mass spectrum interpreted as **698**. Melocelinine therefore has the structure **699**.

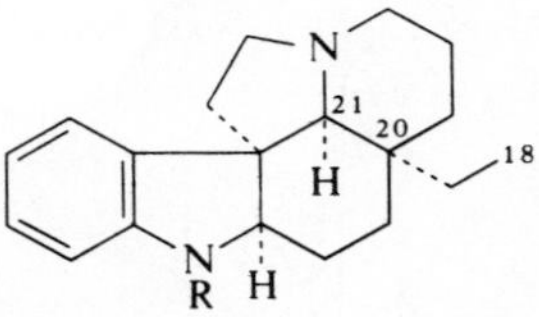

695 R = H
700 R = $COCH_3$

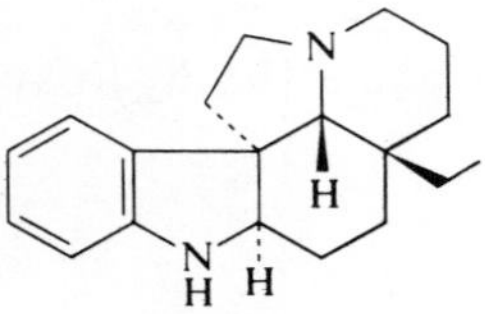

249

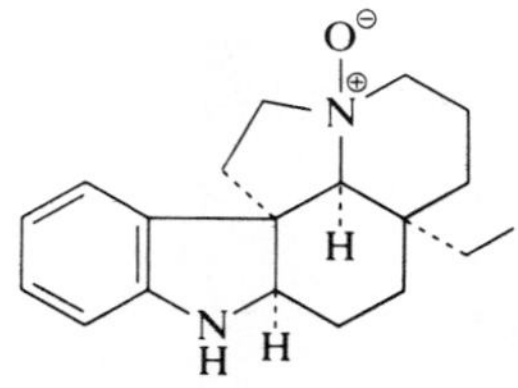

696

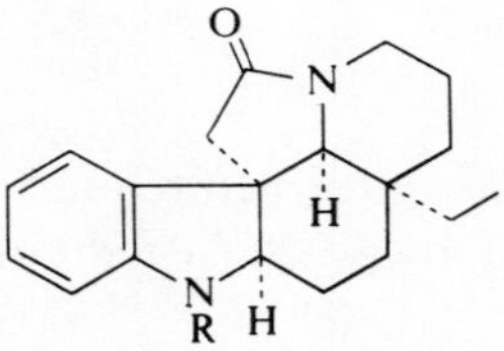

697 R = H
701 R = $COCH_3$

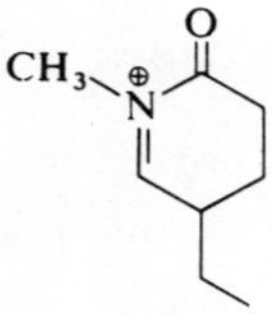

m/e 140
698

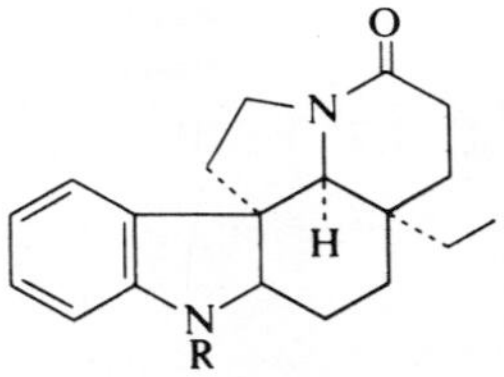

699 R = H
702 R = $COCH_3$

Oxidation of *N*-acetyl-20,21-epi-(+)-aspidospermidine (**700**) with chromium trioxide in pyridine gave a mixture containing both acetylmelocine (**701**) and acetylmelocelinine (**702**). Consequently, melocine (**697**) and melocelinine (**699**) were also assigned the 20,21-epi configuration (*295*).

II.H.1 Pandoline (220) and Epipandoline (221)

Pandoline (**220**) and epipandoline (**221**) have been isolated from *Melodinus polyadenus* (Baillon) Boiteau (*296, 297*) and *Ervatamia obtusiuscula* Mgf. (*298*).

In previous work the C-20 configuration of these alkaloids had been suggested on the basis of chromatographic properties. More recently, two groups have reported on this problem (*297, 299*). Reduction of pandoline with sodium borohydride/glacial acetic acid at 90° for 30 min gave a mixture of 3,7-secopandolines A (**703**) and B (**704**), whereas epipandoline gave only one isomer (**705**). The three compounds were examined by ^{13}C NMR, but the similarity of the chemical shifts could not in itself distinguish between the C-20 configurations. Comparison with (+)-velbanamine (**706**) of established C-20 stereochemistry gave a good chemical correlation. Representative data are shown in Table V. Thus, 3,7-secopandoline A (**703**) and B (**704**) have the C-20 stereochemistry in which the C-20 ethyl group is β. Pandoline and epipandoline are therefore represented by structures **220** and **221**, respectively (*297*).

TABLE V
REPRESENTATIVE CHEMICAL SHIFTS OF PANDOLINE DERIVATIVES

		δ(ppm)					
		C-14	C-15	C-18	C-19	C-20	C-21
3,7-Secopandoline A	(**703**)	30.5	39.5	6.9	32.6	71.2	66.1
3,7-Secopandoline B	(**704**)	30.4	38.3	7.1	33.8	71.3	65.6
(+)-Velbanamine	(**706**)	30.1	40.4	6.9	32.3	71.6	65.8
3,7-Secoepipandoline	(**705**)	29.2	37.7[a]	7.4	35.4	72.6	64.4

[a] Tentative assignment.

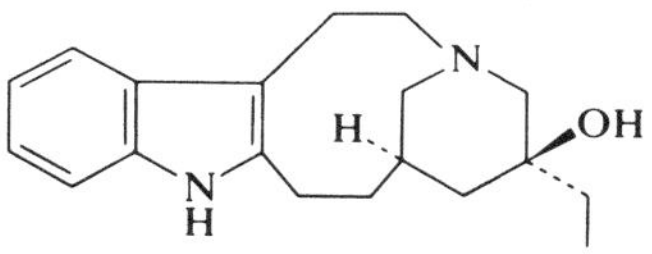

706

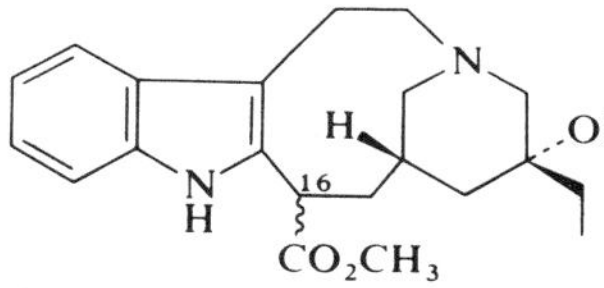

703/704

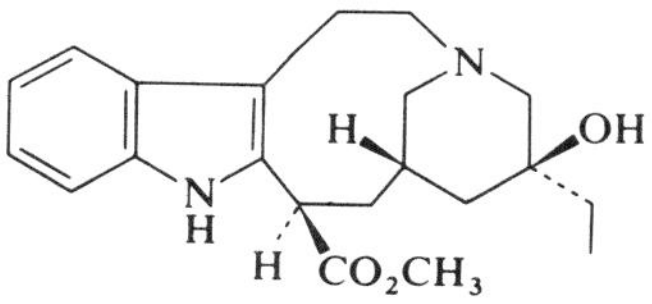

705

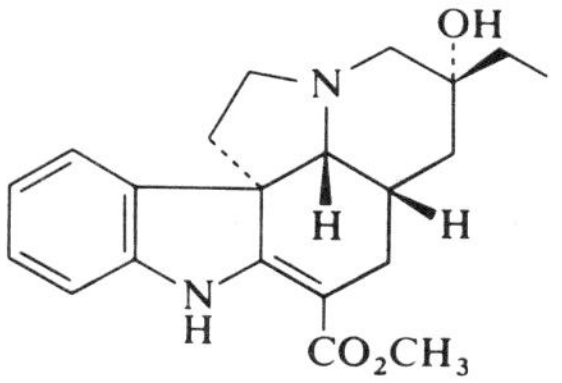

220

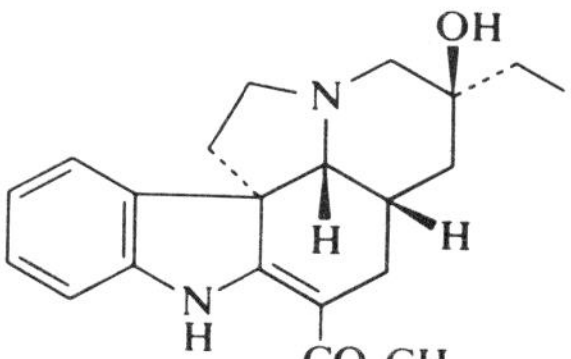

221

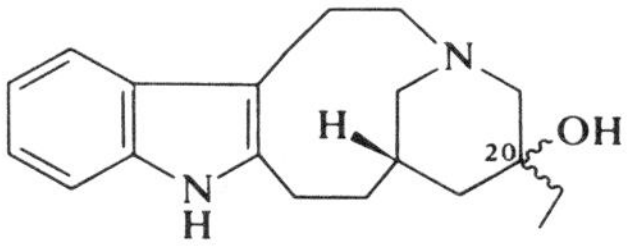

707 β-C_2H_5, α-OH
708 α-C_2H_5, β-OH

A second determination (*299*) was made by conversion of (+)-pandoline (**220**) to (−)-velbanamine (**707**) by catalytic hydrogenation (Pd/C) in trifluoroacetic acid followed by acid-catalyzed decarboxylation to give **707** {mp 137–142°, $[\alpha]_D$ −56° ($CHCl_3$)}. A similar conversion of (+)-20-epipandoline (**221**) led to (−)-isovelbanamine (**708**) {mp 189–193°, $[\alpha]_D$ −21° (MeOH)}.

The two products were identical with authentic samples except for optical rotation. Thus, pandoline (**220**) has the 20*R* configuration in which the hydroxyl group has the α-orientation, and epipandoline (**221**) has the 20*S* configuration.

II.H.2. Pandine (**709**)

A single-crystal X-ray crystallographic determination of pandine indicated that the original gross structure was correct and that the C-20 configuration was *R* (*300*). Pandine is therefore represented by structure **709**.

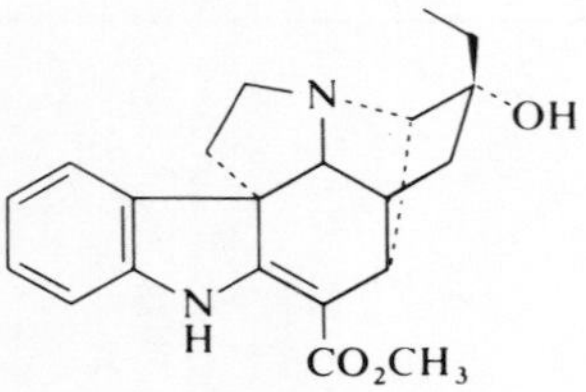

N OH N H CO_2CH_3

709

II.J.6. Tubotaiwine-*N*-Oxide (**273**)

Kingston and co-workers (*301*) have isolated tubotaiwine-*N*-oxide (**273**) as the cytotoxic constituent of the roots of *Tabernaemontana holstii* K. Schum.

II.J.7. Ellipticine (**267**) and Related Compounds

The metabolism of ellipticine (**267**) has been examined in the rat. Although 9-hydroxyellipticine was shown to be the product of metabolism, unlike most aromatic hydroxylations there is no involvement of an arene oxide, and therefore no NIH shift mechanism operates (*302*).

III.A. Rearrangement of Tabersonine (**28**) to Vincamine (**277**)

The rearrangement of tabersonine (**28**) to vincamine (**277**) is an important commercial process. Recently, two (*303, 304*) patents have appeared describing new routes to **277** from **28**.

Oxidative rearrangement of **28** with *m*-chloroperbenzoic acid followed by hydrogenation gave vincamine (**277**) and epivincamine (**299**) in the ratio 1:6.5. Reversal of the reaction sequence and the use of *p*-nitroperbenzoic acid gave **277** and **299** in the ratio 4:1 (*303*).

The oxidative rearrangement of **28** and **2** has also been affected with cupric ion in aqueous dimethylformamide in an oxygen atmosphere at room temperature (*304*). Vincamine (**277**) and 16-epivincamine (**299**) were produced in 8.4% and 12% yields, respectively.

III.L. Chemistry of Vindoline (101)

In spite of the importance of vindoline (**101**), the chemistry of this alkaloid had been little investigated until recently, and was stimulated by a desire to vary the nature of the unit coupling with catharanthine *N*-oxide (*305*).

Deacetyldihydrovindoline (**710**) when heated with *N,N'*-thiocarbonyldiimidazole gave the thiocarbonate **711** in 88% yield, and subsequent treatment with Raney nickel in refluxing THF gave **712**. Catalytic reduction (H_2/Pt) afforded **713** after treatment with potassium *tert*-butoxide. A tertiary hydroxyl group at C-16 was introduced by treatment of **713** with hydrogen peroxide–oxygen in the presence of lithium diisopropylamide, although assignment of the stereochemistry in the product **714** was not possible (Scheme 50). A similar series of reactions in the vindoline series gave **715** (*305*).

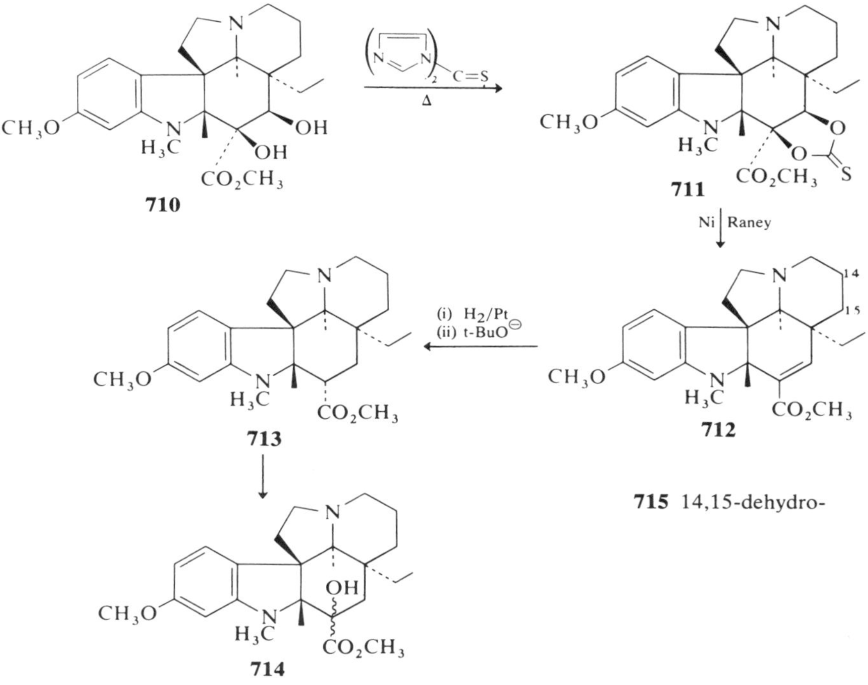

710 → (**711**) → Ni Raney → **712** → (i) H_2/Pt (ii) t-BuO$^{\ominus}$ → **713** → **714**

715 14,15-dehydro-

Scheme 50

A partial review of the synthesis of the *Aspidosperma* alkaloids has been published (*306*).

IV.A.5. Synthesis of Deoxyaspidodispermine (**158**)

Honma and Ban (*307*) have reported a second synthesis of deoxyaspidodispermine (**158**). Crucial in this scheme is the eosine-sensitized stereospecific addition of oxygen to the diene (**716**) in quantitative yield. Catalytic hydrogenation of **717** under pressure gave diol **718**, from which water was eliminated, after basic hydrolysis of the urethane group, to give the indolenine **719**. Lithium aluminum hydride reduction gave the indoline **720** which could be acetylated to deoxyaspidodispermine (**158**) under mild conditions (*307*) (Scheme 51).

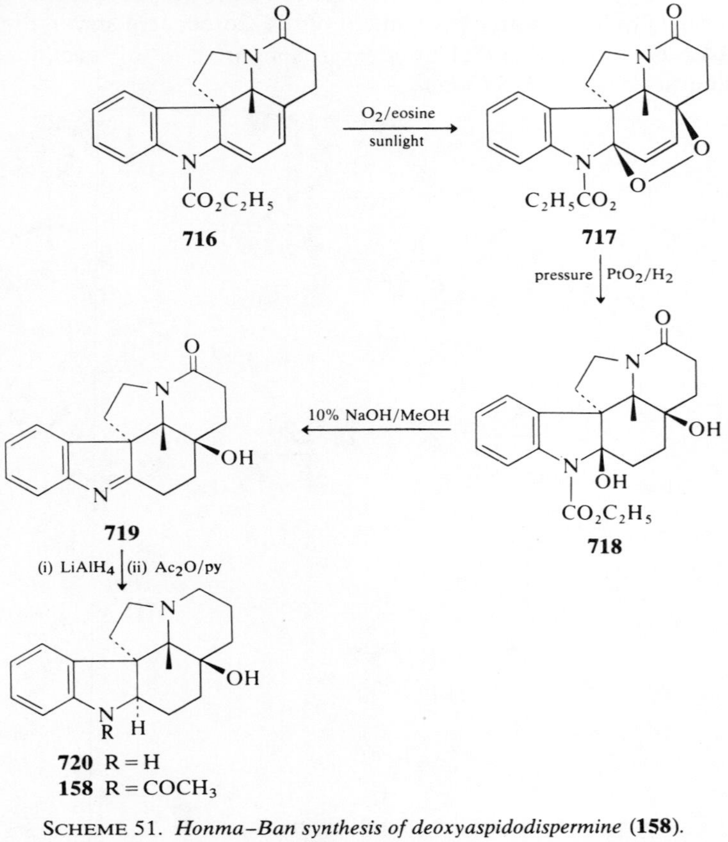

$CO_2C_2H_5$

716

O_2/eosine
sunlight

$C_2H_5CO_2$

717

pressure PtO_2/H_2

OH
OH
$CO_2C_2H_5$

718

10% NaOH/MeOH

OH

719

(i) $LiAlH_4$ (ii) Ac_2O/py

OH
R H

720 R = H
158 R = $COCH_3$

SCHEME 51. *Honma–Ban synthesis of deoxyaspidodispermine* (**158**).

IV.A.7. Synthesis of Aspidoalbidine (**721**)

In 1975, Ban and co-workers reported a synthesis of *N*-acetylaspidoalbidine (**411**) (*211*). More recently, this group has reported (*308*) a synthesis of the parent compound aspidoalbidine (**721**), which is also known as fendleridine.

Lithium aluminum hydride reduction of the aldehyde **447** was previously reported to give the enamine alcohol **450**. Reexamination of the product by NMR spectroscopy indicated that this structure should be revised to **722**. Catalytic hydrogenation gave the amine alcohol **723**, and subsequent oxidation with mercuric acetate in 5% aqueous acetic acid at 75° gave **721** (*308*) (Scheme 52).

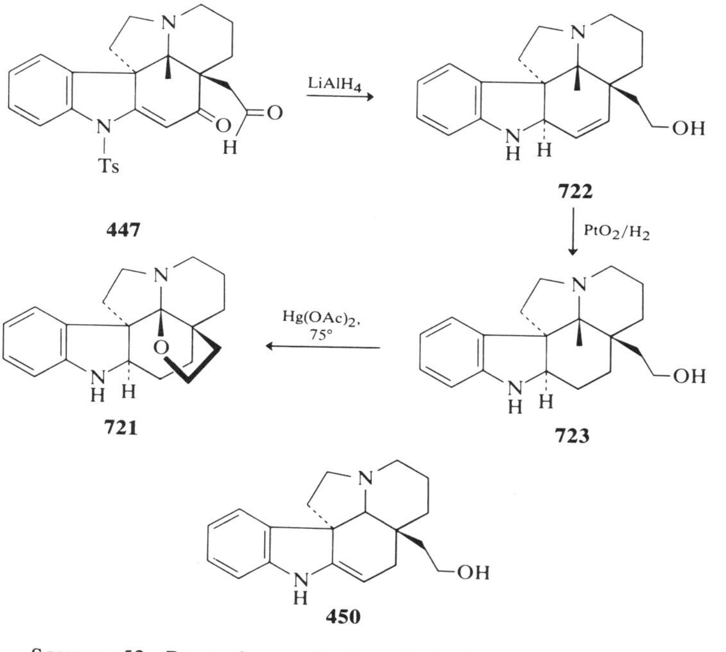

SCHEME 52. *Ban and co-workers' synthesis of aspidoalbidine* (**721**).

IV.A.16. Improved Routes to Intermediates for the Synthesis of Vindoline (**101**) and Vindorosine (**99**)

The tetracyclic indoline (**525**) is a key intermediate in the synthesis of vindoline (**101**), and the corresponding demethoxy analog is important in the formation of vindorosine (**99**). Takano and co-workers have now reported two improved routes to the intermediate **525** (*309*).

The diazoketone **724**, prepared from 6-methoxy-1-methyltryptamine in an unexceptional manner, was treated with trifluoroacetic acid to give the vinylogous amide **725** in 35% yield as a mixture of epimers. Treatment of the mixture with lithium in liquid NH_3 gave **525** in 92% yield.

The second route begins with 1,2-dimethyl-6-methoxytryptamine, which is condensed with diethylethoxymethylenemalonate to give **726**. When this vinylogous urethane was heated with acetic anhydride: acetic acid (3:2), the vinylogous amide **725** was obtained in 52% yield as a mixture of epimers. The corresponding intermediate in the synthesis of vindorosine (**99**) was prepared from 1,2-dimethyltryptamine in 65% yield (*309*).

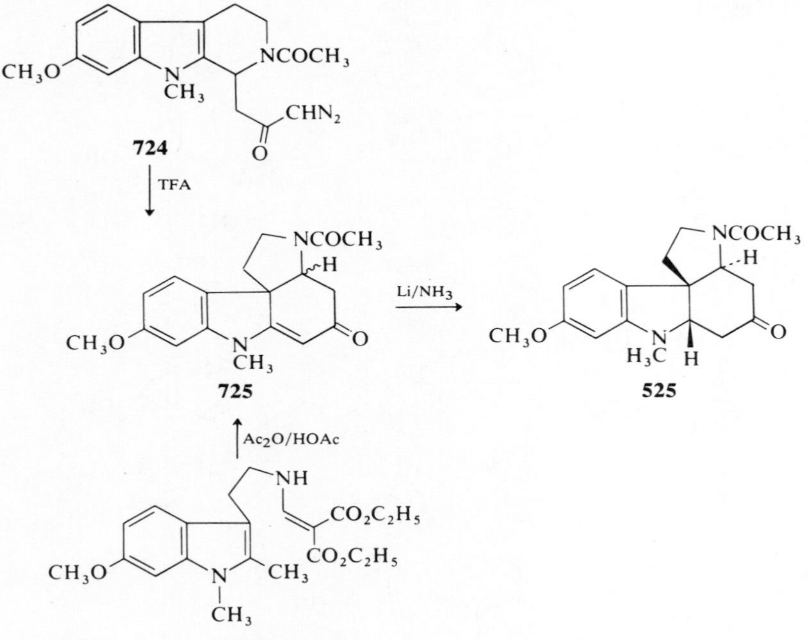

CH_3O $NCOCH_3$ N CH_3 CHN_2 O **724**

TFA

$NCOCH_3$ H CH_3O N CH_3 O **725**

Li/NH_3

$NCOCH_3$ H CH_3O N H_3C H O **525**

$Ac_2O/HOAc$

NH $CO_2C_2H_5$ $CO_2C_2H_5$ CH_3O N CH_3 CH_3 **726**

The novelty of these syntheses merits some explanation. The intermediate **725** is envisaged as being produced from diazoketone **724** by electrophilic attack on C-2 and subsequent rearrangement (Scheme 53). Cyclization of **726** is thought to proceed via the tricyclic amide diester **727**, as shown in Scheme 54.

Ban and co-workers have recently reported (*310*) a total synthesis of **525**, an intermediate in Büchi's synthesis of vindoline (**101**). Previous work (*205*) had succeeded in the synthesis of **417** beginning with 2-hydroxytryptamine (**406**) (Scheme 16). Now, these workers have reported a

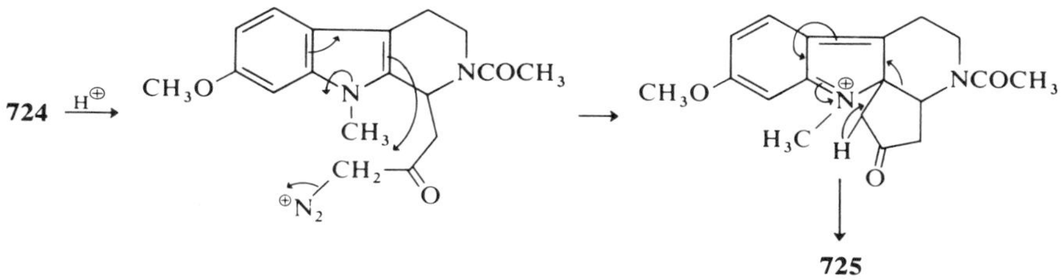

SCHEME 53

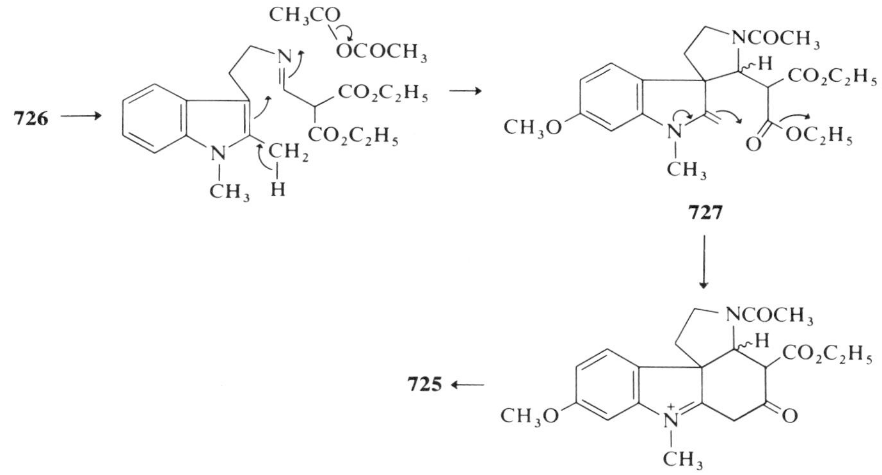

SCHEME 54

synthesis of the analog **728** using 2-hydroxy-6-methoxytryptamine (**729**). The crucial steps are summarized in Scheme 55. Reduction of the vinylogous amide group in **728** to the aminoketone in **525** was accomplished with lithium in liquid ammonia.

IV.D. SYNTHESIS OF ALKALOIDS WITH THE ELLIPTICINE/OLIVACINE SKELETON

A patent (*311*) has appeared on the synthesis of both 6*H*-pyrido[4,3-*b*]carbazole (**637**) and olivacine (**669**), as described previously (*270, 271*).

Two new synthetic approaches to ellipticine have been reported (*312, 313*). One of these (*312*), developed by Jackson and co-workers, does not rely on preformed indoles. The initial steps are formylation of 2,5-dimethyl acetanilide (**730**) and conversion of the product to the amine

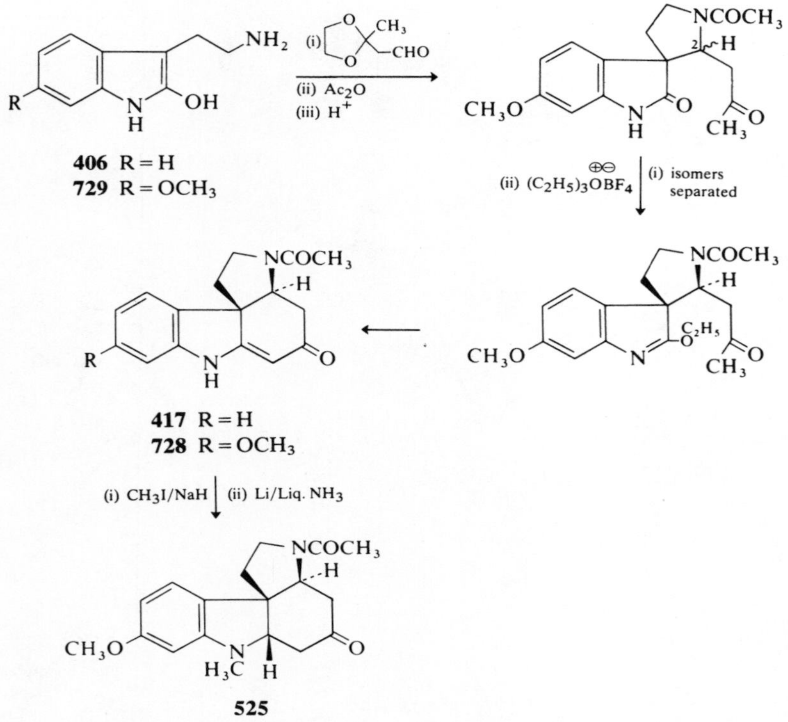

406 R = H
729 R = OCH_3

417 R = H
728 R = OCH_3

525

SCHEME 55. *Ban and co-workers' synthesis of the Büchi intermediate* **525**.

731 via the corresponding imine. N-Tosylation and coupling with *p*-bromoanisole in the presence of potassium carbonate and copper foil (Goldberg reaction) gave the diphenylamine (**732**) (Scheme 56). In principle, thermal dehydrogenation should lead to an *N*-acetyl-9-methoxy-ellipticine derivative.

The carbazole aldehyde **733** has been used in previous syntheses of ellipticine, but Jackson and co-workers (*312*) have reported an improved conversion of this compound to ellipticine (**267**). Condensation with aminoacetaldehydediethylacetal followed by reduction and tosylation afforded the tosylate **734** in 93% yield. Acid hydrolysis gave ellipticine (**267**) in 90% yield (*312*) (Scheme 57).

A quite different synthetic approach has been used by Bergman and Carlsson (*313*). It had been shown previously (*314*) that condensation of 2-ethyl indole (**735**) with 3-acetylpyridine (**736**) in acetic acid gave **737**. This compound has now been successfully transformed to **267**. N-Alkylation with butyl bromide followed by *rapid* heating at 350° for 5 min gave

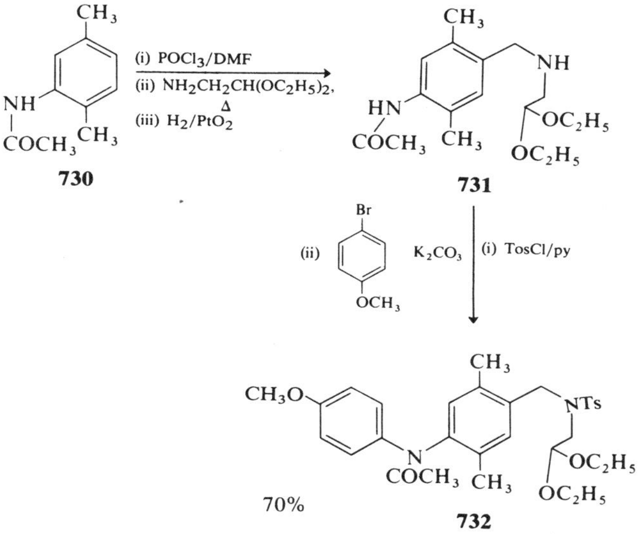

(i) $POCl_3$/DMF
(ii) $NH_2CH_2CH(OC_2H_5)_2$, Δ
(iii) H_2/PtO_2

730 → 731

(i) TosCl/py
(ii) 4-bromoanisole, K_2CO_3

732 70%

SCHEME 56

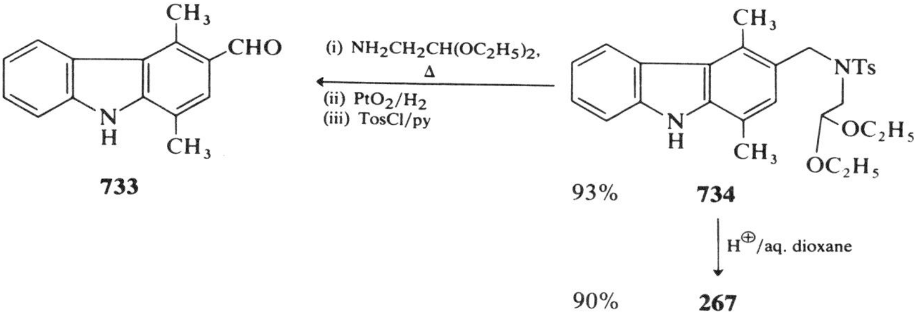

(i) $NH_2CH_2CH(OC_2H_5)_2$, Δ
(ii) PtO_2/H_2
(iii) TosCl/py

733 → 734 93%

$H^{\oplus}$/aq. dioxane

267 90%

SCHEME 57

267 in about 70% overall yield (*313*) (Scheme 58). To date, this is the most efficient synthesis of **267**.

An entirely new approach to the synthesis of ellipticine (**267**) has been reported by Kozikowski and Havan (*315*). Crucial in this synthesis is the Diels–Alder addition of an acrylic acid moiety to an oxazole followed by elimination of the oxygen bridge to give a pyridine. Thus, heating the oxazole **738** in acetic acid–acrylonitrile at 145° gave the indolyl-ethyl pyridine **739** in 16% yield (Scheme 59). The *N*-acetyl derivative of this intermediate had previously been converted to **267** (*266*).

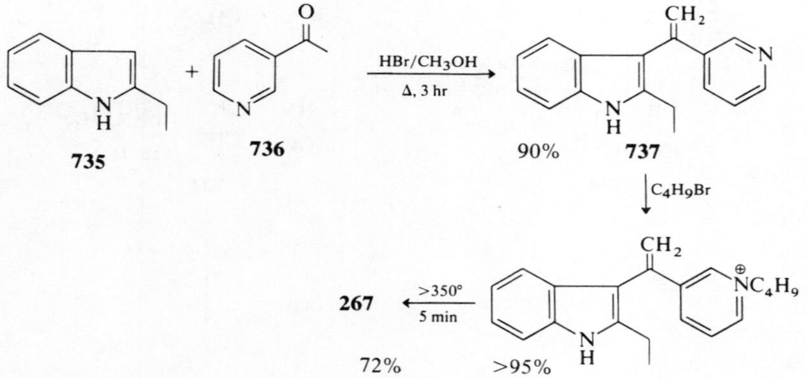

SCHEME 58. *Bergmann–Carlsson synthesis of ellipticine* (**267**).

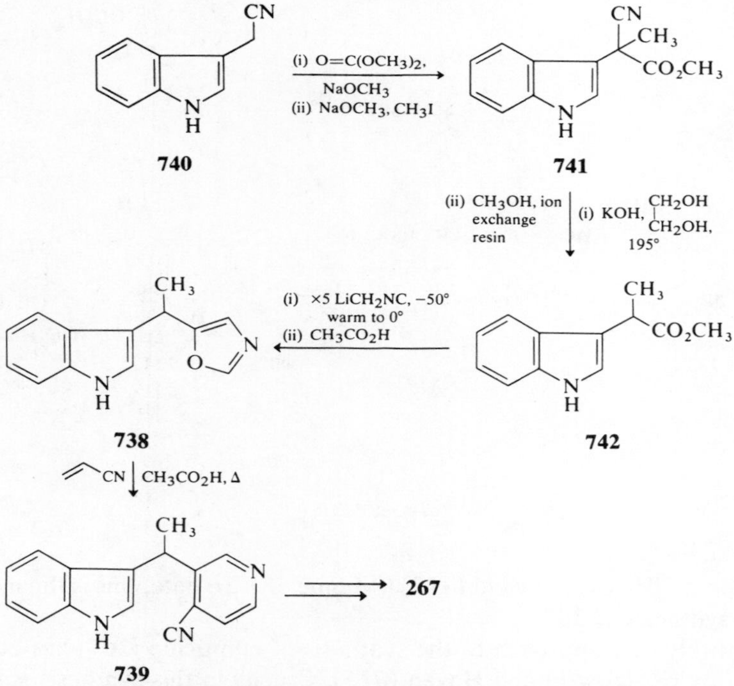

SCHEME 59. *Kozikowski–Havan synthesis of ellipticine* (**267**).

The oxazole was prepared from indole acetonitrile (**740**) by dicarbomethoxylation, hydrolysis, and C-alkylation with sodium methoxide–methyl iodide to afford **741**. Hydrolysis, decarboxylation, and esterification then gave the methyl indolylacetic acid (**742**). Treatment of this ester with the lithium salt of methyl isocyanide gave the oxazole **738** in 80% yield after quenching with acetic acid.

A further modification of methods which have already been described has been discussed by French workers (*316*) for the formation of 9-methoxy ellipticine (**743**). It combines the Borsche (*317*) synthesis of tetrahydrocarbazoles with the Cranwell–Saxton approach (*318*) for the pyridine ring, to produce the carbazole aldehyde **744** (Scheme 60).

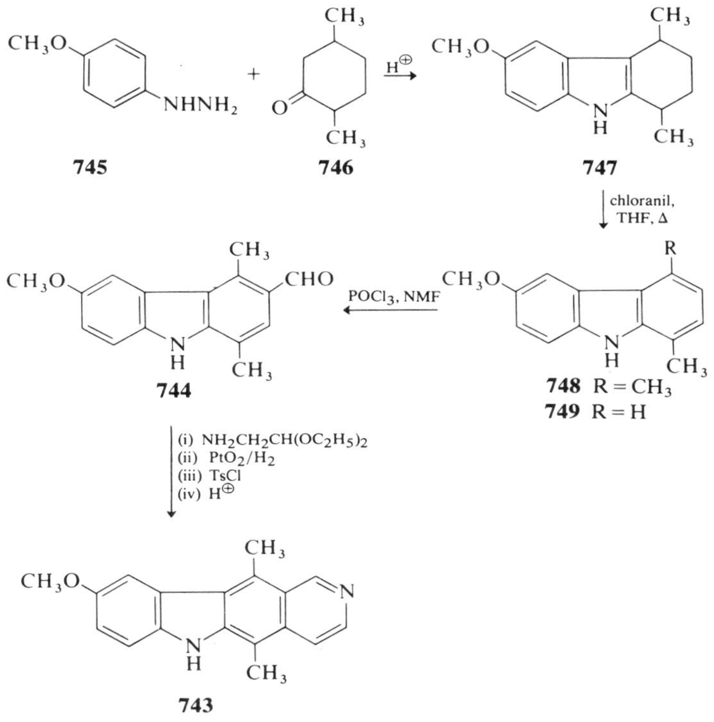

SCHEME 60

Condensation of *p*-*O*-methoxyphenylhydrazine (**745**) with 2,5-dimethylcyclohexanone (**746**) gave mainly the tetrahydrocarbazole **747**, which was dehydrogenated with chloranil in refluxing THF. The resulting carbazole **746** was formylated under Vilsmeier–Haack conditions to afford **744**. Formation of the pyridine ring can then be achieved by condensation with the diethylacetal of aminoacetaldehyde followed by reduction of the imine, N-tosylation, and cyclization to give **743** as described previously (Scheme 57). When the 4-methyl group is not present in the carbazole **748**, i.e., **749**, subsequent cyclization of the tosylamino acetal gave a mixture of both pyrido [4,3-*b*] and [3,4-*c*] carbazoles (*316*).

References

1. I. Kompis, M. Hesse, and H. Schmid, *Lloydia* **34**, 269 (1971).
2. G. A. Cordell, *Lloydia* **37**, 219 (1974).
3. B. Gabetta, *Fitoterapia* **44**, 3 (1973).
4. A. M. Aliev and H. A. Babaev, *Farmatsiya* (*Moscow*) **22**, 82 (1973).
5. J. Le Men and W. I. Taylor, *Experientia* **21**, 508 (1965).
6. W. Döpke and H. Meisel, *Pharmazie* **23**, 521 (1968).
7. D. A. Rakhimov, V. M. Malikov, and S. Yu. Yunusov, *Khim. Prir. Soedin.* **5**, 330 (1969).
8. D. A. Rakhimov, V. M. Malikov, M. R. Yagudaev, and S. Yu. Yunusov, *Khim. Prir. Soedin.* **6**, 226 (1970).
9. B. Pyuskyulev, I. Kompis, I. Ognyanov, and G. Spiteller, *Collect. Czech. Chem. Commun.* **32**, 1289 (1967).
10. W. Döpke, H. Meisel, and G. Spiteller, *Tet. Lett.* 6065 (1968).
11. W. Döpke and H. Meisel, *Tet. Lett.* 749 (1970).
12. W. Döpke, H. Meisel, and E. Gründemann, *Tet. Lett.* 1287 (1971).
13. B. Gabetta, E. M. Martinelli, and G. Mustich, *Fitoterapia* **47**, 6 (1976).
14. M. P. Cava, S. S. Tjoa, Q. A. Ahmed, and A. I. Da Rocha, *J. Org. Chem.* **33**, 1055 (1968).
15. H. Meisel and W. Döpke, *Pharmazie* **26**, 182 (1971).
16. V. M. Malikov and S. Yu. Yunusov, *Khim. Prir. Soedin.* **5**, 65 (1969).
17. M. Gorman, N. Neuss, G. H. Svoboda, A. J. Barnes, and N. J. Cone, *J. Am. Pharm. Assoc., Sci. Ed.* **48**, 256 (1959).
18. V. M. Malikov and S. Yu. Yunusov, *Khim. Prir. Soedin.* **7**, 640 (1971).
19. M. H. Mehri, M. Koch, M. Plat, and P. Potier, *Bull. Soc. Chim. Fr.* 3291 (1972).
20. E. Wenkert, D. W. Cochran, E. W. Hagaman, F. M. Schell, N. Neuss, A. S. Katner, P. Potier, C. Kan, M. Plat, M. Koch, H. Mehri, J. Poisson, N. Kunesch, and Y. Rolland, *J. Am. Chem. Soc.* **95**, 4990 (1973).
21. P. L. Majumder, T. K. Chandra, and B. N. Dinda, *Chem. Ind. (London)* 1032 (1973).
22. G. F. Smith and J. T. Wrobel, *J. Chem. Soc.* 792 (1960).
23. J. R. Chalmers, H. T. Openshaw, and G. F. Smith, *J. Chem. Soc.* 1115 (1957).
24. P. L. Majumder, B. N. Dinda, A. Chatterjee, and B. C. Das, *Tetrahedron* **30**, 2761 (1974).
25. P. L. Majumder and B. N. Dinda, *Phytochemistry* **13**, 645 (1974).
26. H. K. Kim, R. N. Blomster, H. H. S. Fong, and N. R. Farnsworth, *Econ. Bot.* **24**, 42 (1970).
27. A. B. Segelman and N. R. Farnsworth, *J. Pharm. Sci.* **63**, 1419 (1974).
28. G. A. Cordell and N. R. Farnsworth, *J. Pharm. Sci.* **65**, 366 (1976).
29. R. N. Blomster, N. R. Farnsworth, and D. J. Abraham, *Naturwissenschaften* **55**, 298 (1968).
30. N. R. Farnsworth, W. D. Loub, R. N. Blomster, and D. J. Abraham, *Z. Naturforsch., Teil B* **23**, 1061 (1969).
31. D. J. Abraham, N. R. Farnsworth, W. D. Loub, and R. N. Blomster, *J. Org. Chem.* **34**, 1575 (1969).

32. D. A. Rakhimov, V. M. Malikov, and S. Yu. Yunusov, *Khim. Prir. Soedin.* **4**, 331 (1968).
33. P. Potier, A.-M. Bui, B. C. Das, J. Le Men, and P. Boiteau, *Ann. Pharm. Fr.* **26**, 621 (1968).
34. M. B. Patel and J. Poisson, *Bull. Soc. Chim. Fr.* 427 (1966).
35. R. E. Moore and H. Rapaport, *J. Org. Chem.* **38**, 215 (1973).
36. J. Naranjo, M. Hesse, and H. Schmid, *Helv. Chim. Acta* **55**, 1849 (1972).
37. R. Iglesias and L. Diatta, *Rev. CENIC, Cienc. Fis.* **6**, 135 (1975); *CA* **84**, 44502d (1976).
38. R. Iglesias and L. Diatta, *Rev. CENIC, Cienc. Fis.* **6**, 141 (1975); *CA* **84**, 44503e (1976).
39. L. Olivier, F. Quirin, B. C. Das, J. Levy, and J. Le Men, *Ann. Pharm. Fr.* **26**, 105 (1968).
40. L. Olivier, F. Quirin, P. Mauperin, J. Levy, and J. Le Men, *C. R. Acad. Sci., Ser. C* **270**, 1667 (1970).
41. J. D. Medina, J. Hurtado, and R. H. Burnell, *Rev. Latinoam. Quim.* **4**, 73 (1973).
42. R. H. Burnell and J. D. Medina, *Phytochemistry* **7**, 2045 (1968).
43. C. Ferrari, S. McLean, L. Marion, and K. Palmer, *Can. J. Chem.* **41**, 1531 (1963).
44. G. H. Svoboda, M. Gorman, and R. H. Tust, *Lloydia* **27**, 203 (1964).
45. N. R. Farnsworth, H. H. S. Fong, and R. N. Blomster, *Lloydia* **29**, 343 (1966).
46. G. H. Aynilian, N. R. Farnsworth, and J. Trojanek, *Lloydia* **37**, 299 (1974).
47. G. H. Aynilian, S. G. Weiss, G. A. Cordell, D. J. Abraham, F. A. Crane, and N. R. Farnsworth, *J. Pharm. Sci.* **63**, 536 (1974).
48. H. Mehri, M. Koch, M. Plat, and P. Potier, *Ann. Pharm. Fr.* **30**, 643 (1972).
49. R. Z. Andriamialisoa, N. Langlois, and P. Potier, *Tet. Lett.* 163 (1976).
50. A. Ahond, M. M. Janot, N. Langlois, G. Lukacs, P. Potier, P. Rasoanaivo, M. Sangare, N. Neuss, M. Plat, J. Le Men, E. W. Hagaman, and E. Wenkert, *J. Am. Chem. Soc.* **96**, 633 (1974).
51. M. Damak, A. Ahond, and P. Potier, *Tet. Lett.* 167 (1976).
52. N. Langlois and P. Potier, *C. R. Acad. Sci., Ser. C* **273**, 994 (1971).
53. G. H. Aynilian, M. Tin-Wa, N. R. Farnsworth, and M. Gorman, *Tet. Lett.* 89 (1972).
54. G. H. Aynilian, B. Robinson, N. R. Farnsworth, and M. Gorman, *Tet. Lett.* 391 (1972).
55. A. R. Battersby and K. H. Gibson, *Chem. Commun.* 902 (1971).
56. L. Diatta, Y. Langlois, N. Langlois, and P. Potier, *Bull. Soc. Chim. Fr.* 671 (1975).
57. N. Langlois and P. Potier, *C. R. Acad. Sci., Ser. C* **275**, 219 (1972).
58. M. P. Cava, S. K. Talapatra, P. Yates, M. Rosenberger, A. G. Szabo, B. Douglas, R. F. Raffauf, E. C. Shoop, and J. A. Weisbach, *Chem. Ind. (London)* 1875 (1963).
59. M. P. Cava, M. V. Lakshmikantham, S. K. Talapatra, P. Yates, I. D. Rae, M. Rosenberger, A. G. Szabo, B. Douglas, and J. A. Weisbach, *Can. J. Chem.* **51**, 3102 (1973).
60. H. Achenbach, *Z. Naturforsch., Teil B* **22**, 955 (1967).
61. C. Djerassi, A. A. P. G. Archer, T. George, B. Gilbert, and L. D. Antonaccio, *Tetrahedron* **16**, 212 (1961).
62. B. V. Milborrow and C. Djerassi, *J. Chem. Soc. C* 417 (1969).
63. M. Ikeda and C. Djerassi, unpublished results, quoted in Milborrow and Djerassi (*62*).
64. H. H. A. Linde, *Pharm. Acta Helv.* **45**, 248 (1970).
65. M. R. Sharipov, M. Khalmirzaev, V. M. Malikov, and S. Yu. Yunusov, *Khim. Prir. Soedin.* **10**, 422 (1974).
66. J. C. Braekman, C. Hootele, C. Van Moorleghem, M. Kaisin, J. Pecher, L. D. Antonaccio, and B. Gilbert, *Bull. Soc. Chim. Belg.* **78**, 63 (1969).
67. O. Hesse, *Ber.* **10**, 2162 (1877).

68. M. Freund and C. Fauvet, *Ber.* **26**, 1084 (1893).
69. Raymond-Hamet, *C. R. Acad. Sci.* **245**, 2374 (1957).
70. C. Djerassi, B. Gilbert, L. D. Antonaccio, and A. A. P. G. Archer, *Experientia* **16**, 61 (1960).
71. Raymond-Hamet, K. Foley, and M. Shamma, *C. R. Acad. Sci., Ser. D* **265**, 71 (1967).
72. S.-M. Nasirov, V. G. Andrianov, Yu. T. Struchkov, M. R. Yagudaev, V. M. Malikov, and S. Yu. Yunusov, *Khim. Prir. Soedin.* **10**, 811 (1974).
73. C. Riche and C. Pascard-Billy, *Acta Crystallogr., Sect. B* **32**, 1975 (1976).
74. S. M. Nasirov, V. G. Andrianov, Yu. T. Struchkov, and S. Yu. Yunusov, *Khim. Prir. Soedin.* **11**, 197 (1976).
75. N. Abdurakhimova, P. Kh. Yuldashev, and S. Yu. Yunusov, *Khim. Prir. Soedin* **3**, 310 (1967).
76. M. Ikeda and C. Djerassi, *Tet. Lett.* 5837 (1968).
77. L. W. Döpke, H. Meisel, and H.-W. Fehlhaber, *Tet. Lett.* 1701 (1969).
78. H. H. A. Linde, *Helv. Chim. Acta* **48**, 1822 (1965).
79. A. Banerji, P. L. Majumder, and A. Chatterjee, *Phytochemistry* **9**, 1491 (1970).
80. G. A. Cordell, G. N. Smith, and G. F. Smith, unpublished results (1970).
81. K. T. D. De Silva, A. H. Ratcliffe, G. F. Smith, and G. N. Smith, *Tet. Lett.* 913 (1972).
82. D. J. Abraham, R. D. Rosenstein, R. L. Lyon, and H. H. S. Fong, *Tet. Lett.* 909 (1972).
83. R. L. Lyon, H. H. S. Fong, N. R. Farnsworth, and G. H. Svoboda, *J. Pharm. Sci.* **62**, 218 (1973).
84. A. H. Ratcliffe, G. F. Smith, and G. N. Smith, *Tet. Lett.* 5179 (1973).
85. N. Kunesch, C. Miet, M. Troly, and J. Poisson, *Ann. Pharm. Fr.* **26**, 79 (1968).
86. N. Kunesch, J. Poisson, and J. Guilhem, *Bull. Soc. Chim. Fr.* 1919 (1971).
87. N. Kunesch, B. C. Das, and J. Poisson, *Bull. Soc. Chim. Fr.* 2156 (1967).
88. N. Kunesch, B. C. Das, and J. Poisson, *Bull. Soc. Chim. Fr.* 3551 (1967).
89. B. Zsadon and J. Tamás, *Chem. Ind. (London)* 32 (1972).
90. J. Mokry, I. Kompis, M. Shamma, and R. I. Shine, *Chem. Ind. (London)* 1988 (1964).
91. B. Zsadon, J. Tamás, M. Szilasi, and P. Kaposi, *Acta Chim. Acad. Sci. Hung.* **78**, 207 (1973).
92. B. Zsadon, J. Tamás, and M. Szilasi, *Chem. Ind. (London)* 229 (1973).
92a. B. Zsadon, L. Decsei, K. Otta, M. Szilasi, and P. Kaposi, *Acta Pharm. Hung.* **44**, 74 (1974).
93. J. Mokry and I. Kompis, *Naturwissenschaften* **50**, 93 (1963).
94. I. Kompis and J. Mokry, *Collect. Czech. Chem. Commun.* **33**, 4328 (1968).
95. J. Bruneton, A. Bouquet, and A. Cavé, *Phytochemistry* **13**, 1963 (1974).
96. K. Biemann, *Tet. Lett.* 9 (1960).
97. P. Tunmann and D. Wolf, *Z. Naturforsch., Teil B* **24**, 1665 (1969).
98. M. Gorman, N. Neuss, and N. J. Cone, *J. Am. Chem. Soc.* **87**, 93 (1965).
99. M. J. Hoizey, M.-M. Debray, L. Le Men-Olivier, and J. Le Men, *Phytochemistry* **13**, 1995 (1974).
100. J. Le Men, G. Lukacs, L. Le Men-Olivier, J. Lévy, and M. J. Hoizey, *Tet. Lett.* 483 (1974).
101. M.-J. Hoizey, C. Siguat, M.-J. Jacquier, L. Le Men-Olivier, J. Lévy, and J. Le Men, *Tet. Lett.* 1601 (1974).
102. J. Le Men, M. J. Hoizey, G. Lukacs, L. Le Men-Olivier, and J. Lévy, *Tet. Lett.* 3119 (1974).
103. M. Zeches, M.-M. Debray, G. Ledouble, L. Le Men-Olivier, and J. Le Men, *Phytochemistry* **14**, 1122 (1975).
104. M. Gorman, N. Neuss, and N. J. Cone, *J. Am. Chem. Soc.* **87**, 93 (1965).

105. J. P. Kutney, R. T. Brown, E. Piers, and J. R. Hadfield, *J. Am. Chem. Soc.* **92**, 1708 (1970).
106. R. T. Brown, J. S. Hill, G. F. Smith, and K. S. J. Stapleford, *Tetrahedron* **27**, 5217 (1971).
107. I. Chardon-Loriaux and H.-P. Husson, *Tet. Lett.* 1845 (1975).
108. F. Quirin, M.-M. Debray, C. Siguat, P. Thepenier, L. Le Men-Olivier, and J. Le Men, *Phytochemistry* **14**, 812 (1975).
109. F. Khuong-Huu, M. Cesario, J. Guilhem, and R. Goutarel, *Tetrahedron* **32**, 2539 (1976).
110. K. Bernauer, G. Englert, W. Vetter, and E. Weiss, *Helv. Chim. Acta* **52**, 1886 (1969).
111. K. Bernauer, G. Englert, and W. Vetter, *Experientia* **21**, 374 (1965).
112. H. Mehri, M. Plat, and P. Potier, *Ann. Pharm. Fr.* **29**, 291 (1971).
113. M. Plat, M. Hachem-Mehri, M. Koch, U. Scheidegger, and P. Potier, *Tet. Lett.* 3395 (1970).
114. D. W. Thomas, H. K. Schnoes, and K. Biemann, *Experientia* **25**, 678 (1969).
115. D. A. Evans, G. F. Smith, G. N. Smith, and K. S. J. Stapleford, *Chem. Commun.* 859 (1968).
116. P. A. Crooks, B. Robinson, and G. F. Smith, *Chem. Commun.* 1210 (1968).
117. G. A. Cordell, G. N. Smith, and G. F. Smith, *Chem. Commun.* 189 (1970).
118. N. Preaux, M. Koch, M. Plat, and T. Sevenet, *Plant. Med. Phytother.* **8**, 250 (1974).
119. N. Preaux, M. Koch, and M. Plat, *Phytochemistry* **13**, 2607 (1974).
120. J. Bruneton, T. Sevenet, and A. Cavé, *Phytochemistry* **11**, 3073 (1972).
121. W. E. L. Sanchez and K. S. Brown, Jr., *An. Acad. Bras. Cienc.* **43**, 603 (1971).
122. A. J. Gaskell and J. A. Joule, *Tet. Lett.* 77 (1970).
123. M. Pinar, U. Renner, M. Hesse, and H. Schmid, *Helv. Chim. Acta* **55**, 2972 (1972).
124. A. Guggisberg, A. A. Gorman, B. W. Bycroft, and H. Schmid, *Helv. Chim. Acta* **52**, 76 (1969).
125. C. Kan-Fan, G. Massiot, A. Ahond, B. C. Das, H.-P. Husson, P. Potier, A. I. Scott, and C.-C. Wei, *J. Chem. Soc., Chem. Commun.* 164 (1974).
126. G. Massiot, S. K. Kan, P. Gonord, and C. Duret, *J. Am. Chem. Soc.* **97**, 3277 (1975).
126a. C. Kan-Fan, B. C. Das, H.-P. Husson, and P. Potier, *Bull. Soc. Chim. Fr.* 2839 (1974).
127. B. Zsadon, K. H. Otta, and P. Tetenyi, *Acta Chim. Acad. Sci. Hung.* **84**, 71 (1975).
128. J.-M. Panas, A.-M. Morfaux, L. Olivier, and J. Le Men, *Ann. Pharm. Fr.* **30**, 273 (1972).
129. B. Zsadon, M. Szilasi, and P. Kaposi, *Herba Hung.* **13**, 69 (1974); *CA* **83**, 40171b (1975).
130. R. F. Garcia and K. S. Brown, *Phytochemistry* **15**, 1093 (1976).
131. B. Gabetta, E. M. Martinelli, and G. Mustich, *Fitoterapia* **45**, 32 (1974).
132. L. Douzoua, M. Mansour, M.-M. Debray, L. Le Men-Olivier, and J. Le Men, *Phytochemistry* **13**, 1994 (1974).
133. A. Rabaron, M. Plat, and P. Potier, *Phytochemistry* **12**, 2537 (1973).
134. L. Le Men-Olivier, B. Richard, and J. Le Men, *Phytochemistry* **13**, 280 (1974).
135. P. L. Majumder and B. N. Dinda, *J. Indian Chem. Soc.* **51**, 370 (1974).
136. B. Zsadon and P. Kaposi, *Tet. Lett.* 4615 (1976).
137. N. Petotfrére, A. M. Morfaux, M.-M. Debray, L. Le Men-Olivier, and J. Le Men, *Phytochemistry* **14**, 1648 (1975).
138. K. Raj, A. Shoeb, R. S. Kapil, and S. P. Popli, *Phytochemistry* **13**, 1621 (1974).
139. M. Tin-Wa, H. H. S. Fong, R. N. Blomster, and N. R. Farnsworth, *J. Pharm. Sci.* **57**, 2167 (1968).
140. P. Rasoanaivo, N. Langlois, and P. Potier, *Phytochemistry* **11**, 2616 (1972).

141. A. Chatterjee, G. K. Biswas, and A. B. Kundu, *Indian J. Chem.* **11**, 7 (1973).
142. J. G. R. Elferink, J. L. Van Eijk, T. J. Khouw, A. M. H. Van Laer, M. M. A. B. Russel-Gulikers, and O. F. Uffelie, *Pharm. Weekbl.* **103**, 1101 (1968).
143. J. C. Simões, B. Gilbert, W. J. Cretney, M. Hearn, and J. P. Kutney, *Phytochemistry* **15**, 543 (1976).
144. N. Langlois and P. Potier, *Phytochemistry* **11**, 2617 (1972).
145. J. M. Panas, B. Richard, C. Siguat, M.-M. Debray, L. Le Men-Olivier, and J. Le Men, *Phytochemistry* **13**, 1969 (1974).
146. J. R. Knox and J. Slobbe, *Aust. J. Chem.* **28**, 1813 (1975).
147. J. M. Panas, B. Richard, C. Potron, R. S. Razafindrambo, M.-M. Debray, L. Le Men-Olivier, and J. Le Men, *Phytochemistry* **14**, 1120 (1975).
148. J. P. Casson and M. Schmid, *Phytochemistry* **9**, 1353 (1970).
149. P. A. Crooks and B. Robinson, *J. Pharm. Pharmacol.* **22**, 799 (1970).
150. K. N. Kilminster, M. Sainsbury, and B. Webb, *Phytochemistry* **11**, 389 (1972).
151. J. Bruneton and A. Cavé, *Ann. Pharm. Fr.* **30**, 629 (1972).
152. J. Bruneton and A. Cavé, *Phytochemistry* **11**, 846 (1972).
153. J. Bruneton and A. Cavé, *Phytochemistry* **11**, 2618 (1972).
154. C. Kan-Fan, B. C. Das, P. Potier, and M. Schmid, *Phytochemistry* **9**, 1351 (1970).
155. R. H. Burnell and D. D. Casa, *Can. J. Chem.* **45**, 89 (1967).
156. M. Sainsbury and B. Webb, *Phytochemistry* **11**, 2337 (1972).
157. F. J. Abreu, R. Matos, F. Braz, O. R. Gottlieb, F. W. L. Machado, and M. I. L. M. Madruga, *Phytochemistry* **15**, 551 (1976).
158. B. Hwang, J. A. Weisbach, B. Douglas, R. F. Raffauf, M. P. Cava, and K. Bessho, *J. Org. Chem.* **34**, 412 (1969).
159. N. K. Hart, S. R. Johns, and J. A. Lamberton, *Aust. J. Chem.* **25**, 2739 (1972).
160. Y. Rolland, G. Croquelois, N. Kunesch, P. Boiteau, M. DeGray, J. Pecher, and J. Poisson, *Phytochemistry* **12**, 2039 (1973).
161. S. Mamatas-Kalamaras, T. Sévenet, C. Thal, and P. Potier, *Phytochemistry* **14**, 1637 (1975).
162. A. I. Scott, *Bioorg. Chem.* **3**, 398 (1974).
163. P. Mauperin, J. Lévy, and J. Le Men, *Tet. Lett.* 999 (1971).
164. M. F. Kuehne, *J. Am. Chem. Soc.* **86**, 2946 (1964).
165. J. Lévy, P. Maupérin, M. Döé de Mandreville, and L. Le Men, *Tet. Lett.* 1003 (1971).
166. G. Hugel, B. Gourdier, J. Lévy, and J. Le Men, *Tet. Lett.* 1597 (1974).
167. G. Hugel, J. Lévy, and J. Le Men, *Tet. Lett.* 3109 (1974).
168. G. Croquolois, N. Kunesch, and J. Poisson, *Tet. Lett.* 4427 (1974).
169. M.-J. Hoizey, L. Olivier, J. Lévy, and J. Le Men, *Tet. Lett.* 1011 (1971).
170. G. Hugel, J. Lévy, and J. Le Men, *C. R. Acad. Sci., Ser. C* 274, 1350 (1972).
171. J. Lévy, Belg. Pat. 761,628 (1971); *CA* **77**, 19866y (1972).
172. J. Lévy, Belg. Pat. 763,739 (1971); *CA* **76**, 127223z (1972).
173. J. Hannart and J. Lévy, Belg. Pat. 765,427 (1971); *CA* **77**, 19867z (1972).
174. J. Lévy, Ger. Offen. 2,201,795 (1972); *CA* **77**, 152430t (1972).
175. J. Lévy, Belg. Pat. 765,705 (1971); *CA* **77**, 19865x (1972).
176. J. Lévy, Belg. Pat. 818,144 (1974); *CA* **83**, 79457z (1975).
177. J. Lévy, Belg. Pat. 816,692 (1974); *CA* **83**, 97683z (1975).
178. C. Pierron, J. Garnier, J. Lévy, and J. Le Men, *Tet. Lett.* 1007 (1971).
179. W. Klyne, R. J. Swan, B. W. Bycroft, D. Schumann, and H. Schmid, *Helv. Chim. Acta* **48**, 443 (1965).
180. B. W. Bycroft, L. Goldman, and H. Schmid, *Helv. Chim. Acta* **50**, 1193 (1967).
181. J. Lévy, C. Pierron, G. Lukacs, G. Massiot, and J. Le Men, *Tet. Lett.* 669 (1976).

182. W. Hofheinz, P. Schönholzer, and K. Bernauer, *Helv. Chim. Acta* **59**, 1213 (1976).
183. Anonymous, Belg. Pat. 826,066 (1975); *CA* **84**, 135917j (1976).
184. R. Z. Andriamalisoa, L. Diatta, P. Rasoanaivo, N. Langlois, and P. Potier, *Tetrahedron* **31**, 2347 (1975).
185. B. Zsadon and K. Horvath-Otta, *Herba Hung.* **12**, 133 (1973).
186. B. Zsadon, J. Tamas, and M. Szilasi, *Magy. Kem. Foly.* **79**, 341 (1973).
187. T. Gebreyesus and C. Djerassi, *J. Chem. Soc., Perkin Trans I* 849 (1972).
188. N. Neuss, D. S. Fukuda, G. E. Mallett, D. R. Brannon, and L. L. Huckstep, *Helv. Chim. Acta* **56**, 2418 (1973).
189. S. Sakai and N. Shinma, *Chem. Pharm. Bull.* **22**, 3013 (1974).
190. S.-I. Sakai and N. Shinma, *Heterocycles* **4**, 985 (1976).
191. E. Wenkert, *Acc. Chem. Res.* **1**, 78 (1968).
192. R. J. Sundberg, "The Chemistry of Indoles," Academic Press, New York, 1970.
193. J. P. Kutney, *Org. Chem., Ser. One* **9**, 23 (1973).
194. E. Winterfeldt, *Prog. Chem. Org. Nat. Prod.* **31**, 469 (1974).
195. T. Oishi, *Yuki Gosei Kagaku Kyokai Shi* **33**, 239 (1975).
196. Y. Ban, M. Akagi, and T. Oishi, *Tet. Lett.* 2057 (1969).
197. Y. Ban, I. Ijima, I. Inoue, M. Akagi, and T. Oishi, *Tet. Lett.* 2067 (1969).
198. Y. Ban and I. Ijima, *Tet. Lett.* 2523 (1969).
199. N. Sakabe, Y. Sendo, I. Ijima, and Y. Ban, *Tet. Lett.* 2527 (1969).
200. M. Pinar, W. von Philipsborn, W. Vetter, and H. Schmid, *Helv. Chim. Acta* **45**, 2260 (1962).
201. I. Inoue and Y. Ban, *J. Chem. Soc. C* 602 (1970).
202. L. Castedo, J. Harley-Mason, and M. Kaplan, *Chem. Commun.* 1444 (1969).
203. Y. Ban, T. Ohnuma, M. Nagai, Y. Sendo, and T. Olshi, *Tet. Lett.* 5023 (1972).
204. Y. Ban, Y. Sendo, M. Nagai, and T. Oishi, *Tet. Lett.* 5027 (1972).
205. T. Oishi, M. Nagal, and Y. Ban, *Tet. Lett.* 491 (1968).
206. H. K. Schnoes and K. Biemann, *J. Am. Chem. Soc.* **86**, 5693 (1964).
207. T. Ohnuma, T. Oishi, and Y. Ban, *J. Chem. Soc., Chem. Commun.* 301 (1973).
208. T. Ohnuma, K. Seki, T. Oishi, and Y. Ban, *J. Chem. Soc., Chem. Commun.* 296 (1974).
209. K. Seki, T. Ohnuma, T. Oishi, and Y. Ban, *Tet. Lett.* 723 (1975).
210. R. H. Burnell, J. D. Medina, and W. A. Ayer, *Can. J. Chem.* **44**, 28 (1966).
211. Y. Ban, T. Ohnuma, K. Seki, and T. Oishi, *Tet. Lett.* 727 (1975).
212. A. Walser and C. Djerassi, *Helv. Chim. Acta* **48**, 391 (1965).
213. Y. Ban, Y. Honma, and T. Oishi, *Tet. Lett.* 1111 (1976).
214. S. S. Klioze and F. P. Darmory, *J. Org. Chem.* **40**, 1588 (1975).
215. Y. Ban, I. Ijima, I. Inoue, M. Akagi, and T. Oishi, *Tet. Lett.* 2976 (1969).
216. F. E. Ziegler and P. A. Zoretic, *Tet. Lett.* 2639 (1968).
217. F. E. Ziegler, J. A. Kloek, and P. A. Zoretic, *J. Am. Chem. Soc.* **91**, 2342 (1969).
218. F. E. Ziegler and G. B. Bennett, *J. Am. Chem. Soc.* **93**, 5930 (1971).
219. F. E. Ziegler and G. B. Bennett, *J. Am. Chem. Soc.* **95**, 7458 (1973).
220. F. E. Ziegler and G. B. Bennett, *Tet. Lett.* 2545 (1970).
221. S. Takano, S. Hatakeyama, and K. Ogasawara, *J. Am. Chem. Soc.* **98**, 3022 (1976).
222. R. B. Woodward, I. J. Pachter, and M. L. Scheinbaum, *J. Org. Chem.* **36**, 1137 (1971).
223. J. A. Marshall and D. E. Seits, *J. Org. Chem.* **39**, 1814 (1974).
224. S. Takano, M. Hirama, T. Araki, and K. Ogasawara, *J. Am. Chem. Soc.* **98**, 7084 (1976)
225. J. P. Kutney, W. J. Cretney, P. W. Le Quesne, B. McKague, and E. Piers, *J. Am. Chem. Soc.* **92**, 1712 (1970).
226. F. E. Ziegler and E. B. Spitzner, *J. Am. Chem. Soc.* **92**, 3492 (1970).
227. F. E. Ziegler and E. B. Spitzner, *J. Am. Chem. Soc.* **95**, 7146 (1973).

228. G. Büchi, K. E. Matsumoto, and H. Nishimura, *J. Am. Chem. Soc.* **93**, 3299 (1971).
229. M. Ando, G. Büchi, and T. Ohnuma, *J. Am. Chem. Soc.* **97**, 6880 (1975).
230. J. P. Kutney, E. Piers, and R. T. Brown, *J. Am. Chem. Soc.* **92**, 1700 (1970).
231. J. P. Kutney, W. J. Cretney, J. R. Hadfield, E. S. Hall, and V. R. Nelson, *J. Am. Chem. Soc.* **92**, 1704 (1970).
232. J. P. Kutney, N. Abdurahman, C. Gletson, P. W. Le Quesne, E. Piers, and I. Vlatas, *J. Am. Chem. Soc.* **92**, 1727 (1970).
233. J. P. Kutney, K. K. Chan, A. Failli, J. M. Fromson, C. Gletsos, and V. R. Nelson, *J. Am. Chem. Soc.* **90**, 3891 (1968).
234. J. P. Kutney, *Lloydia* **40**, 107 (1977).
235. J.-Y. Laronze, J. Laronze-Fontaine, J. Lévy, and J. Le Men, *Tet. Lett.* 491 (1974).
236. J. Le Men, L. Le Men-Olivier, J. Lévy, M. C. Levy-Appert-Collin and J. Hannart, Ger. Offen. 2,429,691 (1975); *CA* **83**, 10559s (1975).
237. R. V. Stevens, R. K. Mehra, and R. L. Zimmerman, *Chem. Commun.* 877 (1969).
238. R. V. Stevens and M. P. Wentland, *J. Am. Chem. Soc.* **90**, 5880 (1968).
239. C. F. Koelsch, *J. Am. Chem. Soc.* **65**, 2548 (1968).
240. R. V. Stevens, J. M. Fitzpatrick, M. Kaplan, and R. L. Zimmerman, *Chem. Commun.* 857 (1971).
241. J. Harley-Mason and M. Kaplan, *Chem. Commun.* 915 (1967).
242. E. Wenkert, K. G. Dave, C. T. Gnewuch, and P. W. Sprague, *J. Am. Chem. Soc.* **90**, 5251 (1968).
243. E. Wenkert, J. S. Bindra, and B. Chauncy, *Synth. Commun.* **2**, 285 (1972).
244. H.-P. Husson, C. Thal, P. Potier, and E. Wenkert, *J. Org. Chem.* **35**, 442 (1970).
245. H.-P. Husson, C. Thal, P. Potier, and E. Wenkert, *Chem. Commun.* 480 (1970).
246. M. E. Stein, *Diss. Abstr. Int. B* **27**, 3053 (1967).
247. E. E. McEntire, *Diss. Abstr. Int. B* **33**, 1457 (1972).
248. R. L. Zimmerman, *Diss. Abstr. Int. B* **35**, 1588 (1974).
249. J. E. Saxton, A. J. Smith, and G. Lawton, *Tet. Lett.* 4161 (1975).
250. D. D. O'Rell, F. G. H. Lee, and V. Boekelheide, *J. Am. Chem. Soc.* **94**, 3205 (1972).
251. A. Jackson, N. D. V. Wilson, A. J. Gaskell, and J. A. Joulé, *J. Chem. Soc. C*, 2738 (1969), and references therein.
252. L. J. Dolby and H. Biere, *J. Am. Chem. Soc.* **90**, 2699 (1968).
253. L. J. Dolby and H. Biere, *J. Org. Chem.* **35**, 3843 (1970).
254. T. Kametani and T. Suzuki, *J. Chem. Soc. C* 1053 (1971).
255. T. Kametani and T. Suzuki, *J. Org. Chem.* **36**, 1291 (1971).
256. T. Kametani and T. Suzuki, *Chem. Pharm. Bull.* **19**, 1424 (1971).
257. G. Büchi, S. J. Gould, and F. Naef, *J. Am. Chem. Soc.* **93**, 2492 (1971).
258. F. Le Goffic, A. Gouyette, and A. Ahond, *C. R. Acad. Sci., Ser. C* **274**, 2008 (1972).
259. F. Le Goffic, A. Gouyette, and A. Ahond, *Tetrahedron* **29**, 3357 (1973).
260. K. N. Kilminster and M. Sainsbury, *J. Chem. Soc., Perkin Trans. I* 2264 (1972).
261. R. B. Woodward, G. A. Iacobucci, and F. A. Hochstein, *J. Am. Chem. Soc.* **81**, 4434 (1959).
262. Y. Langlois, N. Langlois, and P. Potier, *Tet. Lett.* 955 (1975).
263. R. Besselièvre, C. Thal, H. P. Husson, and P. Potier, *J. Chem. Soc., Chem. Commun.* 90 (1975).
264. M. Sainsbury and B. Webb, *J. Chem. Soc., Perkin Trans. I* 1580 (1974).
265. R. W. Guthrie, A. Brassi, F. A. Mennona, J. G. Mullin, R. W. Kierstad, and E. Grunberg, *J. Med. Chem.* **18**, 755 (1975).
266. M. Sainsbury and R. F. Schinazi, *J. Chem. Soc., Perkin Trans. I* 1155 (1976).
267. S. Suzue, M. Hirobe, and T. Okamoto, *Yakugaku Zasshi* **93**, 1331 (1973).

268. J. P. Kutney and D. S. Grierson, *Heterocycles* **3**, 171 (1975).
269. C. W. Mosher, O. P. Crews, E. M. Acton, and L. Goodman, *J. Med. Chem.* **9**, 237 (1966).
270. T. Kametani, Y. Ichikawa, T. Suzuki, and K. Fukumoto, *Heterocycles* **3**, 401 (1975).
271. T. Kametani, T. Suzuki, Y. Ichikawa, and K. Fukumoto, *J. Chem. Soc., Perkin Trans. I* 2102 (1975).
272. R. Besselièvre and H.-P. Husson, *Tet. Lett.* 1873 (1976).
273. S. J. Martinez and J. A. Joule, *J. Chem. Soc., Chem. Commun.* 818 (1976).
274. J. E. D. Barton and J. Harley-Mason, *Chem. Commun.* 298 (1965).
275. O. Kennar, K. A. Kerr, D. G. Watson, J. K. Fawett, and L. Riva Di Sanseverino, *Chem. Commun.* 1286 (1967).
276. J. M. F. Filho, B. Gilbert, M. Kitagawa, L. A. Paes Leme, and L. J. Durham, *J. Chem. Soc. C* 1260 (1966).
277. C. Kump, J. J. Dugan, and H. Schmid, *Helv. Chim. Acta* **49**, 1237 (1966).
278. B. M. Craven, B. Gilbert, and L. A. Paes Leme, *Chem. Commun.* 955 (1968).
279. B. M. Craven and D. E. Zacharias, *Experientia* **24**, 770 (1968).
280. S. H. Brown, C. Djerassi, and P. G. Simpson, *J. Am. Chem. Soc.* **90**, 2445 (1968).
281. N. C. Ling and C. Djerassi, *Tet. Lett.* 3015 (1970).
282. K. S. Brown and C. Djerassi, *J. Am. Chem. Soc.* **86**, 2451 (1964).
283. J. Kahrl, T. Gebreyesus, and C. Djerassi, *Tet. Lett.* 2527 (1971).
284. I. D. Rae, M. Rosenberger, A. G. Szabo, C. R. Willis, P. Yates, D. E. Zacharias, G. A. Jeffrey, B. Douglas, J. L. Kirkpatrick, and J. A. Weisbach, *J. Am. Chem. Soc.* **89**, 3061 (1967).
285. P. T. Cheng, S. C. Nyburg, F. N. MacLachlan, and P. Yates, *Can. J. Chem.* **54**, 726 (1976).
286. W. E. Oberhänsli, *Helv. Chim. Acta* **52**, 1905 (1969).
287. G. Lukacs, M. De Bellefon, L. Le Men-Olivier, J. Lévy, and J. Le Men, *Tet. Lett.* 487 (1974).
288. P. Yates, F. N. MacLachlan, I. D. Rae, M. Rosenberger, A. G. Szabo, C. R. Willis, M. P. Cava, M. Behforouz, M. V. Lakshmi-Kantham, and W. Zeiger, *J. Am. Chem. Soc.* **95**, 7842 (1973).
289. N. Neuss, H. E. Boaz, J. L. Occolowitz, E. Wenkert, F. M. Schell, P. Potier, C. Kan, M.-M. Plat and M. Plat, *Helv. Chim. Acta* **56**, 2660 (1973).
290. M. Daudon, H. Mehri, M. M. Plat, E. W. Hagaman, F. Schell, and E. Wenkert, *J. Org. Chem.* **40**, 2838 (1975).
291. L. J. Durham, J. N. Shoolery, and C. Djerassi, *Proc. Natl. Acad. Sci. U.S.A.* **71**, 3797 (1974).
292. R. Burton, C. W. M. Grant, and L. D. Hall, *Can. J. Chem.* **50**, 497 (1972).
293. L. D. Hall and C. M. Preston, *Can. J. Chem.* **52**, 829 (1974).
294. A. Chiaroni, C. Riche, L. Diatta, R. Z. Andriamialisoa, N. Langlois, and P. Potier, *Tetrahedron* **32**, 1899 (1976).
295. A. Rabaron and M. Plat, *Plant. Med. Phytother.* **7**, 319 (1973).
296. A. Rabaron, Doctoral Thesis, University of Paris, VI, 1974, quoted in Bruneton *et al.* (*297*).
297. J. Bruneton, A. Cavé, E. W. Hagaman, N. Kunesch, and E. Wenkert, *Tet. Lett.* 3567 (1976).
298. J. Bruneton, T. Sevenet, P. Potier, and A. Cavé, unpublished observations, quoted in Bruneton *et al.* (*297*).
299. J. Le Men, G. Hugel, M. Zeche, M.-J. Hoizey, L. Le Men-Olivier, and J. Lévy, *C. R. Acad. Sci., Ser. C* **283**, 759 (1976).

300. A. Ducruix and C. Pascard, *Acta Crystallogr., Sect. B* **33**, 1990 (1977).
301. D. G. I. Kingston, F. Ionescu, and B. T. Li, *Lloydia* **40**, 215 (1977).
302. V. N. Reinhold and R. J. Bruni, *Biomed. Mass Spectrom.* **3**, 335 (1976).
303. Y. Rolland, G. Lewin, and F. Libot, Ger. Offen. 2,652,165; *CA* **87**, 102510e (1977).
304. S. A. Buskine, Ger. Offen. 2,534,858; *CA* **86**, 106865b (1977).
305. J. P. Kutney, K. K. Chan, W. B. Evans, Y. Fujise, T. Honda, F. K. Klein, and J. P. de Sousa, *Heterocycles* **6**, 435 (1977).
306. E. Wenkert, *Recent Chem. Nat. Prod., Incl. Tobac., Proc. Philip Morris Sci. Symp., 2nd, 1975* p. 15 (1976).
307. Y. Honma and Y. Ban, *Heterocycles* **6**, 129 (1977).
308. Y. Honma, T. Ohnuma, and Y. Ban, *Heterocycles* **5**, 47 (1976).
309. S. Takano, K. Shishido, M. Sato, and K. Ogasawara, *Heterocycles* **6**, 1699 (1977).
310. Y. Ban, Y. Sekine, and T. Oishi, *Tet. Lett.* 151 (1978).
311. T. Kametani, Jpn. Kokai 77/12,916; *CA* **87**, 68519c (1977).
312. A. H. Jackson, P. R. Jenkins, and P. V. R. Shannon, *J. Chem. Soc., Perkin Trans. I* 1698 (1977).
313. J. Bergman and R. Carlsson, *Tet. Lett.* 4663 (1977).
314. J. Bergman and R. Carlsson, *J. Heterocycl. Chem.* **9**, 833 (1972).
315. A. P. Kozikowski and N. M. Havan, *J. Org. Chem.* **42**, 2039 (1977).
316. D. Rousselle, J. Gilbert, and C. Viel, *C. R. Acad. Sci., Ser. C* **284**, 377 (1977).
317. B. Robinson, *Chem. Rev.* **63**, 373 (1973).
318. P. A. Cranwell and J. E. Saxton, *J. Chem. Soc.* 3482 (1962).

—CHAPTER 4—

PAPAVERACEAE ALKALOIDS. II

F. ŠANTAVÝ

Institute of Chemistry, Medical Faculty,
Palacký University, Olomouc, Czechoslovakia

I. Introduction

During the past years the increasing interest and research in the isolation, chemistry, biochemistry, and chemotaxonomy of the Papaveraceae alkaloids had led to the publication of a great number of original

ISBN 0-12-469517-5

communications and summarizing reports. In 1970, information available on these alkaloids was reviewed in the series of *The Alkaloids* (Vol. XII, p. 333), but since then many important papers on this subject have appeared. Therefore, it seems desirable to incorporate the recent advances in this field into a supplementary review. The organization of this chapter is similar to that of the 1970 review. It was, however, necessary to add the isoquinoline alkaloids to the group of benzylisoquinoline alkaloids and to rename the group of tetrahydroprotoberberine and protoberberine alkaloids into a group of alkaloids of the berbine type because the Papaveraceae plants were also found to contain pseudoprotoberberine bases. The narceineimide alkaloids with two nitrogens were classified with the narceine alkaloids, and the group of spirobenzylisoquinoline alkaloids was subdivided into (a) ochotensimine, (b) hypecorine and canadaline and, (c) peshawarine types of alkaloids. Furthermore, it was deemed necessary to add the group of dimeric alkaloids and the group of those alkaloids which occur in the Papaveraceae plants but are not derived from isoquinoline or benzylisoquinoline. New information on the alkaloids in the individual Papaveraceae plants contributes to the chemotaxonomic classification of the individual plant species to the correct tribes and genera. For this reason Section IV on chemotaxonomy, ecology, and callus tissues has been added.

The chemistry and the pharmacology of benzylisoquinoline alkaloids were dealt with in the monograph by Shamma (*1*). He also wrote (*2*) a chapter on the spirobenzylisoquinoline alkaloids for the series *The Alkaloids* and on the isoquinoline alkaloids for the book by Pelletier (*3*). Kametani (*4*) published a two-volume monograph (1968 and 1974) on the isoquinoline alkaloids in which, in addition to the formula, melting point, and optical rotation of the alkaloids, the origin of the plant material studied and the reference numbers indicating the literature on the physical properties, spectroscopy, and the syntheses were given. Each of these two volumes contains a detailed introduction to the chemistry of the individual groups of isoquinoline alkaloids. Kühn, Thomas, and Pfeifer (*5, 6*), and Šantavý (*6a*) have reviewed the Papaveraceae alkaloids. Döpke published a two-volume continuation of the books by H. G. Boit (*Ergebnisse der Alkaloid-Chemie*) on the progress made in the field of alkaloids (including the *Papaver* plant alkaloids) during the years from 1933 to 1968 (*6b*).*

The British Chemical Society has recently begun to publish each year a volume entitled *The Alkaloids* (*10*) which, in always briefer form, records those reports on alkaloid chemistry and biochemistry which have appeared in the preceding year. Farnsworth (*11*) undertook to publish a periodical

* The benzylisoquinoline alkaloids were also found in the *Thalictrum* genus (*Ranunculaceae*) and were dealt with in a review by Schiff and Doskotch (*7*) and in other reports (*8, 9*).

entitled "Pharmacognosy Titles" containing references on the isolation, chemistry, and biology of natural products, including the Papaveraceae alkaloids.

Robinson (*12*) wrote a review on the metabolism and function of alkaloids in plants in which references to morphine, codeine, and thebaine were given. Preininger (*13*) published a comprehensive review including 691 references on the pharmacology and toxicology of the Papaveraceae alkaloids.

In this chapter only the new formulas or those which are necessary for the clarity of the text are given; the formulas of the compounds mentioned in previous chapters of this treatise are not always included.

II. Occurrence, Detection, and Isolation of Alkaloids

Similar to the earlier review published in this series (Vol. XII, p. 333), the Papaveraceae plants were subdivided into tribes and genera as listed in Table I. This shows the source of isolation for the alkaloids.

Table II (*14–293*) is supplementary to the previous tables on Papaveraceae plants and their alkaloids (Vol. IV, p. 79; Vol. IX, p. 44; Vol. X, p. 467; Vol. XII, p. 335). In Table II all the recently investigated plants, the reinvestigated plants, and those for which the already isolated alkaloids have now been examined for structural elucidation are given together with the appropriate references.

Studies of the isolation and the constitution of alkaloids of the Papaveraceae family have led to the investigation of the effect of various types of drying on the quantity of alkaloids in the leaves of *Macleaya* (*Bocconia*) *microcarpa* (*52*), the stability of some *Papaver* alkaloids (in different solvents) to sunlight (*294*), the accumulation of alkaloids in the latex of the Papaveraceae (*243*), the site of origin of the alkaloids in the studied plants (*295, 296*), the different geographical zones (*297, 298*), the time after flowering (*299*), and the seasonal variations in the content of the individual groups of alkaloids, particularly in *P. somniferum* (*300*).

Numerous reports have also been published on the separation of the individual alkaloids by using ion exchangers (*301, 302*), various solvents of different pH (*303*), or multibuffered extraction systems (*102, 103, 304–306*). The isolation of minor bases from extracts of *P. somniferum* (*254, 282, 307–311*) and of *Fumaria officinalis* (*166*) has been reported. The separation of the bases obtained from this plant material was also carried out by countercurrent distribution (*312*) and column chromatography (*313, 314*). The alkaloid coreximine was detected in *P. somniferum* by an isotope dilution method which was based on the biosynthesis of this alkaloid from reticuline (*249*). The isoquinoline type of quaternary alkaloids can be

TABLE I

SURVEY OF THE GENERA OF THE PLANT FAMILY PAPAVERACEAE

Subfamily	Tribe	Genus	Assumed number of species	Isoquinoline	Benzylisoquinoline substituents			Proaporphine
					2	3′	3′,4′	
I. Hypecoideae		*Pteridophyllum*	1	–	–	–	–	–
		Hypecoum	15	–	–	–	–	–
II. Papaveroideae	Platystemoneae	*Meconella*	6	–	–	–	–	–
		Platystemon	57	–	–	–	–	–
	Romneyeae	*Romneya*	2	–	–	–	+	–
	Eschscholtzieae	*Dendromecon*	20	–	–	–	–	–
		Hunnemannia	1	–	–	–	+	–
		Eschscholtzia	123	–	–	–	+	–
	Chelidonieae	*Sanguinaria*	1	–	–	–	–	–
		Eomecon	1	–	–	–	–	–
		Stylophorum	1 (3?)	–	–	–	–	–
		Hylomecon	1 (3?)	–	–	–	–	–
		Chelidonium	1	–	–	–	–	–
		Macleaya	2	–	–	–	–	–
		Bocconia	9	–	–	–	–	–
	Papavereae	*Glaucium*	21	–	–	–	–	–
		Roemeria	6	–	–	–	+	+
		Dicranostigma	3	–	–	–	–	–
		Meconopsis (*Cathcartia*)	45	–	–	–	–	+
		Stylomecon	2	–	–	–	–	–
		Argemone	9	–	–	–	+	–
		Papaver	100	+	–	+	+	+
		Arctomecon	3	–	–	–	–	–
		Canbya	2	–	–	–	–	–
III. Fumarioideae	Corydaleae	*Dicentra*	17	–	–	–	+	–
		Adlumia	1	–	–	–	–	–
		Corydalis	280	+	+	–	+	–
		Roborowskia batalin ?	?	–	–	–	–	–
		Sarcocapnos ?	?	–	–	–	–	–
	Fumarieae	*Rupicapnos* ?	?	–	–	–	–	–
		Fumaria	30	+	–	–	–	–
		Trigonocapnos ?	?	–	–	–	–	–

[a] 13-Methyl- and 13-hydroxy derivatives.

ACCORDING TO ENGLER AND THE GROUPS OF ALKALOIDS FOUND

Types of known alkaloids

Aporphine							Berbine								Benzophenanthridine		Spirobenzylisoquinol			
Aporphine	Dehydroaporphine	Promorphinane	Morphinane	Pavine	Isopavine	Cularine	Tetrahydro-protoberberine	Berberine	1-Hydroxy-protoberberine	Pseudoberberine	Protopine	Phthalidisoquinoline	Narceine	Rhoeadine/Papaverrubine	Reduced	Dehydrogenated	Ochotensimine	Hypecorine	Peshawarine	Dimeric alkaloids
−	−	−	−	−	−	−	−	−	−	−	+	−	−	−	−	−	−	−	−	−
−	−	−	−	−	−	−	+	+	−	−	+	−	−	−	−	+	−	+	+	−
−	−	−	−	−	−	−	−	−	−	−	−	−	−	−	−	−	−	−	−	−
−	−	−	−	−	−	−	−	+	−	−	+	−	−	−	−	+	−	−	−	−
−	−	−	−	−	−	−	−	−	−	−	+	−	−	−	−	−	−	−	−	−
−	−	−	−	−	−	−	−	−	−	−	+	−	−	−	−	−	−	−	−	−
−	−	−	−	−	−	−	+	+	−	−	+	−	−	−	−	+	−	−	−	−
+	−	−	−	+	−	−	+	+	−	−	+	−	−	−	−	+	−	−	−	−
−	−	−	−	−	−	−	−	+	−	−	+	−	−	−	+	+	−	−	−	+
−	−	−	−	−	−	−	−	−	−	−	−	−	−	−	−	−	−	−	−	−
+	−	−	−	−	−	−	+	+	−	−	+	+(?)	−	−	+	+	−	−	−	−
−	−	−	−	−	−	−	+	+	−	−	+	+	−	−	+	+	−	−	−	−
−	−	−	−	−	−	−	+	+	−	−	+	−	−	−	+	+	−	−	−	−
−	−	−	−	−	−	−	−	+	−	−	+	−	−	−	−	+	−	−	−	−
−	−	−	−	−	−	−	+	+	−	−	+	−	−	+	+	+	−	−	−	+
+	+	−	−	−	−	−	+	+	−	−	+	−	−	−	+	+	−	−	−	−
+	+	−	−	−	+	−	+	+	−	−	+	−	−	−	−	−	−	−	−	−
+	−	−	−	−	−	−	−	+	−	−	+	−	−	−	−	+	−	−	−	−
−	−	−	−	−	−	−	+	+	−	−	+	−	−	+	−	+	−	−	−	−
−	−	−	−	−	−	−	−	+	−	−	+	+	−	−	−	−	−	−	−	−
−	−	−	−	+	−	−	+	+	−	−	+	−	−	−	−	+	−	−	−	−
+	+	+	+	−	+	−	+	+	+	+	+	+	+	+	−	+	−	−	−	+
−	−	−	−	−	−	−	−	−	−	−	+	−	−	−	−	−	−	−	−	−
−	−	−	−	−	−	−	−	−	−	−	−	−	−	−	−	−	−	−	−	−
+	−	−	−	−	−	+	−	+	−	−	+	+	−	−	−	+	+	−	−	+
−	−	−	−	−	−	−	−	−	−	−	+	+	−	−	−	−	−	−	−	−
+	+	+	−	−	−	+	+[a]	+[a]	+	+	+	+	+	−	+	+	+	+	−	−
−	−	−	−	−	−	−	−	−	−	−	−	−	−	−	−	−	−	−	−	−
−	−	−	−	−	−	−	−	−	−	−	−	−	−	−	−	−	−	−	−	−
−	−	−	−	−	−	−	−	−	−	−	−	−	−	−	−	−	−	−	−	−
+	−	+	−	−	−	−	+	+	−	−	+	−	+	−	−	−	+	−	−	−
−	−	−	−	−	−	−	−	−	−	−	−	−	−	−	−	−	−	−	−	−

TABLE II

PLANTS OF THE PAPAVERACEAE AND THEIR ALKALOIDS

Plant	Alkaloids	References
Arctomecon californica Torr. & Fremont	Allocryptopine, protopine	*14*
Argemone albiflora Hornem.	Alkaloid AA-1 ($C_{19}H_{21}NO_4$), AA-2 ($C_{20}H_{21}NO_5$), AA-3, allocryptopine, berberine, chelerythrine, norchelerythrine, norsanguinarine, protopine, sanguinarine, (−)-scoulerine, (−)-*β*-scoulerine methohydroxide[a]	*15*
A. albiflora Hornem, ssp. *texana* & Ownb.	Allocryptopine, berberine, coptisine, protopine, sanguinarine	*16*
A. aurantiaca Ownb.	Coptisine, protopine	*17*
A. brevicornuta Ownb.	Berberine, norargemonine	*16*
A. chisosensis Ownb.	Allocryptopine, berberine, protopine	*17*
A. corynbosa Greene, ssp. *arenicola*	Allocryptopine, berberine, protopine, sanguinarine	*16*
A. echinata Ownb.	Berberine, cryptopine	*18*
A. fruticosa Thurber & Gray	Allocryptopine, benzophenanthridine alkaloids, cryptopine, hunnemannine	*18*
A. glauca, var. *glauca*	Allocryptopine, berberine, chelerythrine, protopine, sanguinarine	*19*
A. gracilenta Greene	(−)-Argemonine, (−)-argemonine methohydroxide, (−)-argemonine *N*-oxide, (−)-isonorargemonine, (+)-laudanidine, (−)-munitagine, muramine, (−)-platycerine methohydroxide, protopine, (+)-reticuline	*20*
A. grandiflora Sweet, ssp. *grandiflora*	*α*-Allocryptopine, berberine, (−)-cheilanthifoline, chelerythrine, (+)-codamine, corypalmine, (+)-laudanosine, protopine, sanguinarine	*21*
A. mexicana L.	Allocryptopine, berberine, (−)-cheilanthifoline, chelerythrine, coptisine, cryptopine, dihydrosanguinarine, norchelerythrine, norsanguinarine, protopine, sanguinarine, (−)-*β*-scoulerine methohydroxide, (−)-*α*-stylopine methohydroxide, (−)-*β*-stylopine methohydroxide, 6-acetonyldihydrosanguinarine	*22–27*
A. munita Dur. & Hilg., ssp. *argentea* Ownb.	Allocryptopine, (−)-argemonine, (−)-isonorargemonine, protopine	*28*
A. munita Dur. & Hilg., ssp. *rotundata* (Rydb.) Ownb.	2,9-Dimethoxy-3-hydroxypavinane	*29, 30*
A. ochroleuca Sweet (syn. *A. mexicana*, var. *ochroleuca* (Sweet) Lindl.)	Allocryptopine, berberine, (−)-*α*-canadine methohydroxide, (−)-cheilanthifoline, chelerythrine, coptisine, sanguinarine, (−)-stylopine methohydroxide, (−)-*α*-tetrahydropalmatine methohydroxide	*31*
A. platyceras Link & Otto	Alkaloid AP-1, (−)-argemonine methohydroxide, (−)-*α*-canadine methohydroxide, chelerythrine, corysamine, cyclanoline [=(−)-*α*-scoulerine methohydroxide], magnoflorine, (−)-platycerine methohydroxide, (−)-stylopine methohydroxide	*32–34*
A. platyceras, ssp. *pinnatisecta*	Bisnorargemonine, munitagine	*35*
A. platyceras, ssp. *pleiacantha*	Allocryptopine, berberine, bisnorargemonine, cryptopine, munitagine, protopine	*35*
A. pleiacantha Greene, ssp. *ambigua* Ownb.	Allocryptopine, berberine, bisnorargemonine, cryptopine, munitagine, protopine	*35*
A. polyanthemos (Fedde) Ownb.	Allocryptopine, berberine, chelerythrine, coptisine, norchelerythrine, protopine, sanguinarine, (−)-scoulerine	*17, 36, 37*
A. sanguinea	Allocryptopine, argemonine, berberine, muramine	*17*
A. squarrosa Greene	Berberine, muramine	*38*
A. subfusiformis	Allocryptopine, berberine, chelerythrine, protopine, sanguinarine, three unidentified bases	*39*
A. subfusiformis, ssp. *subfusiformis* Ownb.	Allocryptopine, berberine, chelerythrine, protopine, sanguinarine	*36*

A. subfusiformis, ssp. *subinermis* Ownb.	Allocryptopine, berberine, chelerythrine, protopine, sanguinarine	*36*
A. subintegrifolia Ownb.	Allocryptopine, berberine, protopine	*28*
A. turnerae A. M. Powel	(+)-Armepavine, (−)-tetrahydropalmatine	*16*
Bocconia arborea	1,3-Bis(6-hydrochelerythrinyl)acetone, dihydrosanguinarine, 6-*O*-methyldihydrochelerythrine, oxysanguinarine	*40*
B. cordata Willd.	Alkaloid K_0, allocryptopine, bocconine (= chelirubine), bocconoline, chelerythrine, cryptopine, dehydrocheilanthifoline, dihydrochelerythrine, dihydrosanguinarine, protopine, sanguinarine, two bases mp 180° and 286°	*41–48*
B. frutescens L.	Allocryptopine, berberine, chelerythrine (= *Bocconia* alkaloid P-61), columbamine, coptisine, corysamine, (−)-isocorypalmine, papaverrubine E, protopine, rhoeadine, sanguinarine	*33, 49, 50*
B. laurina	Protopine	*51*
B. microcarpa Maxim.	Allocryptopine, chelerythrine, protopine, sanguinarine, several unidentified bases	*48, 52, 53*
B. pearcei	Unidentified bases	*54*
Chelidonium majus L.	Chelamine (= 10-hydroxychelidonine), chelamidine (= 10-hydroxyhomochelidinine), dihydrochelerythrine, dihydrochelilutine, dihydrochelirubine, dihydrosanguinarine, magnoflorine, oxysanguinarine, sanguinarine, sparteine, stylopine, (−)-α-stylopine methohydroxide, (−)-β-stylopine methohydroxide, substance X (= *N*-demethyldihydroxysanguinarine), alkaloid K_0	*50, 55–63, 223a*
Corydalis ambigua Chem. & Schlecht.	α-Allocryptopine, (+)-base II, (±)-base II, cavidine, (+)-corybulbine (= corydalmine), (+)-corydaline, (+)-corydalmine, dehydrocorydaline, dehydrothalictrifoline, (+)-glaucine, (+)-1-methylcorypalline, noroxyhydrastinine, protopine, (−)-scoulerine, (−)-tetrahydrocolumbamine, (−)-tetrahydrocoptisine, (+)-tetrahydrojatrorrhizine, (±)-tetrahydropalmatine	*64–66*
C. bulbosa (Chinese)	α-Allocryptopine, (+)-corybulbine, (+)-corydaline, dehydrocorydaline, (+)-glaucine, noroxyhydrastinine, protopine, (−)-scoulerine, (−)-tetrahydrocolumbamine, (−)-tetrahydrocoptisine, (+)-tetrahydrojatrorrhizine, tetrahydropalmatine	*64, 67, 68*
C. campulicarpa Hayata	α-Allocryptopine, berberine, ophiocarpine, protopine, seven minor unidentified alkaloids	*69*
C. caseana A. Gray	Caseadine, caseanadine	*70, 71*
C. caucasica	Allocryptopine, chelerythrine, protopine, sanguinarine	*71a*
C. cava (L) Sch. & K. (syn. *C. tuberosa* DC.)	Adlumidiceine, apocavidine, berberine, bulbocapnine, canadine, capnoidine, coptisine, corybulbine, corycavamine, corycavidine, corycavine, corydaline, corydine, corypalmine, corysamine, corytuberine, dehydrocorydaline, dihydroxydimethoxyprotoberberine, domestine, domesticine, glaucine, hydrastinine, isoboldine, isocorybulbine, isocorydine, isocorypalmine, 1,2-methylenedioxy-6a,7-dehydroaporphine-10,11-quinone, 13-methyltetrahydroprotoberberine (= thalictricavine), narcotine, 8-oxocoptisine, palmatine, predicentrine, protopine, reticuline, scoulerine, sinoacutine, stylopine, tetrahydrocoptisine, tetrahydropalmatine	*50, 72–77*
C. claviculata D.C.	Berberine, coptisine, cularicine, protopine, stylopine	*73*
C. decumbens Pers.	Adlumidine, bulbocapnine, protopine, (−)-tetrahydropalmatine	*78*
C. fimbrillifera Korsch.	β-Hydrastine, protopine	*79–81*
C. gigantea Trautv. & Mey.	(−)-Adlumine, (−)-adlumidine, (+)-bicuculline, (−)-cheilanthifoline, dihydrosanguinarine, protopine, sanguinarine, (−)-scoulerine	*119*
C. gortschakovii Schrenk.	(−)-Adlumine, alkaloid mp 156°, (+)-bicuculline, bracteoline, corgoine, corydine, cryptopine, domesticine, gortschakoine, isoboldine, isocorydine, protopine, sendaverine	*82–86*
C. govaniana	*S*-Govadine, *S*-govanine, bicuculline, corygovanine	*87*
C. impatiens	Several unidentified bases	*88*
C. incisa (Thunb.) Pers.	Acetylcorynoline, acetylisocorynoline, alkaloid V, bases TN-4, TN-5, TN-12, TN-16, TN-21, TN-23, (−)-cheilanthifoline, chelidonine, coreximine, corycavine, corydalic acid methyl ester, corydalisol (= TN-14), corydalispirone (= TN-5), corydalmine, corydamine, corynoline, corynoloxine, 11-epicorynoline, (+)-14-epicorynoline (= Base II), *N*-formylisocorynoline, 12-hydroxycorynoline, 6-oxocorynoline, pallidine, protopine, (+)-reticuline, (−)-scoulerine, sinoacutine, (−)-tetrahydrocorysamine	*89–101*
C. intermedia (L.) Merat Nouv.	α-Allocryptopine, alkaloid mp 170°, isoboldine, protopine	*84*
C. Koidzumiana Ohwi (Taiwan)	Allocryptopine, capaurine, cheilanthifoline, corydalidzine, corydaline, (+)-corybulbine, corynoxidine, dihydrosanguinarine, epicorynoxidine, (−)-isocorypalmine, oxysanguinarine, protopine, (+)-reticuline, sanguinarine, (−)-scoulerine, stylopine, tetrahydropalmatine	*102–106*

(*Continued*)

TABLE II (*Continued*)

Plant	Alkaloids	References
Corydalis (Korean)	Berberine, canadine, coptisine, corysamine, norisocorydine, noroxyhydrastinine	*107*
C. ledebouriana	Corledine, ledeborine, ledeboridine, (±)-raddeanine, several unidentified alkaloids	*108–110*
C. leariloba, var. *papilligera*	Corydaline, glaucine, protopine, (−)-scoulerine, (−)-tetrahydrocolumbamine, (−)-tetrahydrocoptisine, tetrahydropalmatine	*78*
C. lutea (L.) D.C.	Adlumidiceine, berberine, bicuculline, coptisine, corypalmine, corysamine, enol lactone of adlumidiceine, isocorydine, isocorypalmine, jatrorrhizine, *N*-methylhydrastine, *N*-methylhydrasteine, ochrobirine, 8-oxocoptisine, oxysanguinarine, protopine, stylopine, tetrahydropalmatine	*73, 111, 112*
C. marshalliana	(−)-Adlumidine, allocryptopine, (+)-bicuculline, (+)-bulbocapnine, corydaline, (+)-corydine, (−)-domesticine, (+)-isoboldine, protopine, alkaloids mp 193°, 241°, and 260°	*71a, 113*
C. nokoensis	Nokoensine	*114*
C. ochotensis Turcz.	Adlumidine, aobamidine, corytenchine, corytenchyrine, dihydrocheilanthifoline, ochotensimine, protopine, raddeanamine, raddeanone, raddeanine, yenhusomidine, yenhusomine	*115–118*
C. ochotensis, var. *raddeana*	Adlumidine, aobamidine, aobamine, bicuculline, cheilanthifoline, dihydrosanguinarine, fumarine, ochotensimine, ochotensine, pallidine, protopine, raddeanamine, raddeanidine, raddeanine, raddeanone, scoulerine, sinoacutine	*120*
C. ochroleuca Koch.	Adlumidine, bicuculline (= Alkaloid F-45), fumarine, fumaramine (= Alkaloid F-46), (+)-glaucine, isocorydine, (−)-scoulerine, (−)-sinoacutine	*76, 121*
C. ophiocarpa Hook. & Thoms.	Adlumine, allocryptopine, berberine, (−)-canadine, coptisine, corypalmine, cryptopine, β-hydroxystylopine, ophiocarpine, protopine, (−)-stylopine	*73, 122*
C. Paczoskii N. Busch.	Corpaine, corydaine	*86, 123, 124*
C. pallida (Thunb.) Pers.	Alkaloid P, alkaloid mp 250°, capauridine, capaurimine, capaurine, corydaline, cryptopine, (+)-isoboldine, (+)-isosalutaridine, kikemanine (= corydalmine), pallidine, protopine, (±)-stylopine, (−)-tetrahydropalmatine	*76, 125–127*
C. pallida Pers, var. *tenuis* Yatabe	Capauridine, capaurimine, capaurine, coptisine, corydalactame (Alkaloid P), (+)-corydaline, corydalmine (= kikemanine), corysamine, (+)-corytuberine, dehydrocapaurine, dehydrocapaurimine, dehydrocorydaline, dehydrocorydalmine (= dehydrokikemanine), dihydrosanguinarine, ginnole, (+)-isoboldine (= aurotensine), oxysanguinarine, pallidine [= (+)-isosalutaridine], palmatine, protopine, sanguinarine, (−)-scoulerine, sinoacutine, (+)-tetrahydrocorysamine, (−)-tetrahydropalmatine, *trans*-3-ethylidine-2-pyrrolidone	*125, 127a–133*
C. persica	Chelerythrine, protopine, sanguinarine	*134*
C. platycarpa Makino	Aurotensine, bicuculline, capaurimine, (−)-capaurine, cheilanthifoline, (+)-corybulbine, corydaline, corysamine, dehydrocapaurimine, dehydrocorybulbine, isocorydine, jatrorrhizine, palmatine, protopine, sanguinarine, stylopine, tetrahydrocolumbamine, tetrahydropalmatine	*89, 135–137*
C. pseudoadunca Popov	(−)-Adlumidine, (+)-bicuculline, coramine, (+)-β-hydrastine, protopine, (−)-scoulerine	*82*
C. racemosa Pers.	Protopine, tetrahydropalmatine	*138*
C. rosea Leyth.	(−)-Adlumidine, (−)-adlumine, protopine	*139*
C. sempervirens (L.) Pers. (syn. *C. glauca* Pursch.)	Adlumiceine, enol lactone of adlumiceine, adlumidiceine, (−)-adlumine, bicucine, bicuculline, capnoidine, coptisine, cryptopine, oxysanguinarine, protopine	*73, 140*
C. sewertzovii Riegel.	Alkaloids mp 189° and 196°, allocryptopine, chelerythrine, corlumine, dihydrosanguinarine, protopine, sanguinarine, severcine	*84, 141*
C. sibirica (L.) Pers.	Adlumidine (= Alkaloid F-16), sibiricine	*76, 88, 142*
C. speciosa Maxim.	α-Allocryptopine, capaurimine, capaurine, corypalline, protopine, tetrahydropalmatine	*106*
C. stewartii Fedde	Coptisine, corycidine, corydicine, domesticine, isoboldine, protopine (= corydine), (+)-tetrahydrocoptisine	*143–145*
C. stricta Steph.	β-Hydrastine, protopine, sanguinarine	*79, 146*
C. tashiroi	Palmatine, protopine, (−)-tetrahydropalmatine	*147*
C. thalictrifolia French.	Adlumidine, adlumine, berberine, cavidine, coptisine, corypalmine, dehydrothalictrifoline, protopine, stylopine, thalictrifoline	*148*

C. vaginans Royle	1-*O*-Methylcorpaine, (+)-ochrobirine, protopine, sanguinarine	*149, 150*
Dicentra canadensis (Goldie) Walp.	Cancentrine, dehydrocancentrines A and B	*151*
D. eximia	Corydine, dicentrine, glaucine, norprotosinomenine	*152*
Dicranostigma franchetianum (Prain) Fedde	Berberine, chelerythrine, isocorydine, magnoflorine, menisperine, protopine	*153*
D. lactucoides Hook f. & Thoms.	Magnoflorine, menisperine [= (+)-isocorydine methohydroxide]	*50*
D. leptopodum (Maxim.) Fedde	Magnoflorine, menisperine	*50*
Eschscholtzia californica Cham.	Allocryptopine, amorphous bases, (−)-bisnorargemonine, californidine, (−)-α-canadine methohydroxide, escholidine (= tetrahydrothalifendine methohydroxide), escholine (= magnoflorine), escholinine [= (+)-romneine methohydroxide], (−)-norargemonine, (−)-α-stylopine methohydroxide	*50, 154–156*
E. douglasii (Hook & Arn.) Walp.	Allocryptopine, amorphous bases, (−)-bisnorargemonine, californidine, (−)-α-canadine methohydroxide, escholidine, escholine, (−)-norargemonine	*155, 156*
E. glauca	Allocryptopine, amorphous bases, (−)-bisnorargemonine, californidine, (−)-α-canadine methohydroxide, escholidine, escholine, norargemonine	*155, 156*
E. lobbii Greene	Allocryptopine, berberine, chelerythrine, chelirubine, coptisine, corydine, corysamine, corytuberine, macarpine, protopine, sanguinarine, (−)-scoulerine	*157*
E. species (*E. oregana* Greene?)	Alkaloid ES-1, ES-2, berberine, californidine, coptisine, corydine, escholamidine, escholamidine methohydroxide, escholamine, escholinine, protopine, scoulerine (−)-α-stylopine methohydroxide	*158*
Fumaria (*Herba Fumariae* Off.?)	Cryptopine, fumariline, fumaritriline, fumaritrine, fumarophycine, *O*-methylfumarophycine, parfumidine, protopine, sinactine, stylopine	*159, 160*
F. densiflora	Alkaloids	*161*
F. indica Pugsley	Amorphous alkaloids, (+)-bicuculline, coptisine, dehydrocheilanthifoline, fumarilicine, fumariline, narceinimide, protopine, (−)-tetrahydrocoptisine	*161–164*
F. kralikii Jord.	(−)-Adlumine, fumarofine, cryptopine, (+)-parfumine, protopine, alkaloid F_K-5	*186*
F. officinalis	Alkaloid F-37 (= fumaricine), alkaloid F-63 (= fumaritine), canadine, cryptopine, fumariline, fumarilicine, fumaroline, fumarophycine, fumoficinaline, *N*-methylhydrastine, *N*-methylhydrasteine, *N*-methyloxohydrasteine, protopine, sanguinarine, scoulerine, sinactine	*74, 76, 111, 165–169, 169a*
F. parviflora Lam.	(+)-Bicuculline, cryptopine, fumaramine, fumaridine, *N*-methylhydrasteine, parfumidine, parfumine, protopine, sanguinarine	*111, 170–174*
F. rostellata	(+)-Adlumine, cryptopine, fumariline, fumaritridine, fumaritrine, fumarostelline, parfumine, protopine, sinactine, stylopine	*160, 175, 176*
F. schleicheri Soyer–Willem	*N*-Methylhydrasteine	*111*
F. vaillantii Loisl.	(−)-Adlumidine, (−)-adlumilantine, (−)-adlumine, (+)-bicuculline, cryptopine, fumaramine, fumaridine (= hydrastine imide), fumvailine, (+)-α-hydrastine, *N*-methylhydrasteine, parfumidine, parfumine, protopine, sanguinarine, vaillantine (= 2,3-didemethylmuramine)	*109, 111, 171, 174, 177–180*
Glaucium contortuplicatum Boiss.	Dicentrine, sinoacutine	*181*
G. corniculatum Curt.	Allocryptopine, (−)-β-canadine methohydroxide, (+)-corydine, heliotrine, protopine, sanguinarine, (−)-β-stylopine methohydroxide	*182, 183*
G. elegans Fisch. & Mey.	Allocryptopine, chelerythrine, (±)-chelidonine (= diphylline), chelirubine, coptisine, corydine, dihydrochelerythrine, glaucine, glauvine, isoboldine, isocorydine, *O*-methylatheroline, protopine, sanguinarine, and quaternary bases of the benzophenanthridine alkaloids	*184, 185*
G. flavum Cr.	Alkaloids I and II, allocryptopine, (−)-aurotensine, bocconoline, bulbocapnine, (−)-chelidonine, chelerythrine, chelirubine, coptisine, (+)-corydine, corytuberine, dehydroglaucine, dehydronorglaucine, dicentrine, (+)-glaucine, glauflavine (= corytuberine), glauvine, 1-hydroxy-2,9,10-trimethoxyaporphine, isoboldine (= aurotensine), (+)-isocorydine, magnoflorine, *O*-methylatheroline, (−)-norchelidonine, protopine, sanguinarine, scoulerine, thaliporphine (= *O*-methylisoboldine), and quaternary salts of alkaloids	*187–199*

(*Continued*)

TABLE II (*Continued*)

Plant	Alkaloids	References
G. flavum Cr., pop. Ghom.	Bulbocapnine, dicentrine, salutaridine	*200*
G. flavum Cr., var. *leiocarpum*	Oxoglaucine (= *O*-methylatheroline)	*179*
G. flavum Cr., var. *vestitum*	Cataline, corunnine, glaucine, pontevedrine, thalicmidine (= thaliporphine)	*198, 201–204*
G. fimbrilligerum Boiss.	Alkaloids mp 160° and 187°, allocryptopine, berberine, chelerythrine, chelidonine, chelirubine, coptisine, (+)-corydine, corytuberine, isoboldine, (+)-isocorydine, (−)-isocorypalmine, magnoflorine, *O*-methylarmepavine, *N*-methylcoclaurine, protopine, sanguinarine	*205–207*
G. grandiflorum	(+)-Glaucine, glauvine, isoboldine, *O*-methylatheroline, protopine, sanguinarine, thalicmidine	*208*
G. leiocarpum Boiss.	Alkaloid K_0, dehydroglaucine, glaucine, protopine	*209*
G. pulchrum Stapf, population Elika	Bulbocapnine, corydine, isocorydine, *N*-methyllindcarpine, protopine	*225*
G. squamigerum Kar. & Kir.	Alkaloids of phenolic and nonphenolic character, allocryptopine, berberine, chelerythrine, coptisine, corydine, protopine, sanguinarine	*210*
G. vitellinum Boiss a. Buhse, pop. Seerjan	Bulbocapnine, dicentrine, glaucine, isocorydine, muramine, protopine, tetrahydropalmatine	*225*
Hunnemannia fumariaefolia Sweet	Scoulerine (= alkaloid HF-1), cyclanoline, escholidine, and other bases	*15, 155*
Hylomecon vernalis Maxim.	Allocryptopine, berberine, canadine, chelerythrine, chelidonine, chelilutine, chelirubine, coptisine, isocorydine, protopine, sanguinarine, stylopine, tetrahydroberberine	*50, 211, 212*
Hypecoum erectum L.	Hypecorine, hypecorinine, protopine	*213, 214*
H. imberbe Sibth. & Sm.	Allocryptopine, chelerythrine, protopine, sanguinarine	*215*
H. parviflorum Kar. & Kir. (Pakistan)	Peshawarine, protopine	*224*
H. pendulum L.	Protopine	*216*
Macleaya (*Bocconia*) *cordata* (tissues culture)	Chelerythrine, sanguinarine	*217, 218*
M. microcarpa	Allocryptopine, chelerythrine, cryptopine, protopine, sanguinarine	*219, 219a*
Meconella oregana, var. *californica*	Protopine	*220*
Meconopsis betonicifolius Franch.	Alkaloid MR-1, allocryptopine, berberine, coptisine, corysamine, cryptopine, isorhoeadine, papaverrubine A, C, D, and E, protopine, rhoeadine	*50, 221*
M. cambrica (L.) Vig.	Alkaloid MC-1($C_{19}H_{21}NO_4$), alkaloids MC-2 and MC-3, berberine, coptisine, corytuberine, magnoflorine, mecambrine, papaverrubine D, roemerine, sanguinarine	*50, 222, 222a*
M. horridula Hook f. & Thoms.	Alkaloids MR-1, MR-2, MR-3, and MR-4, allocryptopine, amurensinine methohydroxide, coptisine, papaverrubine D and E, protopine, rhoeadine (?)	*50, 221*
M. napaulensis DC.	Alkaloids MR-2 and MR-3, allocryptopine, amurensinine, methohydroxide, base mp 212°, coptisine, corysamine, cryptopine, magnoflorine, norharmane alkaloid $C_{13}H_{16}N_2O$, papaverrubine D and E, protopine, rhoeadine, several unidentified alkaloids	*50, 221, 223*
M. paniculata (D. Don) Prain	Alkaloid MR-1, MR-3, allocryptopine, coptisine, corysamine, magnoflorine, papaverrubine D and E, protopine, rhoeadine	*50, 221*
M. robusta Hook. f. & Thoms.	Alkaloids MR-1 and MR-3, allocryptopine, coptisine, corysamine, magnoflorine, papaverrubine D, protopine	*50, 221*
M. rudis Prain	Alkaloids MR-1 ($C_{13}H_{16}N_2O$), MR-2, MR-3, and MR-4, allocryptopine, amurensinine methohydroxide, coptisine, magnoflorine, papaverrubine D and E, protopine, rhoeadine (?)	*50, 221*
M. sinuata Prain	Alkaloids MR-2 and MR-3, allocryptopine, coptisine, papaverrubine D, protopine	*50, 221*

Papaver alboroseum Hulten	Alborine, mecambridine (= oreophyline), papaverrubine D and C	*226*
P. alpinum, ssp. *alpinum*	Alborine, alpinigenine, alpinine, amurensine, amurensinine, amurine, cryptopine, mecambridine, nudaurine, papaverrubine D and C, protopine, sanguinarine	*227*
P. alpinum, ssp. *Kerneri*	Alborine, alpinigenine, alpinine, amurensine, amurensinine, amurine, cryptopine, muramine, mecambridine, nudaurine, papaverrubine D and E, protopine, sanguinarine	*227*
P. alpinum, ssp. *rhaeticum*	Alborine, alpinigenine, alpinine, amurensine, amurensinine, amurine, crytopine (?), mecambridine, muramine (?), nudaurine, papaverrubine D and G, protopine, sanguinarine	*227*
P. alpinum, ssp. *Sendtneri*	Alborine, alpinigenine, alpinine, amurensine, amurensinine, amurine, cryptopine (?), mecambridine, muramine, nudaurine, papaverrubine D and G, protopine, sanguinarine	*227*
P. anomalum Fedde	Alborine, α-allocryptopine, amurensinine, amurine, nudaurine, pavanoline, protopine, reframidine	*226*
P. arenarium M.B.	Macrostomine	*237*
P. armeniacum (L.) DC	Thebaine	*240*
P. bracteatum Lindl.	Alpinigenine, alpinine, bracteoline, codeine, coptisine, floripavidine, 14β-hydroxycodeine, 14β-hydroxycodeinone, isothebaine, mecambridine, *N*-methylcorydaldine, neopine, norcorydine, nuciferine, orientalidine, oripavine, oxysanguinarine, papaverrubine B, D, and E. protopine, salutaridine, thebaine, α-thebaol, two *N*-oxides of thebaine	*228–235c*
P. californicum A. Gray	Coptisine, cryptopine, muramine, papaverrubine A, B, D, and E, protopine, rhoeadine, rhoeaginine	*236*
P. commutatum Fisch. & Mey	Alkaloid Pc-1, coptisine, cryptopine, isocorydine, isorhoeadine, isorhoeagenine, isorhoeagenine-glycoside, papaverrubine A, B, C, D, E, and F, roemerine, (−)-stylopine, methohydroxide (alkaloid R-D)	*236, 237*
P. dubium L.	Alkaloids Rd-B, Rd-C, and Rd-F, oxysanguinarine, protopine	*236*
P. feddei Schwarz	Amurine	*238*
P. fugax Poir.	Aporheine, armepavine, mecambrine, palmatine, papaverrubine B, D, and E, protopine, (+)-remrefidine (isoroemerine metiodide)	*234, 239, 239a*
P. fugax of Turkish Origin	Armepavine, narcotine, rhoeadine, thebaine	*240, 241*
P. glaucum Boiss. & Hauskn.	Coptisine, glaucamine, glaudine, oxysanguinarine, papaverrubine A, B, C, and D	*239*
P. heldreichii Boiss.	Amurine, aporheine, coptisine, glaucine, liriodenine, papaverrubine B	*238, 239*
P. macrostomum Boiss. & Huet.	Macrostomine, sevanine	*242*
P. orientale L.	Alkaloids Or-1, Or-2, bracteine (= orientalinone), (+)-bracteoline, isothebaine, mecambridine, (−)-orientalinone, oripavidine, oripavine, papaverrubine C and D, salutaridine, thebaine	*243–244, 281*
P. pannosum Schwz. (*P. spicatum* Boiss. & Bal., var. *spicatum*)	Amurine, aporphines bases II, III, and IV, dihydronudaurine, glaucine, mecambrine, nudaurine, roemerine, no papaverrubine	*226, 238*
P. pilosum Sibth. & Smith	Amurine	*238*
P. polychaetum Schott & Kotschy	Amorphous alkaloids and protopine	*241, 245*
P. pseudocanescens M. Pop.	Alkaloid PO-5, alborine, amurensine, amurensinine, cryptopine, mecambridine, mecambridine methohydroxide, papaverrubine C, D, and E, protopine, rhoeadine	*226, 246*
P. pseudoorientale	Alkaloids PO-4 and PO-5, aryapavine, bracteoline, macrantaline, macrantoridine, isothebaine, orientalidine, salutaridine	*16a, 243, 247, 281*
P. radicatum Rottb.	β-Allocryptopine, amurensinine, amurine, berberine, cryptopine, *O*-methylthalisopavine, papaverrubine E, protopine, sanguinarine	*248*
P. rhoeas L.	Adlumidiceine, alkaloid R-B, berberine, coptisine (= alkaloid R-U), glaucamine, isorhoeadine (= alkaloid R-A), isorhoeagenine, isorhoeagenine-glycoside (= alkaloid R-C), oxysanguinarine (= alkaloid R-K), papaverrubine A(= alkaloid R-S), papaverrubine C, papaverrubine E (= alkaloid R-M), rhoeadine hydrochloride (= alkaloid R-T), (−)-sinactine, stylopine, (−)-β-stylopine methohydroxide (alkaloid R-D), thebaine (?) norharmane alkaloid (roots)	*50, 158, 236, 249–253*
P. rupifragum Boiss. & Reut.	Coptisine, magnoflorine, papaverrubine A, C, D, and E, protopine, rhoeadine, rhoeageine	*50*

(*Continued*)

TABLE II (*Continued*)

Plant	Alkaloids	References
P. somniferum L.	6-Acteonyldihydrosanguinarine, β-allocryptopine, bases mp 147°, 257°, and 260°, "Bound-morphine," canadine, choline, codeine, coreximine, cryptopine, dihydroprotopine, dihydrosanguinarine, gnoscopine, 16-hydroxythebaine, magnoflorine, 6-methylcodeine, narceine imide, narcotine, normorphine, norsanguinarine (callus tissues), orientaline, 13-oxocryptopine, oxydimorphine, oxysanguinarine, pacodine, palaudine, papaveraldine, papaverine, salutaridine, sanguinarine, stepholidine, thebaine, tetrahydropapaverine, two *N*-oxides of morphine, codeine and thebaine	*233, 249, 254–280, 340*
P. spicatum Boiss. & Bal.	Amurine	*238*
P. strictum Boiss. & Bal.	Amurine	*238*
P. syriacum Boiss. & Blanche	Berberine, coptisine, corysamine, isorhoeadine, (±)-mercambrine, papaverrubine A, D, and E, protopine, rhoeadine, rhoeagenine, (−)-stylopine, (−)-β-stylopine methohydroxide, thebaine	*33, 281*
P. tauricolum Boiss.	Armepavine, mecambrine, palmatine, papaverrubine E, protopine	*239*
P. triniaefolium Boiss.	Aporheine, armepavine, coptisine, mecambrine, oxysanguinarine, papaverrubine B and D, protopine	*234, 282*
P. urbanianum Fedde	(+)-Aporheine, (−)-armepavine, 6a,7-dehydroaporheine (= dehydroroemerine), (−)-mecambrine, muramine, *N*-methyl-6,7-dimethoxytetrahydroisoquinolone-1, palmatine, papaverrubine B, protopine	*283*
Roemeria hybrida (L.) DC	Alkaloid RH-5, coptisine, (−)-isocorypalmine, roehybridine ($C_{31}H_{39}N_3O_5$), roehybrine, roemeridine ($C_{31}H_{39}N_3O_5$)	*284*
R. refracta (Stev.) DC	Coptisine, (+)-mecambrine, (−)-mecambroline, protopine, reframidine, reframine, reframoline, remrefidine, remrefine (= reframine methohydroxide), roemeramine, (−)-roemerine, roemeroline, roemeronine	*285–287*
Romneya coulteri, var. *trichocalyx*	Dihydrosanguinarine, norromneine, romneine, sanguinarine	*288, 289*
Sanguinaria canadensis	α-Allocryptopine, β-allocryptopine, chelerythrine, chelilutine, chelirubine, dihydrosanguilutine, oxysanguinarine, protopine, sanguidimerine (= alkaloid SC-2), sanguilutine, sanguinarine, sanguirubine	*211, 290–293*
Stylomecon heterophylla (Benth) G. Tayl.	Allocryptopine, berberine, coptisine, cryptopine, protopine, stylophylline	*212*
Stylophorum diphyllum (Michx.) Nutt.	(+)-Chelidonine, coptisine, corysamine, protopine, (−)-stylopine	*33*

[a] Methohydroxide or methohalide (methosalt) are trivial names used in European chemical literature for quaternary ammonium salts as for example *N*-methylscoulerinium hydroxide or *N*-methylscoulerinium chloride.

readily separated by column chromatography on polyamide (*315*). Recently, Slavík *et al.* in Czechoslovakia have elaborated on a method for the isolation of quaternary tetrahydroprotoberberine alkaloids (*32, 158, 158a, 183, 221, 223, 246*). Their method is based on the finding that the iodides of these bases can be extracted from the aqueous layer with chloroform. More than 20 quaternary methoalkaloids from various species of the Papaveraceae family have been isolated. Ten of them were new alkaloids which had not been described previously. This method also includes the separation of phenolic and nonphenolic alkaloids.

The methylenedioxy group of alkaloids was identified by color reaction according to Labat and Gaebel (*316–318*) (the ^1H NMR analysis is now widely used for direct quantitative determination of this group). The production of free radicals in some alkaloids of the poppy has been reported (*319*). In recent years, thin-layer chromatography has been used for the qualitative detection and identification of alkaloids of this plant family (*227, 320–323*). The quantitative determination of opium alkaloids was also carried out by using ion exchangers (*324*), Sephadex (*325*), gas chromatography (*326*), and high-speed liquid chromatography (*327, 328*; see also *298, 329*).

The content of morphine in *P. somniferum* of various origins (*276*) and detection of thebaine in the different varieties of *P. bracteatum*, *P. orientale*, and *P. pseudo-orientale* was investigated (*328*). The analysis of these plant species and the chemotaxonomy of the Papaveraceae plants are discussed in Section IV,A.

For the spectral data and physical constants of many *Papaver* alkaloids see Holubek's Atlas, Vols. I–VIII (1965–1973) (*330*).

III. Structures, Syntheses, Biosyntheses, and Chemical and Physicochemical Properties of the Papaveraceae Alkaloids

A. Isoquinoline and Benzylisoquinoline*

Noroxyhydrastinine (**2a**) was isolated from the Korean *Corydalis* (*107*), *N*-methylcorydaldine (**2b**) from *P. bracteatum* (*235*) and *P. urbanianum* (*283*), and (+)-1-methylcorypalline (**1a**) from *C. ambigua* (*66*). The structure of the latter was also confirmed by synthesis of its racemate. Hydrastinine was isolated from *C. cava* (*73*), and *N*-methylcoclaurine (**7c**) from *G. fimbrilligerum* (*206*). Sevanine (**3b**) and macrostomine (**6**) are two new alkaloids of *P. macrostomum* (*242*) whose structures were established on the

* This material is supplementary to *The Alkaloids*, Vol. IV, p. 28; Vol. VII, p. 423; Vol. XII, p. 347.

basis of physicochemical data. The structure of sevanine was also confirmed by its synthesis (*331–333*). From the plant *C. gortschakovii*, the alkaloid corgoine (**8a**) was isolated and then synthesized (*83*, *334*). This plant also gave the alkaloid gortschakoine [1-(4′-methoxy-benzyl)-7-methoxy-8-hydroxytetrahydroisoquinoline] whose UV, IR, and NMR spectra were described (*85*). Norprotosinomenine (**7t**) was isolated from *D. eximia* (*152*) and is assumed to be an intermediate product in the biosynthesis of many benzylisoquinoline alkaloids.

1a $R^4 = H; R^1 = R^2 = R^3 = Me$ 1-Methylcorypalline
1b $R^1 = R^3 = R^4 = Me; R^2 = H$ Salsolidine
1c $R^1 = R^4 = H; R^2 = R^3 = Me$ Corypalline
1d $R^1 = Me; R^2 = R^3 = R^4 = H$ Salsolinol

2a $R^1 = H; R^2 + R^3 = CH_2$ Noroxyhydrastinine
2b $R^1 = R^2 = R^3 = Me$ *N*-Methylcorydaldine
2c $R^1 = R^2 = Me; R^3 = H$ Thalifoline

3a $R^1 = R^2 = R^3 = R^4 = Me$ Papaverine
3b $R^1 = H; R^2 = Me; R^3 + R^4 = CH_2$ Sevanine
3c $R^1 = H; R^2 = R^3 = R^4 = Me$ Pacodine
3d $R^3 = H; R^1 = R^2 = R^4 = Me$ Palaudine

The quaternary alkaloid escholinine (*155*) (from *E. californica*) was shown to be identical with (+)-romneine methohydroxide (**5a**), and its structure was determined from spectral analysis and Hofmann degradation (*154*). The structures of the newly isolated alkaloids escholamine methohydroxide (**4a**) and escholamidine methohydroxide (**4b**) were also elucidated (*158*).*

During the biosynthesis of many alkaloids (isopavine, berberastine, and macarpine), an important role is played by 4-hydroxylaudanosoline (**9a**) which, however, could not be isolated from the plants (*337*).

Studies of the biosynthesis of reticuline (**7f**) have shown (*338, 339*) that the upper half of the alkaloid (rings A and B) can arise from dihy-

* (*R*)-(−)-Armepavine (**7b**) was isolated from *Rhamnus frangula* (Rhamnaceae) (*335*). The tembetarine chloride [methochloride of reticuline (**7f**)] was isolated from *Fagara naranjillo* (Griseb.) (*336*).

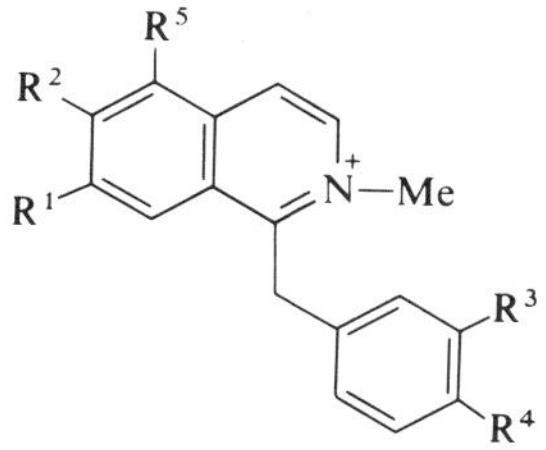

4a $R^1 + R^2 = R^3 + R^4 = O{-}CH_2{-}O$; $R^5 = H$ Escholamine
4b $R^1 + R^2 = O{-}CH_2{-}O$; $R^3 = OMe$; $R^4 = OH$; $R^5 = H$ Escholamidine
4c $R^1 = R^2 = R^4 = R^5 = OMe$; $R^3 = H$ Takatonine

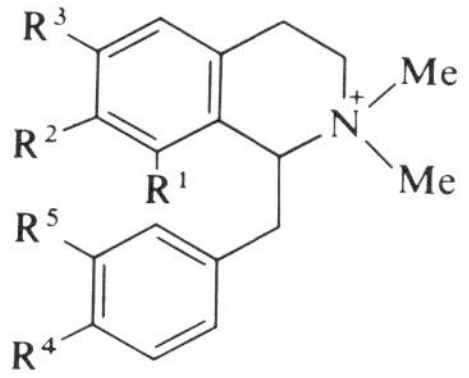

5a $R^1 = H$; $R^2 + R^3 = O{-}CH_2{-}O$; $R^4 = R^5 = OMe$ Romneine methohydroxide (= Escholinine)
5b $R^1 = OH$; $R^2 = R^4 = OMe$; $R^3 = R^5 = H$ Petaline

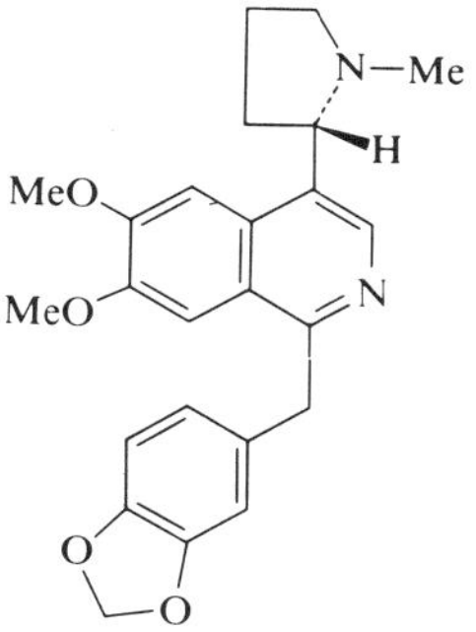

6 Macrostomine

droxyphenylalanine whereas the benzylic portion (ring C) is biosynthesized from *p*-hydroxyphenylalanine which is deaminated to *p*-hydroxyphenylpyruvic acid. *p*-Hydroxyphenylalanine is hydroxylated in the *meta* position, after which condensation of the two halves followed by Mannich reaction led to laudanosoline (**7m**) and reticuline (**7f**) (Scheme 17).

Tracer experiments have shown that in *P. somniferum* papaverine arises from (−)-norreticuline via norlaudanidine and norlaudanosine (*340, 340a*). N-Methylation of tetrahydroisoquinoline bases may occur at several stages

7

7a $R^1 = R^5 = H$; $R^2 = R^3 = OMe$; $R^4 = OH$ *N*-Norarmepavine
7b $R^1 = Me$; $R^2 = R^3 = OMe$; $R^4 = OH$; $R^5 = H$ Armepavine
7c $R^1 = Me$; $R^2 = OMe$; $R^3 = R^4 = OH$; $R^5 = H$ *N*-Methylcoclaurine
7d $R^1 = R^5 = H$; $R^2 = OMe$; $R^3 = R^4 = OH$ Coclaurine
7e $R^1 = R^5 = H$; $R^2 = R^4 = OH$; $R^3 = OMe$ Isococlaurine
7f $R^1 = Me$; $R^2 = R^4 = OMe$; $R^3 = R^5 = OH$ Reticuline
7g $R^1 = Me$; $R^2 = R^3 = R^4 = R^5 = OMe$ Laudanosine
7h $R^1 = Me$; $R^2 = R^3 = R^4 = OMe$; $R^5 = OH$ Laudanidine
7i $R^1 = Me$; $R^2 + R^3 = O{-}CH_2{-}O$; $R^4 = R^5 = OMe$ Romneine
7j $R^1 = H$; $R^2 = R^3 = R^4 = R^5 = OMe$ Tetrahydropapaverine
7k $R^1 = H$; $R^2 + R^3 = O{-}CH_2{-}O$; $R^4 = R^5 = OMe$ *N*-Norromneine
7l $R^1 = Me$; $R^2 = R^3 = R^4 = OMe$; $R^5 = OH$ Laudanine
7m $R^1 = Me$; $R^2 = R^3 = R^4 = R^5 = OH$ Laudanosoline (= Tetrahydropapaveroline)
7n $R^1 = H$; $R^2 = R^3 = R^4 = R^5 = OMe$ *N*-Norlaudanosine
7o $R^1 = Me$; $R^2 = R^5 = OH$; $R^3 = R^4 = OMe$ Protosinomenine
7p $R^1 = H$; $R^2 = R^3 = R^4 = R^5 = OH$ *N*-Norlaudanosoline
7r $R^1 = Me$; $R^2 = R^4 = R^5 = OMe$; $R^3 = OH$ Codamine
7s $R^1 = Me$; $R^2 = R^5 = OMe$; $R^3 = R^4 = OH$ Orientaline
7t $R^1 = H$; $R^2 = R^5 = OH$; $R^3 = R^4 = OMe$ *N*-Norprotosinomenine
7u $R^1 = H$; $R^2 = R^4 = OMe$; $R^3 = R^5 = OH$ *N*-Norreticuline

8a R = H Corgoine
8b R = Me Sendaverine

9a $R^1 = Me$; $R^2 = R^3 = R^4 = R^5 = H$ 4-Hydroxylaudanosoline
9b $R^1 = R^2 = R^4 = Me$; $R^3 = R^5 = H$ 4-Hydroxyreticuline

along the biosynthetic routes, and there is evidence that enzymatic N-demethylation may take place to an appreciable extent (*340*).

Reviews of the synthesis of isoquinoline alkaloids (*341–343*) and of biogenetic types of syntheses (*4, 344, 345*) have been published.

The increasing demand of the pharmaceutical industry for various isoquinoline and benzylisoquinoline derivatives with different electron-donating substituents led to a thorough study of the problem of selective O-demethylation and reetherification or methylenation of the liberated phenolic groups (*346–348*). The alkaloid corypalline (**1c**) can be conveniently prepared from 6,7-dimethoxy-3,4-dihydroisoquinoline (*349*). The first step is partial demethylation of the methoxyl at C-7. It gives rise to the monomethoxy derivative which on hydrogenation affords corypalline (Scheme 1).

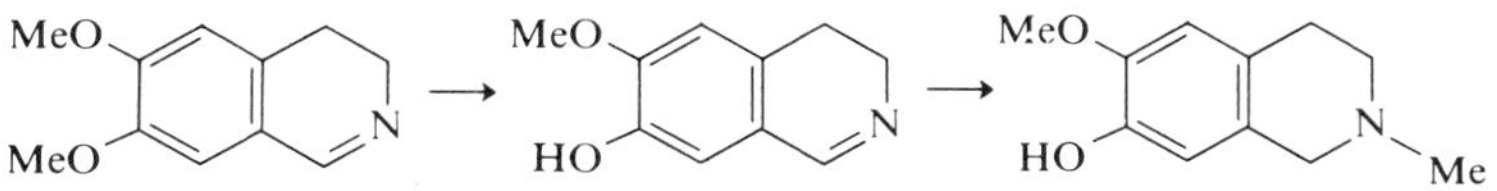

MeO MeO N → MeO HO N → MeO HO N Me

SCHEME 1. *Transformation of 6,7-dimethoxy-3,4-dihydroisoquinoline into corypalline* (**1c**) (*349*).

The Pschorr reaction was described in connection with the synthesis of the papaverine (**3**) derivatives (*350, 351*). The synthesis of petaline (**5b**) was accomplished (*352, 353*). Escholamine (**4a**) and takatonine (**4c**) were synthesized by a modified Pomeranz–Fritsch reaction (*354*). The phenolic oxidation of (*R*)-(−)-*N*-methylcoclaurine (**7c**) and (*S*)-(+)-reticuline (**7f**) with peroxidase proved to be a failure (*355*). The oxidation of reticuline with ferricyanide yielded isoboldine (**24c**) and pallidine (**43b**) and the by-products vanillin and thalifoline (**2c**) (*355*). A new synthesis of 3-oxopapaverine was developed (*356*), and the Eschweiler–Clark method for the synthesis of codamine (**7r**) was modified (*357*). Oxidation of reticuline (**7f**) by enzymatic systems from homogenized *P. rhoeas* in the presence of hydrogen peroxide gave (±)-β-hydroxyreticuline (**10**) (*358*).

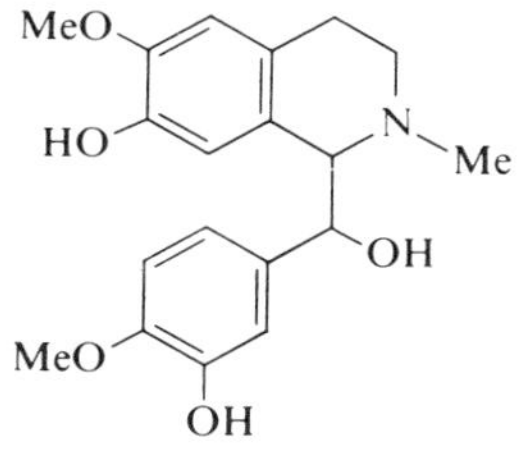

MeO HO N Me OH MeO OH

10 β-Hydroxyreticuline

Treatment of cotarnine, hydrastinine, or other 3,4-dihydroisoquinoline compounds with diazomethane led to the addition of a CH_2 group to the double bond and thus to formation of the aziridine compound. 3,4-Dihydropapaveraldine methiodide with diazomethane gave the benzazepine derivative which permits an expansion of the heterocyclic ring into a seven-membered ring (*359–361*) (see Section III,M). Kametani and Ogasawara used this method to synthesize isopavine (**56**) and rhoeadane (**154**) bases with methylenedioxy groups (*362*).

The racemic laudanosine (**7g**) syntheses were referred to earlier in *The Alkaloids* (Vol. XII, p. 348). From the racemic mixture it is difficult to obtain (*S*)-(+)-laudanosine which is widely distributed in nature. The synthesis of its unnatural (*R*)-form (in poor yields) has been reported (asymmetric reduction of 3,4-dihydropapaverine with lithium butyl-(hydro)dipinan-3α-ylborate) (*363*). Recently, a description of the biogenetic synthesis of these two forms from methyl-L-(+)-3,4-dihydroxyphenylalaninate hydrochloride by condensation with sodium 3-(3,4-dimethoxyphenyl)glycinate at pH 4 and 35° was given. Thus, a diastereoisomeric mixture of the Pictet–Spengler products **11a** and **11b** (ratio

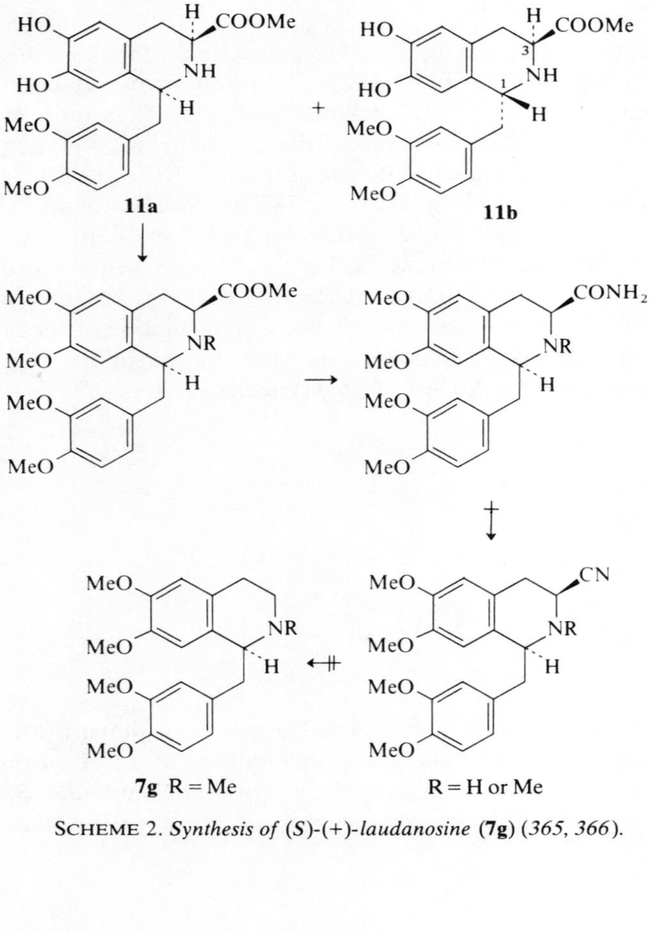

11a + **11b**

7g R = Me

R = H or Me

SCHEME 2. *Synthesis of* (*S*)-(+)-*laudanosine* (**7g**) (*365, 366*).

2.4 : 1) was obtained (see Scheme 2). Those two isomers were separated by chromatography. The predominant isomer, **11a**, has a 1,3-*cis* configuration. By decarboxylation and methylation on nitrogen, it can be converted into (*S*)-(+)-laudanosine (**7g**). Similarly, the stereoisomer **11b** can be converted into (*R*)-(−)-laudanosine. An analogous procedure was developed for (*S*)-(+)-reticuline (**7f**) (*364–366*) (Scheme 2).

On the basis of an analysis of the racemic *O,O'*-dibenzylisococlaurine, the absolute configuration was assigned to (*S*)-(+)- and (*R*)-(−)-isococlaurine (**7e**) and armepavine (**7b**) (*367, 368*). Thus, the two antipodes of (±)-*N*-norromneine (**7k**) were obtained; their absolute configurations were determined by correlation (*288*).

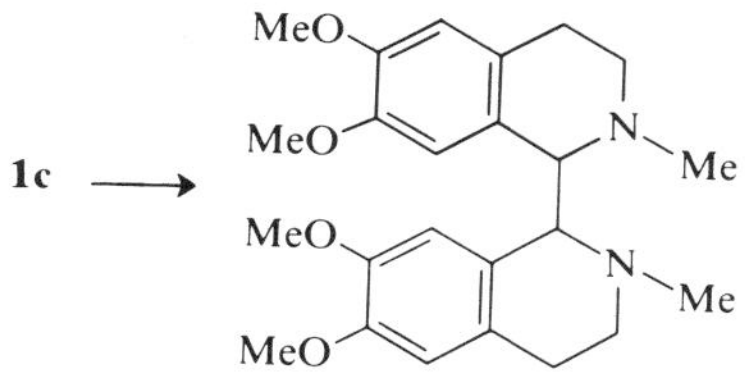

1c ⟶ **12**

Oxidation of corypalline (**1c**) with ferricyanide gave the 1,1-dimer (**12**) (*369, 370*). The mechanism and the effect of trace elements on the autoxidation of papaverine (**3a**) and that of the hydrogenolytic splitting of sendaverine methiodide (**8b**) have also been studied in detail (*370–373*). Knabe studied the rearrangement of dihydroisoquinoline compounds and inferred that the larger substituents on the nitrogen atom sterically hinder the rearrangement (*374*). In the compounds examined, elimination and disproportionation occurred as side reactions. The size of the substituent on the nitrogen atom had no influence on the side reactions (Scheme 3).

The proaporphine, promorphinane, and cularine alkaloids can be reconverted into the benzyltetrahydroisoquinoline bases with sodium in liquid ammonia, a process which permits correlation of the absolute configurations of the benzyltetrahydroisoquinoline, proaporphine, aporphine, promorphinane, and cularine alkaloids (*375*).

Electrolytic oxidation of armepavine (**7b**) and *N*-norarmepavine (**7a**) leads to their fragmentation, to partial carbon–oxygen dimerization, and to the formation of the alkaloid dauricine (**13**) (Scheme 4) (*376*).

Irradiation with sunlight of laudanosine methiodide or methosulfate in methanol or water gives the ring B cleavage products with methoxyl or hydroxyl group at C-9 (*377*).

A new method for the determination of the enantiomeric purity of isoquinoline alkaloids makes use of chiral lanthanide ^{1}H NMR shift reagents

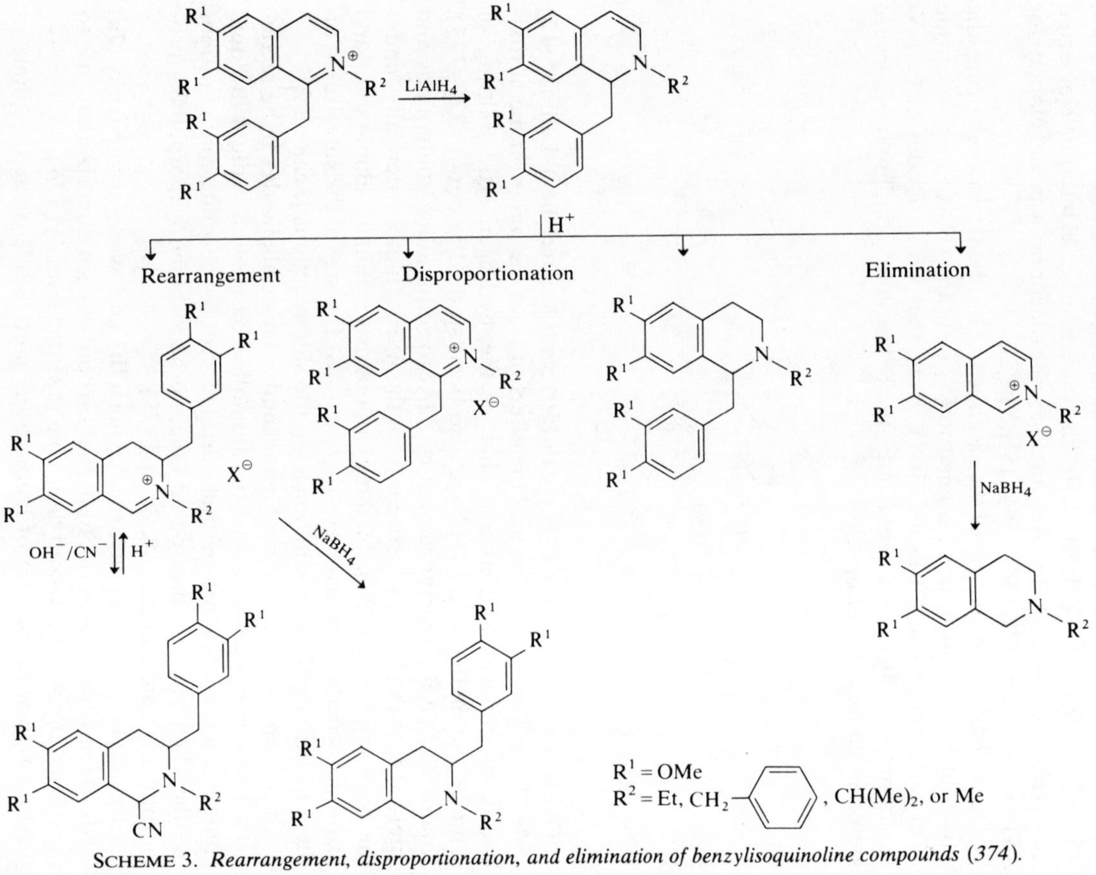

SCHEME 3. *Rearrangement, disproportionation, and elimination of benzylisoquinoline compounds (374).*

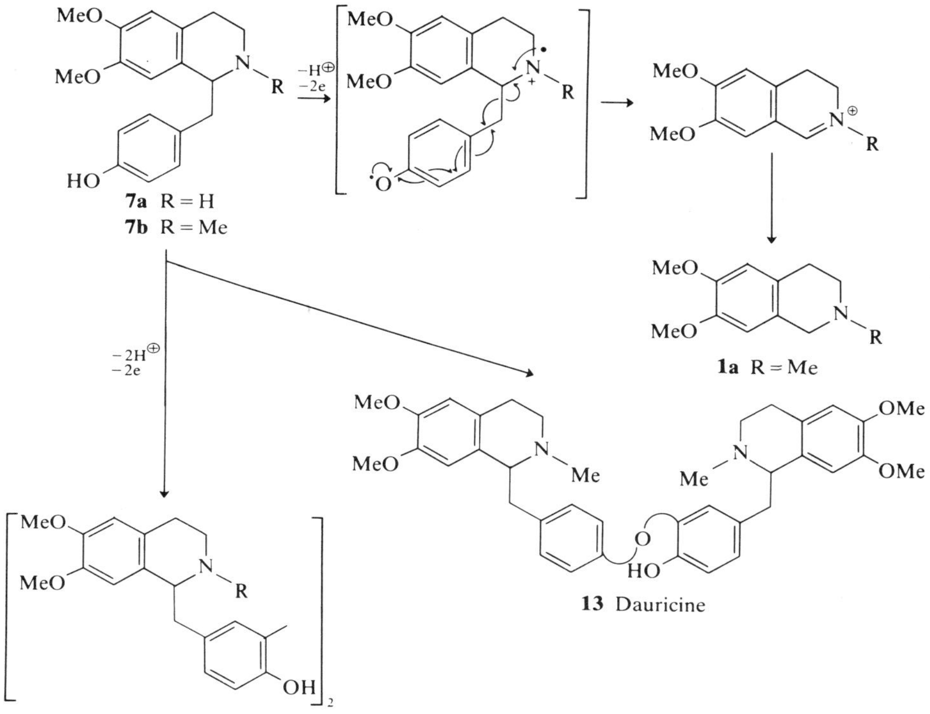

7a R = H
7b R = Me

1a R = Me

13 Dauricine

SCHEME 4. *Electrolytic oxidation of norarmepavine* (**7a**) *or armepavine* (**7b**) (*376*).

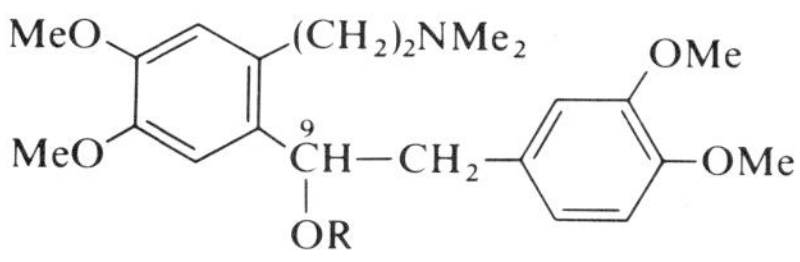

R = H or Me

(*378*). This method was tested on enantiomeric compounds of five alkaloidal mixtures.

Dolejš and Slavík reported the mass spectrometric fragmentation of methines of the benzylisoquinoline alkaloids (*379*). The isoquinoline and 3,4-dihydroisoquinoline bases give characteristic UV spectra on the basis of which this group of substances can easily be identified (*380*). The origin of the three tautomeric forms of hydrastinine and cotarnine was reinvestigated by UV, IR, and ^{1}H NMR spectroscopy and polarography; the presence of the aldehyde form **B** could not be detected (*381*). In an acidic medium the ammonium form **A** prevails and in alkaline medium the carbinol form **C**.

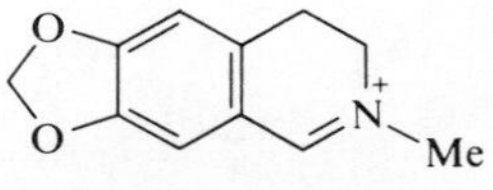

A 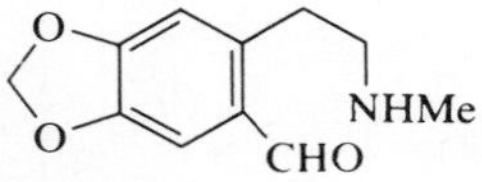 **B** 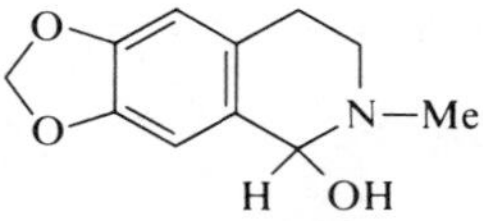 **C**

Independent papers report the isolation and quantitative determination of papaverine and of its oxidation products by thin-layer chromatography (*360, 371*). The stability of papaverine in pharmaceutical products and the origin of its degradation products have been studied (*382*).

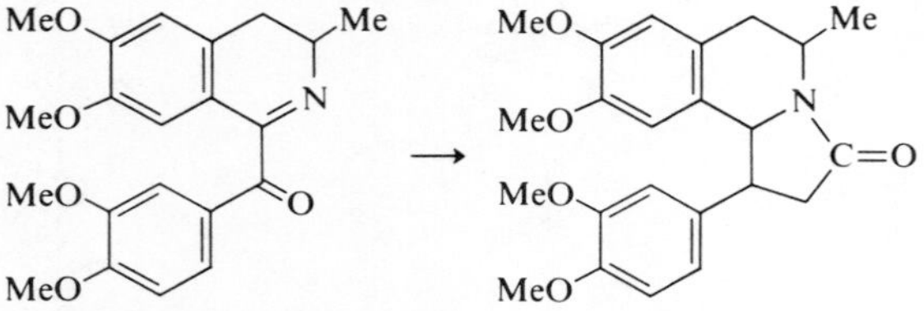

14 → **15**

A study of the benzylisoquinoline bases showed that, on heating with acetic anhydride, 1-(3,4-dimethoxybenzoyl)-6,7-dimethoxy-3-methyl-3,4-dihydroisoquinoline (**14**) gives an intensive green color reaction which is due to compound **15** (*383*). The mechanism of a reaction for the demonstration of the presence of papaverine, giving rise to coralyne (**73**) (Scheme 30), was described (*384*).

Wiegrebe and Röhrbach-Munz (*385*) studied the synthesis and Hofmann degradation of tetrahydroisoquinoline bases which were variously substituted at C-1.

Stock and Shia studied the dissociation constants of corypalline (**1c**) and isocorypalline in acetonitrile (*386*).

The biotransformation of (±)-reticuline (**7f**) into (±)-coreximine (**60a**), norreticuline (**7u**), and scoulerine (**58b**) in the rat, and in homogenized rat liver was demonstrated by tracer experiments with (±)-[*N*-$^{14}CH_3$]reticuline (*354, 387*). Tritium-labeled laudanosine (**7g**) was also incorporated into norlaudanosine (**7n**), tetrahydropalmatine (**58g**), and xylopinine (**60c**) with the same rat liver preparation (*388*). It was found that, in patients with Parkinson's disease, the treatment with dihydroxyphenylalanine gives rise to salsolinol (**1d**) and tetrahydropapaveroline (**7m**) (*389*). Dopamine and acetaldehyde or 3,4-dihydroxyphenylacetaldehyde condense in mammalian tissues to afford 1-substituted tetrahydroisoquinoline bases which might induce a variety of pharmacological reactions (*390*). Such *in vivo* processes can be catalyzed by enzymes to form a single optical isomer which is expected to differ from its antipode in biological activity. To evaluate this concept of "alkaloid formation" in humans, especially in relation to the behavioral changes induced by alcoholism and to other disorders, both optical isomers are necessary. Based on this consideration, the enantiomeric salsolinols (**1d**) and tetrahydropapaverolines (**7m**) were synthesized by O-demethylation of the corresponding isomeric salsolidines (**1b**) and norlaudanosines (**7n**).

B. Proaporphine Group*

Previous knowledge of the proaporphine alkaloids has been summarized in two reviews (*10, 391*) and in the book by Shamma (*1*). These three publications also draw attention to the UV, IR, and NMR spectra, but no assessment of the data is made.

In addition to the common types of proaporphine alkaloids, the alkaloids glauvine (**16**) (*195, 196, 199*) and jolantamine (**17**) (*392*) were isolated whose cyclohexadienone ring (or cyclohexenone ring) forms part of the aporphine skeleton. The structure of glauvine was confirmed (*199*) by its conversion into the aporphine alkaloid *O*-methylatheroline (**18**); that of jolantamine was deduced from the UV and ^{1}H NMR data (*392*). The tetrahydroproaporphine alkaloid roehybrine (**19**) (with a keto group and a methoxyl group on ring D) was isolated (*284*) from the plant *R. hybrida*. (±)-Mecambrine (**20b**) was isolated from *P. syriacum* (*33*).†

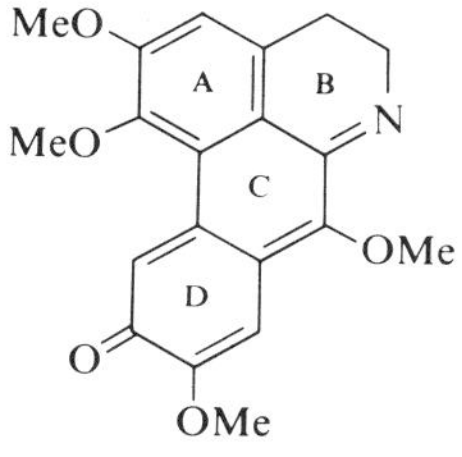

16 Glauvine

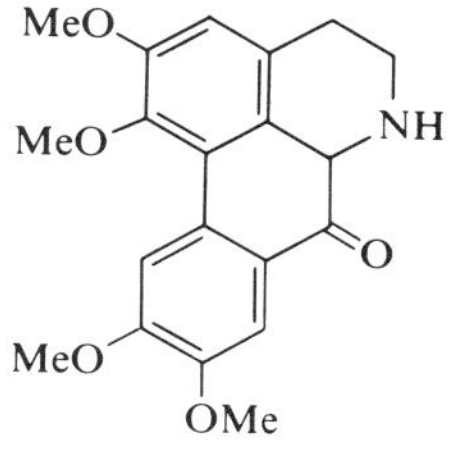

17 Jolantamine

18 *O*-Methylatheroline

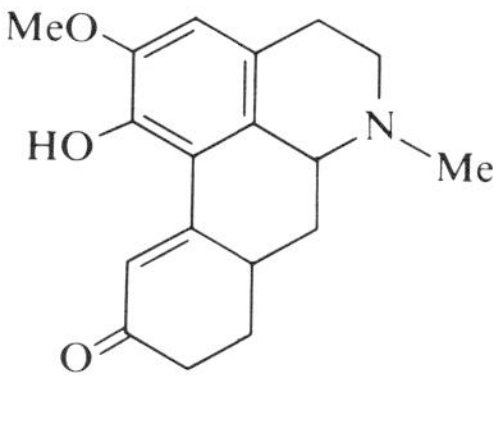

* This material is supplementary to *The Alkaloids*, Vol. XII, p. 349.

† The proaporphine alkaloids of the cyclohexadienone type or those with a partially reduced ring D occur not only in the Papaveraceae but in many other plant families, such as Euphorbiaceae, Lauraceae, Menispermaceae, Monimiaceae, and Nymphaceae (*1, 393*). Their widespread occurrence permits a study of the isolation and the synthesis of these compounds on a large scale.

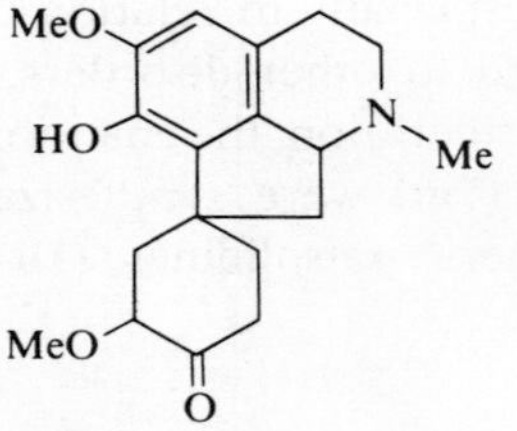

MeO, HO, N, Me, MeO, O

19 Roehybrine

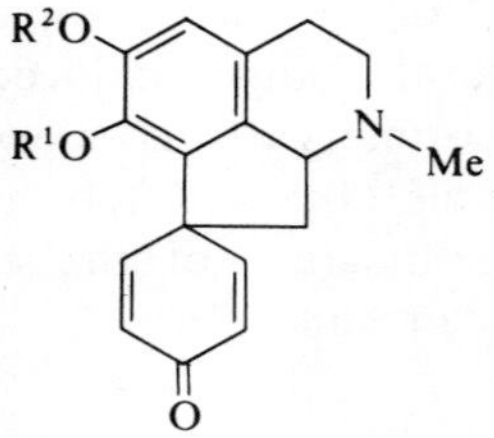

R^2O, R^1O, N, Me, O

20a $R^1 = H$; $R^2 = Me$ Glaziovine
20b $R^1 + R^2 = CH_2$ Mecambrine

In recent years the synthesis of this type of alkaloid (Vol. XII, p. 356) has been studied at length, particularly by phenolic oxidation (*394–399*), electrochemical oxidation (*400–402*), carbon-carbon cleavage, and the photo-Pschorr reaction (*201, 403–409*). In addition, a novel formation of (±)-glaziovine (**20a**) in very low yield through a phenoxynitrenium intermediate (*410, 411)* (Scheme 5), by 8,1′-ring closure of 1-benzylisoquinoline derivatives (*203*), and by spiran ring construction on the basis of cyclopent [*ij*]isoquinoline (*446*), and the synthesis of glaucine (**24h**) via *p*-quinol acetate (Scheme 11) have been reported (*413*).

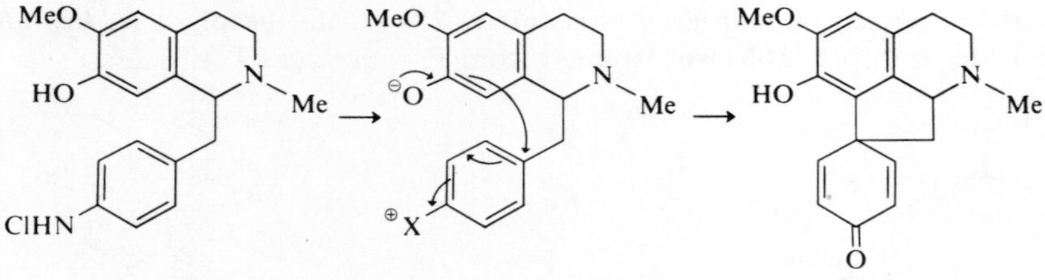

MeO, HO, N, Me, ClHN → MeO, $^{\ominus}O$, N, Me, $^{\oplus}X$ → MeO, HO, N, Me, O

20a

SCHEME 5. *Synthesis of* (±)*-glaziovine* (**20a**) *through a phenoxynitrenium intermediate* (*410, 411*).

Photolysis of the sodium salt of 2-bromo-*N*-ethyl-4′-hydroxybenzylanilide in *N,N*-dimethylformamide gave 2′-ethylspiro(cyclohexa-2,5-diene-1,1′-isoindoline)-3′,4-dione (**21**) in good yield. The reaction was

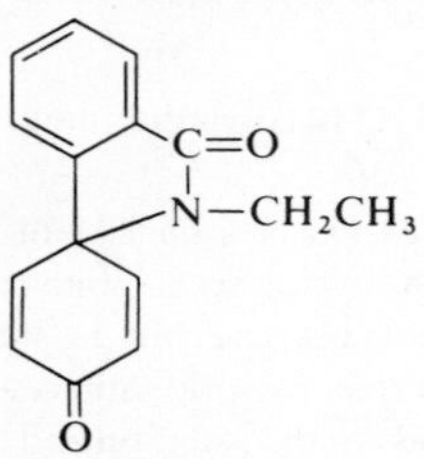

C=O, $N-CH_2CH_3$, O

21

utilized in the syntheses of racemic mecambrine (**20b**), domesticine (**24l**), and amurine (**43c**). It simultaneously yielded the proaporphine, aporphine, and the promorphinane bases (*414*).

In the formation of the aporphine alkaloids, the quinonoid form (cyclohexadienone) may arise not only from ring C but from ring A according to the location of the keto group at C-2 or C-3 (Schemes 6 and 11). Thus, thaliporphine (**24g**) and glaucine (**24h**) have been prepared from codamine (**7r**) (*413*).

The formation of spirodienone products by nonphenolic benzylisoquinoline oxidative anodic coupling (*415*) (yield 90%) and by facile biomimetic syntheses of aporphine and promorphinane alkaloids can proceed along the pathway outlined in Scheme 6 via the morphinanedienone (**43**) (proerythrinandienone), and neoproaporphine (**40**) intermediates. The temperature dependence observed shows that these dienones exist in equilibrium in acid at elevated temperature (*416–417*).

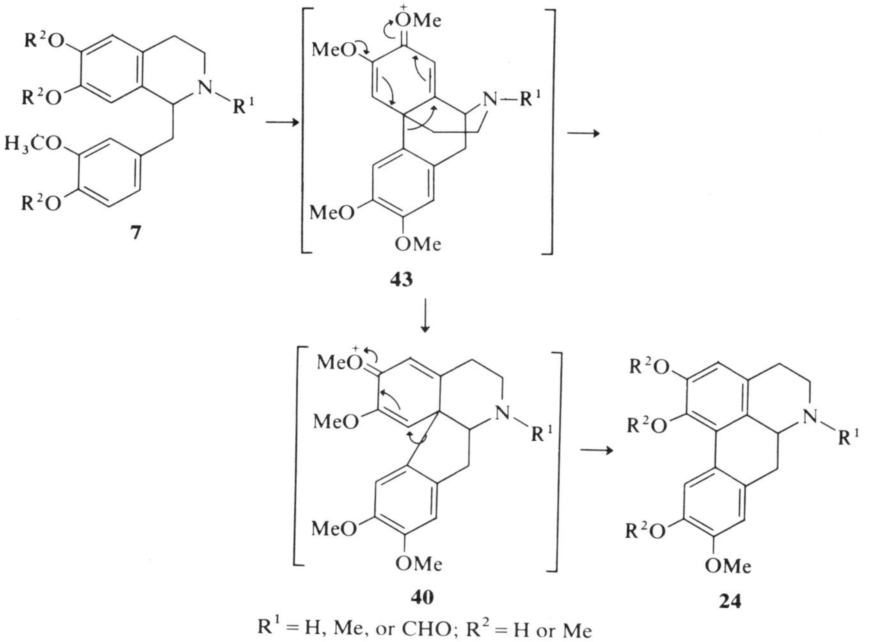

7 → **43** → **40** → **24**

R^1 = H, Me, or CHO; R^2 = H or Me

SCHEME 6. *Formation of the aporphine alkaloids* (**24**) *via the promorphinane* (**43**) *and neoproaporphine* (**40**) *intermediates* (*416*).

Biosynthetic studies have demonstrated that (±)-reticuline is a precursor of the aporphine alkaloids (+)-bulbocapnine, (+)-isoboldine, and (+)-magnoflorine, and these results have been interpreted as indicative of a "direct-coupling" mechanism (*416*). The *in vivo* conversion of reticuline to morphinandienones has also been demonstrated. Therefore, the authors assume that the precursors of the aporphine alkaloids may also be the morphinandienones (promorphinanes) (Scheme 6) (*416*).

The products of the sodium borohydride reduction of (−)-mecambrine (**20b**) and of the catalytic hydrogenation of the same alkaloid and (−)-roemeramine (**22a**) were studied and, on the basis of the ORD curves and ^{1}H NMR spectra, the relative configurations of these two alkaloids and of four stereoisomeric dihydromecambrinols **22b** and **22c** were established (*418*). Casagrande *et al.* (*203*) studied the synthesis of (±)-glaziovine (**20a**) and the stereochemistry of partially reduced proaporphine products (*20, 412, 446*).

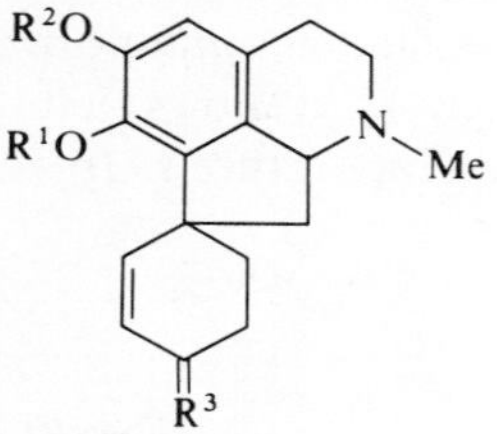

22a $R^1 + R^2 = CH_2$; $R^3 = O$ Roemeramine
22b $R^1 = Me$; $R^2 = H$; $R^3 = H,OH$
22c $R^1 = H$; $R^2 = Me$; $R^3 = H,OH$

Dolejš studied the mass spectrometry of proaporphine alkaloids (*419*). According to him, the fragmentation processes largely depend on the character of the spirane ring and also on the substitution of the aromatic ring. The differences between the UV and IR spectra of proaporphine and promorphinane bases have been reported (*420, 421*). There is a fundamental difference in the UV spectra depending on whether or not the compound carries a methoxyl group on the double bond of the ring D in the α- or α,α'- or in β-position to the keto group, a hydrogen, or one or two methoxyl groups. The methylenedioxy group on the ring A causes an increase in intensity and a bathochromic shift of the secondary band vs. the same band of compounds with two methoxyl or one methoxyl and one hydroxyl group on the ring A. The UV spectra of proaporphine and promorphinane compounds are almost identical. The proaporphine compounds with one hydrogenated double bond in the ring D have also been studied (*420*). The UV spectra of the cyclohexenones differ only slightly from those of the cyclohexadienone compounds. The first primary band of the former is shifted more to the blue end of the spectrum. The measurements show that the substituents of the cyclohexenone or cyclohexadienone ring affect not only the position but also the intensity of the band at about 230 nm and of the aromatic band at about 285 nm.

In spite of the fact that the IR spectra of the cyclohexadienone proaporphine and promorphinane alkaloids have many common characteristic properties in the range between 1700 and 1600 cm^{-1}, some slight differences can be recognized: the proaporphine alkaloids, with a methylenedioxy group instead of two methoxyl groups on the ring A, have a characteristic band at about 1640 cm^{-1} which in $CHCl_3$, KBr, and Nujol is lower than the two *vicinal* bands (at ca. 1665 and 1622 cm^{-1}). This phenomenon is also observed in the promorphinane alkaloids with a methylenedioxy group in the 2,3-positions. In the IR spectra of this type of compounds the strongest bands usually appear at ca. 1660 cm^{-1} ("cross-conjugated dienone system"), ca. 1620 cm^{-1} (double bonds of this system), and ca. 1605 cm^{-1} (aromatic system).

The cyclohexenone alkaloids behave in a similar manner except that absorptions caused by their double bonds are of low intensity and are found at 1620 and 1603 cm^{-1}, and the keto group is responsible for a strong band in the 1690–1675 cm^{-1} region. In the cyclohexadienone as well as the cyclohexenone compounds the presence of a methoxyl group on the double bond in the position α to the keto group causes both a shift of the original band at ca. 1620 cm^{-1} to ca. 1635 cm^{-1} and an increase in its intensity. The position of this band depends on the medium used and on the concentration of the analyzed substances (*421*).

C. Aporphine Group*

Apart from the reports on aporphine alkaloids published elsewhere in this series of *The Alkaloids*, there have appeared two reviews (*393, 422*) and the book by Shamma (*1*). The isolation of the aporphine alkaloids was also carried out by using ion exchangers (*302*).

The new alkaloids corunnine (**27a**), cataline (**28**), and pontevedrine (**29**) were either detected or isolated from *G. flavum* and *G. flavum* var. *vestitum* (*202–204, 423*). *G. flavum* and *G. leiocarpum* were also shown to contain the alkaloids 6,6a-dehydroglaucine (**30**), 6,6a-dehydro-*N*-norglaucine, and thaliporphine (**24g**) (*188, 209*). The alkaloid glauvine (**16**) was isolated from *G. elegans* (*185*), and *N*-methyllindcarpine (**24m**) from *G. pulchrum*, population Elika (*225*). Escholine was found to be identical with magnoflorine (**25h**), and aurotensine with isoboldine (**24c**) (*154, 155, 423a*). *P. urbanianum* yielded 6a,7-dehydroaporheine (**26a**) in addition to the alkaloid (+)-aporheine (**23c**) (*283*). 1,2-Methylenedioxy-6a,7-dehydroaporphine-10,11-quinone (**31a**) was found in *C. cava* (*77*). It arises directly on

* This material is supplementary to *The Alkaloids*, Vol. IV, p. 119; Vol. XII, p. 361; Vol. XIV, p. 225.

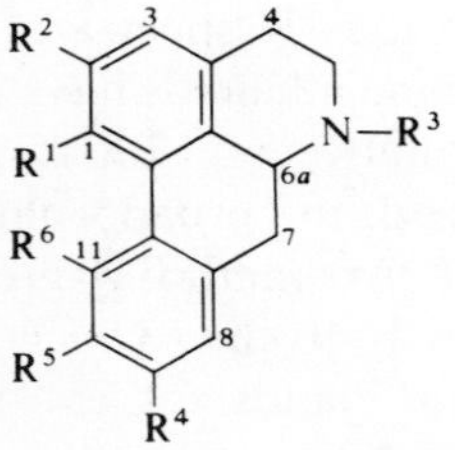

23–25

23a $R^1 = OMe; R^2 = OH; R^3 = R^4 = R^5 = R^6 = H$ Asimilobine
23b $R^1 = R^2 = OMe; R^3 = Me; R^4 = R^5 = R^6 = H$ Nuciferine
23c $R^1 + R^2 = O—CH_2—O; R^3 = Me; R^4 = R^5 = R^6 = H$ Aporheine
23d $R^1 = OH; R^2 = OMe; R^3 = Me; R^4 = R^5 = R^6 = H$ Lirinidine
23e $R^1 + R^2 = O—CH_2—O; R^3 = (Me)_2; R^4 = R^5 = R^6 = H$ Remrefidine

24a $R^1 + R^2 = O—CH_2—O; R^3 = Me; R^4 = R^6 = H; R^5 = OH$ Mecambroline
24b $R^1 = R^5 = OMe; R^2 = R^4 = OH; R^3 = R^6 = H$ Laurolitsine
24c $R^1 = R^4 = OH; R^2 = R^5 = OMe; R^3 = Me; R^6 = H$ Isoboldine (Aurotensine)
24d $R^1 = R^5 = OH; R^2 = R^4 = OMe; R^3 = Me; R^6 = H$ Bracteoline
24e $R^1 = R^5 = OMe; R^2 = R^4 = OH; R^3 = Me; R^6 = H$ Boldine
24f $R^1 + R^2 = O—CH_2—O; R^3 = Me; R^4 = R^5 = OMe; R^6 = H$ Dicentrine
24g $R^1 = OH; R^2 = R^4 = R^5 = OMe; R^3 = Me; R^6 = H$ Thaliporphine
24h $R^1 = R^2 = R^4 = R^5 = OMe; R^3 = Me; R^6 = H$ Glaucine
24i $R^1 = R^4 = R^5 = OMe; R^2 = OH; R^3 = R^6 = H$ Norpredicentrine
24k $R^1 = R^4 = R^5 = OMe; R^2 = OH; R^3 = Me; R^6 = H$ Predicentrine
24l $R^1 = OH; R^2 = OMe; R^3 = Me; R^4 + R^5 = O—CH_2—O; R^6 = H$ Domesticine
24m $R^1 = R^5 = OMe; R^2 = R^6 = OH; R^3 = Me; R^4 = H$ *N*-Methyllindcarpine

25a $R^1 + R^2 = O—CH_2—O; R^3 = R^4 = H; R^5 = OH; R^6 = OMe$ Nandigerine
25b $R^1 = OH; R^2 = R^6 = OMe; R^3 = Me; R^4 = R^5 = H$ Isothebaine
25c $R^1 + R^2 = O—CH_2—O; R^3 = Me; R^4 = H; R^5 = OH; R^6 = OMe$ *N*-Methylnandigerine (= Hernangerine)
25d $R^1 + R^2 = O—CH_2—O; R^3 = Me; R^4 = H; R^5 = OMe; R^6 = OH$ Bulbocapnine
25e $R^1 = R^6 = OH; R^2 = R^5 = OMe; R^3 = Me; R^4 = H$ Corytuberine (= Glauflavine)
25f $R^1 + R^2 = O—CH_2—O; R^3 = Me; R^4 = H; R^5 = R^6 = OMe$ *O*-Methylbulbocapnine
25g $R^1 = R^2 = R^5 = OMe; R^3 = Me; R^4 = H; R^6 = OH$ Isocorydine
25h $R^1 = R^6 = OH; R^2 = R^5 = OMe; R^3 = (Me)_2; R^4 = H$ Magnoflorine (= Escholine)
25i $R^1 = OH; R^2 = R^5 = R^6 = OMe; R^3 = Me; R^4 = H$ Corydine
25k $R^1 = OH; R^2 = R^5 = R^6 = OMe; R^3 = R^4 = H$ Norcorydine
25l $R^1 = R^2 = R^5 = OMe; R^3 = Me; R^4 = H; R^6 = OH$ Norisocorydine
25m $R^1 + R^2 = R^5 + R^6 = O—CH_2—O; R^3 = R^4 = H$ Ovigerine
25n $R^1 + R^2 = R^5 + R^6 = O—CH_2—O; R^3 = Me; R^4 = H$ *N*-Methylovigerine

oxidation of bulbocapnine (**25d**). The oxoaporphine alkaloid *O*-methylatheroline (**35d**) was isolated from the aerial part of *G. elegans*, and the alkaloid liriodenine (**35a**) from *P. heldreichii* (*239*).

(+)-Corydine (**25i**) was isolated from *Thalictrum dioicum* (*424*) and from the bark of *Cinnamomum* (*425*), hernangerine (**25c**) from *Hernandia papuana* (*426*), and boldine (**24e**) and laurolitsine (**24b**) from *Litsea turfosa*

26

26a $R^1 + R^2 = O—CH_2—O$; $R^3 = Me$; $R^4 = R^5 = R^6 = H$ Dehydroaporheine (= Dehydroroemerine)
26b $R^1 = R^2 = OMe$; $R^3 = Me$; $R^4 = R^5 = R^6 = H$ Dehydronuciferine
26c $R^1 + R^2 = O—CH_2—O$; $R^3 = Me$; $R^4 = R^5 = OMe$; $R^6 = H$ Dehydrodicentrine
26d $R^1 + R^2 = O—CH_2—O$; $R^3 = Me$; $R^4 = H$; $R^5 = R^6 = OMe$ Dehydro-*O*-methylbulbocapnine
26e $R^1 = R^2 = R^4 = R^5 = OMe$; $R^3 = Me$; $R^6 = H$ Dehydroglaucine
26f $R^1 = R^2 = R^4 = R^5 = OMe$; $R^3 = R^6 = H$ Dehydronorglaucine

27a R = H Corunnine
27b R = Me Oxoglaucine

28 Cataline

29 Pontevedrine

30

31a R = H
31b R = Me

(*427*). Asimilobine (**23a**) was found in the plants of the family Aristolochiaceae (*428*). Noteworthy is the alkaloid thalphenine (**32**) (Scheme 7), which has five rings and was isolated from *Thalictrum polygamum* (*429*). The presence of this alkaloid indicates that the aporphine bases might also arise by opening of the berbine ring C of the tetrahydroprotoberberine bases and by a new formation of the rings C and E.

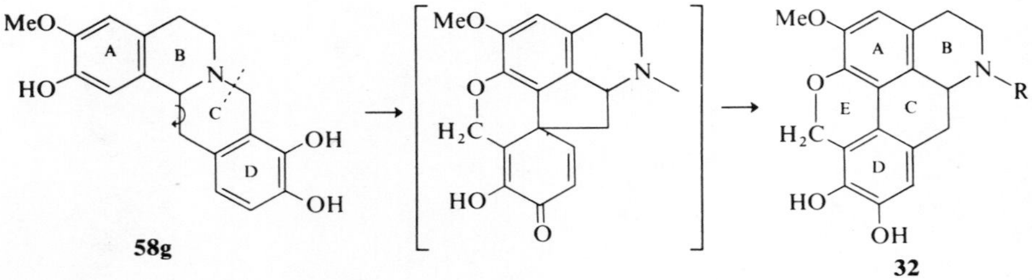

SCHEME 7. *Pathway for the formation of thalphenine* (**32**) *from tetrahydroprotoberberine alkaloids* (**58**) (*429*).

Shamma and Moniot are of the opinion that the methylenedioxy bridge between the rings A and D in thalpenine (**32**) is formed during the biogenesis via the oxonium ion from the methoxyl group (*430*).

Recently, the dehydrogenated aporphine alkaloids and the alkaloids with an additional oxo group (mainly obtained from the callus tissue of *Stephania cepharantha*) have been found in plant material (*431*). There were obtained four intensely fluorescent compounds of the cepharanone (**33**) and the cepharadione (**34**) types.

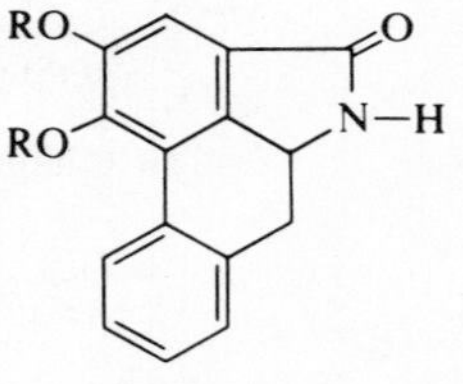

33 Cepharanone 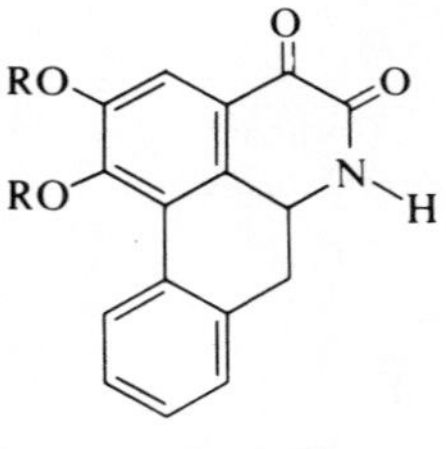 **34** Cepharadione

R = H, Me, or vice versa

The constitutions of 6a,7-dehydroglaucine (**26e**) (*191*), hernandonine (**35c**) (*432*), imenine (**35e**) (*433*), oxonuciferine (**35g**), dicentrinone (**35h**), and lanuginosine (**35b**) have been elucidated (*393, 434*). The latter five oxo bases have not yet been found in the Papaveraceae. Because some of the oxo-aporphine alkaloids have cytotoxic (tumor inhibiting) properties, their further study could be fruitful (*435*).

A ^{1}H NMR study showed that dehydronuciferine (**26b**), dehydrodicentrine (**26c**), and dehydroocopodine initially undergo C-7- and N-protonation in CF_3COOH, the C-protonated immonium salts being formed almost completely under equilibration conditions. Thus, reduction of

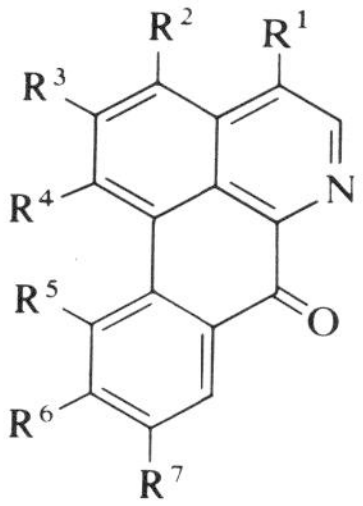

35a $R^1 = R^2 = R^5 = R^6 = R^7 = H$; $R^3 + R^4 = O—CH_2—O$ Liriodenine
35b $R^1 = R^2 = R^5 = R^6 = H$; $R^3 + R^4 = O—CH_2—O$; $R^7 = OMe$ Lanuginosine
35c $R^1 = R^2 = R^7 = H$; $R^3 + R^4 = R^5 + R^6 = O—CH_2—O$ Hernandonine
35d $R^1 = R^2 = R^5 = H$; $R^3 = R^4 = R^6 = R^7 = OMe$ *O*-Methylatheroline
35e $R^1 = R^2 = R^3 = R^4 = OMe$; $R^5 = R^6 = R^7 = H$ Imenine
35f $R^1 = R^5 = H$; $R^2 = OMe$; $R^3 + R^4 = R^6 + R^7 = O—CH_2—O$ Cassamedine
35g $R^1 = R^2 = R^5 = R^6 = R^7 = H$; $R^3 = R^4 = OMe$ Oxonuciferine (Lysicamine)
35h $R^1 = R^2 = R^5 = H$; $R^3 + R^4 = O—CH_2—O$; $R^6 = R^7 = OMe$ Dicentrinone

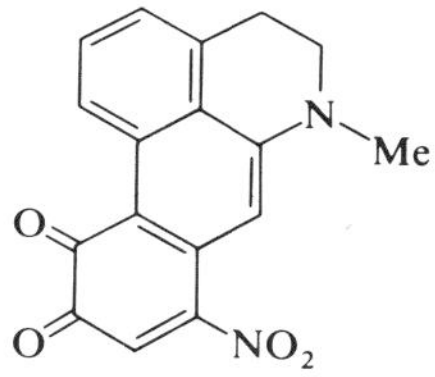

36

dehydronuciferine with amalgamated Zn in DCl–D_2O gave 86% of 6a,7,7-trideuterionuciferine (*436*).

The dehydrogenation of aporphines to the corresponding dehydroaporphines has been accomplished by the use of various chemical oxidants, including permanganate, DDQ, mercuric salts, iodine (*437*), and photochemical oxidation (*193*). Recently, another method has been elaborated: namely, heating the aporphine bases (in refluxing acetonitrile) with 10% palladium on charcoal (*438*). Under these conditions, e.g., nuciferine (**23b**) afforded dehydronuciferine (**26b**) in 90% yield after 15 min reaction time. Kametani *et al.* described the synthesis of 6a,7-didehydroaporphine system by benzyne reactions (*439*). The synthesis of oxoaporphines by photolysis, including the total synthesis of atheroline (**35d**), was also described (*289, 409, 440*).

The generally known Pellagri reaction for identifying morphine by converting it into the blue product was studied again (reviewed in *The Alkaloids*, Vol. XII, p. 361) because many aporphine alkaloids yield a colored spot on oxidation with air–oxygen or iodine (Dragendorff reagent) after paper or thin-layer chromatography. Three other reports (*437, 441–442*) deal with the oxidation of the nonphenolic aporphines by means of iodine to afford the corresponding 6a,7-dehydroaporphines (**26**). In contrast, iodine oxidation of nonphenolic noraporphines proceeds all the way to 7-oxoaporphine. The phenolic aporphine *N*-methylnandigerine (**25c**) is converted in low yield by iodine to the blue 10,11-*o*-quinone (**31b**). This compound is formed as the major product of mercuric chloride oxidation of **23b** and bulbocapnine (**25d**). The 6a,7-dehydroaporphines dehydronuciferine (**26b**) and dehydrodicentrine (**26c**) are oxidized by oxygen at pH 6 to give the corresponding oxoaporphines **35g** and **35h**; **26b** is also rapidly oxidized to **35g** by peracetic acid or by benzoyl peroxide (Scheme 8).

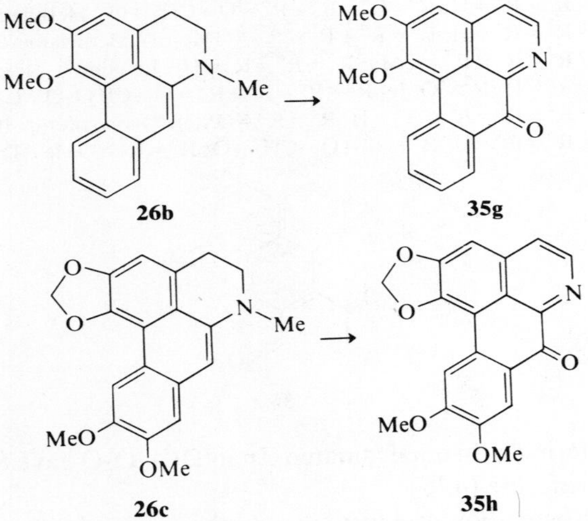

MeO MeO N Me → MeO MeO N O

26b **35g**

O O N Me MeO OMe → O O N O MeO OMe

26c **35h**

SCHEME 8. *Formation of 7-oxoaporphines* (**35**) *from dehydroaporphines* (**26**) (*437*).

It was shown that the oxidation product obtained by treatment of bulbocapnine methyl ether (**25f**) with iodine in ethanol/water is not the tetradehydroaporphinium salt, as reported by Gadamer and Kuntze (*443*), but a dimer of the structure **37** (*441, 441a*) (Scheme 9). Reduction of this compound with zinc in dilute sulfuric acid gives racemic bulbocapnine methyl ether (**25f**), whereas reduction with complex hydrides affords the dimers **38** and **39**.

Aporphine can also be identified by application of the Husemann reaction which takes place on oxidation of this base by nitric acid (*444*). An analysis of the thus obtained red product showed that it is **36**.

The aporphine alkaloids can be N-demethylated and thus converted into *N*-noraporphine alkaloids (retention of the chirality at C-6a) with hydrogen peroxide and following treatment with liquid sulfur dioxide (yield ca. 35%)

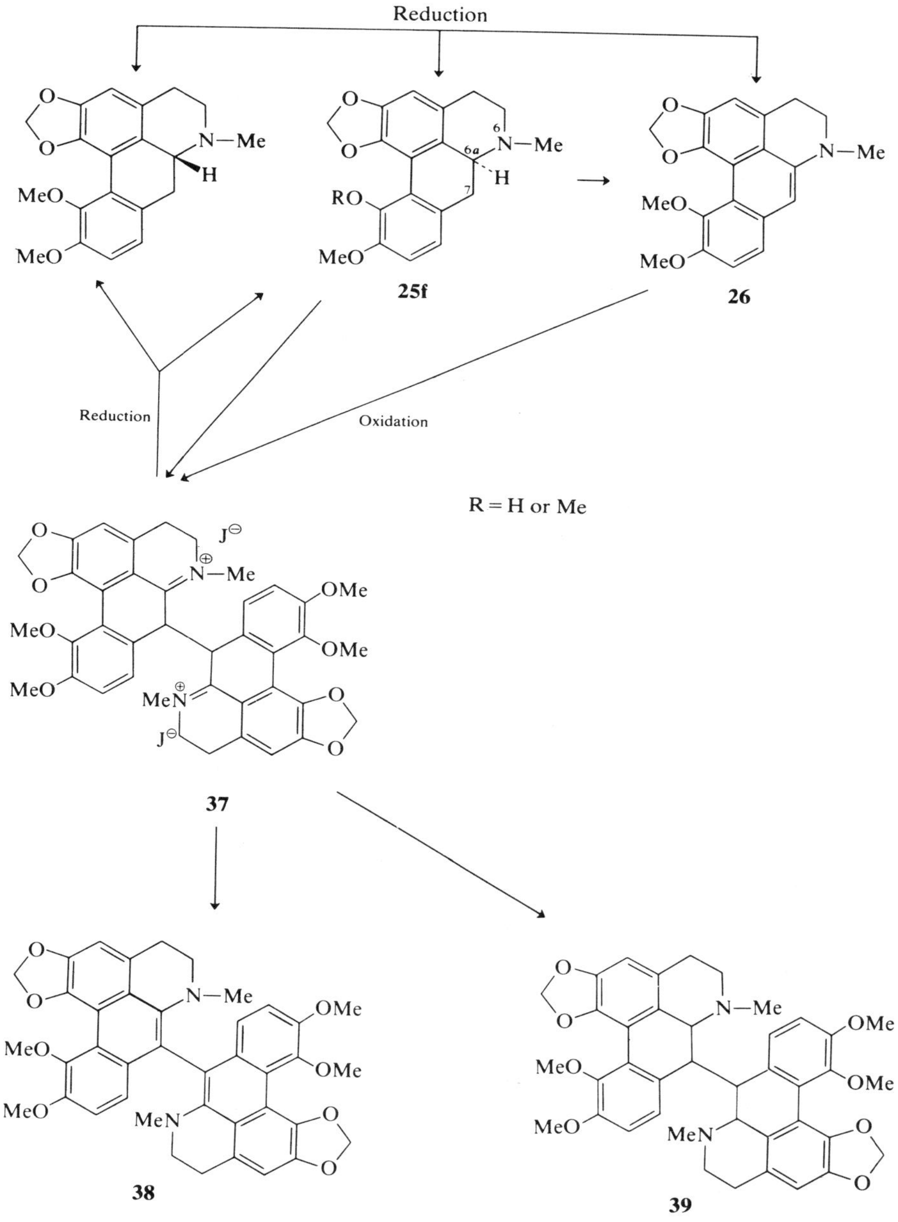

R = H or Me

SCHEME 9. *Oxidation and dehydrogenation products of bulbocapnine methyl ether* (**25f**) (*441*).

(*445*). The by-product is the 6a,7-didehydrobase. The *O*-demethylation of (±)-amuronine, (±)-tetrahydroglaziovine and other proaporphines can be carried out by refluxing in 20% HCl under nitrogen for 24 hr (*446, 446a*). Gereke *et al.* also described selective ether cleavage with boron halide in the aporphine series and thus conversion of (*S*)-bulbocapnine into (*S*)-corytuberine and (*S*)-corydine methyl ether (*441a*). Sodium benzylselenolate in refluxing dimethylformamide has been found to be a superior reagent for the O-demethylation of aryl methyl ethers of nonphenolic aporphine alkaloids. Regioselective demethylation by this method occurs at positions 1, 8, and 11 of the aporphine nucleus; methylenedioxy functions survive the reaction (*447*). Selective cleavage of the methylenedioxy group in *O*-methylbulbocapnine with boron trichloride is discussed in Section III,C (*448*). The ether cleavage of methoxy- and methylenedioxy-substituted isoquinoline has been reviewed (*346*).

Further studies of the synthesis of the aporphine alkaloids were carried out to find simpler procedures and to increase the yields. The most frequently applied methods are the Bischler–Napieralski reaction (*91, 408, 408a, 449, 450*), the Pschorr cyclization (*451, 452*), and the photocyclization of various benzylisoquinoline precursors (*416a, 453, 454*) (Scheme 10).

The alkaloids mecambroline (**24a**) (*65*), bracteoline (**24d**) (*408*), isoboldine (**24c**) (*455*), isothebaine (**25b**) (*456*), predicentrine (**24k**) (*457*), and

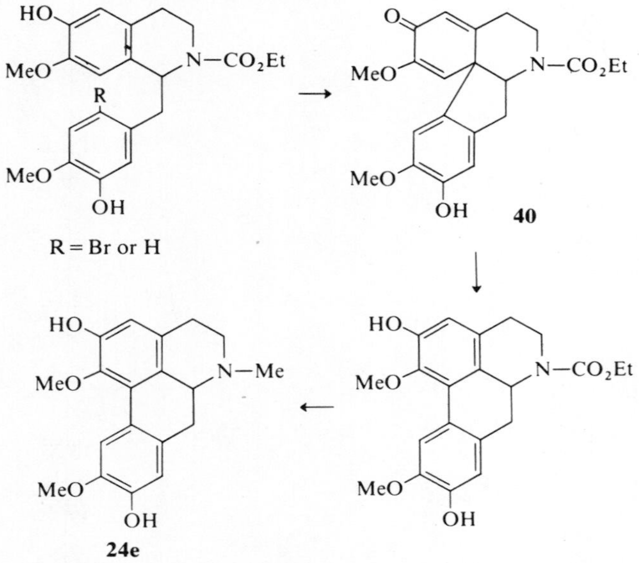

R = Br or H

40

24e

SCHEME 10. *Novel photochemical aporphine synthesis via neoproaporphine* (**40**) (*416a*).

thaliporphine (*439*) were synthesized by the former two methods. The alkaloids nuciferine (**23b**), glaucine (**24h**), boldine (**24e**) (*416a*), norpredicentrine (**24i**) (*453*), and isocorydine (**25g**) (*454*) were obtained by the photochemical route. (±)-Boldine (**24e**) was also prepared by photocyclization of (±)-bromodiphenol to neospirodienone (**40b**), followed by rearrangement to *N*-ethoxycarbonylnorboldine and reduction with lithium aluminum hydride (*416a*) (Scheme 10).

The total syntheses of the less common alkaloids cassamedine (**35f**) (*427*) and *N*-methylovigerine (**25n**) (*458*), and of some nornucipherine derivatives were also achieved (*459*). Thaliporphine (**24g**), domesticine (**24l**), and glaucine (**24h**), can be synthesized via the *p*-quinol acetate route (*415, 460, 461*) (Scheme 11).

The synthesis of the aporphine bases was carried out by using various oxidizing agents. The yields of the final product were only poor when the starting material was laudanosoline (**7n**). Phenol oxidation of (+)-reticuline perchlorate (**7f**) with cuprous chloride and oxygen in pyridine gave (+)-corytuberine (**25e**), (+)-isoboldine (**24c**), and pallidine (**43b**). Under the same reaction conditions, (±)-orientaline perchlorate (**7s**) yielded (±)-orientalinone. Oxidation with cupric chloride and potassium superoxide in pyridine gave rise to similar results (*461a*).

More recently, the preparative value of $VOCl_3$ and VOF_3–TFA in chemical oxidation has been demonstrated (see Section III,B on the proaporphine and promorphinane alkaloids) (*415, 462*). Some other authors used the purified enzyme horseradish peroxidase (*463*). By this method the aporphine base (besides the quaternary dibenzopyrrocoline) is readily obtained from (*S*)-(+)-laudanosoline hydrobromide or from (*R*)-(−)-laudanosoline methiodide with retention of the absolute configuration. The synthesis of 6a,7-dehydroaporphine bases was also carried out by making use of the benzyne reaction (*439*). Reduction of these substances affords the corresponding aporphine bases (*439*). The synthesis of isoquinoline alkaloids by lead tetraacetate oxidation was reviewed by Umezawa and Hoshino (*343*).

The synthesis of deuterium- or tritium-labeled dehydroaporphines has been reported (*436, 464*).

The biosynthesis of the aporphine alkaloids in *P. somniferum* has been studied (*258, 259, 456, 465*). In 1911, Gadamer postulated that the biosynthesis of the aporphine alkaloid glaucine (**24h**) proceeds via the simple benzyltetrahydroisoquinolines (*466*). Later, this hypothesis was experimentally confirmed, but some of the problems remained unresolved owing to the inadequate technique employed then. Recently, some authors have returned to this problem (*152, 258, 259, 467*). By addition of N—CH_3 labeled reticuline to a nutrient solution it was shown that, from the thus

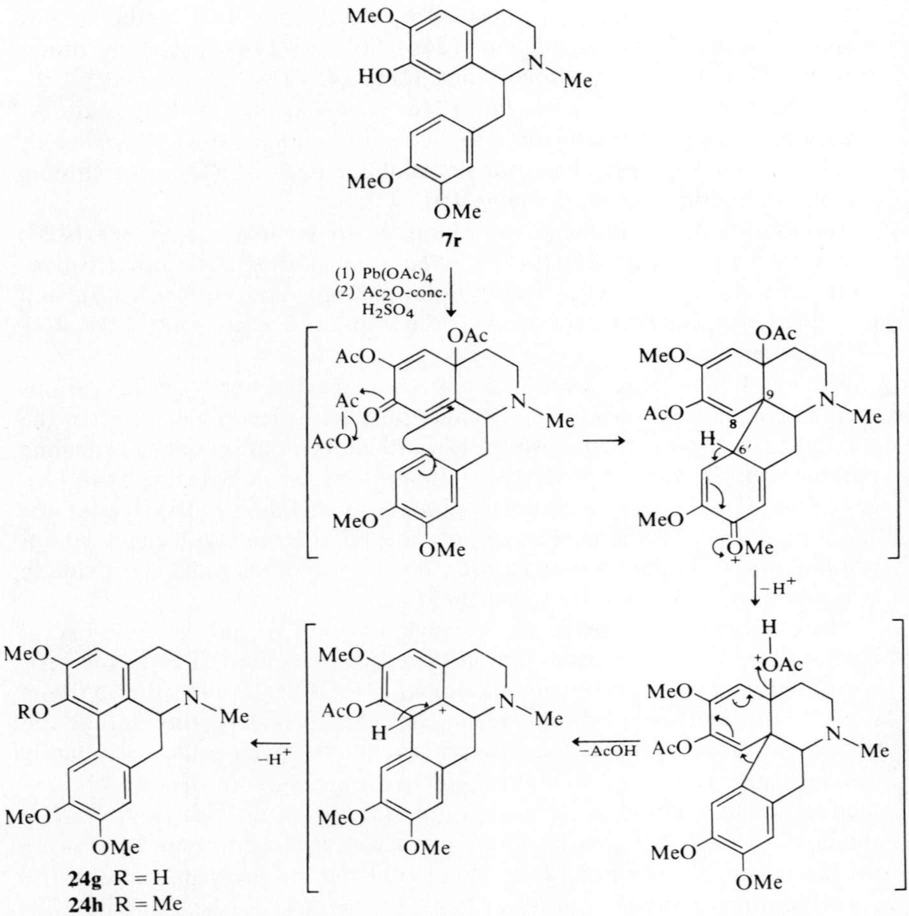

SCHEME 11. *Synthesis of the aporphine alkaloids* (**24**) *via the p-quinol acetate route* (*413*).

labeled precursor, up to 98% of radioactivity passes into bulbocapnine (**25d**) (*467, 468*). A similar experiment with reticuline labeled at several positions revealed the direct diphenyl coupling of reticuline to provide the aporphine skeleton (Scheme 12).

No doubt exists as far as the pathway **A** in Scheme 12 is concerned, though cyclohexadienone could not be demonstrated as an intermediary product (pathway **B**) of this biosynthesis. Experiments with labeled compounds have also shown that, during the biosynthesis of isothebaine (**25b**), the plant utilizes in the first place orientaline (**7s**) and only then the two isomers of orientalinone (*258, 456*). Blaschke's results (*467, 468*) are confirmed by

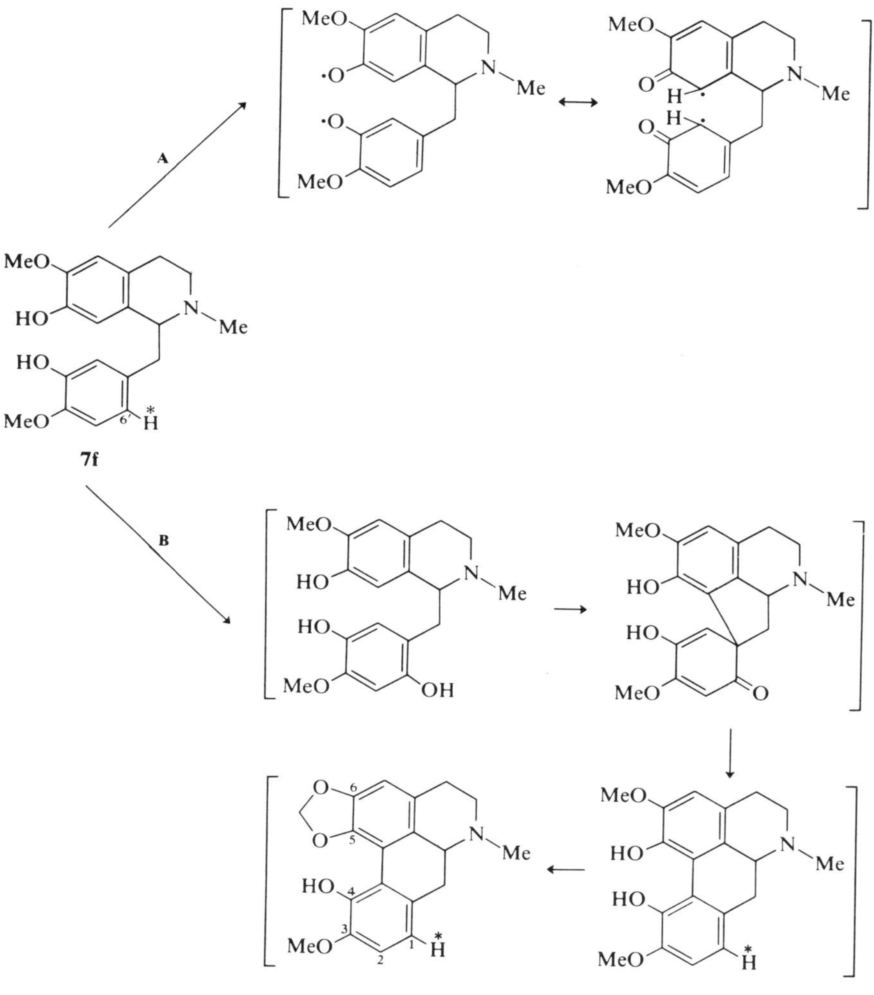

7f

25d Bulbocapnine

25e Corytuberine

SCHEME 12. *Biosynthesis of bulbocapnine* (**25d**) *in Corydalis cava* (*467*).

two papers of Brochmann-Hanssen (*258, 259*). In the first paper the author reports that, when *P. somniferum* was fed specifically labeled (±)-reticuline (**7f**), it was incorporated into isoboldine (**24c**). No incorporation of reticuline could be observed in magnoflorine (**25h**). Because reticuline, isoboldine, and magnoflorine are known to exist in the opium poppy, it may be concluded that aporphines with a 1,2,9,10-substitution pattern [isoboldine type (**24c**)] can be biosynthesized by direct phenol coupling (*ortho–para*)

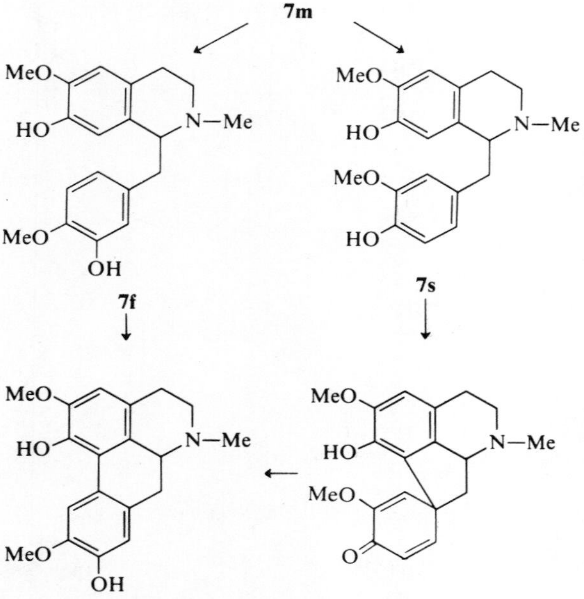

7m

7f **7s**

24c **22d** Orientalinone

SCHEME 13. *Potential pathways for the biosynthesis of orientalinone* (**22d**) *and isoboldine* (**24c**) (*258*).

(Scheme 13); this may not be the case for aporphines with substituents in positions 1, 2, 10, and 11 [corytuberine type (**25e**)]. Because of steric factors these aporphines are more likely to be produced via a dienone proaporphine intermediate followed by a dienone–phenol rearrangement. In the second work, the same authors report that orientaline was detected in the opium poppy by an isotope-dilution method based on its biosynthesis from norlaudanosoline (**7p**) (*258*). Administration of labeled orientaline (**7s**) revealed that in *P. somniferum* this alkaloid was not a precursor of isoboldine (**24c**), and the experimental results provided no evidence for a pathway involving norprotosinomenine (**7t**); this route is, however, very likely for *P. orientale* and *P. bracteatum* (Scheme 13).

Contrary to the above-mentioned conclusions, the plant *Dicentra eximia* elaborates isoboldine and corytuberine types of alkaloids in tracer experiments from norprotosinomenine, probably via the dienones **40a** and **40b** (*152, 467*) (Schemes 6 and 11).

The aporphine alkaloids were dealt with in papers on the UV spectra in neutral and basic solutions (*469*), on IR spectroscopy in the region 1630–1480 cm^{-1} (*470*), on NMR spectroscopy in the (*471–473*), and on the

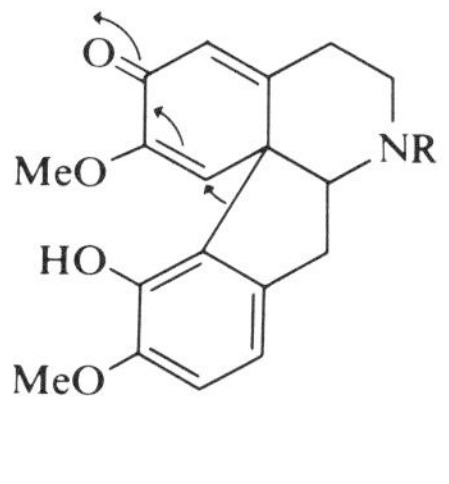

MeO HO MeO NR O

40a

MeO OH NR O

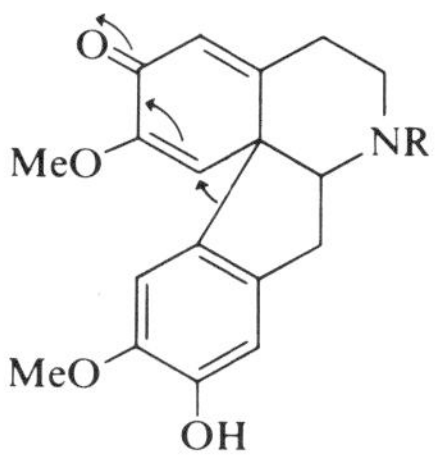

40b
Neoproaporphines

photocolorimetric determination of glaucine (**24h**) in *G. flavum* with tropaeoline (*190*). The stability of this alkaloid in injection solutions has also been studied (*442*).

D. PROMORPHINANE GROUP*

A review of the chemistry of these alkaloids has been written by Stuart (*474*).

Dihydronudaurine (**42**) was isolated from *P. pannosum* (*226*). The plants *C. pallida* var. *tenuis* (Yatabe) (*125*) and *C. incisa* (Pers.) (*475*) yielded the alkaloids sinoacutine (**43a**) and pallidine (**43b**). The acetylated promorphinane alkaloid pavanoline was isolated from *P. anomalum*, but its structure could only be partially elucidated because of the limited quantity available (*226*).

Racemic amurine (**43c**) (*476*), 3,4-dimethoxypromorphinane (**43d**) (*477*), isosalutaridine (**43f**) (*478, 479*), *O*-methylflavinantine (**43g**) (*478*), *O*-demethylsalutaridine (**43e**), and sinoacutine (**43a**) have been synthesized (*480*). Kametani *et al.* (*455, 481–487*) synthesized them either by oxidation with potassium ferricyanide, Ag_2CO_3, electrolytic oxidation, photolytic synthesis, or by the benzyne reaction. All these reactions (yield 1–5%) and

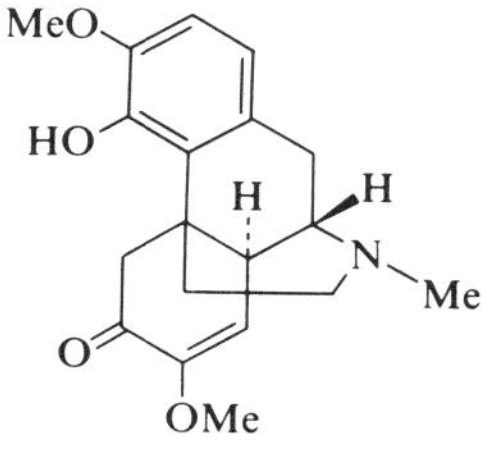

MeO HO H H N Me O OMe

41 Sinomenine

O O H N Me MeO HO H

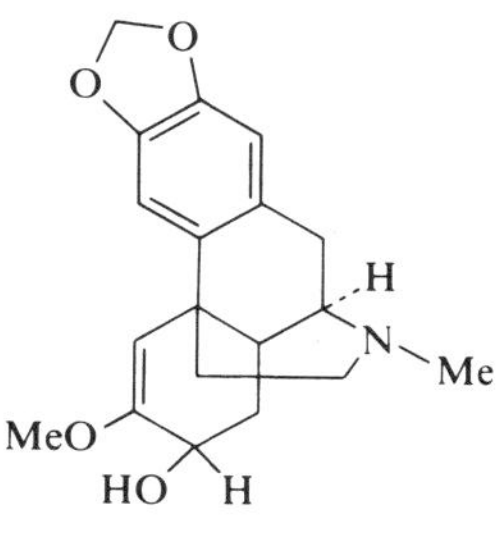

42 Dihydronudaurine

* This material is supplementary to *The Alkaloids*, Vol. XII, p. 362.

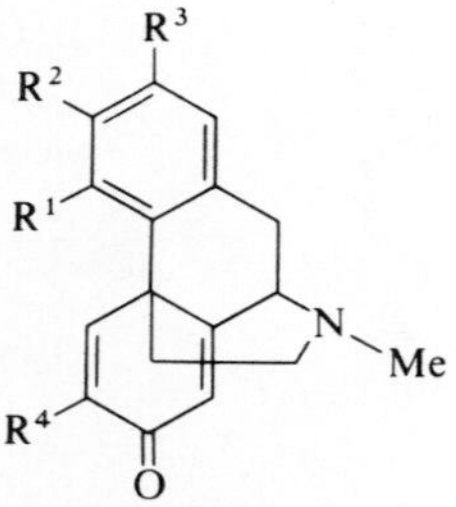

43 —Morphinandienone—Promorphinane

43a $R^1 = OH$; $R^2 = R^4 = OMe$; $R^3 = H$ (*R*)-Salutaridine/(*S*)-Sinoacutine
43b $R^1 = H$; $R^2 = R^4 = OMe$; $R^3 = OH$ Pallidine
43c $R^1 = H$; $R^2 + R^3 = O—CH_2—O$; $R^4 = OMe$ Amurine
43d $R^1 = R^2 = R^4 = OMe$; $R^3 = H$ 3,4-Dimethoxypromorphinane (*O*-Methylsalutaridine)
43e $R^1 = R^4 = OH$; $R^2 = OMe$; $R^3 = H$ 6-*O*-Demethylsalutaridine
43f $R^1 = H$; $R^2 = R^4 = OMe$; $R^3 = OH$ Isosalutaridine
43g $R^1 = H$; $R^2 = R^3 = R^4 = OMe$ *O*-Methylflavinantine
43h $R^1 = H$; $R^2 = OH$; $R^3 = R^4 = OMe$ Flavinantine

those employing the phenol oxidation and the photo-Pschorr reaction have been reviewed (*483*). Synthesis by a modified Pschorr reaction also results in the opening of the methylenedioxy ring in the initial compounds, and this provides a route for further modifications (*488*). Franck and Teez developed a new method for the synthesis of these compounds (*489*). They carried out phenolic oxidation by using $VOCl_3$; the yield of alkaloids was as high as 34%. The synthesis of these promorphinane bases was also accomplished by electrooxidative cyclization (in HBF_4) of variously substituted 1-benzyl-tetrahydroisoquinoline compounds (*402, 415–417, 490, 491*) (yields up to 70–90%). Thus, for example, the racemic alkaloids amurine (**43c**), flavinantine (**43h**), and pallidine (**43b**) were obtained (*492*). The electrochemical oxidation of laudanosine (**7g**) to *O*-methylflavinanthine (**43g**) in MeCN was also studied. Cyclic voltametry, rotating disc voltametry, and preparative electrolysis on isoquinoline alkaloids, amines, and aromatic ethers helped to elucidate the mechanism of the coupling in this reaction (*401*). Schwartz and Mami criticized all the previous methods of phenolic oxidation using $K_3Fe(CN)_6$, MnO_2, $PbAc_4$, Ag_2CO_3 or $VOCl_3$ or VOF_3–TFA. They conclude that the oxidation product of reticuline gives rise to isosalutaridine (**43f**) by *para–para* coupling in addition to the proaporphine/aporphine derivatives (*ortho–para* coupling) which are found in larger quantities (*493*) (Scheme 14).

In order to imitate nature in the synthesis it is necessary first to realize a *ortho–para* coupling of reticuline to yield salutaridine (**43a**). The authors succeeded in that respect by application of a thallium tristrifluoroacetate coupling method at −78° for 3 hr and then at −20° for 12 hr (*493*). During

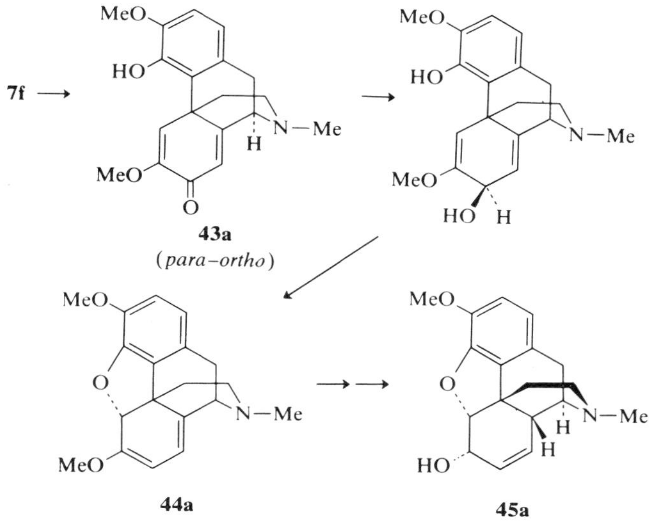

SCHEME 14. *Synthesis of morphinane alkaloids* (**45**) *from R-reticuline* (**7f**) (*493*).

this reaction, isosalutaridine (*para–para*) arises only in small amount. In this manner racemic thebaine (**44a**) can be obtained from racemic *N*-ethoxycarbonylnorreticuline via salutaridine (**43a**) in good yield. Because thebaine can be converted into codeinone (**47a**) and this into codeine (**45a**) and finally into morphine (**45c**), one has a biogenetically patterned synthesis of morphine alkaloids.

Rapoport *et al.* described a simple preparation of 6-*O*-demethylsalutaridine (**43e**) from thebaine with sodium bisulfite in an oxygen atmosphere (*494*). In another synthesis, 14-bromocodeinone (**47b**) in Claisen's alkali yields 6-*O*-demethylsalutaridine (*495*). Methylation of this compound with diazomethane gives rise to *O*-methylsalutaridine (**43d**). The following reaction sequence appears to account for the production of 6-*O*-demethylsalutaridine (*495*) (Scheme 15).

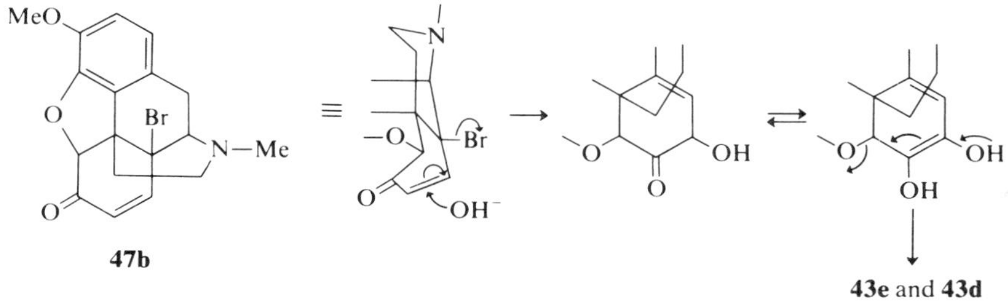

SCHEME 15. *Preparation of 6-O-demethylsalutaridine* (**43e**) *and O-methylsalutaridine* (**43d**) *from 14-bromocodeinone* (**47b**) (*495*).

For the UV and IR spectroscopy of the promorphinane alkaloids see Section III,B.

E. Morphinane Group*

The compounds 6-methylcodeine (**45b**) (*263*), 16-hydroxythebaine (**44b**) (*262*), normorphine (**45d**) (*274*), 14β-hydroxycodeine (**46**), and 14β-hydroxycodeinone (**47c**) (*235*) are new alkaloids of the morphinane type. Normorphine was found in opium with the help of sensitive reactions, and it was established as an active metabolite of morphine in *P. somniferum*. In the *Papaver* species there were always demonstrated two isomeric *N*-oxides of codeine, morphine and thebaine, which were also prepared synthetically (*233*).

The new alkaloid oripavidine, isolated from *P.* orientale, was identified as 3,13-didemethylthebaine (*243a*).

By using ^{14}C-morphine it could be shown that morphine is further degraded to such nonalkaloidal metabolites as thebaol (**50**), which is elaborated directly by the plant from thebaine (*288, 496*).

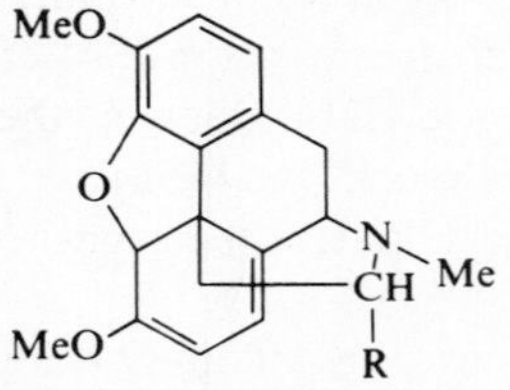

44a R = H Thebaine
44b R = OH 16-Hydroxythebaine

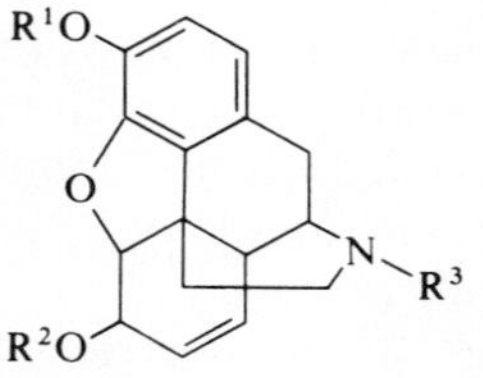

45a $R^1 = R^3 = Me$; $R^2 = H$ Codeine
45b $R^1 = R^2 = R^3 = Me$ 6-Methylcodeine
45c $R^1 = R^2 = H$; $R^3 = Me$ Morphine
45d $R^1 = R^2 = R^3 = H$ Normorphine

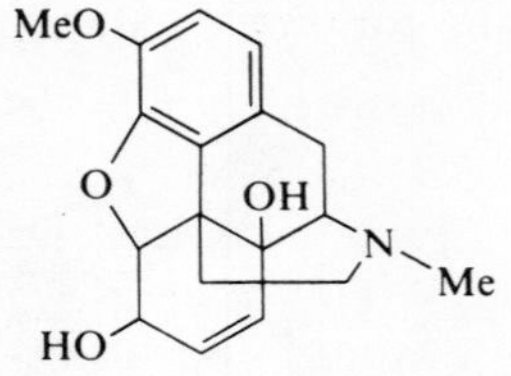

46 14-Hydroxycodeine

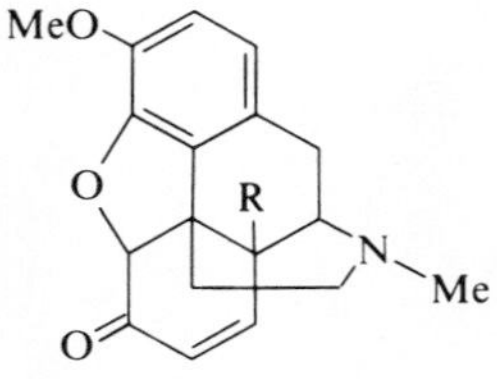

47a R = H Codeinone
47b R = Br 14-Bromcodeinone
47c R = OH 14-Hydroxycodeinone

* This material is supplementary to *The Alkaloids*, Vol. XII, p. 364.

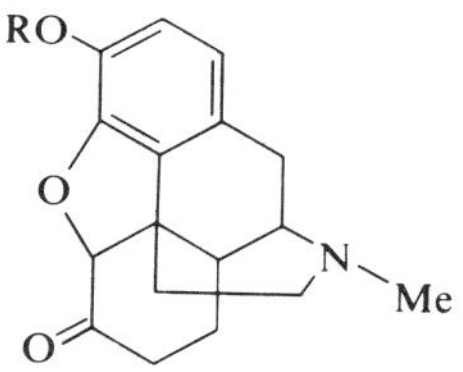

48a R = Me Dihydrocodeinone
48b R = H Dihydromorphinone

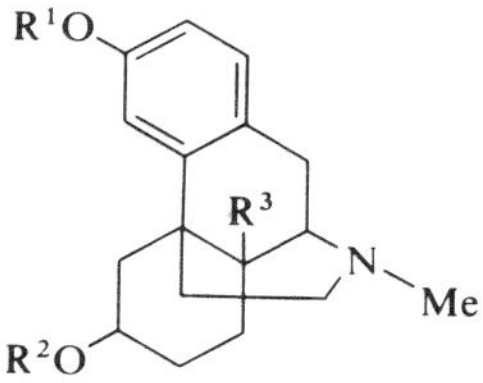

49a $R^1 = Me$; $R^2 = R^3 = H$ Dihydrocodeine
49b $R^1 = Me$; $R^2 = H$; $R^3 = OH$ 14-Hydroxydihydrocodeine
49c $R^1 = R^2 = R^3 = H$ Dihydromorphine

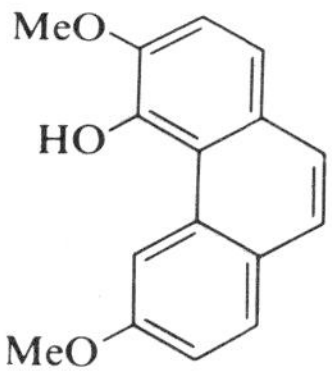

50 Thebaol

A commercially profitable biomimetic synthesis of morphinane alkaloids will require the conversion of reticuline into salutaridine in sufficient yield (Scheme 14) (see Section III,D) and the conversion of salutaridine into thebaine, codeine, and morphine.

Recently, Rapoport *et al.* described a new synthesis of oripavine and thebaine, a practical synthesis of codeine from dihydrothebainone, and the conversion of thebaine to codeine (yield 85% from thebaine) (*497, 498*). The transformation of thebaine into codeine with sulfonyl hydrazine and later with methyl hypobromite was reported by Krausz (yield 70%) (*499*) (Scheme 16).

The synthesis of the morphinane alkaloids from promorphinane bases has also been discussed in Section III,D.

Kametani *et al.* reported the fifth formal total synthesis of morphine and sinomenine (**41**) via racemic sinoacutine (**43a**) and thebaine (*480*). Gaál reviewed the literature on the preparation of dihydrocodeinone (**48a**) and dihydromorphinone (**48b**), and he developed his own procedure for the preparation of these two substances (yield ca. 40–47%) (*500*).

The metabolic pathways in plants and living organisms were studied by means of tritium-labeled morphine (*508, 509*). The use of unnaturally tritium-labeled codeine derivatives showed that *P. somniferum* is able to O-demethylate them into unnatural morphine derivatives (*501*). The

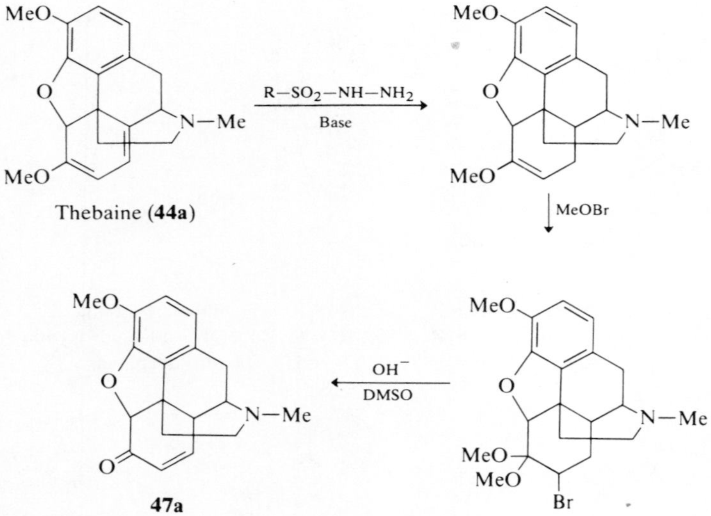

SCHEME 16. *Conversion of thebaine into codeinone* (**47a**) (*499*).

knowledge of the biosynthesis of thebaine, codeine, and morphine has stimulated some authors to confirm the finding of thebaol (**50**) or to find other metabolic intermediate steps, the initial amino acids, or the final products (*493, 501–504*). In *P. somniferum*, radioactive tyrosine is incorporated into different minor alkaloids but not, however, into thebaine and morphine (*505, 506*). On the contrary 3,4-dihydroxyphenylalanine is rapidly incorporated into morphine, thebaine, narcotine, and papaverine (*507*). On incorporation of radioactive phenylalanine, tyrosine (?), and also glycine and urea, radioactive carbon atoms also appear in thebaine, i.e., most of them in the capsules, less in the leaves, and the least in the stalks and roots of *P. bracteatum* Arya II (*298*). The fact that such syntheses take place in the latex indicates that the latter is cytoplasmatic with adequate organelle activity.

Earlier work (in summary, *510*) has defined with considerable precision how the 1-benzylisoquinoline system is converted by opium poppies into the morphine group of alkaloids. Norlaudanosoline (**7p**) was the earliest 1-benzylisoquinoline recognized on this pathway, and it was shown to be built from two aromatic units both derivable from tyrosine; one of them was dopamine. The nature of the second unit was unknown. Recently, the studies of *P. somniferum*, *P. orientale*, and *P. bracteatum* have shown (*339, 511, 511a*) (Scheme 17) that (a) the amino acid (**c**) can act as a specific precursor of morphine (**45c**); only one enantiomer of **c** would be expected to be biologically converted; (b) the aromatic nuclei of both building blocks for the 1-benzylisoquinoline system are dihydroxylated before isoquinoline formation occurs; (c) the dihydroisoquinoline (**d**) may lie on the pathway between the amino acid **c** and norlaudanosoline (**7p**); and (d) externally introduced [2-^{14}C]dopa (**a**) does not significantly label the pool of the keto acid **b** from which the amino acid **c** is presumably built. One possibility is that, in the intact plant, added dopa **a** fails to penetrate to the site of the appropriate aminotransferase.

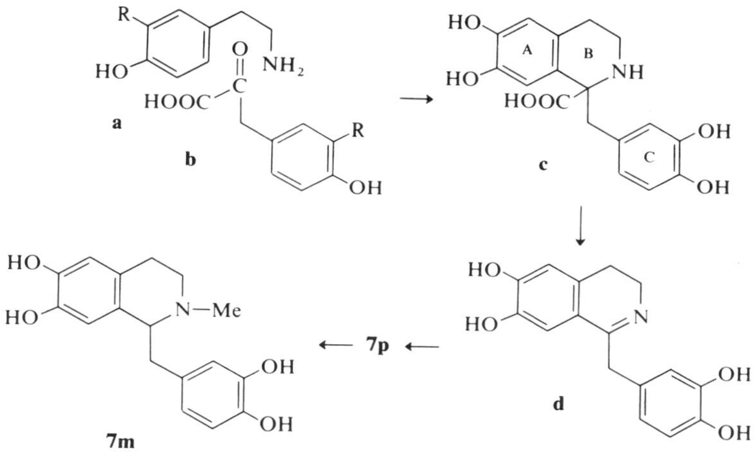

R = H or OH

SCHEME 17. *Biosynthesis of laudanosoline* (**7m**) *in Papaver somniferum* (*339, 511*).

Studies of the development process of the seedling *Papaver* plant showed that of the amino acids, the precursor recognized as tyrosine fluctuates almost in exact parallel with the alkaloid content (*512*). The maximum concentrations of the amino acids (with tyrosine) were studied in three-day-old seedling roots. Michels-Nyomárkay brought this finding into relationship with the higher CO_2 production and thus with the more intensive respiration of the young seedlings of the *Papaver* plant. The seeds contain only traces of alkaloids. The endosperm portion and the embryo of the seed contain only some alkaloids (narceine, thebaine, papaverine, narcotine). Tyrosine could not be detected in any of the mentioned seed parts. The alkaloidal content of the endosperm might form one of the energy sources of the seedling (*512*).

In *P. somniferum*, the alkaloids appear to be stored in the vacuolar sap of the vesicles rather than membrane bound; in this respect, the vesicles behave as normal vacuoles (*268, 298, 513*). However, evidence indicates that the stem latex and vesicles are translocated into the capsule during its rapid expansion after petal fall. During that time, the morphine itself is being synthesized and metabolized in the vesicle (more rapidly in the stem vesicles than in those of the capsule), and the metabolites pass out of the latex into the pericarp, with a significant amount appearing in the ovules. The vesicles are therefore not merely passive accumulators of alkaloids.

Numerous attempts have been made to improve the industrial isolation procedure of opium alkaloids by application of different adsorbents or ion exchangers (*502, 514–520*). Much attention has been paid to the Kabay method for the isolation of opium alkaloids and of the minor nonmorphinane bases from the dry plant *P. somniferum* (*254*).

Ikonomovski studied the conditions of the methylation of morphine into codeine by using phenyltrimethylammonium methoxide (*521*). New developments in photochemistry have also been applied to the morphinane alkaloids: the effects of diffuse sunlight or daylight transform codeine in solution (methanol or ethanol) into methylcodeine and ethylmorphine, and thebaine in organic solvents into 3-*O*-methyl-6-*O*-ethylmorphine and its *N*-oxide, diethylmorphine, and its *N*-oxide (*522*). Thebaine was found to yield codeinone (**47a**) and methylcodeine (**45b**).

An improved N-demethylation of morphine, codeine, and 6,7-benzomorphane was carried out in 90% overall yield (*523*). The demethylation of thebaine to 6-*O*-demethylsalutaridine (**43e**) has been reported (*494*). Attention was also paid to the preparation of tosyloxy and mesyloxy derivatives of some morphinanes, e.g., morphine, codeine, dihydromorphine (**49c**), dihydrocodeine (**49a**), 14-hydroxycodeine (**46**), and 14-hydroxydihydrocodeine (**49b**) (*524*). The rearrangements of all these derivatives as well as the Beckmann and Schmidt rearrangements of 6-oxomorphine bases have also been studied (*525*).

Snatzke *et al.* studied 56 CD curves of morphinane derivatives in the region between 500 and 185 nm (*526*). The sign of the $^1B_{2u}$ band CD can be predicted from the ring chirality. The influence of different substituents in ring C on the Cotton effects is discussed on the basis of the corresponding sector rule. A CD band found for two allyl iodides at 310–300 nm is tentatively assigned a forbidden transition of this chromophore.

Quantum chemical calculations have been reported for morphine and morphinelike opiate narcotics (*527, 528*). An independent ^{1}H NMR spectroscopic study of codeine and isocodeine and of some of their derivatives, respectively, was carried out to confirm their conformation (*529*). The ^{13}C NMR spectra of codeine, its oxidation and reduction products, thebaine, sinomenine (**41**), and 8,14-dihydrothebaine were recorded and interpreted by proton noise decoupling and single-frequency off-resonance methods using spectra of 3-methoxymorphinane and its derivatives as reference data (*530*). Carrol *et al.* also studied the ^{13}C NMR spectra of many morphine derivatives. They showed by comparison of the chemical shifts of the morphine and 14-hydroxymorphine systems to those of the 6,14-*endo*-etheno- and 6,14-*endo*-ethanotetrahydrothebaine systems that the spatial configurations of rings A, B, and D of the two systems were similar (*531*). The acid dissociation constants of morphine in aqueous methanol media were determined by Vysotskaya *et al.* (*532*).

The qualitative and quantitative determinations of morphinane alkaloids or their derivatives in opium were studied by application of different methods (*278, 533, 534*), by differential spectrophotometry (*535, 536*), thin-layer chromatography (*509, 537, 538*), gas chromatography in conjunction with mass spectrometry (*539, 540*), and high-speed liquid chromatography (thebaine) (*327, 534*). A procedure for the quantitative determination of morphine in opium by isotopic dilution has been developed (*541*).

In recent years the qualitative and quantitative determination of morphine and its derivatives in commercially available material or in biological fluids in the human body has become of particular importance in connection with drug addiction (*542*). In humans, morphine is transformed into codeine (*543*).

F. Cularine Group*

With the exception of the dimeric alkaloids, which also contain a cularine skeleton (*The Alkaloids*, Vol. XIV, p. 407 and this Vol., p. 510) no other new alkaloids of this type have been found.

Kametani *et al.* (*544, 545*) and other workers (*546–561*) endeavored to carry out the synthesis of cularine alkaloids by phenolic oxidation (biogenetic type of synthesis) of the corresponding derivatives of laudanosine. The paper (*549*) describes the synthesis of these bases via the 6-ethoxycarbamido-3,4-dihydroisoquinolines, which were converted to 6-aminoisoquinoline. By Ullmann reaction it gives the compound **52** and (±)-cularine (**51**) (Scheme 18). Cularine-type alkaloids were also synthesized by the intramolecular Ullmann reaction of 7,8-disubstituted isoquinoline obtained by the usual Bischler–Napieralski reaction from the phenolic bromoamide (pathway **a**) (*544, 548*). However, in the papers referred to (*557, 558, 561*), the rings A, C, and D were formed first (pathway **b**), and only then was the ring B formed during the synthesis of cularine.

Some authors had erroneously assumed, on the basis of the ORD and CD measurements, that cularine has the (*R*) configuration. This finding was

* This material is supplementary to *The Alkaloids*, Vol. IV, p. 249; Vol. X, p. 463; Vol. XII, p. 368.

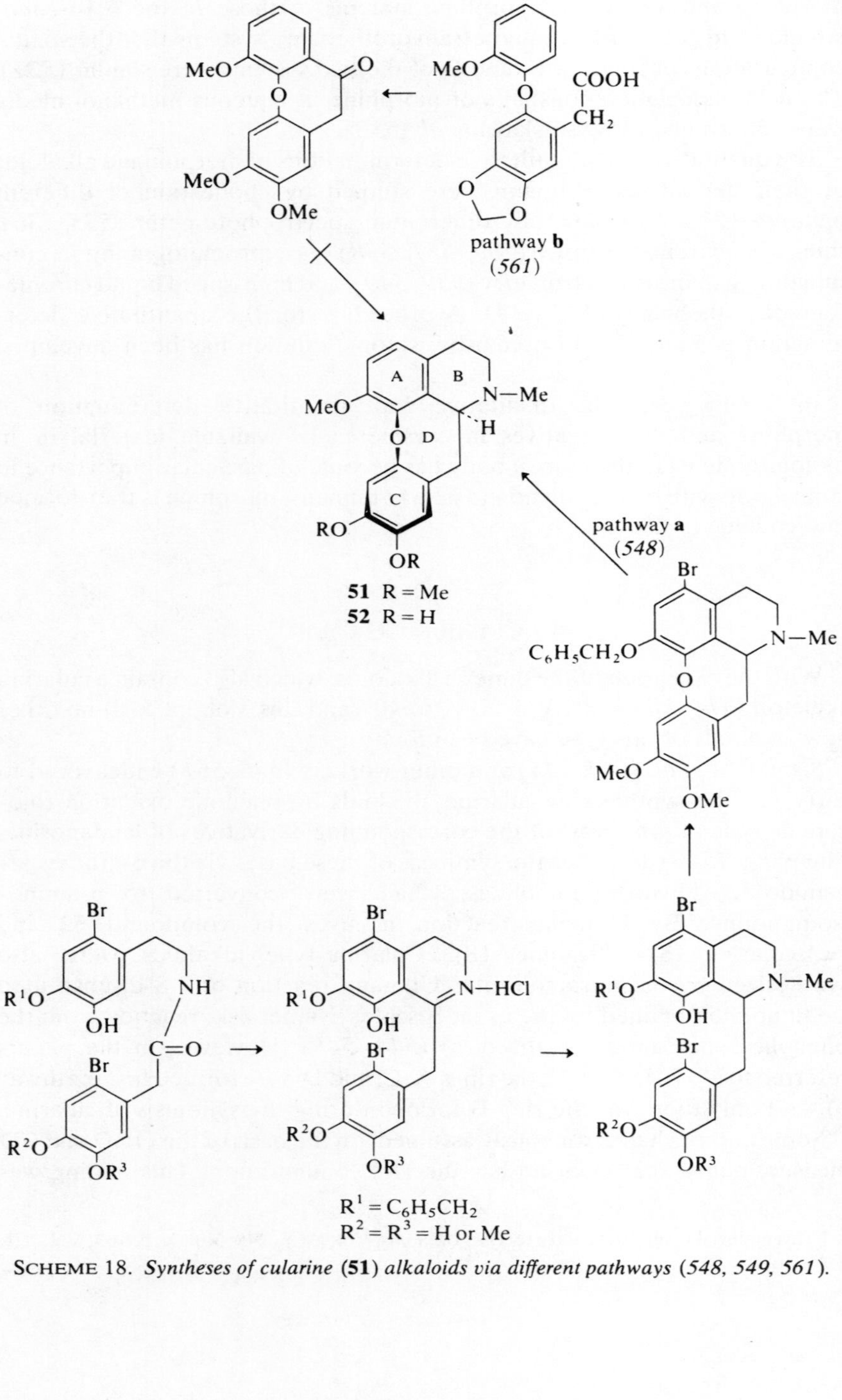

MeO
O
O
MeO
OMe
MeO
O
COOH
CH_2
O
O
pathway **b**
(*561*)
A
B
N—Me
MeO
H
O
D
C
RO
OR
51 R = Me
52 R = H
pathway **a**
(*548*)
Br
$C_6H_5CH_2O$
N—Me
O
MeO
OMe
Br
R^1O
NH
OH
C=O
Br
R^2O
OR^3
Br
R^1O
N—HCl
OH
Br
R^2O
OR^3
Br
R^1O
N—Me
OH
Br
R^2O
OR^3

$R^1 = C_6H_5CH_2$
$R^2 = R^3 = H$ or Me

SCHEME 18. *Syntheses of cularine* (**51**) *alkaloids via different pathways* (*548, 549, 561*).

revised by partial synthesis of (*S*)-1-(2-hydroxy-4,5-dimethoxybenzyl)-2-methyl-7-methoxy-1,2,3,4-tetrahydroisoquinoline (**53**) (which was also obtained by the hydrogenolysis of cularine) from (*S*)-romneine (**7i**) (*562*) (Scheme 19).

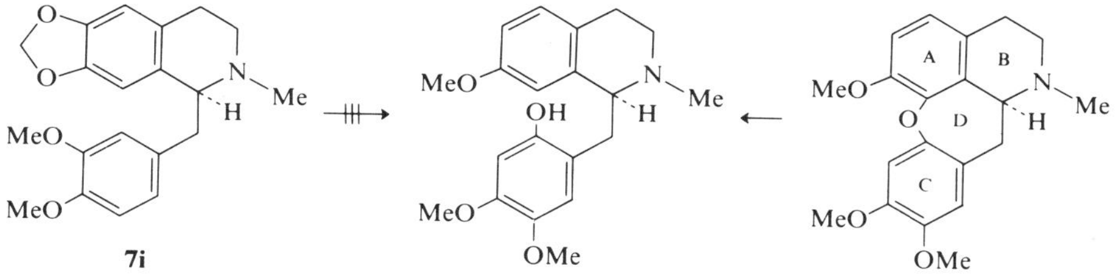

SCHEME 19. *The determination of the absolute configuration of cularine* (**51**) (*562*).

The ORD curve of the product of partial synthesis, the curve of the hydrogenolytic product **53**, and the ORD curve of (*S*)-romneine or (*S*)-laudanosine were identical. Cularine was therefore assigned the (*S*) configuration, which was confirmed by X-ray analysis (*563, 564*).

G. PAVINE GROUP*

The alkaloids argemonine (**54a**), isonorargemonine (**54b**), (−)-2,9-dimethoxy-3-hydroxypavinane (**54c**) (*30*), munitagine (**55a**), and platycerine (**55b**) were isolated from *A. munita* ssp. *rotundata* and *A. gracilenta* (*20*) (Scheme 20). The pavine alkaloids were also found in the plants in the form of their methohydroxides (*34*) and *N*-oxides (*20*). On the basis of the alkaloids found in different *Argemone* species, Stermitz and McMurtrey postulated the theory of the chemotaxonomy and the biogenesis as outlined in Scheme 20 (*20*). The biosynthesis of pavine alkaloids has also been discussed on p. 439.

The two dimeric pavine alkaloids (pennsylpavine and pennsylpavoline) were isolated from *Thalictrum polygamum* (*565*).

The pavine alkaloids bisnorargemonine (**54d**) (*566, 567*), norargemonine (**54e**) (*568*), eschscholtzidine (**54f**) (*569*), platycerine (**55b**) (*570*), munitagine (**55a**) (*571*), and the newly isolated 2,9-dimethoxy-3-hydroxypavinane (**54c**) (*221*) were prepared synthetically. Studies of the reaction conditions of the synthesis led to an increase in the yields of pavinane

* This material is supplementary to *The Alkaloids*, Vol. IV, p. 34; Vol. X, p. 477; Vol. XII, p. 370.

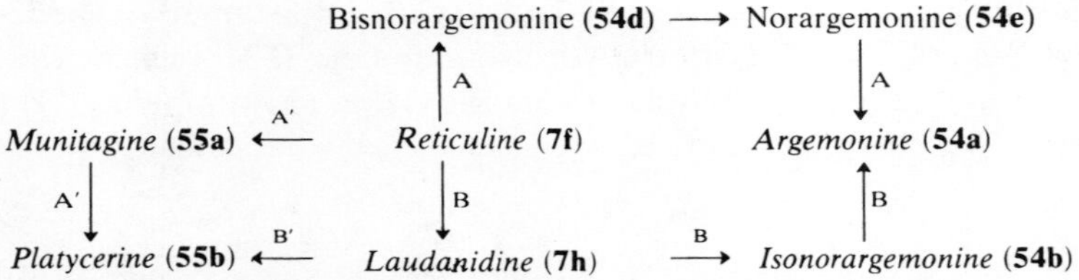

Bisnorargemonine (**54d**) ⟶ Norargemonine (**54e**)

Munitagine (**55a**) ⟵A′ *Reticuline* (**7f**) ⟶A Bisnorargemonine (**54d**); Norargemonine (**54e**) ⟶A *Argemonine* (**54a**)

Munitagine (**55a**) ⟶A′ *Platycerine* (**55b**); *Reticuline* (**7f**) ⟶B *Laudanidine* (**7h**); *Isonorargemonine* (**54b**) ⟶B *Argemonine* (**54a**)

Platycerine (**55b**) ⟵B′ *Laudanidine* (**7h**) ⟶B *Isonorargemonine* (**54b**)

SCHEME 20. *Possible biosynthetic pathways (20) of pavine alkaloids. Alkaloids found in A gracilenta are italicized. Path A, in A. munita and A. hispida; path A′, an alternative cyclization from (+)-reticuline to munitagine in A. gracilenta; path B, in A. gracilenta; path B′, an alternative cyclization from laudanidine to platycerine.*

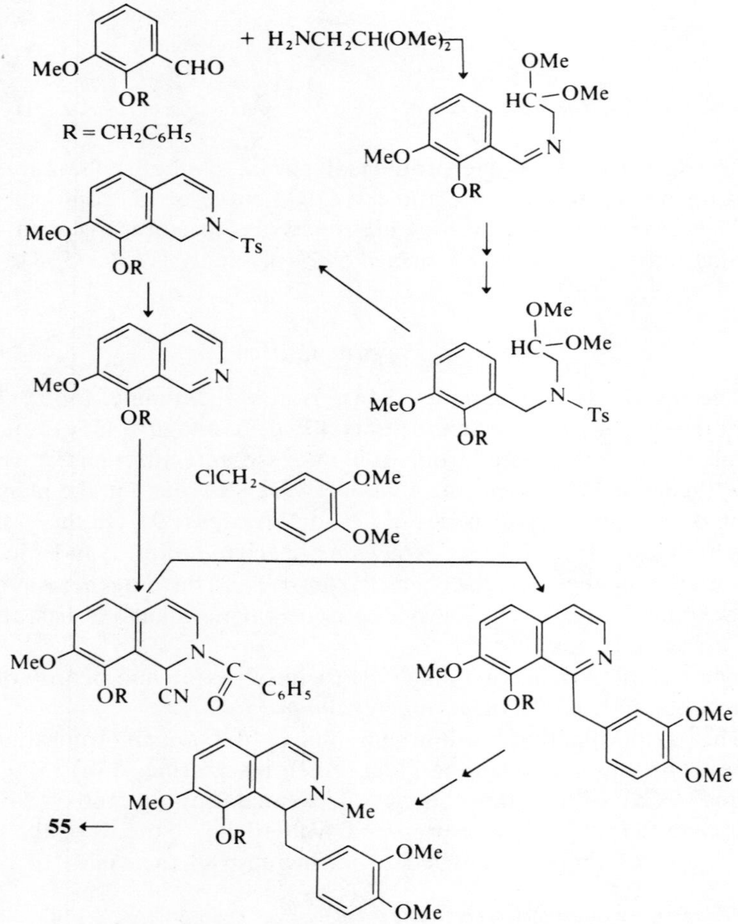

MeO, OR, CHO + $H_2NCH_2CH(OMe)_2$

$R = CH_2C_6H_5$

OMe, HC—OMe, N, MeO, OR

MeO, OR, N—Ts

OMe, HC—OMe, N—Ts, MeO, OR

MeO, OR, N

$ClCH_2$, OMe, OMe

MeO, OR, N, C, C_6H_5, CN, O

MeO, OR, N, OMe, OMe

55 ⟵ MeO, OR, N—Me, OMe, OMe

SCHEME 21. *Synthesis of pavine alkaloids* (**55**) (*570*).

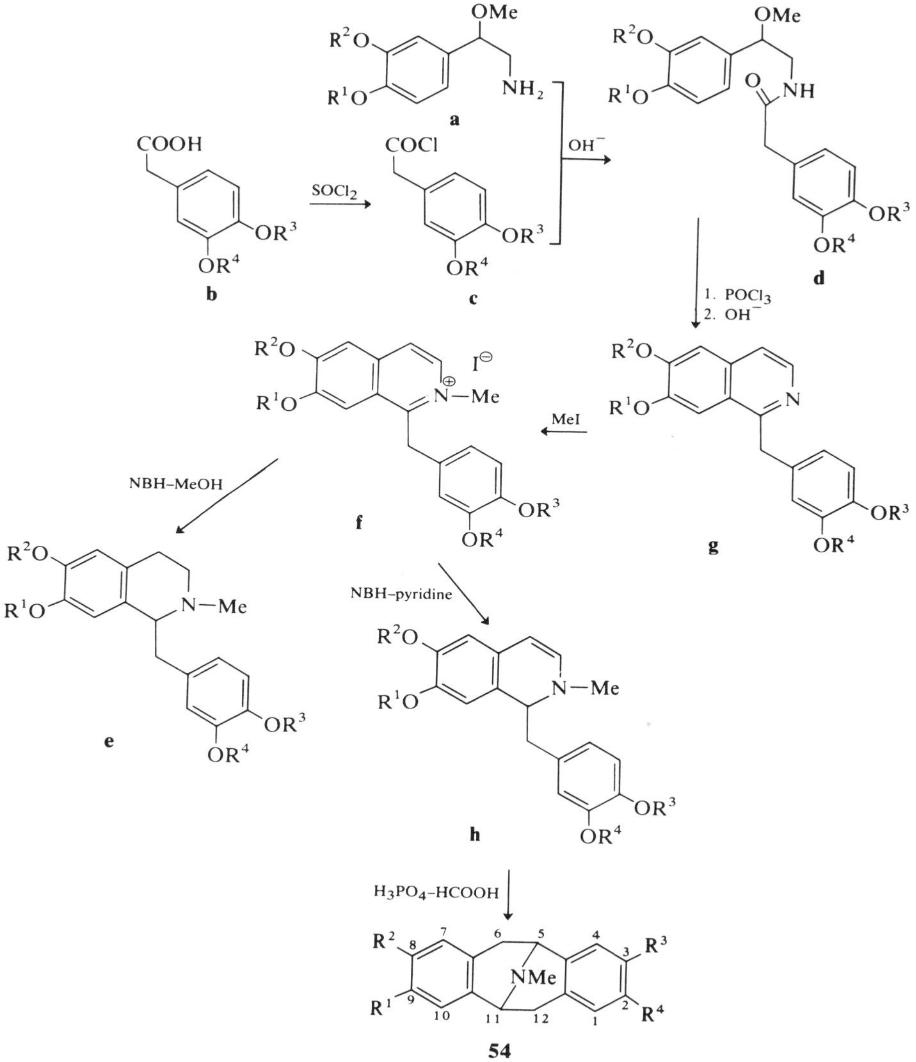

54a $R^1 = R^2 = R^3 = R^4 = OMe$ Argemonine
54b $R^1 = R^2 = R^3 = OMe$; $R^4 = OH$ Isonorargemonine
54c $R^1 = R^4 = OMe$; $R^2 = H$; $R^3 = OH$ 2,9-Dimethoxy-3-hydroxypavine
54d $R^1 = R^4 = OMe$; $R^2 = R^3 = OH$ Bisnorargemonine
54e $R^1 = OH$; $R^2 = R^3 = R^4 = OMe$ Norargemonine
54f $R^1 + R^2 = O{-}CH_2{-}O$; $R^3 = R^4 = OMe$ Eschscholtzidine

SCHEME 22. *A second synthesis of pavine alkaloids (29). For compounds* **a–h**: $R^1 = R^3 = CH_2PH$, $R^2 = R^4 = Me$ *or vice versa.*

alkaloids up to 50% or more: because the cyclization of the 1-benzyl-1,2-dihydroisoquinoline to pavine is unimolecular, the reaction pathway favors high dilution (*572*) (Scheme 21).

The synthesis as well as the NMR spectroscopy of bisnorargemonine (**54d**) and of its isomers were described (*29*) (Scheme 22).

The pavinane alkaloids are usually synthesized by acid-catalyzed rearrangement of *N*-methyl-1,2-dihydro-1-benzylisoquinolines oxygenated at suitable positions. This type of synthesis is applicable to symmetrically substituted *N*-methylpavinanes (e.g., the C-2, C-3, C-8, and C-9 electron-donating oxygen substituents such as argemonine). There was, however, developed an alternative synthesis of this type of alkaloids via a series of reactions which culminate in the Stevens rearrangement to give the tetracyclic *N*-methylpavinane with the correct oxygenation pattern (e.g., the C-3, C-4, C-8, and C-9) (*573*) (Scheme 23) [munitagine type (**55**)].

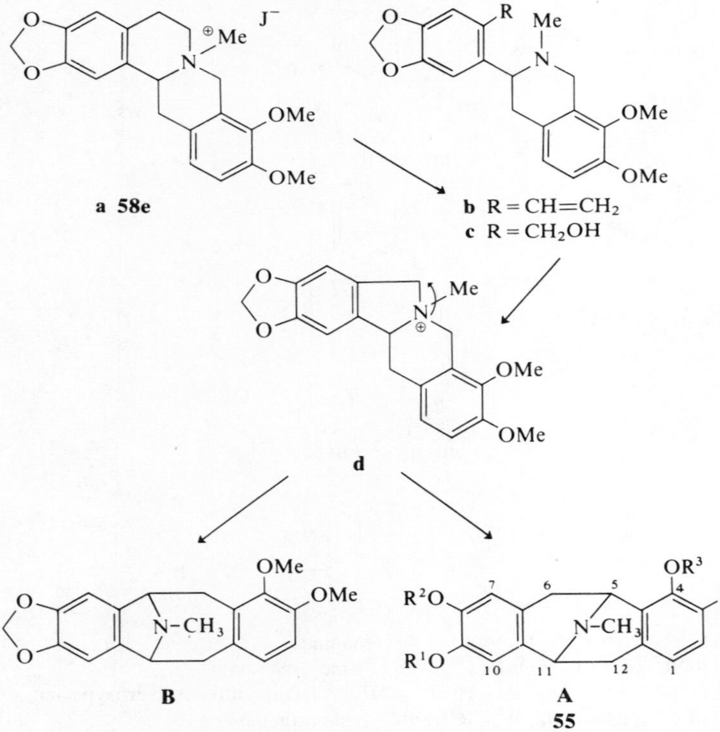

a 58e

b R = CH=CH$_2$
c R = CH$_2$OH

d

B

A
55

55a R^1 = R^3 = H; R^2 = Me Munitagine
55b R^1 = R^2 = Me; R^3 = H Platycerine

SCHEME 23. *Synthesis of pavine alkaloids* (**55**) *from tetrahydroprotoberberinium alkaloids* (**58**) (*573*).

Hofmann degradation of racemic tetrahydroberberine methiodide **a** gave the styrene derivative **b**. Oxidation of **b** with OsO_4–$NaIO_4$ followed by reduction with sodium borohydride afforded the benzyl alcohol derivative **c**. Treatment of **c** with $MeSO_2Cl$ in pyridine led to a cyclic quaternary methomesylate **d**, and stirring of the latter with an excess of phenyllithium in ether at room temperature overnight yielded two types of rearrangement products, **A** (**55**) and **B** (Scheme 23).

H. Isopavine Group*

Since the publication of the last review on this group of alkaloids, only thalisopavine (**56c**) (*417*) and *O*-methylthalisopavine (**56d**) (*248*) have been discovered.

The Soviet workers determined the constitution of roemrefine (=reframine methohydroxide) (**56a**) (*286, 574*), and Dyke and Ellis confirmed the position of the phenolic group on the aromatic ring B in amurensine (**56b**) by a synthesis (*575, 576*). The alkaloids thalisopavine (*417*) and *O*-methylthalisopavine (*248*), isolated from *Thalictrum dasycarpum* (Ranunculaceae), have the same absolute configuration as the other known isopavine alkaloids obtained from Papaveraceae plants.

The first synthesis of the racemic isopavine compounds with methoxyl groups on both aromatic nuclei was reported by Battersby and Yeowell (*577*) and later used for the synthesis of thalisopavine (*417*). A simple and ingenious method for the synthesis of isopavine derivatives with methylenedioxy groups was developed by Dyke (*337, 575, 578–582*). The synthesis requires the formation of 4-hydroxynorlaudanosoline (**9a**), which is probably also a natural precursor of this group of alkaloids. This method does not, however, yield the phenolic isopavines, one of which is reframoline (**56e**). Therefore, another method for their synthesis has been developed (*576*). Instead of the amine, the urethane is prepared from the starting benzyl derivatives; it is then cyclized and saponified into the corresponding alkaloid.

For this synthesis (Scheme 24), deoxypiperoin (**a**) is condensed with the aminoacetaldehyde dialkyl acetal to the imine **b** and then, without isolation, it is reduced with sodium borohydride to the compound **c**. The yield was 60% from the ketone. Treatment of **c** with concentrated hydrochloric acid for five days at room temperature affords an isopavine base hydrochloride in 70% yield. Nine isopavine bases have been prepared in this manner.

Kametani and Ogasawara (*362, 583*) synthesized isopavine derivatives by the conversion of dihydroisoquinoline compounds with diazomethane into

* This material is supplementary to *The Alkaloids*, Vol. X, p. 479; Vol. XII, p. 380.

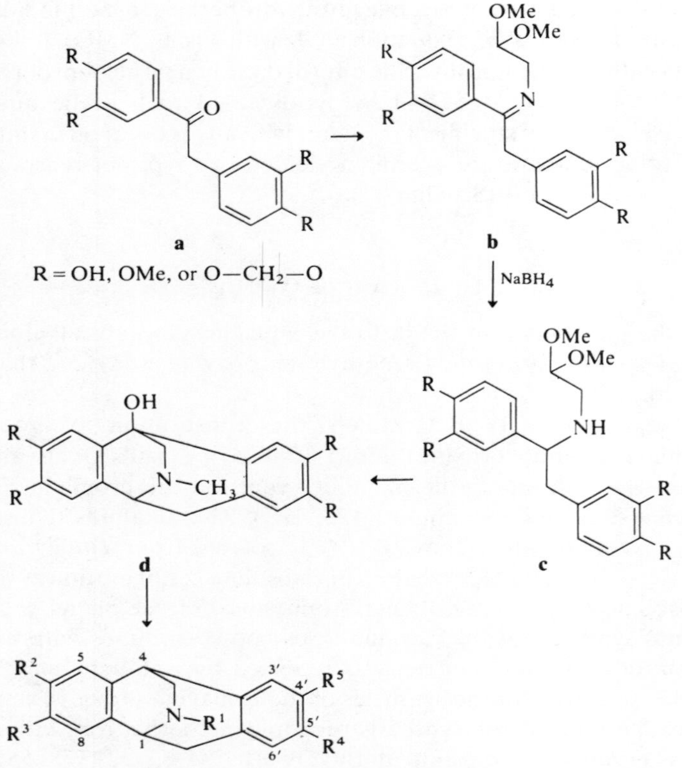

R = OH, OMe, or O—CH$_2$—O

56a R^1 = (Me)$_2$; R^2 + R^3 = O—CH$_2$—O; R^4 = R^5 = OMe Remrefine
56b R^1 = Me; R^2 + R^3 = O—CH$_2$—O; R^4 = OH; R^5 = OMe Amurensine
56c R^1 = Me; R^2 = R^4 = R^5 = OMe; R^3 = OH Thalisopavine
56d R^1 = Me; R^2 = R^3 = R^4 = R^5 = OMe *O*-Methylthalisopavine
56e R^1 = Me; R^2 = OH; R^3 = OMe; R^4 + R^5 = O—CH$_2$—O Reframoline
56f R^1 = Me; R^2 = R^3 = OMe; R^4 + R^5 = O—CH$_2$—O Reframine
56g R^1 = Me; R^2 + R^3 = R^4 + R^5 = O—CH$_2$—O Reframidine

SCHEME 24. *Synthesis of isopavine alkaloids* (*337, 575*).

the aziridinium salt **57a** which can then be converted via benzazepine **57b** in high yield into the isopavine bases using the conditions developed by Dyke (*575, 580*) (Scheme 25).

When the alkaloid carries a phenol group on the aromatic ring A or B, in the position C-1 or C-9, its position can be determined on the basis of the UV spectra of the product of the first step of the Hofmann degradation (*576*).

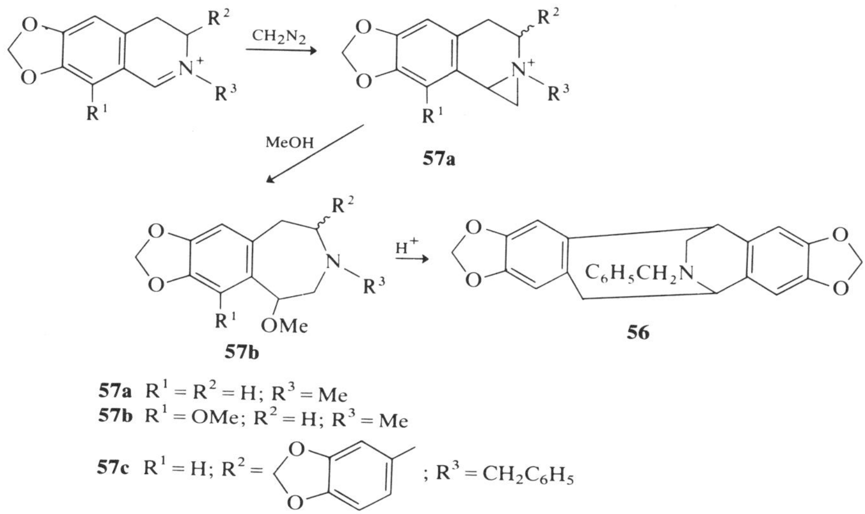

57a $R^1 = R^2 = H$; $R^3 = Me$
57b $R^1 = OMe$; $R^2 = H$; $R^3 = Me$
57c $R^1 = H$; $R^2 =$; $R^3 = CH_2C_6H_5$

SCHEME 25. *Synthesis of isopavine alkaloids* (**56**) *via 3,4-dihydroisoquinoline and aziridinium salts* (**57a**) (*362*).

The absolute configuration of amurensine (**56b**) was determined by Shamma and Nakanishi, who applied the so-called "aromatic chirality method of CD analysis" (*584*).

The fact that the pavine and isopavine alkaloids have the same absolute configuration points to the possibility of a common biogenetic precursor such as the 4-hydroxyreticuline (**9b**) [or the already mentioned 4-hydroxynorlaudanosoline (**9a**)] which, depending upon the plant family or genus, can cyclize directly to an isopavine species or alternatively can undergo dehydration, double-bond isomerization, and intramolecular cyclization to a pavine analog (*584*) (Scheme 26).

Rates of methiodide formation provide a facile means for differentiation between pavine and isopavine bases, because the latter quaternize at a faster rate (*585*).

I. BERBINE (DIBENZO[*a,g*]QUINOLIZIDINE) (PROTOBERBERINE, PSEUDOPROTOBERBERINE, CORYDALINE, AND CORYTENCHIRINE TYPES) GROUP*

Recent chemical progress concerning berberine alkaloids (biosynthesis, bioformation with mammalian tissues, syntheses, reactions including

* This material is supplementary to *The Alkaloids*, Vol. IV, p. 77; Vol. IX, p. 41; Vol. XII, p. 383.

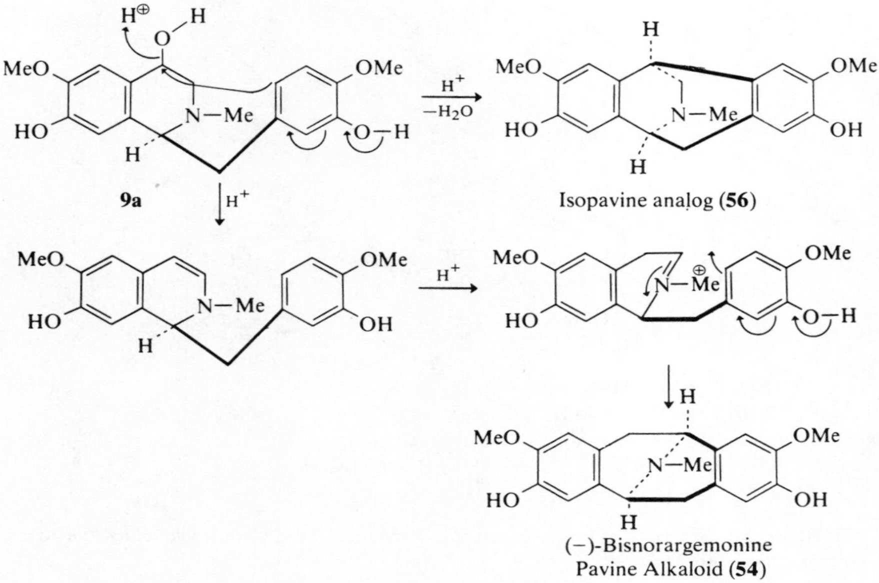

9a

Isopavine analog (**56**)

(−)-Bisnorargemonine
Pavine Alkaloid (**54**)

SCHEME 26. *The possible common pathway of the biosynthesis (and the absolute configuration) of pavine and isopavine alkaloids (3, 584).*

racemization), the new alkaloids, and their stereochemistry have been reviewed (*247, 586*).

A short time back a series of tetrahydroprotoberberine and protoberberine alkaloids (*vide infra*) and their methohydroxides were isolated. Those isolated from plants of the family Papaveraceae can be subdivided into tetrahydroprotoberberine alkaloids of the type **58**, their dehydro derivatives **59**, 13-methyltetrahydroprotoberberine (**68**) (dehydro derivatives **69**), and 13-hydroxytetrahydroprotoberberine (**70**). Alkaloids of the caseanadine type (**62**), the tetrahydropseudoprotoberberine alkaloids (**60**, **64**) (their dehydro derivatives **61**, **63**, and **65** have not been found in nature as yet), the corytenchirine (**83**) type, and **66** with the dehydro derivatives **67** were also isolated. Included in the last group is the alkaloid aryapavine (**66b**) which was recently isolated from *P. pseudoorientale* (*243, 587*). The alkaloids of the types **58** and **59** are subdivided still further according to whether they possess two or three electron-donating oxygen substituents on the ring A.

The tetrahydroprotoberberine alkaloids with three oxygen substituents located in the ring D have been found in *Stephania rotunda* (Menispermaceae) (*588*). These types of alkaloids will probably be helpful in future chemotaxonomy.

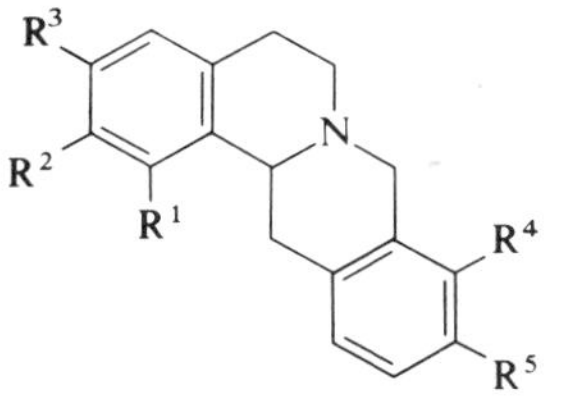

58 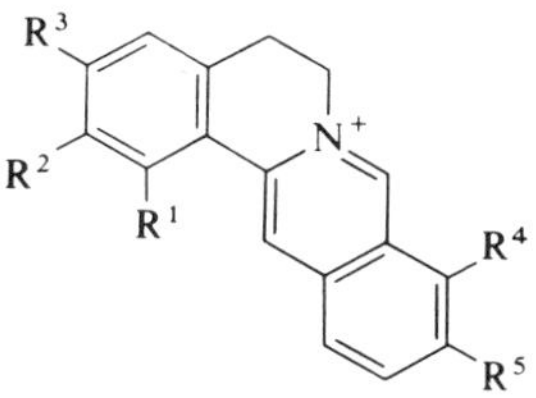 **59**

58a $R^1 = H; R^2 = R^5 = OH; R^3 = R^4 = OMe$ Stepholidine
58b $R^1 = H; R^2 = R^4 = OH; R^3 = R^5 = OMe$ Scoulerine
58c $R^1 = H; R^2 = OH; R^3 = OMe; R^4 + R^5 = O{-}CH_2{-}O$ Cheilanthifoline
58d $R^1 = H; R^2 = R^3 = R^4 = OMe; R^5 = OH$ Kikemanine (= Corydalmine) (Schefferine)
58e $R^1 = H; R^2 + R^3 = O{-}CH_2{-}O; R^4 = R^5 = OMe$ Canadine
58f $R^1 = H; R^2 + R^3 = O{-}CH_2{-}O; R^4 = OMe; R^5 = OH$ Tetrahydrothalifendine
58g $R^1 = H; R^2 = R^3 = R^4 = R^5 = OMe$ Tetrahydropalmatine
58h $R^1 = H; R^2 = R^3 = OMe; R^4 + R^5 = O{-}CH_2{-}O$ Sinactine
58i $R^1 = H; R^2 + R^3 = O{-}CH_2{-}O; R^4 = OH; R^5 = OMe$ Nandinine
58k $R^1 = H; R^2 = R^4 = R^5 = OMe; R^3 = OH$ Corypalmine (Discretinine)
58l $R^1 = H; R^2 = OH; R^3 = R^4 = R^5 = OMe$ Isocorypalmine
58m $R^1 = H; R^2 = R^4 = OMe; R^3 = R^5 = OH$ Discretamine (Aequaline)
58n $R^1 = H; R^2 + R^3 = R^4 + R^5 = O{-}CH_2{-}O$ Stylopine (Tetrahydrocoptisine)
58o $R^1 = R^5 = OH; R^2 = R^3 = R^4 = OMe$ Capaurimine
58p $R^1 = OH; R^2 = R^3 = R^4 = R^5 = OMe$ Capaurine

59a $R^1 = H; R^2 + R^3 = O{-}CH_2{-}O; R^4 = R^5 = OMe$ Berberine
59b $R^1 = H; R^2 = R^3 = OMe; R^4 + R^5 = O{-}CH_2{-}O$ Epiberberine
59c $R^1 = H; R^2 = OH; R^3 = OMe; R^4 + R^5 = O{-}CH_2{-}O$ Dehydrocheilanthifoline
59d $R^1 = H; R^2 = OMe; R^3 = OH; R^4 + R^5 = O{-}CH_2{-}O$ Groenlandicine
59e $R^1 = H; R^2 + R^3 = R^4 + R^5 = O{-}CH_2{-}O$ Coptisine
59f $R^1 = R^5 = OH; R^2 = R^3 = R^4 = OMe$ Dehydrocapaurimine
59g $R^1 = H; R^2 = R^4 = R^5 = OH; R^3 = OMe$

The alkaloid aegualine was found to be identical with discretamine (**58m**), the alkaloid coramine with coreximine (**60a**), and the alkaloid schefferine with (−)-kikemanine (**58d**) (*588a, b, 598, 615*).

The new alkaloid caseanadine (**62a**) was isolated from *C. caseana* (*70, 170*). The alkaloids designated earlier as F-33 (casealutine) and F-35 were named caseamine (**64c**) and caseadine (**64a**) (*589*). They were found to be tetrahydroprotoberberines with a novel substitution pattern (*590*). It has been mentioned in Section II, p. 387 that in *P. somniferum* the presence of the tetrahydropseudoprotoberberine alkaloid coreximine (**60a**) based on its biosynthesis from reticuline) could be demonstrated in the plant by application of an isotope-dilution method (*249*). The alkaloids (*S*)-govanine (**60f**), (*S*)-govadine (**60h**), and corygovanine (**60b**) were isolated from *C. govaniana* and also synthetically prepared (*87*). The alkaloid stepholidine (**58a**) was isolated from opium of Indian origin by chromatography on silica

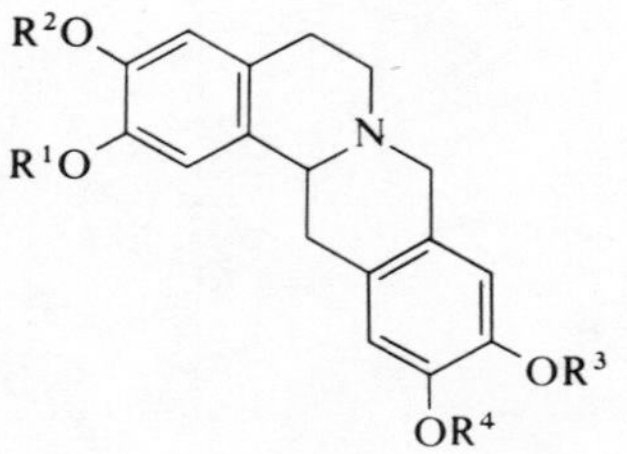

60 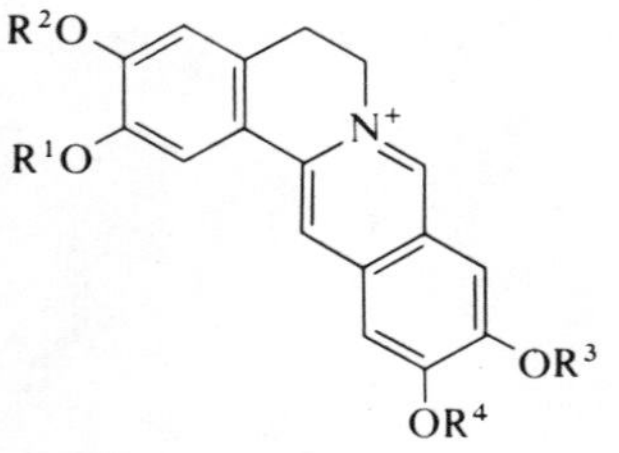 **61**

60a $R^1 = R^4 = H$; $R^2 = R^3 = Me$ Coreximine (Coramine)
60b $R^1 = R^2 = Me$; $R^3 + R^4 = CH_2$ Corygovanine
60c $R^1 = R^2 = R^3 = R^4 = Me$ Xylopinine
60d $R^1 = R^2 = R^3 = Me$; $R^4 = H$ Corytenchine
60e $R^1 = R^2 = R^4 = Me$; $R^3 = H$
60f $R^1 = H$; $R^2 = R^3 = R^4 = Me$ Govanine
60g $R^1 = Me$; $R^2 = H$; $R^3 + R^4 = CH_2$
60h $R^1 = R^3 = H$; $R^2 = R^4 = Me$ Govadine

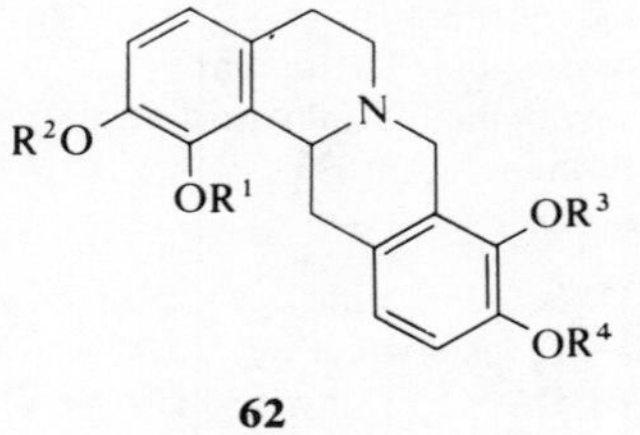

62 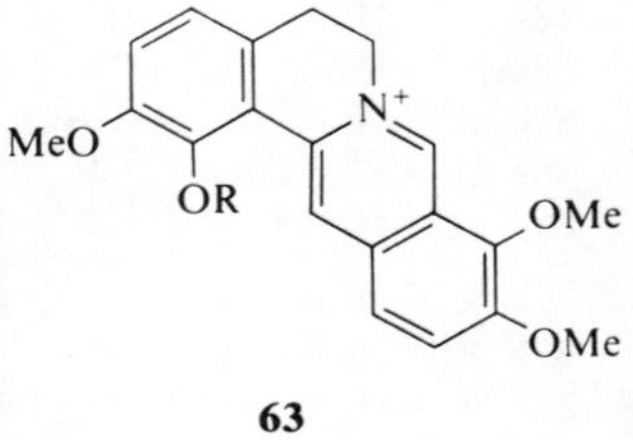 **63**

62a $R^1 = H$; $R^2 = R^3 = R^4 = Me$ Caseanadine
62b $R^1 = R^2 = R^3 = R^4 = Me$ *O*-Methylcaseanadine

gel (*265*). (−)-Canadine (**58e**) was isolated from Kirgiz opium (*257*). The alkaloids HF-1 from *H. fumariaefolia* and (−)-scoulerine (**58b**) were shown to be identical (*15*). The alkaloid coramine (**60a**) was isolated from *C. pseudoadunca* (*82*) and the new alkaloid (−)-kikemanine (**58d**) [= (−)-corydalmine] from *C. pallida* (*127*); it might be identical with the alkaloid F-51 [upon reaction with diazomethane, it generates (−)-tetrahydropalmatine (**58g**)] (*125, 129*). (*S*)-Corytenchine (**60d**) and (*S*)-corytenchirine (**83b**) were isolated from *C. ochotensis* (collected in Taiwan's Central Mountains) (*116, 117*). The presence of corysamine (**69a**) was demonstrated in Korean *Corydalis* (*107*) and in *C. cava* (*77*). (+)-Tetrahydrocorysamine (**68a**) is present in *C. pallida* var. *tenuis* (*131*) and (−)-tetrahydrocorysamine (**68a**) in *C. incisa* Pers. (*95*). The plant *C. cava* (= *C. tuberosa*) was shown to contain, in addition to the alkaloid thalictricavine (**68b**), a new, related alkaloid, epiapocavidine (**68f**), which also belongs to the group of 13-methyltetrahydroprotoberberines (*75*). The plant *C. platy*-

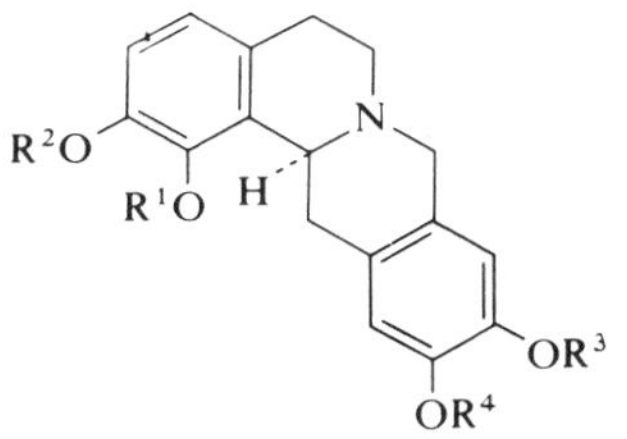

64

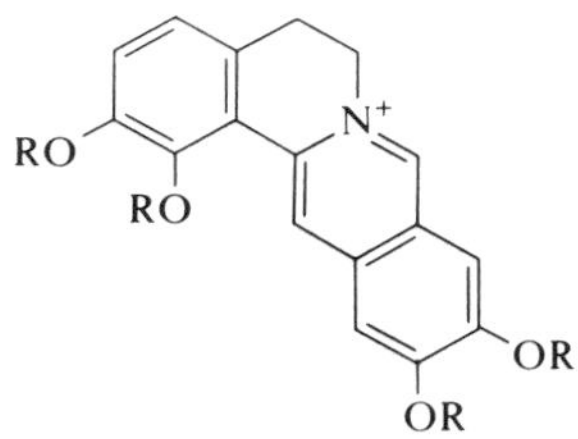

65

64a $R^1 = H$; $R^2 = R^3 = R^4 = Me$ Caseadine
64b $R^1 = R^2 = R^3 = R^4 = Me$ *O*-Methylcaseadine
64c $R^1 = R^3 = Me$; $R^2 = R^4 = H$ or vice versa Caseamine

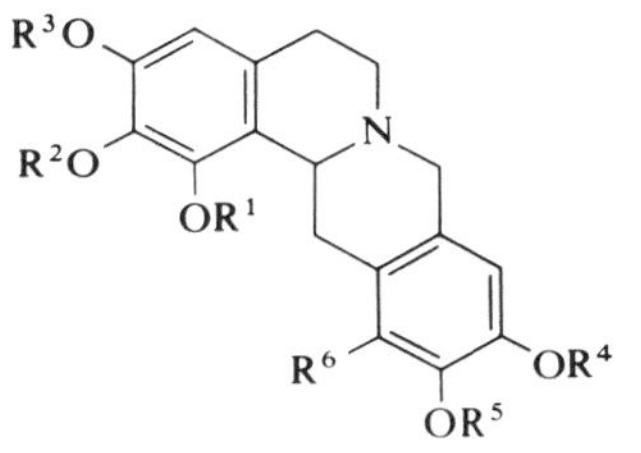

66

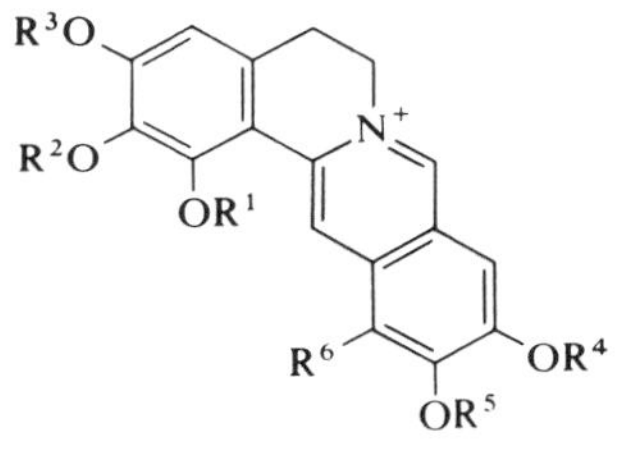

67

66a $R^1 = R^4 = R^5 = Me$; $R^2 + R^3 = CH_2$; $R^6 = CH_2OH$ Mecambridine
66b $R^1 = R^4 = Me$; $R^2 + R^3 = CH_2$; $R^5 = H$; $R^6 = CH_2OH$ Aryapavine
66c $R^1 = R^4 = Me$; $R^2 + R^3 = CH_2$; $R^5 + R^6 = CH_2{-}O{-}CH_2$ Orientalidine

67a $R^1 = R^4 = R^5 = Me$; $R^2 + R^3 = CH_2$; $R^6 = CH_2OH$ Alkaloid PO-5
67b $R^1 = R^4 = Me$; $R^2 + R^3 = CH_2$; $R^5 + R^6 = CH_2{-}O{-}CH_2$ Alkaloid PO-4

carpa was found to contain dehydrocapaurimine (**59f**) (*136*). The isolated and named "base A" from *B. cordata* was established as dehydrocheilanthifoline (**59c**) (*46*). Furthermore, *C. cava* gave an undetermined dihydroxydimethoxyprotoberberine, and *C. ophiocarpa* yielded 13β-hydroxystylopine (**70b**), whose structure was confirmed by synthesis (*122*) (Scheme 33). Dehydrocheilanthifoline (**59c**) was isolated for the first time from *Fumaria* species (*163*).

From *C. koidzumiana*, the alkaloids corynoxidine (*trans*-**72a**) and epicorynoxidine (*cis*-**72b**) have been isolated (*106*). Both alkaloids are *N*-oxides of (−)-tetrahydropalmatine and differ in the fusion of the B/C ring junction, in melting point, and in optical rotation. The plant *C. nokoensis* gave the alkaloid nokoensine, which is an *N*-oxide of the alkaloid capaurine (**58p**) (*114*).

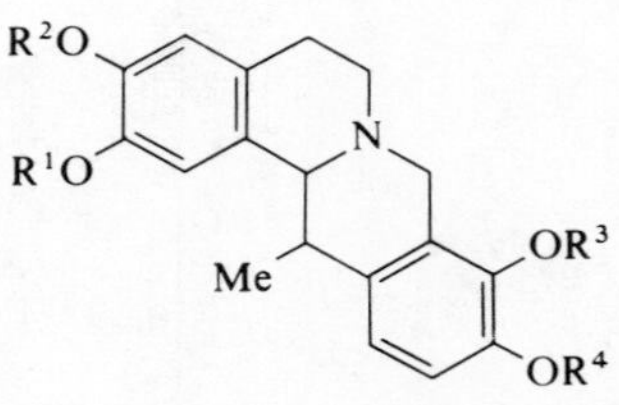

R^2O, R^1O, Me, OR^3, OR^4, N

68

R^2O, R^1O, Me, OR^3, OR^4, N^+

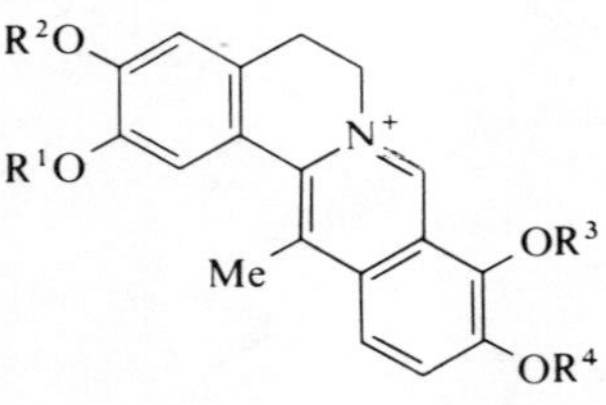

69

68a $R^1+R^2=R^3+R^4=CH_2$ Tetrahydrocorysamine
68b $R^1+R^2=CH_2$; $R^3=R^4=Me$ Thalictricavine and Mesothalictricavine
68c $R^1=R^3=R^4=Me$; $R^2=H$ (+)-Corybulbine [= (+)-Corydalmine]
68d $R^1=R^2=Me$; $R^3+R^4=CH_2$ Cavidine (= Thalictrifoline or Mesocavidine)
68e $R^1=H$; $R^2=Me$; $R^3+R^4=CH_2$ Apocavidine
68f $R^1+R^2=CH_2$; $R^3=Me$; $R^4=H$ Epiapocavidine
68g $R^1=R^3=Me$; $R^2=R^4=H$ Corydalidzine
68h $R^1=R^2=R^3=R^4=Me$ Corydaline and Mesocorydaline

69a $R^1+R^2=R^3+R^4=CH_2$ Corysamine (Worenine)
69b $R^1=R^2=R^3=R^4=Me$ Dehydrocorydaline
69c $R^1=R^3=R^4=Me$; $R^2=H$ Dehydrocorybulbine

The *meso* compounds have their hydrogens at C-13 and C-13a in the *trans* position.

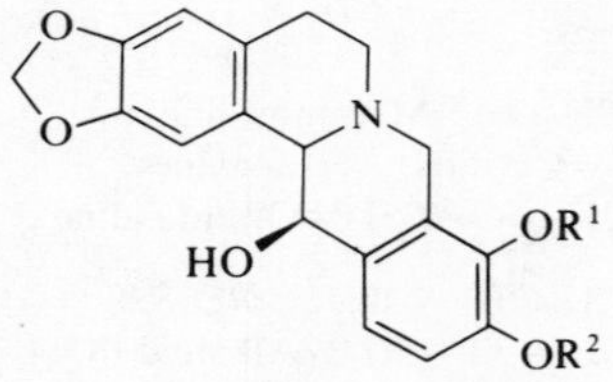

O, O, N, HO, OR^1, OR^2

70

70a Ophiocarpine ($R^1=R^2=Me$)
70b β-Hydroxystylopine ($R^1+R^2=CH_2$)

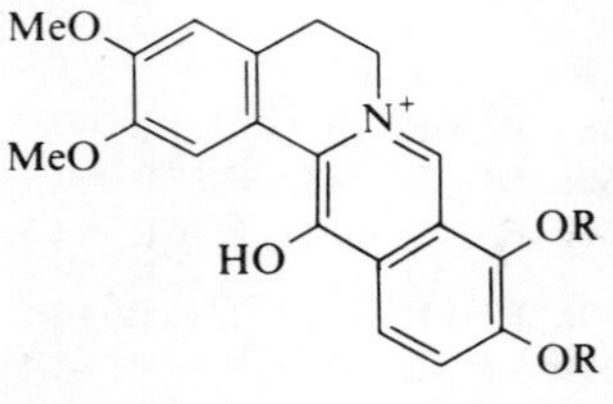

MeO, MeO, N^+, HO, OR, OR

71 13-Hydroxyprotoberberine
(R = H or Me or CH_2)

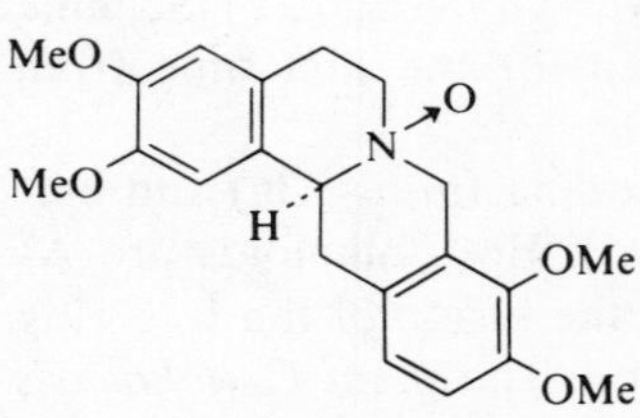

MeO, MeO, N, O, H, OMe, OMe

72a
trans-Corynoxidine

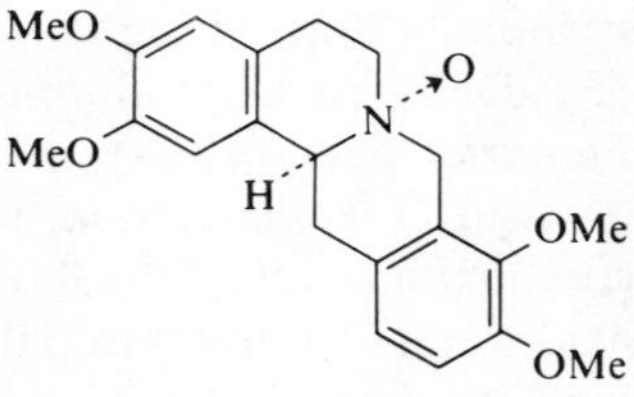

MeO, MeO, N, O, H, OMe, OMe

72b
cis-Corynoxidine
(= Epicorynoxidine)

73

Kametani *et al.* (*591–596*) and Kaneko and Naruto (*131*) reported the structure of capaurine (**58p**) and capaurimine (**58o**). An X-ray analysis showed that this alkaloid of an (*S*) configuration is a *cis*-quinolizidine. The structure of corydalidzine (**68g**), isolated from *C. koidzumiana*, was established by spectroscopic methods and by synthesis of the racemic base (*147*). Discretamine was identified as **58m** (*597*). The alkaloid (+)-corydalmine proved to be identical with the alkaloid (+)-corybulbine (**68c**) (*64*). This finding must, however, await further confirmation by direct comparison with authentic specimens.

The corms of *Stephania glabra* contain, among other alkaloids, the two alkaloids (−)-corydalmine (**58d**) and (−)-stepholidine (**58a**) (*375*). The alkaloid tetrahydrothalifendine (**58f**) was isolated from *Thalictrum fendleri* (*598a*). A revision of the structure of the alkaloids isolated from *Coptis groendlandicum* showed that the alkaloid **A** is coptisine (**59e**); the alkaloid **B** has been identified as groendlandicine (**59d**) (*41, 599*). The UV spectrum of groendlandicine exhibits a significant shift of the longest wavelength band in alkaline medium, which supports the fact that its hydroxyl group is located at C-3.

During the past years, the isolation of a great number of quaternary bases of the already known tetrahydroprotoberberine alkaloids has been reported (Table X) (*32, 158, 158a, 183, 246*). The escholidine perchlorate thus isolated was shown to be 2,3-methylenedioxy-9-methoxy-10-hydroxy-tetrahydroprotoberberinium methoperchlorate (*155*). As shown by ^{13}C NMR spectroscopy, quaternary tetrahydroprotoberberine alkaloids may have the α- and β-form of the conformation.

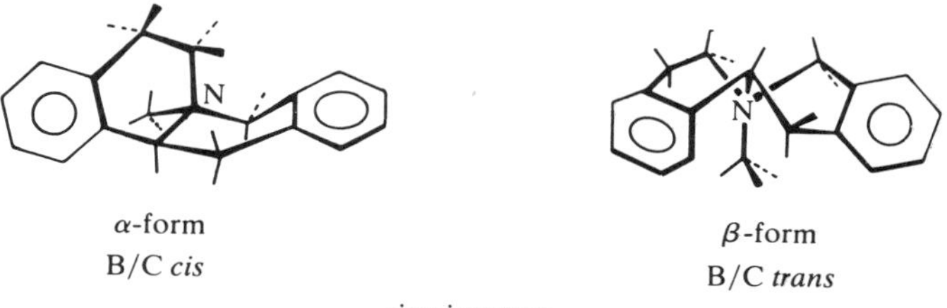

α-form
B/C *cis*

β-form
B/C *trans*

ring juncture

The naturally occurring quaternary tetrahydroprotoberberine types of alkaloids are predominantly in the α-form, except for the β-*l*-canadine methochloride. The stereochemistry of stylopine methohydroxide remains undefined (*600*).

The position of the substituents of the ring D of mecambridine (**66a**), orientalidine (**66c**), and of the alkaloids PO-5 (**67a**) and PO-4 (**67b**) were determined and biogenetically justified (*589, 601*). The structures were also confirmed by total synthesis of racemic mecambridine, orientalidine, and aryapavine (**66b**) (*587, 601–604*).

The tautomeric forms of protoberberine and pseudoprotoberberine compounds were examined by UV, IR, and ^{1}H NMR spectroscopy, and by polarography (*381*). The results showed that, in aqueous alkaline medium, the carbinol form **C** is formed only by the protoberberine compounds and not by the pseudoprotoberberine; the presence of the aldehyde form **B** could not be proven in this group of compounds.

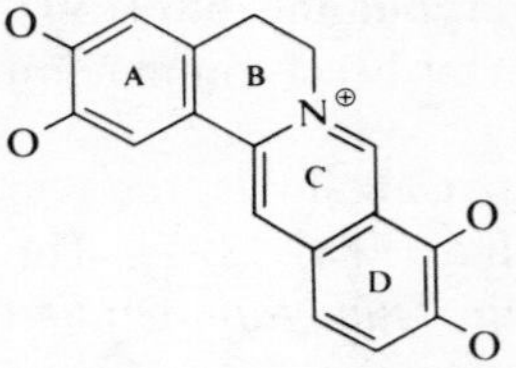

O A B N⊕ O C D O O   O NH CHO O O O   O N H OH O O O

A 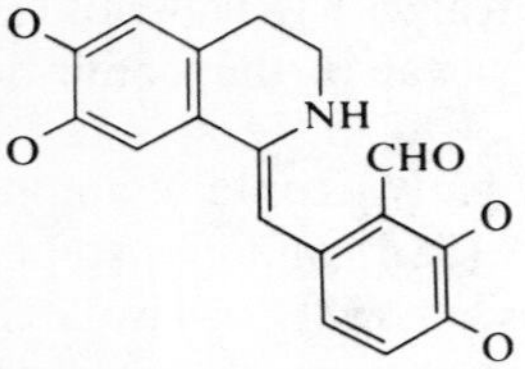 **B** 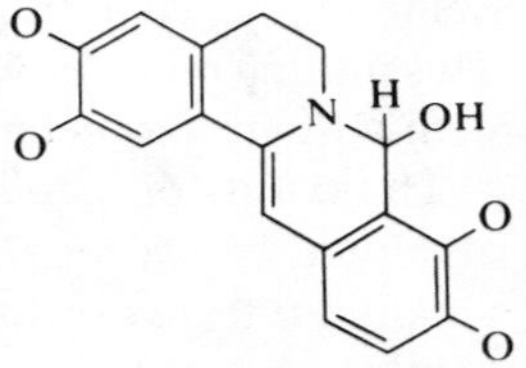 **C**

It was concluded that the conversion of the immonium form **A** into the pseudobase **C** was affected by the polarity of the solvent, the structural type of the compound, and the position and number of electron-donating oxygen substituents of the isoquinoline skeleton. This problem has simultaneously been studied in 3,4-dihydroisoquinoline and cotarnine and in benzophenanthridine (**177**) compounds (Sections III,A and III,N). The equilibrium constant of the pseudobase formation is also dependent on the type of the substituent and its position in the rings A, C, and D (*331*). Habermehl *et al.* studied the disproportionation of pseudobases and reported that in mass spectrometry the pseudobase of berberine undergoes thermal disproportionation into amide and amine which gives rise to an M-14 peak (*605, 606*).

The synthesis of racemic cheilanthifoline (**58c**) was completed (*47*). During the past decade, new syntheses of different pseudoprotoberberine or tetrahydropseudoprotoberberine bases were developed (*607*). For example, (+)-xylopinine (**60c**) was prepared in good yield by addition of substituted cyanobenzocyclobutene (**74a**) (starting material) to 3,4-dimethoxy-*N*-methylphenethylamine or via 3,4-dihydroisoquinoline, and by thermolysis (*608–613*). The by-product was spirobenzylisoquinoline (**198c**) (Scheme 27).

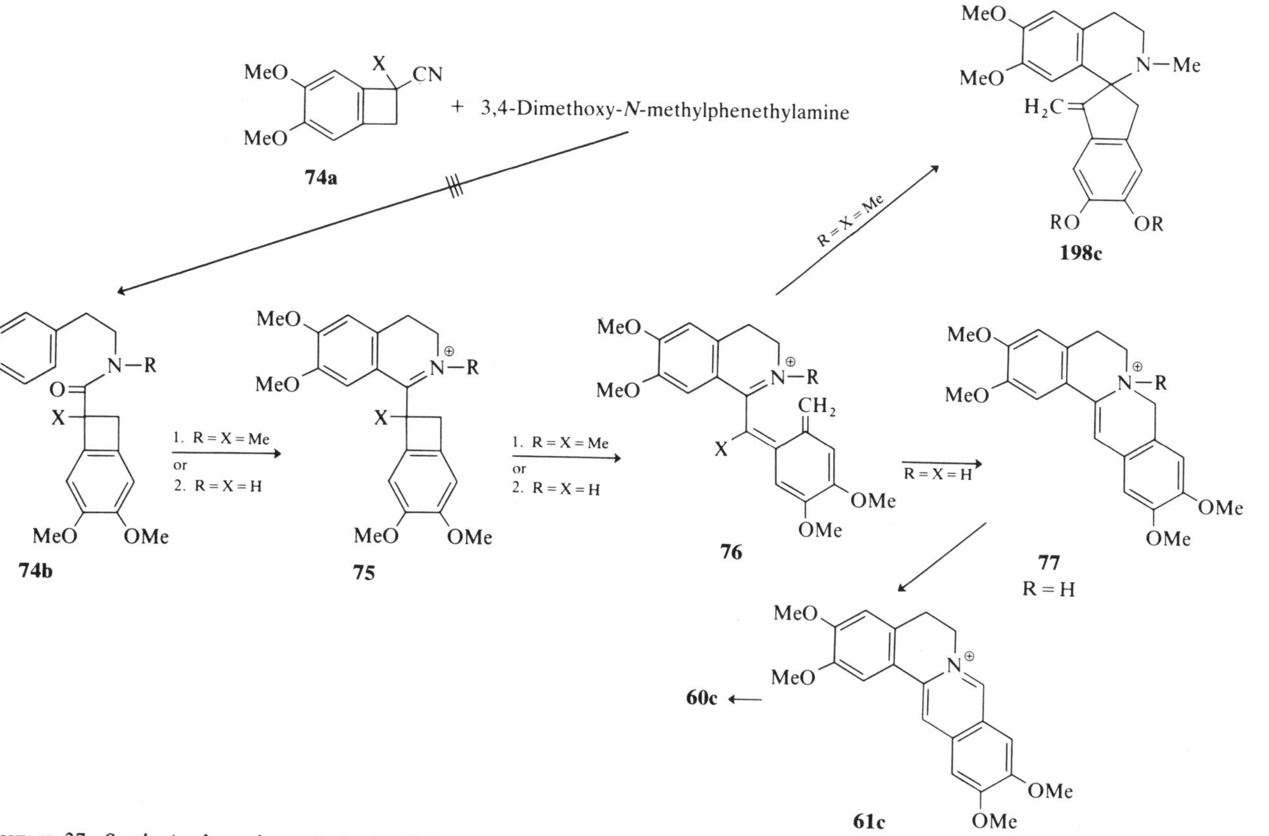

SCHEME 27. *Synthesis of pseudoprotoberberine* (**61**) *and tetrahydropseudoprotoberberine* (**60**) *bases from cyclobutene or cyclobutenyl derivatives* (**74a**, **74b**, *and* **75**) *by thermolysis* (*609*).

Furthermore, the following compounds were synthetically prepared: racemic cheilanthifoline (**58c**) (*47*), kikemanine (**58d**) (*129*), canadine (**58e**), berberine (**59a**) (*590, 614*), tetrahydropalmatine (**58g**) (*475*), sinactine (**58h**), cavidine (**68d**) (*616, 617*), nandinine (**58i**) (*590, 614, 615*), capaurine (**58p**) (*618*), capaurimine (**58o**) (*128, 618a*), xylopinine (**60c**) (*610, 615, 619*), *O*-methylcaseanadine (**62b**) (*70, 620*), thalictricavine (**68b**), and corydaline (**68h**) (*615*). Xylopinine (**60c**) and some other alkaloids were synthesized by benzoylation of 1-alkyl-3,4-dihydroisoquinolines followed by photocyclization. This method provides a useful route to the synthesis of other protoberberine alkaloids (*619*). It is also applicable to the synthesis of cularine (**51**) and spirobenzyltetrahydroisoquinoline alkaloids (**188**). Xylopinine was also synthesized from the corresponding enamide under benzyne reaction conditions (*615*). Kametani *et al.* summarized their findings on the synthesis of these alkaloids and described the formation of protoberberines by debenzylation and photolysis of tetrahydroisoquinolines (*622, 623*). The total stereospecific synthesis of racemic ophiocarpine (**70a**) from the 3,4-dihydroisoquinoline derivative by Mannich cyclization was also described (*624*).

Meise and Zymalkovski reported the formation of the berberine ring system by means of a reaction between homoveratrylamine (**78**) and 3-isochromanone (**79**). In two steps, the yield reached 80% (*625, 626*) (Scheme 28).

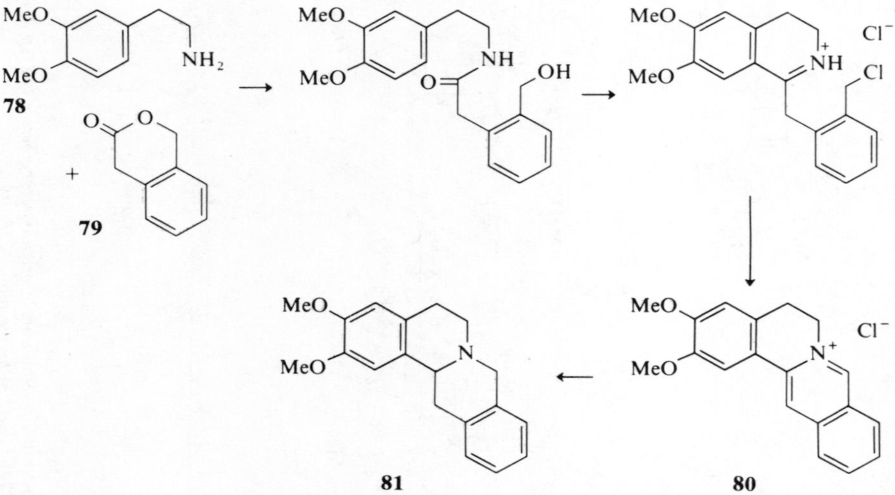

SCHEME 28. *Synthesis of berbine bases* (**80** *and* **81**) *according to Meise and Zymalkowski* (*625, 626*).

In a similar manner the synthesis of tetrahydroprotoberberine derivatives was realized, and biologically active compounds were obtained (*627, 628*).

Brown *et al.* reported the synthesis of many berbine derivatives, isopavine alkaloids, and amurensinine (**56f**) (*581, 629*). They describe some rearrangements of those alkaloids. 3-Hydroxy-2-methoxy-9,10-methylenedioxyberbine (**59d**) and 3-hydroxy-2-methoxy-10,11-methylenedioxyberbine (**60h**) were synthesized through the *N*-formyl derivatives of 6-benzyloxy-1-(6-bromo-3,4-methylenedioxybenzyl)-7-methoxy-1,2,3,4-tetrahydroisoquinoline. One of the compounds thus synthesized was found to be identical with tetrahydro derivatives of groendlandicine (**59d**) (*630*). The syntheses of 13-hydroxytetrahydroprotoberberine (**70**) and 13-hydroxyprotoberberine (**71**) have been described (*122, 631*) (Scheme 29a). The synthesis of a series of dehydrocorydaline derivatives and their UV spectra in ethanolic and alkaline ethanolic media have also been published (*68*).

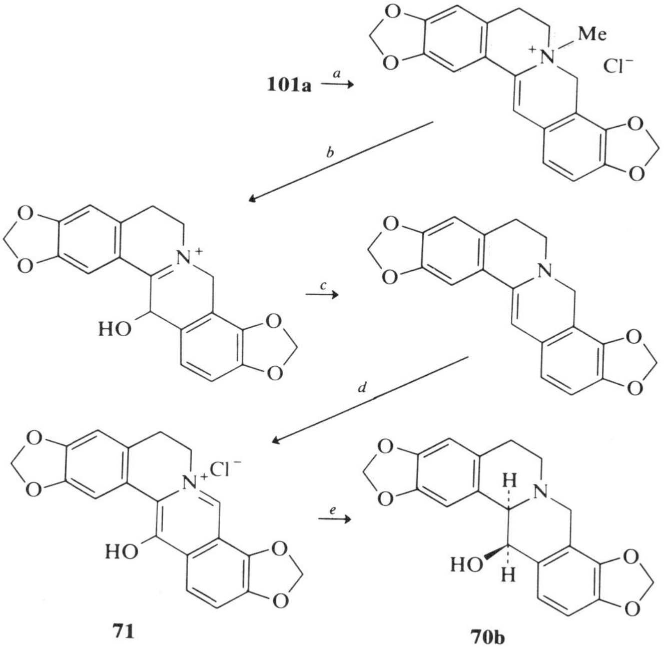

SCHEME 29a. *Synthesis of* (±)-13β-*hydroxystylopine* (**70b**) *from protopine* (**101a**) (*122*). *Path a*, $POCl_3$; *b*, Δ *in vacuo*; *c*, *m-chloroperbenzoic acid*; *d*, O_2; *e*, $NaBH_4$.

Hanaoka *et al.* (*632*) carried out a novel and simple conversion of berberine by photooxygenation to (±)-ophiocarpine (**70a**) and (±)-*epi*-

ophiocarpine, and (±)-α- and (±)-β-phthalidetetrahydroisoquinoline, all of them in excellent yield (Scheme 29b).

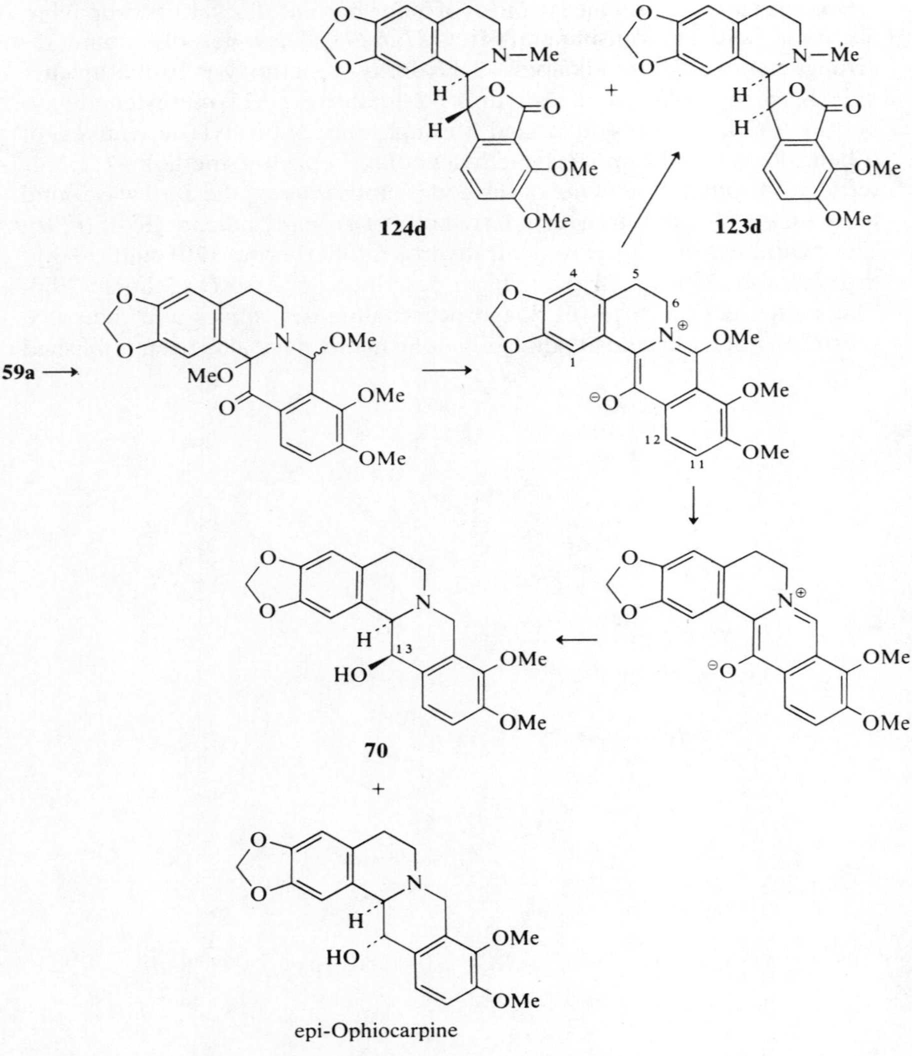

124d

123d

59a →

70

+

epi-Ophiocarpine

SCHEME 29b. *Conversion of berberine* (**59a**) *to racemic ophiocarpine* (**70**) *and hydrastine* (**123** *and* **124**) *derivatives by photooxygenation* (*632*).

The optical antipodes of *N*-methyltetrahydropapaverine were prepared and then converted into the optical antipodes of coralydine (**82b** and **82c**) and *O*-methylcorytenchirine (**83b** and **83c**) (*633*) (Scheme 30). The total

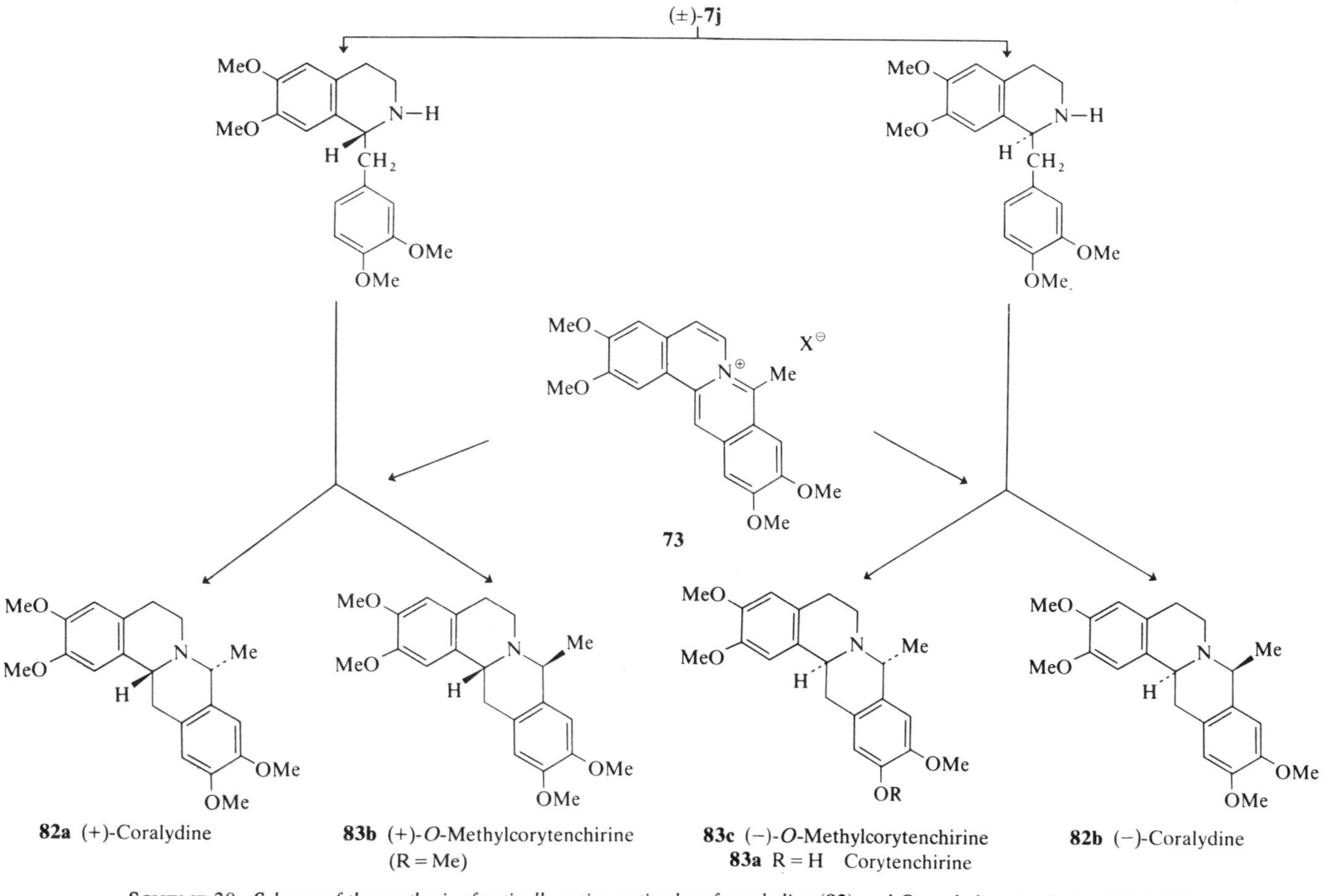

SCHEME 30. *Scheme of the synthesis of optically active antipodes of coralydine* (**82**) *and O-methylcorytenchirine* (**83**) (*633*).

synthesis of racemic corytenchirine (**83a**) was carried out in two ways: (a) irradiation of the (*Z*)-1-benzilidenetetrahydroisoquinoline derivative and (b) the Mannich reaction of 1,2,3,4-tetrahydro-1-(3-hydroxy-4-methoxybenzyl)-6,7-dimethoxyisoquinoline with acetaldehyde (*634*). In that paper the stereochemistry of corytenchirine-type compounds was also discussed. The synthesis and absolute configuration at the chiral centers of 2,3,10,11-tetrahydroxy-8-methylberbine compounds, e.g., (+)-coralydine and (+)-*O*-methylcorytenchirine, were described again (*635*). The absolute configuration of this group of compounds was also derived from X-ray structure

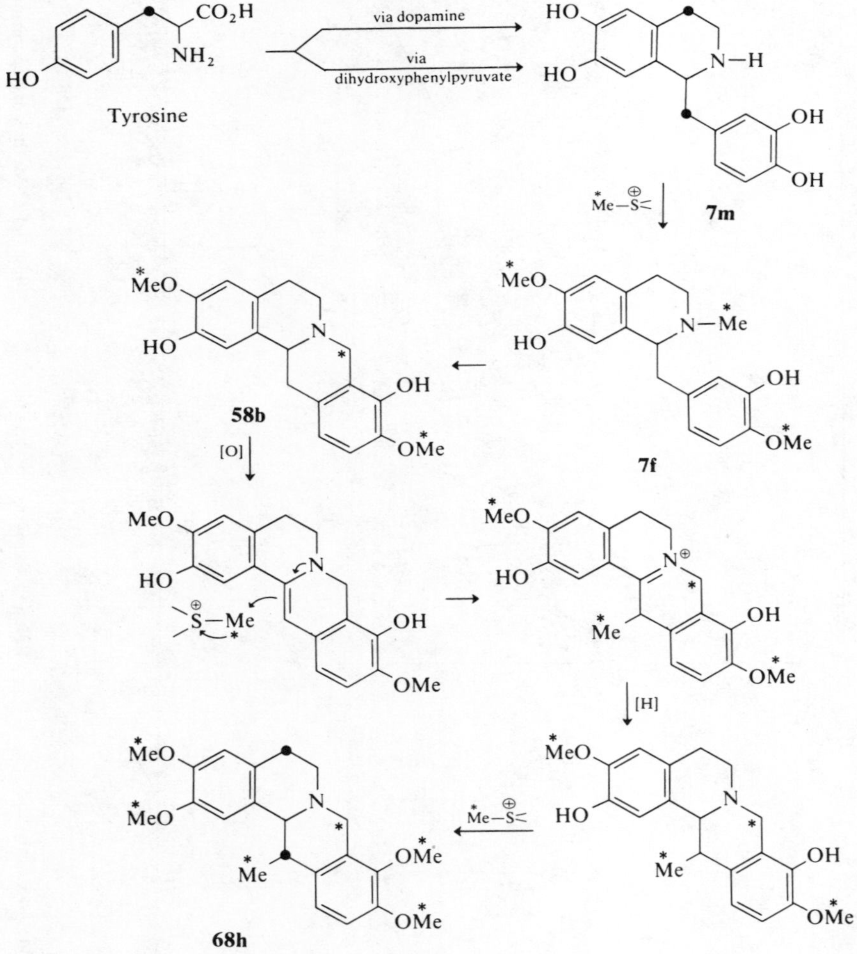

Tyrosine

via dopamine

via dihydroxyphenylpyruvate

7m

7f

58b

68h

SCHEME 31. *Biosynthesis of corydaline* (**68h**) *in Corydalis cava* (*638, 639*).

analysis. Coralyne (**71**) and *O*-methylcorytenchirine (**83**) can also be prepared by Stevens rearrangement of berbine methiodides (*635a*) or by thermolysis of benzocyclobutene (**74**) (*636*) (Scheme 27).

According to Manske, the precursors of corydaline alkaloids (**68**) are already C-methylated on the benzylisoquinoline skeleton before ring C closure takes place (*637*). On the contrary, Blaschke reported that labeled reticuline (**7f**) is converted specifically into corydaline (**68h**) (*638*). The experiments have also shown that the radioactivity from [3-^{14}C]tyrosine and from [methyl-^{14}C]methionine is incorporated nonrandomly into predicted positions of corydaline and ochotensimine in *Corydalis solida* and *C. ochotensis*, respectively (*639*). The methyl group of methionine supplies the C-13 methyl group of corydaline and the exocyclic methylene group of

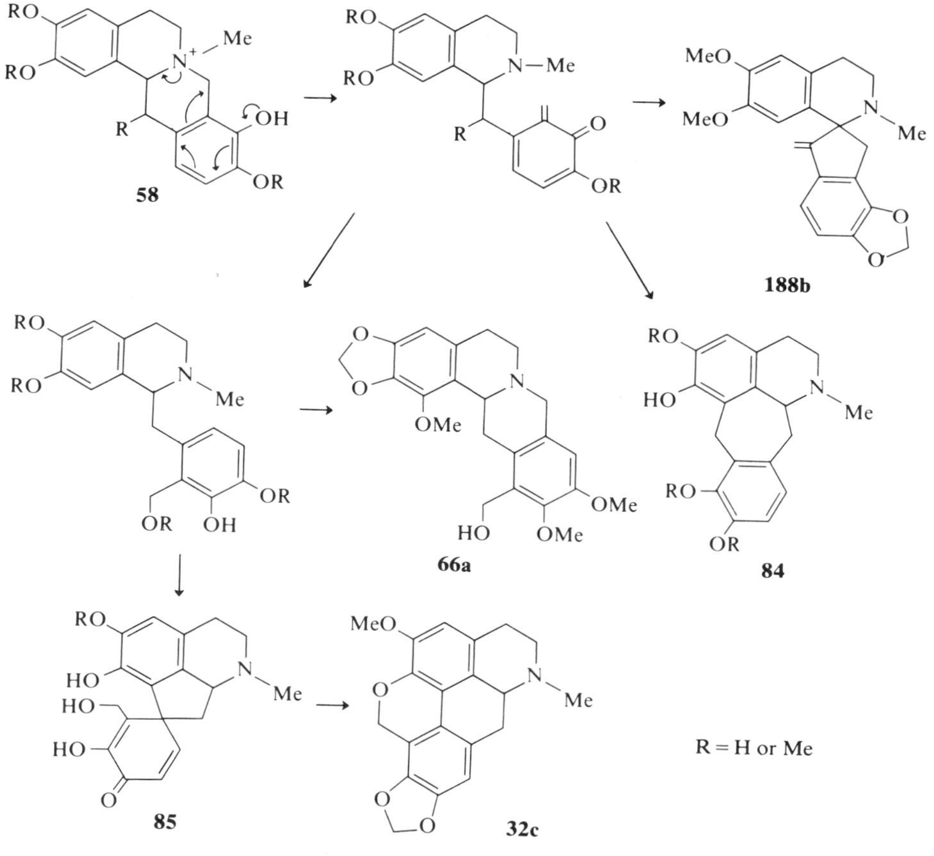

SCHEME 32. *Abnormal Hofmann degradation of phenolic berbinium salts* (**58**) (*609, 609a, 640*).

ochotensimine (**188**), as well as the "*bridge*" carbon atom at the oxocyclic O- and N-attached one-carbon units of each alkaloid (Scheme 31).

Recent studies have shown that the protoberberine and the tetrahydroprotoberberine bases undergo several reactions. The abnormal Hofmann degradation of phenolic tetrahydroprotoberbinium salts (**58**) might serve to explain the origin of mecambridine (**66a**) and orientalidine (**66c**) (*589, 640*) (Scheme 32). In some cases it may even account for the formation of spirobenzyltetrahydroisoquinoline bases (**188**), the base **84**, and those of the *N*-methylthalphenine type (**32**) (*641*). Kametani *et al.* describe the cleavage of tetrahydroberberines with trifluoroacetic anhydride (*642*).

The Oppenauer oxidation of racemic ophiocarpine (**70a**) gave 3-(2′-vinyl-3,4-methylenedioxy)phenyl-11,12-dimethoxy-4-hydroxyisoquinoline (**86**) (*643*). Photolysis of **86** afforded 3,4-dimethoxy-11,12-methylenedioxy-8,9-dihydro[1]benzoxepinol[2,3-*c*]isoquinoline (**87**), whereas photolysis of methyl ether of **86** gave *N*-norchelerythrine (**177c**) (Scheme 33).

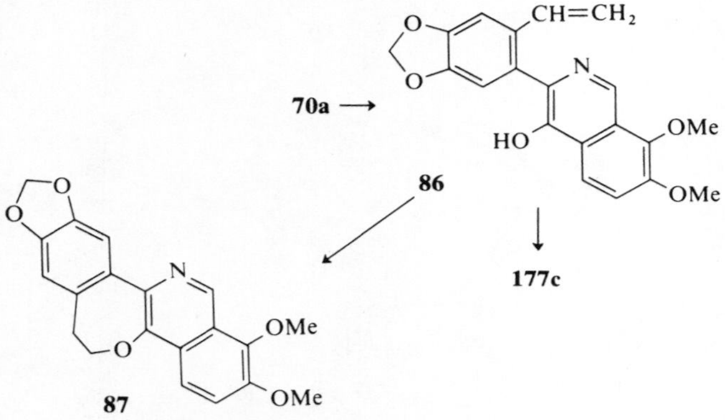

SCHEME 33. *The Oppenauer oxidation of ophiocarpine* (**70a**) (*643*).

Heating 8-oxyberberine (**91**) with concentrated hydrochloric acid converts it into 9-demethyl-8-oxyberberine, and by the same heating of 8-oxyepiberberine (8-oxydehydrosinoacutine) it changes into 9,10-dihydroxy-8-oxyepiberberine (*607, 644*).

The conversion of protoberberine alkaloids (**59**) to benzo[*c*]phenanthridine alkaloids (**177**) was achieved (*643*). Thermal decomposition of the hydrochloride of the tetrahydroprotoberberine methine gave bases **88a** and **88b** (*645*).

The protoberberinium salts were obtained in good yield by dealkylation of their *N*-substituted tetrahydro derivatives and following dehydrogenation with iodine (*646, 647*).

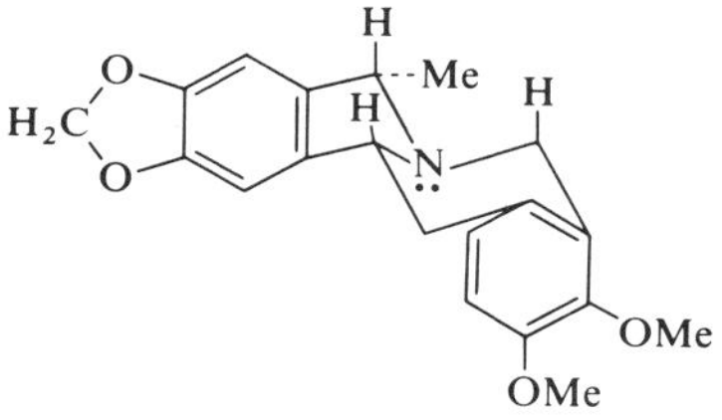

88a

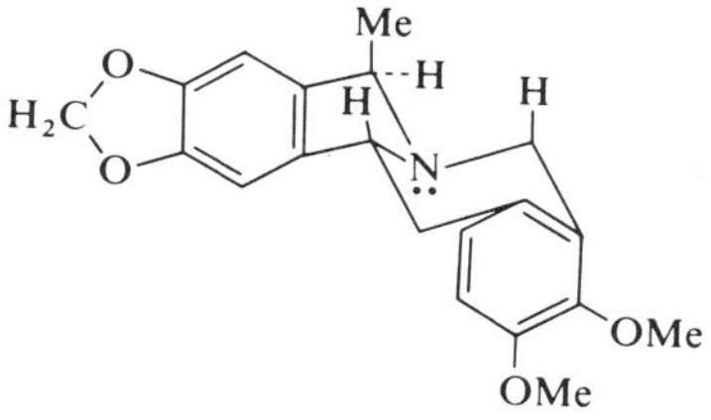

88b

The tetrahydro- or 13-oxotetrahydroprotoberberinium salts can be rearranged (Stevens rearrangement) into spirobenzylisoquinoline or 13-oxospirobenzylisoquinoline alkaloids (**188**) and vice versa. Yields are up to 40% (*635a, 648–650*) (Schemes 62, 63, and 64).

Following photochemical pathways, the enamides **89** afford 8-oxyberbine compounds (**91** and **92**) in a 45–96% yield with elimination of the substituent A in the *ortho* position (*651*) (Scheme 34). This reaction has been extended to form 8-oxyprotoberberine or 13-methyl-8-oxyprotoberberine.

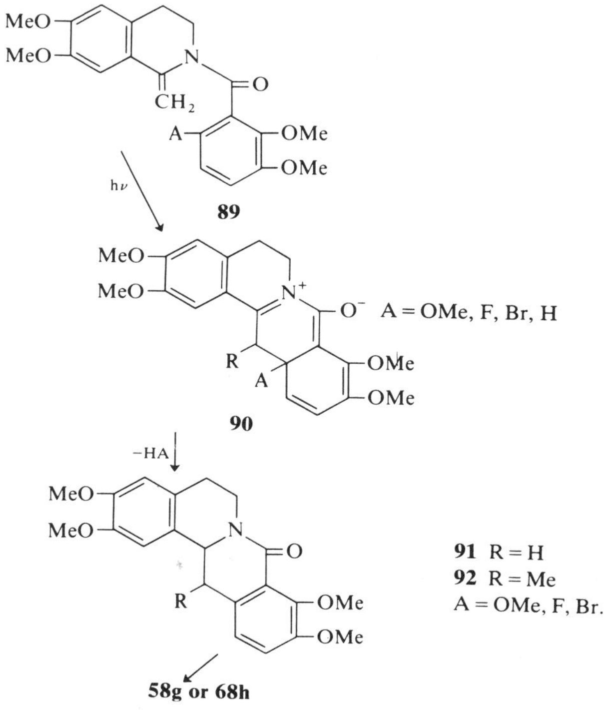

89

A = OMe, F, Br, H

90

91 R = H
92 R = Me
A = OMe, F, Br.

58g or **68h**

SCHEME 34. *Formation of 8-oxyberbines* (**91** *and* **92**) *from 2-aroyl-1-methylene-1,2,3,4-tetrahydroisoquinolines* (**89**) (*651*).

In the presence of phenyllithium or lithium aluminum hydride, (±)-*N*-methyltetrahydroberberinium salts undergo rearrangement to *N*-methyl-7,8-dimethoxy-1,2,3,4-tetrahydroisoquinoline-3-spiro-1′,5′,6′-methylenedioxyindane (**93**) (*652*). On Hofmann's exhaustive methylation this substance yields the compound **94** (Scheme 35).

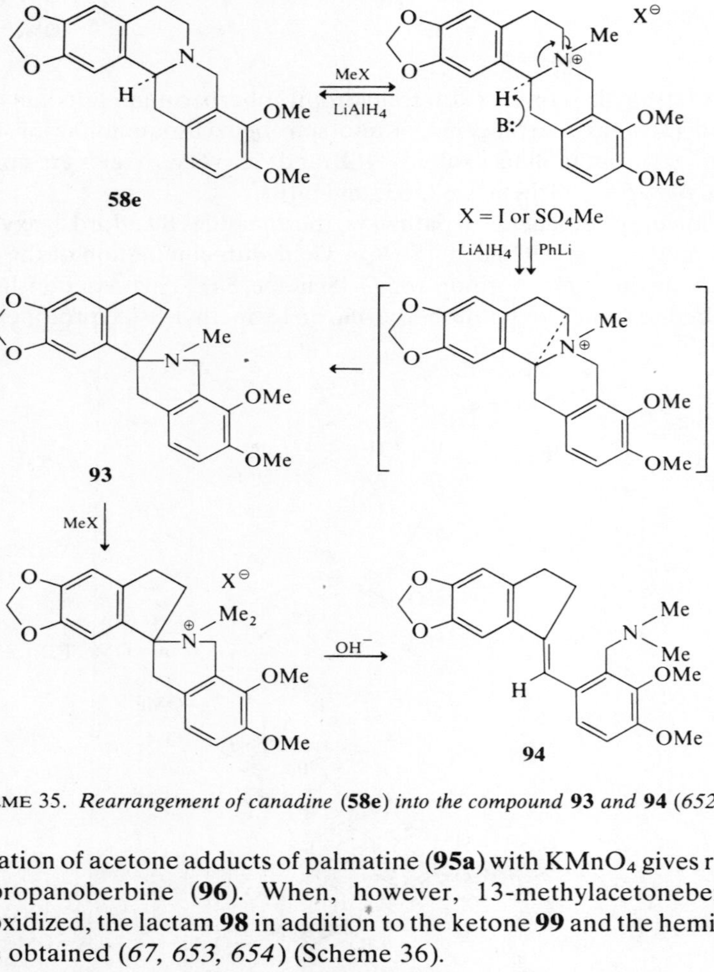

MeX
LiAlH$_4$
58e
X$^{\ominus}$
X = I or SO$_4$Me
LiAlH$_4$ PhLi
93
MeX
X$^{\ominus}$
OH$^-$
94

SCHEME 35. *Rearrangement of canadine* (**58e**) *into the compound* **93** *and* **94** (*652*).

Oxidation of acetone adducts of palmatine (**95a**) with $KMnO_4$ gives rise to 8,13a-propanoberbine (**96**). When, however, 13-methylacetoneberbine (**97**) is oxidized, the lactam **98** in addition to the ketone **99** and the hemiketal **100** are obtained (*67, 653, 654*) (Scheme 36).

In recent years, attention has been paid to the physicochemical properties of the tetrahydroberberine and berberine alkaloids. Three papers have appeared on mass spectrometry of those alkaloids (*655–657*) and their *N*-methohalides (*658*). The mass spectra of 9-methoxytetrahydroprotoberberines show an intense peak which arises by cleavage of the methoxyl

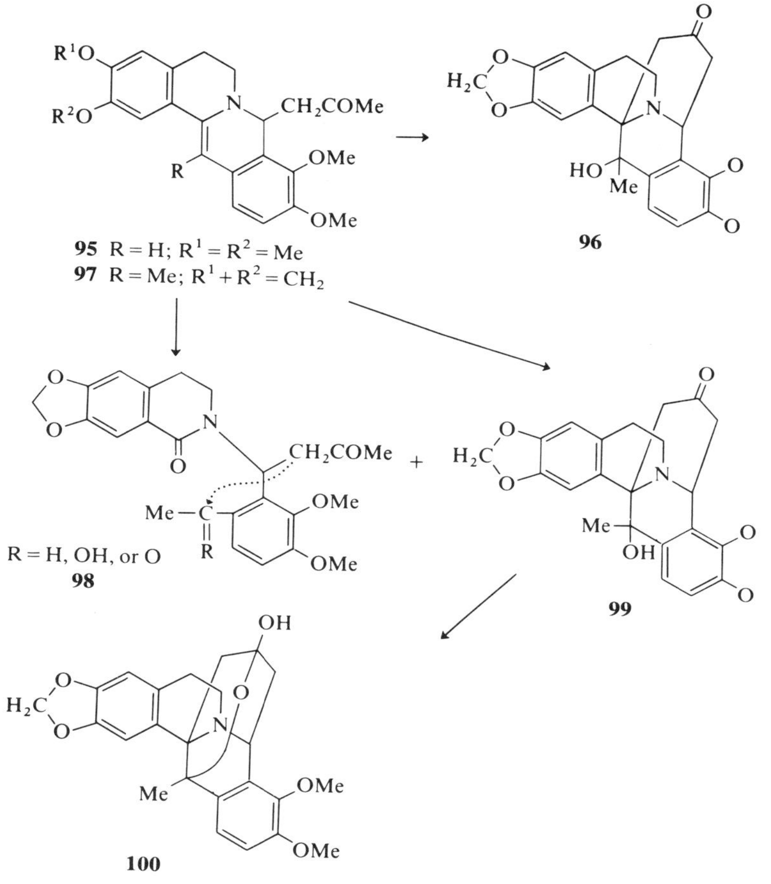

95 R = H; $R^1 = R^2 = Me$
97 R = Me; $R^1 + R^2 = CH_2$

96

R = H, OH, or O
98

99

100

SCHEME 36. *Determination of the structure of the oxidation products of acetoneberberine* (**95**) *and 13-methylacetoneberberine* (**97**) (*653*).

group from the ionized molecule, in contradistinction to the isomers or homologs without such substitution but with methoxyl substituents at other sites (*656*). Location of such a substituent originally present, or produced from a 9-hydroxyl group by treatment with diazomethane, permits differentiation of 9,10- from 10-11-oxygenation patterns in ring D and thus is an important step in the structural elucidation of unknown alkaloids of this class (*588b*).

Ultraviolet spectroscopic studies of the protoberberine and pseudoprotoberberine alkaloids showed that, on the basis of the form of the absorption curve, the alkaloids carrying oxygen substituents at the 9- and 10-positions

(**59**) can easily be distinguished from those with the substituents at C-10 and C-11 (**61**) (*380, 659*). Wiegrebe *et al.* measured the UV spectra of a series of protoberberine compounds (*660*). Another comprehensive work was published on the ^{1}H NMR data of protoberberinium salts in trifluoroacetic acid (*529*). Snatzke *et al.* studied the problem of CD and the rules for the determination of the absolute configuration of tetrahydroprotoberberine bases (*661, 662*).

Shamma *et al.* studied various tetrahydroprotoberberine and tetrahydropseudoprotoberberine alkaloids in respect to the Bohlmann bands, thin-layer chromatography, and the rapidity of the formation of the methiodide (*3, 585*) (Table III). They concluded that there might or might not exist a relationship between the Bohlmann bands in the IR spectrum and the *trans* configuration of the rings B/C. These bands also appear when the ring A carries an oxygen substituent at C-1. According to Shamma, the tetrahydroprotoberberines may exist in any of the three different conformations **a**, **b**, or **c**, depending upon the substitution pattern at C-1 and C-13 (Scheme 37). The alkaloids capaurine (**58p**) and capaurimine (**58r**) exist in the conformation **c** where the rings B/C have a *cis* configuration (half-chair–half-chair).

Takao and Iwasa also studied the stereostructure of tetrahydroprotoberberine alkaloids in the crystalline and liquid phase on the basis of the Bohlmann bands (*663*). They analyzed three groups of alkaloids: group I, alkaloids of the tetrahydropalmatine type (**58g**); group II, alkaloids of the capaurine type (**58p**); and group III, alkaloids of the mesocorydaline type (**68h**). Contrary to Kametani (*594*), they conclude that in groups I and III the preferred conformation present in solution is retained in the crystal state, that is, the group I compounds adopt the B/C-*trans* form and group III compounds the B/C-*cis* form. In the alkaloids of group II, in which the position of the equilibrium is shifted to the B/C-*trans* side rather than that of group III, the crystal contains only one of the configurations.

According to Shamma, the relative configuration of (−)-ophiocarpine (**70a**) at C-13 and C-13a can be deduced from physical measurement (*1*, p. 297). The alkaloid has a wide IR absorption band near 3526 cm^{-1} and is thus strongly hydrogen-bonded, whereas the synthetic 13-epiophiocarpine has only a free hydroxyl peak at 3597 cm^{-1}. Both compounds exhibit strong Bohlmann bands at 2800 cm^{-1}, indicating that rings B and C are *trans* fused. These conclusions are consistent with the ^{1}H NMR spectroscopic findings.

Kametani *et al.* studied the conformation of the tetrahydroprotoberberines and the tetrahydropseudoprotoberberines by ^{13}C NMR spectroscopy (*587*). They conclude that the conformation of the rings B and C can be inferred from the chemical shifts of C-6, C-13, and C-13a, and that it is

TABLE III

DATA FOR TETRAHYDROPROTOBERBERINES AND TETRAHYDROPSEUDOPROTOBERBERINES AS SHOWN IN SCHEME 37

Property	Compound							
	A	**B**	**C**	**D**	**E**	**F**	**G**	**H**
B/C stereochemistry	*trans*	*trans*	*trans*	*trans*	*cis*	*cis*	*cis* (half-chair–half-chair)	*cis* (half-chair–half-chair)
Rate of methiodide formation $\times 10^4$ sec^{-1}	35	53.6	1.1	1.3	273	341	78	85
TLC, R_f (ether)	0.72	0.75	0.84	0.75	0.57	0.40	—	—
Bohlmann bands	Yes	Yes	Yes	Yes	No	No	Yes	Yes
C-13 methyl chemical shift	—	—	0.95	0.93	1.43	1.48	—	—

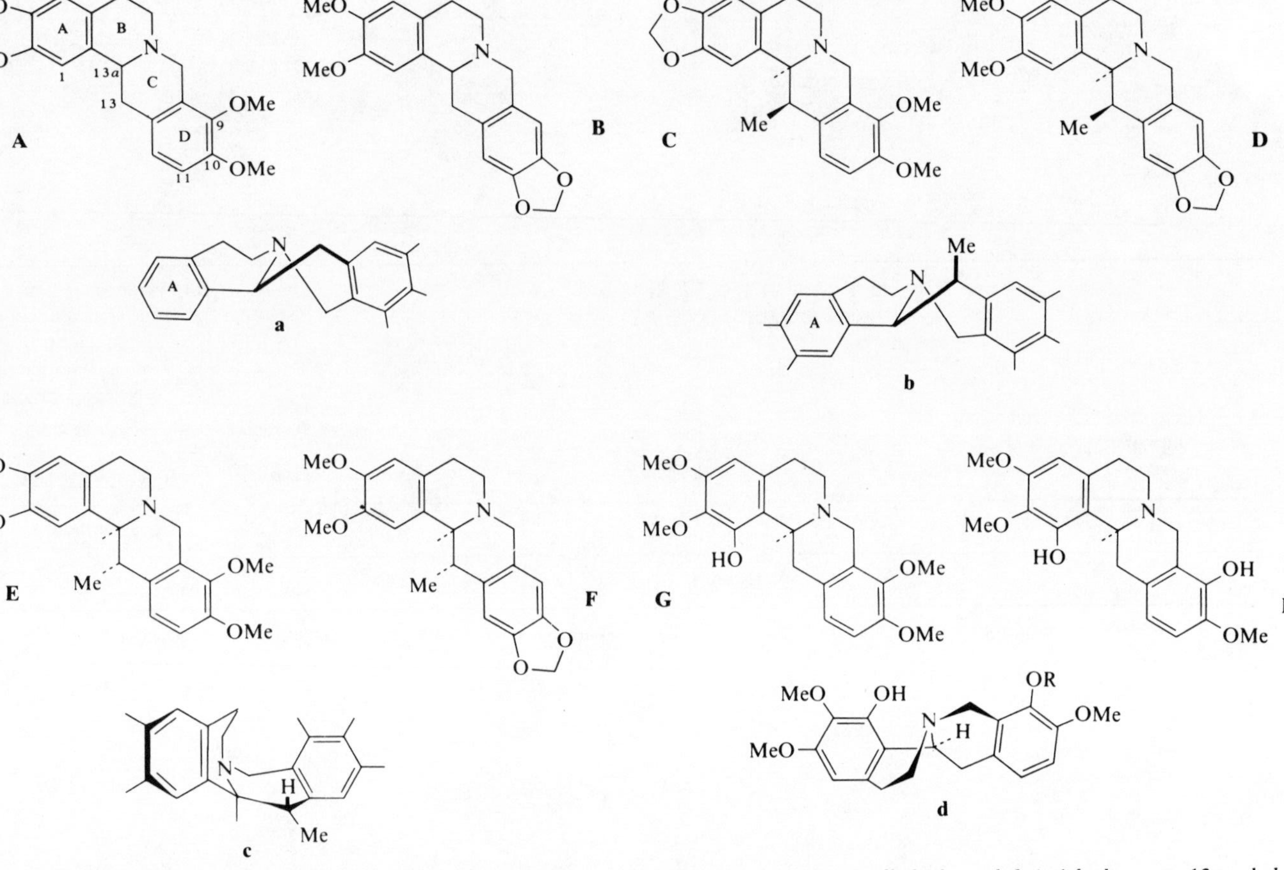

SCHEME 37. *Various conformations of tetrahydroprotoberberine, tetrahydropseudoprotoberberine alkaloids, and their 1-hydroxy- or 13-methyl derivatives. See Table III* (*585*).

possible to determine whether the alkaloid has to be relegated to the protoberberine (**58**) or to the pseudoprotoberberine (**60**) group from the values of the chemical shifts of C-8.

To the alkaloids carrying a methyl group at C-13 (corydaline type **68**) the alkaloids and bases cavidine (**68d**), apocavidine (**68e**), epiapocavidine (**68f**), *meso*-thalictricavine (**68b**) (*75, 664*), and corydalidzine (**68g**) (*640*) have been added. From the coupling constant $J_{H\text{-}13\text{-}H\text{-}13a}$ of these tetrahydro-protoberberine bases, the relative stereochemistries (*cis* or *trans*) of the two hydrogen atoms on the carbon atoms C-13 and C-13a and the conformations of the rings B and C were inferred (*148*). The same conclusions have also been reached by Govindachari *et al.* after a study of the relative configuration of thalictricavine and *meso*-thalictricavine by ^{1}H NMR spectroscopy (*664*). The relative configuration of (±)-coralidine and (±)-O-methylcorytenchirine was also studied by high resolution 270 MHz proton magnetic resonance (*664a*).

A simple method of isolation of berberine has been described, as has its rapid determination in the crude plant material (*665, 666*). Confirmatory evidence has been provided again that, on extraction, the protoberberine bases in chloroform form additive products and the corresponding 8-oxyberberine derivatives (*667*).

Coralyne (**73**) has an antileukemic effect on leukemias L-1210 and P-388 in laboratory animals. Further studies showed that this base conjugates itself with a DNA double helix in aqueous medium (*668*). Dehydrocorydaline (**69b**) has a significant adrenergic neuron-blocking effect (*669*).

Pavelka and Kovář studied 24 derivatives of protoberberine for their interaction with liver alcohol dehydrogenase. The inhibitory power of the tested compounds was correlated with their structures and with some properties following from those structures. The most effective inhibitor of liver alcohol dehydrogenase is 13-ethylberberine, which is bound more firmly to the enzyme at pH 10 than NAD and NADH (*670*).

J. Protopine Group*

Vaillantine (**101c**) was isolated from *F. vaillantii* (*177*). Corydinine, isolated from *C. stewartii* (*143*), was shown to be identical with protopine (**101a**) (*144*). Dihydroprotopine (**102**) was isolated from *P. somniferum* (*279*). Fumaridine and fumaramine possess two nitrogen atoms in their molecules (*171*). Shamma and Moniot have revised these structures and shown that they are of the narceineimide type (**146**) (*671*).

* This material is supplementary to *The Alkaloids*, Vol. IV, p. 147; Vol. X, p. 475; Vol. XII, p. 390.

101a $R^1 = H$; $R^2 + R^3 = R^4 + R^5 = CH_2$ Protopine
101b $R^1 = H$; $R^2 + R^3 = CH_2$; $R^4 = R^5 = Me$ Allocryptopine
101c $R^1 = R^2 = R^3 = H$; $R^4 = R^5 = Me$ Vaillantine
101d $R^1 = OCH_3$; $R^2 = R^3 = Me$; $R^4 + R^5 = CH_2$ 1-Methoxycryptopine
101e $R^1 = OCH_3$; $R^2 + R^3 = CH_2$; $R^4 = R^5 = Me$ 1-Methoxyallocryptopine
101f $R^1 = H$; $R^2 = R^3 = Me$; $R^4 + R^5 = CH_2$ Cryptopine (Thalisopyrine)
101g $R^1 = H$; $R^2 = R^3 = R^4 = R^5 = Me$ Muramine

102 Dihydroprotopine

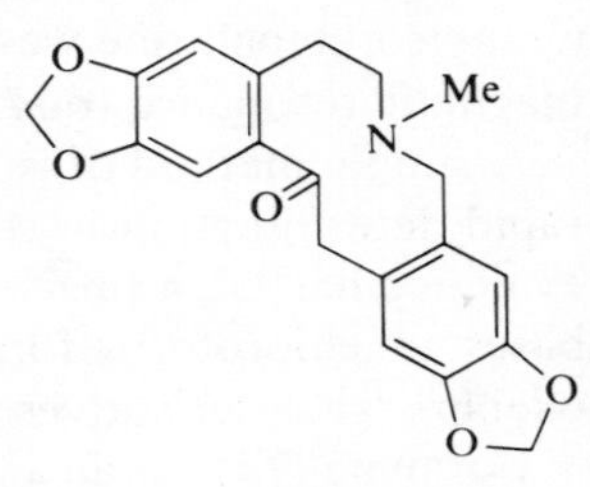

103

The plant *Zanthoxylum conspersipunctatum* Merr. & Perry (Rutaceae) yielded pseudoprotopine (**103**). It was synthesized by using the method of Perkin (*672*).

Allocryptopine (**101b**) was synthesized from tetrahydroprotoberberine methiodide in one step by photooxygenation (*673*). The methosulfate of 13-oxotetrahydroprotoberberine (**160**) also gave 13-oxoallocryptopine (**121a**) and the 13-oxoanalog of allocryptopine (**120b**) (*674*) (Scheme 64).

Teitel *et al.* prepared tetracyclic benzazepineisoindole, i.e., "Schöpf's base VI" (**104a**), from β-hydrastine (*100, 675*). This base is a convenient intermediate for further chemical transformation into the protopine skeleton: Hofmann degradation of the "base VI" yields the compounds **105** (isoindoline structure) and **106**. The 10-membered *trans*- (**108b**) and *cis*-dibenzazecine (**106**) were converted via **107** into α-allocryptopine (**101b**) in poor yield (Scheme 38).

Brown and Dyke studied the products of Hofmann degradation of protopine and of alkaloids related to it (*676, 677*). The structures of γ-anhydrodihydrocryptopine (**112**) and γ-cryptopidene (**113**) are substantiated by their independent preparation from epicryptorubine chloride

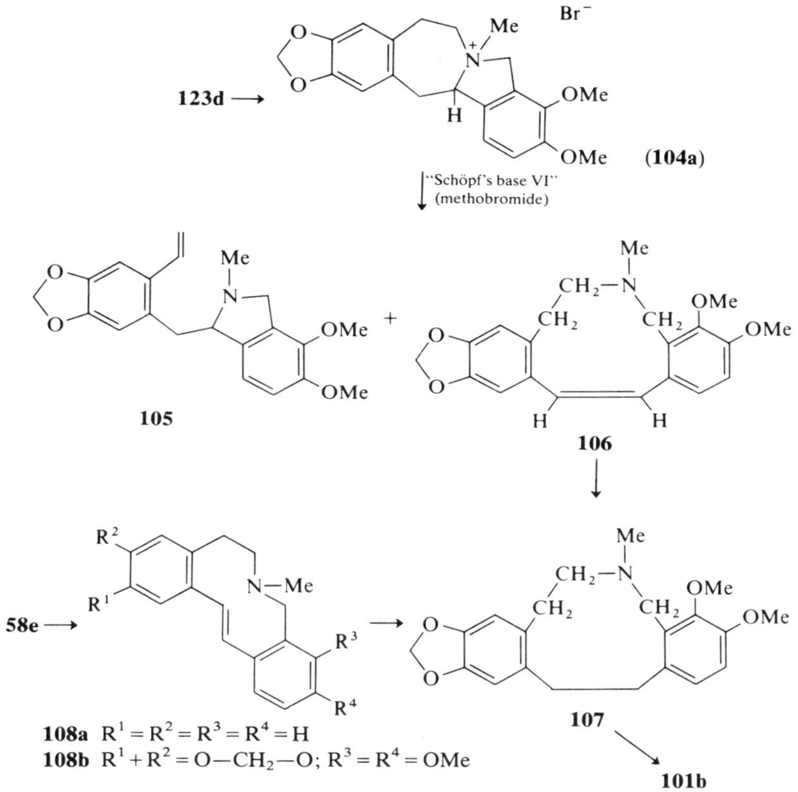

123d → (**104a**) Br^-

"Schöpf's base VI" (methobromide)

105 + **106** → **107** → **101b**

58e → **108a** $R^1 = R^2 = R^3 = R^4 = H$
108b $R^1 + R^2 = O{-}CH_2{-}O$; $R^3 = R^4 = OMe$

SCHEME 38. *Preparation of protopine bases* (**101**) *by Hofmann degradation of the "Base VI"* (**104a**) (*675*).

(ψ-cryptopine chloride) (**109**) by reduction to the tetrahydro base **110** followed by Hofmann degradation via compound **111** (Scheme 39).

Onda *et al.* also studied the degradation products of protopine alkaloids (*678–682*). On the basis of the ^1H NMR spectra, they found that treatment of anhydroprotopine (**114**) with dilute hydrochloric acid affords two stereoisomeric compounds with a spiro-type skeleton (**115** and **116**) (Scheme 40). These authors also showed that irradiation of anhydroprotopine (**114**) and subsequent dehydrogenation give rise to dihydrosanguinarine (**175b**) in good yield and, in the following step, to sanguinarine (**177b**) (Scheme 40).

The compounds **114** and **118** were treated with acetylene dicarboxylate to give the condensation photoproduct **119**; after reduction with lithium aluminum hydride the product **119** was studied by ^1H NMR spectroscopy

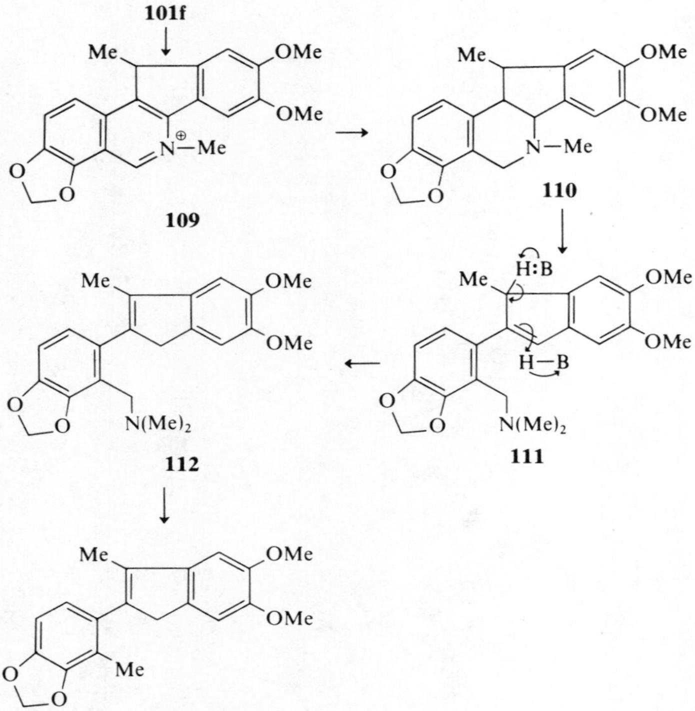

SCHEME 39. *A study of the structure of the products of Hofmann's degradation of cryptopine* (**101f**) (*676, 677*).

and the nuclear Overhauser effect to establish the structure in which the B/C ring juncture is cis and the conformation of the ring B is a flattened boat form (*682*) (Scheme 41).

Although in *P. somniferum* and *Sinomenium acutum* the biosynthesis of the morphinane bases can be carried out with (+)-reticuline as starting material, these bases and also protopine (**101a**) cannot be obtained biosynthetically in satisfactory yield from (+)-tembetarine chloride (reticuline methochloride). Similar results were obtained with *D. spectabilis* (*683*). Important intermediates of the biosynthesis between tetrahydroprotoberberine and protopine alkaloids are the compounds of the dihydroprotopine type (**102**), which were prepared in high yield by partial synthesis with cyanogen bromide (von Braun method) (*684*). The biosynthesis of protopine alkaloids is also discussed in Section III,N.

The protopine alkaloids contain a ten-membered ring with a methylated nitrogen atom and a keto group. This feature is responsible for several

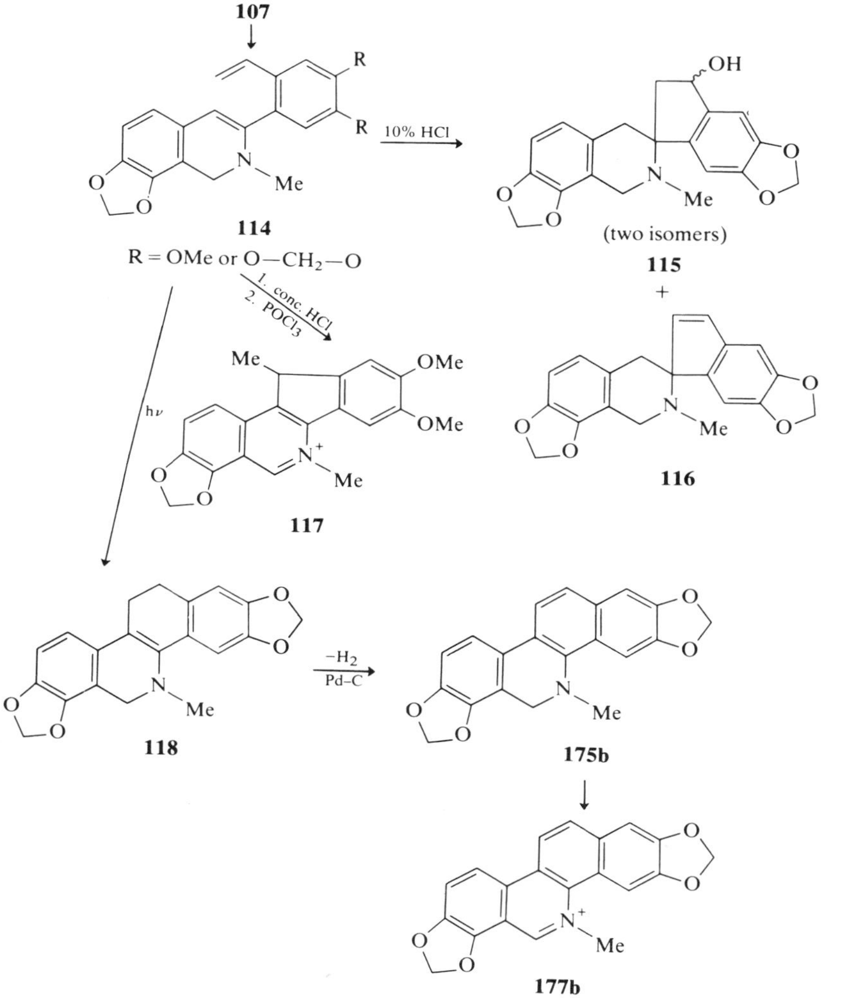

SCHEME 40. *Conversion of alkaloids of the protopine type* (**107**) *into those of the sanguinarine type* (**177**) (*680*).

characteristic chemical reactions and also for the shift of the carbonyl band in the IR spectrum. Therefore, it was interesting to prepare the basic skeleton **120a** of the protopine base and with this in view to establish its physicochemical properties. Bentley and Murray were the first to carry out this synthesis (*685*). The classical method of Perkin (*686*) was also used for that purpose (*687*).

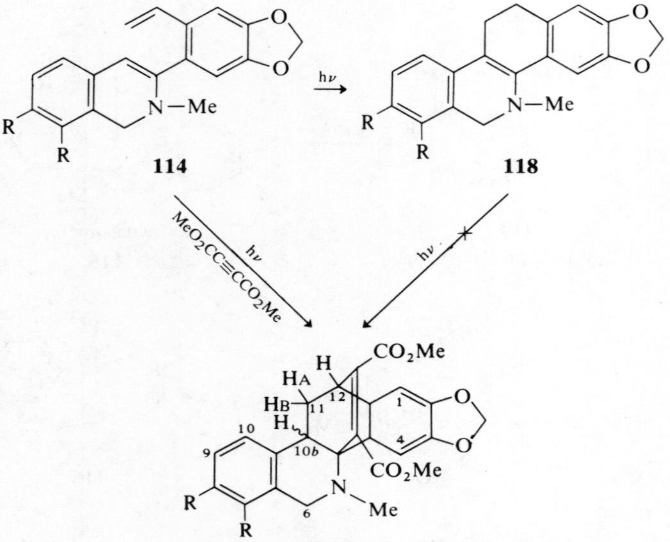

SCHEME 41. *Another transformation of alkaloids of the protopine type (681).*

In the first place, the base **108a**, whose *N*-oxide was quite stable to acid treatment, was prepared. Conversion into the keto base **120a** was achieved only under special conditions. The ^{1}H NMR spectrum of **120a** showed characteristic signals, and its IR spectrum showed a displaced carbonyl band which is typical of protopine alkaloids; its position was due to a transannular interaction with the nitrogen atom. The high stability to acids of the unsubstituted *N*-oxide, when compared to those containing methoxy or methylenedioxy substituents, is indicative of a reduced availability of the π electrons of the stilbenic double bond for the fixation of a proton which is a necessary step for its transformation into **120a**.

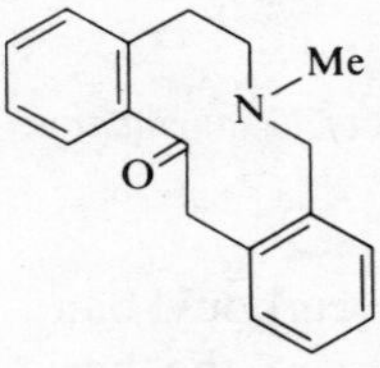

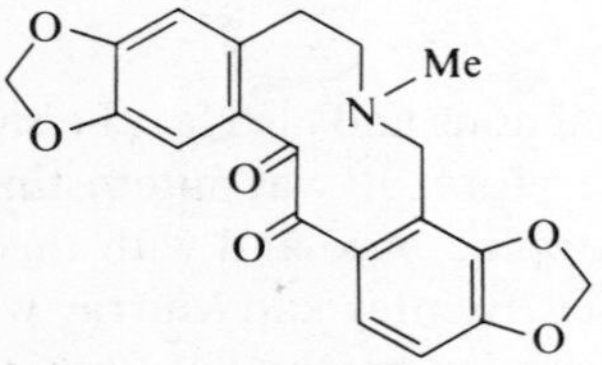

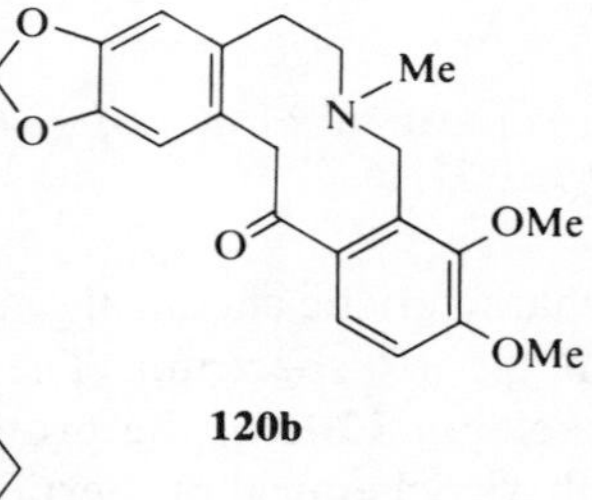

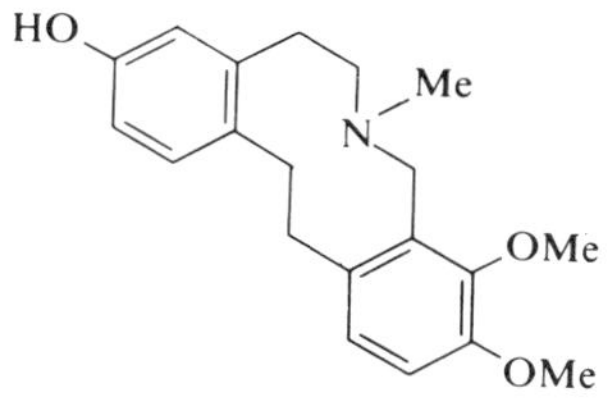

122

The natural abundance ^{13}C NMR spectra of five protopine alkaloids were measured at 22.628 MHz by application of the pulse Fourier transform technique (*688*). Several techniques were employed to make spectral assignments. It was possible to make self-consistent assignments for all the resonances of these alkaloids. Evidence was presented for transannular interaction between the amino group and the carbonyl group at C-14.

Šantavý *et al.* studied the UV spectra to establish whether in 13-oxoprotopine alkaloids (**121**) the interaction of the tertiary nitrogen may also take place with the oxo group at C-13 (*420*). It was found that this interaction takes place only with the keto group at C-14. Such an interaction does take place if only a C-13 keto group (**120b**) is present (*674*). In protopine and cryptopine, the nature of intramolecular N · · · C=O interactions were measured by X-ray crystallography (*689*).

It was reported that the compound **122** might be a unique nonnarcotic analgesic in view of the criteria of structure–pharmacological relations. According to the authors, it has good analgesic properties and does not possess narcotic properties characteristic of morphine (*690*). This has provided an incentive to the elaboration of a series of methods for the synthesis not only of this substance but of those of the berberine and the protopine type which also have analgesic properties (*627, 628*).

K. Phthalidetetrahydroisoquinoline Group*

Corydalis ledebouriana and *C. severtzovii* yielded the new alkaloids corledine (**124g**) and severcine (**124f**) (*141*) (see Table IV). The optical rotation of these two alkaloids shows that they are more likely to belong to the *1R* series than to the *1S* series. (+)-Bicuculline (**125a**) was also isolated from *F. parviflora* (**173**). α-Narcotine (**123h**) was isolated by Gašić and Pergál from European opium by a modification of the method used for its quantitative determination (*270*). At the same time, Trojánek *et al.* isolated this alkaloid from mother liquors obtained after the commercial isolation of

* This material is supplementary to *The Alkaloids*, Vol. IV, p. 168; Vol. VII, p. 433; Vol. IX, p. 117; Vol. XII, p. 397.

TABLE IV
Relative and Absolute Configurations of Phthalidetetrahydroisoquinoline Alkaloids[a]

123 **124** **125** **126**

Structure	Substituents at O atoms: R^1	R^2	R^3	R^4	Nomenclature: **123** *erythro* *1R,9S*	**124** *threo* *1R,9R*	**125** *erythro* *1S,9R*	**126** *threo* *1S,9S*
a	CH_2		CH_2		(−)-Bicuculline	(−)-Capnoidine	(+)-Bicuculline	(+)-Adlumidine
b	Me	Me	CH_2		—	(−)-Adlumine	(+)-Corlumine	(+)-Adlumine
c	H	Me	CH_2		—	—	(+)-Corlumidine	—
d	CH_2		Me	Me	(−)-β-Hydrastine	(−)-α-Hydrastine	—	—
e	Me	Me	Me	Me	Cordrastine II	Cordrastine I	—	—
f	H	Me	Me	Me	—	(−)-Severcine	—	—
g	Me	H	Me	Me	—	(−)-Corledine	—	—
h	CH_2 6-OMe		Me	Me	(−)-α-Narcotine	(−)-β-Narcotine*	(+)-Narcotine	—
i	CH_2 6-OH		Me	Me	(−)-α-Narcotoline	—	—	—

[a] Asterisk denotes an unnatural derivative.

morphine from poppy capsules by using column chromatography on silica gel and thin-layer chromatography on alumina (*272*). Preininger *et al.* isolated bicuculline (**125a**), (−)-adlumine (**124b**), and (−)-capnoidine (**124a**) from *C. sempervirens* (*140*). The alkaloid corydalisol (**204**), which also has 17 carbon atoms, was isolated from *C. incisa* (*92*). It probably arises by reduction of the aldehyde group or carboxyl group derived from the C-8 of the berbine skeleton.

Corledine and severcine, whose phenolic group is located on ring A, yield (−)-adlumine (**124b**) on methylation with diazomethane (*110, 141*).

Narcotine can be split into cotarnine and opianic acid by treatment with dilute sulfuric acid; by oxidation with dilute nitric acid, chromic acid, potassium permanganate; or with mercuric acetate (*691*). Hydrogenolysis (hydrochloric acid or sulfuric acid and zinc) affords hydrocotarnine and meconine. Treatment of hydrocotarnine with sodium in ethanol leads to cleavage of the methoxyl group with the formation of 8-hydroxy-1,2-dihydrohydrastinine. The phthalidetetrahydroisoquinoline alkaloids can be converted into rhoeadane see Section III,M and spirobenzylisoquinoline (see Section III,O,1) precursors (*692*).

A new synthesis of the phthalideisoquinoline alkaloids has recently been announced (*693*). By this method, the phthalideisoquinoline system was synthesized from substituted 2-phenyl-1,3-indandiones as starting material. The key step in this synthesis is the rearrangement of the spirobenzylisoquinoline ring system into the phthalideisoquinoline system. The diastereomeric bases of cordrastines *I* and *II* (**124e** and **123e**) were synthesized by this method, and their relative configurations were established by comparison of their NMR spectra with those of the phthalideisoquinoline alkaloids of known configuration. This synthesis constitutes a new route to this system. It requires only readily available starting material and is widely applicable (Scheme 42) (*693, 693a*).

Another synthesis of phthalideisoquinoline alkaloids is based on 3,4-methylenedioxyphthalide-α-carboxylic acid (starting material piperonal). Condensation with phenethylamine and cyclization (Bischler and Napieralski) yield dehydrophthalideisoquinoline derivatives (**161**—Scheme 65) which on reduction afford ten phthalideisoquinolines and spirobenzylisoquinolines (*693b*). The tosylation of narcotinediol (**129a**) will be shown later in Scheme 49.

Shamma and Georgiev described a simple method* for the preparation of diastereoisomeric phthalidetetrahydroisoquinolines (*694*).

* Note: The presently known natural phthalidetetrahydroisoquinoline alkaloids have the electron-donating substituents of the ring D in the ortho position to the carbonyl group of the ring C. The described synthesis yields, however, compounds with a carbonyl group in the meta position to the electron-donating substituents (*694*).

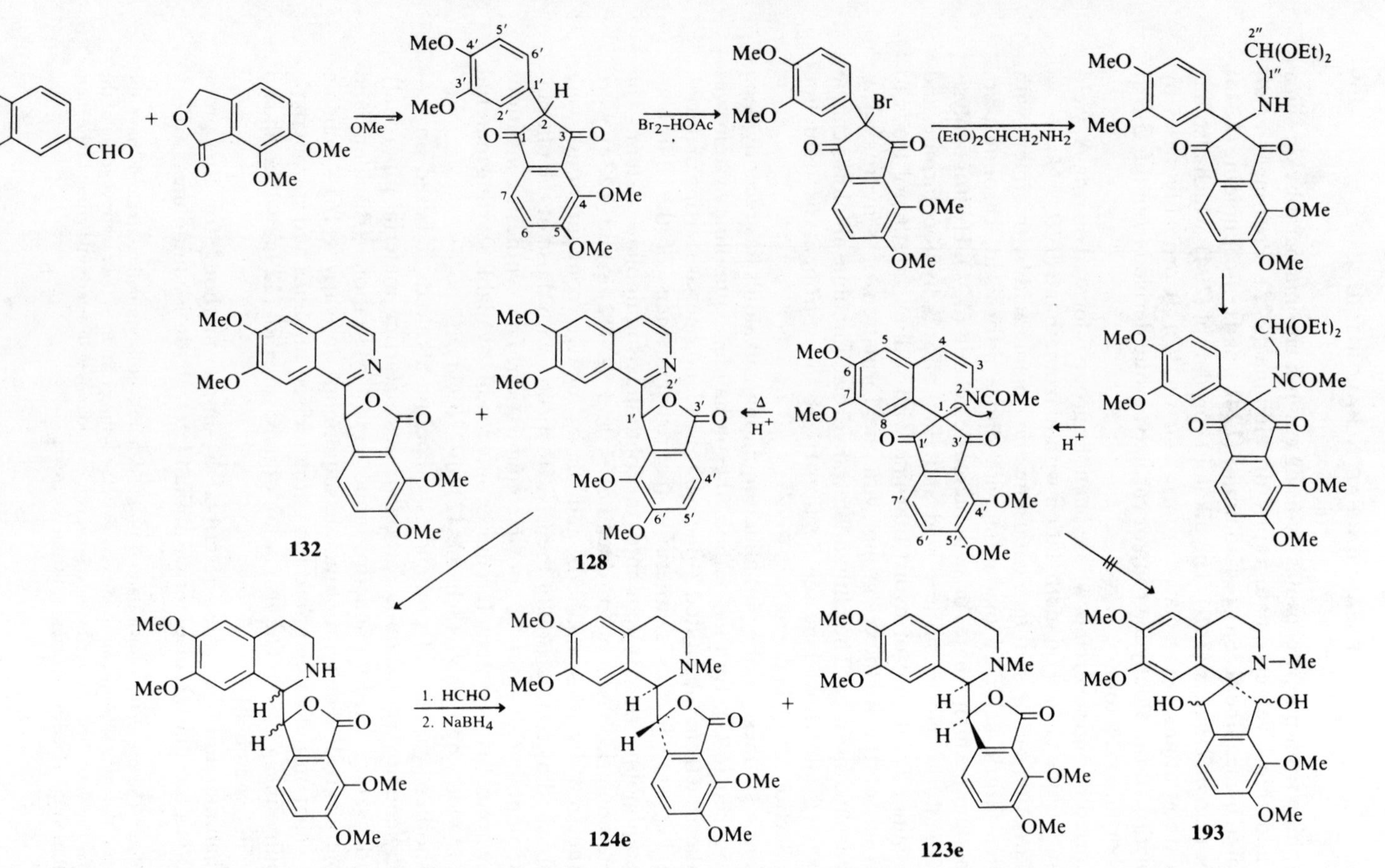

SCHEME 42. *Synthesis of phthalideisoquinoline* (**123** *and* **124**) *and ochrobirine* (**193**) *types of alkaloids from dimethoxybenzaldehyde and dimethoxyphthalide* (*693–693b*).

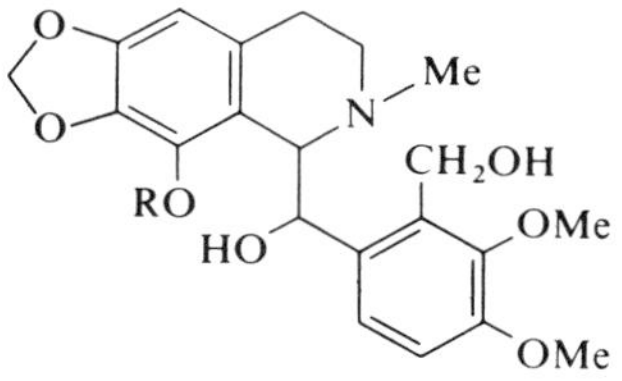

129a R = Me
129b R = H

2′-Hydroxymethylpapaverine (**130b**) is readily available from papaverine (**130a**). Chromium trioxide in AcOH–H_2SO_4 oxidation of 2′-hydroxymethylpapaverine led in 75% yield to the aromatic phthalideisoquinoline **131**. On reduction with Adams catalyst it gave a diastereoisomeric mixture of norphthalidetetrahydroisoquinolines **127a** which could be separated and subsequently individually N-methylated with formaldehyde and reduced with $NaBH_4$ to afford phthalidetetrahydroisoquinoline **127b** (Scheme 43).

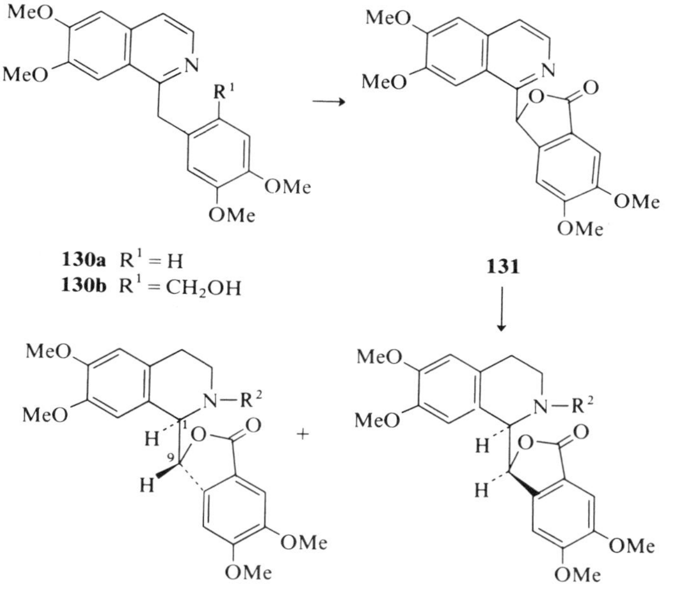

130a $R^1 = H$
130b $R^1 = CH_2OH$

131

127a $R^2 = H$
127b $R^2 = Me$

SCHEME 43. *Synthesis of phthalidetetrahydroisoquinoline bases with erythro and threo configuration at C-1 and C-9 (694).*

On oxidation, hydration, N-methylation, and reduction with $NaBH_4$, berberine gives a mixture of α- and β-hydrastine (*695*). According to these authors, an analogous pathway leads to the formation of (−)-α-narcotine in *P. somniferum*: In this plant the present (−)-scoulerine must suffer oxidation of the N-7 and C-8 bond while the integrity of the C-14 asymmetric center is essentially maintained. (−)-α-Narcotine and (−)-β-hydrastine underwent photochemical racemization and epimerization to give a mixture of the racemic epimers (*696*).

The conformation of the individual diastereoisomers was studied by ^{1}H NMR spectroscopy (*694*). The ^{1}H NMR data show that the preferred conformation is that involving the least steric interaction between the rings A, B and C, D. In the phthalidetetrahydroisoquinolines, the ring C is located in the proximity of the nitrogen atom if that one is secondary. After N-methylation (tertiary nitrogen atom), the ring C is located in the proximity of the ring A and away from the relatively bulky *N*-methyl group. The methoxyl group at C-8 of narcotine (**123h**) affects the resulting conformation of the compound. From the results of the chemical shifts of the ^{1}H NMR at C-1 and C-9 the authors conclude that, in narcotine, the ring C is also located in the proximity of the nitrogen (formula **A**) whereas in

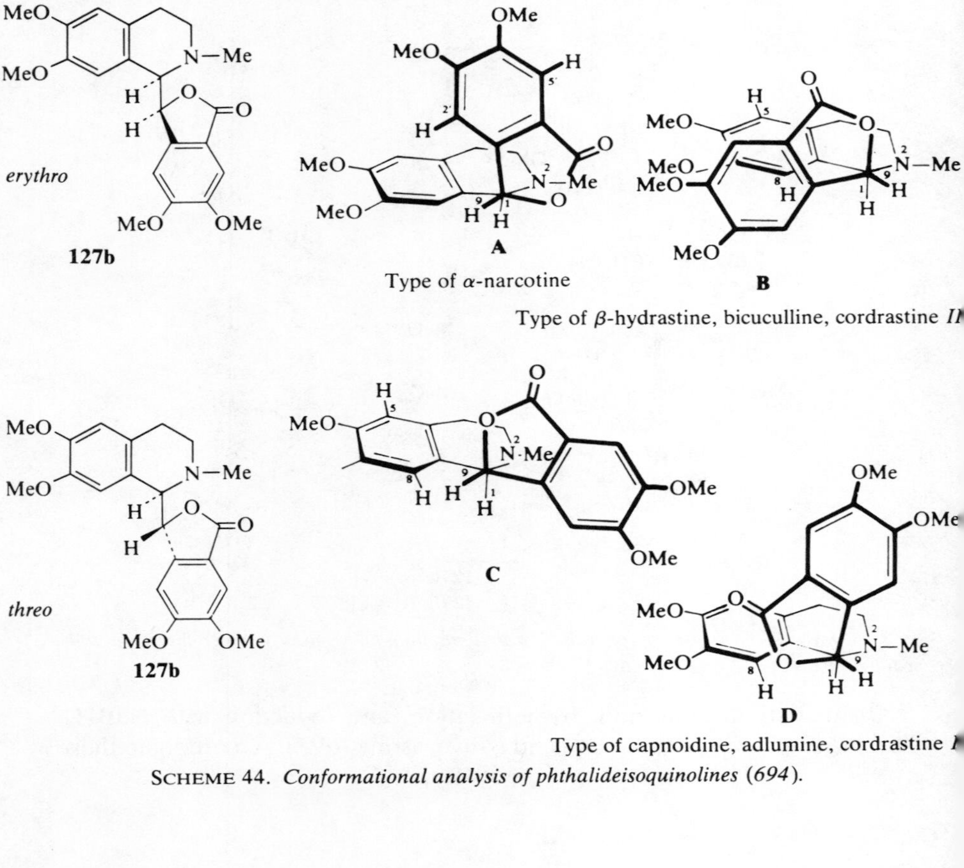

erythro **127b**

A Type of α-narcotine

B Type of β-hydrastine, bicuculline, cordrastine II

threo **127b**

C

D Type of capnoidine, adlumine, cordrastine I

SCHEME 44. *Conformational analysis of phthalideisoquinolines* (*694*).

hydrastine, bicuculline, and cordrastine *II* (*erythro* series) (formula **B**) and in adlumine, capnoidine, and cordrastine *I* (*threo* series) the ring **C** is located in the proximity of the ring **A** (formula **D**) (Scheme 44).

A simple synthesis of racemic cordrastine and hydrastine can be accomplished by condensation of 3,4-dihydro-6,7-dimethoxy-2-methyl- or 3,4-dihydro-6,7-methylenedioxy-2-methylisoquinolinium salts with methyl-6-diazomethyl-2,3-dimethoxybenzoate, derived from 6,7-dimethoxyphthalimide (*697*).

The alkaloids of the narcotine type can also be synthesized from benz[*d*]indeno[1,2-*b*]azepine (**133a**) (*698*) (Scheme 45). Moreover, compound **133a** forms a key substance for the synthesis of the tetrahydroprotoberberine (**58**), protopine (**101**), rhoeadane (**154**), and spirobenzylisoquinoline (**191**) ring skeletons. The compounds **133a** and **133b** arise also by rearrangement from the spirobenzylisoquinoline, protoberberine, and 1-benzoylisoquinoline skeletons. Therefore, it is assumed that even in the plants it plays a key role in the formation and interconversion of the benzylisoquinoline alkaloids with 17 carbon atoms in the skeleton (Scheme 45).

Synthetically, the compound **133** (**c**) (Scheme 46) is easily available from 3,4-dihydropapaveraldine methiodide (**134**) (**a**), which with diazomethane affords first compound **b** and then, by reaction with $POCl_3$ in toluene, the chloro derivative of the compound **c** (besides a small quantity of the formyl derivative of the compound **f**). Oxidation of substance **c** with permanganate in acetone at 0° gives the lactone **d** (**161**). On reduction with $NaBH_4$ it yields the phthalidetetrahydroisoquinoline **e** (**125**) which, according to the ^{1}H NMR spectra, belongs to the *erythro* series (*698*).

The dimerization of narcotoline (**123i**) and its alkyl and aryl derivatives, and the Hofmann degradation of these derivatives were studied by Gorecki and Kubala (*699, 700*). The preparation of the isomers of narcotine was reported (*701–703*).

In view of the attempts of the pharmaceutical industry to obtain derivatives of known alkaloids, Teitel and O'Brien developed methods of gentle saponification of the methoxyl and the methylenedioxy group, respectively, or its complete removal with retention of the optical activity of the skeleton. They found that the optically active dimethoxymethylenedioxy-substituted benzylisoquinoline, phthalide, and tetrahydroprotoberberine can be O-demethylated with boron trichloride and following hydrogenolysis of the bistetrazoyl ether intermediate over Pd–C. During this reaction the compounds retain their absolute configuration (*448*). On the contrary, the aporphine alkaloids are converted into racemates during this procedure. Therefore, the benzylisoquinolines, phthalides, and tetrahydroprotoberberines are far more resistant to hydrogenolysis over Pd–C and dehydrogenation than are the compounds of the aporphine type (*346, 704*).

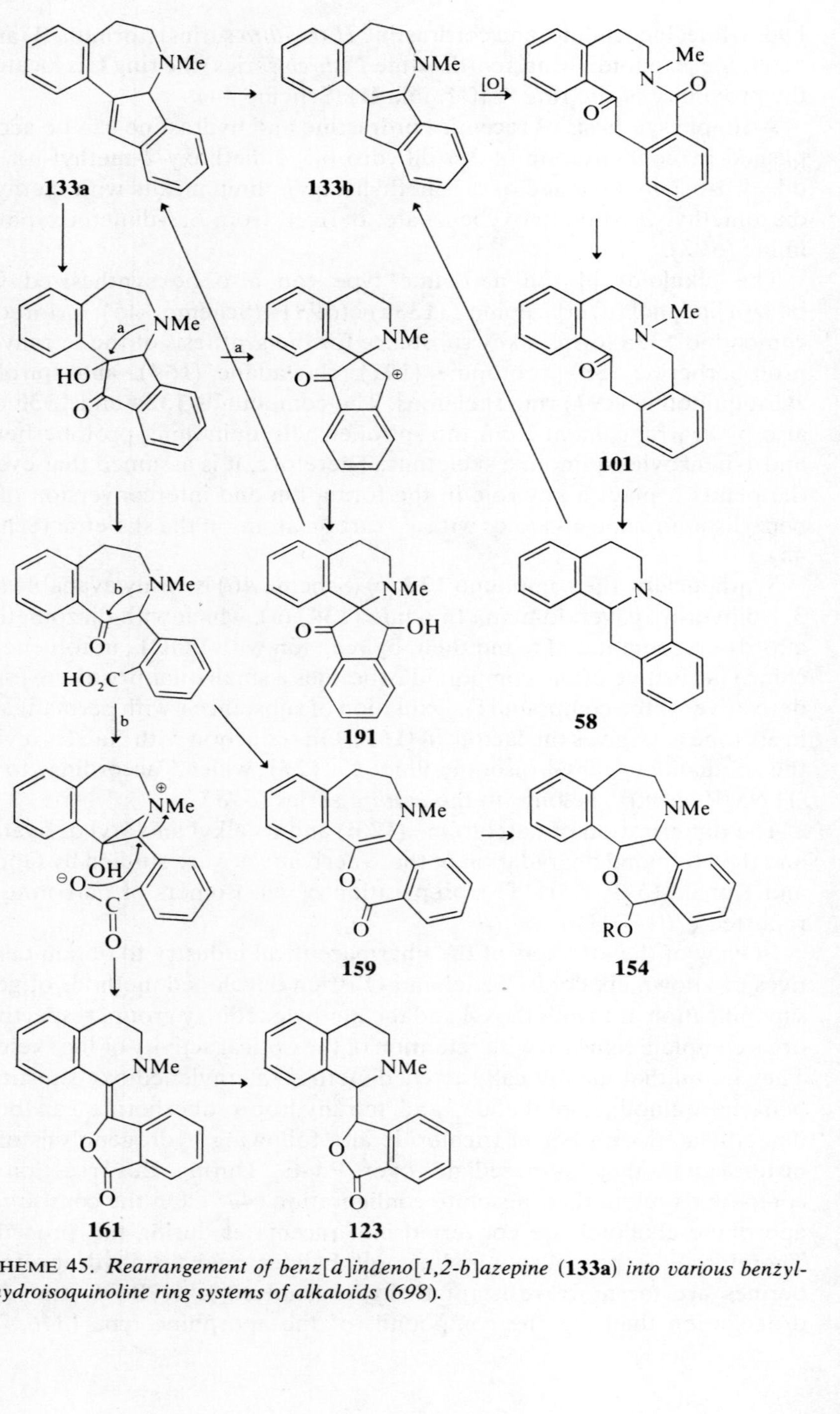

SCHEME 45. *Rearrangement of benz[d]indeno[1,2-b]azepine* (**133a**) *into various benzyltetrahydroisoquinoline ring systems of alkaloids* (*698*).

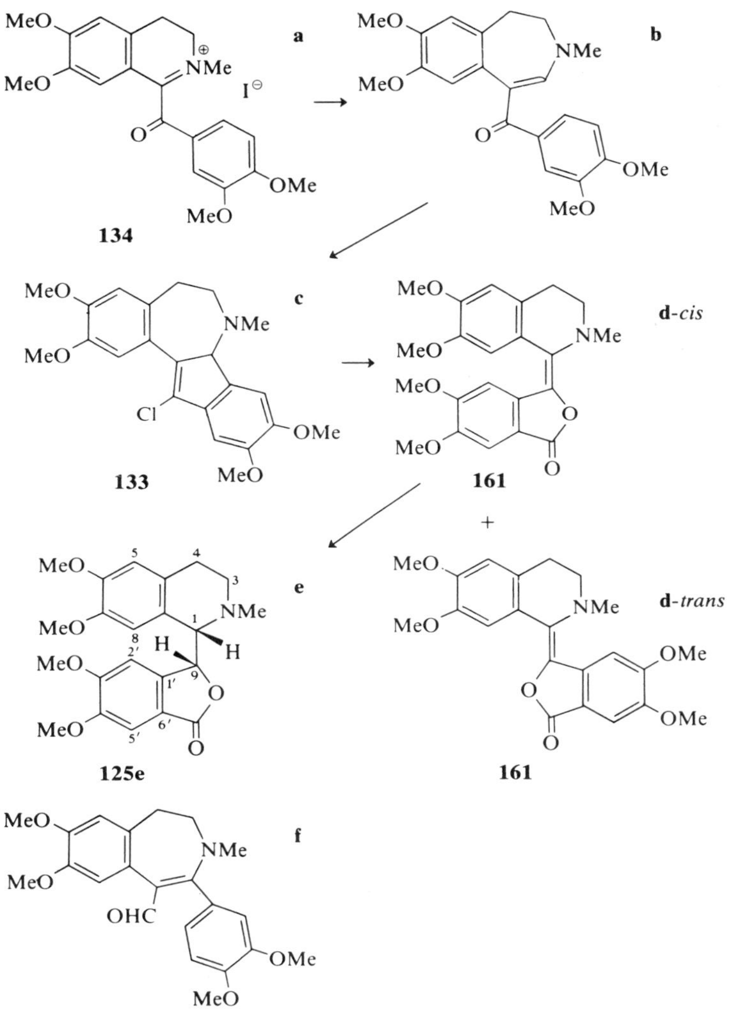

SCHEME 46. *Synthesis of chlorobenz[d]indeno[1,2-d]azepine* (**c**) *and phthalideisoquinoline compounds* (**125**) *from 3,4-dihydropapaveraldine methiodide* (**134**) (*698*).

Upon reaction with diazomethane or dichloromethane, the phenols can be reetherified. Alkaline isomerization affords the corresponding C-9 epimers (Scheme 47) (*704, 705*).

The reduction of phthalidetetrahydroisoquinoline alkaloids was studied in an attempt to duplicate the hemiacetal moiety of rhoeagenine **(154c)** (*706*). The final product depends on the temperature and on the reducing

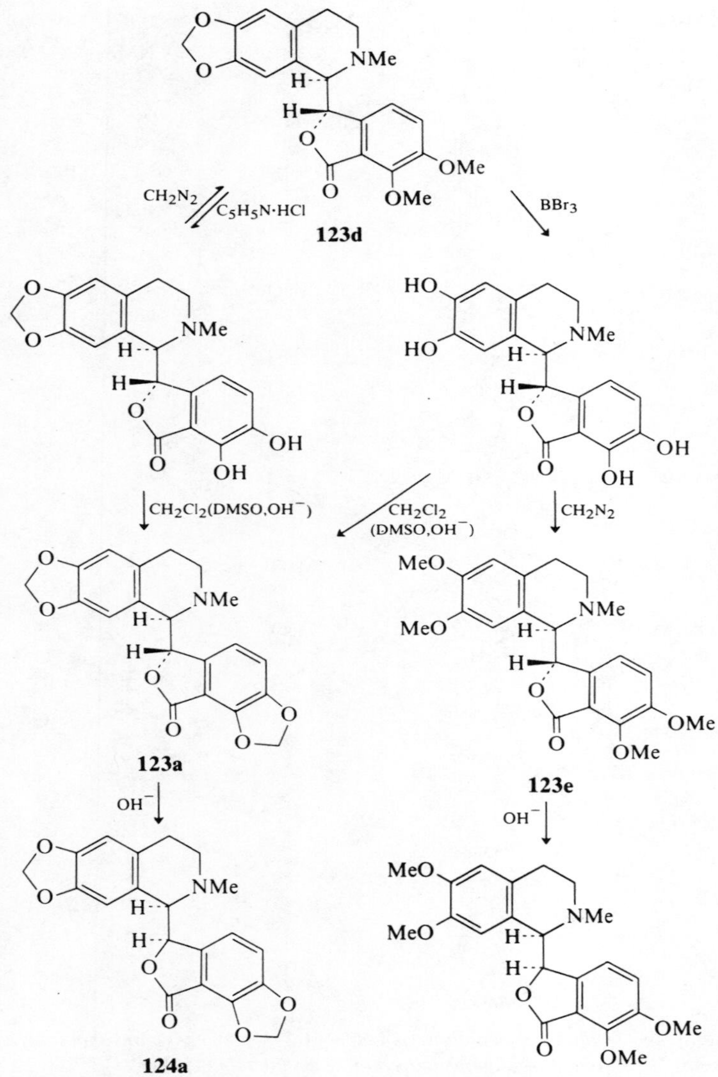

CH_2N_2 ⇌ $C_5H_5N{\cdot}HCl$

123d

BBr_3

$CH_2Cl_2(DMSO,OH^-)$

CH_2Cl_2 $(DMSO,OH^-)$

CH_2N_2

123a

OH^-

123e

OH^-

124a

SCHEME 47. *Interconversions of phthalidetetrahydroisoquinolines* (*704, 705*). DMSO, *Dimethyl sulfoxide.*

agent: Upon reduction with lithium aluminum hydride in ether, narcotine affords narcotinediol (**129a**). Reduction with the same reagent in boiling tetrahydrofuran yields a mixture of narcotinediol and C-8 *O*-demethyl-narcotinediol (**129b**) (Scheme 49). Reduction of narcotine in ether at room temperature with lithium trimethoxyaluminum hydride furnishes narcotinediol and the hemiacetal **135**. Phenylmagnesium bromide reacts with α-narcotine (**123h**) to form two hemiketals, **137a** and **137b** (intermediary product **136**), which are interconvertible in solution via the hydroxy ketone (*707, 708*). With benzylmagnesium bromide, the benzylidene derivative **138** is formed together with hydrocotarnine (Scheme 48).

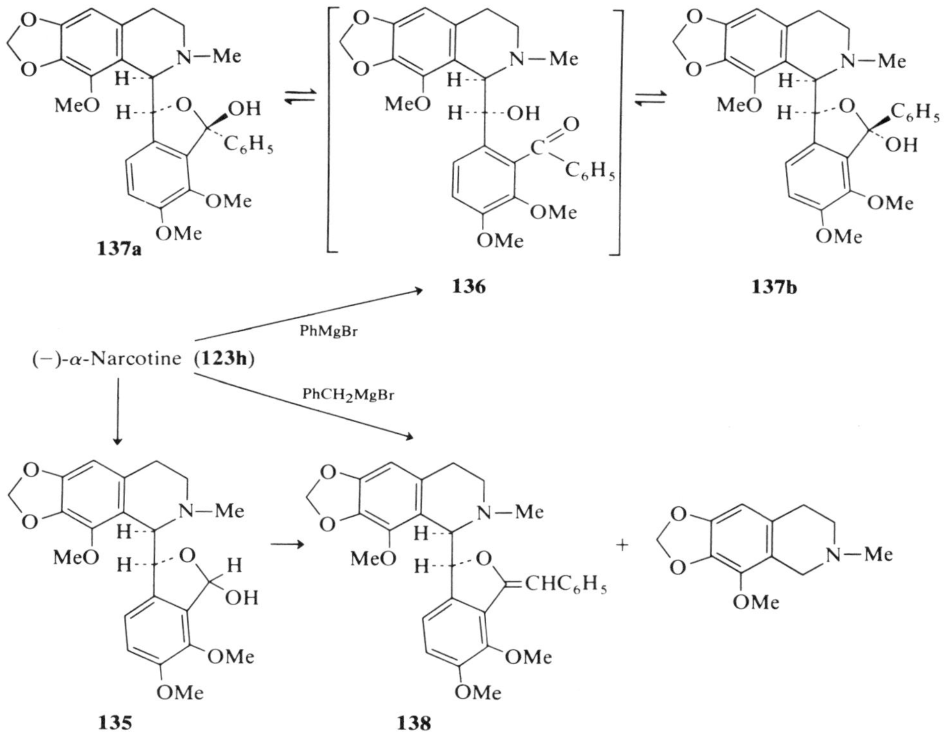

SCHEME 48. *Reaction of α-narcotine* (**123h**) *with Grignard reagents or trimethoxyaluminum hydride* (*706–708*).

Tosylation of α-narcotinediol (**129a**) with p-toluenesulfonyl chloride in pyridine gives the quaternary tosylate **139** (*709*). Boiling in acetic anhydride leads to elimination of the tosyl group to give rise to the compound **140**. On boiling with acetic anhydride in the presence of sodium iodide, the substances **139** and **140** yield the pentacyclic compound **141**. The resulting enamine **140** is acylated in the β-position to the nitrogen atom and, after partial O-demethylation, is cyclized to the compound **141**. Reduction of this compound with $NaBH_4$ results in the formation of the enamine **142**. With methyl iodide it is alkylated in the β-position to the nitrogen atom to give rise to the quaternary salt **143** (Scheme 49).

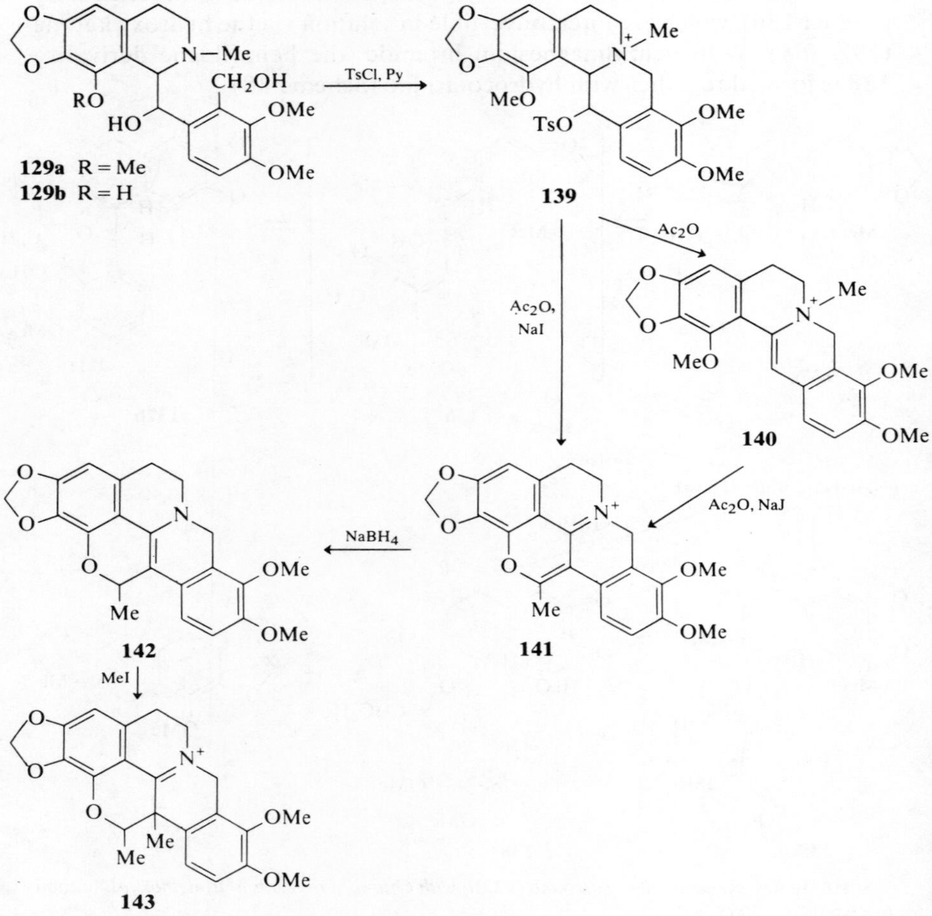

129a R = Me
129b R = H

TsCl, Py

139

Ac_2O

Ac_2O, NaI

140

Ac_2O, NaJ

NaBH$_4$

142

141

MeI

143

SCHEME 49. *Conversion of narcotinediol* (**129a**) *by tosylation and with acetic acid anhydride* (*709*).

The absolute configuration of phthalidetetrahydroisoquinoline alkaloids was studied by circular dichroism (*710, 711*). The earlier conclusions were based on the dissociation constants and ORD data and on some studies of organic-preparation [summarized in Shamma (*1*), p. 364: conversion of (−)-β-hydrastine into (−)-13-epiophiocarpine and of (−)-α-hydrastine into (−)-ophiocarpine (**70a**); these two ophiocarpines yield (−)-canadine on hydrogenolysis]. The absolute configuration of (−)-bicuculline was established by X-ray analysis (*704, 712*). Bicuculline was also studied by molecular orbital methods and proton magnetic resonance (*713*); the conformation of lowest energy for the *N*-methylbicuculline cation were those in which the N-containing ring was in the "half-chair" form **a**, with the phthalide ring pseudoaxial and the *N*-methyl groups staggered (see Fig. 1).

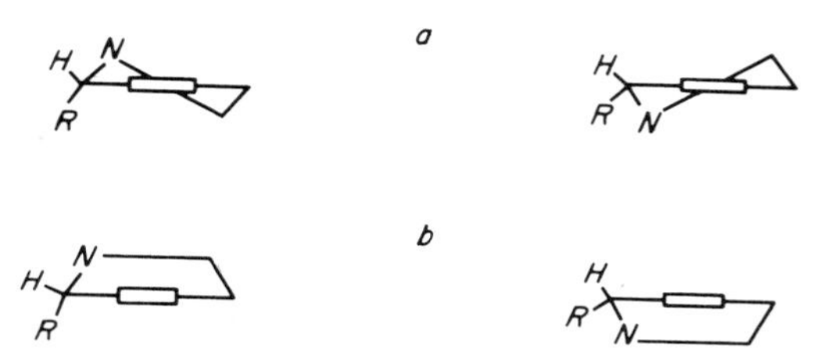

FIG. 1. Forms of the N-containing ring of bicuculline in the molecular orbital calculations. (a) "Half-chair" forms; (b) "half-boat" forms: ring pseudoaxial (left); ring pseudoequatorial (right) (*713*).

Similar results were obtained for the protonated bicuculline cation, in which the proton occupied the equatorial position. These studies were carried out to clarify the effect of the relatively selective antagonists of these two compounds as postsynaptic inhibitors on the mammalian central nervous system.

The UV spectroscopy of phthalidetetrahydroisoquinoline bases have been discussed in Section III,O,1. Stepanenko *et al.* described the spectrophotometric determination of narcotine in plant material by using chloramine B (*714*).

L. NARCEINE AND NARCEINEIMIDE GROUP*

Recently, Preininger *et al.* have isolated two new alkaloids of the narceine group from *Corydalis sempervirens*, namely, adlumidiceine (**144a**) and adlumiceine (**144b**) (*715*). These two alkaloids were isolated in the carboxyl form. The alkaloid *Z*-aobamidine was isolated from *C. ochotensis* var. *raddeana* (*118, 120*), and is an enol lactone of adlumidiceine (**144a**). The related alkaloid bicucullinine (**145b**) was isolated from *C. ochroleuca* and the alkaloid cryptopleurospermine (**145c**) from *Cryptocarya pleurosperma* (Lauraceae) (*121, 716*). From *Corydalis* and *Fumaria* plants were isolated the alkaloids *N*-methylhydrastine (**144c**) and *N*-methyloxohydrasteine (**145a**) (*111*).

Trojánek *et al.* isolated the narceineimide **146a** from poppy capsules (*272*). The extract from this material was subjected to column chromatography on silica gel and thin-layer chromatography. On the basis of chemical

* This material is supplementary to *The Alkaloids*, Vol. IV, p. 147; Vol. XII, p. 397.

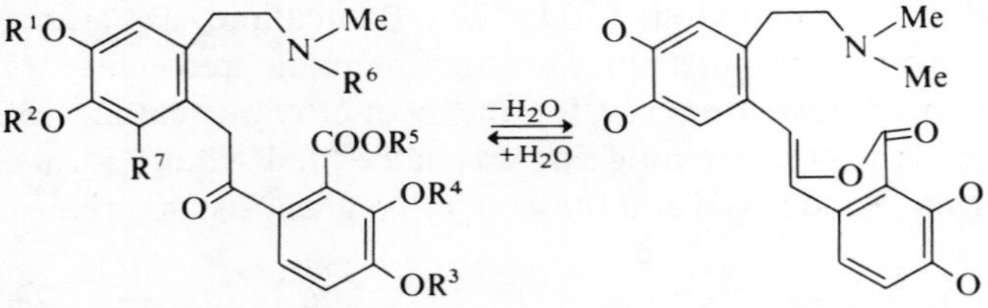

$-H_2O$ / $+H_2O$

144a $R^1+R^2=R^3+R^4=CH_2$; $R^5=R^7=H$; $R^6=Me$ Adlumidiceine
144b $R^1=R^2=R^6=Me$; $R^3+R^4=CH_2$; $R^5=R^7=H$ Adlumiceine
144c $R^1+R^2=CH_2$; $R^3=R^4=R^6=Me$; $R^7=H$ *N*-Methylhydrastine
144d $R^1+R^2=CH_2$; $R^3=R^4=R^6=Me$; $R^5=Et$; $R^7=OMe$ Nornarceine ethyl ester
144e $R^1+R^2=CH_2$; $R^3=R^4=Me$; $R^5=R^6=H$; $R^7=OMe$ Nornarceine
144f $R^1+R^2=CH_2$; $R^3=R^4=R^6=Me$; $R^5=H$; $R^7=OMe$ Narceine
144g $R^1+R^2=R^3+R^4=O{-}CH_2{-}O$; $R^5=R^7=H$; $R^6=Me$ Z-Aobamidine
Aobamidine was isolated as an enol lactone of adlumidiceine (**144a**)
N-Methylhydrastine (**144c**) is also an enol lactone

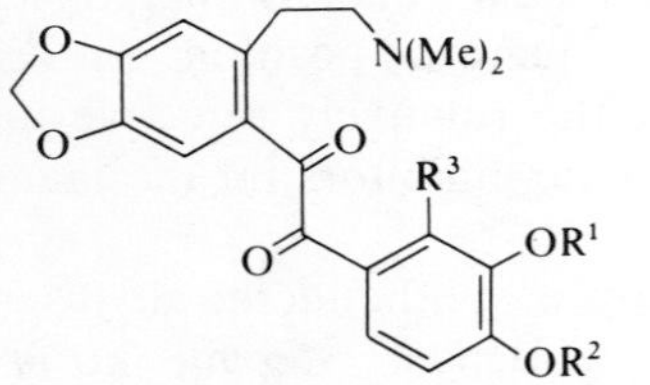

145a $R^1=R^2=Me$; $R^3=COOH$ *N*-Methyloxohydrasteine
145b $R^1+R^2=CH_2$; $R^3=COOH$ Bicucullinine
145c $R^1=R^3=H$; $R^2=Me$ Cryptopleurospermine

degradation, spectral analysis, and a comparison with a synthetic specimen, the structure of narceineimide was established (Scheme 50).

Shamma and Moniot showed that the alkaloids fumaridine and fumaramine do not have the erroneously assigned (*171*) protopine-type structure but the structure derived from narceineimide (**146a**) (*671*). The structure of fumaridine corresponds to the known *N*-methylhydrasteineimide (**146b**), and the structure of fumaramine corresponds to that of the adlumidiceineimide (**146c**) (*121, 717*). Shamma and Moniot also mention in their work that narceineimide compounds may not be true alkaloids, but artifacts formed during work-up.

Trojánek *et al.* found that the thermal decomposition of narceineimide methiodide in the presence of aqueous potassium hydroxide affords a mixture of stereoisomeric narceoneimides (**148**) along with 10,11-methylenedioxy-3,4,12-trimethoxy-7,8-dihydro-5*H*-isoindolo[1,2-*b*][3]benzazepine-5-one (**149**), which is identical with the oxidized red product

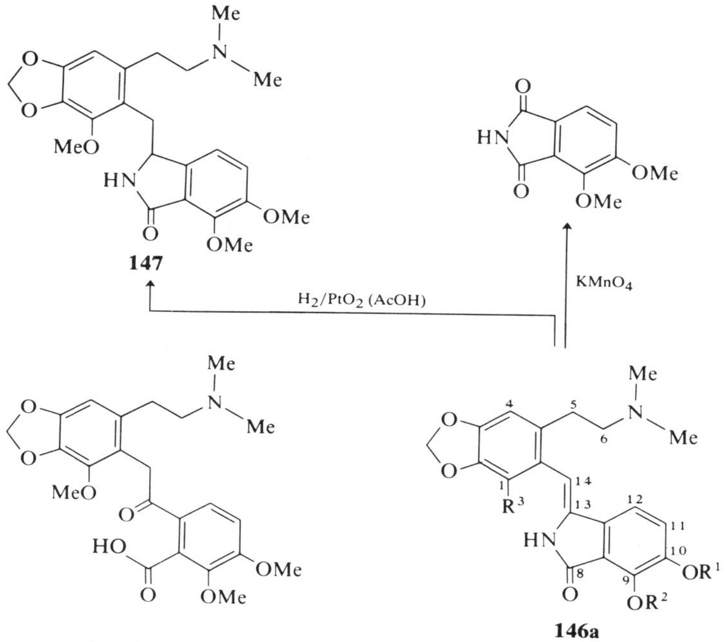

Me
N
Me
O
O
MeO
HN
O
OMe
OMe
147

HN
O
O
OMe
OMe

H_2/PtO_2 (AcOH)

$KMnO_4$

O
O
MeO
O
HO
O
OMe
OMe
Me
N
Me

O
O
R^3
HN
O
OR^2
OR^1
Me
N
Me
146a

146a $R^1 = R^2 = Me$; $R^3 = OMe$ Narceineimide
146b $R^1 = R^2 = Me$; $R^3 = H$ *N*-Methyl hydrasteineimide (Fumaridine)
146c $R^1 + R^2 = CH_2$; $R^3 = H$ Adlumidiceineimide (Fumaramine)

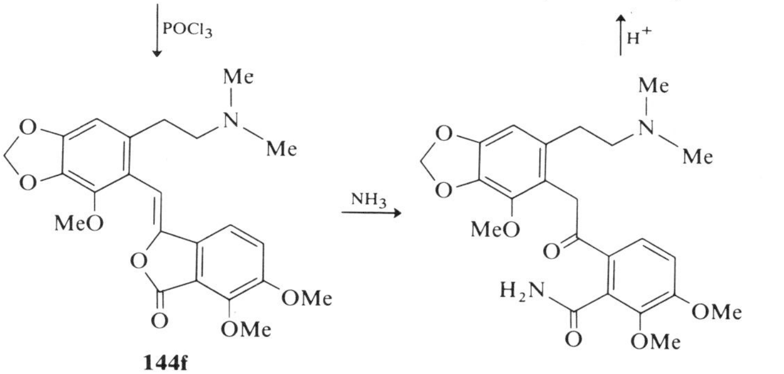

$POCl_3$

H^+

O
O
MeO
O
O
OMe
OMe
Me
N
Me
144f

NH_3

O
O
MeO
O
H_2N
O
OMe
OMe
Me
N
Me

SCHEME 50. *Reactions and synthesis of narceineimide (272).*

obtained from papaverrubines (*718, 719*). During the thermal decomposition of narceineimide methohydroxide, the resulting narceoneimides are accompanied by a lesser amount of the compound **150** and of 7-methyl-9,10-methylenedioxy-3,4,11-trimethoxy-6,7-dihydro-5*H*-isoindolo[1,2-*b*]isoquinol-5-one (**151**) (Scheme 51).

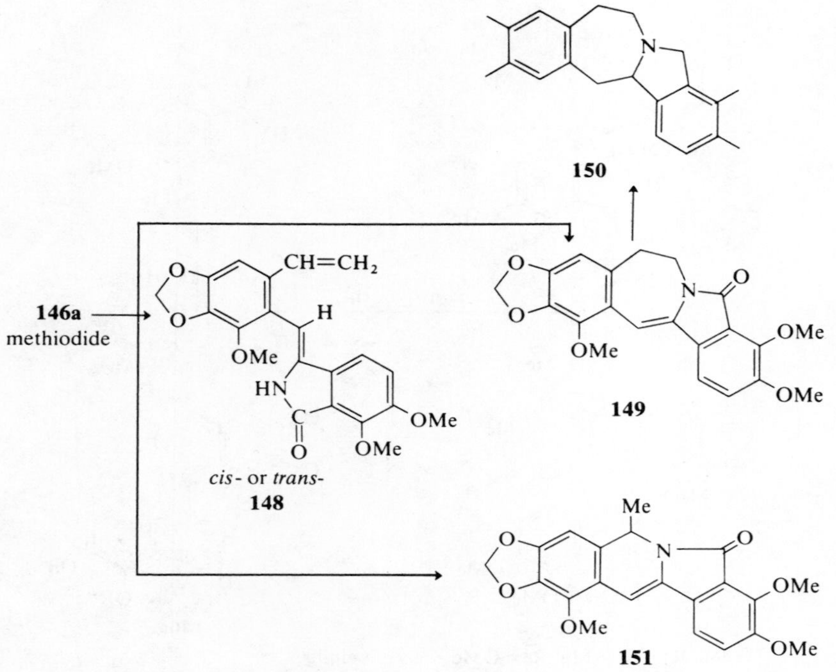

SCHEME 51. *Conversion of narceineimide methiodide* (**146a**) *and determination of its structure* (*718*).

Klötzer *et al.* developed a method to prepare *N*-nornarceine from narcotine (*720, 721*). It consists in conversion of narcotine into its *N*-benzyl bromide which on Hofmann degradation affords *N*-benzylnarcotinium bromide. Catalytic debenzylation yields the nornarceine ethyl ester hydrochloride (**144d**) and nornarceine (**144e**).

M. RHOEADANE GROUP*

Formerly, rhoeadine and papaverrubine alkaloids were known to occur only in the plants of the genus *Papaver*. Recently, Slavík *et al.* have shown that rhoeadine and papaverrubine E are present in *Bocconia frutescens* and papaverrubine D in *Meconopsis betonicifolia*, *M. horridula*, *M. napaulensis*, *M. paniculata*, *M. robusta*, *M. rudis*, and *M. sinuata* (*33, 50*). The genus *Meconopsis* is botanically closely related to the genus *Papaver*, a relation thus confirmed chemotaxonomically (see structures **152–154** in Table V).

* This material is supplementary to *The Alkaloids*, Vol. X, p. 474, Vol. XII, p. 398.

TABLE V

RHOEADANE ALKALOIDS AND THEIR RELATIVE CONFIGURATION AT C-1, C-2, AND C-14

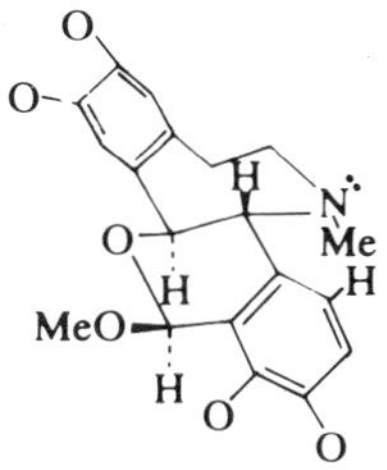

152 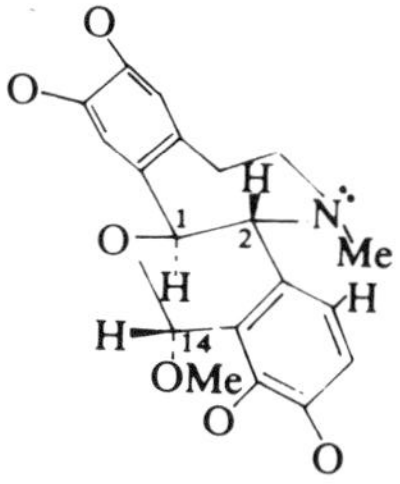 **153** 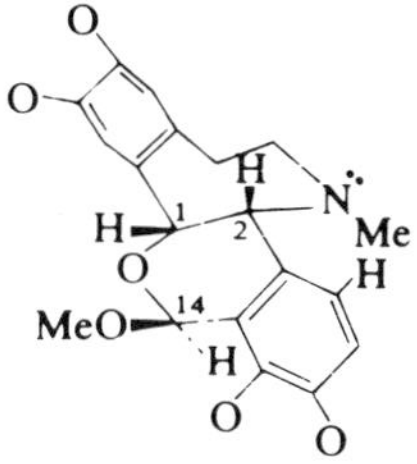 **154**

Structure	Substituents at C and N atoms: 3	7	8	11	12	14	Nomenclature: *1S,2R,14S*	*1S,2R,14R,*	*1R,2R,14S*
a	Me	CH_2		CH_2		Me	Isorhoeadine	*epi*-Isorhoeadine	Rhoeadine
b	Me	CH_2		CH_2		Glucose	Isorhoeagenine–glucoside	—	—
c	Me	CH_2		CH_2		H	Isorhoeagenine	—	Rhoeagenine
d	Me	Me	H	CH_2		Me	—	*N*-Methyl-14-*O*-demethylepiporphyroxine	—
e	Me	Me	Me	CH_2		Me	Glaudine	Epiglaudine	Oreodine
f	Me	Me	Me	CH_2		H	Glaucamine	—	Oreogenine
g	Me	Me	Me	Me	Me	Me	Alpinine	*epi*-Alpinine	—
h	Me	Me	Me	Me	Me	H	—	Alpinigenine	—
i	H	CH_2		CH_2		Me	Papaverrubine A	—	Papaverrubine E
j	H	Me	Me	CH_2		Me	Papaverrubine B	—	Papaverrubine F
k	H	Me	H	CH_2		Me	Papaverrubine D	Papaverrubine C	—
l	H	Me	Me	Me	Me	Me	Papaverrubine G	—	—

The alkaloid alpinigenine (**153h**) was rediscovered in one species of *P. bracteatum* [*P. pseudo-orientale*? (*298*)] which produces mainly thebaine (*232, 722*). The preparation of epialpinine (**153g**) has been described (*232*).

The dimeric alkaloid stepionine was isolated from *Stephania japonica* Miers. (Formosa) (*723*). Reductive fission transforms it into the 3-benzazepine derivative, which is closely related to the rhoeadane alkaloids. It has also been prepared synthetically (*724*).

Rozwadowska and ApSimon again studied the Hofmann degradation of rhoeagenine which was originally used by Späth (*725, 726*). They prepared small quantities of derhoeagenine, epi-derhoeagenine, and of the compounds **155** and **156**, whose structures were elucidated by mass spectrometry and ^{1}H NMR analysis.

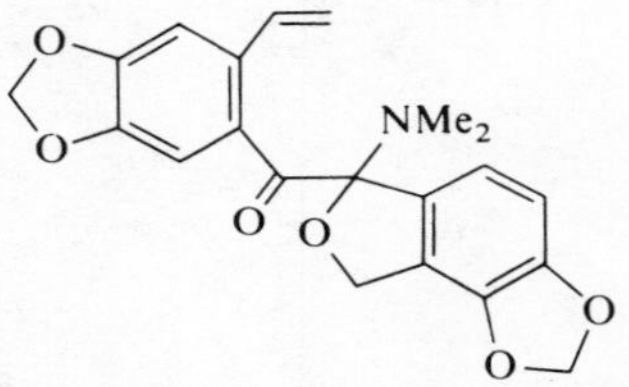

155 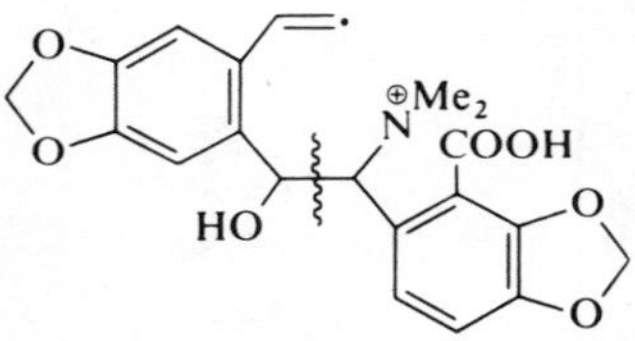 **156**

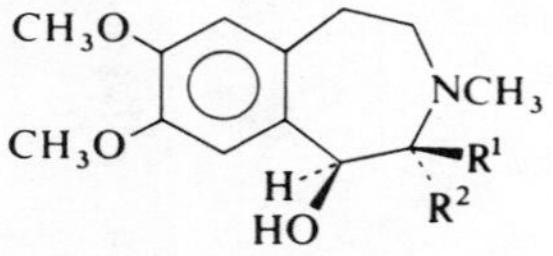

$R^1 = H$, $R^2 =$ phenyl
$R^1 =$ phenyl, $R^2 = H$

157

Inubushi *et al.* carried out the synthesis of *trans* and *cis* isomers of 7,8-dimethoxy-*N*-methyl-2-phenyl-1,2,3,4-tetrahydro-3*H*-3-benzazepin-1-ol (**157**), a compound without electron-donating substituents on the second aromatic ring (*727*). Recently, Shamma and Töke have described an interesting and simple method for the synthesis of rhoeadane skeleton which arises from *N*-benzyl-3,4-dihydroisoquinoline (**a**). Its double bond between C-1 and the quaternary nitrogen permits an opening of the ring B (**b**) and the formation of isorhoeadane (**152**) and 14-oxoisorhoeadane (**159**) (*728*) (Scheme 52).

The photosensitized oxidation of an enaminoketone (**133b**) and its conversion to *cis*-alpinine (**154g**) via **158** and **159** was carried out by Orito, Manske, and Rodrigo (*729*) (Scheme 53).

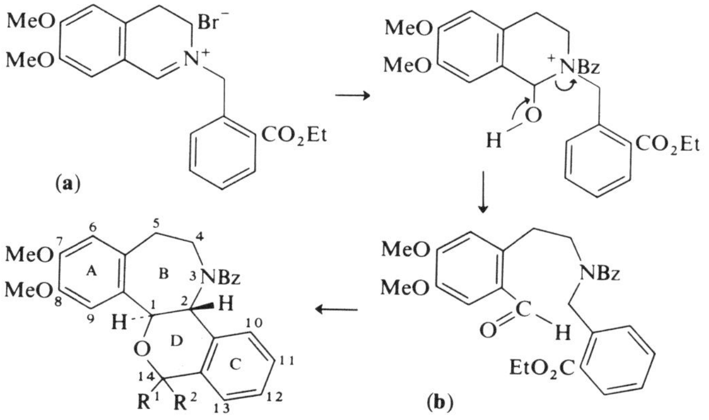

159 $R^1, R^2 = O$
152f $R^1 = H, R^2 = OH$
152e $R^1 = H, R^2 = OMe$

SCHEME 52. *Synthesis of trans-B/D rhoeadane derivatives* (**152**) *according to Shamma and Töke* (*728*).

Nalliah, Manske, and Rodrigo also described the transformation of 13-oxyprotoberberinium salts (**160**) into alkaloids of rhoeadane type (Scheme 54) (*730*). This transformation involves an intermediary benz[*d*]indeno[1,2-*b*]azepine (**133b**) type of compound, as subsequently suggested by Kametani *et al.* (*359, 698*) (Schemes 45, 53, and 55).

A summarizing report on the synthesis of benzazepine compounds and related substances of rhoeadane types was published by Kametani and Fukumoto (*344*).

Independently, Irie *et al.* and Klötzer, Teitel, Blount, and Brossi (*731, 732, 733*) developed two syntheses of rhoeadine and alpinigenine. The first method was based on the rearrangement of the rings B and C in 1,2,3,4-tetrahydro-2-methyl-4′,5′,6,7-bismethylenedioxyisoquinoline-1-spiro-2′-indan-1′-ol into the compound **133d** or **133e** and, finally, to rhoeadine (Scheme 55).

The second method consisted of the conversion of natural phthalidetetrahydroisoquinolines (**123**) into rhoeadane alkaloids (*732, 733*) (Scheme 56).

Recently Prabhakar *et al.* described a simple synthesis of (+)-*cis*-alpinigenine from tetrahydroprotoberberinium hydroxide (methoxyl groups in positions 2, 3, 11, 12) which was converted into a *trans*-azocine intermediate and then into the 13,14-diol. On treatment with periodic acid the latter yielded the corresponding dialdehyde. Photolysis of *o*-aminomethylbenzaldehyde (in butanol, N_2, pyrex filter, and 34°) afforded directly (±)-*cis*-alpinigenine (28% yield) (*733a*).

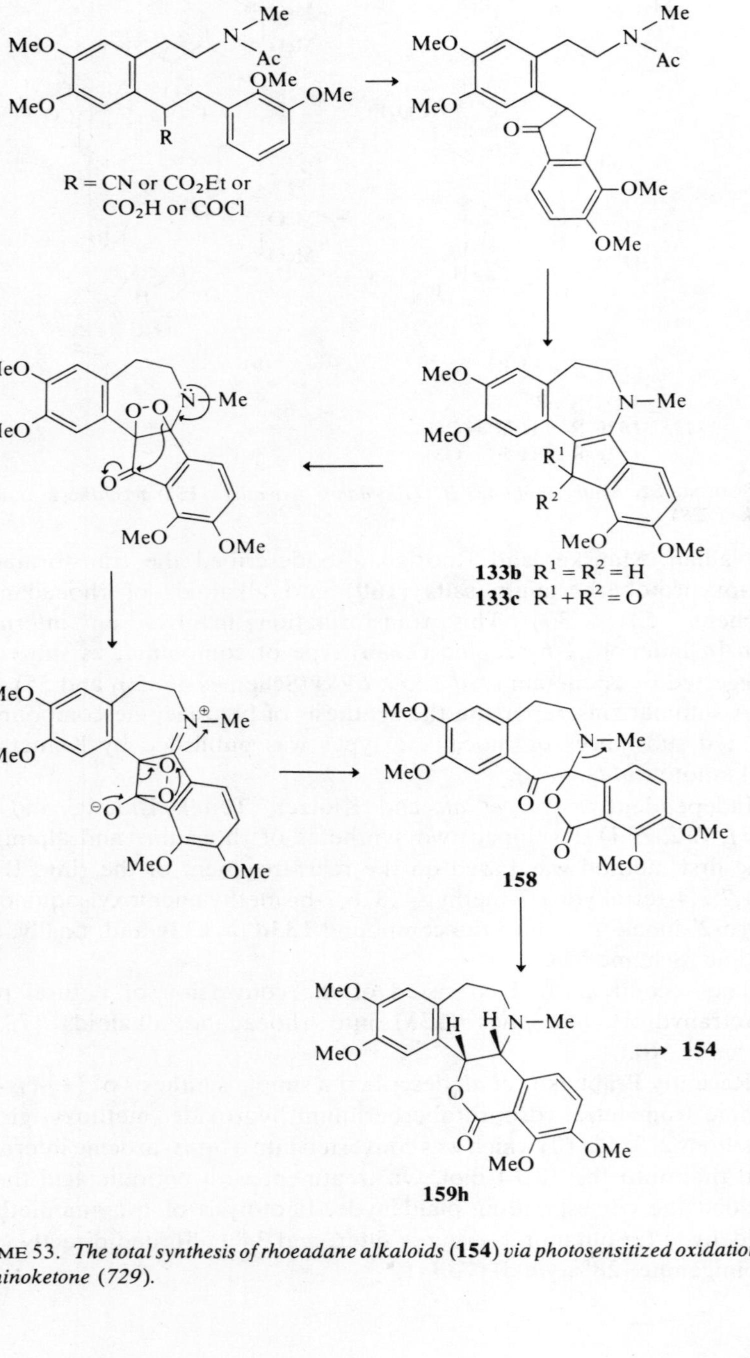

MeO, MeO, N, Me, Ac, OMe, OMe, R

R = CN or CO_2Et or CO_2H or COCl

MeO, MeO, N, Me, Ac, O, OMe, OMe

MeO, MeO, N—Me, R^1, R^2, MeO, OMe

133b $R^1 = R^2 = H$
133c $R^1 + R^2 = O$

MeO, MeO, O–O, N—Me, O, MeO, OMe

MeO, MeO, $N^{\oplus}$—Me, O, O, $^{\ominus}O$, MeO, OMe

MeO, MeO, N—Me, O, O, O, OMe, MeO

158

MeO, MeO, H, H, N—Me, O, O, MeO, OMe → **154**

159h

SCHEME 53. *The total synthesis of rhoeadane alkaloids* (**154**) *via photosensitized oxidation of an enaminoketone* (*729*).

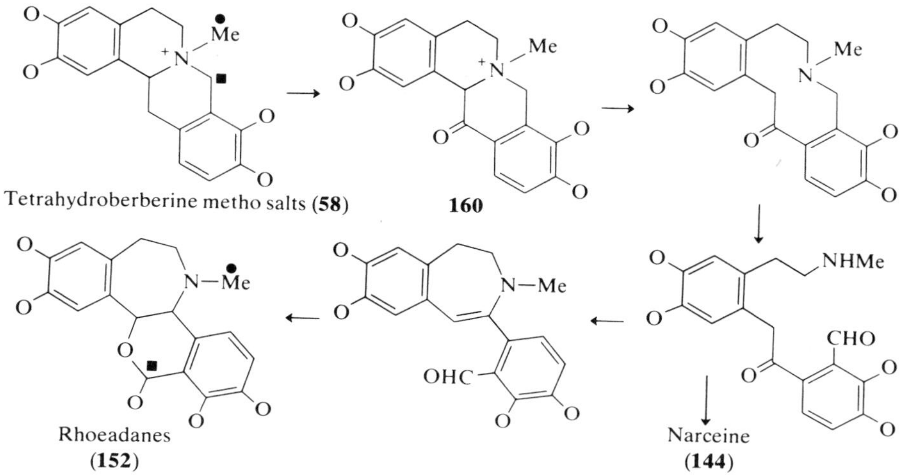

Tetrahydroberberine metho salts (**58**) → **160** → → Narceine (**144**); → Rhoeadanes (**152**)

SCHEME 54. *A suggested pathway for the* in vivo *transformation of tetrahydroprotoberberine metho salts via 13-oxoprotoberberinium salts* (**160**) *into alkaloids of narceine* (**144**) *and rhoeadane types* (**152**) (*730*). ■, ●—*signifies* ^{14}C *labels.*

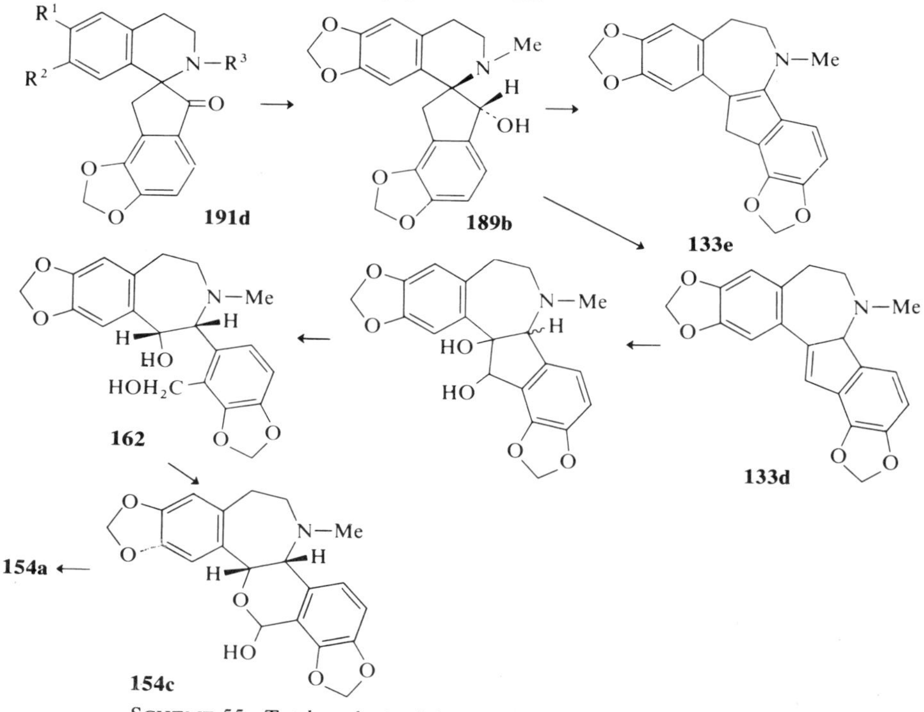

191d → **189b** → **133e**; **189b** → **133d** → → **162** → **154c** → **154a**

SCHEME 55. *Total synthesis of the alkaloid rhoeadine* (**154a**) (*731*).

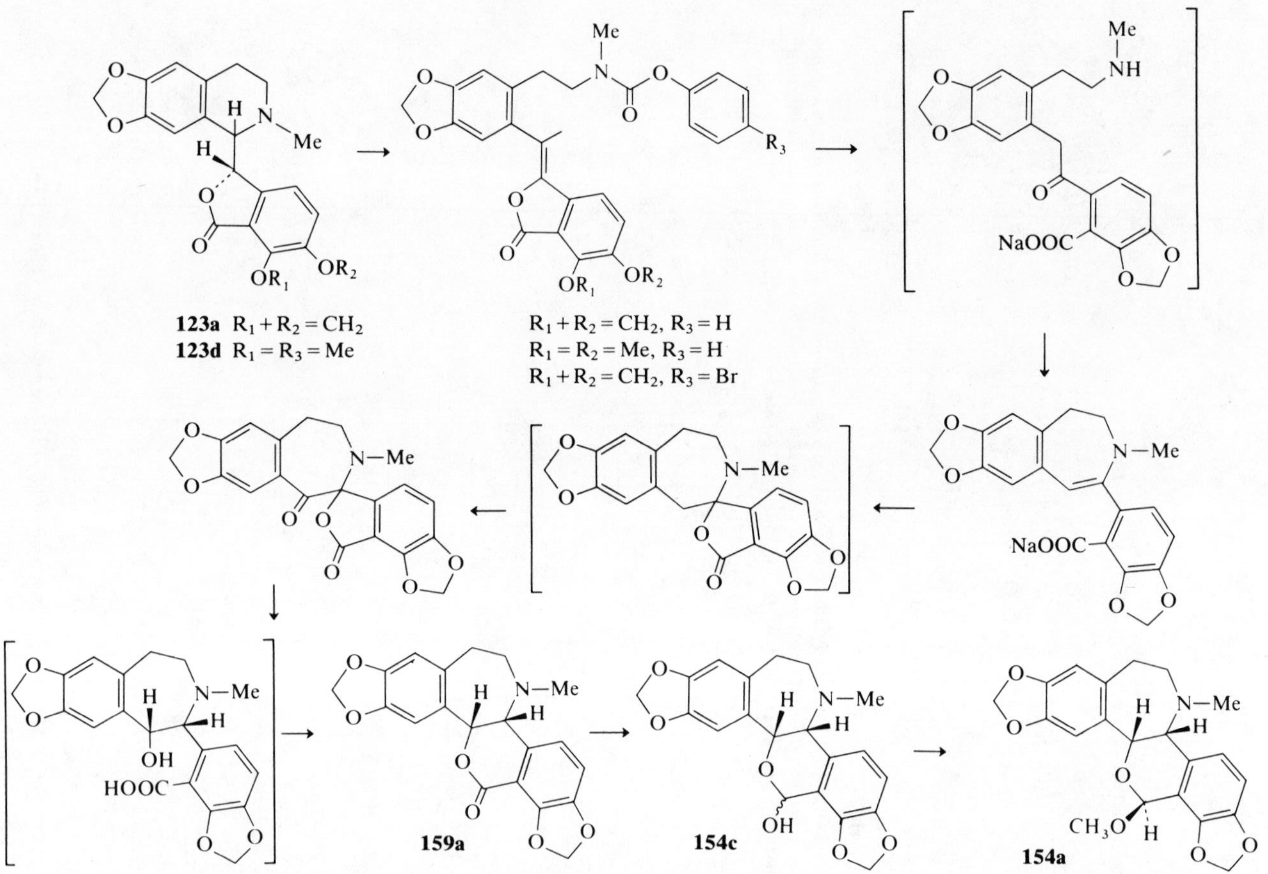

123a $R_1 + R_2 = CH_2$
123d $R_1 = R_3 = Me$

$R_1 + R_2 = CH_2$, $R_3 = H$
$R_1 = R_2 = Me$, $R_3 = H$
$R_1 + R_2 = CH_2$, $R_3 = Br$

159a **154c** **154a**

SCHEME 56. *Conversion of phthalidetetrahydroisoquinoline alkaloids* (**123**) *into alkaloids of rhoeadane type* (**154**) (*733*).

In an earlier paper on papaverrubine derivatives, it was reported that on treatment with dilute acid they were transformed into deep red substances which were assigned the structure of a quasi-dehydrogenated "Schöpf base VI" (**104b**) (*719, 734*). This isoindolo[1,2-*b*][3]benzazepine derivative has now been prepared synthetically at four research centers (*100, 735–737*). Biogenetically interesting are those syntheses which are based on the naturally occurring narceineimide (*737*) or on phthalidetetrahydroisoquinoline alkaloids (*100*). The first synthesis of a papaverrubine-type compound was elaborated by Hohlbrugger and Klötzer (*736*).

Trojánek *et al.* studied the thermal decomposition of 10,11-methylenedioxy-3,4,12-trimethoxy-7,8,13,13a-tetrahydro-5*H*-isoindolo[1,2-*b*]-[3]benzazepine methohydroxide (methoxy analog of Schöpf's base VI) (**104b**) (*738*). There arise three "des" bases. **163a**, **163b**, and **163c**, the first of which is convertible into 1-methoxyallocryptopine (**101e**) (Scheme 57).

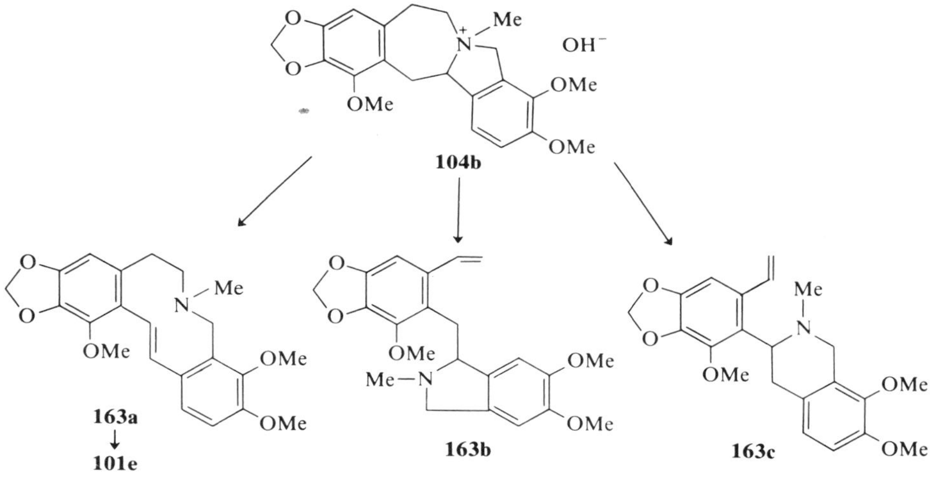

SCHEME 57. *Hofmann degradation of 1-methoxy-"Schöpf's base VI"* (**104b**) *into the compounds* **163a**, **163b**, **163c**, *and* **101e** (*738*).

Battersby *et al.* used (*13S*)- and (*13R*)-labeled scoulerine (**58b**) to study the biosynthesis of rhoeadine. Both of them were incorporated by living *P. rhoeas* plants into rhoeadine. This alkaloid, isolated from plants fed with (*13S*)-labeled scoulerine, had lost 79% of the tritium present in the precursor, whereas the (*13R*)-labeled scoulerine afforded rhoeadine which retained 74% of the original tritium. Bearing in mind the configurational purity of the precursors, these values prove that a stereospecific loss of the pro-*S*-hydrogen occurs from C-13 of scoulerine at some stage of its transformation into rhoeadine. It seems likely from structural considerations that, as before, (−)-scoulerine is converted first into stylopine. Enzymatic hydroxylation at C-13 from the rear side with retention of configuration would remove the pro-*S*-hydrogen to set up the chirality (**a**) for rearrangement possibly via the phosphate (**b**) (Scheme 58); hydroxylation at C-8 is also envisaged. Rearrangement and N-methylation (not necessarily in this order) then provide the correct skeleton from which rhoeadine (**154a**) could arise by obvious steps. Consequently, (+)-rhoeadine is biosynthesized from (−)-scoulerine and not from its (+) enantiomer (*739*).

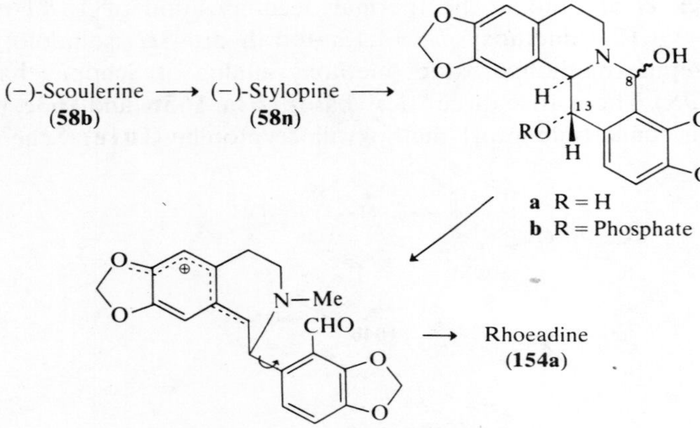

(−)-Scoulerine (**58b**) ⟶ (−)-Stylopine (**58n**) ⟶

a R = H
b R = Phosphate

⟶ Rhoeadine (**154a**)

SCHEME 58. *Pathway of the biosynthesis of rhoeadine* (**154a**) (*739*).

Tani and Tagahara also studied the biosynthesis of rhoeadine and reported that in *P. rhoeas* this alkaloid arises from α-stylopine methochloride (**58n**) via protopine. β-Stylopine methochloride does not participate in this biosynthesis (*740*).

Papaver bracteatum produces alpinigenine (**153h**) during its early stages of development, but this substance is not detectable in older plants (*722*). It has been shown that the leaves of older plants are able to form this alkaloid after being fed tetrahydropalmatine (**58g**). Leaves of *P. orientale*—a species related to *P. bracteatum*, which normally does not produce alpinigenine—do not synthesize this alkaloid after application of tetrahydropalmatine. These results support the assumption that the absence of alpinigenine in old thebaine-producing types of *P. bracteatum* inhibits the biosynthesis in an early stage of development but not the current formation and degradation of this particular alkaloid. Apparently, one of the steps in the biosynthetic pathway is blocked in old plants and a new formation of alpinigenine is triggered by feeding tetrahydropalmatine. Biosynthetic experiments carried out with *P. bracteatum* showed that alpinigenine arises preferentially from

[N-$^{14}CH_3$, 8-^{14}C] DL-tetrahydropalmatine methiodide, but that the intermediate product is the protopine alkaloid muramine (**101g**) (*684, 741–745*). In the resulting alpinigenine C-14 arises from C-8 of tetrahydropalmatine, and the methyl group of the quaternary nitrogen becomes the methyl group of the tertiary nitrogen of alpinigenine. The experiments also showed that tetrahydropalmatine methiodide is converted into alpinigenine far more readily than tetrahydropalmatine (Scheme 54). The initial starting material for tetrahydropalmatine is again

$$\text{DOPA} \longrightarrow \text{norlaudanosoline} \longrightarrow (+)\text{-reticuline.}$$

Rönsch developed a method for the simple degradation of alpinigenine (**153h**). He prepared the oxime **164**, the nitrile **165a**, compound **165b**, and the phthalimide derivative **146**, whose further degradation afforded compounds **166** and **167** (*684, 741, 741a, 744, 745*) (Scheme 59).

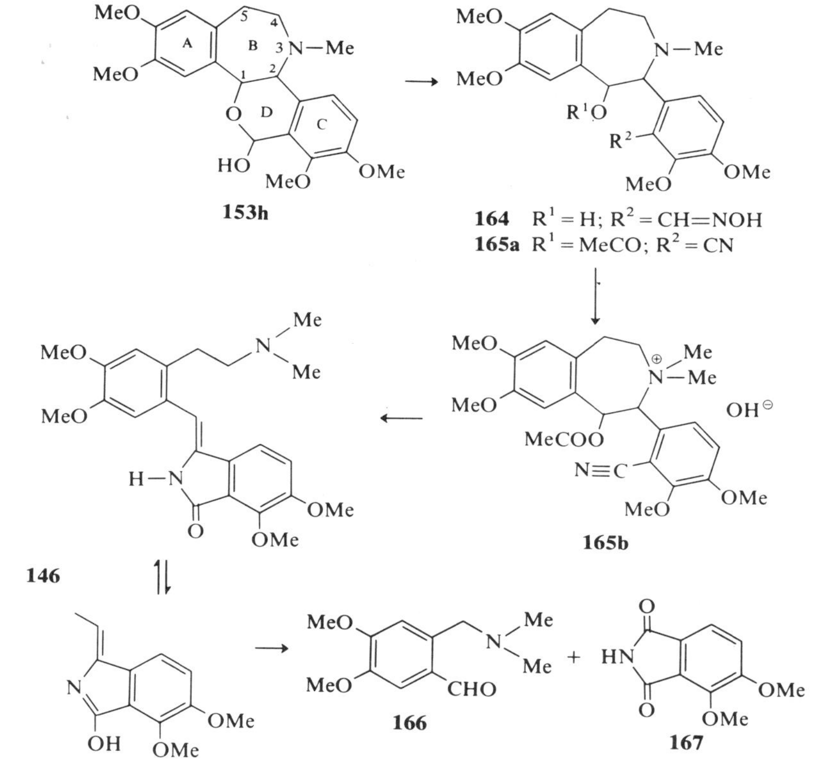

153h

164 $R^1 = H$; $R^2 = CH{=}NOH$
165a $R^1 = MeCO$; $R^2 = CN$

165b

146

166

167

SCHEME 59. *Degradation of rhoeadane alkaloids* (**153**) *for the study of their biogenesis* (*744*).

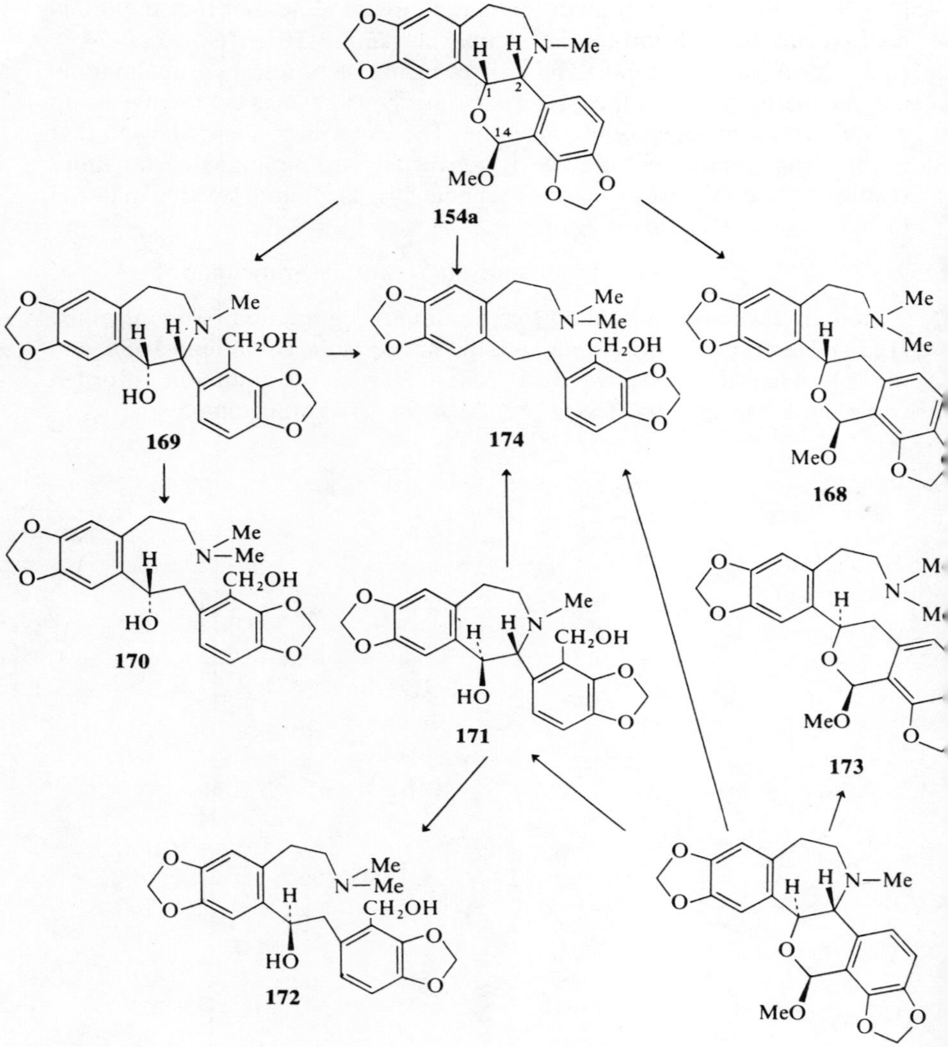

SCHEME 60. *Emde degradation of rhoeadine* (**154a**) *and isorhoeadine* (**152a**) *for the establishment of the absolute configuration at C-1 and C-2* (*698, 748, 749*).

This author applied the Emde degradation of alpinigenine (**153h**) for the elucidation of the absolute configuration of rhoeadine bases **152–154** (*743, 746, 746a*). The Emde degradation of rhoeadine (**154a**) and isorhoeadine (**152a**) was also used by Šimánek *et al.* (*747–749*). Splitting of the methiodides of rhoeadine and isorhoeadine gave, in addition to the optically inactive substance **174**, compounds **168** and **173**, respectively. The optical rotation showed that they are diastereoisomers. The CD spectrum of the substance **168** is, however, a mirror image of that of compound **173**. It was found that the contribution of the methoxyl group at C-14 to the CD in the region between 300 and 270 nm (^1L_b transition) is negligible. Thus, the substances **168** and **173** have the same conformation of the ring D though with opposite helicity (*722*). The Emde degradation of the methiodide of rhoeageninediol (**169**) (*750*) and of isorhoeageninediol (**171**) (*743*) gave the enantiomeric compounds **170** and **172**, respectively. Substance **174** was also a by-product. Because in the compounds **170** and **172** only one center of chirality remained, i.e., at C-1 in native alkaloids, the absolute configuration *1S, 2R, 14S* could be deduced for alkaloids **152** and the configuration *1R, 2R, 14S* for alkaloids **154** (Scheme 60).

The absolute configuration of a whole group of rhoeadine and papaverrubine alkaloids was studied by ORD and CD data (*751, 752*). The Fourier expansion of the endocyclic torsion angle revealed that the seven-membered ring of *N*-methylrhoeadine also shows pseudorotation (*753*). The absolute configuration of the three chiral centers of rhoeagenine methiodide was also determined by Huber on the basis of X-ray diffraction (*754*). Thus, the results obtained from ORD and CD analyses and reported in previous papers on the absolute configuration were confirmed. The relative stereochemistry at C-1 and C-2 was forthcoming from ^{1}H NMR coupling constants, and the relative stereochemistry at C-14 from ^{1}H NMR and kinetic studies (*755*). It was found that treatment of the glaudine type (**152**) with very dilute acid in methanol results in isomerization at C-14 to afford the epiglaudine type (**153**) (Scheme 61). Stronger acid irreversibly epimerizes epiglaudine (rings B/D *trans*) to the oreodine type (B/D *cis*) (**154**), while isomerization takes place again at C-14. Shamma *et al.* applied the CD aromatic chirality method to determine the absolute configuration of (+)-glaudine (**152e**), (+)-epiglaudine (**153e**), and (+)-oreodine (**154e**) (*752*) (Table VI).

N. Benzo[*c*]phenanthridine (α-Naphthaphenanthridine) Group*

Recently, some new alkaloids of this group have been isolated and known alkaloids have been reinvestigated, namely, dihydrochelerythrine (**175a**), dihydrosanguinarine (**175b**), 6-*O*-methyldihydrochelerythrine (**176a**) [the substance arises by crystallization of the chelerythrine base from methanol (= chelerythrine-*ps*-methanolate) (*ps* = pseudo) and therefore is not a naturally occurring substance (*40, 50*)], sanguidimerine (**206a**) and

* This material is supplementary to *The Alkaloids*, Vol. IV, p. 253; Vol. X, p. 485; Vol. XII, p. 333.

TABLE VI
CD DATA OF THE RHOEADANE ALKALOIDS AT ROOM TEMPERATURE (*752*)

Compound	CD extrema (nm) ($\Delta\epsilon$)	
	Split cotton effect[a]	Normal cotton effect[b]
(+)-Glaudine (**152e**)	195 (−5.00), 207 (+7.25)	222 (+5.7), 250 (−2.0), 290 (+5.3)
(+)-Epiglaudine (**153e**)	193 (−7.5), 207 (+9.4)	250 (−1.3), 270 (+5.7)
(+)-Oreodine (**154e**)	197 (−10.0), 207 (+12.5)	238 (−3.00), 288 (+3.9)

[a] In hexane.
[b] In absolute ethanol.

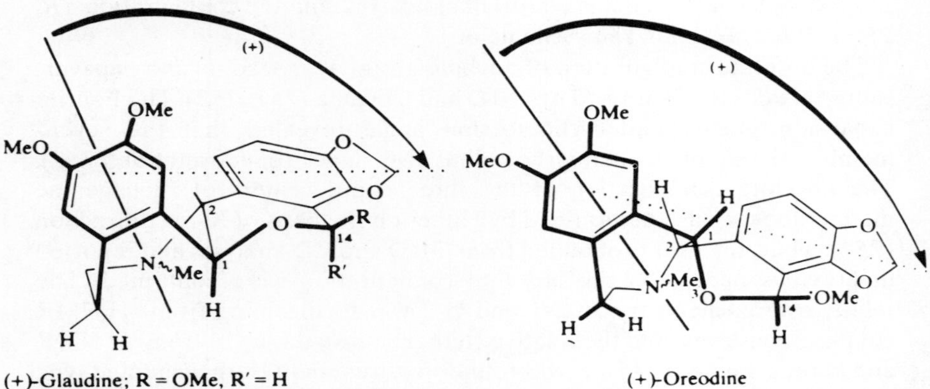

SCHEME 61. *A CD study of the absolute configuration of glaudine* (**152e**), *epiglaudine* (**153e**), *and oreodine* (**154e**) (*752*).

chelerythridimerine (**206b**) (*40, 42, 756*), corynolamine (*757*), bocconoline (**176b**) (*757*), 6-acetonyldihydrosanguinarine (**176c**) (*25*), corynoline (**178a**) (*90, 97, 758–761*), 11-acetylcorynoline (**178b**) (*95*), 12-hydroxy-corynoline (**178c**), 11-epicorynoline (**180**), 6-oxocorynoline (**178d**) (*93, 761*), corydamine (**181a**) (*99, 762*), *N*-formylcorydamine (**181b**) (*762*), corynoloxine (= alkaloid V) (**182**) (*90, 96, 97*), (+)-13-epicorynoline (= isocorynoline) (**179a**) (*95, 98, 763*), acetyl-13-epicorynoline (**179b**) (the absolute configuration of these epicorynoline alkaloids was established by X-ray analysis) (*763*), and the corydalic acid methyl ester (**184**) (*91*). 13-Epicorynoline (**179**) is the first base of this group in which the rings B and C have a *trans* configuration; X-ray analysis showed, however, that it is not a 14- but a 13-epicorynoline. The new alkaloid dihydrosanguilutine (**186**) was isolated from roots of *Sanguinaria canadensis* L. (*291*). Dihydrochelilutine and dihydrochelirubine also belong to this group of alkaloids (*223b*).

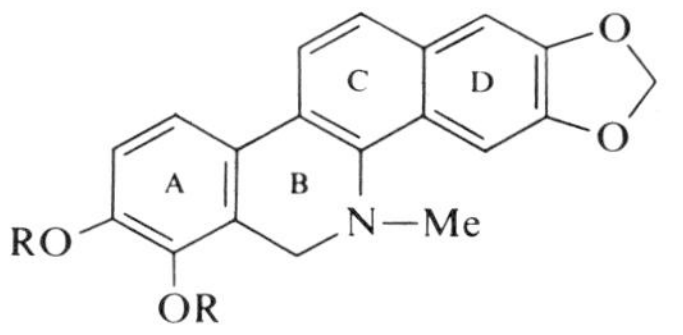

175a $R^1 = R^2 = Me$ Dihydrochelerythrine
175b $R^1 + R^2 = CH_2$ Dihydrosanguinarine

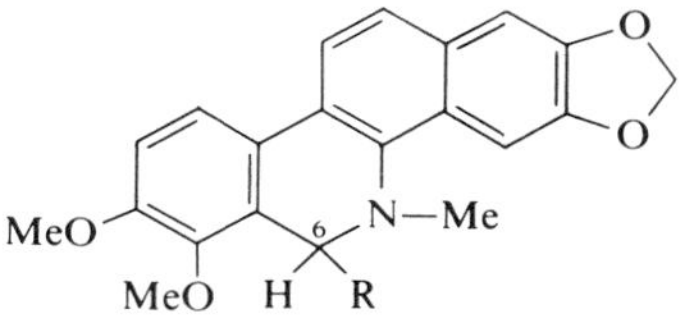

176a $R = OMe$ *O*-Methyldihydrochelerythrine
176b $R = CH_2OH$ Bocconoline
176c $R = CH_2COCH_3$

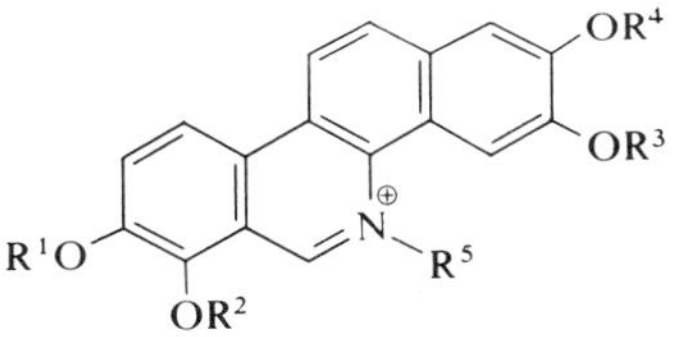

177a $R^1 = R^2 = R^5 = Me$; $R^3 + R^4 = CH_2$ Chelerythrine
177b $R^1 + R^2 = R^3 + R^4 = CH_2$; $R^5 = Me$ Sanguinarine
177c $R^1 = R^2 = Me$; $R^3 + R^4 = CH_2$; $R^5 = H$ Norchelerythrine

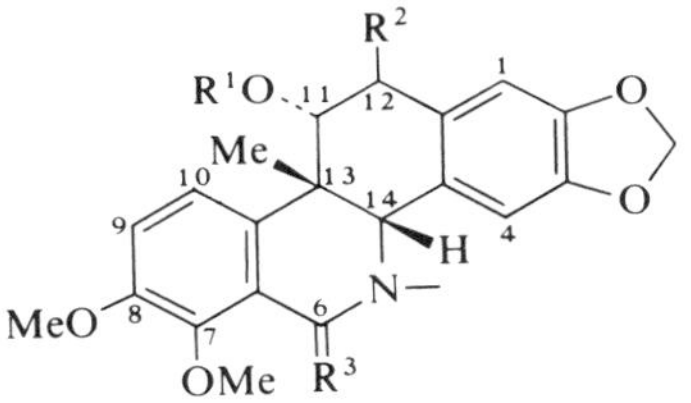

178a $R^1 = R^2 = H$; $R^3 = H_2$ Corynoline
178b $R^1 = Acetyl$; $R^2 = H$; $R^3 = H_2$ *O*-Acetylcorynoline
178c $R^1 = H$; $R^2 = OH$; $R^3 = H_2$ 12-Hydroxycorynoline
178d $R^1 = R^2 = H$; $R^3 = O$ 6-Oxocorynoline

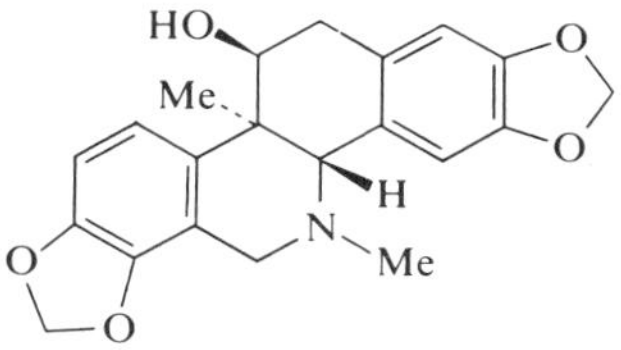

179a 13-Epicorynoline (= Isocorynoline)
179b Acetyl-13-epicorynoline

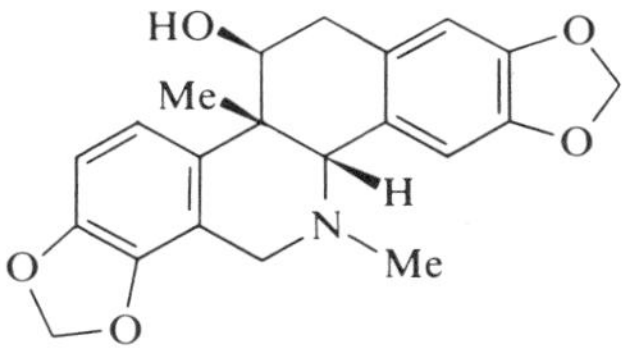

180 11-Epicorynoline

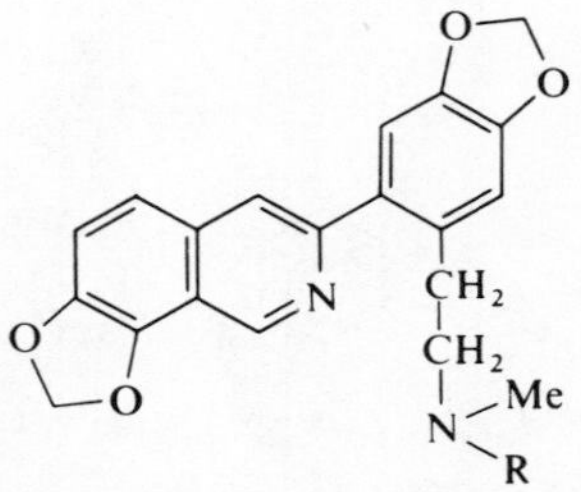

181a R = H Corydamine
181b R = CHO *N*-Formylcorydamine

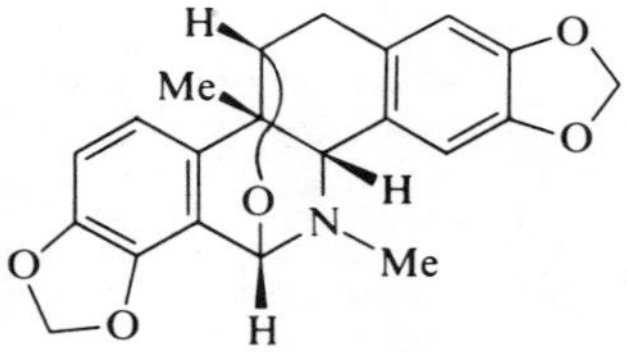

182 Corynoloxine

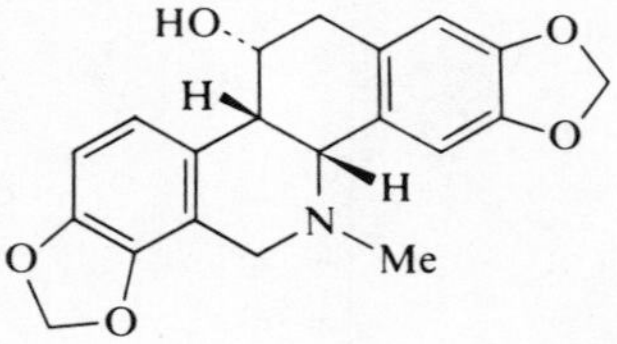

183a Chelidonine

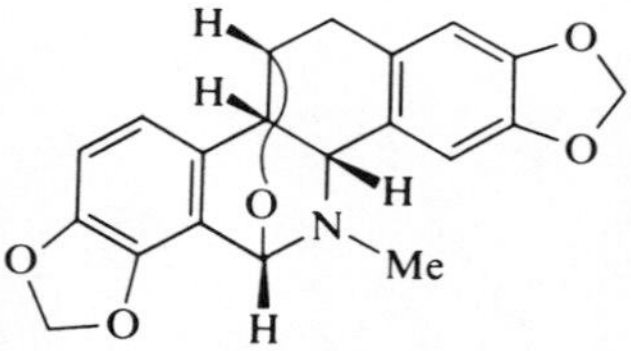

183b Didehydrochelidonine

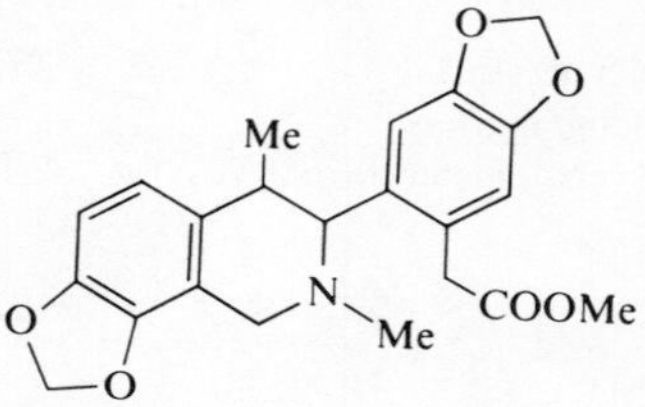

184

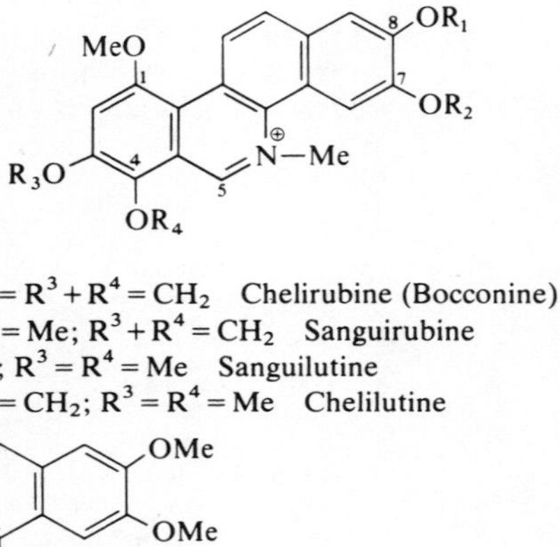

185a $R^1 + R^2 = R^3 + R^4 = CH_2$ Chelirubine (Bocconine)
185b $R^1 = R^2 = Me$; $R^3 + R^4 = CH_2$ Sanguirubine
185c $R^1 = R^2$; $R^3 = R^4 = Me$ Sanguilutine
185d $R^1 + R^2 = CH_2$; $R^3 = R^4 = Me$ Chelilutine

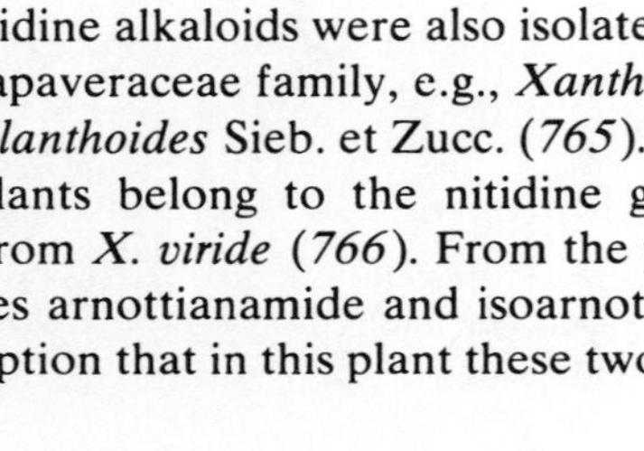

186

The benzophenanthridine alkaloids were also isolated from plants which do not belong to the Papaveraceae family, e.g., *Xanthoxylon tsihanimposa* H. Per. (*764*) and *X. ailanthoides* Sieb. et Zucc. (*765*). However, the bases isolated from those plants belong to the nitidine group. The alkaloid decarine was isolated from *X. viride* (*766*). From the plant *X. cuspidatum* the two isomeric amides arnottianamide and isoarnottianamide were isolated (*767*). The assumption that in this plant these two substances arise by oxidation of chelerythrine and nitidine was confirmed by *in vitro* oxidation of both of them by the novel Baeyer–Villiger-like oxidation of their immonium groups.

Corynoloxine (**182**) has a structure similar to that of didehydrochelidonine (**183b**), which can be obtained by oxidation of chelidonine with permanganate (*768*). Reduction of this substance with sodium borohydride (NBH) gives chelidonine (**183a**). The structures of 12-hydroxycorynoline (**178c**), 11-epicorynoline (**180**), and 6-oxocorynoline (**178d**) were established by correlation of their spectroscopic and chemical data with those of the derivatives of corynoline (**178a**) (*93*). Corynolamine and bocconoline (**176b**) were assigned a structure with a hydroxymethyl group at C-6 (*757*). They form a new type of dihydrobenzo[*c*]phenanthridine alkaloids.

Onda *et al.* also studied some derivatives of corynoline, chelidonine, and sanguinarine, i.e., their reactions with 2,3-dichloro-5,6-dicyanobenzoquinone and with dimethylacetylenedicarboxylate (*679, 679a, 682*).

Corynoline and corynoloxine are converted by Emde or Hofmann degradation into the common compound **187**, which provides evidence confirming the steric and structural relationship of these two substances (*97*).

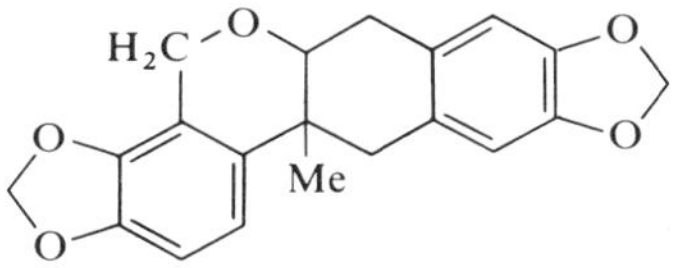

H_2C O O O O O Me

187

The synthesis of 6-oxochelirubine (prepared by photocyclization) showed that the alkaloids chelilutine, sanguirubine, sanguilutine, and chelirubine (bocconine)—independently isolated by Slavík (*45*) and Onda (*44, 769*)—have the fundamental structure **185** (*770, 770a, 770b*).

The syntheses of benzophenanthridine bases were also studied (*132, 771–776*). Much interest was shown in the synthesis of the alkaloids of this group, particularly after their antitumor activity in mouse leukemia L-1210 (LE) and P-388 (PS) as well as some selected antimicrobial activities had been established (*776*). On photocyclization, protopine, tetrahydroprotoberberine, and 13-methyltetrahydroprotoberberine yielded sanguinarine via anhydroprotopine (**114**) (*680, 681, 777, 778*) (Schemes 33 and 40).

Photocyclization of the enamide **a** followed by stereospecific functionalization of the *cis*-lactam **178d** completed the first synthesis of the alkaloid corynoline (**178a**) and 12-hydroxycorynoline (**178c**). Because corynoline has been converted into 6-oxocorynoline (**178d**) and corynoloxine (**182**), this synthesis formally completes the total synthesis of all these alkaloids (*93, 96, 761*):

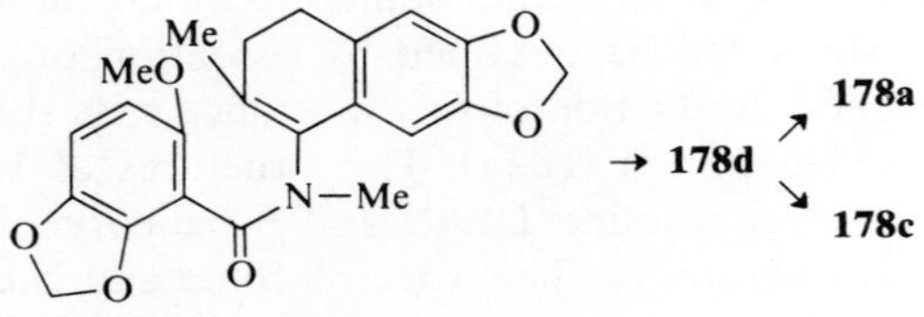

Me
MeO
N—Me
O
O
O
O
O
→ **178d**
↗ **178a**
↘ **178c**

a

Ninomiya *et al.* carried out the total synthesis of avicine, nitidine, and 11-epicorynoline (**180**) (*621, 773*).

In the past decade, much attention of the biochemists and organic chemists was paid to the biosynthesis of both the simple benzylisoquinoline alkaloids and the tetrahydroprotoberberine, phthalideisoquinoline, protopine, rhoeadine, benzophenanthridine, and ochotensane alkaloids. All these groups of natural substances are biosynthesized in a similar manner and therefore are discussed in connection with the benzophenanthridine alkaloids. The final stage was reached on the basis of the results obtained particularly from ^{2}H-, ^{3}H-, ^{13}C-, and ^{14}C-labeled compounds and from enzymatic reactions which take place along with the individual changes in the plants *Chelidonium majus P. somniferum* and *P. bracteatum* (*511a*). (*72, 258, 280, 328, 339, 416, 417, 467, 468, 503, 638, 779, 780*). Much work in that direction was carried out by Battersby and his group (*256, 739, 781–783*) and by Tani *et al.* (*101, 740, 783a*). The results obtained by these two groups can be summarized in Scheme 62 as follows: (a) (*S*)-(+)-Reticuline (**7f**) is the precursor of (*S*)-(−)-stylopine (**58n**), from which all the other tetrahydroprotoberberine and tetrahydropseudoprotoberberine alkaloids of the (*S*) series are derived; (b) C-1 and C-2 of reticuline are unaffected during its conversion into stylopine; (c) C-8 of stylopine is formed by oxidative cyclization involving the *N*-methyl group of reticuline, and this step involves an appreciable isotope effect; (d) the methylenedioxy group on ring D of stylopine arises from the *o*-hydroxymethoxy system of reticuline or scoulerine; (e) the biotransformation of tetrahydroprotoberberine alkaloids into alkaloids of the protopine, rhoeadine, benzophenanthridine, and probably also the phthalideisoquinoline and ochotensane types requires in the first place transformation of the tetrahydroprotoberberine alkaloids into their *N*-methohydrochloride salts; (f) protopine is shown to be stereospecifically formed *in vivo* from (*S*)-(−)-scoulerine of (*S*)-(−)-stylopine methochloride; (g) the benzophenanthridine alkaloids (hydrogenated and

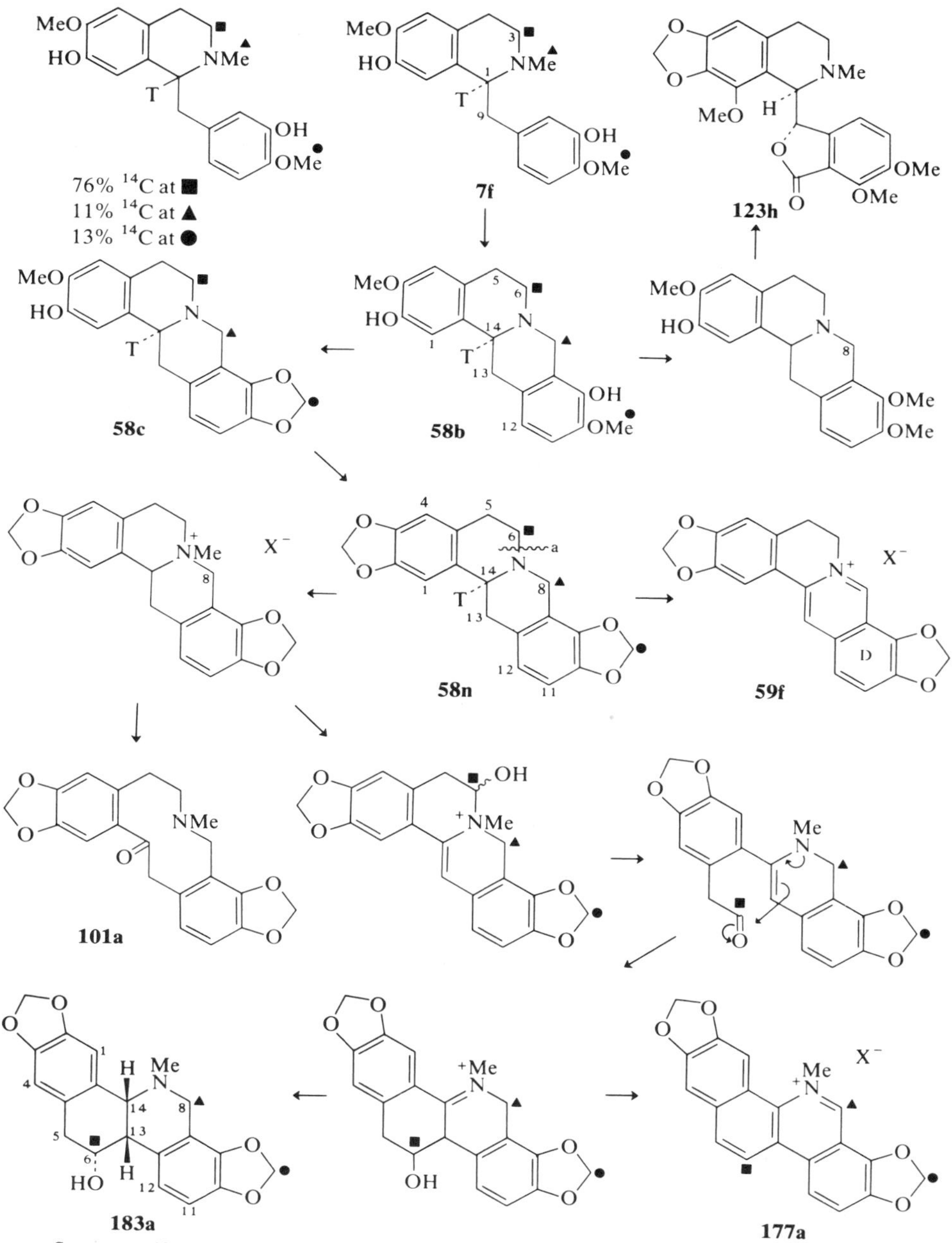

76% ^{14}C at ■
11% ^{14}C at ▲
13% ^{14}C at ●

7f

123h

58c

58b

58n

59f

101a

183a

177a

SCHEME 62. *The biosynthesis of various Papaveraceae alkaloids with labeled reticuline (256).*

dehydrogenated) are stereospecifically biosynthesized from (*S*)-(−)-scoulerine and (*S*)-(−)-stylopine via nandinine (**58i**) (α-*N*-metho salts, see Section III,N), and protopine; (h) (*S*)-(+)-reticuline is incorporated into protopine or into chelidonine (also studied in *Dicentra spectabilis*) 50 times more rapidly than is (*R*)-(−)-reticuline; (i) the 13-methyltetrahydroprotoberberine-type alkaloid was converted via the α-*N*-metho salt into the protopine-type alkaloid, regardless of the configuration of the hydrogens at C-13 and C-13a; (j) in *P. somniferum*, isocorypalmine (**58o**) and canadine (**58e**) participate in the biosynthesis of berberine (**59a**) and narcotine (**123h**); and (k) latex isolated from *P. somniferum* incubated with L-methionine (^{14}C-methyl) and *S*-adenosyl-L-methionine (^{14}C-methyl) was shown to contain methyltransferase enzymes (*784*). Except for narcotine, the major alkaloids reticuline, papaverine, codeine, and thebaine were found to be radioactive with ^{14}C-methyl group. (1) In *P. somniferum* and *P. bracteatum* the precursor of the morphinane alkaloids is not (*S*)-(+)-reticuline but its (*R*)-(−)-enantiomer (*510*).

The absolute configurations of chelidonine, chelidonine acetate, and norchelidonine have been independently studied by application of a kinetic method of partial resolution during esterification with α-phenylbutyric anhydride, by CD analysis (*662*), and by ^{1}H NMR analysis (*785*). The mass spectra and the NMR spectra of chelirubine, chelilutine, sanguinarine, and sanguilutine have been reported (*770, 786*).

A differentiation between the alkaloids of the sanguinarine, dihydrosanguinarine, oxysanguinarine, and the nitidine type is possible on the basis of the UV spectra of their salts in aqueous–alcoholic medium (*380*). This method also shows whether the aromatic ring carries methoxyl or methylenedioxy substituents.

The possible tautomeric forms of alkaloids of the sanguinarine, sanguilutine, and nitidine types were studied by spectral methods (UV, IR, and ^{1}H NMR) and by polarography; the immonium or the aldehyde form in alkaline medium could not be demonstrated (*381*). The equilibrium between sanguinarine acetate and chelerythrine acetate and their corresponding 6-acetoxydihydro tautomers has been studied by ^{1}H NMR and UV spectroscopy (*756, 787*).

The purification of the commercially available sanguinarine (*290*) and a simplified isolation method were described (*787a*). Reports have appeared on thin-layer chromatographic methods of detection of benzophenanthridine alkaloids in *Bocconia frutescens*, *Macleaya microcarpa*, *M. cordata* (*48*), and *Fumaria capreolata* (*788*). A colorimetric method of determining the alkaloids in *B. cordata* has been reported (*789*).

In *Verticillium daphnae*, sanguinarine is detoxicated into dihydrosanguinarine (*790*).

O. Spirobenzylisoquinoline Group

1. Ochotensimine Group*

From the plants of the genus *Corydalis* and *Fumaria* were isolated several new spirobenzyltetrahydroisoquinoline alkaloids of the ochotensimine type (**188**), i.e., corpaine (**189a**) (*86, 123, 124*), 2-*O*-methylcorpaine (**189b**) (*150*), fumaricine (F37) (**189c**), fumaritine (F63) (**189d**) (*175, 791, 792*), fumaritrine (**189e**), fumaritridine (**189f**) (*176*), fumarophycine (**189g**) (*159, 165, 169*), raddeanamine (**189h**) (*118*), *O*-methylfumarophycine (*159, 160*), parfumidine (**190a**) (*160, 170, 179*), parfumine (**190b**) (*108, 172, 173, 175, 179*), fumariline (**190c**) (*160*), sibiricine and corydaine (**191a**) (*124, 793*), yenhusomidine and (±)-raddeanone (**191b**), fumarostelline (**191c**) (*115, 117, 794*), ledeborine (**191f**) (*110*), fumarofine (**192**) (*795*), ochrobirine (**193a**) (*142*), yenhusomine (**193b**) (*115, 794*), raddeanine (**193c**) (*110, 118*), raddeanidine (**193d**) (*118*), and ledeboridine (**193e**) (*110*). The ^{1}H-NMR spectral data of yenhusomidine and raddeanone differed as to the methine hydrogen at C-8 (*127b, 794b*). The same finding was made in corydaine and sibiricine (*794b*).

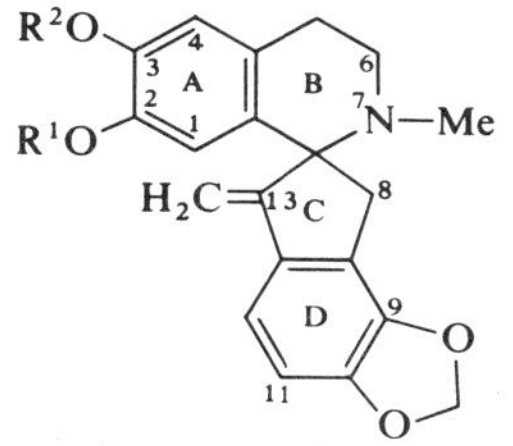

188a $R^1 = CH_3$; $R^2 = H$ Ochotensine
188b $R^1 = R^2 = Me$ Ochotensimine

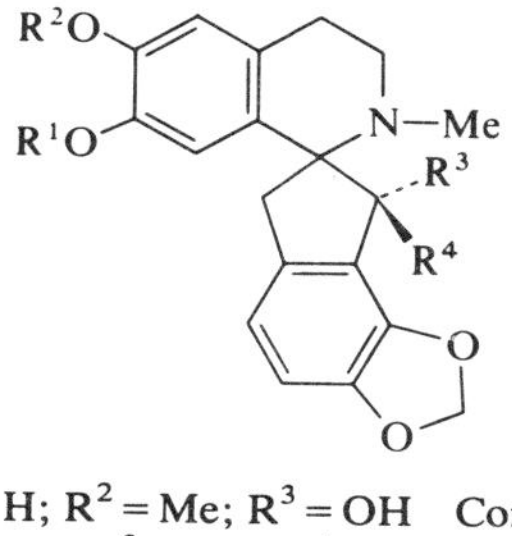

189a $R^1 = R^4 = H$; $R^2 = Me$; $R^3 = OH$ Corpaine
189b $R^1 = R^2 = Me$; $R^3 = OH$; $R^4 = H$ 2-*O*-Methylcorpaine
189c $R^1 = R^2 = Me$; $R^3 = H$; $R^4 = OH$ Fumaricine
189d $R^1 = R^3 = H$; $R^2 = Me$; $R^4 = OH$ Fumaritine
189e $R^1 = R^2 = Me$; $R^3 = H$; $R^4 = OMe$ Fumaritrine
189f $R^1 = H$; $R^2 = Me$; $R^3 + R^4 = H + OMe$ Fumaritridine
189g $R^1 = R^3 = H$; $R^2 = Me$; $R^4 = Ac$ Fumarophycine
189h $R^1 = R^2 = R^3 = Me$; $R^4 = OH$ Raddeanamine

* This material is supplementary to *The Alkaloids*, Vol. XII, p. 420; Vol. XIII, p. 165.

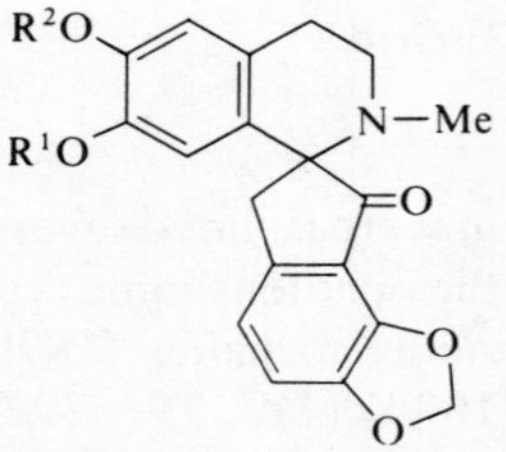

190a $R^1 = R^2 = Me$ Parfumidine
190b $R^1 = Me$; $R^2 = H$ Parfumine
190c $R^1 + R^2 = CH_2$ Fumariline

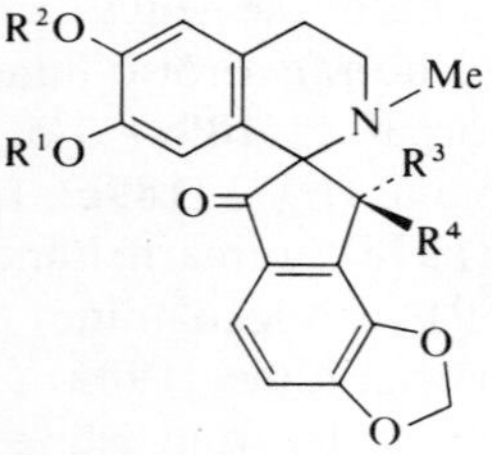

191a $R^1 + R^2 = CH_2$; $R^3 = OH$; $R^4 = H$ / $R^1 + R^2 = CH_2$; $R^3 = H$; $R^4 = OH$ Sibiricine and Corydaine
191b $R^1 = R^2 = Me$; $R^3 = OH$; $R^4 = H$ Yenhusomidine (Raddeanone)
191c $R^1 = H$; $R^2 = Me$; $R^3 + R^4 = H + OH$ Fumarostelline
191d $R^1 = R^2 = Me$; $R^3 = H$; $R^4 = OH$
191e $R^1 = R^2 = Me$; $R^3 + R^4 = O$
191f $R^1 = Me$; $R^2 = R^3 = H$; $R^4 = OH$ Ledeborine

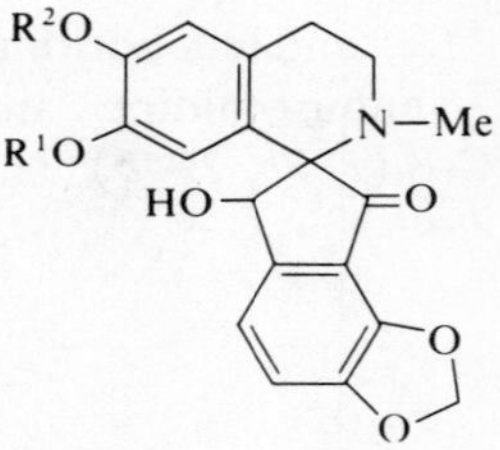

192 $R^1 = H$; $R^2 = Me$ Fumarofine (F-38)

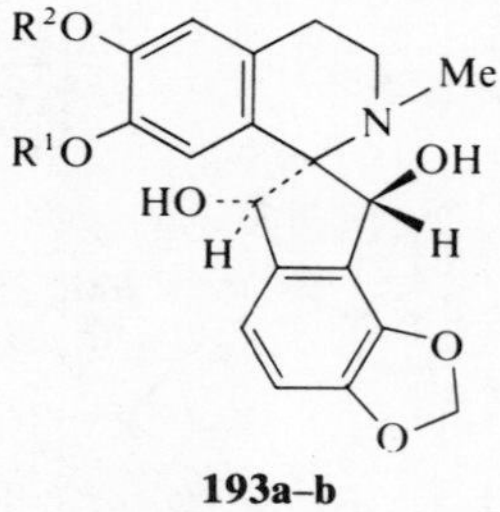

193a–b 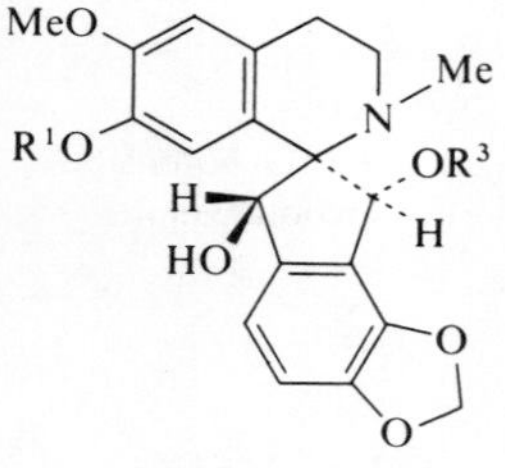 **193c–e**

193a $R^1 + R^2 = CH_2$ Ochrobirine
193b $R^1 = R^2 = Me$ Yenhusomine
193c $R^1 = Me$; $R^3 = H$ Raddeanine
193d $R^1 = Me$; $R^3 = Ac$ Raddeanidine
193e $R^1 = R^3 = H$ Ledeboridine

Shamma and Jones suggested a model for the biogenesis of the spirobenzyltetrahydroisoquinoline alkaloids in *Corydalis* plants (*796, 797*). In the basic medium, the enamine *N*-metho-salt precursor **194** can undergo cleavage to the quinonoid intermediate **195** which by an electrocyclic process forms the spirane **a**. A tautomeric shift then yields the diphenol **b**,

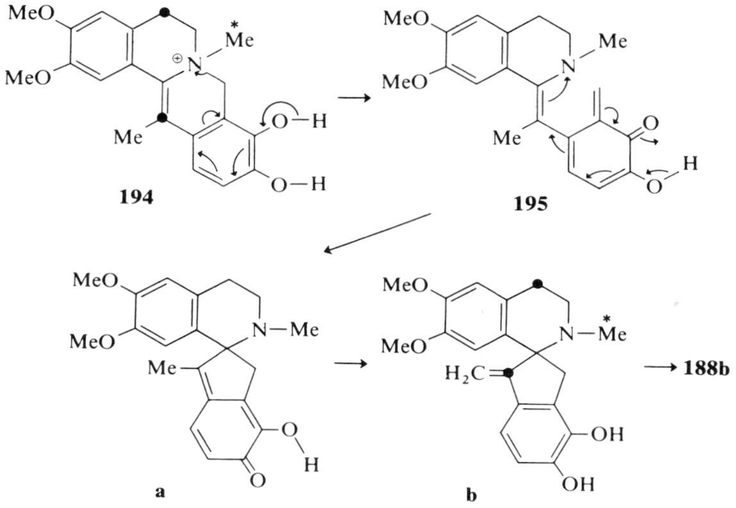

SCHEME 63. *A model for the biosynthesis of the ochotensimine* (**188**) *alkaloids in Corydalis* (*639, 796*) (*see also Scheme* 31).

which is easily convertible to ochotensimine (**188b**) (Scheme 63). The photolytic 13-methyl- (**194**) or 13-oxoprotoberberine (**160**)→spirobenzylisoquinoline rearrangement proceeds in an analogous manner (*649*) (Scheme 64).

The biosynthesis of the spirobenzyltetrahydroisoquinoline alkaloids can proceed from dehydrophthalideisoquinoline alkaloids (**161**) via the aldehyde **195** (Scheme 65). The latter may, however, also arise by the Hantsch tautometry from berberine alkaloids. By that mechanism a group of Canadian authors succeeded in preparing the analogs corydaine and sibiricine (*798, 799*).

The biosynthesis of ochotensimine alkaloids has been dealt with in Section III,N (*639, 796*) (Schemes 31 and 63).

A great number of the spirobenzyltetrahydroisoquinoline alkaloids have already been prepared synthetically: corydaine and sibiricine (*799*), fumaricine (*800*), fumariline (*801*), fumaritine (*791, 801*), isoochotensine (*796, 802*), ochotensimine (*792a, 796, 797, 803, 804*), ochotensine (*805, 806*), ochrobirine (*693a, 796, 807–809*), yenhusomine, and yenhusomidine (*693b, 794, 794a*). The many synthetic routes to the spirobenzyltetrahydroisoquinoline alkaloids can be classified according to the type of intermediates or reactions used in the key step, e.g. (a) indanedione–indanetrione intermediates. These methods include the benzazepine route and are older methods often used (*693, 693a, 698, 792a, 797, 803–809*)

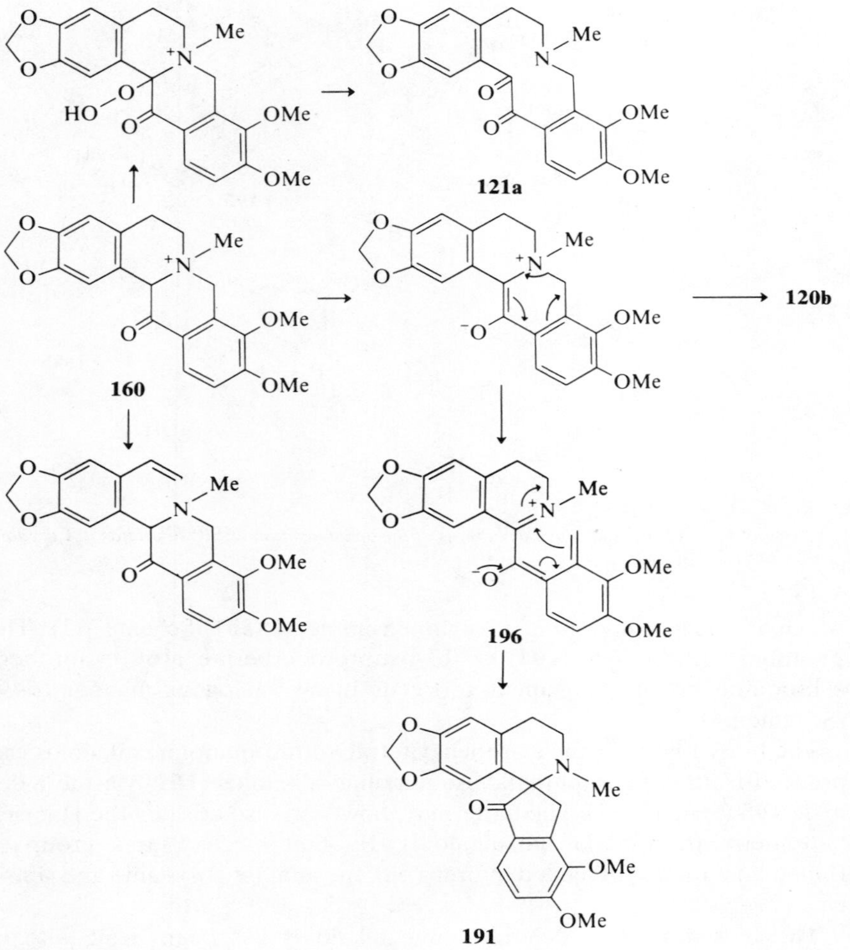

Me +N HO O O OMe OMe O O O O Me N O O OMe OMe **121a**

O O Me +N O OMe OMe **160** → O O +N Me ⁻O OMe OMe → **120b**

O O Me N O OMe OMe O O +N Me ⁻O OMe OMe **196**

O O N Me O OMe OMe **191**

SCHEME 64. *The conversion of a 13-oxoprotoberberinium metho salt to a spirobenzylisoquinoline* (**191**), *13-oxoallocryptopine* (**121a**) *and a 13-oxo analog of allocryptopine* (**120b**) (*649, 674*).

(e.g. Scheme 42); (b) quinonemethide intermediates, including the protoberberine and benzocyclobutene intermediates. The photolytic rearrangement of the protoberberine compounds (*649, 796*) (Schemes 63 and 64) and the thermal rearrangement of the benzocyclobutenyl precursors belong to this group (*609, 674, 796, 810–812*) (Scheme 27); (c) from dehydrophthalideisoquinolines (**161**) (Scheme 46) by intramolecular aldol condensation (*693, 693b, 798*) (Scheme 65); and (d) miscellaneous

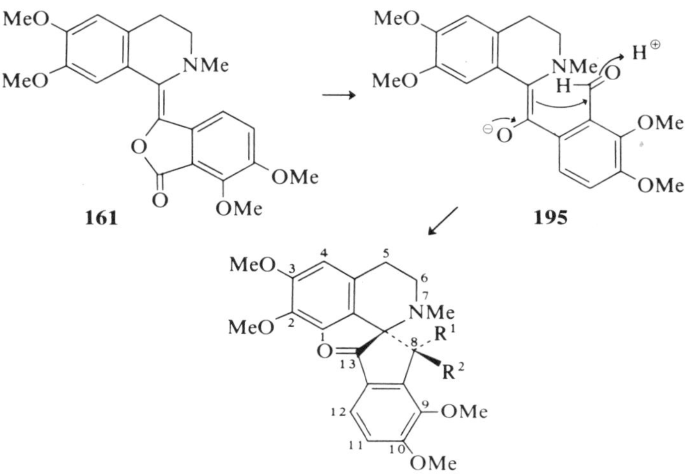

SCHEME 65. *A new synthesis of analogs of sibiricine* (**191a**) [*or corydaine* (**189b**)] *from dehydrocordrastine* (**161**) (*798, 799*).

methods: benzyltetrahydroisoquinoline approach (*802, 805*), abnormal Hofmann degradation of phenolic berbinium salts (*640, 796, 797*) (Scheme 32), and Stevens rearrangement of the *N*-methyltetrahydroberberinium salts with phenyllithium, which is not used for synthesis of actual alkaloids but provides the spirobenzylisoquinoline system (*619, 635a, 650, 652*) (Scheme 66).

Many of these procedures have been described in Vol. XIII of this series and in the book by Shamma (*1*). In this chapter only the papers published after the year 1971 are briefly discussed.

The photo-induced rearrangement can be reversed so that the spirobenzyltetrahydroisoquinoline bases or its ketones afford the berberinium alkaloids and their 8-lactames which on hydrogenation change into the tetrahydroprotoberberine alkaloids (xylopinine yields up to 80%) (*648, 648a*). The spirobenzyltetrahydroisoquinoline alkaloids of the type **191e** can also be retransformed in high yield to benzindanoazepines (**133**) (*cis* and *trans* forms in the ratio 1 : 3) (*813*) or to dibenzocyclopent[*b*]azepine (*797*).

Imai *et al.* synthesized optically active ochotensanes and thus they determined the chirality at C-8a (*814*).

Treatment of (*14S*)-(−)-β- and (*14R*)-(+)-β-canadine methochlorides (**58e**) with organolithiums, lithium aluminum hydride, or sodium methylsulfinyl carbanion in tetrahydrofuran gave (+)-**199** and (−)-2,3-methyl-

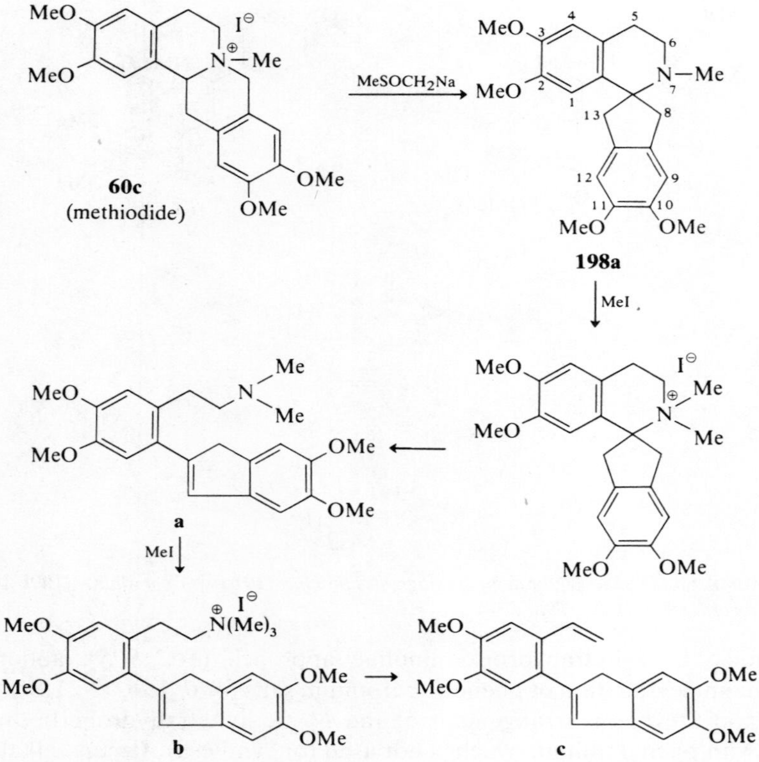

SCHEME 66. *The Stevens rearrangement of tetrahydropseudoprotoberberine methoiodide* (**60c**) *into pseudoochotensane type of skeleton* (**198**) *and determination of its structure via substances* **a**, **b** *and* **c** (*650*).

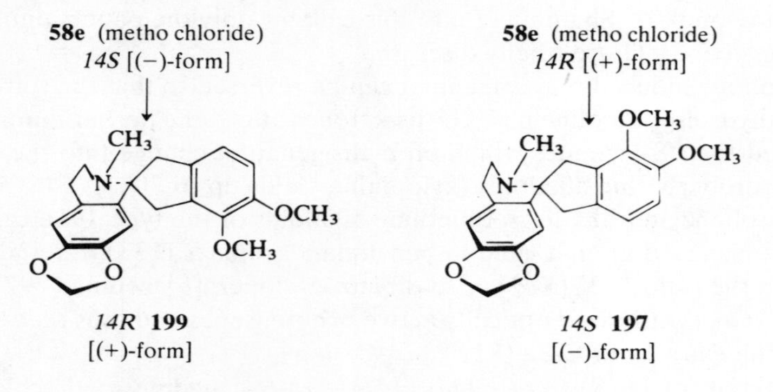

SCHEME 67. *Synthesis of optically active ochotensanes from canadine* (**58e**) *metho chloride* (*814*).

enedioxy-9,10-dimethoxyochotensanes (**197**), respectively (*814*). The structures of those compounds were confirmed by Hofmann degradation and spectral data. The CD spectra of these two compounds showed the Davydov split extrema centered at 284 nm whose first Cotton effects corresponded to the absolute configurations (*814, 815*). Therefore, the absolute configurations should be *14R* for compound **199** and *14S* for **197**.

The structures of the new alkaloids were elucidated by physicochemical methods, particularly by ^{1}H NMR analysis including the Overhauser effect and mass spectrometry (*86, 165, 793, 795*). The mass spectrometry of these bases was also reported in an independent paper together with a summary of the findings made earlier in this field (*816*). These authors attempt to confirm the findings made by mass spectrometric analysis of a series of model compounds. Fragmentation mechanisms are proposed for the formation of the major ions in the spectra of compounds containing one or two oxygen functions in the five-membered ring. The spectra of compounds containing a carbonyl group differ markedly from those with a hydroxyl group in the five-membered ring. The mass spectrometry of the fumariline alkaloids has also been described (*817*).

The position of the long-wavelength band in the UV spectra of ochotensane alkaloids and phthalidetetrahydroisoquinoline alkaloids shows whether the electron-donating oxygen groups of the ring D are located at positions 9,10 (compound **191**), 10,11 (**198b**—pseudoochotensanes); or 11,12 (**190**) in relation to the keto group at C-13 on the ring C (Fig. 2) (*420*).

2. Hypecorine and Canadaline Alkaloids

From *Hypecoum erectum* and *Corydalis incisa* were isolated two alkaloids which, although similar to the spirobenzyltetrahydroisoquinoline or tetrahydroprotoberberine alkaloids, have an additional carbon atom in the benzyl portion of the molecule. Their ring C is spirofused to C-8. The first alkaloids of this type—hypecorine (**200a**) and hypecorinine (**200b**)—were isolated by Yakhontova *et al.* She also assigned to them the correct structure on the basis of degradation reactions and physicochemical data (*214*). The alkaloid hypecorinine was found to be identical with corydalispirone. Its isolation and analysis were independently carried out by Japanese authors (*92, 95*). The structure of this alkaloid was correlated with that of bicucullinediol and adlumidinediol (*92, 95, 118, 214*).

Acid converts hypecorine (**200a**) into compound **201**, which re-forms hypecorine upon basification. Hydrogenation of compound **201** and subsequent acetylation give rise to compound **203**, which also provides confirmatory evidence of the structure of hypecorine (in the paper, a

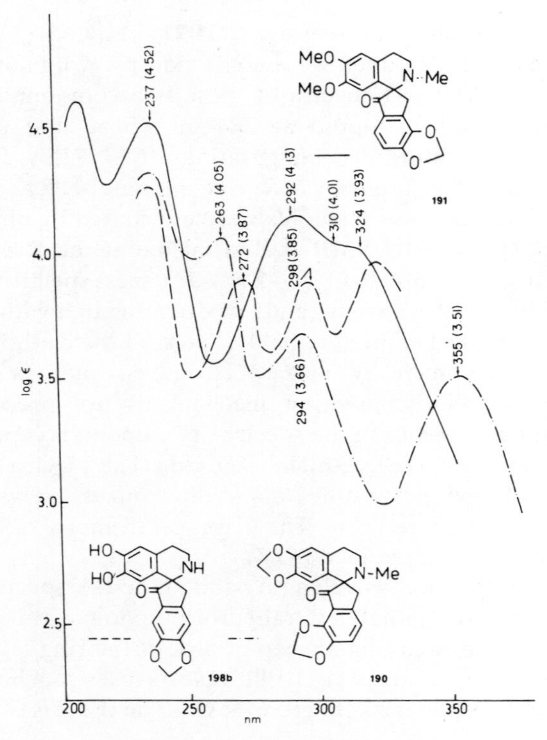

FIG. 2. UV extinction curves (in ethanol) of three spirobenzyltetrahydroisoquinoline bases (**190**, **198**, and **199**) showing the effect of electron-donating substituents of the ring D vs the electron-accepting substituent of the ring C on the position and extinction of the longest wavelength bands (*420*).

TABLE VII

HYPECORINE ALKALOIDS

Compound	Empirical formula	Melting point (°C)	$[\alpha]_D$	References
Hypecorine (**200a**)	$C_{20}H_{19}NO_5$	156	0	*214*
Hypecorinine (**200b**)	$C_{20}H_{17}NO_6$	198	0	*214*
Corydalispirone (**200b**)	$C_{20}H_{17}NO_6$	198	0 ($CHCl_3$)	*92, 95*
Corydalisole (**204**)	$C_{20}H_{21}NO_5$	161	+21.4 ($CHCl_3$)	*92, 95*
Canadaline (**205a**)	$C_{21}H_{23}NO_5$	118	+43 ($CHCl_3$)	*818*
Aobamine (**205b**)	$C_{20}H_{19}NO_5$	Syrup	?	*118*
Macrantaline (**205c**)	$C_{22}H_{27}NO_6$	141	?	*16a*
Macrantoridine (**205d**)	$C_{22}H_{25}NO_7$	114	?	*16a*

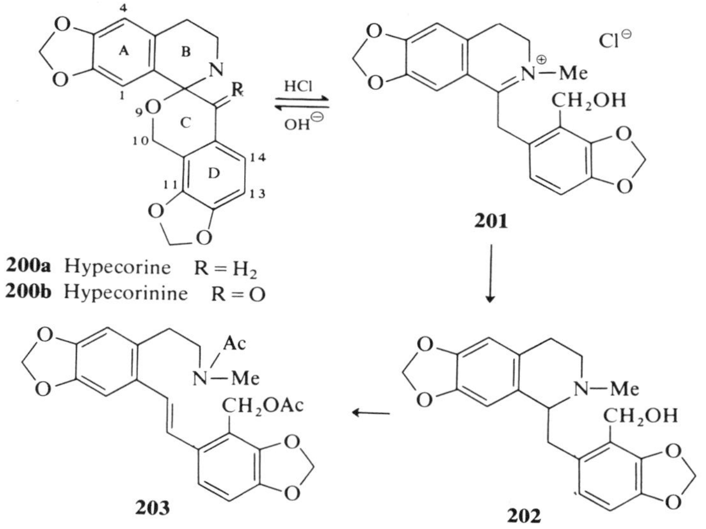

200a Hypecorine R = H_2
200b Hypecorinine R = O

SCHEME 68. *Transformation and degradation of hypecorine* (**200a**) (*819*).

tabulation of the ^{1}H NMR signals of compounds **200a–203** has also been given) (*819*) (Scheme 68).

In *Corydalis incisa* the Japanese authors (*92*) also found corydalisol (**204**), which is related to canadaline (**205a**) (isolated from *Hydrastis canadensis*) (*818*), and aobamine (**205b**) (isolated from *C. ochotensis* var. *raddeana*) (*118*). It could not be determined whether the latter three alkaloids are the precursors in the biosynthesis or degradation products of tetrahydroprotoberberine or spirobenzyltetrahydroisoquinoline alkaloids. Closely related to these alkaloids are macrantaline (**205c**) and macrantoridine (**205d**) which have been found in *P. pseudoorientale* (*16a*).

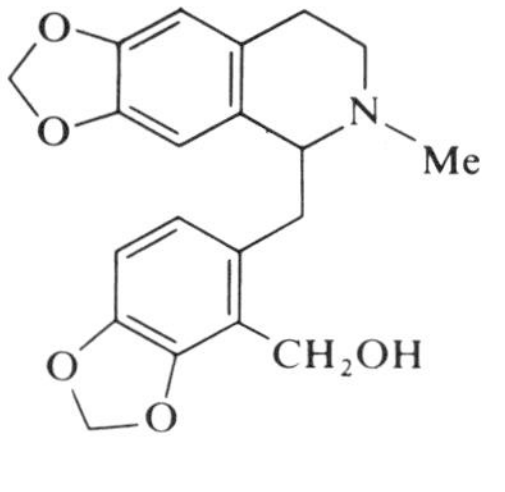

204
Corydalisol

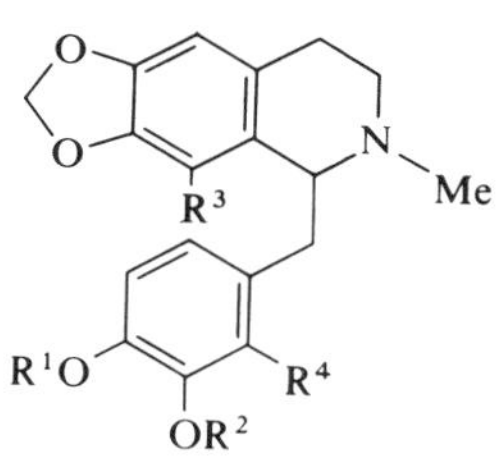

205a $R^1 = R^2 = Me$; $R^3 = H$; $R^4 = CHO$ Canadaline
205b $R^1 + R^2 = CH_2$; $R^3 = H$; $R^4 = CHO$ Aobamine
205c $R^1 = R^2 = Me$; $R^3 = OMe$; $R^4 = CH_2OH$ Macrantaline
205d $R^1 = R^2 = Me$; $R^3 = OMe$; $R^4 = COOH$ Macrantoridine

3. Peshawarine Alkaloids

From the plant *Hypecoum parviflorum* (indigenous to Pakistan) a new alkaloid, peshawarine (**206**), was isolated (*224*) whose structure is related to that of hypecorine, rhoeadine, or narceine type alkaloids. Because this alkaloid has been found only in the plant belonging to the genus *Hypecoum* where the alkaloids of neither the rhoeadine nor the narceine type are present, it is assumed in view of the biogenesis that this new alkaloid is related to the hypecorine alkaloids (**200**).

Peshawarine is actually a de-*N*-methyl-13,14-dihydrooxyrhoeagenine [oxyrhoeagenine (**159a**), Scheme 56], which by action of lithium aluminum hydride gives the de-*N*-methyldihydrorhoeageninediol, identical with

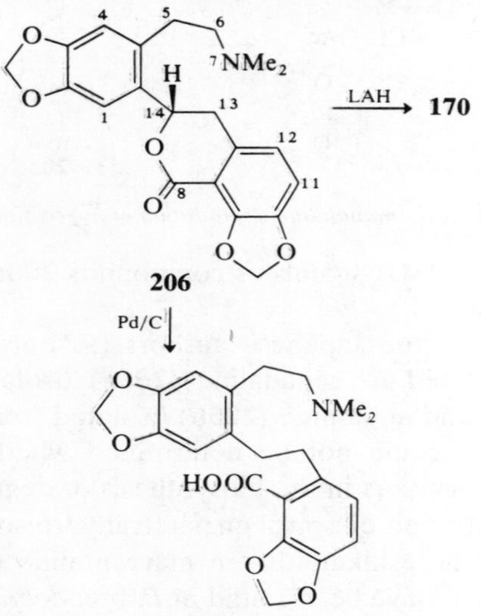

4 5 6 O O 7 NMe$_2$ H 1 14 13 LAH → 170 12 O 11 O 8 O O

206

Pd/C ↓

O O NMe$_2$ HOOC O O

174b

SCHEME 69. *Transformation of peshawarine* (**206**) *into the compounds* **170** *and* **174b** (*821*).

compound **170** (Scheme 60). It has an (*R*) configuration at C-14 which is similar to that of rhoeadine at C-1 (*820*). The racemic form of peshawarine was prepared from rhoeadine (Emde degradation) (*821*) and from coptisine via aobamine (**205b**) (*820*).

A compound related to peshawarine was synthetically prepared from (±)-canadaline (**205a**) (dimethoxy derivative of the ring C) (*224, 820*).

P. Dimeric Alkaloids

From *Dicentra canadensis* Walp. three alkaloids of the dimeric type were isolated: cancentrine (cularine–morphine dimer), which was originally designated by Manske as alkaloid F-22; dehydrocancentrine A; and dehydrocancentrine B (*151, 822–826*). The structure and chemical properties of these three alkaloids were described in *The Alkaloids*, Vol. XIV, p. 407. Recently, Ruchirawat and Somchitman studied the biological model of synthesis of cancentrine from its two moieties (*827*).

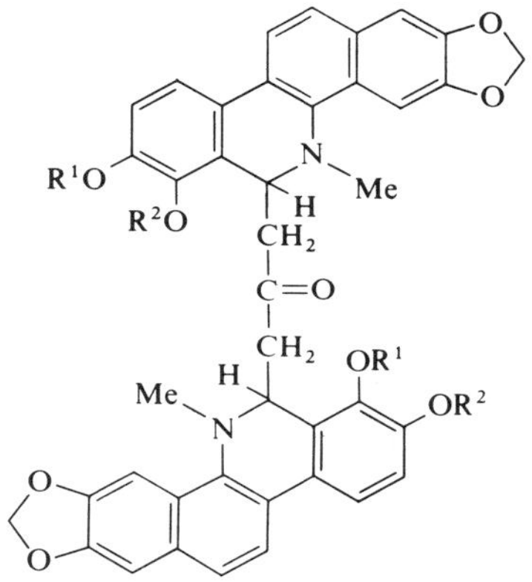

207a $R^1 + R^2 = CH_2$ Sanguidimerine
207b $R^1 = R^2 = Me$ Chelerythridimerine

From the rhizomes of *Sanguinaria canadensis*, the alkaloids chelerythridimerine [1,3-bis(6-hydrochelerythrinyl)acetone] (**207b**) and sanguidimerine (**207a**) have been isolated (*40, 293, 756*). According to these authors, they are not artifacts which have arisen during the handling of the drug but natural products. The UV spectra of these substances are identical with that of dihydrosanguinarine (**175b**). This alkaloid was inactive in the Walker carcinosarcoma 256 and in the P-388 leukemia test system to give ED_{50} of 2.7×10^1 $\mu g/ml$ in the 9-KB carcinoma assay.

The root of *Jatrorrhiza palmata* (Menispermaceae) was found to contain the dimeric bisjatrorrhizine, whose structure has not been definitely established (*423*). *Stephania japonica* gave the dimeric alkaloid benzylisoquinoline-2-phenyl-*s*-homotetrahydroisoquinoline, which is related to the alkaloids of the rhoeadane series **154** (*723, 724*). From *Thalictrum polygamum* (Ranunculaceae) a new group of alkaloids, i.e., the aporphine–pavine dimers, were isolated (*565*). Pennsylpavine, pennsylpavoline, and pennsylpavanamine belong to this group. Shamma *et al.* isolated the two dimeric alkaloids pakistanine and pakistanamine from *Berberis beluchistanica* (Berberidaceae). These are proaporphinebenzylisoquinoline and aporphinebenzylisoquinoline alkaloids (*828*). Tomko found pakistanamine as the major alkaloid in the seeds of *Berberis julianae* Sch. (*829*).

Q. Alkaloids Other than the Isoquinoline or Benzylisoquinoline Group

The plant *Chelidonium majus* L. has long been known to contain the alkaloid (−)-sparteine (**208**) (*830*). Its presence in this plant has been newly confirmed (*59*).

From the plant *Corydalis pallida* Pers. var. *tenuis* Yatabe, *trans*-3-ethylidene-2-pyrrolidone (corydalactame) (alkaloid P) (**209**) was isolated and synthetically prepared (*125, 130, 131*).

Slavík isolated the alkaloid MR-1 from *Meconopsis rudis* Prain. It is an *N*-methyltetrahydronorharmane (6-methoxy-2-methyl-1,2,3,4-tetrahydro-β-carboline) (**210**) (*221*).

The pyrrolizidine alkaloid heliotrine was also isolated from *Glaucium corniculatum* Curt. (*182*).

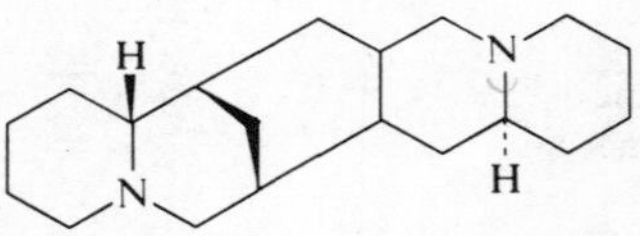

H N N H Me C H N H O MeO N H N Me

208 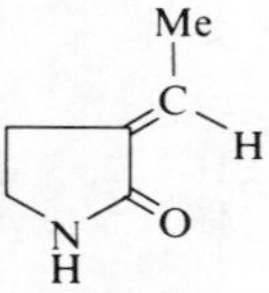 **209** 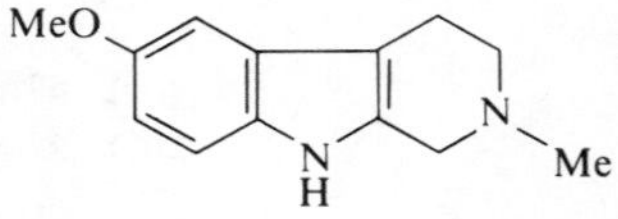 **210**

IV. Chemotaxonomy, Ecology, and Callus Tissues

A. Chemotaxonomy

Hegnauer published a chapter entitled "Chemical Evidence for the Classification of Some Plant Taxa," one section of which he devoted to the plants of the Papaveraceae family. According to him, the classification of this family is still controversial. The chemistry of the constituents suggests that the family Papaveraceae is out of place in the order *Rhoeadales*. Hegnauer assumes that the chemistry and the serology will help to resolve this problem also in the species and genera of this family (*831*). The chemotaxonomy of the plants Papaveraceae, Ranunculaceae, and Berberidaceae was also studied by Tétényi and Vágújfalvi (*832*).

From the viewpoint of botany, taxonomy, and chemotaxonomy, attention was chiefly paid to the plants of the sections *Oxytona* (Macrantha Elk.), *Papaver* (Mecones), among those particularly to the species *P. somniferum*, and to the plant genera *Corydalis*, *Glaucium*, *Fumaria*, and *Argemone* (*vide infra*) (*266*).

Stermitz *et al.* have recently reviewed the chemotaxonomy and the systematization of *Papaver* and *Argemone* plants (*36, 833*). They have also pointed out the characteristic features of the individual genera and sections from the viewpoint of the alkaloids. On the basis of the contained alkaloids, they have tried to differentiate the *A. subfusiformis* taxa from *A. polyanthemos* (Scheme 20). According to these authors, the benzophenanthridine alkaloids (including the *N*-nor derivatives) may not be significant chemotaxonomic markers because they are formed in callus tissues (*269*). The alkaloids contained in the seeds of *A. mexicana* (grown in the U.S.S.R. and in Vietnam) have been studied (*834*).

Slavík and Slavíková showed that the content of alkaloids in *Glaucium fimbrilligerum* is closely related to the species *G. elegans*, *G. oxylobum*, and *G. squamigerum* (*205*). (−)-Isocorypalmin (**58r**) was found not only in *G. fimbrilligerum* but in *Bocconia frutescens* and *Roemeria hybrida* (see Table II). Until now it was only isolated from *P. somniferum* as a minor component. This alkaloid is, however, commonly found in the botanically and biochemically closely related tribe Fumarieae.

Ownbey assumes on the basis of an investigation of the chromosomes that the species *Corydalis aurea* subsp. *aurea*, *C. micrantha* subsp. *micrantha*, and *C. sempervirens* belong to the independent section *Eucroydalis*, which is indigenous to northern America (*835*). Böhm and Günther recommended, on the basis of the presence or absence of amurine (**43c**), that the section *Pilosa* (genus *Papaver*) be divided into two subsections, namely *Pilosa* (*P. pilosum*) and *Pseudo-pilosa* (*P. atlanticum*, *P. latericium*) (*238*).

Preininger and Šantavý studied the chemotaxonomy of the plants of the section *Macrantha* Elkan and *Miltantha* Bernh. (*234*). They conclude that the plants of these two sections contain enzymatic systems which are able to synthesize the (*R*) and the (*S*) series of the individual alkaloids.

Slavík showed that the presence of the rhoeadane alkaloids is not restricted to the plants of the genus *Papaver* but also to those of *Bocconia* and *Meconopsis* (*33, 50, 221*). Dimeric alkaloids were found in the plants of the genus *Bocconia*, *Dicentra*, *Papaver*, and *Sanguinaria* (see Table I).

A new type of spirobenzyltetrahydroisoquinoline alkaloids—hypecorine alkaloids (**200**)—was found in the genus *Hypecoum* which also contains the tetrahydroprotoberberine, protopine, and the benzophenanthridine alkaloids (*214*). This genus also produces the alkaloid peshawarine (**206**), whose structure is related to that of hypecorine (*224*). The presence of cularine alkaloids is restricted to the genera *Dicentra* and *Corydalis*, and the isopavine alkaloids to the genera *Papaver* and *Roemeria*. The pavine alkaloids were found only in the genus *Argemone* (see Table I).

Tétényi reviewed the infraspecific chemical taxa of medicinal plants and among them also those of *P. somniferum* (*300, 832*). The diagram published in Tétényi (*300*) shows the changes in the content of the benzylisoquinoline, phthalideisoquinoline, and morphinane alkaloids during ontogenesis. Histological and electron-microscopical studies revealed that in the germinating plants *P. somniferum* all the alkaloids were formed in the meristem cells (*836*). Vágújfalvi examined the content of the alkaloids in the latex from various sites of four Papaveraceae taxa (*P. somniferum, P. orientale, P. somniferum* × *P. orientale*, and *Chelidonium majus*) (*837*). The results indicate that the participation of the latex vessel system in the alkaloid transport in those species appears to be improbable (*72*).

Thornber compared the presence of 11 benzylisoquinoline alkaloids in the plants of the Papaveraceae, Menispermaceae, Berberidaceae, Magnoliaceae, Ranunculaceae, Tutaceae, Monimiaceae, Annonaceae, Aristolochiaceae, Lauraceae, and Nymphaceae, and he concluded that the alkaloids cularine and morphine (including codeine and thebaine) are present only in the genera *Corydalis* and *Dicentra* (alkaloid cularine) and in the genus *Papaver* (morphinane alkaloids); the hasubanonine and bisbenzylisoquinoline alkaloids do not occur in *Papaver* plants (*838*).

The chemotaxonomical conclusions concerning the sections *Oxytona* (Macrantha), *Orthorhoeades*, *Mecones*, and *Pilosa* have also been discussed (*234*).

B. Ecology

The seasonal variations of the alkaloids and the areas of occurrence of *F. officinalis* and *M. microcarpa* have been studied (*217, 839*). A similar paper was devoted to *P. somniferum* (*513, 840*); an analysis revealed that a general accumulation of alkaloids, especially the hydrophenanthrene group, occurred one day after the flower had opened. This coincides with the peak of the pollination period. At that period, besides the principal opium poppy latex alkaloids (morphine, codeine, thebaine, papaverine, and narcotine), several other Dragendorff positive compounds were detected. The lysosomal property and accumulation of the alkaloids in vacuoles of *Chelidonium majus* have been described (*841*).

Alkaloidal storage, metabolism, and translocation in the vesicles of *P. somniferum* have been reported (*268, 298*). The content of alkaloids in this plant of various ages (no statistical difference was found in the morphine content between the blue- and the white-seeded variety), the accumulation of morphine in several varieties of poppy in different climatic zones (*326, 842*), and the identification of opium-yielding Papavers have also been studied (*843*). The contents of morphine in dried capsules of *P.*

somniferum cultivated in Bulgaria (*277*); recovery of narcotine from opium (*278*); effect of drying on the formation of alkaloid spectra of *Chelidonium majus* (*844*) the alkaloidal storage, metabolism, and translocation in the vesicles of *P. somniferum* (*268, 298,* No. 25); and the optimization of conditions for extracting alkaloids from *F. officinalis* have been reported (*168*). Mura'eva *et al.* studied the *Fumaria* plants for their content of alkaloids in relation to the period of vegetation and the locality of growth (*845*).

Attempts have also been made to cultivate such a type of *P. somniferum* which would contain a large quantity of morphine/codeine alkaloids in the straw (*846, 847*). This material can be adapted agrotechnically for an easier industrial isolation of morphinane alkaloids by the Kabay method.

As already mentioned in Section II the interest in the plants of the section *Oxytona* Bernh. was promoted by the search for new and cheaper sources of thebaine from which codeine derivatives may be prepared (Scheme 14). Among those plants are several varieties of *P. bracteatum* which have been studied thoroughly by Mothes and his school as early as 1963 (*847b*). The dry roots of some of the types of *P. bracteatum* contain up to 0.9% of thebaine. In these plant types thebaine is always accompanied by alpinigenine (*153h*). The plant *P. orientale* is even morphologically closely similar to the plants *P. bracteatum*, and, in some cases, its roots also contain thebaine. It could not, however, be utilized as a medicinal plant because its content of thebaine is low and varies considerably. A working group at the United Nations Secretariat has begun to study systematically the plants of the section *Oxytona* and the content of thebaine present therein. Intensive field studies combined with cytological and chemical analyses of this section have resulted in a reevaluation of the species in this *Oxytona* group (*266, 298, 329, 513*). The section is treated as comprizing three species, *P. bracteatum* (diploid), *P. orientale* (tetraploid), and *P. pseudoorientale* (hexaploid). Different major alkaloids characterize each species, and past confusion about alkaloid chemistry is thus clarified (Table VIII) (*266, 848*).

The quantity of thebaine in *P. bracteatum* depends on the period of ripening and on the part of the plant. The largest quantity was found in the roots (0.5–1.0%/weight of dry drug) and in the poppy heads (0.23–1.40%) (*849*). Böhm studied the content of morphinane alkaloids (thebaine, oripavine, and codeine/morphine) in hybrids obtained by hybridization between *P. somniferum*, *P. bracteatum*, *P. orientale*, and *P. pseudoorientale* and their dependence upon the genetic milieu (*847*).

Recently, a new population, Arya I (northern Teheran) and Arya II (western Iran) of *P. bracteatum* was found. The latter produced thebaine almost exclusively in very high yield. The thebaine content of this new population is 3.5% in dried capsules, more than three times higher the

TABLE VIII

SOME OF THE CRITICAL DIFFERENCES BETWEEN THE THREE SPECIES OF *PAPAVER*[a]

	P. bracteatum	*P. orientale*	*P. pseudoorientale*
Chromosomes (2*n*)	14	28	42
Stomatal length (young leaves)	20–32 μm	32–45 μm	43–60 μm
Cauline leaves on fl. stem	5–7, some on upper third	2–3, rarely on upper third	5–6, some on upper third
Leaves	Pinnatisect, cut almost to midrib; margin dentate	Pinnatisect, not deeply cut; margin dentate	Pinnatisect, deeply cut
Bracts	3–8, some persistent	Absent	Absent or present, small
Calyx	3-valved with thick bristles	2-valved, fine bristles	2- or 3-valved, fine bristles
Buds	Erect, oval to oblong	Pendulous, oval	Erect, oval
Anthers	Dark purple	Yellow or pale violet	Dark, purple
Petals	6, deep red with black blotches at base	4, pale orange with pale blue or white blotches	4–6, orange, to scarlet, black blotches
Pigments of the petal	Do not fade on storage. Distinct chromatographic pattern	Fade on storage. Distinct chromatographic pattern	Fade on storage.
Alkaloids	Thebaine, Alpinigenine	Oripavine, Isothebaine (Thebaine)	Isothebaine, Oripavine Thebaine

[a] Taken mainly from Goldblatt (*266, 848*).

content of previously reported thebaine-producing plants (petals, 1.02; stigmas, 0.63; bracts, 1.81; young leaves, 0.02; stems, 1.03%). The content of the pigments and resinous material in Arya II is very low. Thus, almost pure thebaine could be obtained after one crystallization of extracted material (*851*).

Fairbairn studied the taxonomic, phytochemical, and agricultural problems involved in attempts to convert the wild plant *P. bracteatum* into an agricultural crop to be used as a source material for pharmaceuticals (*266*). In contradistinction to *P. somniferum*, this species contains no morphine and therefore would be unsuitable for the illicit production of heroin. At present, the efforts of the botanist are directed toward the cultivation of *P. bracteatum* varieties which would contain only thebaine mainly in the leaves, so that the roots could be retained for the next crop.

TABLE IX

ALKALOIDS OF PAPAVERACEAE CALLUS TISSUES (*269*)[a]

Subfamily, tribe	Original plants of callus tissues	Type of alkaloids[a]								
		Benzophenanthridine					Protopine	Aporphine	Unknown alkaloids	
		1	2	3	4	5	6	7	CS	QS
Papaveroideae										
Eschscholtzea	*Eschscholtzia californica*	+	+	+	+	+	+	+	+	–
Chelidonieae	*Chelidonium japonicum*	+	+	+	–	–	+	–	+	–
	Macleaya cordata	+	+	+	+	+	+	–	+	–
Papavereae	*Papaver somniferum*	+	+	+	+	–	+	+	+	+
	P. setigerum	+	+	+	+	–	+	+	+	
	P. bracteatum[d]	+	+	+	+	+	+	+	+	+
	P. orientale	+	+	+	+	–	+	+	+	+
	P. rhoeae	+	+	+	+	–	+	+	+	+
Fumarioideae										
Corydaleae	*Dicentra peregrina*	+	+	+	+	–	++	+	+	+
	Corydalis incisa	+	+	+	+	–	+	+	+	–
	C. pallida	+	+	+	+	–	+	+	+	+
Hypecoideae	*Pteridophyllum racemonsum*[b]	+	+	+	–	–	++	+		

[a] *Key*: –, absent; +, present; ++, present (large amount). CS had R_f 0.31 (yellow fluorescence) in $CHCl_3$–MeOH (3:1); QS had R_f 0.3 (yellow fluorescence) in MeOH–H_2O–NH_4OH (8:1:1).

[b] Original plants.

[c] (1) Norsanguinarine, (2) oxysanguinarine, (3) dihydrosanguinarine, (4) sanguinarine, (5) chelirubine, (6) protopine, (7) magnoflorine.

[d] Callus tissues of *P. bracteatum* also form thebaine (*850*).

Because the morphological features of *P. bracteatum* are rather dubious, a rapid semiquantitative spot test was elaborated for the determination of thebaine and for the differentiation of *P. bracteatum* from *P. orientale* and *P. pseudo-orientale* (*852*). A rapid quantitative determination of thebaine in *P. bracteatum* is also carried out by application of high-pressure chromatography (*298*, No. 27).

Recently, Phillipson *et al.* found thebaine, narcotine, rhoeadine, and armepavine with thebaine as the major alkaloid in *P. fugax* of Turkish origin (*240*). The same authors also found thebaine in one race of *P. armeniacum* from Iran. If confirmatory evidence is provided, these plants would represent a new source of thebaine and thus also of codeine.

C. Callus Tissues

The callus tissues from 11 representative species of the Papaveraceae (Table IX) and the redifferentiated platelets from four species were successfully obtained and maintained (*235c, 269*). The alkaloids in the callus tissues and redifferentiated platelets were examined and compared with those of the original plants. All the callus tissues are similar in their alkaloid chemistry and contain benzophenanthridine (particularly norsanguinarine), protopine, and aporphine alkaloids. The platelets have, however, a more specific alkaloid pattern being similar to that of the original plants.

Callus and suspension cultures which were prepared from the seeds of *P. bracteatum*, population Arya II, produce thebaine. All analytical methods confirm the presence of this alkaloid in the eluted fraction and, therefore, indicate that the tissue culture is capable of biosynthesizing thebaine (*255*).

TABLE X
QUATERNARY ALKALOIDS ISOLATED FROM PLANTS OF PAPAVERACEAE FAMILY[a]

A. Benzyltetrahydroisoquinoline	
Escholamine:	*Eschscholtzia sp.* (*E. oregana*?)
Escholamidine:	*E. sp.* (*E. oregana*?)
Escholinine:	*E. californica*
B. Aporphine	
Magnoflorine (Escholine):	*Argemone platyceras, Dicranostigma lactucoides, D. leptopodum, D. franchetianum, Eschscholtzia californica, E. douglasii, E. glauca, Glaucium fimbriligerum, G. flavum, Chelidonium majus, Meconopsis cambrica, M. napaulensis, M. paniculata, M. robusta, M. rudis, Papaver somniferum*

TABLE X (*Continued*)

Menisperine:	*Dicranostigma franchetianum, D. lactucoides, D. leptopodum*
Remrefidine:	*Roemeria refracta*
G. Pavine	
Californidine:	*Eschscholtzia californica, E. douglasii, E. glauca, E. sp. (E. oregana?)*
Argemonine methohydroxide:	*Argemone gracilenta, A. platyceras*
Platycerine:	*A. platyceras*
H. Isopavine	
Amurensinine methohydroxide:	*Meconopsis horridula, M. napaulensis, M. rudis*
Remrefine:	*Roemeria refracta*
I. Tetrahydroprotoberberine[b]	
(−)-α-Stylopine methohydroxide:	*Argemone mexicana, A. ochroleuca, A. platyceras, Eschscholtzia californica, E. sp. (E. oregana?), Chelidonium majus*
(−)-β-Stylopine methohydroxide:	*Argemone mexicana, A. ochroleuca, A. platyceras, Glaucium corniculatum, Chelidonium majus, Papaver commutatum*(?), *P. rhoeas, P. syriacum*
(−)-α-Canadine methohydroxide:	*Argemone ochroleuca, A. platyceras, Eschscholtzia californica, E. douglasii, E. glauca*
(−)-β-Canadine methohydroxide:	*Glaucium corniculatum*
(−)-α-Tetrahydropalmatine methohydroxide:	*Argemone ochroleuca*
(−)-α-Scoulerine methohydroxide: (Cyclanoline)	*Argemone platyceras*
(−)-β-Scoulerine methohydroxide:	*Argemone albiflora, A. mexicana*
Escholidine:	*Eschscholtzia californica, E. douglasii, E. glauca, Hunnemannia fumariaefolia*
Mecambridine methohydroxide:	*Papaver pseudocanescens*

[a] Private communication of Professor J. Slavik.

[b] The quaternary tetrahydroprotoberberine alkaloids (α- and β-cyclanoline) were also found in the plant *Cyclea barbata* (Menispermaceae) (*853*), which indicates their larger distribution and greater significance.

REFERENCES

1. M. Shamma, "The Isoquinoline Alkaloids." Academic Press, New York, 1972.
2. M. Shamma, *in* "The Alkaloids" Vol. XIII, p. 165. Academic Press, New York, 1971.
3. M. Shamma, *in* "Chemistry of the Alkaloids" (S. W. Pelletier, ed.), p. 31. Van Nostrand-Reinhold, Princeton, New Jersey, 1970.
4. T. Kametani, "The Chemistry of the Isoquinoline Alkaloids," Vol. I. Hirokawa Publ. Co., Tokyo, 1968; Vol. II. Kinkodo Publ. Co., Sendai, Japan, 1975.

5. L. Kühn, D. Thomas, and S. Pfeifer, *Wiss. Z. Humboldt-Univ. Berlin, Mathe-Naturwiss. Reihe* **19**, 18 (1970).
6. S. Pfeifer, *Pharmazie* **26**, 328 (1971).
6a. F. Šantavý, *Planta Med.* **19**, 119 (1971).
6b. W. Döpke, "Ergebnisse der Alkaloid-Chemie 1960–1968." Akademie-Verlag, Berlin, 1976.
7. P. L. Schiff, Jr. and R. W. Doskotch, *Lloydia* **33**, 403 (1970).
8. T. R. Govindachari and N. Viswanathan, *J. Sci. Ind. Res.* **31**, 244 (1972); *CA* **77**, 152396m (1972).
9. G. J. Kapadia and M. B. E. Fayez, *J. Pharm. Sci.* **59**, 1699 (1970).
10. J. E. Saxton, or M. F. Grundon, "The Alkaloids," Vols. 1–7, Chem. Soc., London, 1971–1977.
11. N. R. Farnsworth, "Pharmacognosy Titles," Vols. I–VIII. Chicago, Illinois, 1966–1974.
12. T. Robinson, *Science* **184**, 430 (1974).
13. V. Preininger, *in* "The Alkaloids" (R. H. F. Manske, ed.), Vol. XV, p. 207. Academic Press, New York, 1975.
14. F. R. Stermitz and V. P. Muralidharan, *J. Pharm. Sci.* **56**, 762 (1967).
15. K. Haisová, J. Slavík, and L. Dolejš, *Collect. Czech. Chem. Commun.* **38**, 3312 (1973).
16. F. R. Stermitz, D. K. Kim, and K. A. Larson, *Phytochemistry* **12**, 1355 (1973).
16a. G. Sariyar and J. D. Phillipson, *Phytochemistry* **16**, 2009 (1977).
17. F. R. Stermitz, D. E. Nicodem, C. C. Wei, and K. D. McMurtey, *Phytochemistry* **8**, 615 (1969).
18. F. R. Stermitz, R. J. Ito, S. M. Workman, and W. M. Klein, *Phytochemistry* **12**, 381 (1973).
19. F. R. Stermitz, S. M. Workman, and W. M. Klein, *Phytochemistry* **10**, 675 (1971).
20. F. R. Stermitz and K. D. McMurtrey, *J. Org. Chem.* **34**, 555 (1969).
21. M. H. Benn and R. E. Mitchell, *Phytochemistry* **11**, 461 (1972).
22. B. C. Bose, R. Vijayvargiya, A. Q. Saifi, and S. K. Sharma, *J. Pharm Sci.* **52**, 1172 (1963).
23. Bui Thi Yu and D. A. Mura'eva, *Farmatsiya* (*Moscow*) **22** (4), 32 (1973); *CA* **79**, 113201r (1973).
24. W. A. Chelombit'ko, D. A. Mura'eva, and Yu. El Sawi, *Khim. Prir. Soedin.* 208 (1971).
25. W. Doepke, U. Hess, and V. Jimenez, *Z. Chem.* **16**, 54 (1976).
26. K. Haisová and J. Slavík, *Collect. Czech. Chem. Commun.* **40**, 1576 (1975).
27. D. K. Santra and A. N. Saoji, *Curr. Sci.* **40**, 548 (1971); *CA* **75**, 148509s (1971).
28. F. R. Stermitz, J. R. Stermitz, T. A. Zanoni, and J. P. Gillespie, *Phytochemistry* **13**, 1151 (1974).
29. C.-H. Chen and T. O. Soine, *J. Pharm. Sci.* **61**, 55 (1972).
30. R. M. Coomes, J. R. Falck, D. K. Williams, and F. R. Stermitz, *J. Org. Chem.* **38**, 3701 (1973).
31. K. Haisová and J. Slavík, *Collect. Czech. Chem. Commun.* **38**, 2307 (1973).
32. J. Slavík and L. Slavíková, *Collect. Czech. Chem. Commun.* **41**, 285 (1976).
33. J. Slavík and L. Slavíková, *Collect. Czech. Chem. Commun.* **40**, 3206 (1975).
34. J. Slavík, L. Slavíková, and K. Haisová, *Collect. Czech. Chem. Commun.* **38**, 2513 (1973).
35. F. R. Stermitz and R. M. Coomes, *Phytochemistry* **8**, 611 (1969).
36. A. L. Bandoni, F. R. Stermitz, R. V. D. Rondina, and J. D. Coussio, *Phytochemistry* **14**, 1785 (1975).
37. K. Haisová and J. Slavík, *Collect. Czech. Chem. Commun.* **39**, 2491 (1974).
38. F. R. Stermitz, *J. Pharm. Sci.* **56**, 760 (1967).

39. A. L. Bandoni, R. V. D. Rondina, and J. D. Coussio, *Phytochemistry* **12**, 3547 (1972).
40. D. B. MacLean, D. E. F. Gracey, J. K. Saunders, R. Rodrigo, and R. H. F. Manske, *Can. J. Chem.* **47**, 1951 (1969).
41. S. F. Cooper, J. A. Mockle, and F. Šantavý, *Planta Med.* **21**, 313 (1972).
42. K. G. Kiryakov, *Folia Med.* **14**, 75 (1972).
43. K. G. Kiryakov, M. S. Kitova, and A. V. Georgieva, *C.R. Acad. Bulg. Sci.* **20**, 189 (1967); *CA* **67**, 32849h (1967).
44. M. Onda, K. Abe, K. Yonezawa, N. Esumi, and T. Suzuki, *Chem. Pharm. Bull.* **18**, 1435 (1970).
45. J. Slavík and F. Šantavý, *Collect. Czech. Chem. Commun.* **37**, 2804 (1972).
46. N. Takao, Y. Yasumoto, and K. Iwasa, *Yakugaku Zasshi* **93**, 242 (1973).
47. Ch. Tani, S. Takao, H. Endo, and E. Oda, *Yakugaku Zasshi* **93**, 268 (1973).
48. W. Vent, *Acta Bot. Acad. Sci. Hung.* **19**, 385 (1973).
49. C. Tani, and S. Takao, *Yakugaku Zasshi* **87**, 699 (1967).
50. J. Slavík, personal communication.
51. M. R. Falco, J. X. deVries, A. G. de Brovetto, Z. Macció, S. Rebuffo, and I. R. C. Bick, *Tet. Lett.* 1953 (1968).
52. W. A. Chelombit'ko, *Actual. Vopr. Farm.* 29 (1968); *CA* **76**, 76396y (1972).
53. H. Grabarczyk and H. Gertig, *Ann. Pharm. (Poznań)* **8**, 75 (1970); *CA* **74**, 95419h (1971).
54. R. A. Labriola and D. Giacopello, *An. Asoc. Quin. Argent.* **54**, 15 (1966).
55. A. Z. Golubov, T. Sunguryan, I. Z. Bozkhova, and V. B. Chervenkova, *Nauchni Tr. Vissh. Pedagog. Inst., Plovdiv, Mat., Fiz., Khim., Biol.* **6**, 63 (1969); *CA* **71**, 109794d (1969).
56. A. Z. Golubov, T. Sunguryan, V. B. Chervenkova, and I. Z. Bozkhova, *Nauchni Tr. Vissh. Pedagog. Inst., Plovdiv, Mat., Fiz., Khim., Biol.* **8**, 135 (1970); *CA* **75**, 16066k (1971).
57. L. Jusiak, *Acta Pol. Pharm.* **24**, 65 (1967).
58. H. K. Kim, N. R. Farnsworth, R. N. Blomster, and H. H. S. Fong, *J. Pharm. Sci.* **58**, 372 (1969).
59. K. G. Kiryakov and P. Panov, *Med. Probl.* **23**, 123 (1971).
60. R. Lavenir and R. R. Paris, *Ann. Pharm. Fr.* **23**, 307 (1965).
61. J. Porebski and D. Jarosinska, *Diss. Pharm. Pharmacol.* **18**, 625 (1966).
62. M. Turowska, *Herba Pol.* **15**, 18 (1969).
63. L. D. Yakhontova, O. N. Tolkachev, and P. N. Kubal'chich, *Farmatsiya (Moscow)* **22**, 31 (1973); *CA* **78**, 133418z (1973).
64. H. Kaneko and S. Naruto, *J. Org. Chem.* **34**, 2803 (1969).
65. S. Narayanaswami, S. Prabhakar, B. R. Pai, and G. Shanmugasundaram, *Indian J. Chem.* **7**, 755 (1969).
66. S. Naruto and H. Kaneko, *Phytochemistry* **12**, 3008 (1973).
67. J. Iwasa, S. Naruto, and N. Ikeda, *Yakugaku Zasshi* **86**, 534 (1966).
68. S. Naruto and H. Kaneko, *Yakugaku Zasshi* **92**, 1017 (1972).
69. S.-T. Lu, S.-J. Wang, and T.-L. Su, *Yakugaku Zasshi* **91**, 778 (1971); *CA* **75**, 95393r (1971).
70. M. P. Cava, M. V. Lakshmikanthan, and M. J. Mitchell, *J. Org. Chem.* **34**, 2665 (1969).
71. C. K. Yu, D. B. MacLean, R. G. A. Rodrigo, and R. H. F. Manske, *Can. J. Chem.* **49**, 124 (1971).
71a. V. A. Chelombit'ko, *Actual Voprosy Farm.* **2**, 24 (1974); *CA* **84**, 147614 (1976).
72. G. Blaschke, G. Waldheim, M. von Schantz, and P. Peura, *Arch. Pharm. (Weinheim, Ger.)* **307**, 122 (1974).

73. H. Grabarczyk, *Diss. Pharm. Pharmacol.* **20**, 557 (1968).
74. R. H. F. Manske, *Can. J. Chem.* **47**, 1103 (1968).
75. R. H. F. Manske, R. Rodrigo, D. B. MacLean, and L. Baczynskyj, *An. R. Soc. Esp. Fis. Quim.* **68**, 689 (1972).
76. N. M. Mollov, *C.R. Acad. Bulg. Sci.* **20**, 557 (1967).
77. V. Preininger, R. S. Thakur, and F. Šantavý, *J. Pharm. Sci.* **65**, 294 (1976).
78. S. Naruto, K. Namba, and H. Kaneko, *Phytochemistry* **11**, 2642 (1972).
79. Kh. Sh. Baisheva and B. K. Rostotskii, *Tr. Vses. Nauch.-Issled. Inst. Lek. Aromat. Rast.* **10**, 30 (1967); *CA* **68**, 877v (1968).
80. Kh. Sh. Baisheva and B. K. Rostotskii, *Tr. Vses. Nauch.-Issled. Inst. Lek. Aromat. Rast.* **15**, 376 (1969); *CA* **75**, 20711j (1971).
81. Kh. Sh. Baisheva and B. K. Rostotskii, *Lek. Rast.* 376 (1969); *CA* **75**, 148487h (1971).
82. M. U. Ibragimova, M. S. Yunusov, and S. Yu. Yunusov, *Khim. Prir. Soedin.* 438 (1970); 638 (1971); *CA* **74**, 54046r (1971).
83. M. U. Ibragimova, M. S. Yunusov, and S. Yu. Yunusov, *Khim. Prir. Soedin.* 211 (1971); *CA* **75**, 36400t (1971).
84. I. A. N. Israilov, M. U. Ibragimova, M. S. Yunusov, and S. Yu. Yunusov, *Khim. Prir. Soedin.* 612 (1975); *CA* **84**, 86727 (1976).
85. T. Irgashev, I. A. Israilov, N. D. Abdullaev, M. S. Yunusov, and S. Yu. Yunusov, *Khim. Prir. Soedin.* 127 (1977).
86. Kh. Sh. Baisheva, D. A. Fesenko, M. E. Perel'son and B. K. Rostotskii, *Khim. Prir. Soedin.* 574 (1970); *CA* **74**, 50522v (1971).
87. K. Mehra, H. S. Garg, D. S. Bhakuni, and N. M. Khanna, *Indian J. Chem., Sect B.* **14B**, 58, 216, 844 (1976); *CA* **87**, 68513w (1977).
88. V. V. Telyat'eva, *Nauchni Tr., Irkutsk. Gos. Med. Inst.* 60 (1968).
89. T. Kametani, M. Ihara, and T. Honda, *Phytochemistry* **10**, 1881 (1971).
90. S. Naruto, S. Arakawa, and H. Kaneko, *Tet. Lett.* 1705 (1968).
91. G. Nonaka, Y. Kodera, and I. Nishioka, *Chem. Pharm. Bull.* **21**, 1020 (1973).
92. G. Nonaka, and I. Nishioka, *Chem. Pharm. Bull.* **23**, 294 (1975).
93. G. Nonaka and I. Nishioka, *Chem. Pharm. Bull.* **23**, 521 (1975).
94. G. Nonaka and I. Nishioka, *Phytochemistry* **13**, 2620 (1974).
95. G. Nonaka, H. Okabe, I. Nishioka, and N. Takao, *Yakugaku Zasshi* **93**, 87 (1973).
96. N. Takao, *Chem. Pharm. Bull.* **19**, 247 (1971).
97. N. Takao, H.-W. Bersch, and S. Takao, *Chem. Pharm. Bull.* **19**, 259 (1971).
98. N. Takao, H.-W. Bersch, and S. Takao, *Chem. Pharm. Bull.* **21**, 1096 (1973).
99. N. Takao and K. Iwasa, *Chem. Pharm. Bull.* **21**, 1587 (1973).
100. S. Teitel, W. Klötzer, J. Borgese, and A. Brossi, *Can. J. Chem.* **50**, 2022 (1972).
101. C. Tani and K. Tagahara, *Chem. Pharm. Bull.* **22**, 2457 (1974); *Yakugaku Zasshi* **97**, 87 (1977).
102. E. Fujita, K. Bessho, Y. Saeki, M. Ochiai, and K. Fuji, *Lloydia* **34**, 306 (1971).
103. Y. Watanabe, M. Uchiyama, and K. Yasuda, *Yakugaku Zasshi* **77**, 807 (1957).
104. C. Tani, N. Nagakura, and S. Hattori, *Tet. Lett.* 803 (1973); *Chem. Pharm. Bull.* **23**, 313 (1975).
105. C. Tani, N. Nagakura, and S. Hattori, *Yakugaku Zasshi* **94**, 844 (1974); *CA* **81**, 166363 (1974).
106. C. Tani, N. Nagakura, S. Hattori, and N. Masaki, *Chem. Lett.* **10**, 1081 (1975); *CA* **84**, 5211p (1976).
107. H. Kaneko, S. Naruto, and N. Ikeda, *Yakugaku Zasshi* **87**, 1382 (1967); **88**, 235 (1968).
108. I. A. N. Israilov, M. S. Yunusov, N. D. Adullaev, and S. Yu. Yunusov, *Khim. Prir. Soedin.* 536 (1975).

109. G. P. Sheveleva, D. Shargzakov, N. V. Plekhanova, S. T. Aktanova, and A. Sh. Aldasheva, *Soedin. Rast. Kirg.* 41 (1970); *CA* **75**, 148464y (1971).
110. I. A. Israilov, M. S. Yunusov, and S. Yu. Yunusov, *Khim. Prir. Soedin.* 428 (1977); 268 (1975); *CA* **83**, 97661 (1975).
111. P. Forgacs, J. Provost, R. Tiberghieu, G. Buffard, M. Pesson, and J. F. Desconclois, *C.R. Acad. Sci., Ser. D* **276**, 105 (1973); **279**, 855 (1974).
112. V. Preininger, J. Novák, V. Šimánek, and F. Šantavý, *Planta Medica* **33**, 396 (1978).
113. K. G. Kiryakov, I. A. Israilov, and S. Yu. Yunusov, *Khim. Prir. Soedin.* 411 (1974).
114. C. Tani, K. Tagahara, and S. Aratani, *J. Pharm. Soc. Jpn* **96**, 527 (1976).
115. S.-T. Lu, T.-L. Su, T. Kametani, and M. Ihara, *Heterocycles* **3**, 301 (1975).
116. S.-T. Lu, T.-L. Su, T. Kametani, A. Ujiie, and M. Ihara, *Heterocycles* **3**, 459 (1975).
117. S.-T. Lu, T.-L. Su, T. Kametani, A. Ujiie, M. Ihara, and K. Fukumoto, *J. Chem. Soc., Trans. Perkin 1* 63 (1976).
118. T. Kametani, M. Takemura, M. Ihara, and K. Fukumoto, *Heterocycles* **4**, 723 (1976); *J. Chem. Soc., Perkin Trans. 1*, 390 (1977).
119. N. N. Margvelashvili, N. P. Prisjazhnjuk, L. D. Kislov, and O. N. Tolkachev, *Khim. Prir. Soedin.* 832 (1976).
120. T. Kametani, M. Takemura, M. Ihara, and K. Fukumoto, *J. Chem. Soc., Perkin Trans. 1* 390 (1977).
121. R. G. A. Rodrigo, R. H. F. Manske, H. L. Holland, and D. B. MacLean, *Can. J. Chem.* **54**, 471 (1976).
122. P. W. Jeffs and J. D. Scharver, *J. Org. Chem.* **40**, 644 (1975).
123. Kh. Sh. Baisheva, D. A. Fesenko, B. K. Rostotskii, and M. E. Perel'son, *Khim. Prir. Soedin.* 456 (1970).
124. D. A. Fesenko and M. E. Perel'son, *Khim. Prir. Soedin.* 166 (1971); *CA* **75**, 49381n (1971).
125. T. Kametani, M. Ihara, and T. Honda, *Chem. Commun.* 1301 (1969); 1253 (1970); *J. Chem. Soc. C* 1060 (1970).
126. A. F. Polyakova, B. K. Rostotskii, and N. S. Dubinin, *Tr. Alma-At. Gos. Med. Inst.* 456, 458 (1970); *CA* **77**, 85631, 85633w (1972).
127. L. I. Stekol'nikov, *Priroda (Moscow)* 107 (1970); *CA* **74**, 1075t (1971).
127a. C. Tani, N. Nagakura, and S. Hattori, *Yakugaku Zasshi* **95**, 1103 (1975); *CA* **84**, 2251 (1976).
127b. S.-T. Lu, T.-L. Su, T. Kametani, A. Ujiie, M. Ihara, and K. Fukumoto, *J. Chem. Soc. Perkin Trans. 1* 63 (1976).
128. T. Kametani, T. Honda, and M. Ihara, *J. Chem. Soc. C* 2396 (1971).
129. T. Kametani, T. Honda, and M. Ihara, *J. Chem. Soc. C* 3318 (1971).
130. T. Kametani and M. Ihara, *J. Chem. Soc. C* 999 (1971).
131. H. Kaneko and S. Naruto, *Yakugaku Zasshi* **91**, 101 (1971).
132. W. Opolzer and K. Keller, *J. Am. Chem. Soc.* **93**, 3836 (1971).
133. T. Kametani, M. Takemura, and M. Ihara, *Phytochemistry* **15**, 2017 (1976).
134. N. N. Margvelashvili and A. D. Pakaln, *Khim. Prir. Soedin.* 133 (1973); *CA* **78**, 156640e (1973).
135. C. Tani, I. Imanishi, and J. Nishijo, *Yakugaku Zasshi* **90**, 407, 1028 (1970).
136. C. Tani, I. Imanishi, and J. Nishijo, *Yakugaku Zasshi* **90**, 903 (1970); *CA* **73**, 88057q (1970).
137. C. Tani, N. Nagakura, and N. Sugiyama, *Yakugaku Zasshi* **95**, 838 (1975).
138. C.-N. Lin, *Hua Hsueh* 22 (1971); *CA* **75**, 85178k (1971).
139. N. N. Margvelashvili, A. T. Kirjanova, and O. N. Tokachjev, *Khim. Prir. Soedin.* 127 (1972); *CA* **77**, 58825d (1972).

140. V. Preininger, J. Veselý, O. Gašić, V. Šimánek, and L. Dolejš, *Collect. Czech. Chem. Commun.* **40**, 699 (1975).
141. I. A. Israilov, M. S. Yunusov, and S. Yu. Yunusov, *Khim. Prir. Soedin.* 811 (1975); *CA* **84**, 180442 (1976).
142. R. H. F. Manske, R. G. A. Rodrigo, D. B. MacLean, D. E. F. Gracey, and J. K. Saunders, *Can. J. Chem.* **47**, 3589 (1969).
143. M. Ikram, M. H. Hug, and S. A. Warsi, *Pak. J. Sci. Ind. Res.* **9**, 34 (1966); *CA* **67**, 117031k (1967).
144. G. A. Miana, M. Ikram, and S. A. Warsi, *Pak. J. Sci. Ind. Res.* **11**, 337 (1968); *CA* **70**, 97012m (1969).
145. S. Naruto and H. Kaneko, *Phytochemistry* **11**, 2644 (1972).
146. Kh. Sh. Salsheva and B. K. Rostotskii, *Tr. Vses. Nauchno-Issled. Inst. Lek. Aromat. Rast.* **15**, 376 (1969); *CA* **75**, 20711j (1971).
147. S.-T. Lu, C.-N. Lun, and T.-S. Wu, *J. Chin. Chem. Soc.* (*Taipei*) **19**, 41 (1972); *CA* **77**, 123829z (1972).
148. C. K. Yu, D. B. MacLean, R. G. A. Rodrigo, and R. H. F. Manske, *Can. J. Chem.* **48**, 3673 (1970).
149. N. N. Margvelashvili, O. E. Lasskaja, A. T. Kirjanova, and O. N. Tolkačev, *Khim. Prir. Soedin.* 813 (1974).
150. N. N. Margvelashvili, O. E. Lasskaja, A. T. Kirjanova, and O. N. Tolkachjev, *Khim. Prir. Soedin.* 123 (1976).
151. G. R. Clarck, R. H. F. Manske, G. J. Palenik, R. G. A. Rodrigo, D. B. MacLean, L. Baczyjnskyj, D. E. F. Gracey, and J. K. Saunders, *J. Am. Chem. Soc.* **92**, 4998 (1970).
152. A. R. Battersby, J. L. McHugh, J. Staunton, and M. Todd, *Chem. Commun.* 985 (1971).
153. Z. Kowalewski, H. Gertig, and B. Kostka, *Ann. Pharm.* (*Poznan*) **6**, 27 (1967); *CA* **71**, 7811x (1969).
154. J. Slavík and L. Dolejš, *Collect. Czech. Chem. Commun.* **38**, 3514 (1973).
155. J. Slavík, L. Dolejš, and P. Sedmera, *Collect. Czech. Chem. Commun.* **35**, 2597 (1970).
156. J. Slavík and L. Slavíková, *Collect. Czech. Chem. Commun.* **36**, 2067 (1971).
157. J. Slavík, V. Novák, and L. Slavíková, *Collect. Czech. Chem. Commun.* **41**, 2429 (1976).
158. J. Slavík, L. Slavíková, and L. Dolejš, *Collect. Czech. Chem. Commun.* **40**, 1095 (1975).
158a. L. Slavíková and J. Slavík, *Collect. Czech. Chem. Commun.* **31**, 3362 (1966).
159. N. M. Mollov and G. I. Yakimov, *C.R. Acad. Bulg. Sci.* **24**, 1325 (1971); *CA* **76**, 113409c (1972).
160. N. M. Mollov and G. I. Yakimov, *C.R. Acad. Bulg. Sci.* **25**, 59 (1972); *CA* **77**, 85674k (1972).
161. S. J. Smolenski, H. Silinis, and N. R. Farnsworth, *Lloydia* **35**, 1 (1972).
162. V. B. Pandey, B. Dasgupta, and S. Ghosal, *J. Inst. Chem., Calcutta* **46**, Pt. 4, 120 (1974); *CA* **82**, 171263k (1975); *Curr. Sci.* **43**, 748 (1964).
163. V. B. Pandey, A. B. Ray, and B. Dasgupta, *Phytochemistry* **15**, 545 (1976).
164. S. Satish and D. S. Bhakuni, *Phytochemistry* **11**, 2888 (1972).
165. M. Castillo, J. K. Saunders, D. B. MacLean, N. M. Mollov, and G. I. Yakimov, *Can. J. Chem.* **49**, 139 (1971).
166. W. Golkiewicz and T. Wawrzynowicz, *Chromatographie* 356 (1970); *BA* 52, 27542 (1971).
167. J. Hermannsson and F. Sandeberg, *Acta Pharm. Suec.* **10**, 520 (1973); *BA* **58**, 15928 (1974).
168. L. G. Molokhova and B. V. Nazarov, *Farmatsiya* (*Moscow*) **23** (1), 23 (1974); *BA* **59**, 33086 (1975).

169. N. M. Mollov, G. I. Yakimov, and P. P. Panov, *C.R. Acad. Bulg. Sci.* **20**, 557 (1967); *CA* **67**, 117013f (1967).
169a. L. G. Molokhova and B. A. Figurkin, *Trudy Permsk. Gos. Med. Inst.* **118**, 33 (1973); *CA* **83**, 128664 (1975).
170. I. A. N. Israilov, M. S. Yunusov, and S. Yu. Yunusov, *Khim. Prir. Soedin.* 493 (1970).
171. I. A. N. Israilov, M. S. Yunusov, and S. Yu. Yunusov, *Khim. Prir. Soedin.* 588 (1970).
172. I. A. N. Israilov, M. S. Yunusov, and S. Yu. Yunusov, *Dokl. Acad. Nauk SSSR* **189**, 1262 (1969); *CA* **73**, 66767u (1970).
173. K. G. Kiryakov and P. Panov, *Folia Med.* **16**, 101 (1974).
174. S. Susplugas, G. Privat, J. Berlan, and J. P. Sarda, *Trav. Soc. Pharm. Montpellier* **28**, 157 (1968); *CA* **70**, 65127 (1969).
175. K. G. Kiryakov and P. P. Panov, *Dokl. Bolg. Akad. Nauk* **25**, 345 (1972); *CA* **77**, 58795u (1973).
176. N. M. Mollov, K. G. Kiryakov, and G. I. Yakimov, *Phytochemistry* **11**, 2331 (1972).
177. M. U. Ibragimova, I. A. Israilov, M. S. Yunusov, and S. Yu. Yunusov, *Khim. Prir. Soedin.* 476 (1974); *CA* **82**, 28586 (1975).
178. I. A. N. Israilov, M. S. Yunusov, and S. Yu. Yunusov, *Khim. Prir. Soedin.* 194 (1968); *CA* **69**, 57449 (1968).
179. K. G. Kiryakov and P. P. Panov, *Dokl. Bolg. Akad. Nauk* **22**, 1019 (1969); **24**, 1191 (1971); *CA* **72**, 51776 (1970); **76**, 110270q (1972).
180. A. Z. Sadikov, B. Babajev, and T. T. Šakirov, *Khim. Prir. Soedin.* 816 (1974).
181. M. Tin-Wa, N. R. Farnsworth, and K. A. Zirvi, *J. Pharm. Sci.* **65**, 755 (1976).
181. M. Tin-Wa, N. R. Farnsworth, and K. A. Zirvi, *J. Pharm. Sci.* **65**, 755 (1976).
182. K. G. Kiryakov and P. P. Panov, *Farmatsiya (Moscow)* **20**, 45 (1970); *CA* **74**, 83991a (1971).
183. V. Novák, L. Dolejš, and J. Slavík, *Collect. Czech. Chem. Commun.* **37**, 3346 (1972).
184. L. Slavíková, *Collect. Czech. Chem. Commun.* **33**, 635 (1968).
185. L. D. Yakhontova, O. N. Tolkachev, and V. Yu. Baranova, *Khim. Prir. Soedin.* 686 (1973).
186. Kh. Kiryakov and P. Panov, *Dokl. Bolg. Acad. Nauk* **29**, 677 (1976); *CA* **85**, 119633d (1976).
187. L. Bojeva-Ivanova, N. Donev, E. Mermerska, B. Avramova, P. Yoncheva, and S. Stefanov, *Postep Dziedzinie Leku Rosl., Pr. Ref. Dosw. Wygloszone Symp., 1970* 104 (1972); *CA* **78**, 88550y (1973).
188. K. H. B. Dutschewska, A. Orahovats, and N. M. Mollov, *Dokl. Bolg. Akad. Nauk* **26**, 899 (1973).
189. H. Gertig, H. Grabarczyk, Z. Kowalewski, and K. Lewandowska-Rogalla, *Diss. Pharm. Pharmacol.* **17**, 503 (1965); **18**, 375 (1966).
190. H. Gertig, K. Zdzislaw, and K. Nowaczyk, *Herba Pol.* **12**, 111 (1966).
191. K. G. Kiryakov, *Chem. Ind. (London)* 1807 (1968).
192. V. Novák and J. Slavík, *Collect. Czech. Chem. Commun.* **39**, 3352 (1974).
193. A. S. Orahovats, H. B. Dutschewska, and N. M. Mollov, *C.R. Acad. Bulg. Sci.* **26**, 491 (1973).
194. Zh. Stefanov and N. Stoyanov, *Farmatsiya (Sofia)* **20**, 29 (1970); *CA* **74**, 91110c (1971).
195. L. D. Yakhontova, *Khim. Prir. Soedin.* 285 (1967).
196. L. D. Yakhontova, *Khim. Prir. Soedin.* 203 (1968).
197. L. D. Yakhontova, *Farmatsiya (Moscow)* **21**, 25 (1972).
198. L. D. Yakhontova, *Lek. Rast.* 15, 348 (1969); *CA* **75**, 16171r, 20721n (1971).

199. L. D. Yakhontova, V. I. Scheinchenko, and O. N. Tolkachev, *Khim. Prir. Soedin.* 214 (1972); *CA* **77**, 48675r (1972).
200. I. Lalezari, A. Shafiee, and M. Mahjour, *J. Pharm. Sci.* **65**, 923 (1976).
201. T. Kametani, T. Sugahara, H. Sugi, S. Shibuya, and K. Fukumoto, *Chem. Commun.* 724 (1971).
202. I. Ribas, J. Sueiras, and L. Castedo, *Tet. Lett.* 2033 (1972).
203. C. Casagrande, L. Canonica, and G. Severini-Ricca, *J. Chem. Soc., Perkin Trans. 1* 1647 (1975).
204. I. Ribas, J. Sueiras, and L. Castedo, *Tet. Lett.* 3093 (1971).
204a. L. Castedo, R. Suau, and A. Mourino, *Tet. Lett.* 501 (1976).
205. L. Slavíková and J. Slavík, *Collect. Czech. Chem. Commun.* **36**, 2385 (1971).
206. A. K. Yunusov, I. A. Israilov, *Khim. Prir. Soedin.* 538 (1974); *CA* **82**, 14014 (1975).
207. A. K. Yunusov, I. A. Israilov, and M. C. Yunusov, *Khim. Prir. Soedin.* 681 (1973).
208. L. D. Yakhontova, O. N. Tolkachev, and D. A. Pakaln, *Khim. Prir. Soedin.* 684 (1973).
209. K. G. Kiryakov and P. Panov, *C.R. Acad. Bulg. Sci.* **22**, 1019 (1969).
210. L. Slavíková, *Collect. Czech. Chem. Commun.* **31**, 4181 (1966).
211. N. C. Gupta, D. S. Bhakuni, and M. M. Dhar, *Experientia* **26**, 12 (1970).
212. J. Slavík, *Collect. Czech. Chem. Commun.* **32**, 4431 (1967).
213. M. N. Komarova and K. F. Blinova, *Tr. Leningr. Khim.-Farm. Inst.* **26**, 163 (1968).
214. L. D. Yakhontova, M. N. Komarova, M. E. Perelson, K. F. Blinova, and O. N. Tolkachev, *Khim. Prir. Soedin.* 624 (1972); *CA* **78**, 108177n (1973).
215. K. G. Kiryakov and P. Panov, *Med. Probl.* **21**, 91 (1969).
216. T. F. Platanova, P. S. Massagetov, A. D. Kuzovkov, and L. M. Utkin, *Zh. Obshch. Khim.* **26**, 173 (1956).
217. D. A. Muravjeva and V. A. Chelombitko, *Acta Pharm. Jugosl.* **19**, 11 (1969); *CA* **71**, 109759w (1969).
218. D. Neumann, *Kul't. Izol. Organov, Tkanei Kletok Rast., Tr. Vses. Konf., 1st, 1968* 245 (1970); *CA* **74**, 108182c (1971).
219. W. A. Chelombit'ko, *Herba Pol.* **17**, 388 (1971); *CA* **77**, 58870q (1972).
220. F. R. Stermitz and R. M. Coomes, *Phytochemistry* **8**, 513 (1969).
221. J. Slavík and L. Slavíková, *Collect. Czech. Chem. Commun.* **42**, 132 (1977).
222. V. A. Mnatsakanyan and A. R. Mlkrtchyan, *Arm. Khim. Zh.* **19**, 466 (1966); *CA* **66**, 38091f (1967).
222a. S. R. Hemingway and J. D. Phillipson, *J. Pharm. Pharmacol.* **27** (1975), Suppl. 84P.
223. J. Slavík and L. Slavíková, *Collect. Czech. Chem. Commun.* **41**, 3343 (1976).
223a. J. Slavík and L. Slavíková, *Collect. Czech. Chem. Commun.* **42**, 2686 (1977).
224. M. Shamma, A. S. Rothenberg, G. S. Jayatilake, and S. F. Husain, *Heterocycles* **5**, 41 (1976).
225. A. Shafiee, I. Lalezari, and O. Rahimi, *Lloydia* **40**, 352 (1977).
226. S. Pfeifer and D. Thomas, *Pharmazie* **27**, 48 (1972).
227. S. Pfeifer and H. Döhnert, *Pharmazie* **23**, 585 (1968).
228. Y. Aynehchi and S. Jaffarian, *Lloydia* **36**, 427 (1973).
229. P. Cheng and N. J. Doorenbos, *Lloydia* **36**, 440 (1973).
230. J. W. Fairbairn and F. Hakim, *J. Pharm. Pharmacol.* **25**, 353 (1973).
231. F. J. E. M. Küppers, C. A. Salimink, M. Bastart, and M. Paris, *Phytochemistry* **15**, 444 (1976).
232. I. Lalezari, A. Shafiee, and P. Nasseri-Nouri, *J. Pharm. Sci.* **62**, 1718 (1973).
233. J. D. Phillipson, S. S. Handa, and S. W. El-Dabbas, *Phytochemistry* **15**, 1297 (1976).
234. V. Preininger and F. Šantavý, *Pharmazie* **25**, 356 (1970).
235a. H. G. Theuns, J. E. G. van Dam, J. M. Luteijn, and C. A. Salemink, *Phytochemistry* **16**, 753 (1977).

235b. O. N. Denisenko, I. A. Israilov, D. A. Muraveva, and M. S. Yunusov, *Khim. Prir. Soedin.* 547 (1977).
235c. S. Kamimura, M. Akutsu, and M. Nishikawa, *Agric. and Biol. Chem. Japan* **40**, 913 (1976).
236. A. Němečková, F. Šantavý, and D. Walterová, *Collect. Czech. Chem. Commun.* **35**, 1733 (1970).
237. V. A. Mnatsakanyan, M. A. Manushakian, and N. E. Mesropjan, *Khim. Prir. Soedin.* 424 (1977).
238. H. Böhm and K. F. Günther, *Pharmazie* **27**, 125 (1972).
239. V. Preininger, V. Tošnarová, and F. Šantavý, *Planta Med.* **20**, 70 (1971).
239a. M. A. Manushakijan and V. A. Mnacakanjan, *Khim. Prir. Soedin.* 713 (1977).
240. J. D. Phillipson, G. Sariyar, and T. Baytop, *Phytochemistry* **12**, 2431 (1973).
241. S. Pfeifer and L. Kühn, *Pharmazie* **23**, 267 (1968).
242. V. A. Mnatsakanyan, V. Preininger, V. Šimánek, A. Klásek, L. Dolejš, and F. Šantavý, *Tet. Lett.* 851 (1974).
243. A. Shafiee, I. Lalezari, P. Nasseri-Nouri, and R. Asgharian, *J. Pharm. Sci.* **64**, 1570 (1975).
243a. I. A. Israilov, O. N. Denisenko, M. S. Yunusov, S. Yu. Yunusov, and A. D. Muraveva, *Khim. Prir. Soedin.* 714 (1977); *CA* **88**, 170357f (1978).
244. K. Délenk-Heydenreich and S. Pfeifer, *Pharmazie* **24**, 635 (1969).
245. S. Pfeifer and L. Kühn, *Pharmazie* **23**, 199 (1968).
246. V. Novák and J. Slavík, *Collect. Czech. Chem. Commun.* **39**, 883 (1974).
247. L. I. Petlichnaya, *Farm Zh. (Kiev)* **30**, 22 (1975).
248. H. Böhm, L. Dolejš, V. Preininger, F. Šantavý, and V. Šimánek, *Planta Med.* **28**, 210 (1975).
249. E. Brochmann-Hanssen, C. C. Fu, and G. Zanati, *J. Pharm. Sci.* **60**, 873 (1971).
250. B. Drozdz and Z. Kowalewski, *Diss. Pharm. Pharmacol.* **18**, 527 (1966).
251. O. Gašić, V. Preininger, H. Potěšilová, B. Belia, and F. Šantavý, *Bull. Soc. Chim., Beograd* **39**, 499 (1974).
252. W. Golkiewicz, L. Jusiak, and E. Soczewinski, *Diss. Pharm. Pharmacol.* **18**, 485 (1966).
253. L. Jusiak, E. Soczewinski, and A. Waksmundzki, *Diss. Pharm. Pharmacol.* **18**, 479 (1966).
254. S. Szabó, P. Gorecki, and R. Bognár, *Pharmazie* **23**, 719 (1968).
255. A. Shafiee, I. Lalezari, and N. Yassa, *Lloydia* **39**, 380 (1976).
256. A. R. Battersby, J. Staunton, H. R. Wiltshire, R. J. Francis, and R. Southgate, *J. Chem. Soc., Perkin Trans. 1* 1147 (1975).
257. I. A. Bessonova, Z. Š. Fajzutbinova, and S. Yu. Yunusov, *Khim. Prir. Soedin.* 711 (1970); *CA* **74**, 84005u (1971).
257a. H. Koblitz, U. Schumann, H. Boehm, and J. Franke, *Experientia* **31**, 768 (1975).
258. E. Brochmann-Hanssen, C.-H. Chen, H. C. Chiang, C. C. Fu, and H. Nemoto, *J. Pharm. Sci.* **62**, 1291 (1973).
259. E. Brochmann-Hanssen, C. C. Fu, and L. Y. Misconi, *J. Pharm. Sci.* **60**, 1880 (1971).
260. E. Brochmann-Hanssen, C. C. Fu, A. Y. Leung, and G. Zanati, *J. Pharm. Sci.* **60**, 1672 (1971).
261. E. Brochmann-Hanssen, A. Y. Leung, K. Hirai, and G. Zanati, *Planta Med.* **18**, 366 (1970); *CA* **73**, 106333h (1970).
262. E. Brochmann-Hanssen, A. Y. Leung, and W. J. Richter, *J. Org. Chem.* **37**, 1881 (1972).
263. E. Brochmann-Hanssen and B. Nielsen, *J. Pharm. Sci.* **54**, 1393 (1965).
264. E. Brochmann-Hanssen, B. Nielsen, and K. Hirai, *J. Pharm. Sci.* **56**, 754 (1967).
265. E. Brochmann-Hanssen and W. J. Richter, *J. Pharm. Sci.* **64**, 1040 (1975).

266. J. W. Fairbairn, *Conf. on Med. Plants, Marianske Lazne (CSSR), 1975* p. 7 (1975); *Planta Med.* **29**, 26 (1976).
267. J. W. Fairbairn and S. El-Masry, *Phytochemistry* **7**, 181 (1968).
268. J. W. Fairbairn, F. Hakim, and Y. El Kheir, *Phytochemistry* **13**, 1133 (1974).
269. T. Furuya, A. Ikuta, and K. Syono, *Phytochemistry* **11**, 3041 (1972); **13**, 2175 (1974).
270. O. Gašić and M. Pergál, *Hem. Ind.* 20 (1971).
271. O. Gašić, M. Pergál, B. Belia, and R. Sekulić, *Bull. Sci. Jugosl.* Vol. **A15**, 80 (1970).
272. J. Hodková, Z. Veselý, Z. Koblicová, J. Holubek, and J. Trojánek, *Lloydia* **35**, 61 (1972).
273. A. Ikuta, K. Syono, and T. Furuya, *Phytochemistry* **13**, 2175 (1974).
274. R. J. Miller, C. Jollès, and H. Rapoport, *Phytochemistry* **12**, 597 (1973).
275. J. Morice, *J. Ann. Amelior. Plant.* **21**, 465 (1971); *CA* **77**, 2764r (1971).
276. P. Popov, Y. Dimitrov, S. Georgiev, and L. S. Iliev, *Bull. Narc.* **25** (3), 51 (1973).
277. P. Popov, *Farmatsiya (Sofia)* **24**(2), 20 (1974); *CA* **81**, 166307c (1974).
278. V. S. Ramanathan and C. Ramachandran, *Bull. Narc.* **26**(2), 69 (1974); *Chem. Age India* **25**(1), 33 (1974); *CA* **81**, 6234y (1974).
279. Zh. Stefanov, *Tr. Nauchnoizsled. Khim.-Farm. Inst.* 8 (1972).
280. D. Vágújfalvi and P. Tétényi, *Herba Hung.* **6**(2), 221 (1967).
281. J. Slavík and L. Slavíková, *Collect. Czech. Chem. Commun.* **41**, 290 (1976).
282. R. Bognár, Gy. Gaál, P. Kerekes, and S. Szabó, *Pharmazie* **22**, 452 (1967).
283. V. Preininger and V. Tošnarová, *Planta Med.* **23**, 233 (1973).
284. J. Slavík, L. Dolejš, and L. Slavíková, *Collect. Czech. Chem. Commun.* **39**, 888 (1974).
285. S. T. Akramov and S. Yu. Yunusov, *Khim. Prir. Soedin.* 199 (1968); *CA* **69**, 87254g (1968).
286. J. Slavík, L. Slavíková, and L. Dolejš, *Collect. Czech. Chem. Commun.* **33**, 4066 (1968).
287. M. Tomita and Y. Aoyagi, *J. Pharm. Soc. Jpn.* **87**, 1425 (1967).
288. J. Kunitomo, E. Yuge, Y. Nagai, and K. Fujitani, *Chem. Pharm. Bull.* **16**, 364 (1968).
289. F. R. Stermitz, D. K. Kim, and L. Teng, *Phytochemistry* **11**, 2644 (1972).
290. A. Brossi and R. Borer, *Lloydia* **28**, 199 (1965).
291. H. K. Kim and F. R. Stermitz, *Phytochemistry* **14**, 834 (1975).
292. M. Tin-Wa, N. R. Farnsworth, H. H. S. Fong, and J. Trojánek, *Lloydia* **33**, 267 (1970).
293. M. Tin-Wa, H. H. S. Fong, D. J. Abraham, J. Trojánek, and N. R. Farnsworth, *J. Pharm. Sci.* **61**, 1846 (1972).
294. S. Pfeifer, G. Behensen, and L. Kühn, *Pharmazie* **27**, 630 (1972).
295. D. Vágújfalvi, *Acta Bot. Acad. Sci. Hung.* **17**, 217 (1971).
296. D. Vágújfalvi, *Physiol. Alkaloide, Int. Symp., 4th, 1969*, p. 59 (1972). Abh. d. Deutsch. Akad. Wiss. Berlin.
297. Bui Thi Yu and D. A. Muravjeva, *Rastit. Resur.* **9**, 200 (1973); *BA* **57**, 4197 (1974).
298. United Nation Secretariat, Scientific Research on *Papaver bracteatum.* Newly entitled: "Scientific Research on *Papaver* Species as Sources of Codeine, Morphine and Thebaine," ST/SOA/SER. J. Nos. 20, 21, United Nations, New York, 1975; Nos. 22, 23, 24, 25, 26, 27 (1976).
299. M. Tomita, M. Kozuka, H. Ohyabu, and K. Fujitani, *Yakugaku Zasshi* **90**, 82 (1970).
300. P. Tétényi, "Infraspecific Chemical Taxa of Medicinal Plants," pp. 29 and 30. Akadémiai Kiadó, Budapest, 1970.
301. J. H. Knox and J. J. Jurand, *J. Chromatogr.* **82**, 398 (1973).
302. T. T. Shakirov, H. N. Aripov, and S. Yu. Yunusov, *Khim. Prir. Soedin.* 394 (1968).
303. B. I. Shvydkij, V. Z. Kamarenko, and T. S. Dyvzhenko, *Farmatsyia (Moscow)* **20**, 49 (1971); *CA* **75**, 67422f (1971).
304. R. Hakim and J. M. Fujimoto, *J. Pharm. Sci.* **59**, 1783 (1970); *CA* **74**, 24950p (1971).

305. L. Jusiak, *Diss. Pharm. Pharmacol.* **20**, 577 (1968); *BA* **51**, 127622 (1970).
306. O. B. Stepanenko and F. M. Shemyakin, *Farmatsyia (Moscow)* **19**, 37 (1970); *CA* **72**, 136459a (1970).
307. R. Bognár, Gy. Gaál, P. Goreckim, and P. Kerekes, *Herba Hung.* **8**, 195 (1969).
308. R. Bognár, Gy. Gaál, P. Gorecki, and S. Szabó, *Pharmazie* **22**, 525 (1967).
309. Gy. Gaál, *Acta Phys. Chim. Debrecina* 171 (1968); *CA* **71**, 64032k (1969).
310. P. Gorecki and R. Bognár, *Pharmazie* **23**, 590 (1968).
311. J. Tulecki and P. Gorecki, *Ann. Pharm. (Poznan)* 25 (1969); *CA* **72**, 47323j (1970).
312. E. Soczewinski and B. Szabelska, *Diss. Pharm. Pharmacol.* **22**, 243 (1970).
313. A. Z. Golubov and A. T. Venkov, *Nauchni Tr. Vissh. Pedagog. Inst., Plovdiv, Mat., Fiz., Khim., Biol.* **7**, 133 (1969); *CA* **73**, 7133m (1970).
314. V. Sapara, *Cas. Ces. Lek., Ved. Příl.* **63**, 293 (1950).
315. H. Inouye, S. Ueda, and Y. A. Takebe, *J. Chromatogr.* **25**, 167 (1966).
316. T. Kaniewska and B. Borkowski, *Diss. Pharm. Pharmacol.* **20**, 111 (1968).
317. J. A. Labat, *Bull. Soc. Chim. Biol.* **15**, 1344 (1933).
318. G. O. Gaebel, *Arch. Pharm. (Weinheim, Ger.)* **248**, 225 (1910).
319. F. Laterrier and B. Viossat, *C.R. Acad. Sci., Ser. C* **265**, 410 (1967).
320. C. K. Guven and B. Aran, *Eczacilik Bul.* **15**, 28 (1973); *CA* **79**, 97018d (1973).
321. C. K. Guven and N. Guven, *Eczacilik Bul.* **14**, 75 (1972); *CA* **78**, 62218w (1973).
322. S. Kh. Mushinskaya, V. A. Danel'yants, and A. M. Sych, *Postep Dziedzinie Leku Rosl., Pr. Ref. Dosw. Wygloszone Symp.*, 203 (1972); *CA* **78**, 68876f (1973).
323. S. Pfeifer, *J. Chromatogr.* **41**, 127 (1969).
324. P. P. Gladyshev, M. I. Goryaev, and A. N. Baigalieva, *Izv. Akad. Nauk Kaz. SSR, Ser. Khim.* **19**, 57 (1969); *CA* **71**, 64112m (1969).
325. P. P. Gladyshev and M. I. Goryaev, *Izv. Akad. Nauk Kaz. SSR, Ser. Khim.* **19**, 50 (1969); *CA* **71**, 64110j (1969).
326. H. L. Tookey, G. F. Spencer, M. D. Grove, and J. A. Duke, *Bull. Narc.* **27**, No. 4, p. 49 (1975).
327. D. W. Smith, T. H. Beasley, Sr., R. L. Charles, and H. W. Ziegler, *J. Pharm. Sci.* **62**, 1691 (1973).
328. United Nations Secretariat, "Scientific Research on *Papaver bracteatum*," Nos. 8, 15. United Nations, New York 1974. See also refs. 298 and 329.
329. "Scientific Research on *Papaver bracteatum*," Report of the Third Working Group on *P. bracteatum*. United Nations Secretariat, New York, No. 28–31 (1977).
330. J. Holubek, "Spectral Data and Physical Constants of Alkaloids," Vols. I–VIII. Czech. Acad. Sci., Prague, 1965–1973.
331. V. Šimánek, V. Preininger, A. Klásek, and J. Juřina, *Heterocycles* **4**, 1263 (1976).
332. V. Šimánek, V. Preininger, and J. Lasovský, *Collect. Czech. Chem. Commun.* **41**, 1050 (1976).
333. J. Wu, C.-H. Chen, N. A. Shaath, and T. O. Soine, *T'ai-wan Yao Hsueh Tsa Chih* **27**, 105 (1975); *CA* **86**, 106835s (1977).
334. T. Kametani, K. Takahashi, V. L. Chu, and M. Hirata, *Heterocycles* **1**, 247 (1973).
335. M. Pailer and E. Haslinger, *Monatsh.* **103**, 1399 (1972).
336. S. M. Albonico, A. M. Kuck, and V. Deulofeu, *Ann.* **685**, 200 (1965).
337. D. W. Brown, S. F. Dyke, G. Hardy, and M. Sainsbury, *Tet. Lett.* 1515 (1969).
338. Sb. Tewari, D. S. Bhakuni, and R. S. Kapil, *Chem. Commun.* 554 (1975).
339. M. L. Wilson and C. J. Coscia, *J. Am. Chem. Soc.* **97**, 431 (1975).
340. E. Brochmann-Hanssen, C.-H. Chen, C. R. Chen, H.-Ching Chiang, A. Y. Leung, and K. McMurtrey, *J. Chem. Soc., Perkin Trans. 1* 1531 (1975).
340a. H. Uprety, D. S. Bhakuni, and R. S. Kapil, *Phytochemistry* **14**, 1535 (1975).

341. F. D. Popp, *Heterocycles* **1**, 165 (1973).
342. T. Kametani and K. Fukumoto, *Heterocycles* **1**, 129 (1973).
343. B. Umezawa and O. Hoshino, *Heterocycles* **3**, 1005 (1975).
344. T. Kametani and K. Fukumoto, *Heterocycles* **3**, 931 (1975).
345. T. Kametani, K. Fukumoto, and F. Satoh, *Bioorg. Chem.* **3**, 430 (1974).
346. S. Teitel and A. Brossi, *Heterocycles* **1**, 73 (1973).
347. A. Brossi and S. Teitel, *J. Org. Chem.* **35**, 1684 (1970).
348. G. Grethe, V. Toome, H. L. Lee, M. Uskoković, and A. Brossi, *J. Org. Chem.* **33**, 504 (1968).
349. A. Brossi, J. O'Brien, and S. Teitel, *Org. Prep. Proced.* **2**, 281 (1970).
350. M. P. Cava, M. J. Mitchell, and D. T. Hill, *Chem. Commun.* 1601 (1970).
351. M. P. Cava, I. Noguchi, and K. T. Buck, *J. Org. Chem.* **38**, 2394 (1973).
352. G. Grethe, H. L. Lee, M. R. Uskoković, and A. Brossi, *Helv. Chim. Acta* **53**, 874 (1970).
353. T. Kametani, T. Kobari, K. Fukumoto, and M. Fujihara, *J. Chem. Soc. C* 1796 (1971).
354. A. J. Birch, A. H. Jackson, and P. V. R. Shannon, *J. Chem. Soc., Perkin Trans. 2* 2190 (1974).
355. T. Kametani, K. Fukumoto, K. Kigasawa, and K. Wakisaka, *Chem. Pharm. Bull.* **19**, 714 (1971).
356. G. N. Dorofeenko and V. G. Korobkova, *Zh. Obshch. Khim.* **40**, 249 (1970); *CA* **73**, 15043x (1970).
357. T. Kametani, T. Terui, H. Agui, and K. Fukumoto, *J. Heterocycl. Chem.* **5**, 753 (1968).
358. T. Kametani, S. Takano, and T. Kobari, *J. Chem. Soc. C* 1030 (1971).
359. T. Kametani, S. Hirata, F. Satoh, and K. Fukumoto, *J. Chem. Soc., Perkin Trans. 1* 2509 (1974).
360. H. O. Bernhard and V. Snieckus, *Tetrahedron* **27**, 2091 (1971).
361. B. Goeber, S. Pfeifer, V. Hanuš, and G. Engelhardt, *Arch. Pharm.* (*Weinheim, Ger.*) **301**, 763 (1968).
362. T. Kametani and K. Ogasawara, *Chem. Pharm. Bull.* **21**, 893 (1973).
363. J. F. Archer, D. R. Boyd, W. R. Jackson, M. F. Grundon, and W. A. Khan, *J. Chem. Soc. C* 2560 (1971).
364. M. Konda, T. Shioiri, and Sh.-i. Yamada, *Chem. Pharm. Bull.* **23**, 1025 (1975).
365. M. Konda, T. Shioiri, and Sh.-i. Yamada, *Chem. Pharm. Bull.* **23**, 1063 (1975).
365a. M. Konda, T. Oh-ishi, and Sh.-i. Yamada, *Chem. Pharm. Bull.* **25**, 69 (1977).
366. Sh.-i. Yamada, M. Konda, and T. Shioiri, *Tet. Lett.* 2215 (1972).
367. D. S. Bhakuni, S. Satish, and M. M. Dhar, *Tetrahedron* **28**, 1093 (1972).
368. D. G. Farber and A. Giacomazzi, *An. Assoc. Quim. Argent.* **58**, 133 (1970).
369. B. Umezawa, O. Hoshino, H. Hara, and J. Sakakihara, *Chem. Pharm. Bull.* **16**, 381 (1968).
370. J. B. Bobbitt, K. H. Weisgraber, A. S. Steinfeld, and S. G. Weiss, *J. Org. Chem.* **35**, 2884 (1970).
371. G. M. Habashy and N. A. Farid, *Talanta* **20**, 699 (1973).
372. E. Pawelczyk and T. Hermann, *Diss. Pharm. Pharmacol.* **21**, 267, 347 (1969).
373. B. Olesch and H. Böhm, *Arch. Pharm.* (*Weinheim, Ger.*) **305**, 222 (1972).
374. J. Knabe and H. Powilleit, *Arch. Pharm.* (*Weinheim, Ger.*) **304**, 52 (1971).
375. M. P. Cava, K. Nomura, S. K. Talapatra, M. J. Mitchell, R. H. Schlessinger, K. T. Buck, J. L. Beal, B. Douglas, R. F. Raffauf, and J. A. Weisbach, *J. Org. Chem.* **33**, 2785 (1968).
376. J. M. Bobbitt and R. C. Hallcher, *Chem. Commun.* 543 (1971).
377. J. B. Brunner and Le Van Thuc, *Chem. Ind.* (*London*) 453 (1976).
378. N. A. Shaath and T. O. Soine, *J. Org. Chem.* **40**, 1987 (1975).

379. L. Dolejš and J. Slavík, *Org. Mass Spectrom.* **7**, 775 (1973).
380. L. Hruban, F. Šantavý, and S. Hegerová, *Collect. Czech. Chem. Commun.* **35**, 3420 (1970).
381. V. Šimánek, V. Preininger, S. Hegerová, and F. Šantavý, *Collect. Czech. Chem. Commun.* **37**, 2746 (1972).
382. S. Pfeifer, G. Behnsen, L. Kühn, and R. Kraft, *Pharmazie* **27**, 734 (1972).
383. H. Budzikiewicz and U. Krüger, *Ann.* **737**, 119 (1970).
384. W. Awe, H. Halpaap, and O. Hertel, *Arzneim.-Forsch.* **10**, 936 (1960).
385. W. Wiegrebe and B. Rohrbach-Munz, *Helv. Chim. Acta* **58**, 1825 (1975).
386. J. T. Stock and G. A. Shia, *Anal. Chim. Acta* **84**, 211 (1976); *CA* **85**, 160374h (1976).
387. T. Kametani, M. Takemura, K. Takahashi, M. Ihara, and K. Fukumoto, *Heterocycles* **3**, 139 (1975).
388. T. Kametani, M. Takemura, M. Ihara, K. Takahashi, and K. Fukumoto, *J. Am. Chem. Soc.* **98**, 1956 (1976).
389. M. Sandler, S. B. Carter, K. R. Hunter, and G. M. Stern, *Nature* (*London*) **241**, 439 (1973).
390. S. Teitel, J. O'Brien, and A. Brossi, *J. Med. Chem.* **15**, 845 (1972).
391. K. L. Stuart and M. P. Cava, *Chem. Rev.* **68**, 321 (1968).
392. M. K. Yusupov, D. A. Abdullaeva, Kh. A. Aslanov, and A. S. Sadykov, *Dokl. Akad. Nauk SSSR* **208**, 1123 (1973); *CA* **78**, 136482b (1973).
393. H. Guinaudeau, M. Leboeuf, and A. Cave, *Lloydia* **38**, 275 (1975).
394. W. V. Curran, *Chem. Commun.* 478 (1971).
395. S. Ishiwata and K. Itakura, *Chem. Pharm. Bull.* **18**, 416 (1970).
396. S. Ishiwata and K. Itakura, *Chem. Pharm. Bull.* **18**, 1224 (1970).
397. S. Ishiwata and K. Itakura, *Chem. Pharm. Bull.* **18**, 1841 (1970).
398. S. Ishiwata, K. Itakura, and K. Misawa, *Chem. Pharm. Bull.* **18**, 1219 (1970).
399. T. Kametani and I. Noguchi, *J. Chem. Soc. C* 502 (1969).
400. J. M. Bobbitt, *Heterocycles* **1**, 181 (1973).
401. J. Y. Becker, L. L. Miller, and F. R. Stermitz, *J. Electroanal. Chem. Interfacial Electrochem.* **68**, 181 (1976).
402. J. M. Bobbitt, I. Noguchi, R. S. Ware, K. Ng. Chiong, and S. J. Huang, *J. Org. Chem.* **40**, 2924 (1975).
403. Z. Horii, Y. Nakashita, and C. Iwata, *Tet. Lett.* 1167 (1971).
404. T. Kametani, S. Shibuya, T. Nakano, and K. Fukumoto, *J. Chem. Soc. C* 3818 (1971).
405. T. Kametani, T. Sugahara, H. Sugi, S. Shibuya, and K. Fukumoto, *Tetrahedron* **27**, 5993 (1971).
406. T. Kametani, H. Sugi, S. Shibuya, and K. Fukumoto, *Chem. Ind.* (*London*) 818 (1971).
407. T. Kametani, H. Sugi, S. Shibuya, and K. Fukumoto, *Chem. Pharm. Bull.* **19**, 1513 (1971).
408. P. Kerekes, K. D. Heydenreich, and S. Pfeifer, *Tet. Lett.* 2483 (1970); *Ber.* **105**, 609 (1972).
408a. J. L. Neumeyer, M. McCarthy, K. K. Weinhardt, and P. L. Levins, *Tet. Lett.* 3107 (1967); *J. Org. Chem.* **33**, 2890 (1968).
409. S. M. Kupchan and P. F. O'Brien, *Chem. Commun.* 916 (1973).
410. T. Kametani, K. Takahashi, K. Ogasawara, and K. Fukumoto, *Tet. Lett.* 4219 (1973).
411. T. Kametani, K. Takahashi, K. Ogasawara, Chu Van Loc, and K. Fukumoto, *Collect. Czech. Chem. Commun.* **40**, 712 (1975).
412. C. Casagrande, L. Canonica, and G. Severini-Ricca, *J. Chem. Soc., Perkin Trans. 1* 1659 (1975).
413. O. Hoshino, T. Toshioka, and B. Umezawa, *Chem. Pharm. Bull.* **22**, 1302 (1974).

414. Z. Horii, Sh. Uchida, Y. Nakashita, E. Tsuchida, and C. Iwata, *Chem. Pharm. Bull.* **22**, 583 (1974).
415. S. M. Kupchan, V. Kameswaran, J. T. Lynn, D. K. Williams, and A. J. Liepa, *J. Am. Chem. Soc.* **97**, 5623 (1975).
416. S. M. Kupchan and Ch.-K. Kim, *J. Org. Chem.* **41**, 3210 (1976).
416a. S. M. Kupchan, Ch.-K. Kim, and K. Miyana, *Chem. Commun.* 91 (1976).
417. S. M. Kupchan and A. Yoshitake, *J. Org. Chem.* **34**, 1062 (1969).
418. J. Slavík, P. Sedmera, and K. Bláha, *Collect. Czech. Chem. Commun.* **35**, 1558 (1970).
419. L. Dolejš, *Collect. Czech. Chem. Commun.* **39**, 571 (1974).
420. F. Šantavý, L. Hruban, V. Šimánek, and D. Walterová, *Collect. Czech. Chem. Commun.* **35**, 2418 (1970).
421. S. Dvořáčková, L. Hruban, V. Preininger, and F. Šantavý, *Heterocycles* **3**, 575 (1975).
422. M. P. Cava and A. Venkateswarlu, *Annu. Rep. Med. Chem.* 331 (1969).
423. M. L. Carvalhas, *J. Chem. Soc., Perkin Trans. 1* 327 (1972).
423a. J. Slavík, *Collect. Czech. Chem. Commun.* **33**, 323 (1968).
424. M. Shamma and S. S. Salgar, *Phytochemistry* **12**, 1505 (1973).
425. E. Gellert and R. E. Summons, *Aust. J. Chem.* **23**, 2095 (1970).
426. F. N. Lahey and K. F. Mak, *Aust. J. Chem.* **24**, 671 (1971).
427. M. P. Cava and S. S. Libsch, *J. Org. Chem.* **39**, 577 (1974).
428. K. Ito, H. Furukawa, M. Haruna, and M. Sato, *Yakugaku Zasshi* **90**, 1163 (1970).
429. M. Shamma, J. L. Moniot, S. Y. Yao, and J. A. Stanko, *Chem. Commun.* 408 (1972).
430. M. Shamma and J. L. Moniot, *Heterocycles* **3**, 297 (1975).
431. M. Akasu, H. Itokawa, and M. Fujita, *Tet. Lett.* 3609 (1974).
432. K. Ito and H. Furukawa, *Tet. Lett.* 3023 (1970).
433. M. P. Cava, M. D. Glick, R. E. Cook, M. Srinivasan, J. Kunitomo, and A. I. daRocha, *Chem. Commun.* 1217 (1969).
434. S. K. Talapatra, A. Patra, and B. Talapatra, *Chem. Ind. (London)* 1056 (1969).
435. P. E. Sonnet and M. Jacobson, *J. Pharm. Sci.* **60**, 1254 (1971).
436. A. Venkateswarlu and M. P. Cava, *Tetrahedron* **32**, 2079 (1976).
437. M. P. Cava, A. Venkateswarlu, M. Srinivasan, and D. L. Edie, *Tetrahedron* **28**, 4299 (1972).
438. M. P. Cava, D. L. Edie, and J. M. Saá, *J. Org. Chem.* **40**, 3601 (1975).
439. T. Kametani, S. Shibuya, and S. Kano, *J. Chem. Soc., Perkin Trans. 1* 1212 (1973).
440. T. Kametani, R. Nitadori, H. Terasawa, K. Takahashi, and M. Ihara, *Heterocycles* **3**, 821 (1975).
441. M. Gerecke, R. Borer, and A. Brossi, *Helv. Chim. Acta* **58**, 185 (1975).
441a. M. Gerecke, R. Borer, and A. Brossi, *Helv. Chim. Acta* **59**, 2551 (1976).
442. J. Lutomski and J. Wisniewski, *Herba Pol.* **19**, 318 (1974); **20**, 40 (1975).
443. J. Gadamer and F. Kunze, *Arch. Pharm. (Weinheim, Ger.)* **249**, 598 (1911).
444. K. Rehse, *Arch. Pharm. (Weinheim, Ger.)* **305**, 625 (1972).
445. M. P. Cava and M. Srinavasan, *J. Org. Chem.* **37**, 330 (1972).
446. C. Casagrande, L. Canonica, and G. Severini-Ricca, *J. Chem. Soc., Perkin Trans. 1* 1652 (1975).
446a. J. S. Bindra and A. Grodski, *J. Org. Chem.* **42**, 910 (1977).
447. R. Ahmad, J. M. Saá, and M. P. Cava, *J. Org. Chem.* **42**, 1228 (1977).
448. S. Teitel and J. P. O'Brien, *Heterocycles* **5**, 85 (1976).
449. S. Ishiwata and K. Itakura, *Chem. Pharm. Bull.* **17**, 2256 (1969).
450. S. Ishiwata and K. Itakura, *Chem. Pharm. Bull.* **17**, 2261 (1969).
451. J. Gadamer, M. Oberlin, and A. Schoeler, *Arch. Pharm. (Weinheim, Ger.)* **263**, 81 (1925).

452. T. Kametani, S. Shibuya, R. Charubala, M. S. Premila, and B. R. Pai, *Heterocycles* **3**, 439 (1975).
453. M. Premila and B. R. Pai, *Indian J. Chem.* **13**, 13 (1975).
454. T. Kametani, T. Sugahara, and K. Fukumoto, *Tetrahedron* **27**, 5367 (1971).
455. T. Kametani, A. Kozuka, and K. Fukumoto, *J. Chem. Soc. C* 1021 (1971).
456. A. R. Battersby, T. J. Brockson, and R. Ramage, *J. Chem. Soc. D* 464 (1969).
457. T. Kametani and S. Shibuya, *Heterocycles* **3**, 439 (1975).
458. M. P. Cava and M. Srinavasan, *Tetrahedron* **26**, 4649 (1970).
459. R. J. Vavrek, J. G. Cannon, and R. V. Smith, *J. Pharm. Sci.* **59**, 823 (1970).
460. O. Hoshino, H. Hara, N. Serizawa, and B. Umezawa, *Chem. Pharm. Bull.* **23**, 2048 (1975).
461. H. Hara, O. Hoshino, and B. Umezawa, *Chem. Pharm. Bull.* **24**, 262, 1921 (1976).
461a. T. Kametani, Y. Satoh, M. Takemura, Y. Ohta, M. Ihara, and K. Fukumoto, *Heterocycles* **5**, 175 (1976).
462. M. A. Schwartz, R. A. Holton, and S. W. Scott, *J. Am. Chem. Soc.* **91**, 2800 (1969); M. A. Schwartz, *Synth. Commun.* **3**, 33 (1973).
463. A. Brossi, A. Ramel, J. O'Brien, and S. Teitel, *Chem. Pharm. Bull.* **21**, 1839 (1973).
464. J. Z. Ginos, A. LoMonte, G. C. Cotzias, A. K. Bose, and R. J. Brambilla, *J. Am. Chem. Soc.* **95**, 2991 (1973).
465. E. Brochmann-Hanssen, C.-H. Chen, H. C. Chiang, and K. McMurtrey, *Chem. Commun.* 1269 (1972).
466. J. Gadamer, *Arch. Pharm.* (*Weinheim, Ger.*) **249**, 498 (1911).
467. G. Blaschke, *Arch. Pharm.* (*Weinheim, Ger.*) **303**, 358 (1970).
468. G. Blaschke, *Arch. Pharm.* (*Weinheim, Ger.*) **301**, 432 (1968).
469. M. Shamma and S. Y. Yao, *J. Org. Chem.* **36**, 3253 (1971).
470. E. L. Kristallovich, M. R. Jagudagev, E. F. Ismailov, and S. Yu. Yunusov, *Khim. Prir. Soedin.* 646 (1973).
471. W. H. Baarschers, R. R. Andt, K. Pachler, J. A. Weisbach, and B. Douglas, *J. Chem. Soc.* 4778 (1964).
472. S. R. Johns, J. A. Lamberton, and A. A. Sioumis, *Chem. Commun.* 480 (1966).
473. M. Shamma and J. L. Moniot, *Experienta* **32**, 282 (1976).
474. K. L. Stuart, *Chem. Rev.* **71**, 47 (1971).
475. T. Kametani and S. Kaneda, *Yakugaku Zasshi* **87**, 1070 (1967).
476. T. Kametani, K. Fukumoto, and T. Sugahara, *Tet. Lett.* 5459 (1968).
477. T. Kametani, K. Fukumoto, F. Satoh, and H. Yagi, *Chem. Commun.* 1398 (1968).
478. T. Kametani, K. Fukumoto, A. Kozuka, H. Yagi, and M. Koizumi, *J. Chem. Soc. C* 2034 (1969).
479. T. Kametani, M. Koizumi, and K. Fukumoto, *Chem. Pharm. Bull.* **17**, 2245 (1969).
480. T. Kametani, M. Ihara, K. Fukumoto, and H. Yagi, *J. Chem. Soc. C* 2030 (1969).
481. T. Kametani and I. Noguchi, *J. Chem. Soc. C* 447 (1968).
482. T. Kametani, R. Charubala, M. Ihara, M. Koizumi, K. Takahashi, and K. Fukumoto, *J. Chem. Soc. C* 3315 (1971).
483. T. Kametani and K. Fukumoto, *J. Heterocycl. Chem.* **8**, 341 (1971).
484. T. Kametani, S. Shibuya, K. Kigasawa, M. Hiiragi, and O. Kusama, *J. Chem. Soc. C* 2712 (1971).
484a. T. Kametani, T. Sugahara, and K. Fukumoto, *Chem. Pharm. Bull.* **22**, 966 (1974).
485. T. Kametani, S. Shibuya, H. Sugi, O. Kusama, and K. Fukumoto, *J. Chem. Soc. C* 2446 (1971).
486. T. Kametani, K. Shizhido, and S. Takano, *J. Heterocycl. Chem.* **12**, 305 (1975); *CA* **83**, 79433p (1975).

487. T. Kametani, H. Sugi, S. Shibuya, and K. Fukumoto, *Tetrahedron* **27**, 5379 (1971).
488. T. Kametani, T. Sugahara, and K. Fukumoto, *Chem. Ind.* (*London*) 833 (1969).
489. B. Franck and V. Teetz, *Angew. Chem.* **83**, 409 (1971).
490. J. R. Falck, L. L. Miller, and F. R. Stermitz, *Tetrahedron* **30**, 931 (1974).
491. L. L. Miller, F. R. Stermitz, and J. R. Falck, *J. Am. Chem. Soc.* **93**, 5941 (1971).
492. E. Kotani and S. Tobinaga, *Tet. Lett.* 4759 (1973).
493. M. A. Schwartz and I. S. Mami, *J. Am. Chem. Soc.* **97**, 1239 (1975).
494. L. F. Bjeldanes and H. Rapoport, *J. Org. Chem.* **37**, 1453 (1972).
495. D. E. Rearick and M. Gates, *Tet. Lett.* 507 (1970).
496. J. Reisch, M. Gombos, K. Szendrei, and I. Novák, *Arch. Pharm.* (*Weinheim, Ger.*) **307**, 814 (1974).
497. D. D. Weller and H. Rapoport, *J. Med. Chem.* **19**, 1171 (1976).
498. R. B. Barber and H. Rapoport, *J. Med. Chem.* **18**, 1074 (1975); **19**, 1175 (1976).
499. F. Krausz, "Scientific Research on *Papaver bracteatum*," ST/SOA/SER.J/27, No. 23, Annex III. United Nations Secretariat, United Nations, New York, 1976.
500. Gy. D. Gaál, *Acta Phys. Chim. Debrecina* **8**, 39 (1962).
501. G. W. Kirby, S. R. Massey, and P. Steinreich, *J. Chem. Soc., Perkin Trans. 1* 1642 (1972).
502. V. Kamedulski, B. Dimov, and Iv. Tonev, *Farmatsiya* (*Sofia*) **22**, 34 (1972); *CA* **77**, 130524z (1972).
503. H. L. Parker, G. Blaschke, and H. Rapoport, *J. Am. Chem. Soc.* **94**, 1276 (1972).
504. M. E. Su, *Korean J. Pharmacogn.* **3**(1), 31 (1972); *BA* **56**, 39543 (1973).
505. J. W. Fairbairn, J. M. Palmer, and A. Paterson, *Phytochemistry* **7**, 2117 (1968).
506. G. J. Kapadia, G. S. Rao, E. Leete, M. B. E. Fayez, Y. N. Vaishnav, and H. M. Fales, *J. Am. Chem. Soc.* **92**, 6943 (1970).
507. J. W. Fairbairn, M. Djoté, and A. Paterson, *Phytochemistry* **7**, 2111 (1968).
508. J. Fishman, B. I. Norton, and W. Hembriee, *J. Labelled Compd.* **9**, 563 (1973); *CA* **80**, 37340p (1974).
509. D. J. Doedens and R. B. Forney, *J. Chromatogr.* **100**, 225 (1974).
510. G. Blaschke, *Mitt. Dtsch. Pharm. Ges.* **39**, 225 (1969).
511. A. R. Battersby, C. R. F. Jones, and R. Kazlauskas, *Tet. Lett.* 1873 (1975).
511a. C. C. Hodges, J. S. Horn, and H. Rapoport, *Phytochemistry* **16**, 1939 (1977).
512. K. Michels-Nyomárkay, *Herba Hung.* **9**, 43, 85 (1970).
513. J. W. Fairbairn, *J. Pharm. Pharmacol.* **25**, 113P (1973).
514. P. P. Gladyshev, M. I. Goryaev, and I. Y. Sadchikov, *Izv. Akad. Nauk Kaz. SSR, Ser. Khim.* **18**, 69 (1968).
515. A. Gyeresi and G. Racz, *Pharmazie* **28**, 271 (1973).
516. F. Kavka, J. Trojánek, and Z. Čekan, *Pharmazie* **20**, 220 (1965).
517. F. Kavka, J. Trojánek, and Z. Čekan, *Pharmazie* **20**, 429 (1965).
518. S. K. Mushinskaya, Yu. V. Shostenko, E. S. Vysotskaya, and N. G. Bozhko, *Khim. Farm. Zh.* **6**(12), 34 (1972); *CA* **78**, 75812k (1973).
519. A. F. Rubstov and E. M. Salomatin, *Farmatsiya* (*Moscow*) **22**(4), 54 (1973); *CA* **80**, 512v (1973).
520. J. Trojánek, F. Kavka, J. Vít, and Z. Čekan, *Pharmazie* **20**, 172 (1965).
521. K. Ikonomovski, *Acta Pharm. Jugosl.* **23**, 169 (1973); *CA* **79**, 137324j (1973).
522. S. Pfeifer, G. Behnsen, and L. Kühn, *Pharmazie* **27**, 648 (1972).
523. K. C. Rice, *J. Org. Chem.* **40**, 1850 (1975).
524. S. Makleit, *Acta Phys. Chim. Debrecina* **18**, 265 (1972–1973).
525. R. Bognár, S. Makleit, L. Radics, and I. Seki, *Org. Prep. Proced. Int.* **5**, 49 (1973).
526. R. Bognár, Gy. Gaál, P. Kerekes, A. Lévi, S. Makleit, F. Snatzke, and G. Snatzke, *Collect. Czech. Chem. Commun.* **40**, 670 (1975).

527. Z. Dinya, S. Makleit, R. Bognár, and P. Jékel, *Acta Chim. Acad. Sci. Hung.* **71**, 125 (1972).
528. G. H. Loew and D. S. Berkowitz, *J. Med. Chem.* **18**, 656 (1975).
529. A. E. Jacobson, H. J. C. Yeh, and L. J. Sargent, *Org. Magn. Reson.* **4**, 875 (1972).
530. Y. Terui, K. Tori, S. Maeda, and J. A. Pople, *Tet. Lett.* 2853 (1975).
531. F. I. Carroll, C. G. Moreland, G. A. Brine, and J. A. Kepler, *J. Org. Chem.* **41**, 996 (1976).
532. E. S. Vysotskaya, S. Kh. Mushinskaya, and Yu. V. Shostenko, *Zh. Fiz. Khim.* **49**(7), 1859 (1975); *CA* **83**, 206460x (1975).
533. V. A. Danel'yants, M. S. Kh. Mushinskaya, and Yu. V. Shostenko, *Khim. Farm. Zh.* **7**, 47 (1973); *CA* **78**, 164150k (1973).
534. J. H. Knox and J. Jurand, *J. Chromatogr.* **87**, 95 (1974).
535. V. A. Danel'yants and Yu. V. Shostenko, *Farm. Zh.* **27**(6), 48 (1972); *BA* **56**, 38840 (1973).
536. R. V. D. Rondina, A. L. Bandoni, and J. D. Coussio, *J. Pharm. Sci.* **62**, 502 (1973).
537. V. A. Danel'yants and Yu. V. Shostenko, *Farm. Zh.* **28**(3), 79 (1973); *CA* **79**, 97025d (1973).
538. H. H. Loh, *J. Chromatogr.* **76**, 505 (1973).
539. A. Bechtel, *Chromatographia* **5**, 404 (1972).
540. R. M. Smith, *J. Forensic Sci.* **18**(4), 327 (1973).
541. E. Brochmann-Hanssen, *J. Pharm. Sci.* **61**, 1118 (1972).
542. P. Popov, *Bull. Narc.* **25**, 51 (1973).
543. U. Borner and S. Abbott, *Experientia* **29**, 180 (1973).
544. T. Kametani, K. Fukumoto, and M. Fujihara, *Bioorg. Chem.* **1**, 40 (1971).
545. T. Kametani, T. Kikuchi, and K. Fukumoto, *Chem. Pharm. Bull.* **16**, 103 (1968).
546. H.-Ch. Hsu, T. Kikuchi, S. Aoyagi, and H. Iida, *Yakugaku Zasshi* **92**, 1030 (1972).
547. H. Iida, H.-Ch. Hsu, T. Kikuchi, and K. Kawano, *Yakugaku Zasshi* **92**, 1242 (1972).
548. H. Iida, H.-Ch. Hsu, and T. Kikuchi, *Chem. Pharm. Bull.* **21**, 1001 (1973).
549. S. Ishiwata, T. Fujii, N. Miyaji, N. Satoh, and K. Itakura, *Chem. Pharm. Bull.* **18**, 1850 (1970).
550. A. H. Jackson and G. W. Stewart, *Chem. Commun.* 149 (1971).
551. A. H. Jackson and G. W. Stewart, *Tet. Lett.* 4941 (1971); *J. Chem. Soc. D* 149 (1971).
552. A. H. Jackson, G. W. Stewart, G. A. Charnock, and J. A. Martin, *J. Chem. Soc., Perkin Trans. 1* 1911 (1974).
553. T. Kametani, M. Fujihara, and K. Fukumoto, *Chem. Commun.* 352 (1971).
554. T. Kametani, K. Fukumoto, and M. Fujihara, *Chem. Commun.* 352 (1971).
555. T. Kametani, K. Fukumoto, and M. Fujihara, *Chem. Pharm. Bull.* **20**, 1800 (1972).
556. T. Kametani, K. Fukumoto, K. Kigasawa, M. Hiiragi, and H. Ishimaru, *Heterocycles* **3**, 311 (1975).
557. T. Kametani, H. Iida, and C. Kibayashi, *J. Heterocycl. Chem.* **6**, 61 (1969).
558. T. Kametani, H. Iida, and C. Kibayashi, *J. Heterocycl. Chem.* **7**, 339 (1970).
559. T. Kametani, H. Iida, T. Kikuchi, M. Mizushima, and K. Fukumoto, *Chem. Pharm. Bull.* **17**, 709 (1969).
560. T. Kametani, T. Sugahara, and K. Fukumoto, *Yakugaku Zasshi* **89**, 610 (1969).
561. I. Noguchi and D. B. MacLean, *Can. J. Chem.* **53**, 125 (1975).
562. J. I. Kunitomo, K. Morimoto, K. Yamamoto, Y. Yoshikawa, K. Azuma, and K. Fuitane, *Chem. Pharm. Bull.* **19**, 2197 (1971).
563. T. Kametani, T. Honda, H. Shimanouchi, and Y. Sasada, *Chem. Commun.* 1072 (1972).
564. H. Shimanouchi, Y. Sasada, T. Honda, and T. Kametani, *J. Chem. Soc., Perkin Trans. 2* 1226 (1973).
565. M. Shamma and J. L. Moniot, *J. Am. Chem. Soc.* **96**, 3338 (1974).

566. Ch.-H. Chen, T. O. Soine, and K.-H. Lee, *J. Pharm. Sci.* **59**, 1529 (1970).
567. Ch.-H. Chen, T. O. Soine, and K.-H. Lee, *J. Pharm. Sci.* **60**, 1634 (1971).
568. K.-H. Lee and T. O. Soine, *J. Pharm. Sci.* **57**, 1922 (1968).
569. M. Premila and B. R. Pai, *Indian J. Chem.* **11**, 1084 (1973).
570. F. R. Stermitz and D. K. Williams, *J. Org. Chem.* **38**, 1761 (1973).
571. F. R. Stermitz, D. K. Williams, S. Natarajan, M. S. Premila, and B. R. Pai, *Indian J. Chem.* **12**, 1249 (1974); *CA* **82**, 171260g (1975).
572. D. A. Walsh and R. E. Lyle, *Tet. Lett.* 3849 (1973).
573. K. Ito, H. Furukawa, T. Iida, K. H. Lee, and T. O. Soine, *Chem. Commun.* 1037 (1974).
574. M. S. Yunusov, S. T. Akramov, and S. Yu. Yunusov, *Khim. Prir. Soedin.* 225 (1968).
575. S. F. Dyke and A. C. Ellis, *Tetrahedron* **27**, 3803 (1971); *Phytochemistry* **11**, 867 (1972).
576. S. F. Dyke, A. C. Ellis, R. G. Kinsman, and A. W. C. White, *Tetrahedron* **30**, 1193 (1974).
577. A. R. Battersby and D. A. Yeowell, *J. Chem. Soc.* 1988 (1958).
578. D. W. Brown, S. F. Dyke, and M. Sainsbury, *Tetrahedron* **25**, 101 (1969).
579. D. W. Brown, S. F. Dyke, G. Hardy, and M. Sainsbury, *Tet. Lett.* 2609 (1968).
580. S. F. Dyke and A. C. Ellis, *Tetrahedron* **28**, 3999 (1972).
581. M. Sainsbury, D. W. Brown, S. F. Dyke, and G. Hardy, *Tetrahedron* **25**, 1881 (1969).
582. M. Sainsbury, S. F. Dyke, D. W. Brown, and R. G. Kinsman, *Tetrahedron* **26**, 5265 (1970).
583. T. Kametani, S. Hirata, and K. Ogasawara, *J. Chem. Soc., Perkin Trans. 1* 1466 (1973).
584. M. Shamma, J. L. Moniot, Wan Kit Chan, and K. Nakanishi, *Tet. Lett.* 3425 (1971).
585. M. Shamma, C. D. Jones, and J. A. Weiss, *Tetrahedron* **25**, 4347 (1969).
586. T. Kametani, M. Ihara, and T. Honda, *Heterocycles* **4**, 483 (1976).
587. T. Kametani, A. Ujiie, M. Ihara, K. Fukumoto, and H. Koizumi, *Heterocycles* **3**, 371 (1975).
588. T. Kametani and I. Noguchi, *Yakugaku Zasshi* **89**, 721 (1969).
588a. Hsüch-Ching Chiang and E. Brochmann-Hanssen, *J. Org. Chem.* **42**, 3190 (1977).
588b. E. Brochmann-Hanssen and Hsüch-Ching Chiang, *J. Org. Chem.* **42**, 3588 (1977).
589. V. Preininger, V. Šimánek, and F. Šantavý, *Tet. Lett.* 2109 (1969).
590. T. Kametani, I. Noguchi, K. Saito, and S. Kaneda, *J. Chem. Soc. C* 2036 (1969).
591. T. Kametani, K. Fukumoto, H. Yagi, H. Iida, and T. Kikuchi, *J. Chem. Soc. C* 1178 (1968).
592. T. Kametani and M. Ihara, *J. Chem. Soc. D* 1241 (1970).
593. T. Kametani, M. Ihara, and T. Honda, *J. Chem. Soc. C* 2342 (1970).
594. T. Kametani, M. Ihara, T. Honda, H. Shimanouchi, and Y. Sasada, *J. Chem. Soc. C* 2541 (1971).
595. T. Kametani, M. Ihara, Y. Kitahara, C. Kabuto, H. Shimanouchi, and Y. Sasada, *Chem. Commun.* 1241 (1970).
596. T. Kametani, K. Wakisaka, T. Kikuchi, M. Ihara, H. Shimanouchi, and Y. Sasada, *Tet. Lett.* 627 (1969).
597. W. J. Richter and E. Brochmann-Hanssen, *Helv. Chim. Acta* **58**, 209 (1975).
598. E. Gellert and R. Rudzats, *Aust. J. Chem.* **25**, 2477 (1972).
598a. M. Shamma and Sr. M. A. Podczasy, *Tetrahedron* **27**, 727 (1971).
599. K. Jewers, A. H. Manchanda, and P. N. Jenkins, *J. Chem. Soc., Perkin Trans. 2* 1393 (1972).
600. Ken-ichi Yoshikawa, I. Morishima, Jun-ichi Kunitomo, M. Ju-Ichi, and Y. Yoshida, *Chem. Lett.* 961 (1975).
601. T. Kametani, M. Takemura, K. Takahashi, M. Takeshita, M. Ihara, and K. Fukumoto, *J. Chem. Soc., Perkin Trans. 1* 1012 (1975).

602. T. Kametani, A. Ujiie, and K. Fukumoto, *J. Chem. Soc., Perkin Trans. 1* 1954 (1974).
603. T. Kametani, A. Ujiie, and K. Fukumoto, *Heterocycles* **2**, 55 (1974).
604. T. Kametani, A. Ujiie, M. Ihara, and K. Fukumoto, *Heterocycles* **3**, 143 (1975); *J. Chem. Soc., Perkin Trans. 1* 1822 (1975).
605. G. Habermehl and J. Schunck, *Ann.* **750**, 128 (1971).
606. G. Habermehl, J. Schunck, and G. Schaden, *Ann.* **742**, 138 (1970).
607. T. R. Govindachari, K. Nagarajan, R. Charubala, and B. R. Pai, *Indian J. Chem.* **8**, 763 (1970); *CA* **73**, 131186v (1970).
608. T. Kametani, T. Kato, and K. Fukumoto, *Tetrahedron* **30**, 1043 (1974).
609. T. Kametani, K. Ogasawara, and T. Takahashi, *Chem. Commun.* 675 (1972).
609a. T. Kametani, M. Takemura, and K. Fukumoto, *Heterocycles* **2**, 433 (1974).
610. T. Kametani, K. Ogasawara, and T. Takahashi, *Tetrahedron* **29**, 73 (1973).
611. T. Kametani, Y. Katoh, and K. Fukumoto, *J. Chem. Soc., Perkin Trans. 1* 1712 (1974).
612. T. Kametani, T. Takahashi, T. Honda, K. Ogasawara, and K. Fukumoto, *J. Org. Chem.* **39**, 447 (1974).
613. T. Kametani, T. Takahashi, K. Ogasawara, and K. Fukumoto, *Tetrahedron* **30**, 1047 (1974).
614. T. Kametani, K. Fukumoto, T. Terui, K. Yamaki, and E. Taguchi, *J. Chem. Soc. C* 2709 (1971).
615. T. Kametani, T. Sugai, Y. Shoji, T. Honda, F. Satoh, and K. Fukumoto, *J. Chem. Soc., Perkin Trans. 1* 1151 (1977).
616. I. Ninomiya, T. Naito, and H. Takasugi, *J. Chem. Soc., Perkin Trans. 1* 1720, 1791 (1975).
617. I. Ninomiya, H. Takasugi, and T. Naito, *Heterocycles* **1**, 17 (1973).
618. T. Kametani, H. Iida, T. Kikuchi, T. Honda, and M. Ihara, *J. Heterocycl. Chem.* **7**, 491 (1970).
618a. T. Kametani and K. Ohkubo, *Yakugaku Zasshi* **89**, 279 (1969).
619. T. Kametani, T. Honda, T. Sugai, and K. Fukumoto, *Heterocycles* **4**, 927 (1976).
620. S. Ishiwata and K. Itakura, *Chem. Pharm. Bull.* **18**, 1846 (1970).
621. I. Ninomiya and T. Naito, *Chem. Commun.* 137 (1973); *J. Chem. Soc., Perkin Trans. 1* 762 (1975).
622. T. Kametani and K. Fukumoto, *Heterocycles* **3**, 29 (1975).
623. T. Kametani, K. Fukumoto, M. Ihara, M. Takemura, H. Matsumoto, B. R. Pai, K. Nagarajan, M. S. Premila, and H. Suguna, *Heterocycles* **3**, 811 (1975).
624. T. Kametani, H. Matsumoto, Y. Satoh, H. Nemoto, and T. Fukuomoto, *J. Chem. Soc., Perkin Trans. 1* 376 (1977).
625. W. Meise and F. Zymalkowski, *Arch. Pharm.* (*Weinheim, Ger.*) **304**, 182 (1971).
626. W. Meise and F. Zymalkowski, *Tet. Lett.* 1475 (1969); *Arch. Pharm.* (*Weinheim, Ger.*) **304**, 175 (1971).
627. W. Nagata, H. Itazaki, K. Okada, T. Wakabayashi, K. Shibata, and N. Tokutake, *Chem. Pharm. Bull.* **23**, 2867 (1975).
628. W. Nagata, K. Okada, H. Itazaki, and S. Uyeo, *Chem. Pharm. Bull.* **23**, 2878 (1975).
629. S. F. Dyke, D. W. Brown, M. Sainsbury, and G. Hardy, *Tetrahedron* **27**, 3495 (1971).
630. H. Sugana and B. R. Pai, *Collect. Czech. Chem. Commun.* **41**, 1219 (1976).
631. T. Takemoto and Y. Kondo, *Yakugaku Zasshi* **82**, 1413 (1962).
632. M. Hanaoka, C. Mukai, and Y. Arata, *Heterocycles* **6**, 895 (1977).
633. H. Bruderer, J. Metzger, and A. Brossi, *Helv. Chim. Acta* **58**, 1719 (1975).
634. T. Kametani, A. Ujiie, M. Ihara, K. Fukumoto, and Sh.-T. Lu, *J. Chem. Soc., Perkin Trans. 1* 1218 (1976).
635. H. Bruderer, J. Metzger, A. Brossi, and J. J. Daly, *Helv. Chim. Acta* **59**, 2793 (1976).

635a. T. Kametani, A. Ujiie, Sh. Py. Huang, M. Ihara, and K. Fukumoto, *J. Chem. Soc., Perkin Trans.* **1**, 394 (1977).
636. T. Kametani, C. Ohtsuka, H. Nemoto, and K. Fukumoto, *Chem. Pharm. Bull.* **24**, 2525 (1976).
637. R. H. F. Manske, "The Alkaloids," Vol. IV, p. 1. Academic Press, New York, 1954.
638. G. Blaschke, *Arch. Pharm.* **301**, 439 (1968).
639. H. L. Holland, M. Castillio, D. B. MacLean, and I. D. Spenser, *Can. J. Chem.* **52**, 2818 (1974).
640. T. Kametani, M. Takemura, K. Fukumoto, T. Terui, and A. Kozuka, *J. Chem. Soc., Perkin Trans. 1* 2678 (1974).
641. T. Kametani, M. Takemura, K. Takahashi, M. Takeshita, M. Ihara, and K. Fukumoto, *Heterocycles* **2**, 653 (1974).
642. T. Kametani, S. Shibuya, S. Hirata, and K. Fukumoto, *Chem. Pharm. Bull.* **20**, 2570 (1972).
643. V. Šmula, R. H. F. Manske, and R. Rodrigo, *Can. J. Chem.* **50**, 1544 (1972).
644. T. R. Govindachari, K. Nagarajan, R. Charubala, and B. R. Pai, *Indian J. Chem.* **8**, 766 (1970).
645. Ch. Tani, S. Takao, and M. Sugiura, *Yakugaku Zasshi* **90**, 1012 (1970).
646. T. Kametani, K. Kigasawa, M. Hiiragi, N. Wagatsuma, and K. Wakisaka, *Tet. Lett.* 635 (1969).
647. T. Kametani, E. Taguchi, K. Yamaki, A. Kozuka, and T. Terui, *Chem. Pharm. Bull.* **21**, 1124 (1973).
648. H. Irie, K. Akagi, S. Tani, K. Yabusaki, and H. Yamane, *Chem. Pharm. Bull.* **21**, 855 (1973).
648a. T. Kametani, S. P. Huang, A. Ujiie, M. Ihara, and K. Fukumoto, *Heterocycles* **4**, 1223 (1976).
648b. D. Greenslade and R. Ramage, *Tetrahedron* **33**, 927 (1977).
649. B. Nalliah, R. H. F. Manske, R. Rodrigo, and D. B. MacLean, *Tet. Lett.* 2795 (1973).
650. S. Kano, T. Yokomatsu, E. Komiyama, S. Tokita, Y. Takahagi, and S. Shibuya, *Chem. Pharm. Bull.* **23**, 1171 (1975).
651. G. R. Lenz, *J. Org. Chem.* **39**, 2839, 2846 (1974).
652. Y. Kondo, T. Takemoto, and K. Kondo, *Heterocycles* **2**, 659 (1974).
653. S. Naruto, H. Nishimura, and H. Kaneko, *Tet. Lett.* 2127 (1972).
654. S. Naruto, H. Nishimura, and H. Kaneko, *Chem. Pharm. Bull.* **23**, 1271, 1276, 1565 (1975).
655. L. Dolejš and J. Slavík, *Org. Mass. Spectrom.* **3**, 141 (1970).
656. W. J. Richter and E. Brochmann-Hanssen, *Helv. Chim. Acta* **58**, 203 (1975).
657. A. S. Yunusov, Ya. V. Rashkers, M. U. Ibragimova, and S. Yu. Yunusov, *Khim. Prir. Soedin.* 380 (1971).
658. Ch. Tani, S. Takao, and K. Tagahara, *Yakugaku Zasshi* **93**, 197 (1973).
659. M. Shamma, M. J. Hillman, and C. D. Jones, *Chem. Rev.* **69**, 779 (1969).
660. W. Wiegrebe, D. Sasse, H. Reinhart, and L. Faber, *Z. Naturforsch. Teil B* **25**, 1408 (1970).
661. G. Snatzke and P. C. Ho, *Tetrahedron* **27**, 3645 (1971).
662. G. Snatzke, J. Hrbek, Jr., L. Hruban, A. Horeau, and F. Šantavý, *Tetrahedron* **26**, 5013 (1970).
663. N. Takao and K. Iwasa, *Chem. Pharm. Bull.* **24**, 3185 (1976).
664. T. R. Govindachari, K. Nagarajan, R. Charubala, B. R. Pai, and P. S. Subramanian, *Indian J. Chem.* **8**, 769 (1970).
664a. D. Tourwe, G. van Binst, and T. Kametani, *Org. Mag. Reson.* **9**, 341 (1977).

665. M. M. Tadžibajev, I. N. Zatorskaja, K. L. Lutfullin, and T. T. Šarikov, *Khim. Prir. Soedin.* 48 (1974).
666. K. Kimura, Y. Novo, and S. Honda, *Shoyakugaku Zasshi* **26**(2), 141 (1972); *CA* **79**, 35176v (1973).
667. G. A. Miana, *Phytochemistry* **12**, 1822 (1973).
668. K. Yuen Zee-Cheng and C. C. Cheng, *J. Pharm. Sci.* **62**, 1572 (1973).
669. K. Kurahashi and M. Fujiwara, *Can. J. Physiol. Pharmacol* **54**, 287 (1976).
670. S. Pavelka and J. Kovář, *Collect. Czech. Chem. Commun.* **40**, 753 (1975).
671. M. Shamma and J. L. Moniot, *Chem. Commun.* 89 (1975).
672. R. M. Sotelo and D. Giacopello, *Aust. J. Chem.* **25**, 385 (1972).
673. M. Hanaoka, Ch. Mukai, and Y. Arata, *Heterocycles* **4**, 1685 (1976).
674. B. Nalliah, R. H. F. Manske, and R. Rodrigo, *Tet. Lett.* 1765 (1974).
675. S. Teitel, J. Borgese, and A. Brossi, *Helv. Chim. Acta* **56**, 554 (1973).
676. D. W. Brown and S. F. Dyke, *Tet. Lett.* 2605 (1968).
677. S. F. Dyke and D. W. Brown, *Tetrahedron* **25**, 5375 (1969).
678. M. Onda, K. Abe, and K. Yonezawa, *Chem. Pharm. Bull.* **16**, 2005 (1968).
679. M. Onda, M. Gotch, and J. Okada, *Chem. Pharm. Bull.* **23**, 1561 (1975).
679a. M. Onda, K. Yuasa, and J. Okada, *Chem. Pharm. Bull.* **22**, 2365 (1974).
680. M. Onda, K. Yonezawa, and K. Abe, *Chem. Pharm. Bull.* **17** 404, 2565 (1969).
681. M. Onda, K. Yonezawa, and K. Abe, *Chem. Pharm. Bull.* **19**, 31 (1971).
682. M. Onda, K. Yonezawa, K. Abe, H. Toyama, and T. Suzuki, *Chem. Pharm. Bull.* **19**, 317 (1971).
683. D. H. R. Barton, R. B. Boar, and D. A. Widdowson, *J. Chem. Soc. C* 807 (1969).
684. H. Rönsch, *J. Prakt. Chem.* **314**, 382 (1972).
685. K. W. Bentley and A. W. Murray, *J. Chem. Soc.* 2497 (1963).
686. R. D. Haworth and W. H. Perkin, *J. Chem. Soc.* 845, 1764, 1769 (1926).
687. A. L. Margni, D. Giacopello, and V. Deulofeu, *J. Chem. Soc. C* 2578 (1970).
688. T. T. Nakashima and G. E. Maciel, *Org. Magn. Reson.* **5**, 9 (1973).
689. S. R. Hall and F. R. Ahmed, *Acta Crystallogr., Sect. B* **24**, 337, 346 (1968).
690. Y. Sawa and S. Maeda, Japan Unexamined Patent 49-41386, 49-41387 (1974); see Nagata *et al.* (*627*).
691. W. Wiegrebe, E. Rosel, W. D. Sasse, and H. Keppel, *Arch. Pharm.* **302**, 22 (1969).
692. H. L. Holland, M. Curcumelli-Rodostamo, and D. B. MacLean, *Can. J. Chem.* **54**, 1472 (1976).
693. V. Šmula, N. E. Cundasawmy, H. L. Holland, and D. B. MacLean, *Can. J. Chem.* **51**, 3287 (1973).
693a. N. E. Cundasawmy and D. B. MacLean, *Can. J. Chem.* **50**, 3028 (1972).
693b. B. C. Nalliash, D. B. MacLean, R. G. A. Rodrigo, and R. H. F. Manske, *Can. J. Chem.* **55**, 922 (1977).
694. M. Shamma and V. St. Georgiev, *Tet. Lett.* 2339 (1974); *Tetrahedron* **32**, 211 (1976).
695. J. L. Moniot and M. Shamma, *J. Am. Chem. Soc.* **98**, 6714 (1976).
696. T. Kametani, H. Inoue, T. Honda, T. Sugahara, and K. Fukumoto, *J. Chem. Soc., Perkin Trans. 1* 374 (1977).
697. T. Kametani, T. Honda, H. Inoue, and K. Fukumoto, *J. Chem. Soc., Perkin Trans. 1* 1221 (1976).
698. T. Kametani, S. Hirata, M. Ihara, and K. Fukumoto, *Heterocycles* **3**, 405 (1975).
699. P. Gorecki, *Herba Pol.* **19**, 352, 357, 365 (1973).
700. P. Gorecki and T. Kubala, *Herba Pol.* **19**, 152, 195 (1973).
701. Gy. Gaál, P. Kerekes, and R. Bognár, *J. Prakt. Chem.* **313**, 935 (1971).
702. Gy. Gaál, P. Kerekes, P. Gorecki, and R. Bognár, *Pharmazie* **26**, 431 (1971).

703. P. Kerekes and R. Bognár. *J. Prakt. Chem.* **313**, 923 (1971).
704. S. Teitel, J. P. O'Brien, and A. Brossi, *J. Org. Chem.* **37**, 1879 (1972).
705. S. Teitel and J. P. O'Brien, *J. Org. Chem.* **41**, 1657 (1976).
706. M. Shamma and J. A. Weiss, unpublished results: see Shamma (*1*).
707. A. R. Battersby, M. Hirt, D. J. McCaldin, R. Southgate, and J. Staunton, *J. Chem. Soc. C* 2163 (1968).
708. M. M. Janot and H. Pourat, *Bull. Soc. Chim. Fr.* Series 6, 823 (1955).
709. V. Šimánek and A. Klásek, *Tet. Lett.* 4133 (1971); *Collect. Czech. Chem. Commun.* **38**, 1614 (1973).
710. T. Kikuchi and T. Nishinaga, *Tet. Lett.* 2519 (1969).
711. G. Snatzke, G. Wollenberg, J. Hrbek, Jr., K. Bláha, F. Šantavý, W. Klyne, and R. J. Swan, *Tetrahedron* **25**, 5059 (1969).
712. C. Gorinsky and D. S. Moss, *J. Cryst. Mol. Struct.* **3**, 299 (1973).
713. P. R. Andrews and G. A. R. Johnston, *Nature* (*London*) *New Biol.* **243**, 29 (1973).
714. O. B. Stepanenko, *Farmatsiya* (*Moscow*) **22**(4), 73 (1973).
715. V. Preininger, V. Šimánek, O. Gašić, F. Šantavý, and L. Dolejš, *Phytochemistry* **12**, 2513 (1973).
716. S. R. Johns, J. A. Lamberton, A. A. Sioumis, and R. I. Willing, *Aust. J. Chem.* **23**, 353 (1970).
717. M. Freund, *Ann.* **271**, 360 (1892).
718. J. Trojánek, Z. Koblicová, Z. Veselý, V. Suchan, and J. Holubek, *Collect. Czech. Chem. Commun.* **40**, 681 (1975).
719. D. Walterová and F. Šantavý, *Collect. Czech. Chem. Commun.* **33**, 1623 (1968).
720. W. Klötzer and W. E. Oberhaensli, *Helv. Chim. Acta* **56**, 2107 (1973).
721. W. Klötzer, S. Teitel, and A. Brossi, *Monatsh.* **103**, 1210 (1972).
722. H. Böhm, *Biochem. Physiol. Pflanz.* **162**, 474 (1971).
723. T. Ibuka, T. Konoshima, and Y. Inubushi, *Chem. Pharm. Bull.* **23**, 114 (1975).
724. T. Ibuka, T. Konoshima, and Y. Inubushi, *Chem. Pharm. Bull.* **23**, 133 (1975).
725. M. D. Rozwadowska and J. W. ApSimon, *Tetrahedron* **28**, 4125 (1972).
726. E. Späth, L. Schmid, and H. Sternberg, *Monatsh.* **68**, 33 (1936).
727. Y. Inubushi, T. Harayama, and K. Takeshima, *Chem. Pharm. Bull.* **20**, 689 (1972).
728. M. Shamma and L. Töke, *Chem. Commun.* 740 (1973); *Tetrahedron* **31**, 1991 (1975).
729. K. Orito, R. H. F. Manske, and R. Rodrigo, *J. Am. Chem. Soc.* **96**, 1944 (1974).
730. B. Nalliah, R. H. F. Manske, and R. Rodrigo, *Tet. Lett.* 2853 (1974).
731. H. Irie, S. Tani, and H. Yamane, *Chem. Commun.* 1713 (1970); *J. Chem. Soc., Perkin Trans. 1* 2986 (1972).
732. W. Klötzer, S. Teitel, J. F. Blount, and A. Brossi, *J. Am. Chem. Soc.* **93**, 4321 (1971); *Monatsh.* **103**, 435 (1972).
733. W. Klötzer, S. Teitel, and A. Brossi, *Helv. Chim. Acta* **54**, 2057 (1971); **55**, 2228 (1972).
733a. S. Prabhakar and Ana M. Lobo, *Chem. Commun.* 419 (1977).
734. C. Schöpf and M. Schweickert, *Ber.* **98**, 2566 (1965).
735. H. O. Bernhard and V. Snieckus, *Tet. Lett.* 4867 (1971).
736. R. Hohlbrugger and W. Klötzer, *Ber.* **107**, 3457 (1974).
737. Z. Veselý, J. Holubek, and J. Trojánek, *Chem. Ind.* (*London*) 478 (1973).
738. Z. Veselý, J. Holubek, H. Kopecká, and J. Trojánek, *Collect. Czech. Chem. Commun.* **40**, 1403 (1975).
739. A. R. Battersby and J. Staunton, *Tetrahedron* **30**, 1707 (1974).
740. C. Tani and K. Tagahara, *Yakugaku Zasshi* **97**, 93 (1977).
741. H. Rönsch, *Eur. J. Biochem.* **28**, 123 (1972).

741a. H. Rönsch, *Phytochemistry* **16**, 691 (1977).
742. H. Rönsch, *Mitt. Chem. Ges. DDR* **19**, 75 (1972).
743. H. Rönsch and H. Böhm, 4. Intern. Symp. Biochemie unde Physiologie d. Alkaloide, Halle (Saale), 25. bis 28. Juni 1969. Band b. *Abh. Dtsch. Akad. Wiss. Berlin* 287 (1972).
744. H. Rönsch, *Tet. Lett.* 5121 (1969).
745. H. Böhm and H. Rönsch, *Z. Naturforsch., Teil B* **23**, 1553 (1968).
746. H. Rönsch. *Tet. Lett.* 4431 (1972).
746a. H. Rönsch, A. Guggisberg, M. Hesse, and H. Schmid, *Helv. Chim. Acta* **60**, 2402 (1977).
747. V. Šimánek, L. Hruban, V. Preininger, A. Němečková, and A. Klásek, *Collect. Czech. Chem. Commun.* **40**, 705 (1975).
748. V. Šimánek, A. Klásek, L. Hruban, V. Preininger, and F. Šantavý, *Tet. Lett.* 2171 (1974).
749. V. Šimánek, A. Klásek, and F. Šantavý, *Tet. Lett.* 1779 (1973).
750. H. Böhm and H. Rönsch, *Z. Naturforsch., Teil B* **23**, 1552 (1968).
751. J. Hrbek Jr., L. Hruban, V. Šimánek, F. Šantavý, and G. Snatzke, *Collect. Czech. Chem. Commun.* **38**, 2799 (1973).
752. M. Shamma, J. L. Moniot, W. K. Chan, and K. Nakanishi, *Tet. Lett.* 4207 (1971).
753. W. M. J. Flapper and C. Romers, *Tetrahedron* **31**, 1705 (1975).
754. C. S. Huber, *Acta Crystallogr., Sect B* **26**, 373 (1970); **28**, 982 (1972).
755. M. Shamma and J. A. Weiss, *Chem. Commun.* 212 (1968).
756. M. Tin-Wa, H. K. Kim, H. H. S. Fong, N. R. Farnsworth, J. Trojánek, and D. J. Abraham, *Lloydia* **35**, 87 (1972).
757. H. Ishii, K. Hosoya, and N. Takao, *Tet. Lett.* 2429 (1971).
758. Ch. Tani, N. Takao, and K. Tagahara, *Yakugaku Zasshi* **84**, 1217 (1964).
759. T. Kametani, *J. Chem. Soc., Perkin Trans. 2* 1605 (1973).
760. T. Kametani, T. Honda, M. Ihara, H. Shimanouchi, and Y. Sasada, *Tet. Lett.* 3729 (1972); *J. Chem. Soc., Perkin Trans. 2* 1605 (1973).
761. I. Ninomiya, O. Yamamoto, and T. Naito, *Chem. Commun.* 437 (1976).
762. G. Nonaka and I. Nishioka, *Chem. Pharm. Bull.* **21**, 1410 (1973).
763. N. Takao, M. Kamigauchi, K. Iwasa, K. I. Tomita, T. Fujiwara, and A. Wakahara, *Tet. Lett.* 805 (1974).
764. N. Weber, *Ber.* **106**, 3769 (1973).
765. H. Ishii and T. Komaki, *Yakugaku Zasshi* **86**, 631 (1966).
766. P. G. Waterman, *Phytochemistry* **14**, 843 (1975).
767. H. Ishii, T. Ishikawa, Sh.-T. Lu, and Ih.-Sh. Chen, *Tet. Lett.* 1203 (1976).
768. M. H. Benn and R. E. Mitchell, *Can. J. Chem.* **47**, 3701 (1969).
769. M. Onda, K. Takiguchi, M. Hirakura, H. Fukushima, M. Okagawa, and F. Navi, *Nippon Nogei Kagaku Kaishi* **39**, 168 (1965).
770. H. Ishii, K. Harada, T. Ishida, E. Ueda, and K. Nakajima, *Tet. Lett.* 319 (1975).
770a. H. Ishii, T. Ishikawa, Yu.-Ich. Ichikawa, and M. Sakamoto, *Chem. Pharm. Bull.* **25**, 3120 (1977).
770b. S. V. Kessar, Y. P. Gupta, K. Dhingra, G. S. Sharma, and S. Narula, *Tet. Lett.* 1459 (1977).
771. I. Ninomiya, O. Yamamoto, and T. Naito, *Heterocycles* **4**, 743 (1976).
772. S. V. Kessar, G. Singh, and P. Balakrishnan, *Tet. Lett.* 2269 (1974); *Indian J. Chem.* **12**(3), 323 (1974).
773. I. Ninomiya, O. Yamamoto, and T. Naito, *Heterocycles* **5**, 67 (1976).
774. T. Kametani, K. Kigawawa, M. Hiiragi, O. Kusama, E. Hayashi, and H. Ishimaru, *Chem. Pharm. Bull.* **19**, 1150 (1971).

775. M. Sainsbury, S. F. Dyke, and B. J. Moon, *J. Chem. Soc. C* 1797 (1970).
776. F. R. Stermitz, J. P. Gillespie, L. G. Amoros, R. Romero, T. A. Stermitz, K. A. Larson, S. Earl, and J. E. Ogg, *J. Med. Chem.* **18**, 708 (1975).
777. M. Onda, K. Yuasa, J. Okada, K. Kataoka-Yonezawa, and K. Abe, *Chem. Pharm. Bull.* **21**, 1333 (1973).
778. M. Onda and K. Kawakami, *Chem. Pharm. Bull.* **20**, 1484 (1972).
779. A. Jindra, *Folia Pharm.* **5**, 819 (1969); *CA* **70**, 57633d (1969).
780. K. Kuzminska, *Wiad. Bot.* **13**, 249 (1969); *CA* **72**, 97362s (1970).
781. A. R. Battersby, R. J. Francis, M. Hirst, E. A. Ruveda, and J. Staunton, *J. Chem. Soc., Perkin Trans. 1* 1140 (1975).
782. A. R. Battersby, J. Staunton, H. R. Wiltshire, R. J. Francis, and R. Southgate, *J. Chem. Soc. C* 1147 (1975).
783. A. R. Battersby, J. Staunton, H. R. Wiltshire, B. J. Bircher, and C. Fuganti, *J. Chem. Soc., Perkin Trans. 1* 1162 (1975).
783a. N. Takao, K. Iwasa, M. Kamigauchi, and M. Sugiura, *Chem. Pharm. Bull.* **24**, 2859 (1976).
784. M. D. Antoun and M. F. Roberts, *Planta Med.* **28**, 6 (1975).
785. C. Y. Chen and D. B. MacLean, *Can. J. Chem.* **45**, 3001 (1967).
786. J. Slavík, L. Dolejš, V. Hanuš, and A. D. Cross, *Collect. Czech. Chem. Commun.* **33**, 1619 (1968).
787. O. N. Tolkashev and O. E. Lasskaja, *Khim. Prir. Soedin.* 741 (1974); *CA* **82**, 171254 (1975).
787a. R. D. Sipanovic, R. C. Howell, and A. A. Bell, *J. Heterocycl. Chem.* **9**, 1453 (1972).
788. J. Susplugas, S. El Nouri, V. Massa, and P. Susplugas, *Trav. Soc. Pharm. Montpellier* **34**, 115 (1974); *CA* **82**, 28500 (1975).
789. G. A. Maslova, *Khim. Prir. Soedin.* 261 (1974).
790. C. R. Howell, R. D. Stipanovic, and A. A. Bell, *Pestic. Biochem. Physiol.* **2**, 364 (1972); *BA* **55**, 62341 (1973).
791. D. B. MacLean, R. A. Bell, J. K. Saunders, C.-Y. Chen, and R. H. F. Manske, *Can. J. Chem.* **47**, 3593 (1969).
792. M. Shamma, *Annu. Rep. Med. Chem.* 323 (1969).
792a. S. McLean, M.-S. Lin, and J. Whelan, *Can. J. Chem.* **48**, 948 (1970).
793. R. H. F. Manske, R. Rodrigo, D. B. MacLean, D. E. F. Gracey, and J. K. Saunders, *Can. J. Chem.* **47**, 3585 (1969).
794. H. Irie, A. Kitagawa, A. Kuno, J. Tanaka, and N. Yokotani, *Heterocycles* **4**, 1083 (1976).
794a. S. MacLean and D. Dime, *Can. J. Chem.* **55**, 924 (1977).
795. C. K. Yu, J. K. Saunders, D. B. MacLean, and R. H. F. Manske, *Can. J. Chem.* **49**, 3020 (1971).
796. M. Shamma and C. D. Jones, *J. Am. Chem. Soc.* **91**, 4009 (1969); **92**, 4943 (1970).
797. M. Shamma and J. F. Nugent, *Tet. Lett.* 2625 (1970); *Chem. Commun.* 1642 (1971).
798. H. L. Holland, D. B. MacLean, R. G. A. Rodrigo, and R. H. F. Manske, *Tet. Lett.* 4323 (1975).
799. B. C. Nalliah, D. B. MacLean, R. G. A. Rodrigo, and R. H. F. Manske, *Can. J. Chem.* **55**, 922 (1977).
800. T. Kishimoto and S. Uyeo, *J. Chem. Soc. C* 2600 (1969).
801. T. Kashimoto and S. Uyeo, *J. Chem. Soc. C* 1644 (1971).
802. T. Kametani, S. Takano, S. Hibino, and T. Terui, *J. Heterocycl. Chem.* **6**, 49 (1969).
803. S. O. de Silva, K. Orito, R. H. F. Manske, and R. Rodrigo, *Tet. Lett.* 3243 (1974).
804. S. McLean, M.-S. Lin, and J. Whelan, *Tet. Lett.* 2425 (1968).

805. H. Irie, T. Kishimoto, and S. Uyeo, *J. Chem. Soc. C* 3051 (1968).
806. R. B. Kelly and B. A. Beckett, *Can. J. Chem.* **47**, 2501 (1969).
807. S. McLean and J. Whelan, *Can. J. Chem.* **51**, 2457 (1973).
808. T. Kametani, S. Hibino, and S. Takano, *J. Chem. Soc. D* 925 (1971); *J. Chem. Soc., Perkin Trans. 1* 391 (1972).
809. B. Nalliah, Q. A. Ahmed, R. H. F. Manske, and R. Rodrigo, *Can. J. Chem.* **50**, 1819 (1972).
810. T. Kametani, T. Takahashi, and K. Ogasawara, *J. Chem. Soc., Perkin Trans. 1* 1464 (1973).
811. T. Kametani, Y. Hirai, F. Satoh, K. Ogasawara, and F. Fukumoto, *Chem. Pharm. Bull.* **21**, 907 (1973).
812. T. Kametani, Y. Hirai, H. Takeda, M. Kajiwara, T. Takahashi, F. Satoh, and K. Fukumoto, *Heterocycles* **2**, 339 (1974).
813. T. Kametani, S. Hirata, S. Hibino, H. Nemoto, M. Ihara, and K. Fukumoto, *Heterocycles* **3**, 151 (1975).
814. J. Imai, Y. Kondo, and T. Takemoto, *Heterocycles* **3**, 467 (1975).
815. M. Shamma, J. L. Moniot, R. H. F. Manske, W.-K. Chan, and K. Nakanishi, *Chem. Commun.* 310 (1972).
816. C. K. Yu and D. B. MacLean, *Can. J. Chem.* **49**, 3025 (1971).
817. A. Kato, K. Akagi, H. Irie, and S. Uyeo, *Yakugaku Zasshi* **95**, 1058 (1975); *CA* **84**, 5210n (1976).
818. J. Gleye, A. Ahond, and E. Stanislas, *Phytochemistry* **13**, 675 (1974).
819. L. D. Yakhontova, M. N. Komarova, O. N. Tolkachev, and M. E. Perel'son, *Khim. Prir. Soedin.* 491 (1976).
820. M. Shamma, A. S. Rothenberg, and S. F. Hussain, *Heterocycles* **6**, 707 (1977).
821. V. Šimánek, V. Preininger, F. Šantavý, and L. Dolejš, *Heterocycles* **6**, 711 (1977).
822. D. B. MacLean, L. Baczynskyj, R. Rodrigo, and R. H. F. Manske, *Can. J. Chem.* **50**, 862 (1972).
823. R. H. F. Manske, *Can. J. Res., Sect. B* **7**, 258 (1932); **16**, 81 (1938).
824. R. Rodrigo, *in* "The Alkaloids" (R. H. F. Manske, ed.), Vol. XIV, p. 407. Academic Press, New York, 1973.
825. R. Rodrigo, R. H. F. Manske, L. Baczynskyj, J. K. Saunders, and D. B. MacLean, *Can. J. Chem.* **50**, 853 (1972).
826. R. Rodrigo, R. H. F. Manske, V. Šmula, D. B. MacLean, and L. Baczynskyj, *Can. J. Chem.* **50**, 3900 (1972).
827. S. Ruchirawat and V. Somchitman, *Tet. Lett.* 4159 (1976).
828. M. Shamma, J. L. Moniot, S. Y. Yao, G. A. Miana, and M. Ikram, *J. Am. Chem. Soc.* **95**, 5742 (1973).
829. J. Tomko, *Chem. Zvesti* **29**, 265 (1975); **30**, 225 (1976).
830. E. Späth and F. Kuffner, *Ber.* **64**, 1127 (1931).
831. R. Hegnauer, *in* "Perspectives in Phytochemistry" (J. B. Harborne and T. Swain, eds.), p. 121. Academic Press, New York, 1969.
832. P. Tétényi and D. Vágújfalvi, *Herba Hung.* **6**, 123 (1967).
833. F. R. Stermitz, *Recent Adv. Phytochem.* **1**, 161 (1968).
834. Thi Yu Bui, *Farmatsiya (Moscow)* **23**(2), 36 (1974); *CA* **81**, 60863r (1974).
835. G. B. Ownbey, *Am. Midl. Nat.* **45**, 184 (1951).
836. S. Sárkány, K. Michels-Nyomárkay, and G. Verzár-Petri, *Pharmazie* **25**, 625 (1970).
837. D. Vágújfalvi, *Bot. Kozl.* **57**, 113 (1970).
838. C. W. Thornber, *Phytochemistry* **9**, 157 (1970).
839. W. O. Nagel, *Diss. Abstr. Int. B* **33**(6), 2473 (1972); *CA* **78**, 94139n (1973).

840. Y. M. ElKheir, *Planta Med.* **27**, 275 (1975).
841. P. Matile, B. Jans, and R. Rickenbacher, *Biochem. Physiol. Pflanz.* **161**, 447 (1970); *CA* **74**, 72773e (1971).
842. V. M. Malinina and R. M. Ivanova, *Dokl. Vses. Akad. Skh. Nauk* **2**, 19 (1974); *CA* **81**, 60831d (1974).
843. P. C. Maiti, *J. Indian Acad. Forensic Sci.* **12**(1), 19 (1973); *CA* **80**, 14156w (1974).
844. K. Michels-Nyomárkay, *Ann. Univ. Sci. Budap. Rolando Eotvos Nominatae, Sect. Biol.* (13), 127 (1971); *CA* **80**, 130517m (1974).
845. D. A. Mura'eva and B. A. Figurkin, *Biol. Nauki* (*Moscow*) **15**, 87 (1972); *CA* **78**, 13760h (1973).
846. K. R. Khanna and U. P. Singh, *Planta Med.* **28**, 92 (1975).
847. H. Böhm, *Planta Med.* **19**, 93 (1971).
847b. D. Neubauer and K. Mothes, *Planta Med.* **11**, 387 (1963).
848. P. Goldblatt, *Ann. Mo. Bot. Gard.* **61**, 264 (1974).
849. C. B. Coffman, C. E. Bare, and W. A. Gentner, *Bull. Narc.* **27**(3), 41 (1975).
850. S. Kamimura, M. Akutsu, and M. Nishikawa, *Agric. Biol. Chem.* (*Jpn.*) **40**, 907 and 913 (1976).
851. I. Lalezari, P. Nasseri, and R. Asgharian, *J. Pharm. Sci.* **63**, 1331 (1974).
852. P. G. Vincent, C. E. Bare, and W. A. Gentner, *Lloydia* **39**, 76 (1976).
853. H.-J. Martin, P. Pachaly, and F. Zymalkowski, *Arch. Pharm.* (*Weinheim, Ger.*) **310**, 314 (1977).

Note added in proof. Wiegrebe *et al.* used isotope dilution analysis to study the absolute configuration of (+)-laudanosine, (+)-6′-hydroxymethyllaudanosine, and 3-phenylisochromanderivative. The latter is related to peshawarine (**206**) [H. M. Stephan, G. Langer, and W. Wiegrebe, *Pharm. Acta Helv.* **51**, 164 (1976)].

Shamma *et al.* also described a simple conversion of berberine in high yield (photooxidation of oxyberberine) into (±)-β-hydrastine (**123**) [M. Shamma, D. M. Hindenlang, Tai-Teh Wu, and J. L. Moniot, *Tet. Lett.* 4285 (1977)].

—CHAPTER 5—

MONOTERPENE ALKALOID GLYCOSIDES

R. S. KAPIL

Central Drug Research Institute
Lucknow, India

AND

R. T. BROWN

The University
Manchester, England

ISBN 0-12-469517-5

I. Introduction

The last 15 years have seen a considerable growth of interest in the various aspects of natural products. Traditional isolation and structure elucidation are no longer the sole preoccupations of natural product chemists. More and more emphasis is now being laid on the synthesis, properties, and function of organic molecules in the living system. The study of the biosynthesis of the *Ipecacuanha* and terpenoid indole alkaloids is an example of the changing trends. Initial identification of the basic building units allowed hypotheses on intermediates and reaction mechanisms to be constructed and tested by tracer studies. As more details of the pathways were revealed, chemists were induced to search for the novel structures predicted by theory. Most of the monoterpene alkaloid glycosides of the *Ipecacuanha* and indole series which were discovered recently as the result of such investigations are reviewed in this chapter.

II. Occurrence

Rewarding sources of monoterpene alkaloid glycosides have been the Asiatic tree *Adina rubescens* Hemsl. and related Rubiaceous plants: *A. cordifolia* (Roxb.) Hook. f. ex Brandis (*1, 2*), *Nauclea latifolia* Sm. (*3*), *N. diderrichii* Merrill (*4*), and *Anthocephalus cadamba* Miq. (*5*). The heartwood and bark of the first has yielded so far no less than nine glycoalkaloids (*6*) besides a new series of novel carboxy terpenoid indole alkaloids (*7*). Other rich sources of these glycosides are the Apocynaceous species, namely, *Rhazya orientalis* A.DC. (*8*) and *R. stricta* Decne. (*9*). Monoterpene alkaloid glycosides are now established precursors for the biosynthesis of *Ipecacuanha*, terpenoid indole, and *Camptotheca* alkaloids, and their isolation from alkaloid-producing plants involves the use of sophisticated methods of isolation and the study of the proper time and state for collection of the plant material.

III. The Dopamine Derivatives

A. IPECOSIDE

Ipecoside, a nitrogen-containing glycoside, was isolated in 1952 from the roots of *Cephaelis ipecacuanha* (Brot.) A. Rich. (Rubiaceae) (*10*). However, serious attempts toward the elucidation of its structure were started only in the early 1960's in the laboratories of Professors Janot and Levisalles in France and of Professor Battersby in England. The elucidation of its structure as **1** or **2** immediately classified it as a novel compound, and ipecoside became the progenitor of a new series of monoterpenoid alkaloidal glycosides. As it turned out, this discovery had far-reaching implications for our understanding of the biosynthesis of *Ipecacuanha* and terpenoid indole alkaloids.

Earlier work established the correct molecular formula of ipecoside as $C_{27}H_{35}NO_{12}$ {mp 175°; $[\alpha]_D -185°$ (MeOH)} and showed that it contains two phenolic *o*-dihydroxy groups. On methylation with diazomethane, ipecoside gave *O,O*-dimethylipecoside (**3** or **4**). On treatment with acetyl chloride, the former afforded a hexaacetate, whereas the latter furnished a tetraacetyl derivative (*11*).

Subsequent collaborative work by British and French chemists confirmed these findings. The UV spectrum, λ_{max} 238, and 285 nm, indicated the presence of a β-alkoxyacrylate and a catechol system, respectively. The former was supported by the IR absorption at 1690 cm^{-1}. The band at 1630 cm^{-1} indicated that ipecoside was an amide. Catalytic hydrogenation gave dihydroipecoside, which showed unchanged UV and IR spectra, indicating that the double bond was not conjugated with carbonyl or phenolic groups. Kuhn–Roth oxidation of ipecoside furnished acetic acid, whereas dihydroipecoside afforded a mixture of acetic and propionic acids, indicating the presence of a vinyl group in **1** or **2**.

A clue to the structure of ipecoside was obtained from the NMR and mass spectra of the glycoalkaloid and many of its derivatives. The NMR spectrum of ipecoside hexatrimethylsilyl ether showed the presence of two aromatic protons, four olefinic protons, one methoxycarbonyl, and one *N*-acetyl function. In the mass spectrum of ipecoside no molecular ion could be detected, but important characteristic peaks appeared at *m/e* 220, 206, 178, 164, and 43, assigned to the cleavage pattern of substituted *N*-acetyl-1-alkyl-1,2,3,4-tetrahydroisoquinolines.

Mild acid hydrolysis of ipecoside furnished D-glucose, while vigorous conditions also gave acetic acid. On enzymatic hydrolysis with β-glucosidase, ipecoside afforded the aglucone (**5** or **6**); on similar treatment, *O,O*-dimethyldihydroipecoside yielded the aglucone (**7** or **8**) which with acid isomerized to the more stable diastereoisomeric form (**9** or **10**). Jones oxidation of **9** or **10** furnished the enol δ-lactones **11** or **12** and **13** or **14**, both of which showed carbonyl absorption at 1773 cm^{-1} in the IR spectra. The NMR spectrum of **13** or **14** was very informative in that practically every proton could be identified. Moreover, it was possible to show by double resonance experiments the coupling of each proton to its neighbor over the sequence C-5, C-6, C-7, C-2, C-3, and C-4. The combined evidence thus indicated structure **1** or **2** for ipecoside.

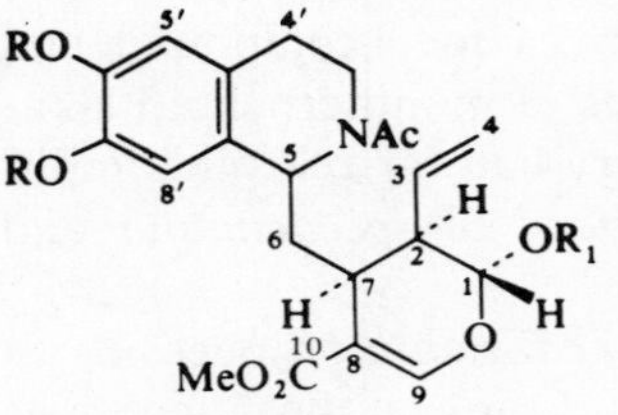

1 R = H, R_1 = β-D-Glu, 5α-H
2 R = H, R_1 = β-D-Glu, 5β-H
3 R = Me, R_1 = β-D-Glu, 5α-H
4 R = Me, R_1 = β-D-Glu, 5β-H
5 R = R_1 = H, 5α-H
6 R = R_1 = H, 5β-H
7 R = Me, R_1 = H, 5α-H, 3,4-dihydro
8 R = Me, R_1 = H, 5β-H, 3,4-dihydro

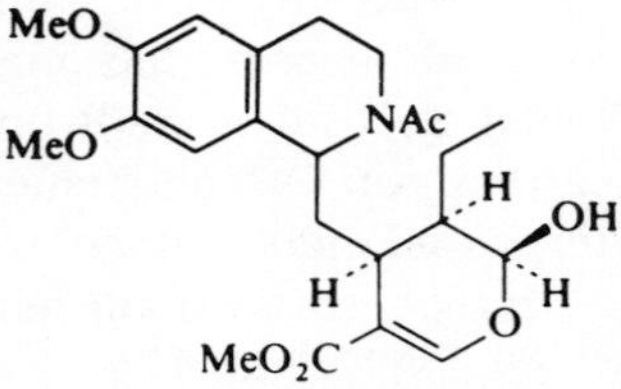

9 5α-H
10 5β-H

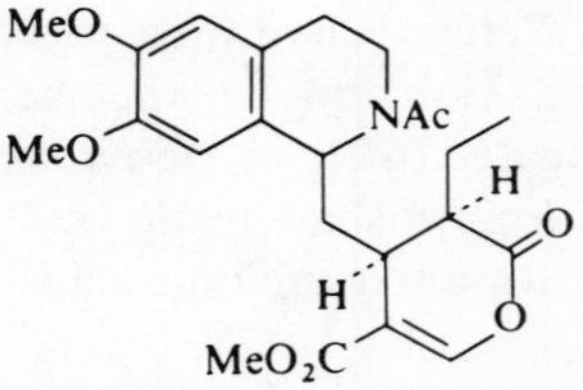

11 5α-H
12 5β-H

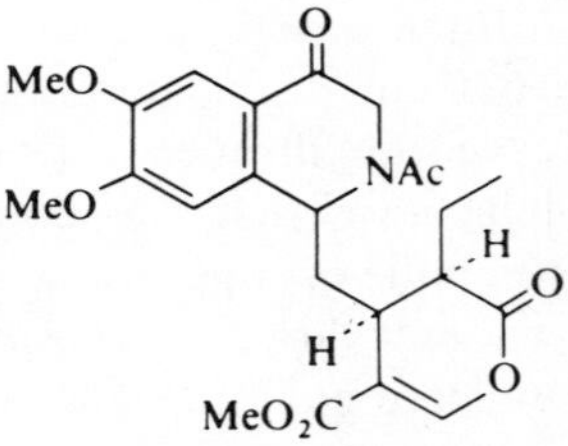

13 5α-H
14 5β-H

The absolute stereochemistry of ipecoside was determined by correlation studies with dihydroprotoemetine. Because the yield of the product in each step was low and a complex mixture of compounds was formed during the reaction, recourse was taken to the use of radioactive material. Vigorous acid hydrolysis of [*O*-methyl-^{3}H]*O*,*O*-dimethyldihydroipecoside furnished **15** or **16** in equilibrium with **17** or **18**, and also several isomeric benzoquinolizidines. From this, after sodium borohydride reduction, (−)-[*O*-methyl-^{3}H]dihydroprotoemetine (**19**) of known absolute configuration (*12*) was isolated.

O,O-Dimethyldihydroipecoside

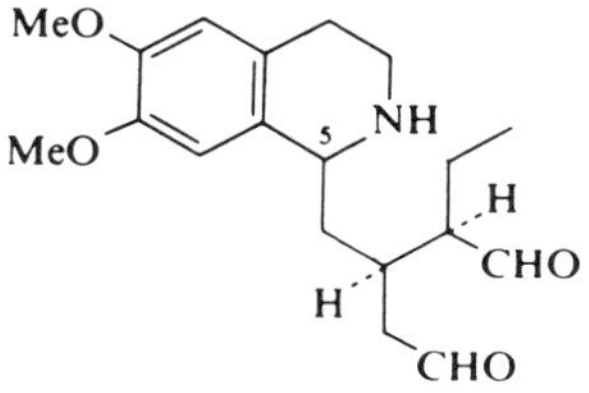

15 5α-H
16 5β-H

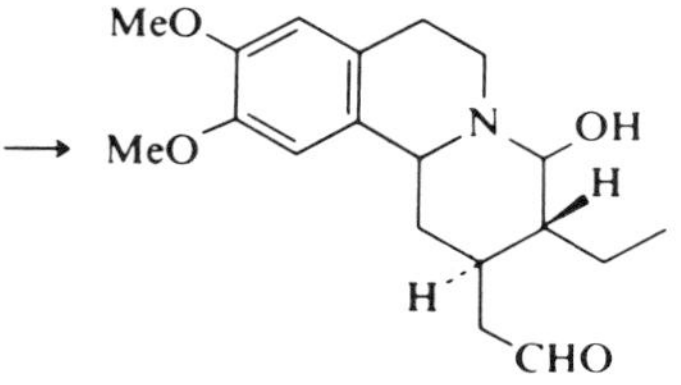

17 5α-H
18 5β-H

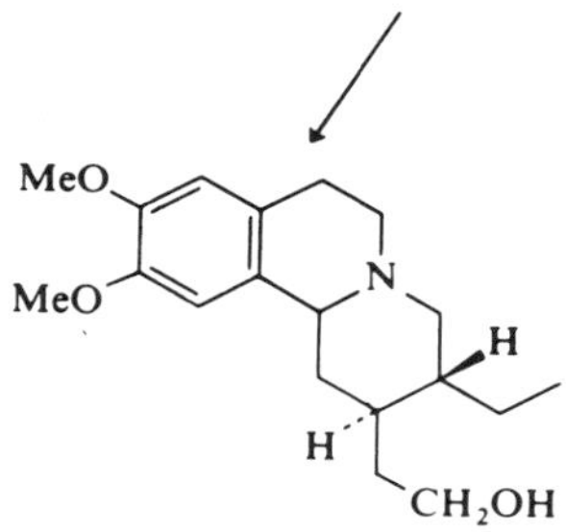

19 5α-H
20 5β-H

When these stereochemical results were applied to vincoside, the indolic analog of deacetylipecoside and its C-3 epimer, strictosidine, which had been correlated chemically with corynantheine and antirhine (Section IV,A), conflicting results were obtained. They cast serious doubt on the validity of the previous assignment and indicated that inversion of C-5 must have occurred at some stage in the chemical operations. An independent X-ray crystallographic study of *O,O*-dimethylipecoside was therefore undertaken. The results obtained confirmed in detail the structure and absolute stereochemistry derived chemically, except that the hydrogen atom at C-5 should have a β configuration (*13*).

The structure assigned to ipecoside (**2**) has been confirmed by partial synthesis. Dopamine (**21**), on condensation with secologanin (**22**) at pH 5.0, afforded a mixture of deacetylipecoside (**23**) and deacetylisoipecoside (**24**) separated by countercurrent distribution. There was a preponderance of the 5β isomer in the reaction. Acetylation of deacetylipecoside followed by Zemplen O-deacetylation furnished ipecoside identical with the natural material (*14*).

In a subsequent modification pure, hexa-*O*-acetylipecoside was readily obtained by acetylation of the crude condensation product of dopamine with secologanin followed by column chromatography over silica gel (*14*).

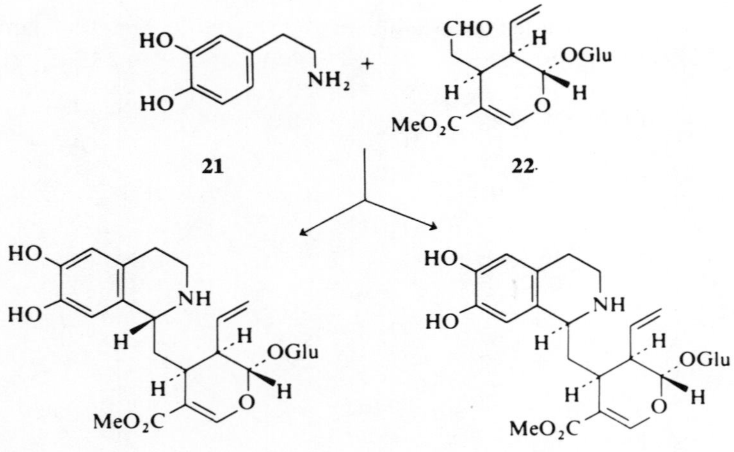

B. ALANGISIDE

Alangiside {$C_{25}H_{31}NO_{10}$; $[\alpha]_D - 105°$ (MeOH)} was isolated from the roots, leaves, and fruit of *Alangium lamarckii* Thw. (Alangiaceae). However, the best source was found to be the unripe fruit. The UV spectrum (bathochromic shift in alkali) was similar to that of ipecoside. The IR spectrum at 1650 cm^{-1} indicated the presence of an amide function. The NMR spectrum showed the presence of an aromatic methoxyl group, a vinyl grouping, two uncoupled aromatic protons, and a doublet at τ2.61 assigned to the —CO$\overset{|}{C}$=CH—O moiety.

β-Glucosidase cleaved alangiside to yield D-glucose and the aglucone **27**. The mass spectrum of alangiside showed a weak M^+ ion at m/e 505 and a medium peak at m/e 343. The base peak in the spectrum appeared at m/e 274, and there were characteristic fragments of isoquinoline ions at m/e 178 (**30**), 177 (**31**), and 176 (**32**). The ion at m/e 343 corresponded to the loss of $C_6H_{11}O_5$ (with a H transfer) giving the aglucone **27** which in turn lost —COCH=CHO— to yield the base peak.

Acetylation of alangiside with acetic anhydride furnished a pentaacetate. Methylation with diazomethane gave *O*-methylalangiside, while catalytic hydrogenation afforded a dihydro derivative containing a *C*-ethyl group (NMR). On acetylation, the former yielded a tetraacetate; the latter furnished the pentaacetyl derivative. This evidence in conjunction with biogenetic considerations supports structure **28** for *O*-methylalangiside.

Correlation with deacetylipecoside (**23**) was therefore carried out. Treatment of **23** with aqueous sodium carbonate gave the lactam **29**, which with diazomethane afforded the ether **28** identical with *O*-methylalangiside. Alangiside could therefore be assigned structure **25** or **26** (*15*).

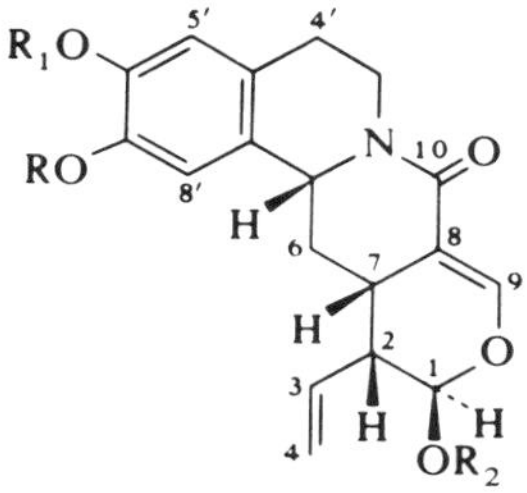

25 $R = H, R_1 = Me, R_2 = \beta$-D-Glu
26 $R = Me, R_1 = H, R_2 = \beta$-D-Glu
27 $R = R_2 = H, R_1 = Me$
28 $R = R_1 = Me, R_2 = \beta$-D-Glu
29 $R = R_1 = H, R_2 = \beta$-D-Glu

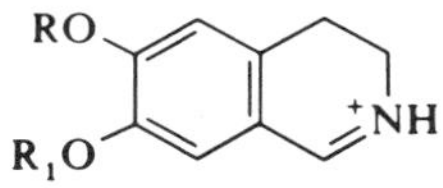

m/e 178

30 $R = H, R_1 = Me$
or $R = Me, R_1 = H$

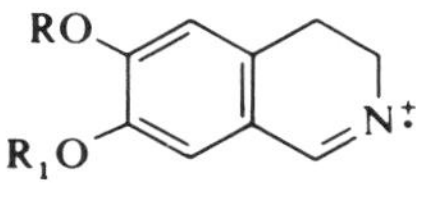

m/e 177

31 $R = H, R_1 = Me$
or $R = Me, R_1 = H$

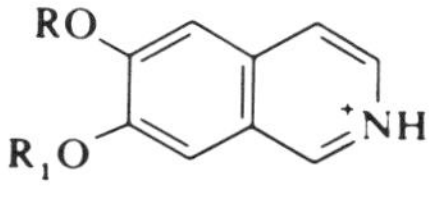

m/e 176

32 $R = H, R_1 = Me$
or $R = Me, R_1 = H$

To distinguish between these structures, a synthesis of alangiside was undertaken. Attempted condensation of 4-hydroxy-3-methoxyphenylethylamine with secologanin under a variety of conditions was not successful. However, 3-hydroxy-4-methoxyphenylethylamine condensed smoothly with secologanin at pH 5.2 to afford **33** and **34**. No attempts were made to obtain the individual epimers in a pure form, and the crude mixture was treated with a solution of sodium carbonate when mainly the 5β epimer (**26**) which differed from alangiside was obtained, and this on treatment with diazomethane afforded *O*-methylalangiside. The phenolic hydroxyl of alangiside was thus located at C-7′, and structure **25** is a complete representation of the molecule (*16*).

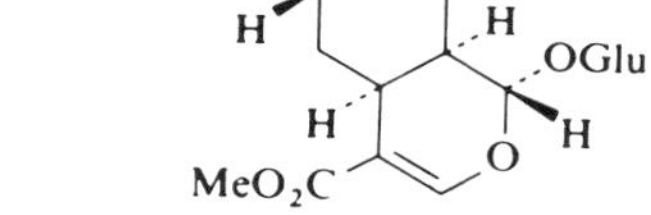

33

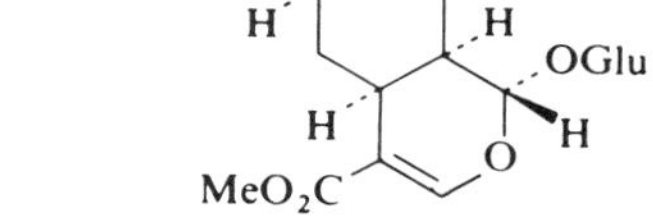

34

IV. The Tryptamine Derivatives

A. VINCOSIDE AND STRICTOSIDINE

As a result of biosynthetic studies on the monoterpenoid indole alkaloids, which in particular established loganin as the key intermediate, it was predicted that the crucial protoalkaloid should be **35** (undefined stereochemistry) formed from the hypothetical secologanin and tryptamine (*17*). This argument received support from the structures of ipecoside (*vide supra*) and cordifoline (*vide infra*). Subsequently, secologanin was discovered (*14, 18*) and its structure established by correlation with ipecoside (*14*). Both C-3 epimers, **35** and **36**, were then synthesized by condensation with tryptamine, and their presence in *Vinca rosea* Linn. (Apocynaceae) demonstrated by radiochemical dilution analysis and by direct isolation (*19*). However, only the major product, vincoside, functioned as an *in vivo* precursor for representatives of the *Corynanthé*, *Aspidosperma*, and *Iboga* alkaloids in *V. rosea*, while the minor product, isovincoside,* was not significantly incorporated (*19, 20*).

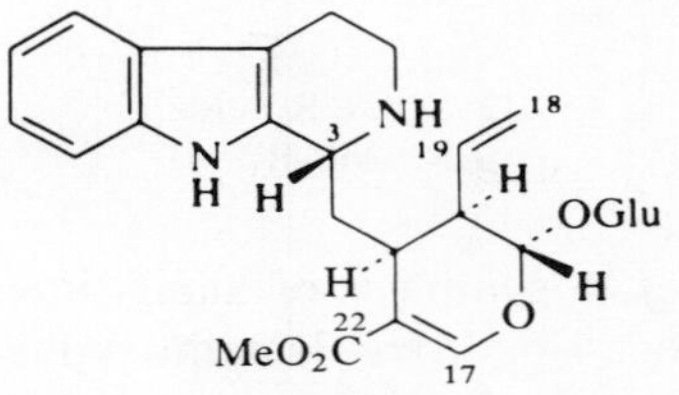

35 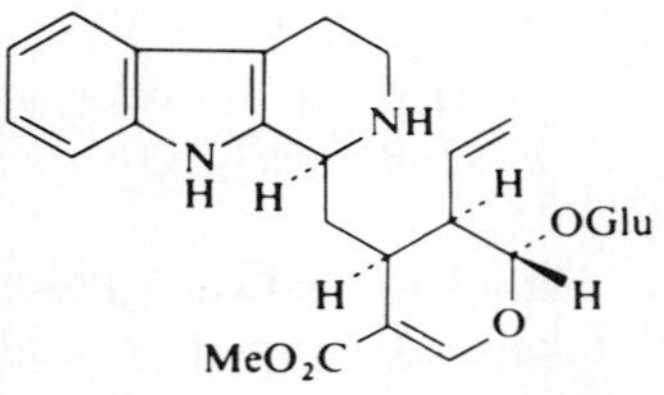 **36**

Although vincoside was unstable as the free base it could be isolated as the hydrochloride {$[\alpha]_D - 112°$ (MeOH)}. Acetylation gave the pentaacetate {$[\alpha]_D - 126°$ ($CHCl_3$)} whose spectral properties were in accord with the anticipated structure (**38** or **40**); the IR spectrum showed indolic NH, acetate, α,β-unsaturated ester, and amide bands; the UV absorption corresponded to the sum of indolic and β-alkoxyacrylate chromophores; and NMR signals could be assigned to the vinyl, methoxycarbonyl, and five acetyl groups, as well as H-17 and aromatic protons. The mass spectral fragmentation pattern was highly characteristic, with significant ions corresponding to cleavages *p, q, r,* and *s* (**41**). Zemplen O-deacetylation afforded *N*-acetylvincoside (**37** or **39**) {mp 180–181°; $[\alpha]_D - 170°$ (MeOH)}, which was also found to occur naturally in *V. rosea*. In a similar manner, the C-3 epimer afforded a hydrochloride {$[\alpha]_D - 143°$ (MeOH)}, a pentaacetate

* Isovincoside was subsequently found to be identical with strictosidine isolated from *Rhazya* species. We will use the latter name to avoid confusion.

(**38** or **40**) {[α]$_D$ $-75°$ ($CHCl_3$)}, and a crystalline acetamide {mp 172–173°; [α]$_D$ $-107°$ (MeOH)} which, as expected, had spectral properties similar to the vincoside derivatives.

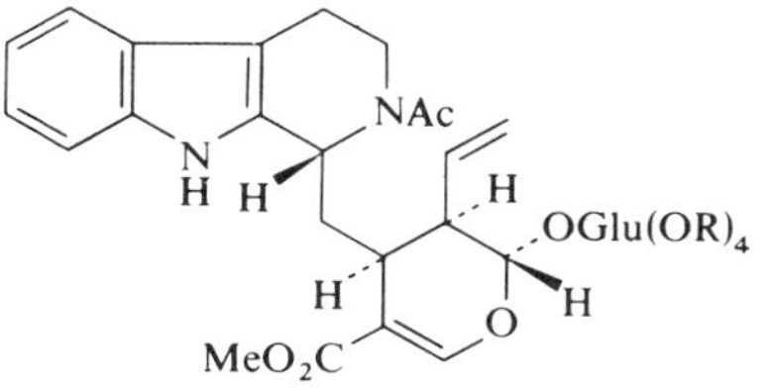

37 R = H
38 R = Ac

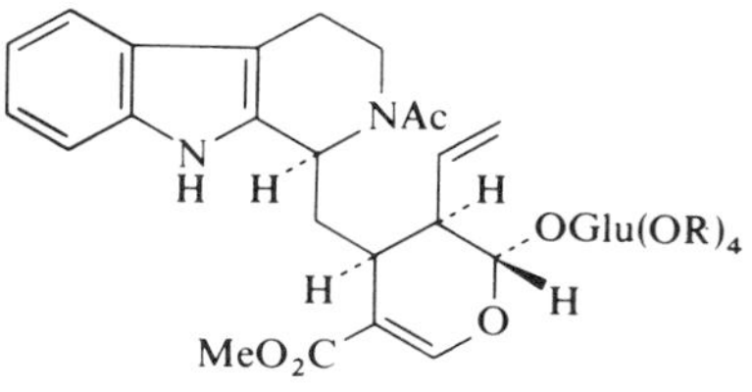

39 R = H
40 R = Ac

In a contemporary investigation on *R. stricta*, the glucoalkaloid strictosidine was isolated for which the gross structure **35** (undefined stereochemistry) was independently deduced (*21*). The UV, IR, and NMR spectra of the base and the derived pentaacetate were, as above, invaluable in detecting the functional groups. Essential structural information came from the mass spectra of the pentaacetyl (**41**) and pentatrimethylsilyl (**42**) derivatives which exhibited major ions attributable to cleavages *p*, *q*, *r*, and *s*. Comparison with the corresponding derivatives of cordifoline (q.v.) revealed a markedly similar fragmentation in accord with their common secoiridoid moiety, and in particular the occurrence of an ion at *m/e* 165 [167 (**44**) in the dihydro series] which could similarly be attributed to the pyrylium ion **43** formed by cleavages *r* and *t*.

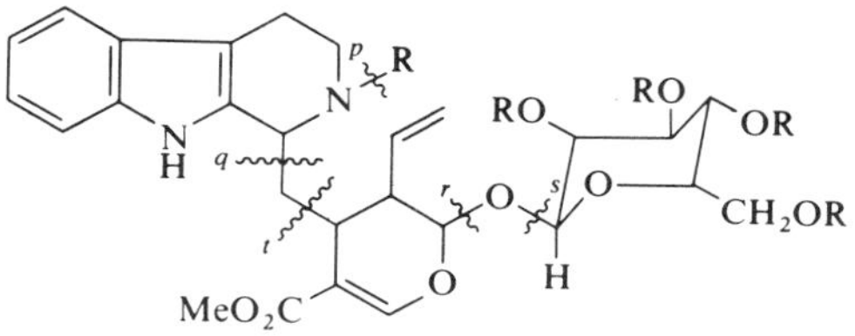

41 R = Ac
42 R = $SiMe_3$

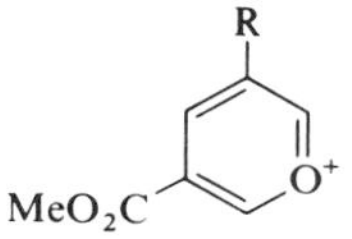

43 R = Vinyl
44 R = Et

An important observation was that strictosidine could readily be transformed into vallesiachotamine (**48**), a process which was rationalized as resulting from the hydrolysis of the glycoside, ring-opening, condensation of N-4 with C-17, and prototropic rearrangement to the conjugated aldehyde, though not necessarily in that sequence (Scheme 1). Because the last process generates two geometrical isomers, vallesiachotamine is actually a mixture, thus affording an explanation of the anomalous NMR behavior reported previously (*22*). Treatment of strictosidine with methanolic hydrochloride converted it to a novel N-1 cyclized structure, nacyline {$C_{21}H_{22}N_2O_3$; [α]$_D$ $+220°$ (MeOH)} (**49**) (*23*).

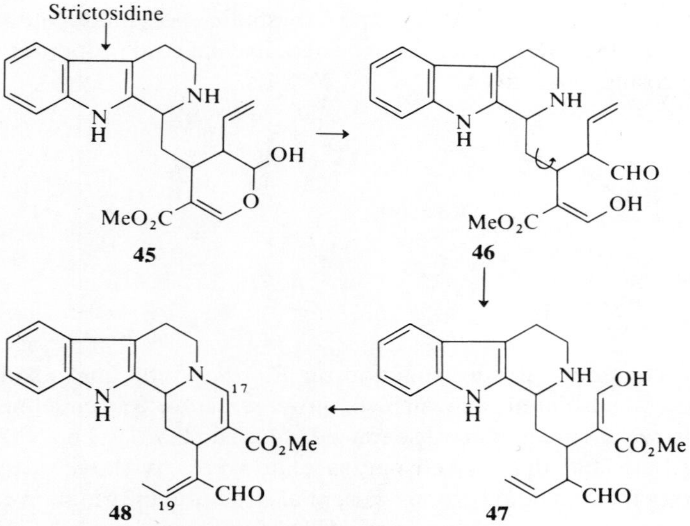

SCHEME 1

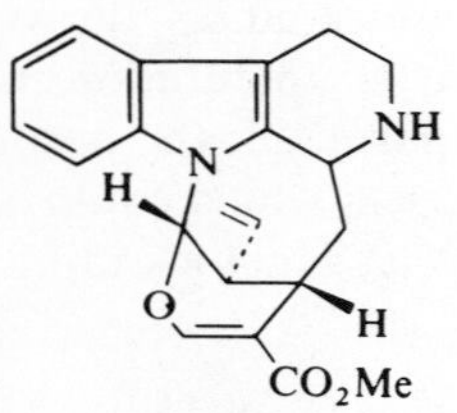

The difference in stability of vincoside and strictosidine as the free bases was quite marked. Aqueous sodium carbonate at room temperature rapidly converted the former into vincoside lactam (**50** or **52**) {mp 201–202°; $[\alpha]_D - 118°$ (MeOH)}, whereas the latter remains unaffected. In order to obtain strictosidine lactam (strictosamide) (**50** or **52**) {$[\alpha]_D - 75°$ (MeOH)} it was found necessary to heat the reaction mixture at 70° for a considerably longer time. Since then, both these transformation products have been isolated from natural sources—vincoside lactam from *A. rubescens* (*9*) and strictosidine lactam from *R. stricta* (*24*) and *N. latifolia* (*3*). One notable difference between the epimers was that strictosidine lactam tetraacetate (**51** or **53**) displayed one acetate signal in its NMR spectrum at anomalously high field ($\tau 8.78$), a characteristic not shared by the vincoside derivative.

Because the absolute stereochemistry of secologanin has been established, it follows from the partial syntheses that the configuration at every asymmetric center in vincoside and strictosidine is known except that

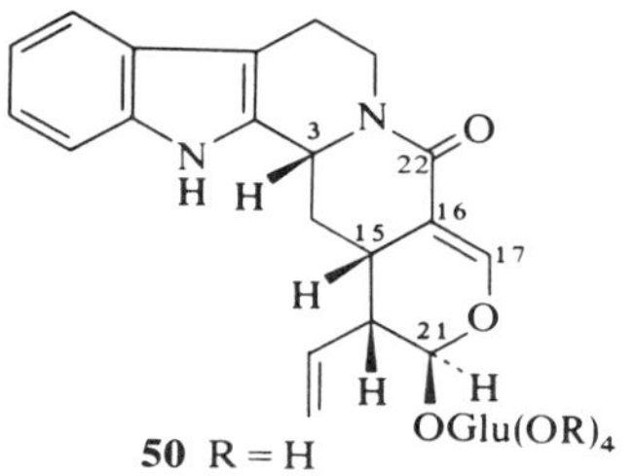

50 R = H
51 R = Ac

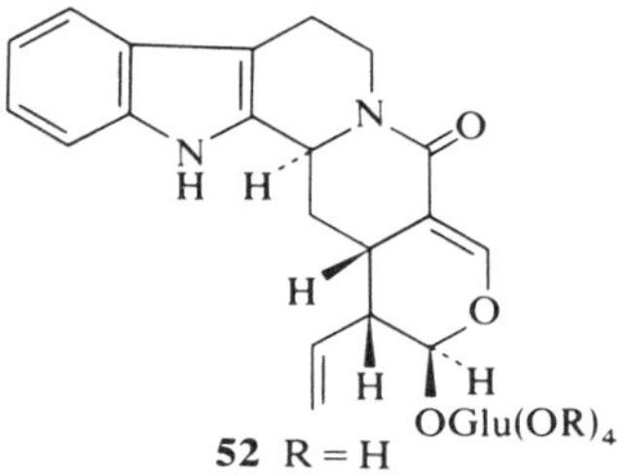

52 R = H
53 R = Ac

at C-3 (*14*). Comparison of the molecular rotation differences resulting from N-acetylation in the two epimeric series with those of ipecoside and isoipecoside derivatives demonstrated that C-3 in vincoside had the same configuration as the corresponding center (C-5) in ipecoside, whose complete stereochemistry had apparently been established as **1** by correlation with (−)-dihydroprotoemetine (*19*). It was thus concluded that vincoside had the 3α configuration (**36**) and strictosidine the 3β configuration (**35**). Further, this appeared to be in agreement with biosynthetic results where vincoside, but not strictosidine, was incorporated into *Corynanthé* alkaloids of 3α configuration with retention of H-3 (*19, 20*). Subsequently, however, some doubt arose about these assignments, and as a result of independent investigations by four groups, conclusive evidence has been obtained which reversed the original assignments and established vincoside as the 3β isomer (**35**) and strictosidine as the 3α isomer (**36**) (*9, 24, 25*).

First, correlations of strictosidine with dihydroantirhine of known stereochemistry were made by two routes (Scheme 2) (*24, 26*). One exploited the ready formation of vallesiachotamine from strictosidine noted above, dihydrostrictosidine (**54**) giving a tetrahydrovallesiachotamine (**55**) which after hydrolysis and decarboxylation was reduced to dihydroantirhine (**56**); the other involved conversions of **55** and dihydrostrictosidine lactam (**57**) to a common diol (**58**). Only the more important intermediates are summarized in Scheme 2, but at every stage the complex array of C-3, C-16, and C-20 isomers formed were meticulously separated, distinguished, and correlated by spectroscopic and equilibration studies. Particular care was taken to check whether inversion at C-3 had occurred at any stage, and to this end a small amount of dihydrovincoside lactam (**60**) containing a tritium label was put through a parallel reaction sequence. Dilution analysis indicated that the resulting diol (**61**) differed at C-3 from the diols obtained from strictosidine lactam—a result supporting the absence of C-3 epimerization in this correlation.

Second, a chemical and chiroptical correlation was made between vincoside, strictosidine, and corynantheine (**66**), an alkaloid whose absolute

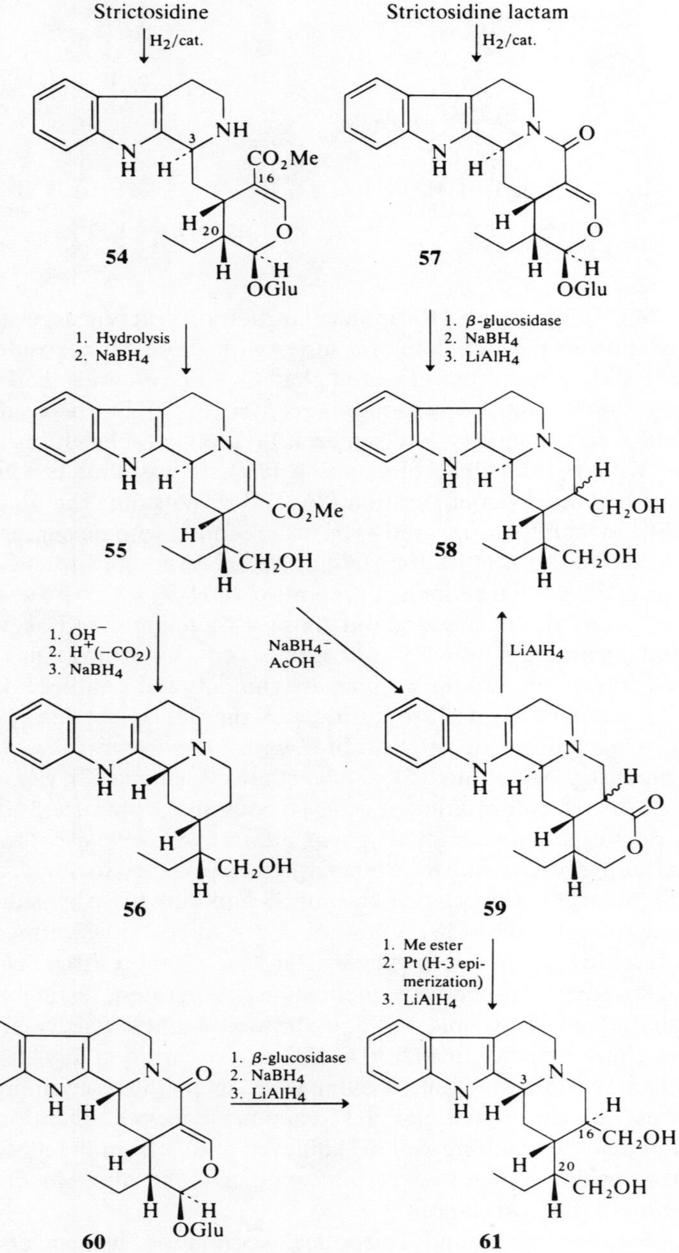

Strictosidine

H_2/cat.

54

1. Hydrolysis
2. $NaBH_4$

55

1. OH^-
2. $H^+(-CO_2)$
3. $NaBH_4$

56

Strictosidine lactam

H_2/cat.

57

1. β-glucosidase
2. $NaBH_4$
3. $LiAlH_4$

58

$NaBH_4^-$
AcOH

$LiAlH_4$

59

1. Me ester
2. Pt (H-3 epimerization)
3. $LiAlH_4$

60

1. β-glucosidase
2. $NaBH_4$
3. $LiAlH_4$

61

SCHEME 2

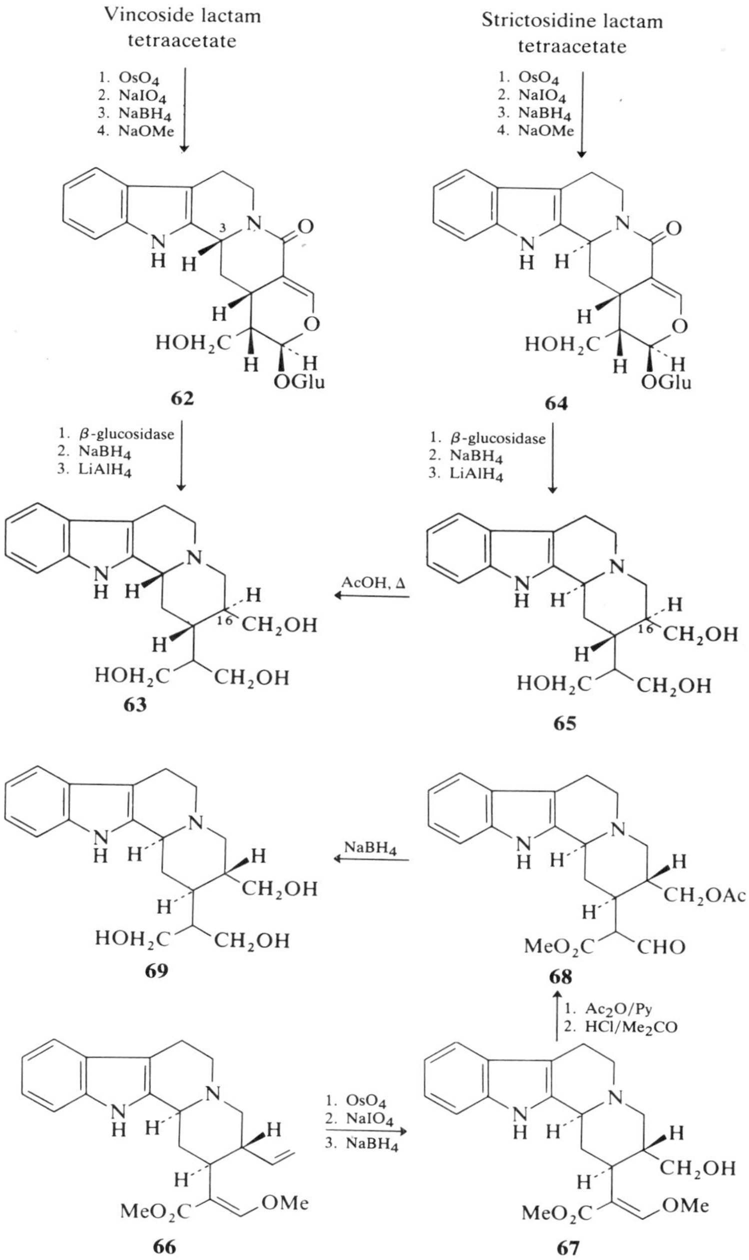

Vincoside lactam tetraacetate

1. OsO_4
2. $NaIO_4$
3. $NaBH_4$
4. NaOMe

62

1. β-glucosidase
2. $NaBH_4$
3. $LiAlH_4$

63

Strictosidine lactam tetraacetate

1. OsO_4
2. $NaIO_4$
3. $NaBH_4$
4. NaOMe

64

1. β-glucosidase
2. $NaBH_4$
3. $LiAlH_4$

65

AcOH, Δ

68

NaBH$_4$

69

1. Ac_2O/Py
2. HCl/Me_2CO

66

1. OsO_4
2. $NaIO_4$
3. $NaBH_4$

67

SCHEME 3

configuration had been rigorously established through a primary standard (*9, 27*). The crucial observation was that, depending on their C-3 stereochemistry, either vincoside or strictosidine should be degraded to triol **63**, which could be compared with an enantiomeric triol (**69**), derivable from corynantheine (Scheme 3). Cleavage of C-18 from vincoside lactam tetraacetate followed by hydrolysis of the glycoside and reduction gave the norcorynanetriol **63** (mp 240–242°) and its C-16 epimer. The ORD curves of both these compounds exhibited strong negative Cotton effects between 244 and 298 nm and constituted very strong evidence for a 3β configuration for vincoside by analogy with standard tetracyclic tetrahydro-β-carboline alkaloids (*28*). Subsequent cleavage of C-18 from corynantheine, hydrolysis, and reduction afforded 3α,15α,20β,18-norcorynane-17,19,22-triol (**69**) (mp 240–242°), a substance shown to be enantiomeric with the major product obtained from vincoside which therefore had 3β stereochemistry. To guard against the possibility that H-3 had been epimerized, strictosidine lactam tetraacetate was taken through an identical reaction sequence to give a pair of triols (**65** and its C-16 isomer) which were readily shown to differ from the previous pair. Both gave positive Cotton effects in accordance with a 3α configuration and on heating in acetic acid were largely epimerized to the vincoside derivatives.

Third, the X-ray structure determination of *O,O*-dimethylipecoside established the stereochemistry of H-5 in ipecoside as β rather than α; as a consequence of the correlation via molecular rotation differences, the corresponding H-3 in vincoside must have the β configuration (*13*).

Finally, X-ray crystallographic studies of *N*-4-*p*-bromobenzylvincoside tetraacetate also confirmed 3β configuration in vincoside (*29*).

B. Strictosidinic Acid

An amorphous material found in *R. orientalis* which gave strictosidine on methylation is presumably the corresponding acid (**70**) (*8*).

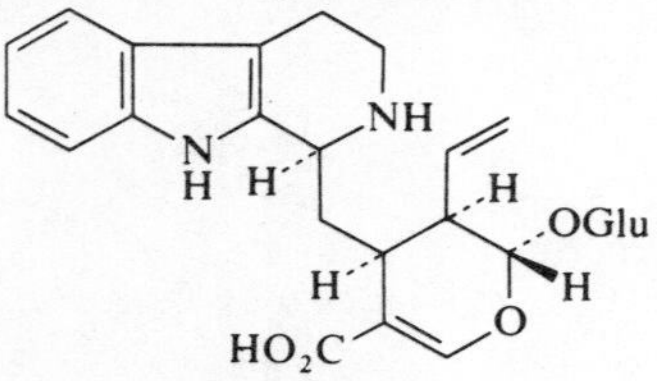

70

C. 5-Oxostrictosidine

Extensive fractionation of the polar material from the heartwood of *A. rubescens* yielded a neutral glycoalkaloid characterized as the pentaacetate derivative {$C_{37}H_{42}N_2O_{15}$; $[\alpha]_D -20°$ ($CHCl_3$)}. Its structure was deduced as 5-oxostrictosidine (**71**) on the basis of spectral data (UV, IR, NMR, CD, and mass) and chemical transformations. Particularly informative was the cleavage of the lactam ring in **72** which, on treatment with sodium methoxide in methanol followed by brief reactylation, furnished **73** (*6*).

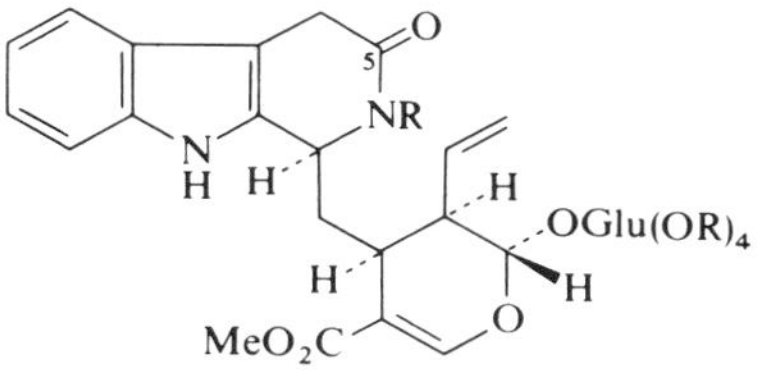

71 R = H
72 R = Ac

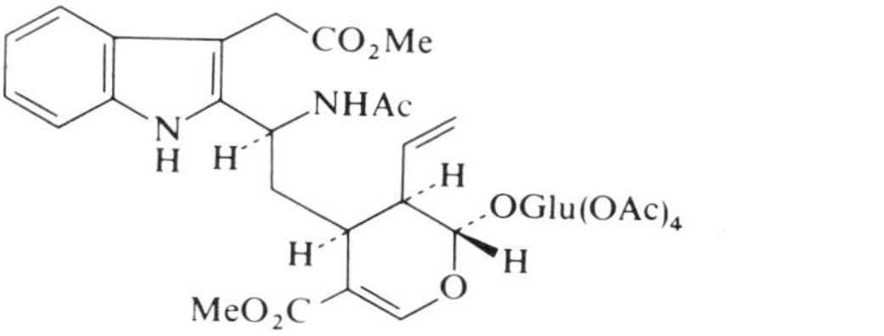

73

D. Lyaloside

A new β-carboline glycoalkaloid, lyaloside {$C_{27}H_{30}N_2O_9$; mp 168–169°; $[\alpha]_D -202°$ (MeOH)}, has recently been isolated from the root bark of *Pauridiantha lyalli* (Baker) Bremek (Rubiaceae). It has been formulated as (**74**) on the basis of spectroscopic evidence. The coupling constant (J = 5.0 Hz) between protons at C_{20} and C_{21} has been interpreted for the α configuration of proton at C_{21} (*30*).

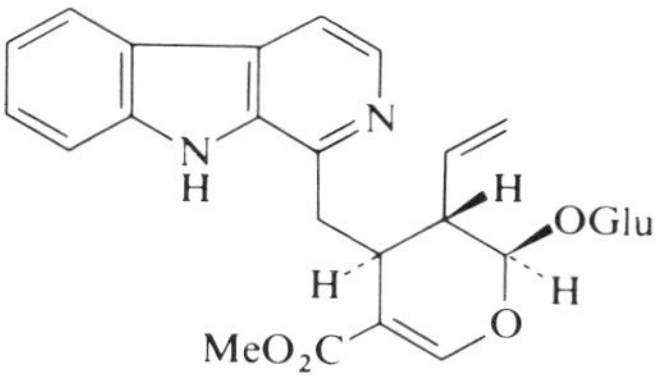

74

E. Palinine

Countercurrent separation of the alkaloids of *Palicourea alpina* (Sw.) DC. (Rubiaceae) furnished palinine {$C_{27}H_{32}N_2O_{10}$; mp 166.5–168°; $[\alpha]_D -252.3°$ (MeOH)}. The UV spectrum, λ_{max}(log ϵ): 236.5 (4.62), 290.5 (4.20), and 394 (3.61) nm, with a bathochromic shift in acid strongly supported the presence of β-carboline chromophore in the glycoalkaloid. Palinine on acetylation with pyridine and acetic anhydride gave the pentaacetate **77** which on purification with silica gel afforded the tetraacetate **76**. The NMR spectrum of the latter showed close resemblance to deoxycordifoline derivatives, and the high-resolution mass spectrum indicated a facile loss of water. The fragmentation pattern was reminiscent of cordifoline pentaacetate, leading to structure **75** for palinine (*31*).

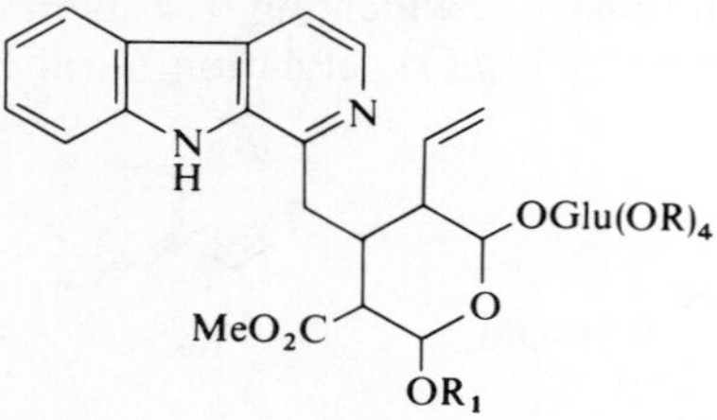

75 R = R_1 = H
76 R = Ac, R_1 = H
77 R = R_1 = Ac

F. Rubescine

From *A. rubescens* a neutral glycoalkaloid, rubescine {$C_{35}H_{36}N_2O_{11}$; $[\alpha]_D - 121°$ (MeOH)}, has been isolated which yielded a crystalline pentaacetate {mp 158–161°; $[\alpha]_D - 79°$ (MeOH)}. Treatment with diazomethane afforded dimethylrubescine, which in turn formed a triacetate. On catalytic hydrogenation two moles of hydrogen were absorbed to give tetrahydrorubescine {$[\alpha]_D - 79°$ (MeOH)}. In addition to indole and β-alkoxyacrylamide chromophores, the UV and NMR spectra coupled with the chemical behavior, indicated the presence of a *trans*-dihydroxycinnamate ester. Examination of the mass spectral fragmentations showed this to be located on a hexoside moiety (cleavages *p*, *q*, and *r* in **78**), and subsequent deacetylation of rubescine with sodium methoxide afforded vincoside lactam and methyl caffeate. At 220 MHz the protons of the glucose unit in 18,19,2″,3″-tetrahydrorubescine could be distinguished, and decoupling experiments showed that H-3′ had been shifted to lower field by acylation. Consequently, rubescine was assigned the 3′-caffeoylvincoside lactam structure **78** (*32*).

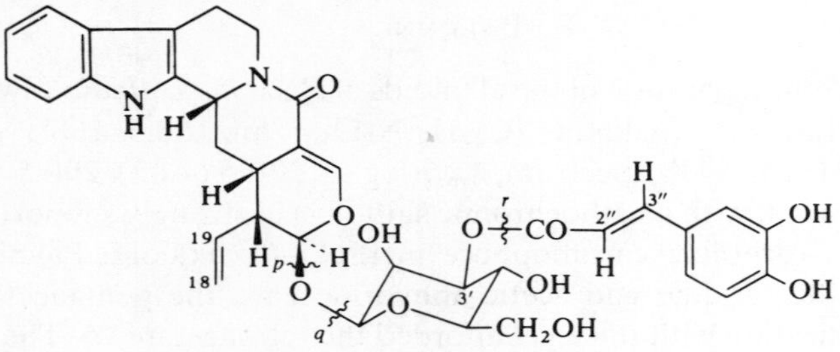

78

G. 10β-D-Glucosyloxyvincoside Lactam

Another lactam from *A. rubescens* was characterized as the octaacetate {$C_{48}H_{56}N_2O_{12}$; $[\alpha]_D -86°$ ($CHCl_3$)}. Its UV spectrum [λ_{max} 231, 273 (inf.), 294 (inf.), and 310 (inf.) nm] corresponded to the sum of β-alkoxyacrylamide and 5-alkoxyindole chromophores. Two successive losses of 330 mass units ($C_{14}H_{18}O_9$) in the mass spectrum from the molecular ion at *m/e* 1012 suggested that it was a bishexoside. The NMR spectrum confirmed these features and also indicated three olefinic protons, which were presumably in a vinyl group, as catalytic hydrogenation added only one mole of hydrogen. After deacetylation, both hexose units were removed from the dihydro derivative by β-glucosidase (thus establishing their identity) to give a dihydroaglucone ($C_{20}H_{22}N_2O_4$). In neutral solution this had a UV spectrum similar to the starting material, but on addition of alkali there was a shift in the absorption maximum to 326 nm. At this stage, a possible relationship to vincoside lactam was suspected which was strengthened when both were shown from CD spectra to have the same 3β configuration. Final confirmation of the structure was obtained by partial synthesis. Condensation of secologanin tetraacetate and serotonin (5-hydroxytryptamine) in glacial acetic acid at 100° afforded 10-hydroxyvincoside lactam tetraacetate (**79**) directly. Catalytic hydrogenation, deacetylation, and hydrolysis with β-glucosidase gave a dihydroaglucone (**82**) identical with that obtained from the bisglucoside which must therefore be 10β-D-glucosyloxyvincoside lactam (**80**). The octaacetate could be assigned the corresponding structure **81** (*33*).

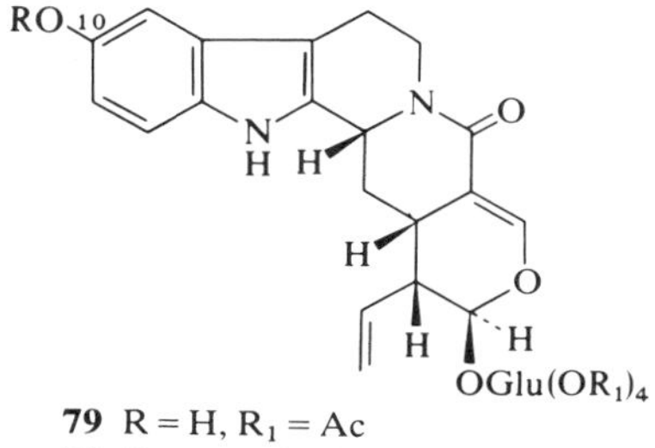

79 R = H, R_1 = Ac
80 R = β-D-Glu$(OH)_4$, R_1 = H
81 R = β-D-Glu$(OAc)_4$, R_1 = Ac

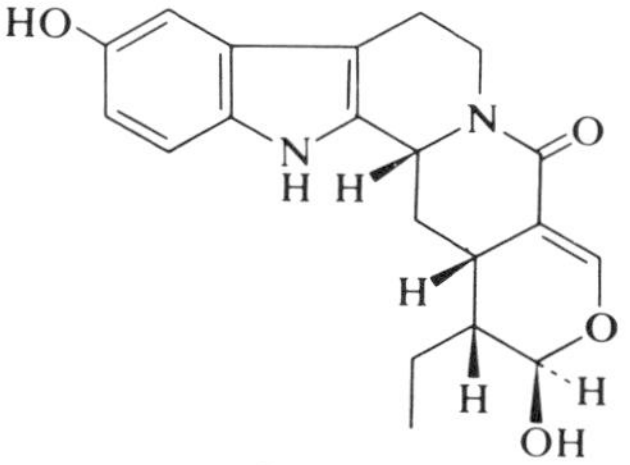

82

H. Cadambine, 3α- and 3β-Dihydrocadambine, and 3α- and 3β-Isodihydrocadambine

A combination of gel-permeation and ion-exchange chromatography on an extract of the heartwood of *A. cadamba* led to the isolation of three new indole glycosides, cadambine {$C_{27}H_{32}N_2O_{10}$; mp 210–211°; $[\alpha]_D -71°$

(MeOH)}, dihydrocadambine {$C_{27}H_{34}N_2O_{10}$; $[\alpha]_D -135°$ (MeOH)}, and isodihydrocadambine {pentaacetate; $C_{37}H_{44}N_2O_{15}$; $[\alpha]_D -66°$ (MeOH)}. On addition of acid, the UV chromophore of cadambine changed from an indole to a 3,4-dihydro-β-carbolinium type (**83**), but it reverted to an indole on basification. It was reduced by sodium borohydride in methanol to 3α-dihydrocadambine and in acetic acid to a mixture of 3α- and 3β-dihydro derivatives. Further chemical transformations with full spectral analysis (UV, IR, NMR, CD, and mass) have led to the assignment of structure and absolute stereochemistry **84** to cadambine, **85** to dihydrocadambine, and **87** to isodihydrocadambine (*34*).

Two of them are thus tryptamine analogs of rubenine (q.v.) with similar seven-membered rings involving bonding between N-4 and C-18, the third containing an equally unusual N-4–C-19 bond. 3α-Dihydrocadambine has also been isolated from *N. diderrichii* (*4*). More recent examination of *A. cadamba* leaves has afforded several more glycoalkaloids, including strictosidine and two vincoside derivatives—3β-dihydrocadambine (**86**) {pentaacetate; $[\alpha]_D -69°$ ($CHCl_3$)} and 3β-isodihydrocadambine (**88**) {pentaacetate; mp 170–171°; $[\alpha]_D \pm 0°$ ($CHCl_3$)} (*5*).

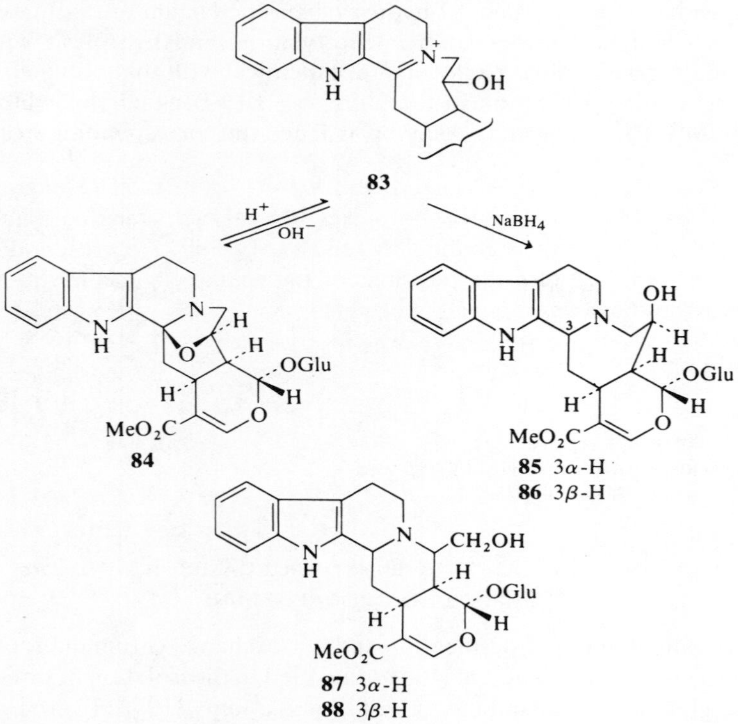

83

H^+ / OH^- ; $NaBH_4$

84

85 3α-H
86 3β-H

87 3α-H
88 3β-H

V. The Tryptophan Derivatives

In principle there appears no reason why a condensation product of secologanin with tryptophan instead of tryptamine should not occur in nature to give a new series of carboxy terpenoid indole alkaloids parallel to the known tryptamine derivatives. A few representatives have been isolated in recent years, and probably more will follow with the adoption of isolation techniques suited to amino acids.

A. Cordifoline

After ipecoside, cordifoline was the second alkaloid derived from secologanin to be discovered and gave the first indication of the existence of the common secoiridoid moiety in the indole series (*1*). It was isolated from the heartwood of *A. cordifolia* as the pentaacetate ($C_{36}H_{40}N_2O_{18}$; mp 142–144°) and was shown to contain carboxyl, phenolic, and olefinic groups by the following reactions. Methylation with diazomethane furnished methylcordifoline pentaacetate {$[\alpha]_D - 185°$ (MeOH)} which could be hydrogenated to methyldihydrocordifoline pentaacetate. Hydrolysis of cordifoline pentaacetate followed by methylation and reacetylation afforded dimethylcordifoline tetraacetate. The other features of the molecule were largely elucidated by spectral data.

A hexoside was indicated by the presence in all the mass spectra of a strong peak at m/e 331 ($C_{14}H_{19}O_9$) owing to the oxonium ion **91** and related ions at m/e 169, 127, and 109 formed by loss of acetic acid and ketene. Complementary ions were found at M − 331 and M − 347 corresponding to cleavages *p* and *q*. From the positions of the acetate signals in the NMR spectrum the sugar was inferred to be glucose, and this was confirmed when the free glycoside was hydrolyzed by β-glucosidase (*35*).

The β-carboline partial structure was represented in the mass spectra by major ion (**93**) formed by the favored cleavage *r* and hydrogen transfer. Further information about this system was supplied by the NMR spectrum of the pentaacetate, which showed phenolic acetate, indolic NH, and four aromatic protons. One of the latter was a singlet at τ1.27, attributed to H-6, and spin decoupling of the others indicated that the acetoxy group must be at either C-10 or C-11. Comparison of the UV spectrum of cordifoline pentaacetate [λ_{max}(log ϵ): 239 (4.66), 271 (4.60), 305 (inf.) (3.82), 338 (3.61), and 350 (3.61) nm] with that of harman-3-carboxylic acid (**94**) revealed a marked correspondence if one postulated the presence of a β-alkoxyacrylate function in the former and constructed a summation spectrum. This established the basic carboxy-β-carboline chromophore, and eventually the phenolic hydroxy group was located at C-10 by reference to adifoline (q.v.).

The remainder of the molecule was represented by a mass spectral peak at *m/e* 165 ($C_9H_9O_3$) corresponding to the pyrylium ion **43**, which shifted to *m/e* 167 (**44**) after hydrogenation. A vinyl group was indicated by three olefinic protons in the NMR spectra which disappeared after hydrogenation and were replaced by a methyl absorption; that the latter was due to an ethyl group was confirmed when Kuhn–Roth oxidation of methyldihydrocordifoline gave propionic acid. The presence of the system $MeO_2C\overset{|}{C}{=}CH{-}O$ was shown by IR bands at 1680 and 1635 cm^{-1}, UV absorption in the 240 nm region, and NMR signals at $\tau 2.51$ ($-CO\overset{|}{C}{=}CH{-}O$) and $\tau 6.15$ (OMe).

Most importantly, in the NMR spectrum of cordifoline pentaacetate it proved possible to locate the hydrogens around ring D and to confirm the assignments by spin decoupling. Thus H-20 ($\tau 7.42$) was coupled to H-15 ($\tau 6.75$), H-19 ($\tau 4.25$), and H-21 ($\tau 4.61$); H-15 was further coupled to the C-14 methylene protons ($\tau 6.30$ and 6.70) and also showed a small allylic interaction with H-17 ($\tau 2.51$). Further, the observed coupling constants between H-20 and H-15 (~3.5 Hz) and H-20 and H-21 (5.0 Hz) could be compared with values calculated from dihedral angles in Dreiding models and indicated their relative stereochemistry. These data led to structure **90** for cordifoline pentaacetate. Because cordifoline also gave a pentapropionate, it must correspond to the unacylated glucoside **89** (*1*).

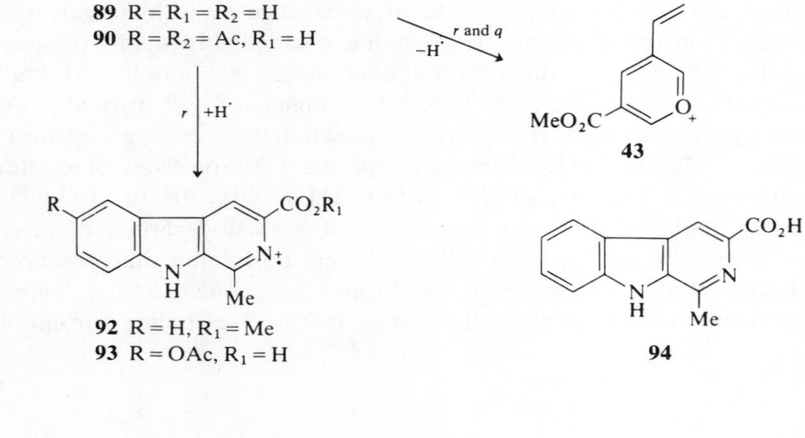

89 $R = R_1 = R_2 = H$
90 $R = R_2 = Ac, R_1 = H$

91

43

92 $R = H, R_1 = Me$
93 $R = OAc, R_1 = H$

94

B. Deoxycordifoline

In a later examination of *A. cordifolia*, Merlini and Nasini, after methylation and acetylation of the alkaloidal constituents, isolated another β-carboline glycoside {mp 91–93°; $[\alpha]_D -144°$ ($CHCl_3$); $\lambda_{max}(\log \epsilon)$; 237 (4.51), 265 (inf.) (4.47), 273 (4.52), 307 (3.88), and 344 (3.71) nm}. Characteristic mass spectral fragments at m/e 421, 405, 331, and 165 indicated that the noncarboline moiety was the same as that in cordifoline pentaacetate. The molecular ion at m/e 752 and a major peak at m/e 240 corresponding to **92** suggested the structure of methyl-10-deoxycordifoline tetraacetate (**96**). As in the case of cordifoline, this was substantiated by analysis of the NMR spectrum with the aid of spin-decoupling experiments which allowed assignments to almost all the protons and their relative stereochemistry.

Eventually, the proposed structure was confirmed by partial synthesis which also provided the absolute configuration of both deoxycordifoline and cordifoline. Condensation of methyl D-tryptophanate with secologanin tetraacetate gave a mixture of the C-3 epimers of methyl 5β-tetrahydrodeoxycordifoline tetraacetate (**100** or **101**) which was aromatized by 5-chlorobenzotriazole to methyldeoxycordifoline tetraacetate (**96**) identical with the natural product. In particular, their CD spectra ($\theta_{307} + 7.2 \times 10^3$, $\theta_{270} - 2.5 \times 10^4$, and $\theta_{239} + 2.0 \times 10^4$) were superimposable and virtually identical with that of methylcordifoline pentaacetate showing the same absolute configuration. Deoxycordifoline was therefore assigned structure **95** (*2*) which has recently been isolated as a crystalline product {mp 178–180°; $[\alpha]_D -212°$ (H_2O)} from *A. rubescens* (*36*).

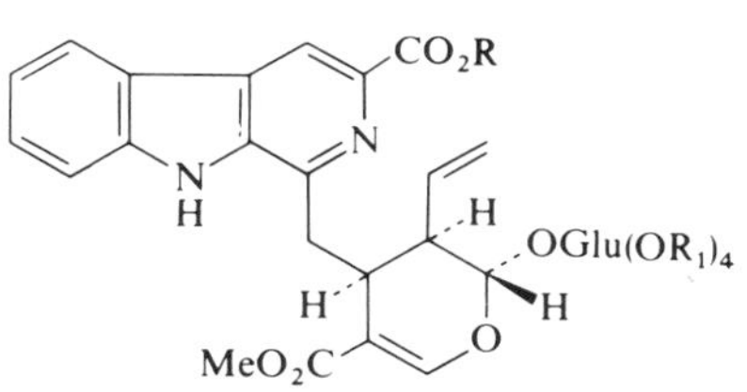

95 $R = R_1 = H$
96 $R = Me$, $R_1 = Ac$

C. Deoxycordifoline Lactam

From the heartwood of *A. rubescens*, Brown and Fraser obtained a glycoalkaloid characterized as the tetraacetate {$C_{37}H_{43}N_2O_{15}$; mp 165–168°; $[\alpha]_D -135°$ (MeOH)} and its methyl ester {$[\alpha]_D -103°$ (MeOH)}. Although the UV spectrum [$\lambda_{max}(\log \epsilon)$: 245 (4.51), 287 (4.13), 306 (3.73), 326 (4.02), and 337 (4.07) nm] did not correspond to any known system, addition of alkali caused a rapid irreversible change to a 3-carboxy-β-carboline chromophore. Subsequent treatment of the methyl ester with sodium methoxide in methanol followed by reacetylation afforded methyldeoxycordifoline tetraacetate. This reaction with alkali was reminiscent of the facile cleavage of *N*-acyl indoles; moreover, no indolic NH was apparent in the IR or NMR spectra, so the only compatible structure is that of deoxycordifoline *N*-1-lactam (**97**). Support for this was provided by mass spectra where, in addition to the typical glycosidic and β-carboline fragmentations, a series of ions arising from cleavages *p*, *q*, and *r* could be discerned; in particular, an intense peak at *m/e* 234 given by the acid was attributed to ion **98**. Ultimately, a synthesis of the previously unknown 2-methoxycarbonyl-4,5-dihydrocanthin-6-one (**99**) provided a reasonable model for the UV chromophore, after allowing for the absence of a β-alkoxyacrylamide system, and confirmed the proposed structure (*37*).

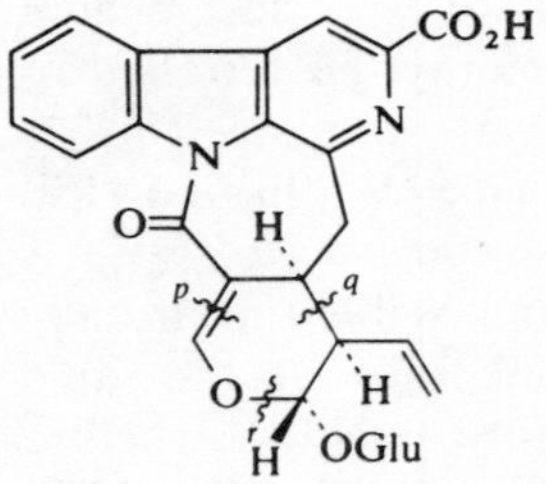

CO_2H N N O H p q O H r H OGlu

97

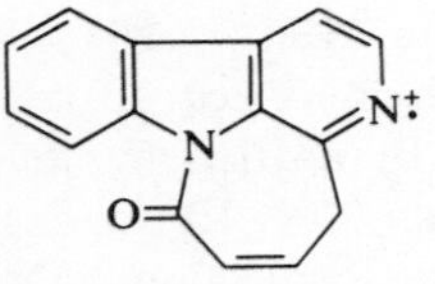

N N⁺ O

m/e 234

98

CO_2Me N N O

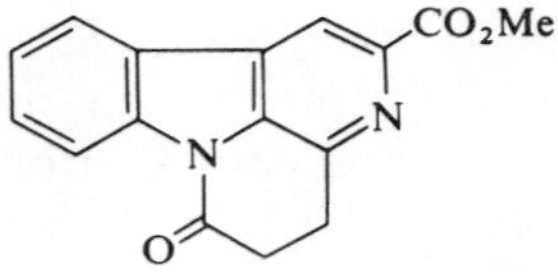

99

D. 3α,5α-TETRAHYDRODEOXYCORDIFOLINE

An obvious inference from the structures of cordifoline and deoxycordifoline was that their biogenetic precursor would have to be a tetrahydrodeoxycordifoline (TDC) (**104** and/or **105**) formed by condensation between L-tryptophan and secologanin. One of these, the 3α isomer, has subsequently been found in *R. orientalis* (*8*) and in *A. rubescens* (*38*), but the other has so far eluded isolation as a natural product.

Extraction of *R. orientalis* roots yielded an amino acid glycoside {mp 232°; $[\alpha]_D -280°$ (MeOH)}. It formed ester and acetate derivatives which could be converted to the corresponding dihydro series. Examination by UV, IR, NMR, and mass spectra indicated the gross structure of a TDC (carboxystrictosidine). An amorphous material which cooccurred corresponded to the dicarboxylic acid **106**, as both gave the same ester with diazomethane.

Equilibration studies on model compounds **109** showed that in those with H-3 and H-5 cis the latter proton appeared as a low-field NMR signal at ca.

$\tau 4.0$ but not in the trans series. Because the methyl pentaacetate derivative of the TDC showed just such a signal, whereas its C-5 epimer obtained by treatment with base did not, the former must be the cis isomer, either $3\alpha,5\alpha$ or $3\beta,5\beta$. To distinguish between these possibilities and also to establish the complete structure and stereochemistry, all four diastereoisomers (**102**, **103**, **107**, and **108**) were synthesized from D- or L-tryptophan and secologanin. The naturally occurring compound corresponded to the cis isomer from L- tryptophan and hence was assigned $3\alpha,5\alpha$-TDC (**104**) structure (*8*).

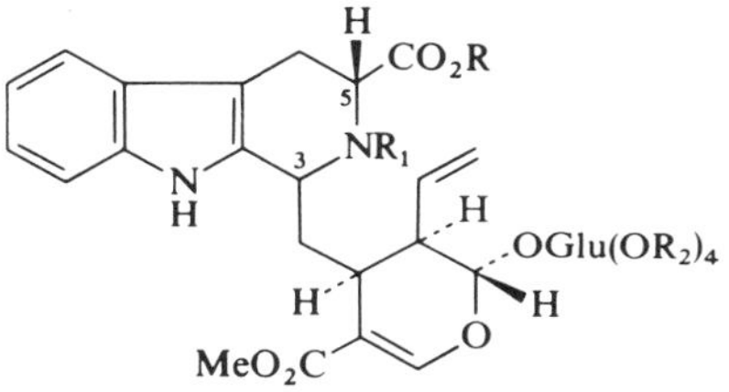

100 $R = Me, R_1 = H, R_2 = Ac, 3\alpha$-H
101 $R = Me, R_1 = H, R_2 = Ac, 3\beta$-H
102 $R = R_1 = R_2 = H, 3\alpha$-H
103 $R = R_1 = R_2 = H, 3\beta$-H

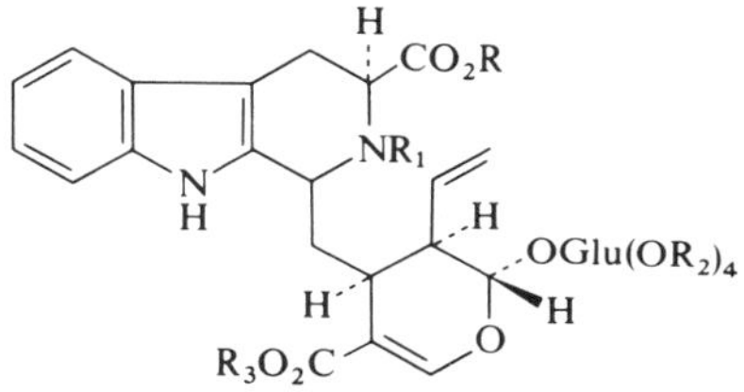

104 $R = R_1 = R_2 = H, R_3 = Me, 3\alpha$-H
105 $R = R_1 = R_2 = H, R_3 = Me, 3\beta$-H
106 $R = R_1 = R_2 = R_3 = H, 3\alpha$-H
107 $R = R_3 = Me, R_1 = R_2 = Ac, 3\alpha$-H
108 $R = R_3 = Me, R_1 = R_2 = Ac, 3\beta$-H

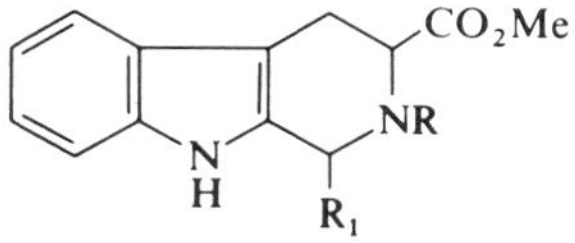

109 $R = H$ or Ac, R_1 = Alkyl

E. $3\alpha,5\alpha$- and $3\beta,5\alpha$-Tetrahydrodeoxycordifoline Lactams

Although $3\beta,5\alpha$-TDC has not yet been found in nature, the derived lactam **111** was detected in *A. rubescens* together with the $3\alpha,5\alpha$-TDC lactam, **110**. Fractionation of a methanolic extract by gel-permeation chromatography followed by acetylation afforded two isomeric acids ($C_{35}H_{38}N_2O_{14}$) which methylated to ester A {$[\alpha]_D - 84°$ ($CHCl_3$)} and ester B {$[\alpha]_D - 79°$ ($CHCl_3$)}. Both contained indole, β-alkoxyacrylamide, and hexoside tetraacetate systems (UV, IR, NMR, and mass spectra). The identical mass spectra overall corresponded to that expected for a vincoside or strictosidine lactam substituted by a methoxycarbonyl group, and this impression was reinforced by the NMR spectra, ester A showing strong similarity to the latter and ester B to the former. The major structural features were then established by partial synthesis. Condensation of methyl L-tryptophanate and secologanin tetraacetate gave two 5α-TDC epimers which, on treatment with base followed by methylation and acetylation, afforded two lactams (**112** and **113**) identical with esters A and B. A parallel sequence using D-tryptophan confirmed that no epimerization of H-5 had occurred during these reactions. The last unknown feature was the stereochemistry at C-3, which was readily assigned from CD spectra: ester A had a positive Cotton effect between 250 and 300 nm and was therefore identified as methyl $3\alpha,5\alpha$-TDC lactam tetraacetate (**112**), whereas ester B had a negative Cotton effect, corresponding to the $3\beta,5\alpha$ isomer (**113**). Finally, one distinguishing feature in the NMR spectrum of 3β isomer was that the H-5 signal appeared at low field ($\tau 3.90$) as a result of deshielding by the lactam carbonyl which cannot occur with the other isomer (*37*).

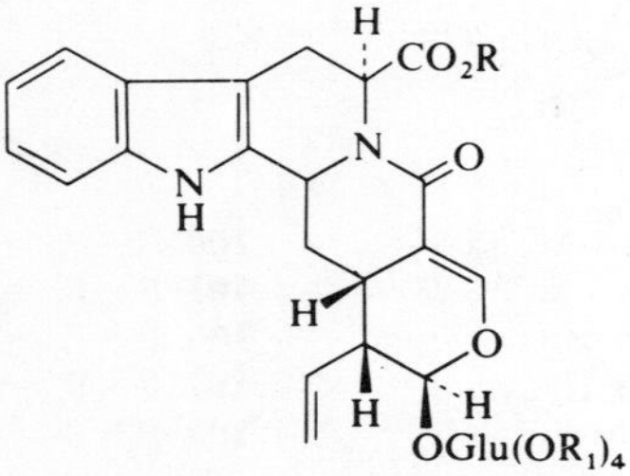

110 $R = R_1 = H$, 3α-H
111 $R = R_1 = H$, 3β-H
112 $R = Me$, $R_1 = Ac$, 3α-H
113 $R = Me$, $R_1 = Ac$, 3β-H

F. Rubenine

Rubenine $\{C_{28}H_{32}N_2O_{11}$; $[\alpha]_D -42°$ (MeOH)$\}$ was isolated from the heartwood of *A. rubescens* and further characterized as the tetraacetate $\{[\alpha]_D -46°$ ($CHCl_3$)$\}$ and its picrolonate (mp 154–156°).

Spectral data readily demonstrated the presence of indole, methyl β-alkoxyacrylate, and hexoside functions. Ions in the mass spectrum of the acetate at m/e 183, 182, 169, and 168 (rather than the more common series m/e 184, 183, 170, and 169) indicated a tetrahydro-β-carboline with a substituent α to N-4; however, the most distinctive feature was the ready loss of CO and CO_2 from the molecular ion which was attributed to the breakdown of a lactone. Accordingly, treatment with sodium methoxide opened the lactone ring, and reactylation gave a methyl ester pentaacetate (**114**) in which simultaneous addition of methanol to the alkoxyacrylate had occurred, as shown by loss of the UV chromophore at 238 nm and the appearance of two new signals in the NMR spectrum.

Hydrolysis of rubenine with β-glucosidase furnished D-glucose and the aglucone **116**. The IR spectrum (1740 cm^{-1}) of the latter indicated the presence of a δ-lactone. Furthermore, the newly liberated hydroxyl had to be linked to the β-alkoxyacrylate via a hemiacetal system, because addition of alkali now produced a marked enhancement in the UV absorption at ca. 275 nm attributable to an ionized β-hydroxyacrylate chromophore.

OH
MeO_2C O
OH^- / H^+
CHO
$\bar{O}$
MeO_2C

Prolonged treatment of the aglucone **116** in methanol with diazomethane afforded a methyl acetal (**117**) and a minor methanolysis product **118**, neither of which showed a base shift in UV spectrum. Reduction of **116** with sodium borohydride furnished the diol **119**.

At this stage, two plausible structures could be advanced for rubenine **115** and **120**—the latter seeming more likely from examination of molecular models. However, it was not supported by the mass spectra. For example, the expected ready loss of CH_2OAc from the position α to N-4 in the corresponding ring-opened derivative **121** was not observed. Eventually, conclusive evidence excluding structure **120** was obtained from NMR spectra, which showed that the lactone was derived from a secondary rather than a primary alcohol.

Detailed examination of the spin-decoupled NMR spectra enabled assignments to be made for practically all the protons and also provided information about their relative stereochemistry. Thus, in rubenine, H-20 (τ7.80) was found to be coupled with H-21 (τ4.65 trans a-a), H-15 (τ6.80 cis a-e), and H-19 (τ4.85 cis a-e); H-19 was further strongly coupled to H_b-18 (τ6.25 trans e-e) but weakly coupled to H_a-18 (τ6.40 cis a-e); H_b-14 (τ8.20) had strong interactions with both H-3 (τ5.67 trans a-a) and H-15 (trans a-a), and the latter had only a weak interaction with H_a-14 (τ8.70 cis a-e); again, H-5 (τ5.85) had a substantial coupling with H_a-6 (τ6.85 cis a-e) and a small coupling with H_b-6 (τ6.50 trans a-e). The above chemical and spectral data could only be satisfied by structure and relative stereochemistry **115** for rubenine. Finally, the CD spectrum established an α orientation for H-3, and because of the rigid hexacyclic structure also proved the absolute configuration of the molecule to be as shown. Rubenine is thus a derivative of 3α,5α-TDC with a novel N-4 → C-18 bond in a seven-membered ring. Its immediate precursor could well be the 18,19-epoxide in which nucleophilic attack by N-4 at C-18 is followed by lactonization of the newly generated alcohol (*39*).

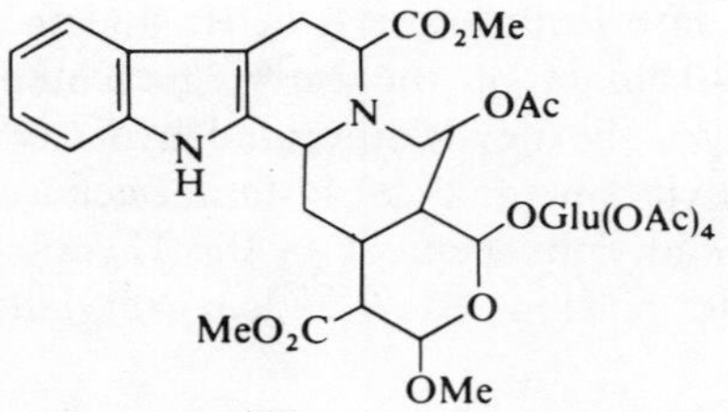

114

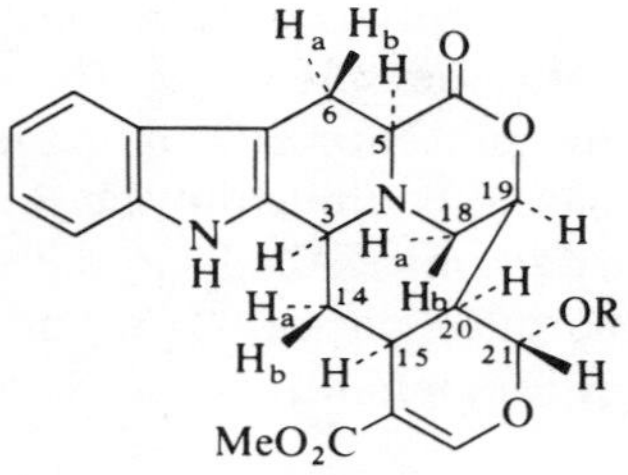

115 R = β-D-Glu
116 R = H
117 R = Me

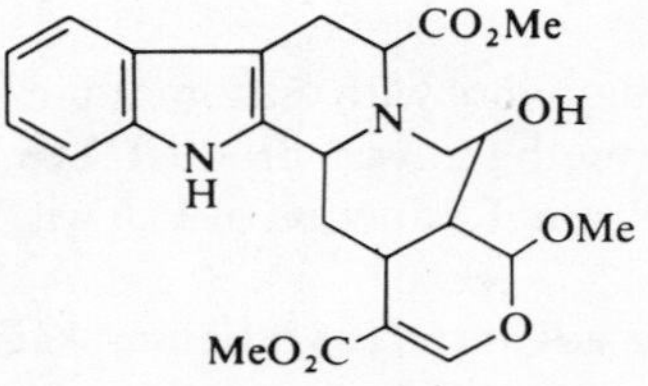

118

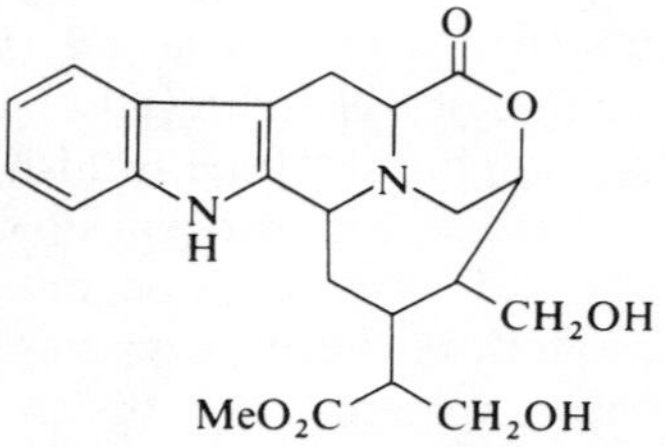

119

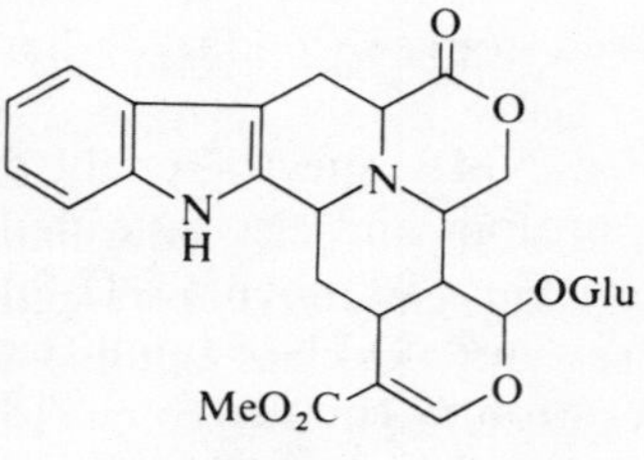

120

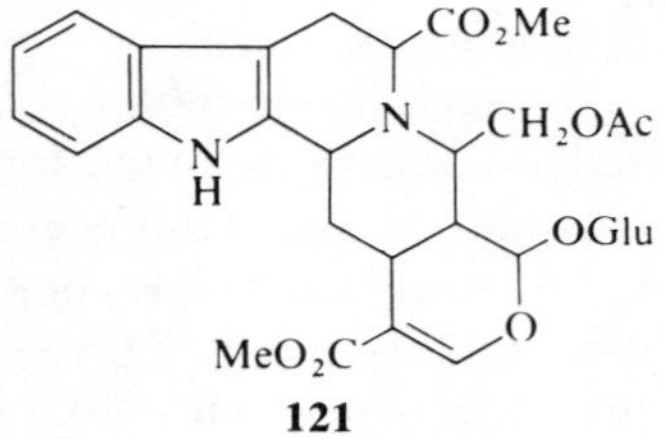

121

G. Macrolidine

Continued investigation on the heartwood of *A. rubescens* resulted in the isolation of macrolidine as the tetraacetate {$C_{36}H_{40}N_2O_{14}$; $[\alpha]_D + 2°$ ($CHCl_3$)}. The major features of **123** were determined from UV, IR, NMR, and mass spectra and conversion into $3\alpha,5\alpha$-tetrahydrodeoxycordifoline pentaacetate (q.v.) by cleavage of the lactone ring with sodium methoxide followed by reacetylation. Catalytic hydrogenation and Zemplen deacetylation of **123** afforded *N*-4-acetyldihydromacrolidine (**124**) {$[\alpha]_D + 39.5°$ (MeOH)}, for which virtually every proton could be assigned in the NMR spectrum. The secondary hydrogen of the sugar at C-2′, C-3′, and C-4′ all appeared in the high-field region $\tau 6.15$–6.60, compared with $\tau 4.70$–5.10 for the tetraacetate; hence signals at $\tau 5.40$ (*dd*, $J = 12.0$ and 2.5 Hz) and 6.06 (*dd*, $J = 12.0$ and ca. 8.0 Hz) were attributed to the 6′-methylene group moved downfield by acylation from ca. $\tau 6.50$ in the alcohol. All these data were compatible with the explanation that only the primary hydroxyl group of the sugar was acylated and hence involved in the lactone ring formation, leading to structure **122** for macrolidine (*40*).

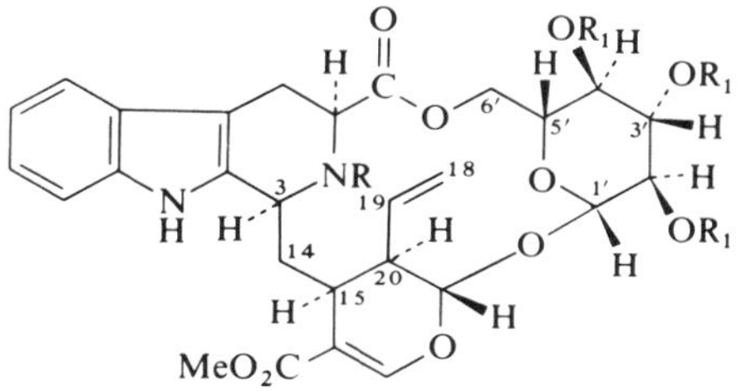

122 $R = R_1 = H$
123 $R = R_1 = Ac$
124 $R = Ac$, $R_1 = H$, 18,19-dihydro

H. Adifoline

Although adifoline and deoxyadifoline are not glycoalkaloids, they have been included in this review because of their obvious derivation from the aglucones of cordifoline and deoxycordifoline.

As previously reported in Volume VIII of this series (*41*), adifoline (mp > 350°) was isolated from *A. cordifolia* by King and co-workers and shown to be a β-carboline derivative (*42*). Subsequent work has enabled a complete structure (**126**) to be assigned for this base (*43*). On prolonged treatment with diazomethane, adifoline ($C_{22}H_{20}N_2O_7$) furnished trimethyladifoline, whereas acetylation gave a diacetate convertible to a monomethyl derivative; these reactions indicated a carboxy and two phenolic or enolic functions. A secondary methyl doublet and a methoxyl singlet at $\tau 8.19$ and 6.16, respectively, were apparent in the NMR spectrum, the latter being assigned to a methoxycarbonyl group because it could be removed by hydrolysis.

Reduction of adifoline with sodium borohydride gave a product which had lost UV (238 nm) and IR (1640, 1613 cm^{-1}) absorptions characteristic of a β-alkoxyacrylate ester, subsequent methylation and acetylation yielding dimethyltetrahydroadifoline diacetate. On the other hand, the same chromophore was unaffected in trimethyladifoline. These reactions were consistent with the partial structures shown below.

$$\text{MeO}_2\text{C}\text{-pyranol (OH)} \rightleftharpoons \text{MeO}_2\text{C}\text{-CHO, OH} \xrightarrow{CH_2N_2} \text{MeO}_2\text{C}\text{-CHO, OMe}$$

$$\downarrow NaBH_4 \qquad\qquad \downarrow NaBH_4$$

$$\text{MeO}_2\text{C}\text{-}CH_2OH\text{, }CH_2OH \qquad\qquad \text{MeO}_2\text{C}\text{-}CH_2OH\text{, OMe}$$

The UV, NMR, and mass spectra of adifoline and its derivatives revealed a phenolic carboxy-β-carboline unit identical with that of cordifoline (q.v.); treatment of trimethyladifoline with lithium aluminum hydride reduced both esters and changed the UV spectrum to one identical with that of 7-methoxyharman, thus locating the position of phenolic hydroxyl at C-10.

Adifoline could be readily dehydrated by acid to anhydroadifoline (**127**), which was not affected by sodium borohydride. This result indicated that elimination of one mole of water from the hemiacetal had occurred to give a pyran derivative; this was supported by UV and NMR spectra and by the presence of a strong pyrylium ion **43** at m/e 165 in the mass spectrum formed by favorable cleavages p and q. Methylation with diazomethane gave dimethylanhydroadifoline (**128**), which provided an important clue to the structure, because NH and OH absorptions were absent from its IR and NMR spectra, implying that N-1 was bonded to a carbon chain.

At this stage, analysis of the spin-decoupled NMR spectra of adifoline and anhydroadifoline at 100 and 220 MHz led to structures **126** and **127**, respectively. In the latter, the 18-methyl group (τ8.15) was coupled to H-19 (τ5.14), which was allylically coupled to H-21 (τ2.56); in turn, H-21 was broadened by a slight interaction with H-15 (τ6.03). In addition to a small coupling with H-17 (τ1.94), H-15 had a substantial coupling with only one (τ6.58) of the 14-methylene protons, the lack of coupling to the other (τ5.44) being consistent with a dihedral angle of ca. 90°. Finally, the methylene group showed the expected large geminal coupling. The additional key feature in the NMR spectrum of adifoline was H-20 (τ7.30). Because it was coupled to H-15 (τ6.63), H-19 (τ5.50), H-21 (τ3.23), it led inevitably to structure **126**. Moreover, the observed coupling constants enabled estimates to be made of dihedral angles which indicated the stereochemistry to be as depicted for adifoline.

I. Deoxyadifoline

Following a process of methylation and acetylation Merlini and Nasini isolated from *A. cordifolia* a new β-carboline alkaloid {$C_{25}H_{24}N_2O_7$;

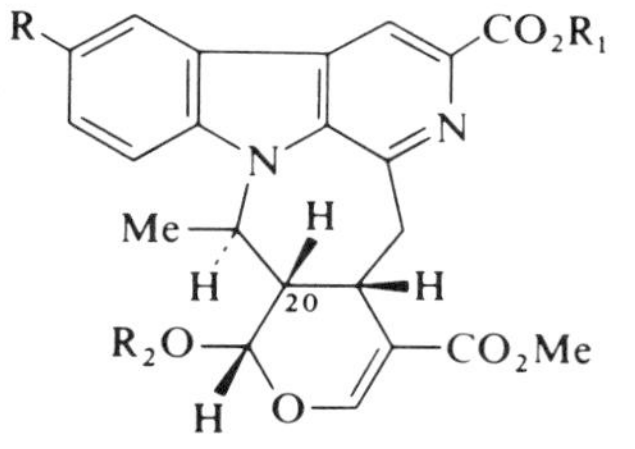

125 R = H, R_1 = Me, R_2 = Ac
126 R = OH, R_1 = R_2 = H

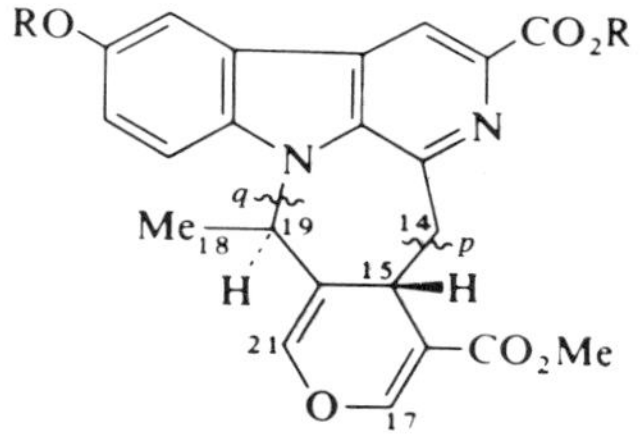

127 R = H
128 R = Me

mp 253–255°; $[\alpha]_D$ 560° ($CHCl_3$); M^+ 464}. The UV spectrum [λ_{max}(log ϵ): 235 (4.45), 263 (4.33), 284 (inf.) (4.00), 330 (3.73), and 342 (3.71) nm] was indicative of an unsubstituted 3-carboxy-β-carboline chromophore which was supported by a strong peak at m/e 240 in the mass spectrum. An acetate and two methoxyl groups appeared in the NMR spectrum, and this also confirmed the lack of substitution in the benzene ring. More importantly, the absence of an NH signal again signified that N-1 was common to two rings. From a complete analysis of the spin-decoupled NMR spectrum, the gross structure of a methyl-10-deoxyadifoline acetate (**125**) was proposed (*2*). However, $J_{15,20}$ in the deoxy compound is 9.0 Hz, much larger than that for adifoline (6.0 Hz) and suggesting a trans diaxial relationship more compatible with the stereochemistry of **129** than with **125**. The occurrence of two diastereomeric series would not be unexpected, for the seven-membered ring may well be formed from a rearranged deoxycordifoline or cordifoline aglucone (**130** or **131**) by conjugate addition of N-1 to C-19, which could take place from either side (*37*).

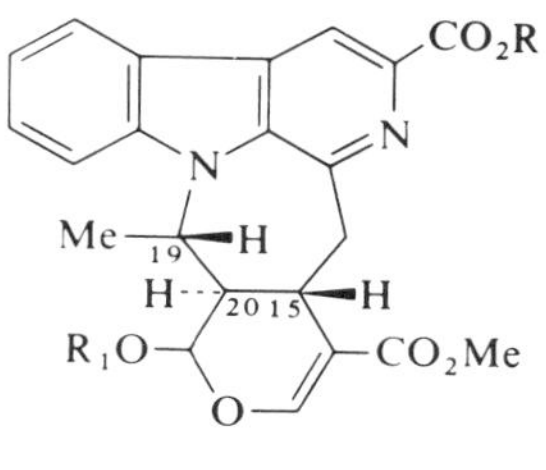

129 R = Me, R_1 = Ac

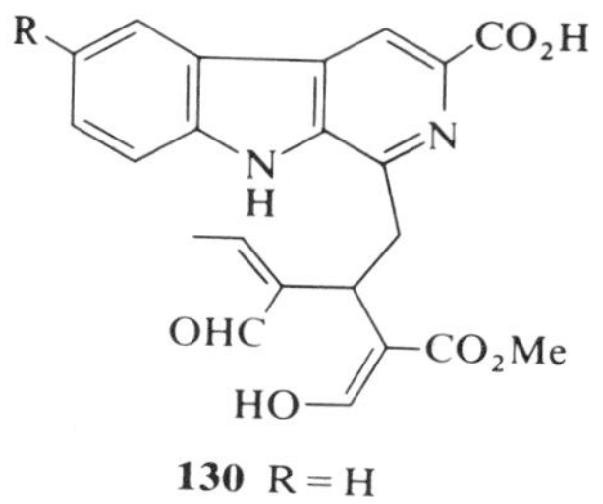

130 R = H
131 R = OH

VI. Biosynthesis

A. Ipecoside

It was originally proposed on the basis of structural and stereochemical relationships that the C_{9-10} unit of *Ipecacuanha* and terpenoid indole alkaloids might have a common biosynthetic origin (*44, 45*). Striking support for this hypothesis has been provided by recent work on the biosynthesis of ipecoside (**2**), cephaeline (**132**), and emetine (**133**) in *C. ipecacuanha.*

Feeding experiments with [2-^{14}C]geraniol (**144**) and [*O*-methyl-^{3}H, 2-^{14}C]loganin (**147**) showed that the C_{10} unit of ipecoside and C_9 unit of cephaeline was of monoterpenoid origin (*46*). Further conclusive evidence that these alkaloids were biosynthesized in a manner analogous to vincoside and terpenoid indole alkaloids was obtained by feeding [*O*-methyl-^{3}H, 6-$^{3}H_2$]secologanin (**22**) and [3′-^{14}C]deacetylipecoside (**23**). Both of these served as specific precursors for ipecoside and cephaeline (*47*).

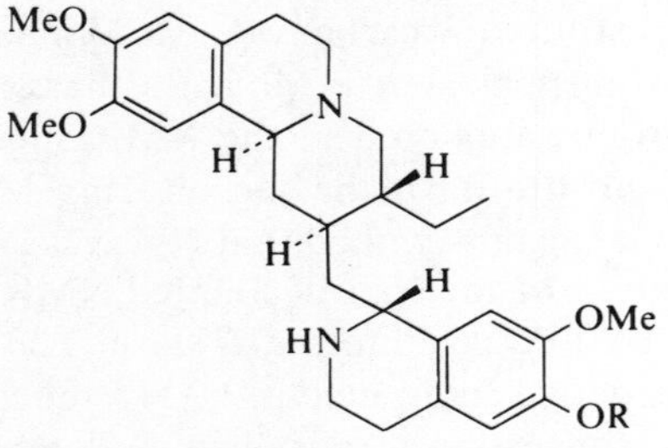

132 R = H
133 R = Me

B. Vincoside and Strictosidine

In accordance with biosynthetic predictions, feeding experiments with [Ar-^{3}H] tryptophan and [*O*-methyl-^{3}H]loganin in *V. rosea* showed good incorporations of both into strictosidine (*48*) and of the latter into vincoside (*19*).

VII. The Biosynthetic Role of Monoterpenoid Glycoalkaloids

A. Terpenoid Indole Alkaloids

One of the most fascinating problems in natural product biosynthesis during the last decade was the origin of the C_{9-10} unit found in *Ipecacuanha*, indole, and *Cinchona* alkaloids. In the initial phases of the work it was proved by tracer studies that this unit did not arise from phenylalanine (*45*), tyrosine (*45*), shikimic acid (*49*), malonate (*45*), acetate (*45, 50, 51*), formate (*45*), or methionine (*52*). The role of glycine has also been investigated (*53*).

Subsequently, in an independent investigation, Battersby, Scott, Arigoni and their respective co-workers showed by feeding experiments on *V. rosea* that the C_{9-10} unit as represented by the *Corynanthé–Strychnos* (**136**), *Aspidosperma* (**137**), and *Iboga* (**138**) skeletons in ajmalicine (**139a**), akuammicine (**141**), vindoline (**142**), and catharanthine (**143**), respectively, was derived from two molecules of mevalonic acid (**134**) (*54–56*). It was further

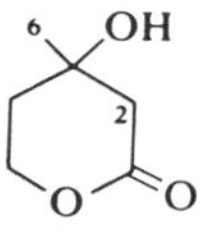

134 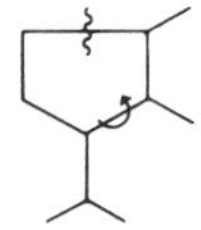 **135** 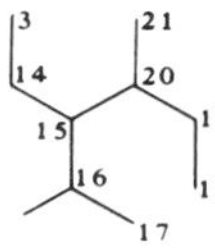 **136** 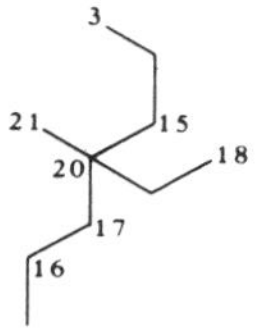 **137** 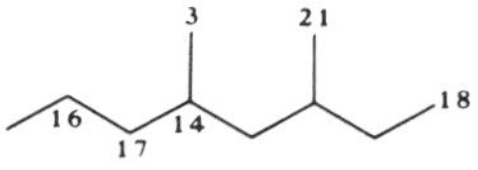 **138**

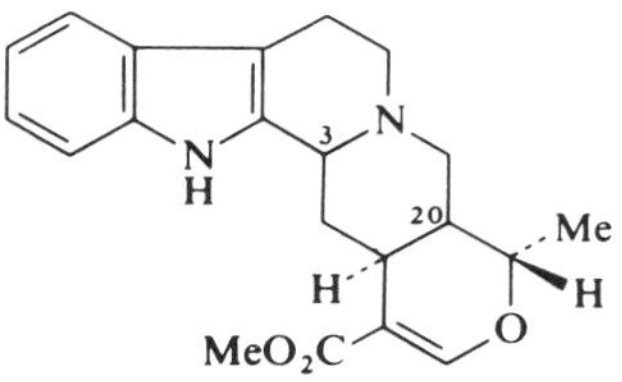

139a 3α-H, 20β-H
139b 3α-H, 20α-H
140a 3β-H, 20β-H
140b 3β-H, 20α-H

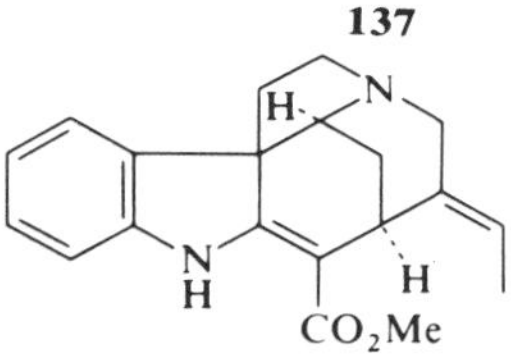

141

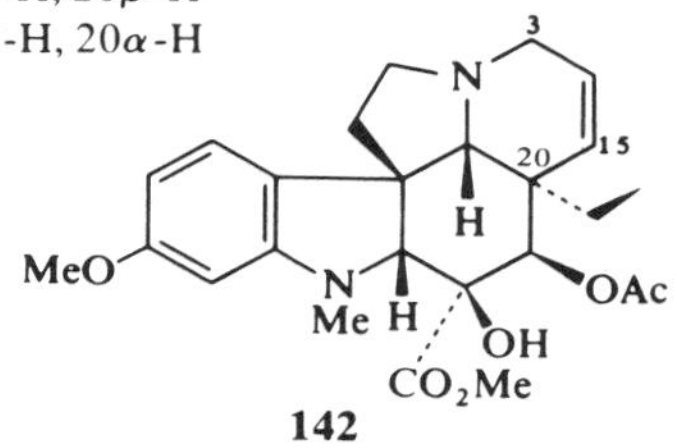

142 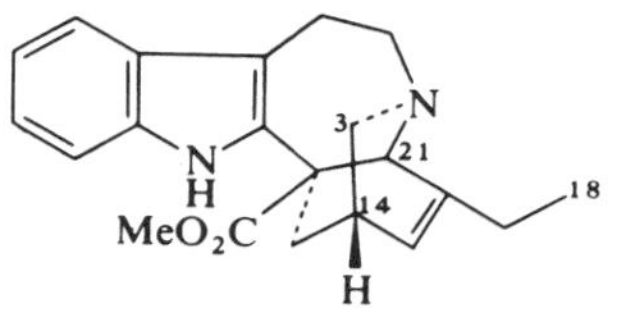 **143**

observed that at some stage of indole alkaloid biosynthesis the C-2 and C-6 of one mevalonate unit had become equivalent—a situation previously encountered in the biosynthesis of plumeride (*57*). This paved the way for vigorous efforts in this direction. Soon after, it was established by radiochemical work that the sequence for the biosynthesis of indole alkaloids was through geraniol (**144**) (*58–61*), 10-hydroxygeraniol (**145**) (*62, 63*), deoxyloganin (**146**) (*64*), loganin (**147**) (*65, 66*), and secologanin (**22**) (*67*) (Scheme 4). The cis isomers (nerol and its 10-hydroxy derivative) were found to be more efficient precursors. Two important aspects which require further exploration in this sequence are (a) the mechanism of the cyclization step of 10-hydroxygeraniol to a cyclopentane monoterpene derivative, and (b) the cleavage mechanism of loganin to secologanin whether or not it

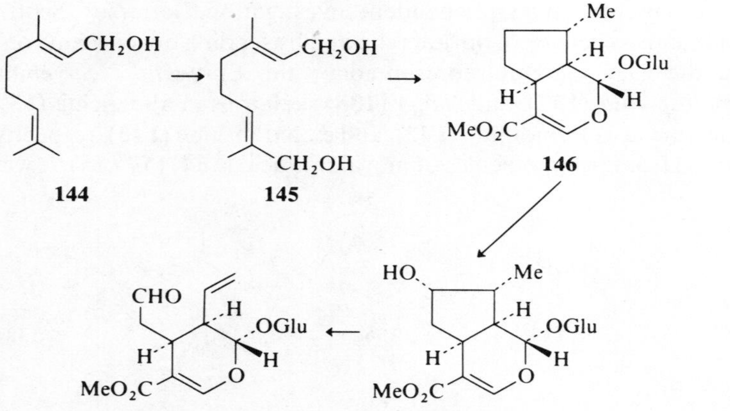

SCHEME 4

proceeds via the hypothetical 4-hydroxyloganin (**148**) which has recently been synthesized (*68*). It will be of considerable interest to see the role of **148** in the biosynthesis of terpenoid indole alkaloids. The reported incorporation of [6-^{3}H]sweroside (**149**) into the indole alkaloids probably proceeds via secologanic acid (**150**) (*69*).

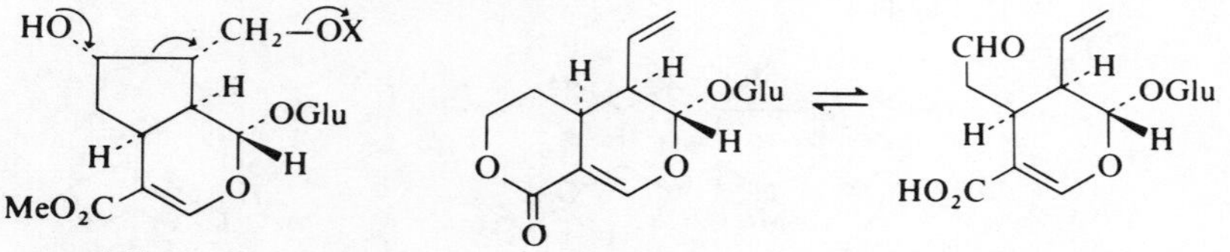

On condensation with tryptamine, secologanin furnished vincoside and strictosidine. When both these epimers were fed separately to *V. rosea*, apparently only vincoside acted as a biological precursor for representative members of the *Corynanthé*, *Iboga*, and *Aspidosperma* alkaloids. Before the C-3 stereochemistry of vincoside was known, the formation of the *Corynanthé* skeleton seemed straightforward: enzymatic removal of the glucose and opening of the aglucone to give the dialdehyde **151** would be followed by cyclization to the immonium salt **152** from which ajmalicine, corynantheine (**153**) and geissoschizine (**154**) could be plausibly derived. However, when the configuration of H-3 was established as β, an intriguing mechanistic problem arose, because vincoside served as sole precursor for ajmalicine and other alkaloids of 3α configuration, the inversion occurring with retention of H-3. One explanation invoked a hypothetical amino acetal intermediate, mancunine (**155**), formed by direct cyclization of N-4 to C-21 (*70*). Opening of the ether bridge would generate the immonium species **156**, which could invert H-3 to give **152** by reversible C-3 → N-4 cleavage of **157** (Scheme 5). *In vitro* support for these suggestions has come from

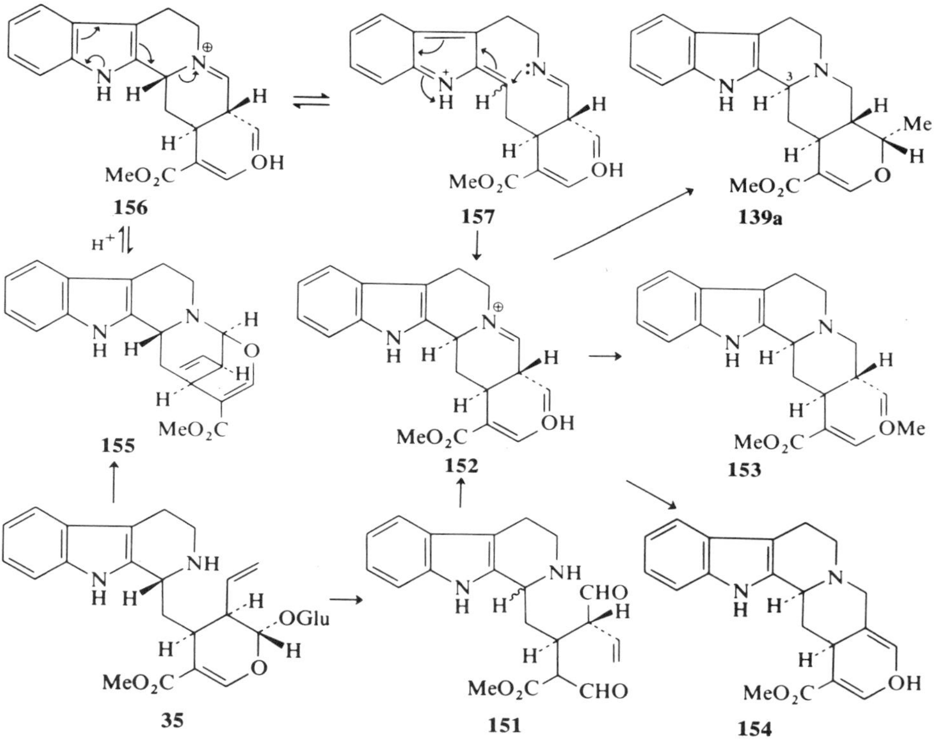

SCHEME 5

synthesis of 18,19-dihydromancunine (**158**) from a vincoside derivative and its reduction to the *Corynanthé* system **159** (*71*). Furthermore, N^1-methylvincoside has been converted via an analog of **156** into N^1-methyldihydrogeissoschizine (**160**) and N^1-methyltetrahydroalstonine (**161**), the inversion of C-3 occurring spontaneously with retention of the hydrogen as required (*72*).

However, very recently in a series of biomimetic experiments it has been established that under virtually physiological conditions vincoside is transformed into akuammigine (**140b**) and strictosidine into tetrahydroalstonine (**139b**) and ajmalicine *i.e.* the precursor for the 3α *Corynanthé* alkaloids was the 3α glycoalkaloid and hence no inversion at C-3 was required (*73*). Concurrent biosynthetic work on *V. rosea* it was found that strictosidine was incorporated into ajmalicine and tetrahydroalstonine, and also acted as a general precursor for vindoline and catharanthine (*74, 75*). Furthermore vincoside was apparently not incorporated into these alkaloids by *V. rosea* cultures (*74*). However, in view of its biomimetic transformation into akuammigine it would seem to be a likely *in vivo* precursor for the 3β *Corynanthé* alkaloids.

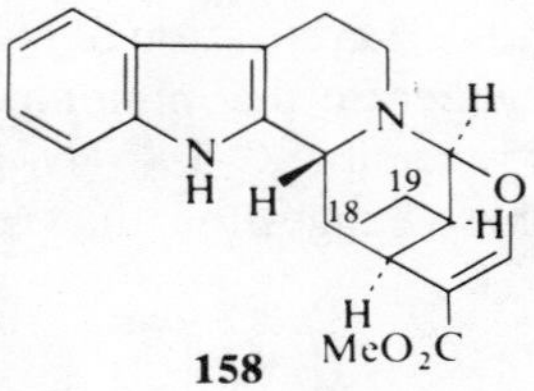

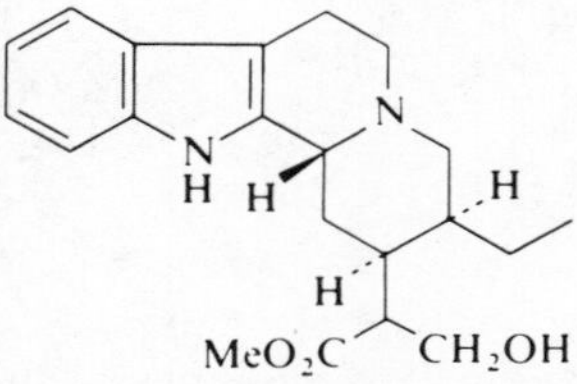

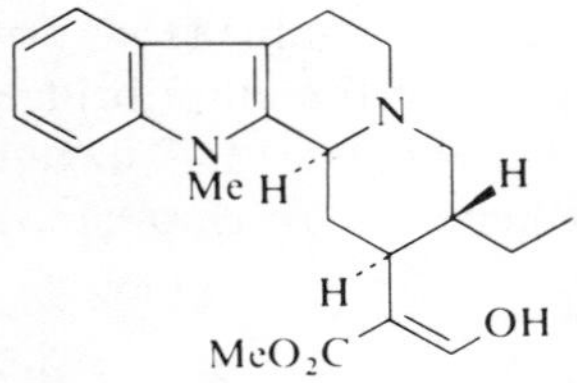

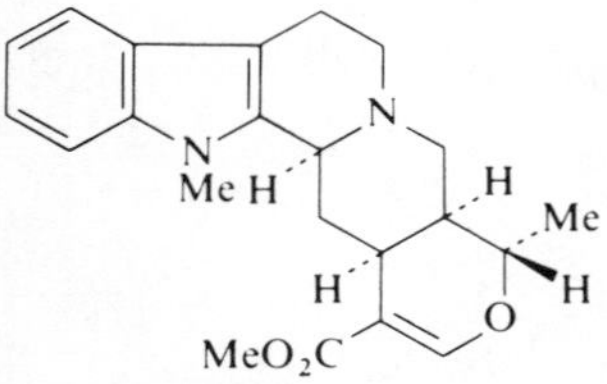

Further progress in this field was achieved by the following: (a) sequence studies on the alkaloid formation in very young seedlings of *V. rosea* (isolation and characterization); (b) double labeling of building blocks (to gain further insight on the mechanisms involved); and (c) steady-state labeling with tryptophan and sequential autoradiography of those alkaloids whose radioprofile is sufficiently dynamic to suggest that a true intermediate is involved. In sequence studies, *Corynanthé* systems appeared before any detectable amounts of *Aspidosperma* and *Iboga* alkaloids (*76*). The feeding experiments with [*O*-methyl-^{3}H,Ar-^{3}H]geissoschizine (**154**) showed it to act as a precursor of *Corynanthé-Strychnos*, *Aspidosperma*, and *Iboga* alkaloids (*77, 78*). Further work led to the isolation of geissoschizine oxindole (**162**) (*79*) and preakuammicine (**163**) (*80*) from *V. rosea* as natural products. The latter was found to be an efficient precursor of akuammicine and vindoline (*79*). The sodium borohydride reduction of **163** furnished stemmadenine (**164**) (*80*), which was found to be the key progenitor of *Aspidosperma* and *Iboga* alkaloids (*76*).

It has been proposed that the incorporation of stemmadenine in vindoline and catharanthine proceeds via dehydrosecondine (**165**). A Diels–Alder

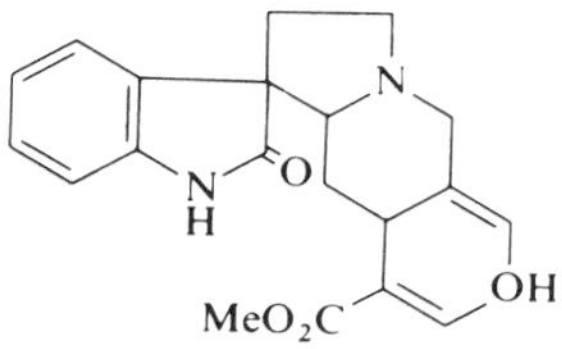

162

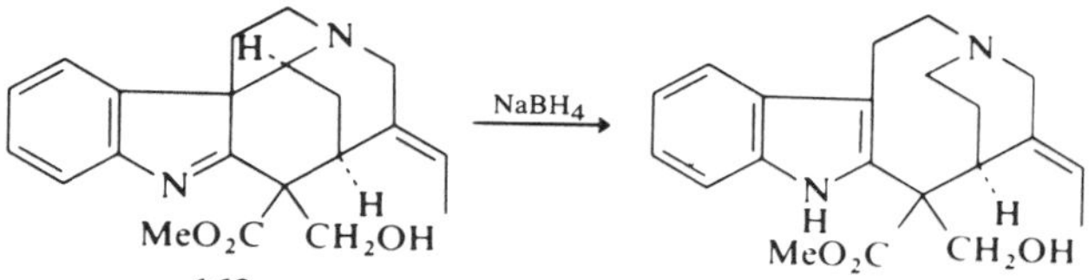

163 $\xrightarrow{NaBH_4}$ 164

reaction as shown in the two indicated ways (Scheme 6) would lead to the formation of tabersonine (**166**) and catharanthine (**143**), respectively. The essential correctness of these ideas has been confirmed by feeding experiments with [*O*-methyl-^{3}H,11-^{14}C] and [Ar-^{3}H]tabersonine to *V. rosea*

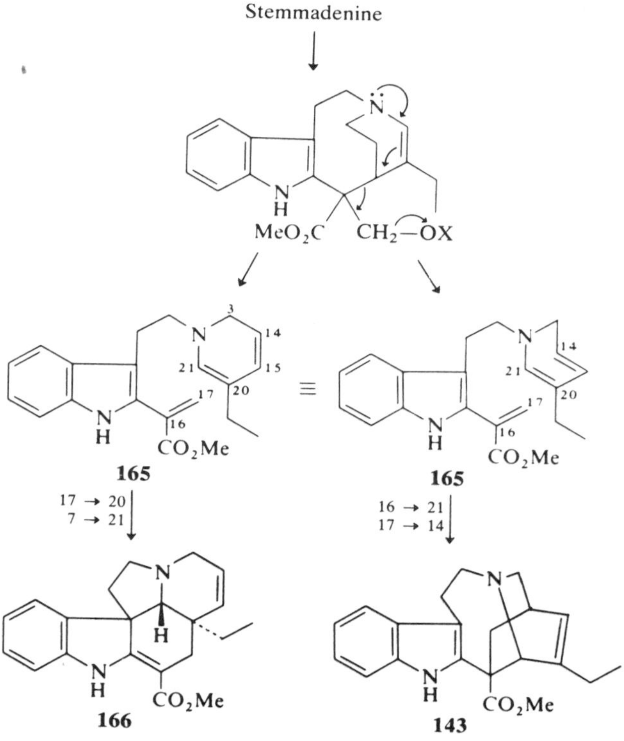

Stemmadenine

165 ≡ 165

17 → 20, 7 → 21 | 16 → 21, 17 → 14

166 | 143

SCHEME 6

when it was found to be incorporated in vindoline and catharanthine (*76*). Although dehydrosecodine has not been detected in nature, many of its derivatives, such as 15,20-dihydro- and 16,17,15,20–tetrahydrosecodines (*81*) and the related dimeric presecamines (*82*), have been isolated from *R. stricta*. *In vitro* analogies to these biotransformations have recently been described (*83*).

Ajmalicine (**139a**) and 3-isoajmalicine (**140a**) were both converted to oxindole alkaloids, mitraphylline (**167**) and isomitraphylline (**168**) in *Mitragyna parvifolia* (Roxb.) Korth. (Rubiaceae) (*84*). Feeding experiments on *Vinca minor* Linn. (Apocynaceae) had shown that vincamine (**169**) is biosynthesized via geissoschizine, stemmadenine, secodine, and tabersonine (*85*). Apparicine (**170**), which lacks one carbon atom of the tryptamine side chain, appears to be derived in *Aspidosperma pyricollum* Müell. Arg. (Apocynaceae) from stemmadenine (*86*).

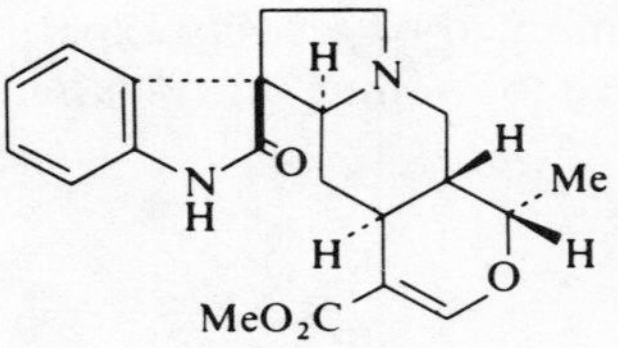

H N H Me N O H H O MeO_2C
167

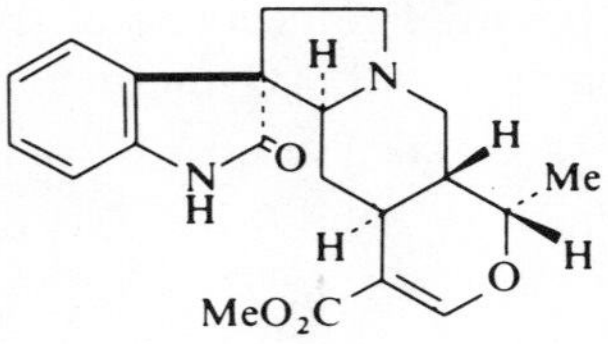

H N H Me N O H H O MeO_2C
168

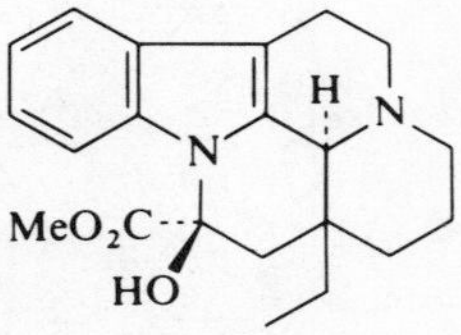

H N N MeO_2C HO
169

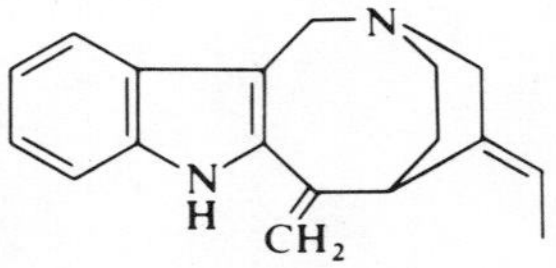

N N H CH_2
170

The biosynthesis of strychnine (**172**) in *Strychnos nux-vomica* Linn. (Loganiaceae) has been reinvestigated. In contrast to earlier short-term experiments when insignificant or no incorporation was obtained by feeding the Wieland–Gumlich aldehyde (**171**) (*87*), prolonged contact feedings have shown **172** to be derived from geissoschizine and the Wieland–Gumlich aldehyde (*88*). The *Corynanthé* alkaloids yohimbine (**173**), rauwolscine (**174**), and corynanthine (**175**) have also been found to be biosynthesized via the strictosidine pathway in *Rauwolfia canescens* Linn. (Apocynaceae) (*89*). A cell-free system which gave high incorporations of [2-^{14}C]tryptamine and [*O*-methyl-^{3}H]secologanin in the biosynthesis of geissoschizine and ajmalicine from seedlings and tissue cultures of *V. rosea* has recently been described (*90*).

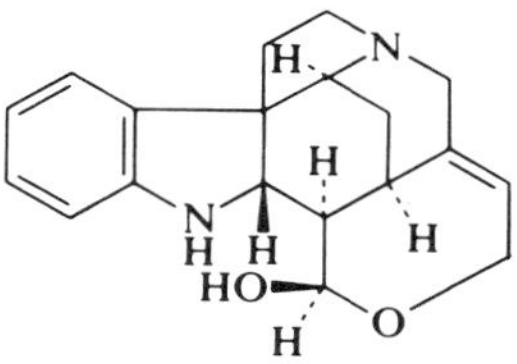

171

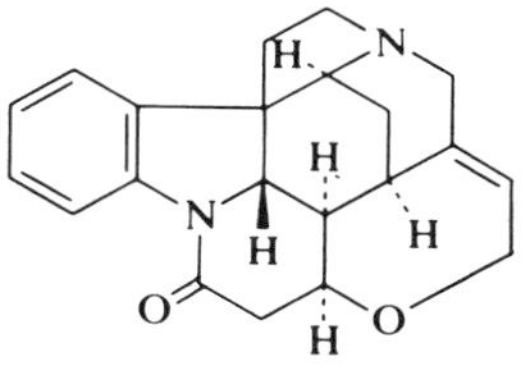

172

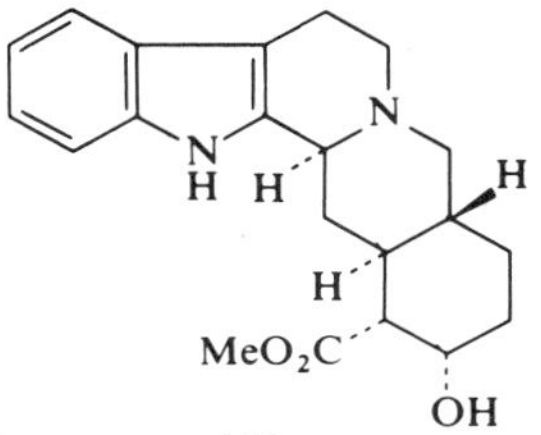

173

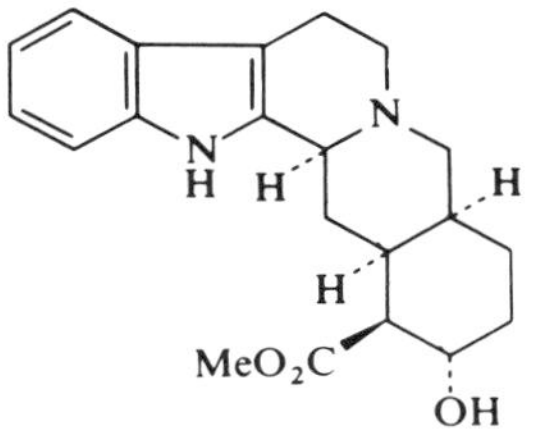

174

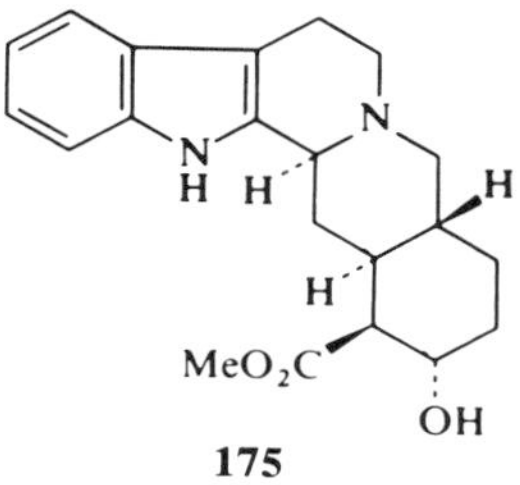

175

B. *CINCHONA* ALKALOIDS

The *Cinchona* alkaloids have long been thought to be derived from indolic precursors (*91*), and recent work on the biosynthesis of quinine (**183**), cinchonine (**186**), and cinchonidine (**182**) in *Cinchona ledgeriana* Moens ex Trimen (Rubiaceae) has confirmed these assumptions. Tracer work has shown that the quinoline ring and C_9 unit of quinine originate from tryptophan (*92*) and geraniol (*93, 94*), respectively. Loganin, vincoside, corynantheal (**176**), and cinchonaminal (**177**) were shown as the key intermediates in the biosynthesis of quinine, cinchonine, and cinchonidine.

Feeding experiments with doubly labeled [1-$^3H_2$, 1-^{14}C]tryptamine showed that 50% of the tritium at C-1 of tryptamine was lost during the formation of *Cinchona* alkaloids—a result which indicates that the cleavage between C-5 and N-4 in corynantheal is a stereospecific process. The ring expansion to quinoline must occur via the indolenine (**178**) and the recyclization of **179** to furnish cinchonidinone (**180**), which was found by radiochemical dilution analysis to be a natural product in *C. ledgeriana*. The specific incorporation of [11-$^3H_2$]cinchonidinone in cinchonine and quinine and the incorporation of [11-$^3H_2$]cinchonidine in cinchonidinone is best understood by the sequence **180** → **186** and its reversal **186** → **182**, as shown in Scheme 7 (*95*).

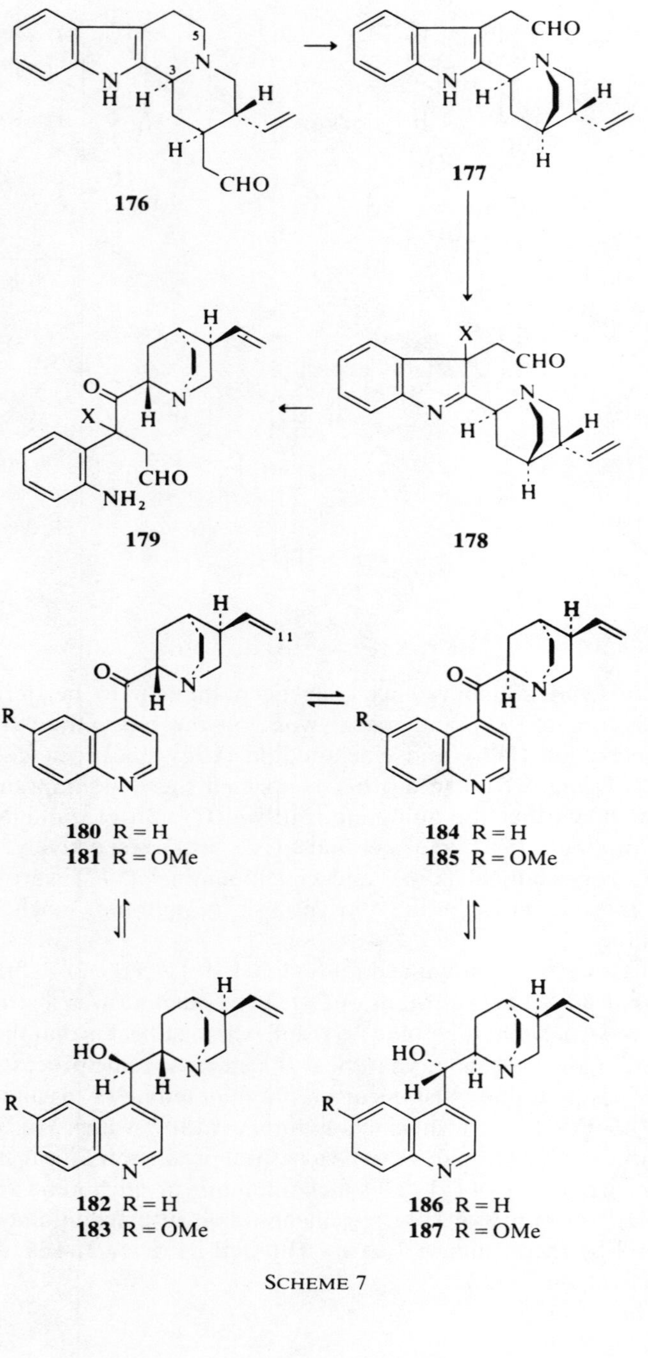

5
N
3
N
H
H
H
H
CHO
176
CHO
N
N
H
H
H
H
177
X
CHO
N
N
H
H
H
178
H
O
N
H
X
CHO
NH_2
179
H
11
O
N
H
R
N
180 R = H
181 R = OMe
H
O
N
H
R
N
184 R = H
185 R = OMe
H
HO
H
H
N
R
N
182 R = H
183 R = OMe
H
HO
H
H
N
R
N
186 R = H
187 R = OMe

SCHEME 7

C. *Camptotheca* Alkaloids

Much interest has recently been generated in *Camptotheca* alkaloids because of their reported high anticancer activity (*96*). These bases bear close structural resemblance to terpenoid indole alkaloids; therefore, it was suggested by Wenkert and co-workers (*97*) that camptothecin (**188**) was probably a masked indole alkaloid of the corynantheidine type. The isolation of vincoside and strictosidine lactams (*9*) provided even closer analogies. Even before *in vivo* testing of the hypothesis, Winterfeldt and his colleagues (*98*) were able to achieve a biogenetically oriented total synthesis of camptothecin involving the oxidative cleavage of an indole and cyclization to a quinolone derivative which was subsequently converted to the required quinoline. Feeding experiments on *Mappia foetida* Miers (Icacinaceae) and on *Camptotheca acuminata* Decne. (Nyssaceae) established that camptothecin is derived from tryptophan and secologanin, and is indeed a terpenoid indole alkaloid (*99, 100*). Furthermore, it was found that the lactam of strictosidine rather than of vincoside is the *in vivo* precursor to camptothecin, in contrast to all other terpenoid indole alkaloids so far investigated (*100*), a distinction also observed during *in vitro* conversions (*101*).

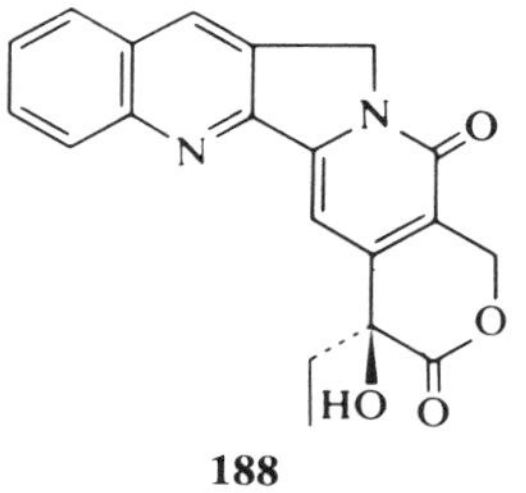

188

D. Carboxy Terpenoid Indole Alkaloids

The isolation of cordifoline and related bases led to the prediction of the existence of a common precursor derived by condensation between tryptophan and secologanin. This tetrahydrodeoxycordifoline (TDC) could be (a) a biogenetic cul-de-sac (b) an alternative intermediate to vincoside for some terpenoid indole alkaloids, or (c) the progenitor of a novel range of acidic terpenoid indole alkaloids in which the carboxy group of tryptophan was retained. The last possibility stimulated a search for such alkaloids which was amply rewarded by the discovery in *A. rubescens* of adirubine (*7*), 5α-carboxytetrahydroalstonine (*102*), and 5α-carboxycorynanthine (*103*), shown to have the tryptophan-based *Corynanthé* type structures **189**, **190**, and **191**, respectively. Surely, with the adoption of modern techniques for the isolation of these bases, a rewarding investigation could be the characterization of carboxy representatives of *Aspidosperma* and *Iboga* alkaloids.

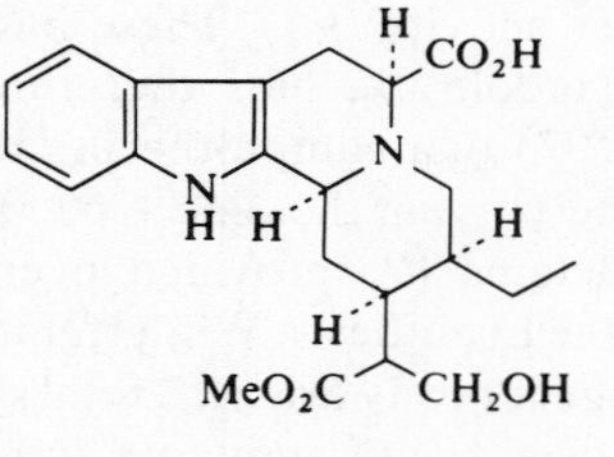

189 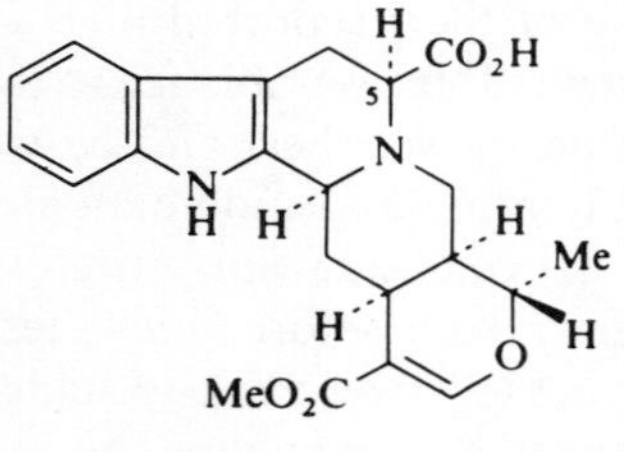 **190**

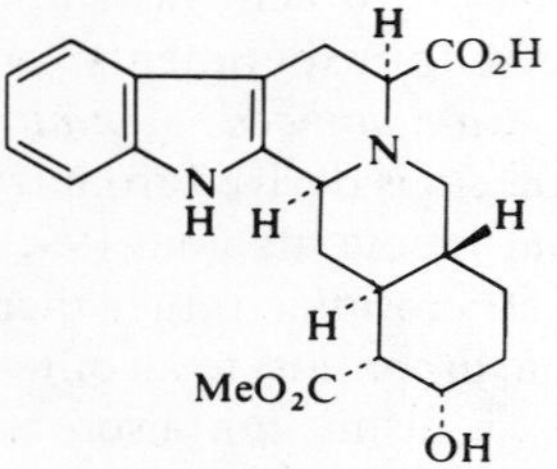

191

VIII. Addendum

1. Extraction of the leaves of *Strychnos decussata* (Pappe) Gil. (Loganiaceae) with methanol followed by column chromatography over silica gel furnished a novel indole glycoside {$C_{26}H_{34}N_2O_7$; mp 201°; $[\alpha]_D -134°$ (MeOH)}. Based on spectral analysis (UV, NMR, and mass), it has been assigned the pentacyclic structure (**192**) (undefined stereochemistry) (*104*).

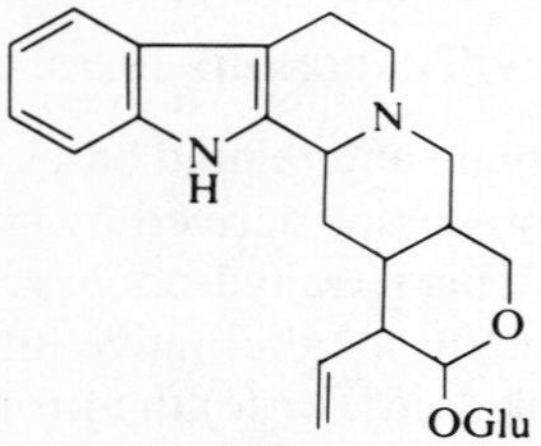

192

2. From the leaves of *P. lyalli* a new glycoalkaloid pauridianthoside {$C_{27}H_{28}N_2O_{10}$; mp 208–210°; $[\alpha]_D -264°$ (MeOH)}, has been isolated. The UV spectrum [$\lambda_{max}(\log \epsilon)$: 218 (4.60), 286 (4.20), and 384 (3.82) nm] indicated the presence of a carbonyl group conjugated with the harman nucleus. The mass spectrum showed the expected molecular ion at *m/e* 540,

and the fragmentation pattern was characteristic of 14-oxolyaloside structure (**193**), which was further confirmed by a detailed study of the 250 MHz NMR spectrum of its tetraacetate {mp 143–145°; $[\alpha]_D - 175°$ ($CHCl_3$)} (*105*).

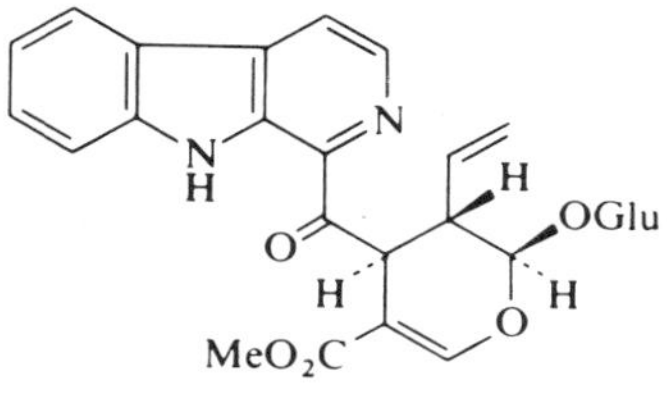

193

3. 20,21-Didehydroajmalicine (cathenamine) (**194**) has been identified as a key intermediate in the enzymatic production of ajmalicine, 19-epiajmalicine (**195**), and tetrahydroalstonine when tryptamine and secologanin were incubated with cell suspension cultures of *V. rosea* (*106*).

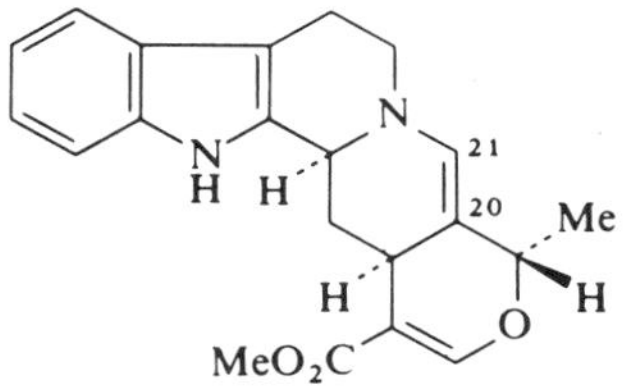

194 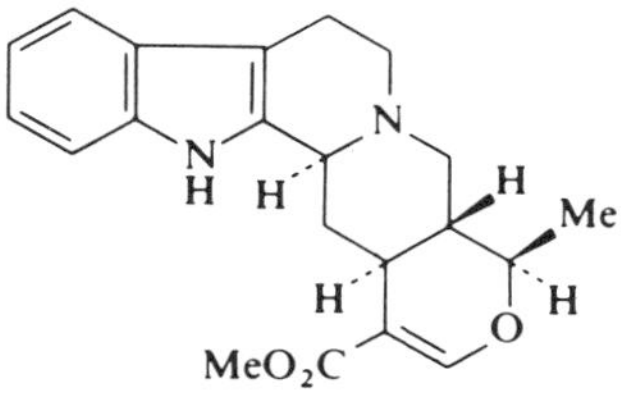 **195**

4. Quite recently Zenk and co-workers have shown that strictosidine is also a biological precursor for terpenoid indole alkaloids of 3β configuration. The biosynthetic conversion proceeds with loss of hydrogen at C-3, while it is retained in the formation of 3α, *Corynanthé* alkaloids (*107*).

REFERENCES

1. R. T. Brown and L. R. Row, *Chem. Commun.* 453 (1967).
2. L. Merlini and G. Nasini, *Gazz. Chim. Ital.* **98**, 974 (1968).
3. R. T. Brown, unpublished observations; F. Hotellier, P. Delaveau, and J. L. Pousset, *Plantes médicinales phytotherapie*, **11**, 106 (1977).
4. G. I. Dimitrienko, D. G. Murray, and S. McLean, *Tet. Lett.* 1961 (1974).
5. R. T. Brown and C. L. Chapple, *Tet. Lett.* 2723 (1976).
6. R. T. Brown and A. A. Charalombides, *Experientia* **31**, 505 (1975).
7. R. T. Brown, C. L. Chapple, and G. K. Lee, *J.C.S., Chem. Commun.* 1007 (1972).
8. K. T. D. De Silva, D. King, and G. N. Smith, *Chem. Commun.* 908 (1971).
9. W. P. Blackstock, R. T. Brown, and G. K. Lee, *Chem. Commun.* 910 (1971).
10. P. Bellet, *Ann. Pharm. Fr.* **10**, 81 (1952).

11. P. Bellet, *Ann. Pharm. Fr.* **12**, 466 (1954).
12. A. R. Battersby, B. Gregory, H. Spencer, J. C. Turner, M.-M. Janot, P. Potier, P. François, and J. Levisalles, *Chem. Commun.* 219 (1967).
13. O. Kennard, P. J. Roberts, N. W. Isaacs, F. H. Allen, W. D. S. Motherwell, K. H. Gibson, and A. R. Battersby, *Chem. Commun.* 899 (1971).
14. A. R. Battersby, A. R. Burnett, and P. G. Parsons, *Chem. Commun.* 1280 (1968); *J. Chem. Soc. C* 1187 (1969).
15. R. S. Kapil, A. Shoeb, S. P. Popli, A. R. Burnett, G. D. Knowles, and A. R. Battersby, *Chem. Commun.* 904 (1971).
16. A. Shoeb, K. Raj, R. S. Kapil, and S. P. Popli, *J. Chem. Soc. Perkin Trans. 1* 1245 (1975).
17. A. R. Battersby, *Pure Appl. Chem.* **14**, 117 (1967).
18. I. Souzu and H. M. Mitsuhashi, *Tet. Lett.* 191 (1970).
19. A. R. Battersby, A. R. Burnett, and P. G. Parsons, *Chem. Commun.* 1282 (1968); *J. Chem. Soc. C* 1193 (1969).
20. A. R. Battersby, A. R. Burnett, E. S. Hall, and P. G. Parsons, *Chem. Commun.* 1582 (1968).
21. G. N. Smith, *Chem. Commun.* 912 (1968).
22. C. Djerassi, H. J. Monteiro, A. Wasler, and L. J. Durham, *J. Am. Chem. Soc.* **88**, 792 (1966).
23. R. T. Brown, C. L. Chapple, and A. Ghashford, *J.C.S. Chem. Commun.* 295 (1975).
24. K. T. D. De Silva, G. N. Smith, and K. E. H. Warren, *Chem. Commun.* 905 (1971).
25. M. Pinar, M. Haraoka, M. Hesse, and H. Schmid, *Helv. Chim. Acta* **54**, 15 (1971); A. R. Battersby, *Alkaloids (London)* **1**, 31 (1970); A. I. Scott, M. B. Staytor, P. B. Reichardt, and J. G. Sweeny, *Bioorg. Chem.* **1**, 157 (1971).
26. S. R. Johns, J. A. Lamberton, and J. L. Occolowitz, *Aust. J. Chem.* **20**, 1463 (1967); Y. K. Sowa and H. Matsumara, *Tetrahedron* **25**, 5319 (1969).
27. E. Wenkert and N. V. Bringi, *J. Am. Chem. Soc.* **81**, 1474 (1959); E. Ochiai and M. Ishikawa, *Chem. Pharm. Bull.* **7**, 256 (1959).
28. W. Klyne, R. J. Swan, N. J. Dastoor, A. A. Gorman, and H. Schmid, *Helv. Chim. Acta* **50**, 115 (1966); C. M. Lee, W. F. Trager, and A. H. Beckett, *Tetrahedron* **23**, 375 (1967).
29. K. C. Mattes, C. R. Hutchinson, J. P. Springer, and J. Clardy, *J. Am. Chem. Soc.* **97**, 6270 (1975).
30. J. Levesque, J. L. Pousset, and A. Cavé, *C. R. Acad. Sci., Ser. C* **280**, 593 (1975).
31. K. L. Stuart and R. B. Woo-Ming, *Tet. Lett.* 3853 (1974).
32. W. P. Blackstock and R. T. Brown, *Tet. Lett.* 3727 (1971).
33. R. T. Brown and W. P. Blackstock, *Tet. Lett.* 3063 (1972).
34. R. T. Brown and S. B. Fraser, *Tet. Lett.* 1957 (1974); R. T. Brown, S. B. Fraser, and J. Banerji, *ibid.* 3335.
35. R. T. Brown, unpublished observations.
36. R. T. Brown and B. F. Warambwa, *Phytochemistry* **17** (1978).
37. R. T. Brown and S. B. Fraser, *Tet. Lett.* 841 (1973).
38. W. P. Blackstock, R. T. Brown, C. L. Chapple, and S. B. Fraser, *J.C.S., Chem. Commun.* 1006 (1972).
39. R. T. Brown and A. A. Charalambides, *J.C.S., Chem. Commun.* 765 (1973).
40. R. T. Brown and A. A. Charalambides, *J.C.S., Chem. Commun.* 553 (1974).
41. J. E. Saxton, *in* "The Alkaloids" (R. H. F. Manske, ed.), Vol. VIII, Chapter 5. Academic Press, New York, 1965.
42. A. D. Cross, F. E. King, and T. J. King, *J. Chem. Soc.* 2714 (1961).
43. R. T. Brown, K. V. J. Rao, P. V. S. Rao, and L. R. Row, *Chem. Commun.* 350 (1968).
44. E. E. van Tamelen, P. E. Aldrich, and T. J. Katz, *J. Am. Chem. Soc.* **79**, 6426 (1957).

45. A. R. Battersby, R. Binks, W. Lawrie, G. V. Parry, and B. R. Webster, *Proc. Chem. Soc., London* 369 (1963); *J. Chem. Soc.* 7459 (1965).
46. A. R. Battersby and B. Gregory, *Chem. Commun.* 134 (1968).
47. A. R. Battersby and R. J. Parry, *Chem. Commun.* 901 (1971).
48. R. T. Brown, G. N. Smith, and K. S. J. Stapleford, *Tet. Lett.* 4349 (1968).
49. K. Stolle, D. Gröger, and K. Mothes, *Chem. Ind. (London)* 2065 (1965).
50. H. Goeggel and D. Arigoni, *Experientia* **21**, 369 (1965).
51. E. Leete, A. Ahmad, and I. Kempis, *J. Am. Chem. Soc.* **87**, 4168 (1965).
52. D. H. R. Barton, G. W. Kirby, R. H. Prager, and E. M. Wilson, *J. Chem. Soc.* 3900 (1965).
53. J. P. Kutney, J. F. Beck, V. R. Nelson, K. L. Stuart, and A. K. Bose, *J. Am. Chem. Soc.* **92**, 2174 (1970).
54. F. McCapra, T. Money, A. I. Scott, and I. G. Wright, *Chem. Commun.* 537 (1965).
55. H. Goeggel and D. Arigoni, *Chem. Commun.* 538 (1965).
56. A. R. Battersby, R. T. Brown, R. S. Kapil, A. O. Plunkett, and J. B. Taylor, *Chem. Commun.* 46 (1966).
57. D. A. Yeowell and H. Schmid, *Experientia* **20**, 250 (1964).
58. A. R. Battersby, R. T. Brown, J. A. Knight, J. A. Martin, and A. O. Plunkett, *Chem. Commun.* 346 (1966).
59. P. Loew, H. Goeggel, and D. Arigoni, *Chem. Commun.* 347 (1966).
60. E. S. Hall, F. McCapra, T. Money, K. Fukumoto, J. R. Hanson, B. S. Mootoo, G. T. Phillips, and A. I. Scott, *Chem. Commun.* 348 (1966); T. Money, I. G. Wright, F. McCapra, E. S. Hall, and A. I. Scott, *J. Am. Chem. Soc.* **90**, 4144 (1968).
61. E. Leete and S. Ueda, *Tet. Lett.* 4915 (1966).
62. S. Escher, P. Loew, and D. Arigoni, *Chem. Commun.* 823 (1970).
63. A. R. Battersby, S. H. Brown, and T. G. Payne, *Chem. Commun.* 827 (1970).
64. A. R. Battersby, A. R. Burnett, and P. G. Parsons, *Chem. Commun.* 826 (1970).
65. A. R. Battersby, R. S. Kapil, J. A. Martin, and L. Mo, *Chem. Commun.* 133 (1968).
66. P. Loew and D. Arigoni, *Chem. Commun.* 137 (1968).
67. A. R. Battersby, A. R. Burnett, and P. G. Parsons, *Chem. Commun.* 1280 (1968); *J. Chem. Soc. C* 1187 (1969).
68. L.-F. Tietze, *Angew. Chem., Int. Ed. Engl.* **12**, 853 (1973); *J. Am. Chem. Soc.* **96**, 946 (1974).
69. R. Guarnaccia and C. J. Coscia, *J. Am. Chem. Soc.* **93**, 6320 (1971).
70. R. T. Brown and C. L. Chapple, *J.C.S., Chem. Commun.* 886 (1973).
71. R. T. Brown, C. L. Chapple, and A. A. Charalambides, *J.C.S., Chem. Commun.* 756 (1974).
72. R. T. Brown, C. L. Chapple, R. Platt, and H. Spencer, *J.C.S., Chem. Commun.* 929 (1974).
73. R. T. Brown, J. Leonard, and S. K. Sleigh, *J.C.S., Chem. Commun.* 636 (1977).
74. J. Stöckigt and M. H. Zenk, *J.C.S., Chem. Commun.* 646 (1977); A. I. Scott, S.-L. Lee, P. De Capite, M. G. Culver, and C. R. Hutchinson, *Heterocycles* **7**, 979 (1977).
75. R. T. Brown, J. Leonard, and S. K. Sleigh, *Phytochemistry* **17**, 899 (1978).
76. A. A. Qureshi and A. I. Scott, *Chem. Commun.* 948 (1968).
77. A. I. Scott, P. C. Cherry, and A. A. Qureshi, *J. Am. Chem. Soc.* **91**, 4932 (1969).
78. A. R. Battersby and E. S. Hall, *Chem. Commun.* 793 (1969).
79. A. I. Scott and S. I. Heimberger, unpublished observations.
80. A. I. Scott and A. A. Qureshi, *J. Am. Chem. Soc.* **91**, 5874 (1969).
81. G. A. Cordell, G. F. Smith, and G. N. Smith, *Chem. Commun.* 189 (1970).
82. G. A. Cordell, G. F. Smith, and G. N. Smith, *Chem. Commun.* 191 (1970).
83. A. I. Scott and C. C. Wei, *J. Am. Chem. Soc.* **94**, 8263, 8264, 8266 (1972).

84. E. J. Shellard, K. Sarpong, and P. J. Houghton, *J. Pharm. Pharmacol.* **23**, 244S (1971); E. J. Shellard and P. J. Houghton, *Planta Med.* **21**, 16 (1972).
85. J. P. Kutney, J. F. Beck, V. R. Nelson, and R. S. Sood, *J. Am. Chem. Soc.* **93**, 255 (1971).
86. J. P. Kutney, V. R. Nelson, and D. C. Wigfield, *J. Am. Chem. Soc.* **91**, 4278, 4279 (1969).
87. Ch. Schlatter, E. E. Waldner, H. Schmid, W. Maier, and D. Gröger, *Helv. Chim. Acta* **52**, 776 (1969).
88. S. I. Heimberger and A. I. Scott, *J.C.S., Chem. Commun.* 217 (1973).
89. A. S. Mulla, R. S. Kapil, and S. P. Popli, unpublished observations.
90. A. I. Scott and S.-L. Lee, *J. Am. Chem. Soc.* **97**, 6906 (1975).
91. R. Goutarel, M.-M. Janot, V. Prelog, and W. I. Taylor, *Helv. Chim. Acta* **33**, 150 (1950).
92. N. Kowanko and E. Leete, *J. Am. Chem. Soc.* **84**, 4919 (1962).
93. A. R. Battersby, R. T. Brown, R. S. Kapil, J. A. Knight, J. A. Martin, and A. O. Plunkett, *Chem. Commun.* 888 (1966).
94. E. Leete and J. N. Wemple, *J. Am. Chem. Soc.* **88**, 4743 (1966); **91**, 2698 (1969).
95. A. R. Battersby and R. J. Parry, *Chem. Commun.* 30, 31 (1971).
96. J. L. Hartwell and B. J. Abbott, *Adv. Pharmacol. Chemother.* **7**, 117 (1969).
97. E. Wenkert, K. G. Dave, R. G. Lewis, and P. W. Sprague, *J. Am. Chem. Soc.* **89**, 6741 (1967).
98. E. Winterfeldt, T. Korth, D. Pike, and M. Boch, *Angew. Chem.* **84**, 265 (1972); M. Boch, T. Korth, J. M. Nelke, D. Pike, H. Radunz, and E. Winterfeldt, *Ber.* **105**, 2126 (1972).
99. A. I. Scott and S. P. Popli, unpublished observations (1972).
100. C. R. Hutchinson, A. H. Heckendorf, P. E. Daddona, E. Hagaman, and E. Wenkert, *J. Am. Chem. Soc.* **96**, 5609 (1974); A. H. Heckendorf and C. R. Hutchinson, *Tet. Lett.* 4153 (1977).
101. C. R. Hutchinson, G. J. O'Loughlin, R. T. Brown, and S. B. Fraser, *J.C.S., Chem. Commun.* 928 (1974).
102. R. T. Brown and A. A. Charalambides, *Tet. Lett.* 1649 (1974).
103. R. T. Brown and A. A. Charalambides, *Tet. Lett.* 3429 (1974).
104. A. Petitjean, P. Rasoanaivo, and J. M. Razafintsalama, *Phytochemistry* **16**, 154 (1977).
105. J. Levesque, J. L. Pousset, and A. Cavé, *Fitoterapia* **48**, 5 (1977).
106. J. Stöckigt, H. P. Husson, C. Kan-fan, and M. H. Zenk, *J.C.S., Chem. Commun.* 164 (1977).
107. M. Rueffer, N. Nagakura, and M. H. Zenk, *Tet. Lett.* 1593 (1978).

SUBJECT INDEX

D

E

F

G

H

I

N

O

P

Q

R

S

T

U

V